4th International Conference on Advances in Solidification Processes (ICASP-4)

IOP Conference Series: Materials Science and Engineering
Volume 117

Windsor, United Kingdom
8 - 11 July 2014

ISBN: 978-1-5108-2232-0
ISSN: 1757-8981

Printed from e-media with permission by:

Curran Associates, Inc.
57 Morehouse Lane
Red Hook, NY 12571

Some format issues inherent in the e-media version may also appear in this print version.

Printed by Curran Associates, Inc. (2016)

For permission requests, please contact the Institute of Physics at the address below.

Institute of Physics
Dirac House, Temple Back
Bristol BS1 6BE UK

Phone: 44 1 17 929 7481
Fax: 44 1 17 920 0979

techtracking@iop.org

Additional copies of this publication are available from:

Curran Associates, Inc.
57 Morehouse Lane
Red Hook, NY 12571 USA
Phone: 845-758-0400
Fax: 845-758-2633
Email: curran@proceedings.com
Web: www.proceedings.com

Table of contents

Volume 117

4th International Conference on Advances in Solidification Processes (ICASP-4)
8–11 July 2014, Windsor, UK

Accepted papers received: 22 February 2016
Published online: 31 March 2016

Preface

Dendritic Growth

Intermetallic Formation

Eutectic, Peritectic and Monotectic Solidification

Microstructure Formation

Advanced Solidification Processing

Solidification Under External Fields

Rapid Solidification

Defect Formation

Preface

This volume of IOP Conference Series - Materials Science and Engineering represents a selection of the papers presented at the fourth International Conference on Advances in Solidification Processes held in Old Windsor (UK) on 8-11 July 2014 (ICASP-4). It followed previous meetings in the series held in Stockholm in 2005, Graz in 2008 and Aachen/Rolduc in 2011.

ICASP-4 was hosted by BCAST, a Specialist Research Institute of Brunel University London, UK. BCAST is the lead partner of the EPSRC Future LiME Hub (formerly the EPSRC Centre – LiME), a national centre of excellence in liquid metal engineering supported by the UK Engineering and Physical Sciences Research Council. BCAST and the EPSRC Future LiME Hub have a balanced portfolio of fundamental and user-led research in the areas of solidification science and the development of new alloys and technologies for closed-loop recycling, to support sustainability of the metallurgical industry.

ICASP is a triennial conference that focuses on recent advances in the science and technologies of solidification and casting of all metals including metal matrix composites. Topics include nucleation, growth and microstructure formation; shape and bulk casting techniques and rapid solidification; and the formation of defects. In particular, ICASP-4 celebrated the 60th anniversary of "constitutional supercooling", which is widely regarded as the beginning of modern quantitative solidification science. For that purpose the conference staged 16 plenary lectures delivered by key researchers in our field.

As well as the 16 plenary lectures the ICASP-4 programme featured 20 invited lectures, 109 oral presentations delivered in three parallel sessions, and 66 poster presentations in a dedicated poster session. The programme also included the inaugural presentation of the John Hunt Medal to Professor Rohit Trivedi and his delivery of the John Hunt Memorial Lecture.

The editors would like to thank the authors for their contributions to this volume and for their cooperation during the editorial process. We are grateful to the reviewers for their constructive criticism. On behalf of the ICASP-4 Organising Committee, we would like to thank members of the International Scientific Committee for promoting the conference. We would also like to thank the EPSRC, the IOM3 (Institute of Materials, Minerals and Mining), MICRESS, the staff of the Beaumont Estate, Old Windsor, the staff and students of BCAST, and colleagues at Brunel University, for their support of the conference.

Professor Z. Fan
Dr I.C. Stone

ICASP
4
2014
4th International Conference on Advances in Solidification Processes
Beaumont Estate, Old Windsor, UK, 8th - 11th July 2014

IOP Conf. Series: Materials Science and Engineering **117** (2016) 011002 doi:10.1088/1757-899X/117/1/011002

ICASP-4 Committees

Organising Committee

Professor Zhongyun Fan	*BCAST, Brunel University London, UK*
Professor Wilfried Kurz	*EPFL, Switzerland*
Professor Andrew Mullis	*University of Leeds, UK*
Dr Christopher Gourlay	*Imperial College, UK*
Dr Ian Stone	*BCAST, Brunel University London, UK*
Dr Markus Köhler	*BCAST, Brunel University London, UK*
Mr Matthew Ralph	*BCAST, Brunel University London, UK*
Mr Thomas Ryan	*Brunel University London, UK*

Scientific Committee

Professor Lars Arnberg	*NTNU, Norway*
Professor Livio Battezzati	*Università di Torino, Italy*
Professor Christoph Beckermann	*University of Iowa, USA*
Professor Bernard Billia	*IM2NP, Aix-Marseille University, France*
Professor David Browne	*University College Dublin, Ireland*
Professor Andreas Bührig-Polaczek	*RWTH Aachen University, Germany*
Professor Madhusudan Chakraborty	*Indian Institute of Technology, India*
Professor Kamanio Chattopadhyay	*Indian Institute of Science, India*
Professor Hervé Combeau	*École des Mines de Nantes, France*
Professor Arne K Dahle	*University of Queensland, Australia*
Professor Chuan Dong	*Dalian University of Technology, China*
Professor Yong Du	*Central South University, China*
Professor Zhongyun Fan	*BCAST, Brunel University London, UK*
Professor Yves Fautrelle	*SIMAP, France*
Professor Ludo Froyen	*University of Leuven, Belgium*
Dr Charles André Gandin	*CNRS, MINES Paris Tech, France*
Dr Christopher Gourlay	*Imperial College, UK*
Professor László Gránásy	*Wigner Research Centre for Physics, Hungary*
Professor Jacques Lacaze	*Ensiacet University, France*
Professor Andreas Ludwig	*University of Leoben, Austria*
Professor Andrew Mullis	*University of Leeds, UK*
Professor B S Murty	*Indian Institute of Technology, India*
Professor Michel Rappaz	*EPFL, Switzerland*
Professor Lorenz Ratke	*DLR, Germany*
Professor Markus Rettenmayr	*Friedrich Schiller University, Germany*
Professor András Roósz	*University of Miskolc, Hungary*
Professor Rainer Schmid-Fetzer	*TU Clausthal, Germany*
Professor Georg Schmitz	*ACCESS, Aachen, Germany*
Professor Peter Schumacher	*University of Leoben, Austria*
Professor David J St John	*University of Queensland, Australia*
Professor Niels Tiedje	*Technical University of Denmark, Denmark*
Professor Peter Voorhees	*Northwestern University, USA*
Dr Nils Warnken	*University of Birmingham, UK*
Professor Gerhard Zimmermann	*ACCESS, Aachen, Germany*

Peer review statement

All papers published in this volume of *IOP Conference Series: Materials Science and Engineering* have been peer reviewed through processes administered by the proceedings Editors. Reviews were conducted by expert referees to the professional and scientific standards expected of a proceedings journal published by IOP Publishing.

The effect of the melt thermal gradient on the size of the constitutionally supercooled zone

A Prasad[1,a], L Yuan[2,b], P D Lee[2,c], M Easton[3,d] and D StJohn[1,e]
[1] Centre for Advanced Materials Processing and Manufacturing, The University of Queensland, Brisbane, Australia
[2] The Manchester X-ray Imaging Facility, School of Materials, The University of Manchester, Oxford Rd., Manchester, M13 9PL, UK
[3] School of Aerospace, Mechanical and Manufacturing Engineering, RMIT University, Melbourne, Australia

Email: [a]a.prasad3@uq.edu.au, [b]lang.yuan@manchester.ac.uk, [c]peter.lee@manchester.ac.uk, [d]mark.easton@rmit.edu.au, [e]d.stjohn@uq.edu.au

Abstract. Recent verification of the analytical Interdependence model by a numerical solidification model (μMatIC) confirmed the critical role of constitutional supercooling (CS) in achieving sufficient undercooling to trigger successful nucleation events. The location of the maximum amount of CS (ΔT_{CSmax}) is some distance from the interface of the previously growing grain and this distance contributes to the final as-cast grain size. The effect of the thermal gradient, *G*, on the size of the CS zone (*CSZ*) was neglected in that work. However, *G* is expected to affect the size of the *CSZ* (i.e. the length of the *CSZ*, x'_{CSZ}, and the location of ΔT_{CSmax}, x'_{CSmax}). This investigation assesses the effect of *G* on x'_{CSZ} and x'_{CSmax}. A range of *G* values is introduced into both the analytical and the numerical models to obtain a correlation between the value of *G* and the dimensions of the CSZ. The result of a test case from the analytical model shows that x'_{CSmax} initially decreases rapidly and then decreases gradually approaching zero at very high values of *G*. Independent of the analytical model, the results from the numerical model replicate the trend obtained from the analytical model.

1. Introduction

The Interdependence model, which can predict the grain size of a casting, states that a new grain is nucleated when the amount of constitutional supercooling (ΔT_{CS}) of an already growing grain is large enough to activate an inoculant particle, which requires a certain minimum undercooling [1-4]. Equation 1 shows the key distances of the Interdependence model for a near-zero thermal gradient at large distances from the advancing solid-liquid (S-L) interface (for example T_{Al} in figure 1).

$$d_{gs} = x_{cs} + x'_{dl} + x_{sd} \tag{1}$$

Here d_{gs} is the predicted grain size, x_{cs} is the growth radius of the already growing grain, x'_{dl} the solute diffusion length ahead of the growing grain and x_{sd} the separation distance of the inoculant particles. In the original model [4] the expression for x_{cs} contains a '*z*' term which sets the additional CS required for subsequent nucleation events where *z* is a function of the thermal gradient present in the system. However, no detailed analysis of the thermal gradient has been undertaken and *z* assumes a pre-set user-defined value. The most appropriate value of *z* can only be determined following a detailed analysis of the effect of thermal gradient on the CS zone (CSZ). Here the size of the CSZ is described as the length x'_{CSZ} ahead of the S-L interface where ΔT_{CS} is positive, as well as the position, x'_{CSmax}, of ΔT_{CSmax} within this length. Note that x'_{CSZ} is not equal to x'_{dl}, the diffusion length as defined previously [4], in figure 1. x'_{dl} is independent of G and thus x'_{CSZ} can be smaller or larger than x'_{dl} depending on the value of G. The prime symbol is used because x' is the distance from the S-L interface whereas x (for example x_{CS} in figure 1) is measured from the point of nucleation [4].

As a first step the Interdependence model is used to qualitatively study the effect of G on the size of CSZ. The qualitative analysis from the analytical model is compared with the results from the μMatIC numerical model [5, 6]. Some preliminary results from this analysis are presented in this paper.

Figure 1 shows the equilibrium temperature, T_E, and the actual melt temperature, T_A, ahead of a growing grain. ΔT_{CS} is given by (T_E - T_A) and T_E changes along the length ahead of the growing grain governed by the solute diffusion from the growing S-L interface. In the case of zero thermal gradient, given by line T_{A1} in figure 1, ΔT_{CS} is simply T_E - T_{A1}. Note that when $T_A = 0$ (i.e. T_{A1}) x'_{CSZ} is infinite. The point of maximum supercooling, ΔT_{CSmax}, is given by the distance from the S-L interface to where T_E first attains its largest value. However, a thermal gradient is usually present in castings from the periphery to the thermal centre. The effect of the thermal gradient is usually studied in directional solidification experiments and numerical models. Within these, there are limited studies on the effect of thermal gradient on the nucleation of new grains. Dong-Lee [7] and Badillo-Beckermann [8] briefly mentioned the role of G in nucleation ahead of columnar grains in their study of the columnar to equiaxed transition (CET) and highlighted the importance of the influence of G on CS.

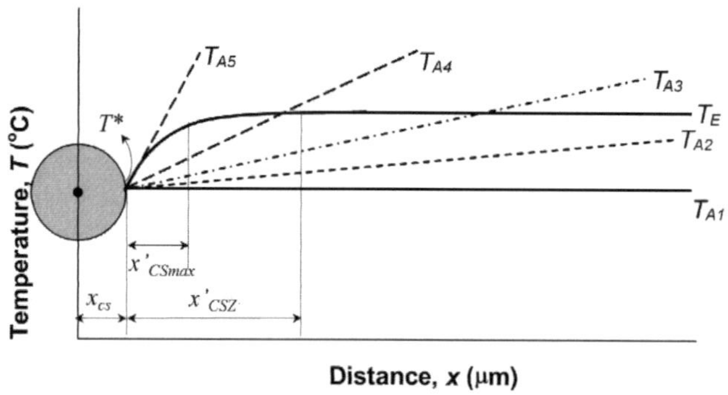

Figure 1. Schematic representation of the CSZ as described by the Interdependence model with varying degrees of thermal gradient, G. The approximate lengths of x'_{CSZ} (where ΔT_{CS} is positive), and x'_{CSmax} (corresponding to ΔT_{CSmax}) for T_{A4} are shown in the figure.

Assuming that the inoculant particle separation distance, x_{sd}, does not change in equation 1, the variation of the nucleation distance of new grains becomes a function of the sum of the first two terms only. Figure 1 shows this scenario conceptually for non-zero thermal gradients. The schematic representation shows that increasing G would decrease x'_{CSZ} approaching 0 at high G values. x'_{CSZ} is the total distance from the interface where ΔT_{CS} is still positive. The changing shape of $T_E(x)$ near the interface suggests that between the S-L interface and x'_{CSZ} there exists a distance, x'_{CSmax}, where ΔT_{CS} is at its maximum value ΔT_{CSmax}. Both x'_{CSZ} and x'_{CSmax} are shown schematically for T_{A4} in figure 1. To verify this, the effect of thermal gradient is studied in two ways. An analytical expression is developed which relates x'_{CSZ} and x'_{CSmax} to the thermal gradient, G. A numerical model, μMatIC [5, 6], is also used to simulate the solidification of a grain in the presence of a thermal gradient and x'_{CSZ} is estimated. Finally, the results from the analytical and μMatIC models are compared.

2. Analytical solution

Equation 2 shows the general form describing the equilibrium temperature, $T_E(x)$. This was derived by combining the expressions for T_E and $C_l(x)$ [4]. In the case of zero thermal gradient ΔT_{CSmax} is equal to T_{liq} (the liquidus temperature of the alloy) minus T^* (the temperature at the S-L interface) at some distance from the interface where the solute concentration attains the alloy composition C_0. In the case of non-zero thermal gradient, the length of positive ΔT_{CS}, i.e. x'_{CSZ}, is equal to T_E - T_A, if T_A is known. However, similar to T_E, T_A changes with distance, the general expression for which is given in equation 3. In these two equations, C_0 is the alloy composition, m is the slope of the liquidus line of the alloy system in question, $C_l^*(x')$ is the solute concentration at the interface and G represents the general thermal gradient term. By determining $T_E(x') - T_A(x')$, it can be shown that ΔT_{CS} is a function of distance, $\Delta T_{CS}(x')$, as shown in Equation 4:

$$T_E(x') = T_{liq} + m\left\{(C_l^*(x') - C_0)e^{-\frac{V}{D}x'}\right\} \tag{2}$$

$$T_A(x') = T^* + Gx' \tag{3}$$

$$\Delta T_{CS}(x') = T_E(x') - T_A(x') = T_{liq} + m\left\{(C_l^*(x') - C_0)e^{-\frac{V}{D}x'}\right\} - T^* - Gx' \tag{4}$$

By changing the $[m(C_l^*(x') - C_0)]$ term in equation 4 to (T^* - T_{liq}), the following is obtained:

$$\Delta T_{CS}(x') = T_E(x') - T_A(x') = \left\{(T^* - T_{liq})\left(e^{-\frac{V}{D}x'} - 1\right)\right\} - Gx' \tag{5}$$

Equation 5 states that for any value of G, ΔT_{CS} is zero when x' is 0 (i.e. at the S-L interface) and for $G = 0$, ΔT_{CS} is $(T_{liq} - T^*)$ from x'_{dl} to ∞. For $G > 0$, $\Delta T_{CS} = 0$ at some x' for a given value of G. x'_{CSZ} can now be evaluated for values of G (>0) and is given by the distance between the points where $\Delta T_{CS} = 0$. Furthermore, ΔT_{CSmax} corresponding to x'_{CSmax} lies between the two extremes where $\Delta T_{CS} = 0$. x'_{CSmax} can be evaluated by mathematically maximizing the expression for $\Delta T_{CS}(x')$ in equation 5 with respect to the distance x'. This is shown in equation 6. Note that T^* has been assumed to be constant.

$$x'_{CSmax} = -\frac{D}{V}\ln\left\{\left(\frac{GD}{V}\right)\cdot\frac{1}{T_{liq} - T^*}\right\} \tag{6}$$

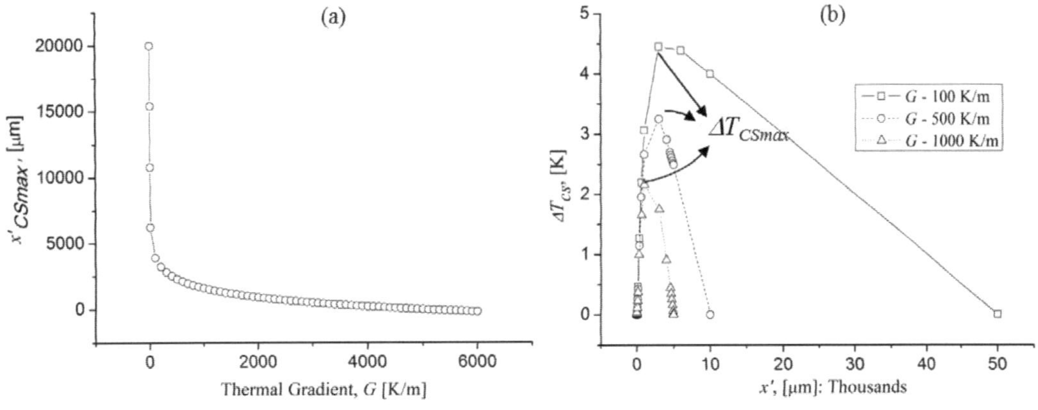

Figure 2. (a) x'_{CSmax}, the location of ΔT_{CSmax}, shown as a function of thermal gradient. At high G values, x'_{CSmax} approaches zero. (b) ΔT_{CS} (= $T_E(x') - T_A(x')$) as a function of distance, x', for different G. As G increases x'_{CSZ}, the total distance of positive ΔT_{CS}, and ΔT_{CSmax} decrease.

Equation 5 was used to plot the change in ΔT_{CS} with distance for a given thermal gradient G, and equation 6 was used to assess the effect of thermal gradient on the distance at which ΔT_{CSmax} is attained. Table 1 shows the values used. Figure 2a shows x'_{CSmax} corresponding to ΔT_{CSmax} as a function of the thermal gradient, G. As the thermal gradient increases, x'_{CSmax} decreases exponentially due to the logarithmic term in equation 6: faster at low G values then decreasing gradually at higher G. At high G values, x'_{CSmax} tends to 0. Figure 2b shows the variation in ΔT_{CS} with distance x' from the S-L interface and confirms the observations made in the previous paragraph. The peak value of ΔT_{CS} within x'_{CSZ} is ΔT_{CSmax}. As G increases x'_{CSZ} decreases and so does ΔT_{CSmax}.

3. Numerical solution: The μMatIC model

The numerical model was used to simulate a centrally placed single grain growing in a 5mm x 5mm computational domain with a left to right thermal gradient G (i.e. the temperature increases from left to right). Details of the model and its simulations are provided elsewhere [5-7, 9-11]. The alloy system modelled was Al-7%Si with an initial temperature of 887 K (614°C) at the left hand side. Five simulations were performed over a range of G, as given in table 1. The simulation time for each run was 86 s with a cooling rate of 0.3 K/s (calculated in the model as $G.V$ where V is the pulling velocity in a directional solidification experiment). The time chosen was to ensure that the right hand side of the computational domain reached a temperature lower than the alloy liquidus, T_{liq}. Following the simulation the grain size and the solute diffusion length ahead of the growing grain was estimated at 25, 50, 75 and 86 s.

Table 1. Values used in equations 5 and 6. Note that the analytical model was used as a test case only for the purposes of this paper. The G values used in the numerical model are also shown.

Item	Description	Value	Units
T_{liq}	Liquidus of the alloy system	913	[K]
T*	S-L interface temperature	908	[K]
G (analytical)	Thermal gradient	~0 – 10000	[K/m]
D (analytical)	Solute diffusion in the liquid	2e-9	[m^2/s]
v (analytical)	Interface growth rate	2e-6	[m/s]
G (numerical)	Thermal gradient in μMatIC	0, 100, 1000, 2000, 5000	[K/m]

* Note that the absolute values of T_{liq} and T* are not important in equation 5.

Figure 3 shows the solutal profile ahead of the growing grain which shows an exponential decay with distance. Close to the interface, between G = 100 to 2000 K/m, the solute concentration is high, which decreases with increasing G values, i.e. $C_{G=1000} < C_{G=100}$. At large distances, profiles from all G values attain the alloy composition of 7%. The solute concentration at any point was translated to an equilibrium liquidus temperature, T_E, at that location. Similar to equation 4, the ΔT_{CS} related to the actual temperature can be obtained by subtracting the location dependent thermal gradient $T_A(x)$ from the CS governed equilibrium liquidus temperature at a given location. This is shown in figure 3b for G = 5000 K/m at t = 50 s. T_E, T_A and ΔT_{CS} are shown in the plot. The solute concentration is denoted as C_E-C_0. These operations were carried out for all the numerical simulation results.

At G = 5000 K/m the concentration ahead of the interface is low. Figure 4b shows the time varying grain growth for G = 5000 K/m. It shows that at t = 50 s grain growth is in its very initial stages and, hence, the interfacial solute concentration is much lower than that for other G values. On the other hand, at G = 0, the concentration is very high adjacent to the S-L interface which then exponentially decreases. Figure 4c shows the growth of a grain at t = 50 s with increasing G from left to right. At G = 0 the growth is much more advanced than at higher thermal gradients. The reduced growth with increasing thermal gradient can be explained by the higher temperatures close to the interface when G is high. These higher temperatures ahead of the S-L interface also significantly reduce the growth of the dendrite along the $+x'$ direction.

The ΔT_{CS} profile ahead of the growing grain in figure 3b follows that of the analytical model with a convex-up shape. See figure 1 for a schematic representation and figure 2 for analytical results.

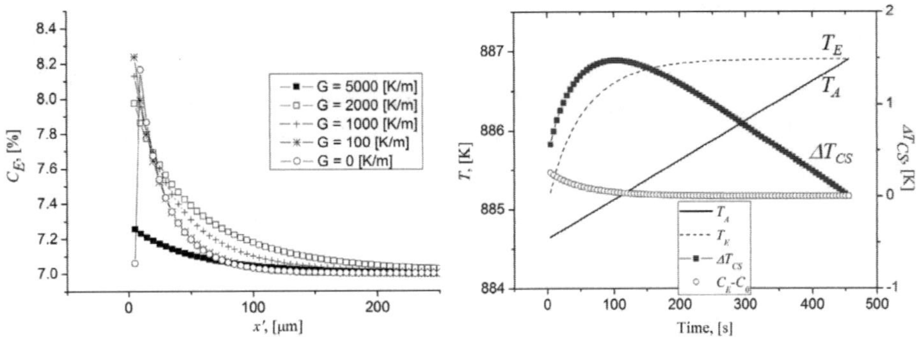

Figure 3. (a) Plot of the solute profile ahead of a growing grain for different G values at t = 50 s. (b) The corresponding ΔT_{CS} over distance x', evaluated as T_E - T_A.

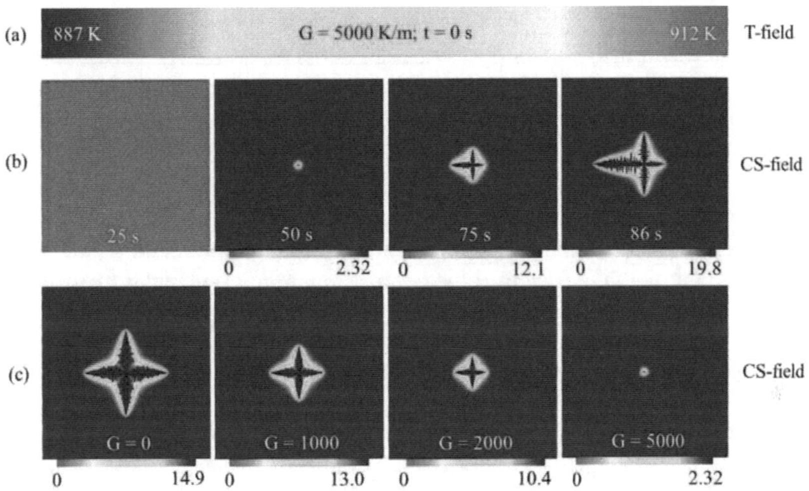

Figure 4. (a) The temperature field at t = 0 s for G = 5000 K/m. (b) Grain growth with increasing time for G = 5000 K/m shown in a CS field. (c) The dendrite size at t = 50 s (also in CS-field) for increasing values of G from left to right. As G increases the time required for a given amount of grain growth increases. In both figures (b) and (c), the higher temperatures on the right hand side reduce the amount of growth in the x' direction. The scale bar shows the ΔT_{CS} values ahead of the growing grains. As t increases or G decreases ΔT_{CSmax} increases.

4. Discussion and conclusions

x'_{CSmax} calculated from the numerical simulations as explained above is plotted for different G values in figure 5a. This is done for two times, t = 40 s and 50 s. As G increases, x'_{CSmax} decreases. At G = 5000 K/m, the result for t = 40 s does not show any value as no solidification had commenced by that time. At 2000 K/m the x'_{CSmax} values for the two times are almost the same. At lower G values (1000 K/m or lower), there is already significant growth by t = 40 s (cf figure 4 which shows lower G values produce larger dendrites).

Comparison between the analytical and numerical results is presented in Figure 5b. It shows that the general trend of the analytical and the numerical models are similar with good agreement between the two at high G values. With increasing G the x'_{CSmax} corresponding to ΔT_{CSmax} decreases exponentially. Moreover, the numerical model shows a decrease in ΔT_{CS} with increasing *G*. Thus, the conceptual representation in figure 1 is confirmed by both approaches. The analytical solution may overestimate x'_{CSmax} particularly at low *G* values, possibly due to the assumption of a constant interfacial temperature T^*. This effect is confirmed by plotting the data for a lower difference between $T_{liq.}$- T^* (i.e. 2K in equation 5) in figure 5b, which shows a better agreement with the numerical solution compared to $T_{liq.}$-T^* = 5 K. Moreover, with grain growth, the S-L interface moves and its position changes with respect to x'. Thus, further development needs to calculate changes in T^* rather than assuming it to be a constant as is done here.

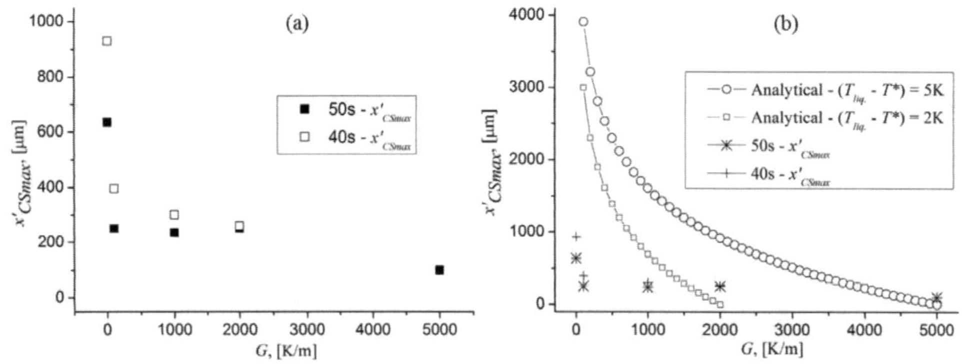

Figure 5. (a) x'_{CSmax} corresponding to ΔT_{CSmax} at t = 40s and 50s from the numerical model, evaluated as T_E - T_A. (b) Comparison of analytical and numerical models. The general trend is similar although the analytical model shows much higher values of x'_{CSmax}.

The decrease in x'_{CSZ} and x'_{CSmax} represents a decrease in the size of the *CSZ* with increase in the thermal gradient. This is due to a greater increase in the temperature ahead of the growing interface. Moreover, a very steep gradient (see T_{A5} in figure 1) may result in no nucleation ahead of the interface. This is seen as $x'_{CSmax} \rightarrow 0$ in the analytical solution in figure 2b. On the other hand, the results from the numerical model in figure 4 show that the degree of supercooling, ΔT_{CS}, may also be a function of the time of solidification. Depending upon the amount of grain growth, x'_{CSmax} may not be fully developed (e.g. *G* = 5000 K/m, *t* = 50s) and, therefore, the supercooling achieved may be insufficient. Indeed, the analysis has shown that an increase in the amount of grain growth results in a proportional increase in x'_{CSmax}. The implication of this is that the potent particle distribution may assume a critical role in the subsequent nucleation events. From qualitative analysis, in general, a system with a steeper gradient with fully developed growing equiaxed grains may require a higher density of potent particles to trigger sufficient nucleation for a fine equiaxed grain structure.

Acknowledgements
DStJ and AP acknowledge the support of ARC Discovery Grant DP120101672 and the Centre for Advanced Materials Processing and Manufacturing (AMPAM) at UQ. PDL, LY acknowledge the Manchester X-ray Imaging Facility and the Research Complex at Harwell, funded in part by the EPSRC (EP/I02249X/1). Financial support from the ExoMet Project, which is co-funded by the European Commission in the 7th Framework Programme (contract FP7-NMP3-LA-2012-280421), by the European Space Agency and by the individual partner organisations is also acknowledged.

References
[1] Easton M A and StJohn D H *Acta Materialia* 2001 **49** 1867
[2] Easton M and StJohn D *Metallurgical and Materials Transactions* A 2005 **36A** 1911
[3] Qian M, Cao P, Easton M A, McDonald S D and StJohn D H *Acta Materialia* 2010 **58** 3262
[4] StJohn D H, Qian M, Easton M A and Cao P *Acta Materialia* 2011 **59** 4907
[5] Wang W, Lee P D and McLean M *Acta Materialia* 2003 **51** 2971

[6] Yuan L and Lee P D *Modelling and Simulation in Materials Science and Engineering* 2010 **18**
[7] Dong H B and Lee P D *Acta Materialia* 2005 **53** 659
[8] Badillo A and Beckermann C *Acta Materialia* 2006 **54** 2015
[9] Prasad A, Yuan L, Lee P D and StJohn D H *Acta Materialia* 2013 **61** 5914
[10] Atwood R C and Lee P D *Acta Materialia* 2003 **51** 5447
[11] Atwood R C, Sridhar S and Lee P D *Scripta Materialia* 1999 **41** 1255

IOP Conf. Series: Materials Science and Engineering **117** (2016) 012002 doi:10.1088/1757-899X/117/1/012002

The interface of heterogeneous nucleation on single crystal substrates

L Yang, M Xia[a] and J Li
School of Materials Science and Engineering, Shanghai Jiao Tong University, China

E-mail: [a] mingxu.xia@sjtu.edu.cn

Abstract. Under controlled nucleation process was achieved by solidifying a high purity Al droplet on a single crystal Al_2O_3 substrate in a high vacuum chamber. The following X-Ray Diffraction (XRD) analysis and measured undercooling prove that the nucleation was triggered by the substrate. Various lattice mismatches between new crystal and substrate (C/S) were obtained through this approach. Combining XRD patterns and high resolution transmission electron microscope analysis we found that the morphology of interface was affected by lattice misfit. An epitaxial layer was found at C/S interface with larger lattice misfit, as in Al(100)//Al_2O_3(0001) system. Further experiments on introduced alloying element, Sb, into liquid Al shows a suppressed epitaxial layer of Al. Chemical reaction between liquid and substrate also contributes to the formation of the interface. The nucleation of Al on the MgO substrates was actually nucleated on $MgAl_2O_4$, chemical reaction product of Al and MgO, rather than MgO.

1. Introduction

Nucleation has been focused on for about 200 years since the pioneering investigation of Fahrenheit [1] on the solidification tendency of supercooled water. However, due to the limitation on time resolution of examine approaches [2], there was no direct observation results on nucleation process so far until the study on the nucleation of protein apoferritin [3]. A few relevant studies on the ordering of liquid structure adjacent to liquid-solid (L/S) interface were believed to be able to disclose how the nucleation starts. For instance, S.H. Oh et al. [4] provided the evidence for the ordering of liquid Al atoms adjacent to L/S interface by knocking oxygen atoms out of α-Al_2O_3 wafer. Lee and Kim [5] proved the ordering structure adjacent to Al_2O_3 contains oxygen rather than being a pure Al layer. Gandman et al. [6] pioneered the research to the ordering degree of the layer which varies as a function of adjacent crystal planes. However, even though the ordering structure was proved to be adjacent to S/L interface, there is no solid evidence for nucleation starting from the ordering structure.

In most of the cases, the studies on nucleation focus on its kinetics, thermodynamics and even numeric or analogue simulations [7]. P. Schumacher et al. [8,9] investigated the crystallization of amorphous Al alloy on TiB_2 substrates in order to demonstrate the heterogeneous nucleation of Al alloy. They found a layer of Al_3Ti at Al/TiB_2 interface. Based on their results, it was well recognized now that the heterogeneous nucleation of Al catalyzed by TiB_2 is through the formation of Al_3Ti layer. But high resolution transmission electron microscopy (HRTEM) analysis shows there are only one or two layers of Ti-contained structure adjacent to the interface. Thus, it is hardly to identify atomic structure of the layer. Fan [10] proposed an hypothesis assuming that heterogeneous nucleation of liquid metal initiates from epitaxial growth, where astrained layer will develop before new crystal forms. The hypothesis is in line with previous L/S interface results. The strained layer takes atomic structure from substrate, called pseudomorphic solid (PS) layer. The strain energy of PS layer is closely related to the lattice misfit between new crystal and substrate (C/S) and will be released through the development of misfit dislocations or planar defects. After the strain energy released, the strained layer normally vanishes. So there is no evidence for the existence of PS layer based on the hypothesis.

Since the strain energy of epitaxial layer is related to the lattice misfit of C/S interface, we will enlarge the lattice misfit deliberately in this study so that the strained layer can be reserved after nucleation given the strain energy large enough to overcome the energy released by misfit dislocation. Alloying element effect and chemical reaction effect are also concerned as factors affecting nucleation interface in this study.

2. Experimental

In order to focus on lattice misfit effect on nucleation, the substrates with different crystal planes (0001), (10-10), (11-20), (1-102) are cut from one single crystal α-Al_2O_3. High purity Al (99.999wt%) was purified using glass fluxing method to remove potential heterogeneous nuclei from liquid Al. Intrinsic undercooling of pure Al was examined on levitation apparatus in a high vacuum chamber to reduce the interference of the container, which is in turn to be compared with the undercooling of liquid Al solidified on substrates. By the comparison, we can suggest whether the nucleation is triggered by the substrates or not. X-ray diffraction (XRD) analysis was employed to detect the nucleation surface of new crystal from the bottom of the sample where nucleation was triggered by the substrates. The lattice misfit calculation follows Bramfitt equation [11]. Parameters of new crystal for calculation were taken from the plane with preferred orientation. The details can be referred to ref.[12]. Furthermore, alloying element, Sb, was added into liquid Al with 1wt% to examine the alloying element effect on the nucleation interface. Single crystal MgO were used to replace Al_2O_3 substrates to demonstrate chemical reaction effect.

Nucleation behaviour of the liquid metal was detected through high speed camera. Undercooling was recorded using a pyrometer with an accuracy of ± 1 K and a time resolution of 50 μs. The thin foils for transmission electronic microscopy (TEM) and HRTEM were prepared from cut slices perpendicular to C/S interface with a thickness less than 80 μm before ion beam thinned using a Gatan precision ion polishing system (PIPS) at 5.0 kV and an incident angle of 4-6°. A Tecnai G2 F20 S-twin TEM instrument was used for TEM, HRTEM and energy dispersive X-ray spectroscopy (EDS) analyses with an accelerating voltage of 200 kV.

3. Results

3.1. Substrate interfered heterogeneous nucleation

Figure 1 shows typical undercooling measured from levitation experiment and from the sample solidified on single crystal substrates. It reveals that the undercooling of liquid Al solidified on the substrate was smaller than the levitation sample. The latter has a 44.4 °C after glass fluxing treatment, while the sample solidified on single crystal substrate only has 23.8 °C undercooling. The difference between these two experiments indicates that the substrate interfered the nucleation of liquid Al. Furthermore, XRD patterns of the samples solidified on different substrates also prove solidification was affected by the substrate (as shown in figure 2). The preferred orientation detected on nucleation surface varies with the exposed plane of the substrate.

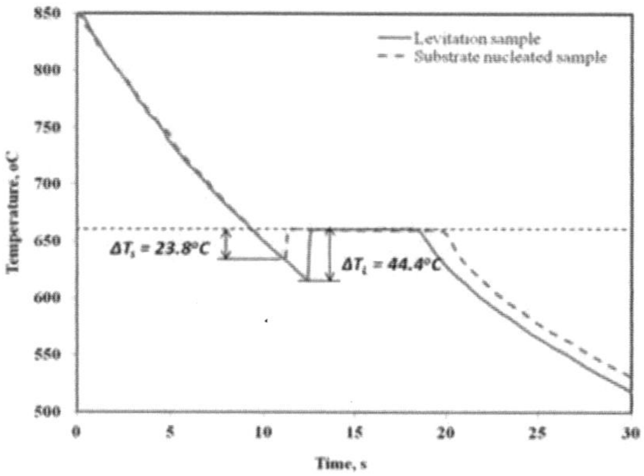

Figure 1. Cooling curves of levitation sample and the sample solidified on the substrate showing the nucleation of liquid Al was affected by the substrate.

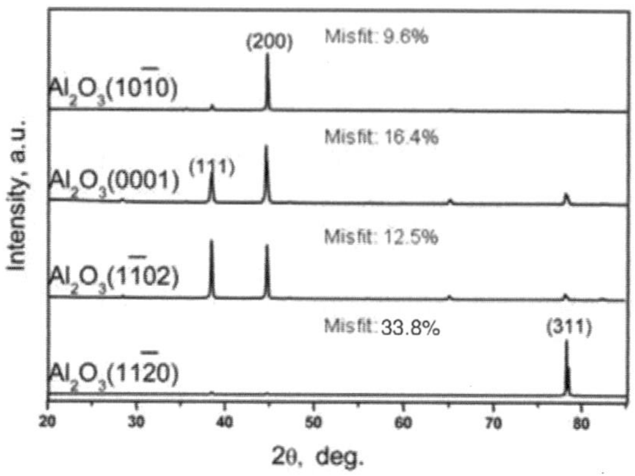

Figure 2. XRD patterns of the nucleation surface of the sample solidified on different substrates. The relevant lattice misfits obtained from XRD results are list as well.

3.2. Interface with various lattice misfits

Due to the substrates effect on the preferred orientation of nucleation surface, parameters for lattice misfit calculation were selected from the crystal plane with preferred orientation. The calculated misfits are listed in figure 2. The interface between Al crystal and Al_2O_3 substrates with different crystal planes were presented in figure 3. For the samples with larger lattice misfit, for instance Al(100) solidified on Al_2O_3(0001), there is a dark plate growing from the interface towards Al side while for the sample with smaller lattice misfit, there is only a distinctive interface without any perturbation. Misfit also affects the shape of the plate. For the interface with larger misfit, the plate is wider and flat while with smaller misfit, the plate is thicker but slightly narrow like a platform standing on the interface. HRTEM analysis on the plate shows that there is large amount of planar defects in the plate (figure 4). Select area electron diffraction (SAED) at the interface shows both Al_2O_3 pattern and an unknown phase as indexed. Combining both atomic lattice image and SAED patterns in figure 4, the unknown phase could be a micro-twin structure. EDS result proves the plate is composed of pure Al.

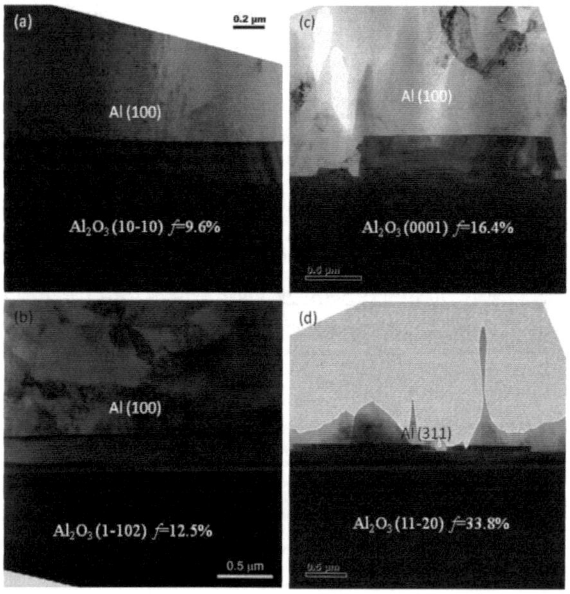

Figure 3. The C/S interfaces with different lattice misfit (a) Al(100) // Al_2O_3(10-10), (b) Al(100) // Al_2O_3(1-102), (c) Al(100) // Al_2O_3(0001), (d) Al(311) // Al_2O_3(11-20) showing the larger lattice misfit will produce epitaxial layer.

IOP Conf. Series: Materials Science and Engineering **117** (2016) 012002 doi:10.1088/1757-899X/117/1/012002

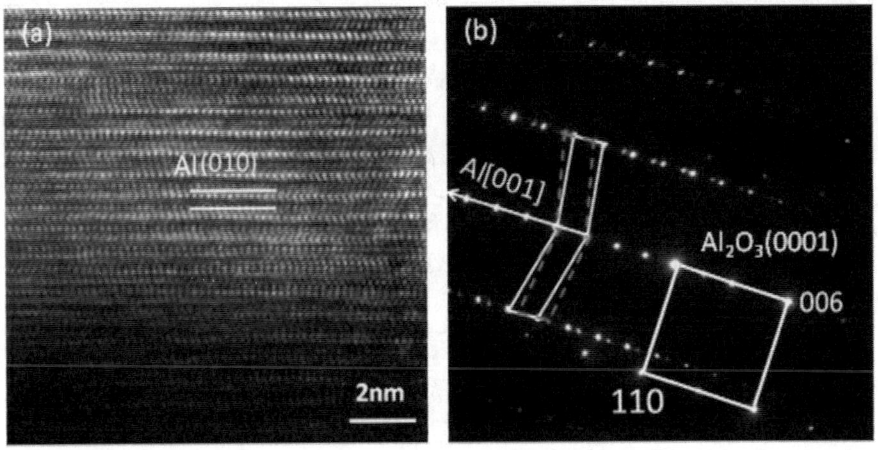

Figure 4. (a) HRTEM image and (b) SAED patterns of epitaxial layer show large amount of planar defects in the layer, which is composed of Al micro-twins.

3.3. Interface with alloying element

1wt% Sb was added into pure Al to investigate the alloying element effect on the nucleation behaviour of Al alloy. XRD patterns of the nucleation surface were presented in figure 5. Comparing with XRD pattern of pure Al, the peaks changed after adding 1wt% Sb. Al (311) peak detected on Al_2O_3 (11-20) substrate for pure Al was replaced by Al (111) and Al (220) with a few AlSb peaks accompanying.

But on Al_2O_3 (0001) substrate, the strongest peak was altered to Al (220) only comparing with (111) and (200) without Sb addition. Thus, the alloying element does modify nucleation behaviour of liquid Al. Apart from the preferred orientation of nucleation surface, the epitaxial layer adjacent to the C/S interface was also suppressed as shown in figure 6.

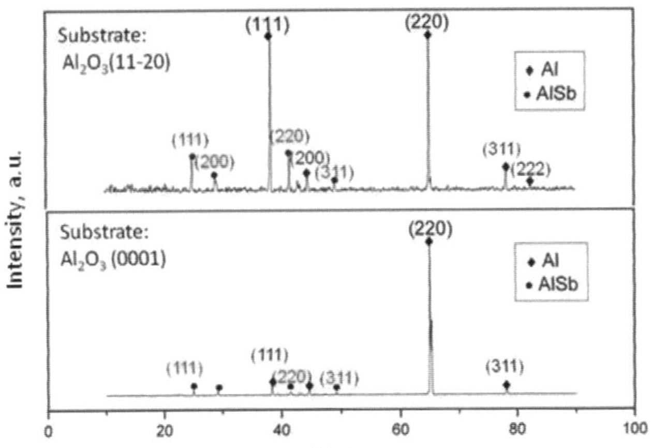

Figure 5. XRD patterns of Al-1wt%Sb nucleated on Al_2O_3 (11-20) and Al_2O_3 (0001) substrates

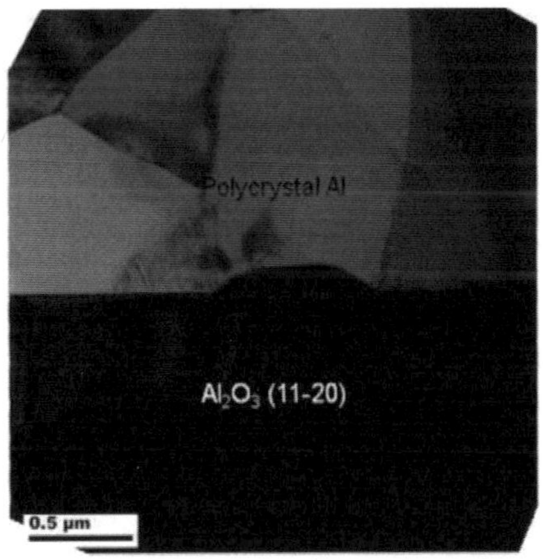

Figure 6. TEM picture of the Al/Al_2O_3 interface showing a suppressed perturbation occurs adjacent to the interface and Al side shows typical poly-crystal feature.

3.4. Interface with chemical reaction

Due to the chemical reactivity of liquid Al on MgO substrate, single crystal MgO was selected as substrates to investigate the effect of chemical reaction on heterogeneous nucleation. The preferred orientation of nucleation surface is presented in figure 7. Apart from an additional Al (220) peak on MgO (111) substrate, Al (111) peak was found as a preferred orientation in all three samples. The relevant misfit for each sample is marked in figure 7. Undercooling on different substrates are relatively equal within the margin of error (figure 8). It does suggest that "misfit effect" may not affect the nucleation of Al on MgO. Further investigation on C/S interface found a Zigzag interface replacing the original interface between Al and MgO as shown in figure 9. HRTEM and SAED pattern analysis prove this zigzag layer is $MgAl_2O_4$ with (111) plane exposed to Al. Thus, whichever the single crystal MgO substrate was selected in the study, the nucleation only occurs on $MgAl_2O_4$ substrate with (111) exposed plane, which subsequently lead to an average undercooling in figure 8.

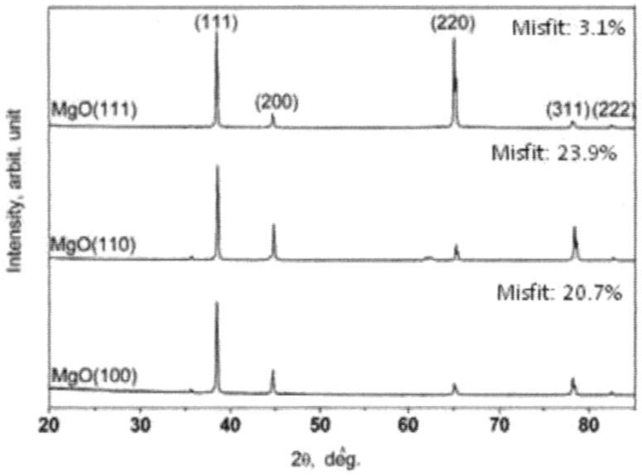

Figure 7. XRD patterns of nucleation surface of pure Al nucleated on different MgO substrates.

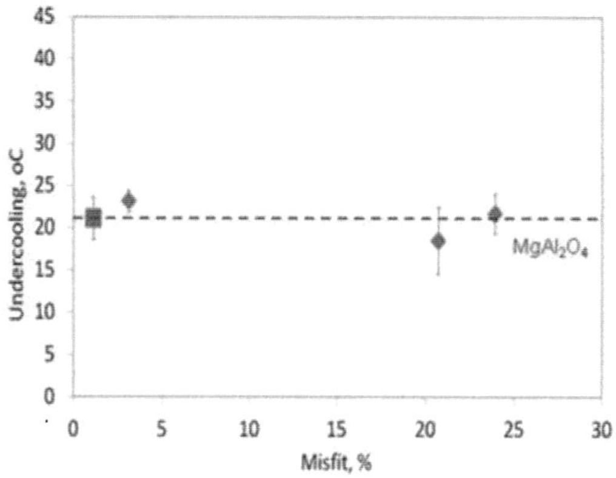

Figure 8. Undercooling of Al nucleated on different MgO substrates showing lattice misfit Al/MgO does not affect undercooling of Al, which actually affected by $MgAl_2O_4$ substrate.

4. Discussion

4.1. Formation of epitaxial layer

The dissipation of strain energy in PS layer is the key factor affecting whether PS layer vanishing or not after nucleation. In this study, large lattice misfits (16.4% and 33.8%) were deliberately created by introducing single crystal substrate Al_2O_3 into Al nucleation process. The expected strain energy is no less than 574 Jm^{-2} for Al_2O_3 (11-20) substrate, which can be calculated according to [13] while based on dislocation theory [14]. The maximum energy that can be dissipated through introduced misfit dislocation is only about 0.964 Jm^{-2}, much less than the energy need to be dissipated. The rest of strain energy will be reserved in the layer and forming planar defects, as micro-twins in this study. For lower lattice misfit system, the strain energy is much less and will be able to dissipate through introducing lattice dislocation, as predicted by Fan [10], et al.

4.2. Type of heterogeneous nucleation interface

However, due to the interference of the interface reaction, the morphology of the interface also varies. With low lattice misfit, the interface of C/S is distinctively clean, without any perturbation. Further enlarging misfit of C/S, perturbation will be found due to the undissipated strained atomic layer, as so called epitaxial layer in this study. Lattice misfit also affect the morphology of epitaxial layer as shown in figure 3 c and d. Apart from lattice misfit, alloying element and chemical reactive substrate also take some effect on heterogeneous nucleation interface. Alloying element affects the preferred orientation of new crystals. But due to the new formed intermetallics, perturbation or epitaxial layer can be suppressed as shown in figure 6. In addition, chemical reaction changes the morphology of the interface. For instance, MgO reacted with Al forming $MgAl_2O_4$ layer between Al and MgO which replaced MgO as new substrate. In general, there are 4 typical interfaces presented in this study:

- Interface without perturbation with smaller C/S misfit.
- Interface with epitaxial layer with larger C/S misfit.
- Interface with suppressed perturbation after adding alloying element.
- Eroded interface by chemical reaction.

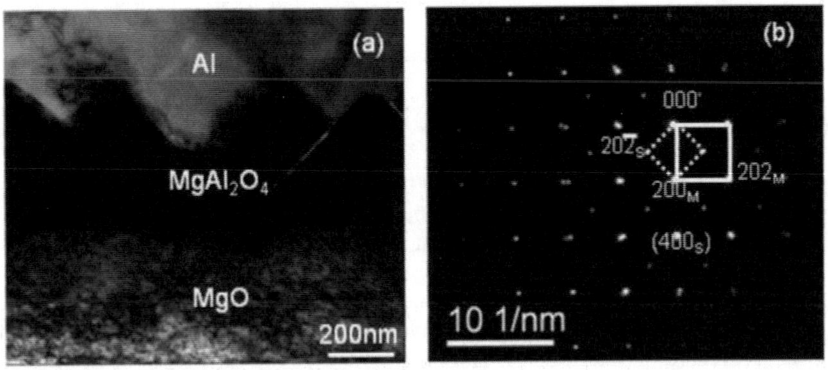

Figure 9. (a) TEM image of Al nucleated on MgO substrate showing a zigzag interface produced by chemical reaction. (b) Indexed SAED patterns of the layer show the pattern presenting $MgAl_2O_4$ feature.

5. Conclusions

Various heterogeneous nucleation interfaces were demonstrated in this study. The effect of C/S misfit on nucleation interface was investigated through a controlled nucleation process, which proves epitaxial layer can be reserved after solidification by using substrate with larger lattice misfit with new crystal. After adding alloying element, epitaxial growth will be suppressed since the introduced alloy element may change preferred orientation of the new crystals. Chemical reaction is another factor playing an important role on the interface of heterogeneous nucleation. In Al-MgO system, chemical reaction product, $MgAl_2O_4$ replaced MgO as new substrate.

Acknowledgement
This work is funded by NSFC (No.51174134), Shanghai STC (No. 11JC1405900) and NBRPC (No.2011CB012900). Authors would also like to thank the useful discussions and valuable advice given from Prof. Fan, Dr. Y. Wang and Dr. W. Yang at BCAST, Brunel University, UK.

References

[1] Fahrenheit D G 1724 *Phil. Trans. Roy. Soc.* **39** 78
[2] Ziese F, Maret G and Gasser U 2013 *J. Phys.: Condens. Matter.* **25** 375105
[3] Tau S –T and Vekilov P G 2000 *Nature* **406** 494
[4] Oh SH, Kauffmann Y, Scheu, Kaplan W D and Ruhle M 2005 *Science* **310** 661
[5] Lee S B and Kim YM 2011 *Acta Mater.* **59** 1383
[6] Gandman M, Kauffmann Y, Koch CT and Kaplan WD 2013 *Phys. Rev. Lett.* **110** 086106
[7] Kelton K F and Greer A L 2010 *Nucleation in condensed matter-Applications in materials and biology* (Amsterdam: Elsevier)
[8] Schumacher P and McKay B J 2003 *J. Non-Cryst Solids* **317** 123
[9] Schumacher P and Greer A L 1994 *Mater. Sci. Eng. A* **178** 309

[10] Fan Z 2013 *Metall. Mater. Trans.* **44A** 14091418
[11] Bramfitt B L 1970 *Metall. Trans.* **1** 1987
[12] Yang L, Xia M, Li J G, 2014 *Mater Lett*, Accepted.
[13] Ayers J E 2007 *Heteroepitaxy of semiconductors* (Boca Raton: Taylor & Francis Group)
[14] Hull D, Bacon D J 1984 *Introduction to dislocations, 3rd ed.* (Oxford: Pergamon press)

Validated thermodynamic prediction of AlP and eutectic (Si) solidification sequence in Al-Si cast alloys

S M Liang and R Schmid-Fetzer[a]
Institute of Metallurgy, Clausthal University of Technology, D-38678 Clausthal-Zellerfeld, Germany

Email: [a]schmid-fetzer@tu-clausthal.de

Abstract. The eutectic microstructure in hypoeutectic Al-Si cast alloys is strongly influenced by AlP particles which are potent nuclei for the eutectic (Si) phase. The solidification sequence of AlP and (Si) phases is, thus, crucial for the nucleation of eutectic silicon with marked impact on its morphology. This study presents this interdependence between Si- and P-compositions, relevant for Al-Si cast alloys, on the solidification sequence of AlP and (Si). These data are predicted from a series of thermodynamic calculations. The predictions are based on a self-consistent thermodynamic description of the Al-Si-P ternary alloy system developed recently. They are validated by independent experimental studies on microstructure and undercooling in hypoeutectic Al-Si alloys. A constrained Scheil solidification simulation technique is applied to predict the undercooling under clean heterogeneous nucleation conditions, validated by dedicated experimental observations on entrained droplets. These specific undercooling values may be very large and their quantitative dependence on Si and P content of the Al alloy is presented.

1. Introduction

Chemical modification, with addition of Sr and/or Na, is a common process performed in Al-Si cast alloys to improve mechanical properties, especially tensile elongation. Trace additions of Sr (about a few hundred ppm) to hypoeutectic Al-Si alloys can results in a transformation of the eutectic silicon morphology from a coarse plate-like structure to a well refined fibrous structure [1-3]. However, the inhomogeneous modification with some coarse eutectic silicon can often be observed either in sand cast or permanent mould cast Al-Si alloys after Sr modification as shown in figure 1. A better understanding of the nucleation of Al-Si eutectic is essential for solving this problem. Many dedicated experimental studies have been carried out on the nucleation of Al-Si alloys, and it is generally accepted that the coarse eutectic Si nucleates on AlP particles [4, 5].

In our previous work [6], using a well developed thermodynamic description of the Al-Si-P ternary system, a precipitation map has been developed and the threshold value of P in Al-Si alloys has been determined. The calculated results not only agree with the microstructure and thermal analysis results of casting alloys but also with the dedicated entrained droplet experiments. In this work, we shall illustrate the careful comparison between the entrained droplet experimental and constrained Scheil simulation for Al-Si-P to facilitate understanding of the nucleation of eutectic Si.

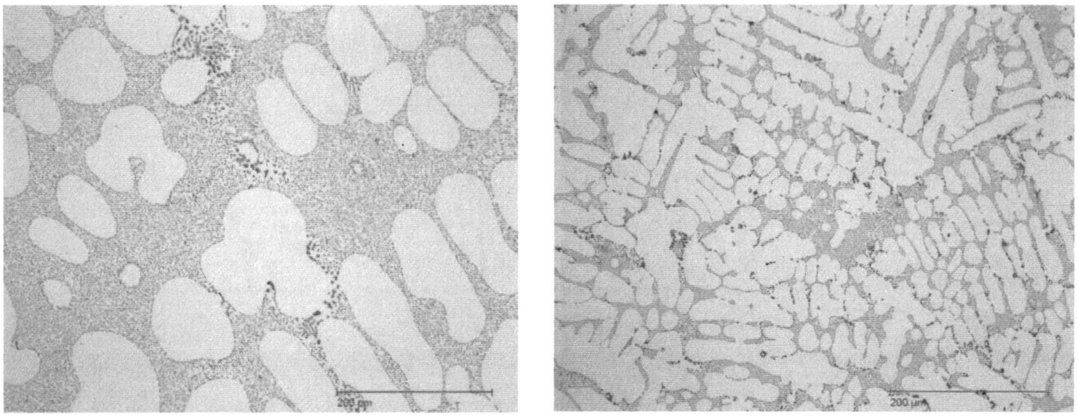

(a) (b)

Figure 1. Inhomogeneous modification of Sr in Al-Si alloys (a) sand cast and (b) permanent mould cast.

2. Entrained droplet experiment and constrained Scheil simulation

Experimental studies of the nucleation of solidification are complicated by the presence of extraneous impurities. An entrained droplet technique, firstly developed by Wang and Smith [7], has proved successful in isolating extraneous impurities and thus achieving clean heterogeneous nucleation conditions. The entrained droplets technique includes two steps: Firstly, an alloy is rapidly solidified by melt spinning to produce a bimodal microstructure consisting of finely dispersed low melting second phase particles embedded in a higher melting point matrix. Subsequently, the alloy is carefully heated to just melt the second phase (droplets) while the matrix remains solid, and then slowly cooled in a thermal analyzer to solidify the entrained droplets. Extraneous impurities are segregated into an insignificant number of droplets, leaving clean nucleation conditions for a sufficiently large number of droplets entrained in the matrix. This technique has been successfully applied to Al-Si alloys [8-13].

A bimodal microstructure of Al-Si alloys can be obtained by melt spinning with some Al-Si eutectic droplets embedded in the Al matrix. The droplet solidification kinetics is studied by differential scanning calorimetry (DSC). The DSC spectrum of the cooling curve of the bimodal microstructure often presents two exothermal events. From the two exothermal onset temperatures, two eutectic nucleation temperatures, T_{N1} and T_{N2}, are determined. Then two values of eutectic nucleation undercooling, ΔT_1 and ΔT_2, are defined as the difference between the Al–Si eutectic equilibrium temperature, 577 °C, and the experimentally observed T_{N1} and T_{N2}, respectively. The first peak refers to the solidification of the melt droplets located at grain boundaries normally with small undercooling, due to the large grain boundary fraction and enrichment of extraneous impurities. The second peak refers to the solidification of entrained droplets, in which the undercooling is largely associated with the nucleation ability of the heterogeneous nucleation sites under clean conditions.

The Scheil-Gulliver model [14], assuming equilibrium in the remaining liquid phase and blocked diffusion in the solidified solid phases, is often used as a simplified simulation of the solidification and phase formation of real cast alloys. During standard Scheil simulation, all phases are allowed to form without consideration of the nucleation barrier of different phases. The solid phase will immediately precipitate from the liquid provided its thermodynamic stability conditions are met. However, the formation of (Si) phase inside the entrained droplets of Al-Si alloys is largely constrained, due to lack of extraneous impurities for the heterogeneous nucleation of eutectic (Si) phase. An advanced "constrained Scheil" simulation is performed here to simulate the solidification of entrained droplets of Al-Si alloys. In constrained Scheil simulation the formation of (Si) phase is suspended based on the experimental fact that very large undercooling, well below the formation of (Si) under standard Scheil conditions, is observed. That is reflected in the constrained Scheil simulation by allowing the atoms of the droplet to distribute into any solid phase except (Si) under the local (metastable) equilibrium at the liquid/solid interface. This simulation continues with growing undercooling below the stable eutectic (Si) temperature until the residual liquid composition in the droplet hits the saturation limit to precipitate another solid phase. In the present case this is AlP, which then triggers the immediate growth of (Si) at large undercooling, thus terminating the constrained Scheil simulation. Therefore, this constrained simulation is useful for the Al-Si-P alloys with P content below the threshold limit so that AlP does not form before the stable eutectic (Si). All thermodynamic calculations in the present work were performed using the integrated software package Pandat (www.computherm.com) [15] with the well developed thermodynamic description of Al-Si-P ternary system [6].

3. Comparison of calculated results with experimental data of Al-Si-P ternary alloys

Ho and Cantor [10] carried out a series of entrained droplets experiments for Al-3Si alloys with different levels of P. Three typical different levels of P are considered in this work for comparison with thermodynamic predictions, namely alloy 1: Al-3% Si-35ppm P; alloy 2: Al-3% Si-2ppm P and alloy 3: Al-3% Si-0.25ppm P. Mass% and ppm = μg/g is used throughout this paper. These three levels of P in the Al-3Si alloys represent three types of solidification sequence, showing different morphology of eutectic Si and undercooling values. The ΔT_1 value is consistently observed at about 3 K in all alloys. This small undercooling is attributed to the droplets at grain boundaries with sufficient nucleation sites, also from impurities enriched in the eutectic melt at grain boundaries. However, the ΔT_2 value, corresponding to the eutectic nucleation in "clean" Al-Si-P droplets in the (Al) matrix, range from about 7–9 K to 60 K depending on the content of P. This dependency can be well explained by the constrained Scheil simulation.

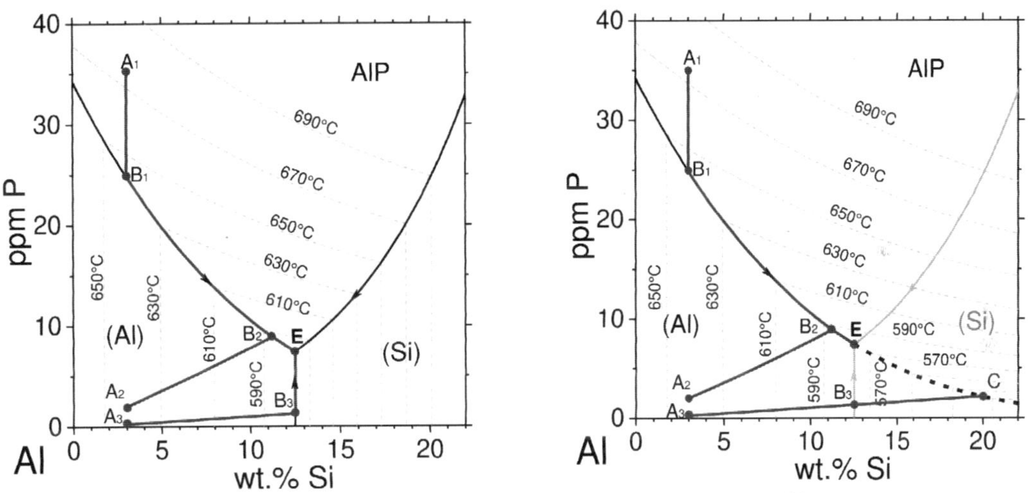

(a) (b)

Figure 2. The solidification paths of the three alloys investigated in the work [10]: (a) standard Scheil solidification; (b) constrained Scheil solidification with suspended (Si) phase.

Figure 2(a) illustrates the liquidus projection of the Al-Si-P ternary system together with the standard Scheil solidification paths of the three alloys in [10]. All three alloys follow the solidification path of A → B → E without considering the nucleation barrier of (Si) phases. Clearly, the solidification sequences of the alloys differ with P content increasing from 0.25 ppm P to 2 ppm P and 35 ppm P. The solidification path of alloy 1 (Al-3Si-35ppm P) is along $A_1 \rightarrow B_1 \rightarrow E$. The AlP phase is the primary phase precipitating before the (Al) phase, and the solidification sequence is L → AlP → AlP + (Al) → AlP + (Al) + (Si). The solidification ends at ternary eutectic L → AlP + (Al) + (Si). The solidification path of alloy 2 (Al-3Si-2ppm P) is along $A_2 \rightarrow B_2 \rightarrow E$, with solidification sequence

L → (Al) → (Al) + AlP → (Al) + AlP + (Si). The AlP phase precipitates after the (Al) phase but before the (Si) phase. The solidification of alloy 3 (Al-3Si-0.25ppm P) follows the path $A_3 \rightarrow B_3 \rightarrow E$, with solidification sequence L → (Al) → (Al) + (Si) → (Al) + (Si) + AlP. The AlP phase can only form at the last stage of solidification at ternary eutectic L → AlP + (Al) + (Si) after the monovariant L → (Al) + (Si) eutectic.

If no effective potential nuclei are available in the melt for (Si) phase, then the formation of (Si) phase would be constrained. Figure 2(b) represents the solidification paths of the alloys under constrained Scheil simulation. In alloy 1 and alloy 2, the AlP phase precipitates before the start solidification of eutectic (Si), and can provide nucleation nuclei for eutectic (Si) phase. Therefore, solidification paths of alloy 1 and alloy 2 under the constrained Scheil simulation are the same as those under the standard Scheil simulation. On the other hand, for alloy 3, with 0.25 ppm P, no AlP precipitates before eutectic (Si) to facilitate the nucleation of Si, thus the formation of (Si) is suspended. The monovariant line L → (Al) + AlP will extend to lower temperature into the region of the suspended (Si). Thus, the solidification path under constrained Scheil simulation of alloy 3 is following $A_3 \rightarrow B_3 \rightarrow C$, rather than $A_3 \rightarrow B_3 \rightarrow E$. At point C, the melt is saturated with AlP which may now precipitate in the undercooled melt.

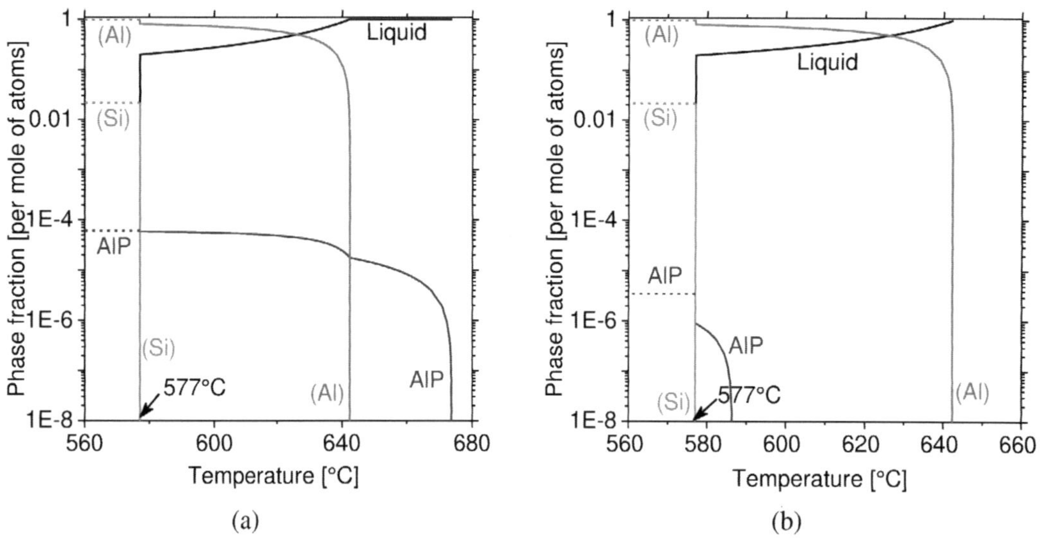

Figure 3. Phase fraction evolution of alloys during Scheil solidification simulation: (a) alloy 1: Al-3% Si-35 ppm P alloy; (b) alloy 2: Al-3% Si-2 ppm P alloy. The AlP phase precipitates before (Al) in Al-3% Si-35 ppm P alloy, and after (Al) but before eutectic (Si) in Al-3% Si-2 ppm P alloy.

Figure 3 represents the phase evolution of alloys 1 and 2 during the standard Scheil simulation, which is identical to the constrained Scheil simulation. The AlP phase precipitates before the (Si) phase and no undercooling of eutectic (Si) can be predicted in these two alloys, which is formally set to zero. The early precipitated AlP phase can provide effective nucleation sites for eutectic (Si), and generate coarse eutectic (Si) morphology. The DSC measured undercooling of alloy 1 and 2 are in the narrow range of 7–9 K, which is due to a kinetic nucleation effect that is beyond the scope of the thermodynamic treatment.

For alloy 3, the solidification sequence under standard Scheil simulation is L → (Al) → (Al) + (Si) → (Al) + (Si) + AlP, shown by the solid lines in figure 4. In the constrained Scheil simulation the (Si) phase is suspended and solidification proceeds along the dashed lines in figure 4 until the formation of

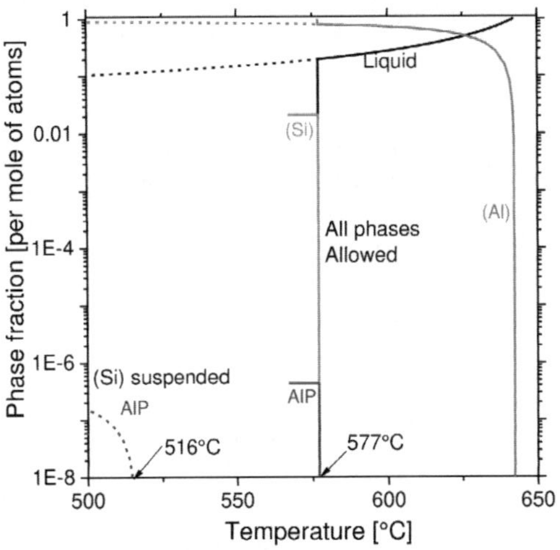

Figure 4. Scheil solidification simulation of alloy Al-3wt.% Si-0.25ppm P; the standard simulation (all phases are allowed to form, solid lines) is complemented by the constrained simulation, suspending the (Si) phase (dashed lines).

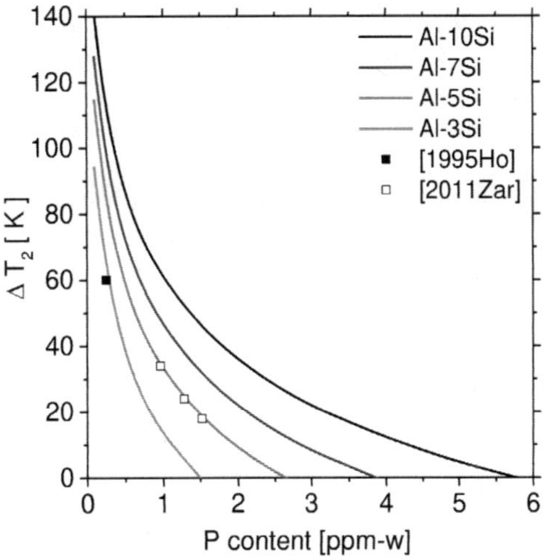

Figure 5. The interdependence of undercooling, ΔT_2, and P content in Al-Si alloys predicted by constrained Scheil simulation. For the experimental data on Al-3Si [10] and Al-5Si alloys [12] the calculation was done as detailed in the text.

Table 1. Nucleation undercooling, ΔT_2, of eutectic (Si) observed in entrained droplet experiments compared with the results predicted by the present thermodynamic calculation. Input values for the calculation are indicated by italic font and output values by bold font.

Alloy (wt.%)	P content (ppm)		ΔT_2 (K)		Exp. Ref
	Exp.	Calc.[(a)]	Exp.	Calc.[(b)]	
Al-5Si (HP)	< 2	**1.0**	34	*34*	[12]
Al-5Si (LP)	< 2	**1.3**	24	*24*	[12]
Al-5Si (HP) -P	< 2+0.5	**1.5**	18	*18*	[12]
Al-5Si (HP) -P	3	*3*	2	**(0)** [(c)]	[13]
Al-5Si (HP) -P	5	*5*	0	**(0)**	[13]
Al-3Si (UHP)	0.25	*0.25*	60	**61**	[10, 11]
Al-3Si (LP)	2	*2*	9	**(0)**	[10, 11]
Al-3Si (UHP) -P	35	*35*	7	**(0)**	[10, 11]
Al-3.58Si (UHP)	<1	**0.3**	63	*63*	[8]
Al-2.90Si (HP)	<1	**0.5**	37	*37*	[8]
Al-3.45Si (LP)	2	*2*	9	**(0)**	[8]

(a) If a fixed experimental P content was reported it was applied as input value (italic font) for the calculation of ΔT_2 (bold font). If a range was reported the calculation was done inversely.
(b) For the inverse calculation the experimental value of ΔT_2 (italic font) is used as input value and the P content (bold font) is calculated.
(c) Zero (0) indicates that AlP forms before (Si) under normal Scheil conditions, therefore the constrained Scheil simulation cannot provide undercooling below the stable eutectic at 577°C. The P content is above the threshold value in these alloys.

AlP in the undercooled melt at 516 °C occurs. This AlP occurrence triggers the formation of eutectic (Si), thus, the predicted undercooling of (Si) is ΔT_2 = 61 K. That value is shown in table 1 as calculated result, marked by bold font, where the given alloy composition used as input value for the calculation is marked by italic font. The analogous calculation (ΔT_2 obtained from given P content) results in the formal value of zero because the P content is above the threshold value and AlP forms before (Si) under normal Scheil conditions as detailed in figure 3. Indeed, a small value of ΔT_2, not higher than 9 K, is experimentally observed. For the other five alloys in table 1 a larger uncertainty is reported for the P content and, therefore, the inverse calculation was performed, P content obtained from ΔT_2. These calculated P contents not only agree reasonably well with the reported experimental range for the alloys. For example, Zarif et al. [12] report a range (upper limit) of the P content in their three Al-5Si alloys. The measured ΔT_2 values for their high purity (HP) — low purity (LP) — and P added alloys are 34, 24, and 18 K, respectively. The calculated P content shows a reasonable trend of 1.0 ppm — 1.3 ppm — 1.5 ppm. More importantly, as clearly stated by [12], the most accurate value is the difference of 0.5 ppm P added deliberately to the Al-5Si (HP) alloy to produce the Al–5Si (HP)–P alloy. The calculated difference is also 0.5 ppm P (exactly 0.56 ppm) and that indicates perfect agreement with the experimental observation. The same realistic trend is also shown for the ultra high purity (UHP) and HP alloys of [8] with calculated values of 0.3 and 0.5 ppm P, respectively. All entrained droplet results are collected in table 1, they all support the present thermodynamic approach. Based on this validated method the relationship between P content and the value of ΔT_2 is calculated for a range of Al-Si alloys with 3, 5, 7, and 10 % Si as shown in figure 5. It is noted that the determination of P content in Al alloys in the ppm range is difficult and associated with some error bar that also applies to the calculated values.

4. Conclusion

Normal and constrained Scheil simulation has been successfully applied to simulate the solidification of Al-Si-P alloys. For P content above a threshold value the solidification sequence switches so that AlP is formed before eutectic silicon, resulting in coarse eutectic morphology and small undercooling.

The P threshold value is strongly dependent on the Si content of hypoeutectic Al-Si alloys and may reach up to 7.4 ppm P for the eutectic Al-12.5Si alloy. The quantitative information on the P content is therefore important and the often provided purity information "below 10 ppm P" is insufficient.

For alloys with purity below the P threshold value very large undercooling may occur under clean nucleation conditions, such as found in entrained droplets. This undercooling, ΔT_2, can be quantitatively predicted by the constrained Scheil simulation approach, indicating that the formation of (Si) may be suppressed until the metastable saturation with AlP is encountered, which subsequently triggers the formation of eutectic silicon on this potent nucleant. The value of ΔT_2 depends strongly on both the P and Si content of the alloy as demonstrated in figure 5.

References

[1] Heusler L, Schneider W 1997 *AFS Transactions* **105** 915
[2] Dahle A *et al.* 2001 *Metall. Mater. Trans. A* **32** 949
[3] Cho Y H *et al.* 2008 *Metall. Mater. Trans. A* **39** 2435
[4] Nogita K *et al.* 2004 *J. Electron Microsc.* **53** 361
[5] Dahle A K *et al.* 2005 *Mater. Sci. Eng. A* **413-414** 243
[6] Liang S-M, Schmid-Fetzer R 2014 *Acta Mater.* **72** 41
[7] Wang C-C, Smith C S 1950 *AIME Trans.* **188** 136
[8] Zhang D L, Cantor B 1993 *Metall. Trans. A* **24** 1195
[9] Cantor B 1994 *Mater. Sci. Eng. A* **178** 225
[10] Ho C R, Cantor B 1995 *Acta Metall. Mater.* **43** 3231
[11] Ho C R, Cantor B 1995 *J. Mater. Sci.* **30** 1912
[12] Zarif M *et al.* 2011 *Metall. Mater. Trans. A* **42** 1684
[13] Li J H *et al.* 2014 *Acta Mater.* **72** 80

[14] Scheil E 1942 *Z. Metallkd.* **34** 70
[15] Cao W *et al.* 2009 *Calphad* **33** 328

Property enhancement by grain refinement of zinc-aluminium foundry alloys

W K Krajewski[1a], A L Greer[2], G Piwowarski[1], P K Krajewski[1]
[1]AGH University of Science and Technology, Faculty of Foundry Engineering, 23 Reymonta Street, 30-059 Krakow, Poland
[2]University of Cambridge, Department of Materials Science & Metallurgy, 27 Charles Babbage Road, Cambridge CB3 0FS, UK

Email: [a]krajwit@agh.edu.pl

Abstract. Development of cast alloys with good mechanical properties and involving less energy consumption during their melting is one of the key demands of today's industry. Zinc foundry alloys of high and medium Al content, i.e. Zn-(15-30) wt.% Al and Zn-(8-12) wt.% Al, can satisfy these requirements. The present paper summarizes the work [1–9] on improving properties of sand-cast ZnAl10 (Zn-10 wt.% Al) and ZnAl25 (Zn-25 wt. % Al) alloys by melt inoculation. Special attention was devoted to improving ductility, whilst preserving high damping properties at the same time. The composition and structural modification of medium- and high-aluminium zinc alloys influence their strength, tribological properties and structural stability. In a series of studies, Zn – (10-12) wt. % Al and Zn – (25-26) wt.% Al - (1-2.5) wt.% Cu alloys have been doped with different levels of added Ti. The melted alloys were inoculated with ZnTi-based refiners and it was observed that the dendritic structure is significantly finer already after addition of 50 - 100 ppm Ti to the melted alloys. The alloy's structure and mechanical properties have been studied using: SEM (scanning electron microscopy), LM (light microscopy), dilatometry, pin-on-disc wear, and tensile strength measurements. Grain refinement leads to significant improvement of ductility in the binary high-aluminium Zn-(25-27) Al alloys while in the medium-aluminium alloys the effect is rather weak. In the ternary alloys Zn-26Al-Cu, replacing a part of Cu with Ti allows dimensional changes to be reduced while preserving good tribological properties. Furthermore, the high initial damping properties were nearly entirely preserved after inoculation. The results obtained allow us to characterize grain refinement of the examined high-aluminium zinc alloys as a promising process leading to the improvement of their properties. At the same time, using low melting ZnTi-based master alloys makes it possible to avoid the excessive melt overheating needed for TiCAl or TiBAl refiners and reduces the possibility of gas pick-up and material loss.

1. Introduction

Grain refinement of non-ferrous cast alloys - mainly those based on Al, is a common practice, which allows fine microstructures to be obtained with increased ductility.

Grain-refinement can be also used for zinc-aluminium alloys with increased Al content of 20-30 wt.% Al [1 – 4]. It should be noted that AlZn-based cast alloys are numbered among the structural materials of improved damping capacity. It is well known that high-aluminium zinc alloys, for example ZA-27 alloy, fall in the category of HiDAlloys (high-damping alloys). It has also been noted that high-zinc aluminium alloys show increased damping. Both groups, i.e. high-aluminium zinc and high-zinc aluminium alloys solidify naturally with a coarse structure and using the refinement process allows highly refined structures to be obtained [5–7]. On the other hand, many structural materials are required to have improved damping properties. However, damping capacity and a fine grain structure are believed to be contradictory factors [6]. The high-aluminium zinc alloys also display naturally low friction, which allows for their use in bushings and bearings. The bearing characteristics of ZA27 alloy are comparable with those of standard industrial bearing alloys, such as tin-lead bronze. This is due to the presence of the bearing ε-$CuZn_4$ phase in a soft α(Al) matrix. The ε-$CuZn_4$ metastable phase takes part in the so-called four-phase reaction to form the stable T'-Al_5Cu_4Zn phase. The ε-$CuZn_4$ phase is present in the structure of high-aluminium zinc for a long time after casting, while the four-phase reaction can lead to an increase of volume of the Zn-Al-Cu alloys, by as much as 4.5% [8]. A partial or total replacement of Cu with Ti, introduced with the Al-12Ti master alloy, leads to the in-situ formation of $Ti(Al,Zn)_3$ particles in the microstructure of Zn-26Al-Cu alloys [5]. It should be also noted that the Zn-Al alloys with increased Al content show structural instability of the solid solution of zinc in aluminium α', which is the main component of their microstructure. The α' solid solution also undergoes phase transformations, which are accompanied by, among other effects, changes in dimension. However, the latter transformation is usually completed in a relatively short time after casting, i.e. in ~1 month [8-9]. The present paper summarizes work on the development of high-aluminium zinc cast alloys on the basis of joint investigations performed over the last ten years at AGH University of Science and Technology – Faculty of Foundry Engineering and at University of Cambridge, UK – Department of Materials Science and Metallurgy.

2. Experimental

The alloys Zn–(25-26)%Al-(1-2.5)%Cu-(0-1.5)%Ti and Zn-(10-12)% Al and master alloys Al-33%Cu, Zn-4%Ti and Al-12%Ti were melted from elements of minimum purity 99.9 % (all compositions in wt.%). Melting was performed in an electric furnace with an argon protective atmosphere, in a clay-graphite crucible of 2-litre capacity (the details of melting and casting are given in Refs [1–3]). Optical light metallography (LM) was performed using a Leica DM IRM microscope. Scanning electron microscopy (SEM) was performed on unetched samples with a Philips XL30 microscope equipped with an energy dispersive X-ray EDX spectrometer (Link-Isis). Wear-resistance pin-on-disc (T01M device, Poland) investigations were performed using sample rods ∅8 x 24 mm. Dry sliding wear tests were performed against a rotating steel disc of 50 HRC, at a load giving 0.8 MPa pressure and at a sliding speed of about 0.7 m/s, for a total sliding distance of 10 km. Dilatometry was performed using samples ∅5 x 35 mm, which were homogenized in air in an annealing furnace at 370°C for 48 h, and then quenched into water at room temperature. During the first 48 hours after quenching, measurements were made using a DI-105 dilatometer; thereafter the length changes of the samples were manually measured using a screw-micrometer of accuracy of 0.001 mm. Damping coefficient tests were carried out using the ultrasonic Krautkramer measurement set, model USLT 2000. The entire study was conducted for 1 MHz longitudinal waves, using the MK1S mini-transducer with head diameter of 10 mm. To determine the attenuation intensity, the echo method (*pulse-echo method*) was adopted, using the internal software of the Krautkramer USLT 2000 device. Prior to testing, the device was calibrated with a pure-zinc cast sample.

3. Results and discussion

3.1. High-Al zinc alloys

Grain refinement of the high-aluminium molten alloys by the Zn-4Ti master alloy is very effective, as can be clearly seen in figure 1 (a) to (f). At the same time elongation increases while tensile strength remains basically preserved (figure 2). It should be noted that the Zn-Ti based master alloys have a mass density very close to the inoculated Zn-Al alloys and that they dissolve very quickly already at temperatures of 500°C. This allows zinc alloys to be treated without the detrimental overheating that is required when using Al-Ti based refiners.

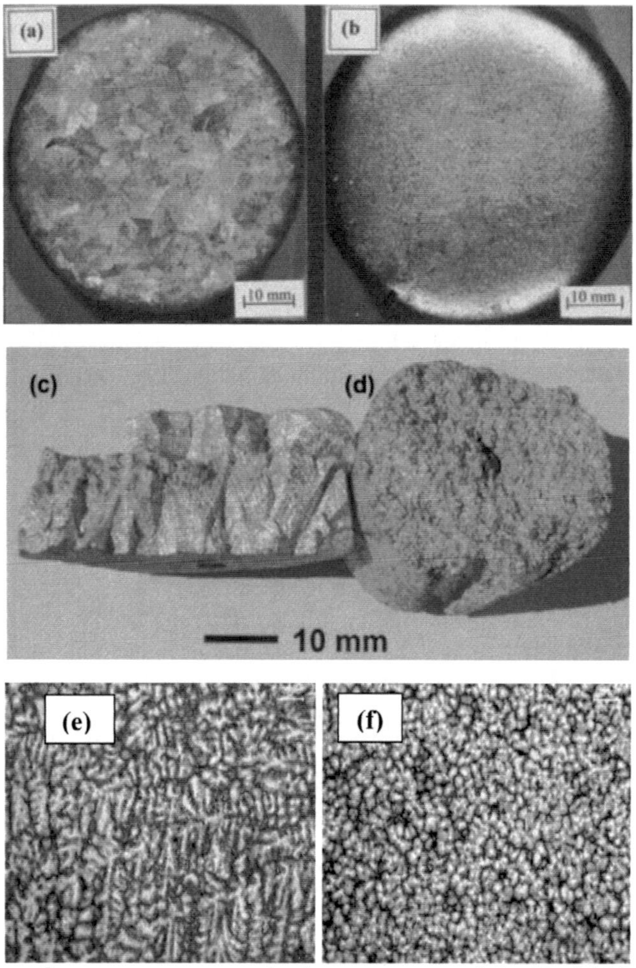

Figure 1. LM pictures of macrostructures and microstructures of the Zn-25Al alloy. (a), (c) and (e) initial alloy; (b), (d) and (f) alloy inoculated with Zn-Ti MA – 0.04 wt.% Ti; (a) and (b) surfaces of the samples; (c) and (d) crystalline fractures; (e) and (f) ground, polished and etched surfaces of a section [3, 6, 10].

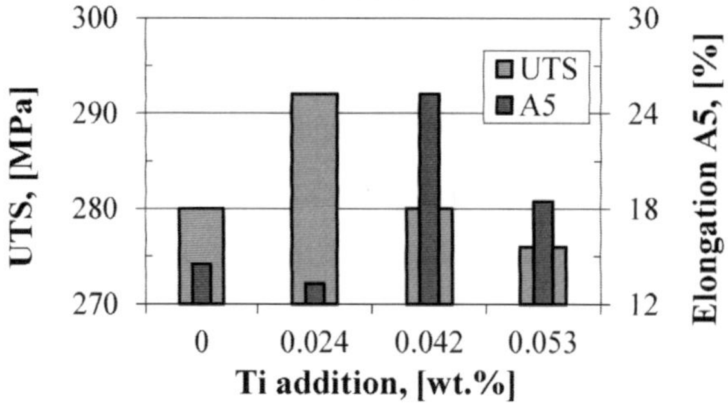

Figure 2. Tensile strength and elongation of the Zn-25Al alloy inoculated with Zn-4Ti master alloy [10].

As noted in [10] the addition of Ti exceeding 250 ppm caused increased elongation. On the other hand the increased Ti addition of 400 ppm caused decreasing UTS. Taking this into consideration it can be concluded that optimal Ti addition should be in a range of 200 – 300 ppm Ti. However, this should be confirmed in detailed examinations.

As mentioned above, adding copper to the high-aluminium alloys leads to good tribological properties on one hand, but also causes detrimental structural instability on the other. Figures 3 and 4 show that a partial replacement of Cu with Ti significantly decreases the structural instability while preserving the good tribological properties.

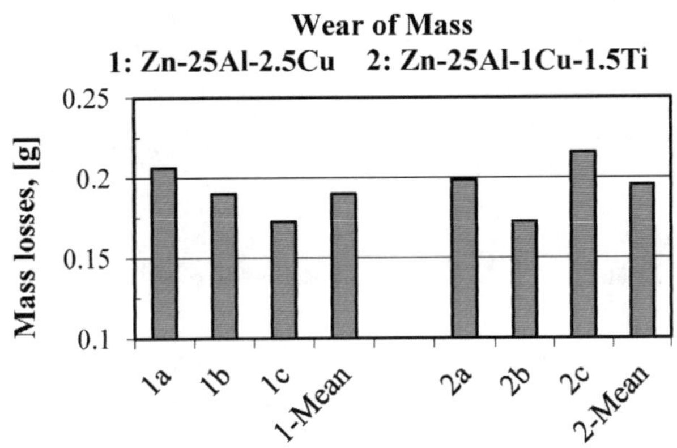

Figure 3. Mass losses of the samples after 10 km distance in the sliding wear test. The samples were aged naturally for 5 years after casting [10, 11].

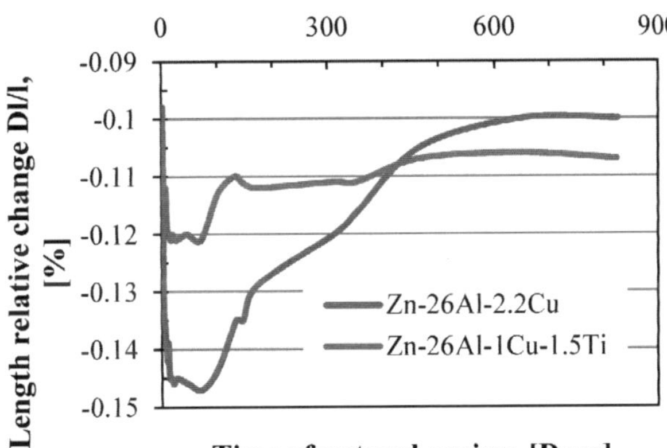

Figure 4. Relative changes of the sample length for the examined alloys during natural ageing after solution treatment and quenching into a supersaturated state.

3.2. Medium-Al zinc alloys

Figures 5–7 show the effect of grain refinement on structure and selected properties of the medium-aluminium zinc alloys represented by Zn-(10-12) wt.% Al binary alloys. During these investigations the alloys were inoculated with the Zn-Ti based master alloys. From figure 5 it appears that the inoculation causes significant refinement of the α(Al) dendrites. However, the structure refinement causes much lower, i.e. ~10%, increase of the ductility measured by elongation – figure 6 in comparison with the high-aluminium alloys elongation changes – figure 2. Further comparing figures 2 and 5 shows that tensile strength of the inoculated alloys is maintained in both the examined high- and medium-aluminum alloys. Figure 7 shows the influence of the performed inoculation (and structural refinement) on the damping properties measured by ultrasonic-wave attenuation. The results collected in Figure 7 show that the attenuation coefficient decreases by about 25% in comparison to the initial, unmodified alloy. However, the inoculated alloys belong to the group of the high-damping alloys, with the attenuation coefficient still being above 100 dB/m.

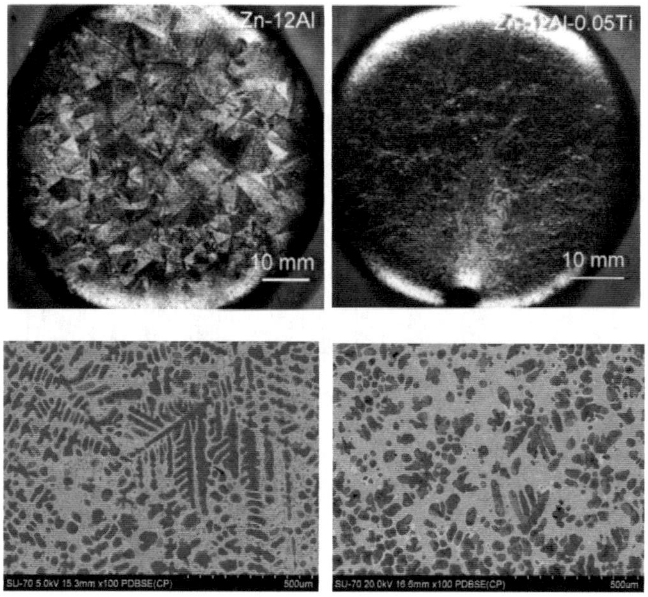

Figure 5. Macrostructures and microstructures of the initial (left-hand side) and inoculated (right-hand side) Zn-12Al-0.05Ti alloy modified with Zn-Ti master alloy [2, 11].

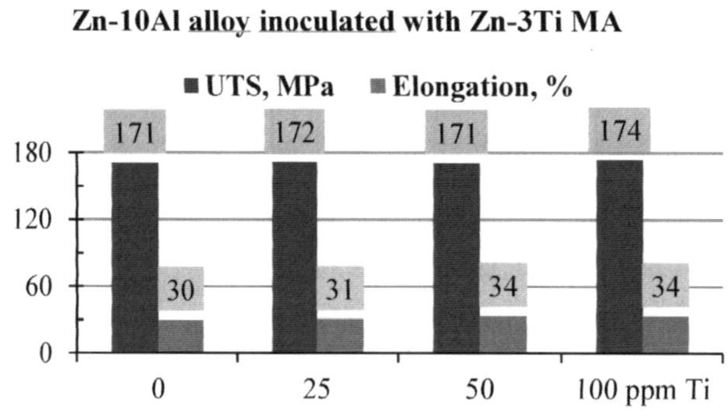

Figure 6. Tensile properties of Zn-10Al alloy inoculated with small Ti addition introduced with Zn-3Ti master alloy.

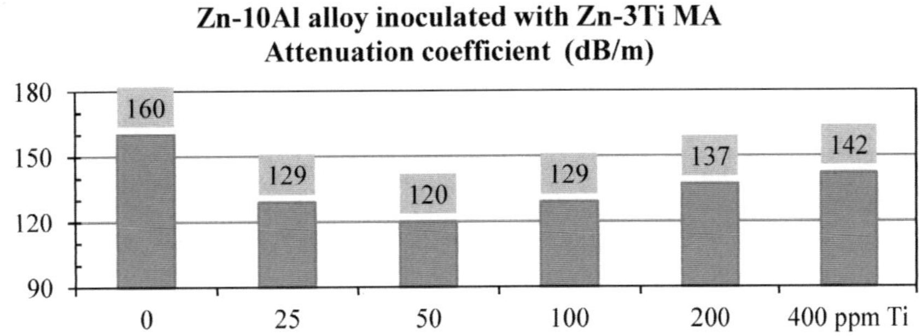

Figure 7. Influence of small Ti addition introduced with Zn-3Ti master alloy on the damping properties of Zn-10Al alloy.

4. Conclusion

Based on the studies described above, the following conclusions can be drawn:

Grain refinement of the examined high-aluminium zinc alloys is a promising process leading to improvement of their properties. At the same time, using the low-melting-point ZnTi-based master alloys avoids the excessive melt overheating needed for the TiCAl or TiBAl refiners and reduces the possibility of gas pick-up and material loss. In the ternary alloys Zn-26Al-Cu, partially replacing Cu with Ti allows the dimensional changes to be reduced while preserving good tribological properties.

Grain refinement of the medium-aluminium binary zinc alloys is less effective, considering increase of ductility measured by elongation changes. It was also noted that grain refinement decreased the attenuation coefficient of 1 MHz ultrasonic waves by about 25%. Taking this into account, it is concluded that further studies of other property changes, e.g. creep and tribological properties, would be desirable.

Acknowledgements

The authors are grateful to the Polish Science National Centre for financial support under grant No. UMO-2012/05/B/ST8/01564 and to the University of Cambridge, Department of Materials Science and Metallurgy for provision of laboratory facilities.

References

[1] W. Krajewski, *Archives of Metallurgy*, **44** (1999), 51-61
[2] W. Krajewski, *Metallurgy and Foundry Engineering MaFE*, **40** (2000), 143-146
[3] W. K. Krajewski, A.L. Greer, T.E. Quested, W. Wolczynski, *Solidification and Crystallization*, chapter 16 (pp.137-149). Euromat 2003 papers. Viley-VCH, 2004, Ed. D.M. Herlach
[4] W.K. Krajewski W.K., A.L. Greer, *Materials Science Forum*, **508** (2008), 281-286
[5] W.K. Krajewski, *Metallurgia Italiana*, No. 11-12 (2006), 27-30
[6] W.K. Krajewski, A.L. Greer, J. Zych, J. Buras, *Foundry Trade Journal*, **180** (2007), No. 3642, April 2007, 97-100
[7] W.K. Krajewski, J. Buras, M. Zurakowski, A.L. Greer, *Archives of Metallurgy and Materials*, **54** (2009), 335-338
[8] W.K. Krajewski, P.L. Zak, J. Orava, A.L. Greer, P.K. Krajewski, *Archives of Foundry Engineering*, **12** (2012) 61-66
[9] W.K. Krajewski, A.L. Greer, J. Orava, P.K. Krajewski, E. Tyrala, *International Foundry Research – Giessereiforschung*, **65** (2013), No. 2, 26-29
[10] W.K. Krajewski, A.L. Greer, P.K. Krajewski, *Archives of Metallurgy and Materials*, **58** (2013), 859-861
[11] W.K. Krajewski, A.L. Greer, P.K. Krajewski, G. Piwowarski, *Proceedings of 71th World Foundry Congress*, May 2014, Bilbao, Spain

IOP Conf. Series: Materials Science and Engineering **117** (2016) 012005 doi:10.1088/1757-899X/117/1/012005

The critical effect of Fe on the grain refinement of aluminium via Al-5Ti-1B addition

Y Zhang[a] and N Ma
School of Materials Science and Engineering, Shanghai Jiao Tong University, 800, Dongchuan Road, Shanghai, 200240, P.R. China

Email: [a]robertzyj@sjtu.edu.cn

Abstract. The influence of Fe on the nucleation potency of TiB_2 particles was investigated by employing grain refinement of high purity aluminium in this study. Experiment results showed that without Fe, high purity aluminium cannot be refined by 0.8wt.% addition of Al-5Ti-1B. However, high purity aluminium containing 0.08wt.% Fe can be refined effectively by 0.2wt.% addition of Al-5Ti-1B, its grain size was about 206μm in diameter. Fine equiaxed grains of about 153μm in diameter can be obtained for high purity aluminium containing 0.08wt.% Fe and 0.006wt.% Ti. Grain refinement mechanism should include nucleation and dendrite remelting and multiplication. Both nucleation and dendrite remelting and multiplication played essential role on the grain refinement of aluminium. The effect of Fe was linked to increase active nuclei by segregation on TiB_2 surface and then to promote nucleation of α-Al. The performance of Ti was to enhance the dendrite remelting and multiplication by forming small conglomeration zones of Ti atoms concentrated around TiB_2 particles.

1. Introduction

Addition of Al-Ti-B mater alloys into molten aluminium has become a standard and effective practice in aluminium foundries worldwide in order to enhance mechanical properties, reduce ingot cracking, improve feeding, improve the homogeneity of the cast structure by obtaining fine, equiaxed grains after solidification. To understand the mechanism of grain refinement and to promote the development of Al-Ti-B refiner, during the last several decades, various theories had been proposed to explain the experimental observations. There is a friendly consensus that the excess Ti and TiB_2 particles are necessary and important for commercial pure aluminium (CPAl) obtaining a good grain refinement result [1-7].

Ti was considered as the strongest element to act as growth restriction factor (GRF). Usually, Q is referred to as the growth restriction factor, and is a measure of the growth-restricting effect of solute elements on the growth of the solid-liquid interface in the absence of solute interactions [3, 8, 9]. It is defined as $mc_0(k-1)$, where m is the slope of the liquidus, C_0 is the concentration of the solute in the alloy, and k is the equilibrium partition coefficient between the solid and liquid at the growing interface. Empirically, the average grain size can be described as a liner function of 1/Q for various Al alloy. The effects of solute content on grain refinement in an isothermal melt studied by Men and Fan [10] indicate that 1/3 power law can describe the experimental data more accurately than a linear relationship.

For excess Ti, another important role is that it can change the interface between TiB_2 and Al by forming $TiAl_3$ interfacial layer. Shumacher and Greer [11-13] revealed that a thin $TiAl_3$ layer about

3nm in thickness was formed between the TiB_2 and the α-Al by observation of $Al_{85}Y_8Ni_5Co_2$ metal glass inoculated with an Al-Ti-B master alloy. Mohanty and Gruzleski [14] reported that the formation of interfacial $TiAl_3$ layer at the TiB_2/Al interface can nucleate α-Al, in which the interface was confirmed by electron probe micro-analytical techniques (EPMA) and the nucleation of α-Al was proved by observation of boride particle within equiaxed grains.

In principle, the mechanism of grain refinement is quite straight forward [14]. When the added Al-Ti-B master alloy dissolves, numerous potent heterogeneous nuclei (TiB_2 with $TiAl_3$ layer) are released and dispersed into the melt, and subsequently a large number of these sites act as active nucleants during solidification [6, 14].

Our previous work testified that 0.2% addition of Al-5Ti-1B cannot refine high purity aluminium (HPAl) effectively [15, 16]. According to the dominant grain refinement mechanism, 0.2% addition of Al-5Ti-1B must refine HPAl effectively. But the experiment results against the prediction of dominant grain refinement mechanism. Therefore, we should reconsider the grain refinement mechanisms and try to explain the new phenomena reasonably. Furthermore, clearly the understanding of grain refinement mechanisms can promote further development of this kind of grain refiner.

For grain refinement, another very important method is to refine aluminium by physical method, including electromagnetic vibration [17-19], and pulse magneto-oscillation technique [20]. The principle of this kind of grain refinement mechanism is to achieve grain multiplication by promoting collapse of dendrite arms, dendrite fragmentation caused by vibration or stirring. Besides, another technique to achieve grain refinement using dendrite multiplication mechanism was made by Wannasin [21], in which gas bubbles were employed to agitate a molten metal during solidification. During this process, the "mother dendrites" created at the cold surfaces are fragmented resulting in grain multiplication. The physical method of grain refinement indicates that the dendrite remelting and multiplication plays an important role in the refinement process.

In the present study, to reveal the essence of grain refinement mechanism of aluminium properly, the effect of Fe, Si and Ti on the grain refinement of HPAl was investigated. According to the experiment results, an improved grain refinement mechanism was proposed.

2. Experiment

Al-5Ti-1B master alloys (LSM, UK) were employed for studying grain refinement of HPAl (>99.99wt.%Al, hereafter, all compositions are denoted in wt%). For studying the behaviour of Fe and Ti on the grain refinement process of HPAl, pure Fe and Al-10Ti were used for the present purpose. TP-1 testing method [22] was employed to evaluate the grain refinement performance of different refiner on aluminium. In each case the melt was maintained at 720±5°C and slow stir was maintained to disperse refiner uniformity and keep the particles suspended in the melt for about 30 seconds with a graphite rod after addition was completed melting. The melt was held for 10 minutes and then cast into TP-1 mould. Each cone is sectioned perpendicular to its axis 38mm from its base for macro-analysis, ground mechanically on 200 grit and 400 grit water-proof adhesive papers and then etched in reagent (45 mL HCl, 15 mL HNO_3, 15 mL HF, 25 mL H_2O) to reveal the macro-structural feature. Normally, the sample for assessment of grain refinement was taken from the center of sample (center grain). For TP-1 sample, the solidification is occurred from the periphery of TP-1 mould to the center. Therefore, to assess the grain refinement properly, the periphery of TP-1 sample (periphery grain) was used to observe its microstructure evolution.

3. Results and discussion

Figure 1 is the grain structure of HPAl, HPAl refined by 0.2% addition of Al-5Ti-1B, and HPAl refined by 0.8% addition of Al-5Ti-1B, respectively. The average grain size is about 1000 μm, 500 μm and 350 μm for without refinement, 0.2% addition and 0.8% addition of refiner of HPAl. Compared to good refinement performance of Al-5Ti-1B refiner to CPAl, obviously, no effective refinement was obtained for HPAl even with 0.8% addition of refiner. In this case, very high excess Ti (0.0224wt% Ti for 0.8% addition of refiner) and a sufficient number of TiB_2 nuclei (four times to 0.2% addition of refiner to HPAl) were added into the melt. The question is why there was no effective grain refinement was obtained, even though there were enough potent nucleation sites and very a high value of Q.

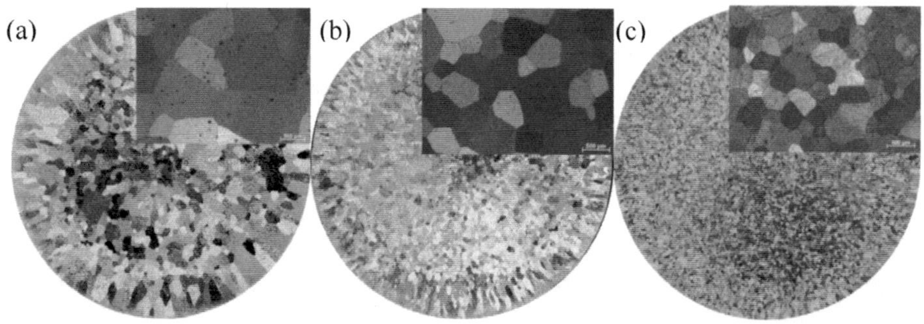

Figure 1. Grain structure of HPAl (inset is center grain structure), (a) without refinement, (b) refined by 0.2% Al-5Ti-1B, (c) refined by 0.8% Al-5Ti-1B.

It should be noticed that the big difference between CPAl and HPAl is that the CPAl contents 0.08% Fe, 0.03%Si and 0.006%Ti [15, 16]. The grain refinement of HPAl by 0.8% addition of refiner indicates that the value of Q caused by solute Ti does not play dominant role on grain refinement process, and excess Ti cannot make TiB_2 particles act as active nucleation sites. There should be another significant premise nucleation mechanism to make TiB_2 particles and solute Ti active.

According to the difference of composition between CPAl and HPAl, the impurities added into HPAl were 0.08% Fe and/or 0.06% Ti. It can be seen from figure 2 that the center grain size of HPAl containing 0.06% Ti refined by 0.2% addition of Al-5Ti-1B was about 450μm and was at the same level to HPAl refined by 0.2% addition of refiner. Also, very coarse column periphery grains were obtained. In this case, the excess Ti within melt was equal to CPAl refined by 0.2% addition of Al-5Ti-1B, its value is 0.0116%. If Q supplied by excess Ti plays important role on crystal growth, better grain refinement should be achieved. This demonstrates that the value of Q has very little effect on the grain refinement of aluminium. Figure 3 presents the grain size of HPAl containing 0.08% Fe refined by 0.2% addition of refiner. Fine equiaxed grains of about 206μm in diameter were obtained at the sample center (figure 3a). It should be noticed that fine column grains at the periphery of TP-1 sample were observed (figure 3b). Compared to the coarse column periphery grains of HPAl refined by 0.2% addition of Al-5Ti-1B (figure 1b), the change of periphery grain structure indicates that Fe can promote nucleation of TiB_2 particles with excess Ti.

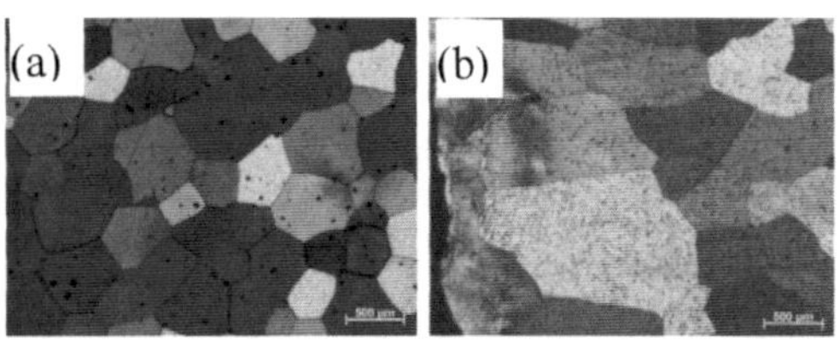

Figure 2. (a) Center grain and (b) periphery grain morphology of HPAl containing 0.006% Ti refined by 0.2% Al-5Ti-1B

Figure 3. (a) Center grain and (b) periphery grain morphology of HPAl containing 0.08% Fe refined by 0.2% Al-5Ti-1B

The experiment results mentioned above show that Fe has strong influence on the grain structure of HPAl after refinement than that of Ti. These testified that Fe plays significant role on the grain refinement of aluminium.

The reciprocal effect of Ti and Fe on the grain refinement of HPAl was shown in figure 4. It can be seen that the periphery column structure disappeared completely (figure 4b) and there were very fine equiaxed grains at the center of about 153μm in diameter (figure 4a). In view of the coarse grain structure for HPAl with 0.2% addition of Al-5Ti-1B, the better grain refinement results of HPAl containing Fe and Ti refined by 0.2% addition of refiner verify that the effect of excess Ti on the grain refinement is to enhance nucleation efficiency by dendrite melting and multiplication other than to promote nucleation potency of TiB_2 particles.

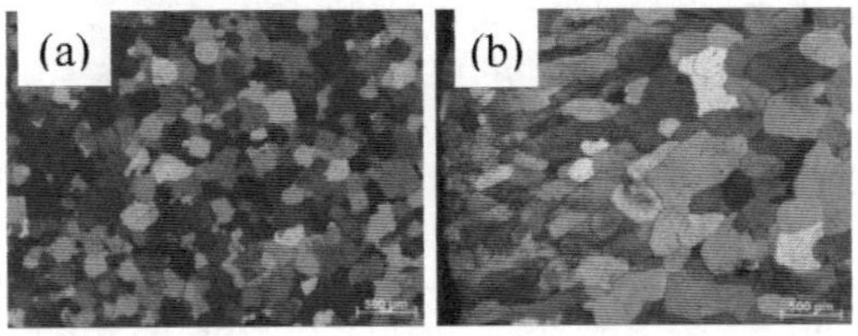

Figure 4. (a) Center grain and (b) periphery grain morphology of HPAl containing 0.08% Fe and 0.006% Ti refined by 0.2%Al-5Ti-1B.

Based on present experiment results, the grain refinement of aluminium should include nucleation, dendrite remelting and multiplication and solute Fe plays a significant role on the nucleation process. McKay's [23] observation on TiB_2/Al interface showed that Fe and Si were very easily segregated on TiB_2/Al interface. Combination present results and McKay's report, the conclusion of the promotion of the interface segregation on the potency of TiB_2 acted as active nucleation site can be drawn. This is why HPAl containing 0.08% Fe can be refined by 0.2% Al-5Ti-1B, while HPAl cannot be refined effectively. The periphery grain structure of HPAl containing 0.08% Fe after refinement was a convinced proof for our speculation (figure 3b).

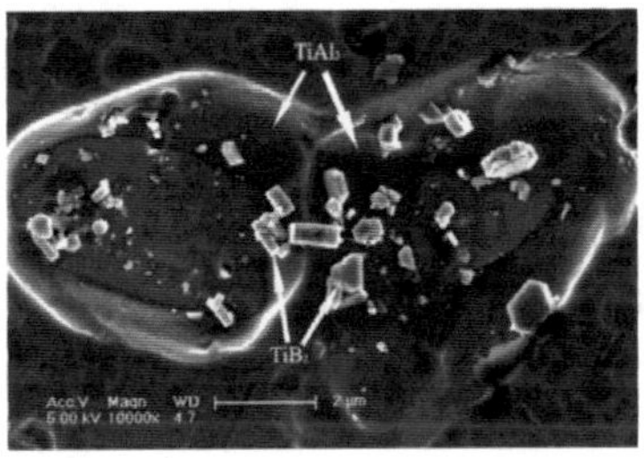

Figure 5. Microstructure of Al-5Ti-1B.

Careful investigation on the microstructure of Al-5Ti-1B grain refiner shows that TiB_2 particles were always found on the surface of $TiAl_3$ phase (figure 5). This was attributed to the good orientation relationship between TiB_2 and $TiAl_3$ [11-13]. Once solidification occurred, TiB_2 can act as the substrate of $TiAl_3$ phase. When refiner was added into melt aluminium, bulk $TiAl_3$ phase was melt quickly [2, 8]. But it does not distributed uniformly within melt due to the restriction effect of TiB_2. As a result, the Ti atoms were concentrated around TiB_2 particles to form many small conglomeration zones. During solidification process of refined aluminium, some of TiB_2 particles were act as active nuclei due to the segregation of Fe, and the other TiB_2 particles carried on Ti atoms zones to promote the dendrite remelting and multiplication. For HPAl refined by Al-5Ti-1B refiner, there are not enough active TiB_2 nuclei due to the absence of Fe segregation. This is why 0.2%, 0.8% addition of Al-5Ti-1B refiner cannot refine HPAl effectively while 0.2% addition of Al-5Ti-1B has good refinement performance on HPAl containing 0.08% Fe. For addition of TiB_2 and excess Ti into aluminium separately, the Ti atoms conglomeration zones (concentration fluctuations) are still exist within melt, but their size is much bigger than that of addition of Al-5Ti-1B, due to the absence of restriction of

TiB_2 particle. Then, there is indeed limited effect on the promotion of dendrite remelting and multiplication. With prolonging holding time or elevating holding temperature of refined aluminium melt, the small conglomeration zones of Ti will be decreased and disappear eventually. As a result, atoms of Ti tend to distribute uniformity within melt. This is the reason that the fading mechanism of grain refinement occurred at long-holding time and high temperature.

Compared to microstructure of refined HPAl with Fe and Ti, the effect of elements on the refinement process was revealed. Fe can promote the nucleation of TiB_2, and Ti can enhance refinement performance by the powerful ability of promoting fine dendrite remelting and multiplication. The different content of Ti on grain refinement performance of refined HPAl shows that there is a critical value for conglomeration zone of Ti atoms to promote the fine dendrite remelting.

Using this theory, poisoning effect of Zr, Mn and Cr on the Al-Ti-B refiner can be described reasonably. Normally, Zr, Mn and Cr were considered to form intermetallic compound and then to weaken the grain refinement by decreasing the Q value. In present study, the value of Q was testified that it was not the critical factor to refinement of aluminium. According to present experiment results, even with very high value of Q and plenty of TiB_2 particles, effective grain refinement did still not occur for HPAl. From this point of view, the essence of poisoning was attributed to insufficient active TiB_2 particles deduced by lack of Fe segregation, in which the formation of compound between Zr, Mn, Cr and Fe occurred. If more concentrate of Fe was added into melt, the grain refinement would be recovered. It was reported that the higher iron content can reduce the poisoning effect of zirconium [24]. This is well agreed with grain refinement mechanism proposed in present study.

4. Conclusion

In this investigation, the performance of Fe, Ti and TiB_2 on the grain refinement of aluminium was studied by employing Al-5Ti-1B refiner. According to the experiments results, Fe plays significant role on the grain refinement of aluminium and it can enhance nucleation potency by segregation onto TiB_2 surface. The grain refinement process should include nucleation, dendrite remelting and multiplication. Both of them play an important role on the grain refinement of aluminium. The effect of excess Ti was to form many small conglomeration zones and to promote the dendrite remelting and multiplication.

Acknowledgments

Authors would like to acknowledge Dr Brian McKay at BCAST, Brunel University for his kind discussion and assistance.

References

[1] McCartney D G 1989 *Int. Mater. Rev.* **34** 247
[2] Bäckerud L, Gustafson P and Johnsson M 1991 *Aluminium* **67** 910
[3] Easton M and StJohn D 1999 *Metall. Mater. Trans. A* **30** 1613
[4] Easton M and StJohn D 1999 *Metall. Mater. Trans. A* **30** 1625
[5] Murty B S, Kori S A and Chakraborty M 2002 *Int. Mater. Rev.* **47** 3
[6] Quested T E 2004 *Mater. Sci. Technol.* **20** 1357
[7] Sigworth G K 1996 *Scripta Mater.* **34** 919
[8] Jones G P and Pearson J 1976 *Metall. Trans.* **7B** 223
[9] Easton M A and StJohn D 2001 *Acta Mater.* **49** 1867
[10] Men H and Fan Z 2011 *Acta Mater* 2011 **59** 2704
[11] Schumacher P and Greer A L 1994 *Mater. Sci. Eng. A* **178** 309
[12] Schumacher P and Greer A L 1994 *Mater. Sci. Eng. A* **181-182** 1335
[13] Schumacher P, Greer A L, Worth J and Evans P V 1998 *Mater. Sci. Technol.* **14** 394
[14] Mohanty P S and Gruzleski J E 1995 *Acta Metall. Mater.* **43** 2001.
[15] Zhang Y J, Wang H W, Ma N H and Li XF 2005 *Mater. Lett.* **59** 3398
[16] Zhang Y J, Ma N H, McKay B, Li X F and Wang H W 2012 *Advanced Materials Research.*

538-541 2264
[17] Radjai A and Miwa K 2000 *Metall. Mater. Trans. A* **31** 755
[18] Yoshiki Mizutani YO, Kenji Miwa, Kazuo Yasue, Takuya Tamura and Yasuji Sakaguchi 2004 *Mater. Trans.* **45** 1944
[19] Mizutani Y, Tamura T and Miwa K *Mater. Sci. Eng. A* **413-414** 205
[20] Gong Y Y, Luo J, Jing J X, Xia Z Q and Zhai Q J 2008 *Mater. Sci. Eng. A* **497** 147
[21] Wannasin J, Martinez R A and Flemings M C 2006 *Scripta Mater.* **55** 115
[22] Standard test procedure for aluminium alloy grain refiners (TP-1) 1990 *The Aluminium Association* (Washington, DC)
[23] McKay B 2001 *Heterogeneous nucleation in Al-Si alloys* (UK: University of Oxford)
[24] Bunn A M, Schumacher P, Kearns M A, Boothroyd C B and Greer A L 1999 *Mater. Sci. Technol.* **15** 1115

IOP Conf. Series: Materials Science and Engineering **117** (2016) 012006 doi:10.1088/1757-899X/117/1/012006

Heterogeneous nucleation of entrained eutectic Si in high purity melt spun Al-Si alloys investigated by entrained droplet technique and DSC

J H Li[1a], M Albu[2], T H Ludwig[3], F Hofer[2], L Arnberg[4], P Schumacher[1,5]
[1] Institute of Casting Research, Montanuniversität Leoben, A8700 Leoben, Austria
[2] Institute for Electron Microscopy and Nanoanalysis (FELMI), Graz University of Technology, Center for Electron Microscopy Graz, A-8100 Graz, Austria
[3] Hydro Aluminium, Årdal Metal Plant, N-6882 Øvre Årdal, Norway
[4] Department of Materials Science and Engineering, Norwegian University of Science and Technology, N-7491 Trondheim, Norway
[5] Austrian Foundry Research Institute, A8700 Leoben, Austria

Email: [a]jie-hua.li@hotmail.com

Abstract. Entrained droplet technique and DSC analyses were employed to investigate the influence of trace elements of Sr, Eu and P on the heterogeneous nucleation of entrained eutectic Si in high purity melt spun Al-5wt.% Si alloys. Sr and Eu addition was found to exert negative effects on the nucleation process, while an increased undercooling was observed. This can be attributed to the formation of phosphide compounds having a lower free energy and hence may preferentially form compared to AlP. Only a trace P addition was found to have a profound effect on the nucleation process. The nucleation kinetics is discussed on the basis of the classical nucleation theory and the free growth model, respectively. The estimated AlP patch size was found to be sufficient for the free growth of Si to occur within the droplets, which strongly indicates that the nucleation of Si on an AlP patch or AlP particle is a limiting step for free growth. The maximum nucleation site density within one droplet is directly related to the size distribution of AlP particles or AlP patches for Si nucleation, but is independent of the cooling rates. Although the nucleation conditions were optimized in entrained droplet experiments, the observed mechanisms are also valid at moderate cooling conditions, such as in shape casting.

1. Introduction

Al-Si based alloys are important casting alloys that constitute ~ 90 % of all shape castings [1]. Modification of eutectic Si is of great necessity to improve the mechanical properties [2]. Nevertheless, the modification mechanism, with respect to nucleation and growth of eutectic Si, is still a matter of debate. Generally, not only higher nucleation undercoolings are observed but also growth temperatures are significantly decreased as measured by thermal analysis, suggesting that nucleation is depressed and subsequent growth is also hindered [3].

With respect to the growth, well-accepted twin plane re-entrant edge (TPRE), poisoning of the TPRE [4] or impurity induced twinning (IIT) mechanisms [3] have been reported to be valid under certain conditions. Therefore, the growth of eutectic Si is not a focus of our present investigation. Instead, the main focus is on the nucleation of eutectic Si.

With respect to nucleation, a much more detailed research is required to elucidate the nucleation kinetics of eutectic Si. It is well-known that nucleation is notoriously difficult to be investigated in conventional solidification conditions because of the following facts: (i) the inherent presence of impurities, (ii) the very short nucleation time, and (iii) the very small scale of size. Despite these difficulties, heterogeneous nucleation has been investigated by the novel entrained droplet technique [5], which was first suggested by Wang and Smith [6] to study heterogeneous nucleation. The potential of this technique was recognized and developed further by Cantor and co-workers [5, 7, 8, 9] who employed rapid solidification to produce micrometer to nanometer size droplets, thereby improving the reproducibility of nucleation undercooling by up to 0.2 °C. Ho and Cantor [5] studied high purity Al-Si alloys containing traces of P using the entrained droplet technique and found that just 0.25-2 ppm P is sufficient to form AlP, which could act as a nucleation site for eutectic Si verifying the results by Crosley and Mondolfo [10] and Flood and Hunt [11].

In this paper, the entrained droplet technique and DSC analyses were employed to investigate the influence of trace elements of Sr, Na, Eu and P on the nucleation of entrained eutectic Si in high purity melt spun Al-5wt.% Si based alloys. The nucleation kinetics are discussed on the basis of the classical nucleation theory [12] and the free growth model [13, 14], respectively.

2. Experimental methods

Al-5 Si alloys (wt. %, used through the paper, in case not specified otherwise) with controlled additions of Sr, Eu and P were manufactured. The experimental details about sample preparation and composition measurement and DSC measurement can be found in [15-17].

The DSC results were reproducible within 0.2 °C for several subsequent runs on one sample. In order to elucidate the nucleation kinetics of entrained eutectic Si, a series of DSC experiments with different cooling rates were performed for different Al-5Si based alloys (high purity (5N Al and 5N Si) and medium purity (5N Al and 4N Si)).

3. Results

Figure 1a shows the DSC thermogram of high purity Al-5Si alloy without any addition (Sr, Eu) but with a low concentration of 0.4 ppm P only (5N Al + 5N Si), resulting from the intrinsic remaining P content of the high purity Al (5N). The first sharp exotherm A (grain boundary eutectic peak) occurred with an onset temperature of 575.5 °C, which is 1.5 °C less than the equilibrium eutectic solidification temperature, whilst the small exotherm B (entrained droplet peak) occurred with an onset temperature of 544.5 °C. Undercooling is defined here as the difference between the onset temperatures of the grain boundary eutectic peak and the entrained droplet peak. The undercooling (ΔT) was measured to be ~ 31.0 °C (for 5N Al + 5N Si) and ~ 20.5 °C (for 5N Al + 4N Si), respectively. The undercooling decreases significantly from 31.0 °C to 20.5 °C when a lower grade Si (4N) is used, which has been associated by Ho and Cantor [5] and Li and Schumacher [15] as an effect of P contamination of the Si used. During reheating four times in DSC, exotherm B becomes less significant, as shown in figure 2, indicating that no significant P is available for the nucleation of eutectic Si after reheating.

Figure 1b shows the DSC solidification exotherms of high purity Al-5Si alloy with 3 ppm P addition. With the presence of 3 ppm P, the formation of a shoulder in the eutectic droplet peak (exotherm B) was observed to be just after the solidification of the grain boundary eutectic exotherm A. Deconvolution of the peak gives a rough estimated onset temperature of the entrained droplet peak of ~575 °C [15], which is approximately identical to the grain boundary eutectic onset temperature, strongly indicating that the presence of P, even with a trace level, can significantly enhance the nucleation of eutectic Si and thereby reduce the undercooling of eutectic Si.

Figure 1c shows the DSC solidification exotherms of high purity Al-5Si alloy with 100 ppm Sr addition. Exotherm B occurred with an onset temperature of 526 °C with a measured undercooling of about 49.5 °C, indicating that the nucleation of the entrained droplets started at a very high undercooling. Furthermore, a new peak emerged on the DSC trace with the onset temperature of 566.5 °C, which may represent the precipitation of the highly undercooled Al_2Si_2Sr phase. It should be noted here that the presence of Al_2Si_2Sr phase is expected above the eutectic reaction in equilibrium conditions. However, because of the high purity elements used here, no significant nucleation sites for the Al_2Si_2Sr phase appear to be present and thereby delay the formation of Al_2Si_2Sr phase.

Figure 1d shows the DSC solidification exotherms of high purity Al-5Si alloy with 50 ppm Sr and 1 ppm P addition. A decreased undercooling of 38 °C was observed, which can be attributed to the presence of 1 ppm P, again indicating that P concentrations and thereby AlP nucleation site are extremely vital to the nucleation of eutectic Si.

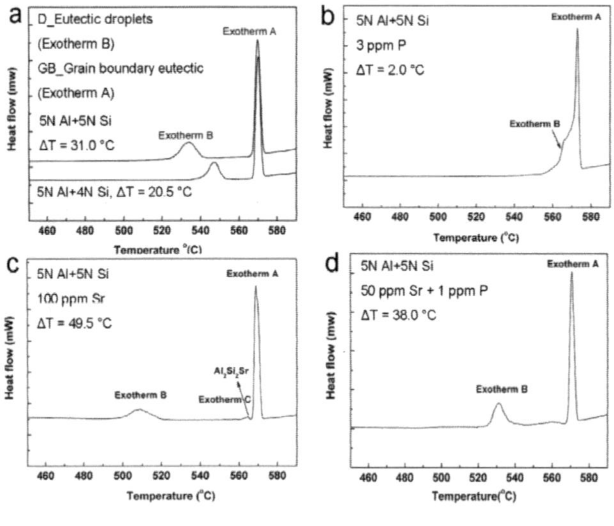

Figure 1. DSC solidification exotherms of Al-5Si alloys without (a) and with 3 ppm P (b), 100 ppm Sr (c), 50 ppm Sr and 1 ppm P (d), at a cooling rate of 10 °C / min.

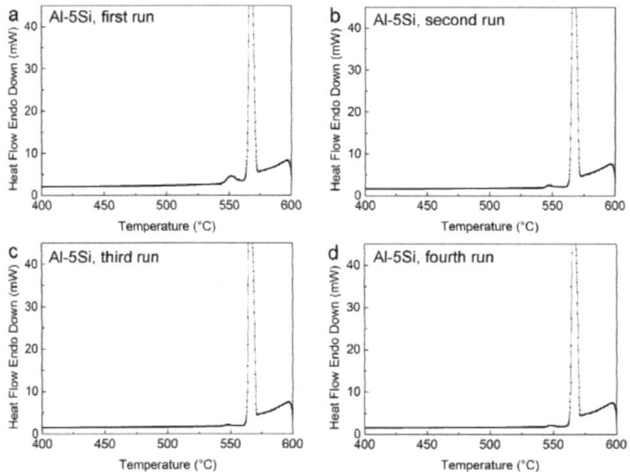

Figure 2. DSC solidification exotherms of Al-5Si alloys at a cooling rate of 10 °C / min. During reheating four times in DSC, exotherm B becomes less significant; indicating that less P is available for the nucleation of eutectic Si.

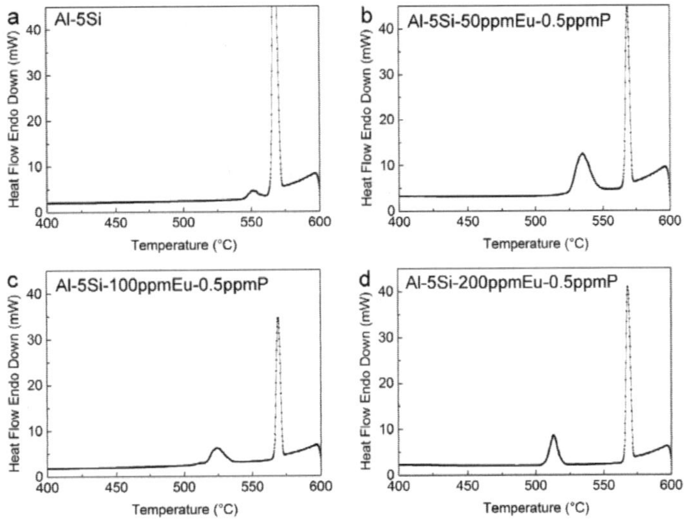

Figure 3. DSC solidification exotherms of Al-5Si alloys without (a) and with 50 ppm Eu and 0.5 ppm P (b), 100 ppm Eu and 0.5 ppm P (c), 200 ppm Eu and 0.5 ppm P (d), at a cooling rate of 10 °C / min.

Similar to the Sr addition (figure 1c), an increased undercooling was also observed in the case of Eu addition, which can be related to modification caused by Eu [19]. Figure 3 shows the DSC solidification exotherms of high purity Al-5Si alloys with 50 ppm, 100 ppm, 200 ppm Eu addition together with 0.5 ppm P. With increasing Eu contents from 50 ppm, 100 ppm to 200 ppm, the onset temperature of exotherm B decreases from 544 °C, 532 °C to 517 °C, respectively. The measured undercooling of entrained eutectic droplet is about 31.5 °C, 43.5 °C, and 58.5 °C. Indeed, an increased undercooling, even compared with 100 ppm Sr, was observed. However, the presence of 3 ppm P, even with 200 ppm Eu addition, results in an absence of exotherm B, as shown in figure 4b, highlighting the importance of AlP for the nucleation of eutectic Si.

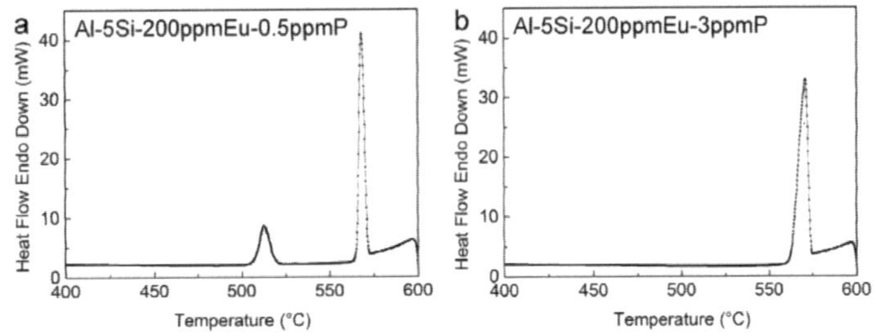

Figure 4. DSC solidification exotherms of Al-5Si alloys with 200 ppm Eu and 0.5 ppm P (a), 200 ppm Eu and 3 ppm P (b) at a cooling rate of 10 °C / min. The presence of 3 ppm P results into an absence of exotherm B.

Only a trace P addition (i.e. 3 ppm) has a profound effect on the nucleation of entrained eutectic droplets (figure 1d, figure 4). Either/or a decreased undercooling was observed with the presence of P. However, Sr and Eu addition depletes or consumes the available AlP nucleation site. An increased undercooling was observed, which can be attributed to the formation of phosphide compounds having a lower free energy and hence form preferentially compared to AlP [15].

In order to understand the observed undercooling of Al-Si entrained droplets and thereby the nucleation kinetics, classic nucleation theory [20], as described by Kim et al [21], was firstly used. As reported in our previous research [15], a wetting angle of about 16° for the nucleation of Si was evaluated for high purity Al-5Si alloy, indicating that the acting substrate for Si is still highly effective. However, the resulting number of nucleation sites (-12.44 for high purity Al-5Si alloy) is far smaller than 1 nucleation site [15], indicating that the physical description of the classical nucleation theory based on a continuum approach does not hold for a small contact angle below 30°. Clearly, a better model is required to elucidate the nucleation kinetic of Si.

Free growth model is subsequently used to describe the nucleation kinetics of entrained eutectic Si in Al-Si based alloys. Similar to the nucleation and free growth of Al on active TiB_2 particles, the size of AlP patches or AlP particles and their size distribution is proposed to become a rate limiting step [22, 23]. Free growth away from the AlP patches or AlP particles occurs when the critical size of Si nucleation sites is identical to the equivalent size of the AlP patches or AlP particles [15]. Furthermore, the free growth model does not imply a constant nucleation rate. At a given temperature interval, a number of AlP patches or AlP particles will fulfil the free growth criterion from which free growth of Si can occur. Applying the approach by Kim et al [21] for droplet nucleation kinetics and using the free growth criterion, yields

$$\ln\left(-\frac{\Delta T-\Delta T_0}{\Delta T_\sigma}\right)+\frac{1}{2}\left(\frac{\Delta T-\Delta T_0}{\Delta T_\sigma}\right)^2=\ln\left(\frac{n_{max}}{\sqrt{2\pi}}\right) \quad (1)$$

From equation (1), it is clear that the maximum number density of nucleation sites (n_{max}) within one droplet is directly related to the size distribution of AlP particles or AlP patches for Si nucleation, but it is independently of the cooling rates. On the other hand, the size distribution of AlP particles or AlP patches depends on the available P content in a given droplet. Thus, the limiting step for the growth of Si on AlP patches can be regarded as the free growth criterion. The maximum number density of nucleation sites (n_{max}) is dependent on the size distribution of droplets and P content. In the case of high purity Al-5Si alloy, the calculated maximum nucleation site value (n_{max}) is about 6.77×10^2.

4. Conclusions

Entrained droplet technique and DSC analyses were employed to investigate the heterogeneous nucleation of entrained eutectic Si in high purity melt spun Al-5wt.% Si alloys. Sr and Eu addition exerts negative effects on the nucleation process, while an increased undercooling was observed. This can be attributed to the formation of phosphide compounds having a lower free energy, which preferentially formed compared to AlP. Only a trace P addition has a profound effect on the Si nucleation.

References

[1] Davis JR et al., editor. 1998 ASM handbook, vol.15. Casting, third ed. (Metals Park, OH: ASM international) pp.751-753.
[2] Pacz A. 1921 U.S Patent No. 1387900;.
[3] Lu Shu-Zu and Hellawell A 1987 *Met. Trans.* A **18** 1721-1733.
[4] Day MG and Hellawell A 1968 *Proc. R Soc. Lond* A **305** 473-491.
[5] Ho CR and Cantor B 1995 *Acta Metall. Mater.* **43** 3231-3246.
[6] Wang CC and Smith CS 1950 *J. of Metals Trans. AIME.* **188** 136-138.
[7] Ho CR, Cantor B 1995 *J. Mater. Sci.* **20** 1912-1920.
[8] Moore KI, Zhang DL and Cantor B 1990 *Acta Metall. Mater.* **38** 1327-1342.
[9] Cantor B and O'Reilly KAQ 1995 *Acta Metall. Mater.* **43** 405-417.
[10] Crosley PB and Mondolfo LF 1966 *Modern Castings* **49** 63.
[11] Flood SC and Hunt JD 1981 *Metal Science* **15** 287-294.
[12] Cho YH, Lee HC, Oh KH and Dahle AK 2008 *Metall. Mater. Trans. A.* **10** 2435.
[13] Quested TE, Greer AL 2005 *Acta Mater.* **53** 4643-4653.
[14] Quested TE, Greer AL 2005 *Acta Mater.* **53** 2683-2692.

[15] Li JH, Zarif M, Albu M, McKay B, Hofer F and Schumacher P 2014 *Acta Mater.* **72** 80-98.
[16] Li JH, Zarif M, Dehm G and Schumacher P 2012 *PHIL. Mag.* **92** 3789-3805.
[17] Zarif M, McKay B and Schumacher P 2011 *Metall. Mater. Trans. A.* **41** 1684.
[18] Li JH, Barrirero J, Engstler M, Mücklich F and Schumacher P 2014 *Metall. Mater. Trans. A.* unpublished.
[19] Li JH, Wang XD, Ludwig TH, Tsunekawa Y, Arnberg L, Jiang JZ and Schumacher P 2014 *Acta Mater.* unpublished.
[20] Turnbull D and Vonnegut B 1952 *Ind. Eng. Chem.* **44** 1292.
[21] Kim WT, Zhang D L and Cantor B 1991 *Metall. Trans. A* **22** 2487.
[22] Greer AL 2003 *Phil. Trans. R. Soc. Lond. A* **361** 479-495.
[23] Bunn AM, Schumacher P, Kearns MA, Boothroyd CB and Greer AL 1999 *Mater. Sci. Technol.* **15** 1115-1123.

IOP Conf. Series: Materials Science and Engineering **117** (2016) 012007 doi:10.1088/1757-899X/117/1/012007

Phase-field modelling of β(Ti) solidification in Ti-45at.%Al : columnar dendrite growth at various gravity levels

A Viardin, R Berger, L Sturz, M Apel and U Hecht

ACCESS e.V., Materials and Processes, Intzestr. 5, D-52072 Aachen, Germany

E-mail: `a.viardin@access-technology.de`

Abstract. The effect of solutal convection on the solidification of γ titanium aluminides, specifically on β(Ti) dendrite growth, is not well known. With the aim of supporting directional solidification experiments under hyper-gravity using a large diameter centrifuge, 2D-phase field simulations of β(Ti) dendrite growth have been performed for the binary alloy Ti-45at.%Al and various gravity scenarios. Both, the direction and magnitude of the gravity vector were varied systematically in order to reveal the subtle interplay between the convective flow pattern and mushy zone characteristics. In this presentation, gravity effects are discussed for early dendrite growth. For selected cases the evolution on longer timescales is also analyse of and oscillatory modes leading to dynamically stable steady state growth are outlined. In a dedicated simulation series forced flow is superimposed, as to mimic thermally driven fluid flow expected to establish on the macroscopic scale (sample size) in the centrifugal experiments. Above a certain threshold this flow turns dominant and precludes solutally driven convective effects.

1. Introduction

The motivation of this study is to reveal convective effects on the characteristics of the mushy zone during directional solidification of bcc β(Ti) dendrites in hyper-gravity conditions. Such conditions commonly occur during centrifugal casting of near-net shape parts from titanium aluminides (e.g. Ti-45at.%Al) but may be explored more systematically by means of solidification experiments inside ESAs large diameter centrifuge (LDC). In order to guide experiment design and evaluation phase, field simulations were performed for different hyper-g scenarios, focusing on columnar growth of β(Ti) dendrites as function of the gravity vector $\vec{g}$. Both, the magnitude and direction of $\vec{g}$ were varied within experimentally feasible limits. This means that $\vec{g}$ is either parallel or anti-parallel to the direction of solidification ranging from 0 to $+20\vec{g}$ (parallel) and from 0 to -$20\vec{g}$ (anti-parallel) respectively. In a dedicated simulation series the effect of additional lateral forced flow was investigated for the case of -$17\vec{g}$. Previous work on gravity effects on the solidification behavior of an Al-Cu alloy has been reported based on simulations [1, 2, 3] and based on experiments for a Pb-Sn alloy [4]. These studies showed that gravity impacts on the characteristics of the mushy zone; the primary dendrite arm spacing decrease for specific fluid flow configurations that promote growth of tertiary arms while inhibiting growth of primary dendrite tips through advection of solute.

2. Model description

The phase field model coupled to melt flow that was used in the present work is described in [1, 5]. The simulations are restricted to two dimensions and to early β(Ti) dendrite growth, e.g. to time periods that are just long enough to allow reaching steady state growth. The partial differential equations for melt flow are solved on a grid with grid spacing twice bigger than that for concentration and phase field, a method suggested by Beckermann [6]. In order to calculate the advection term, the flow velocity is projected on the finer grid using linear interpolation. This method allows speeding up the computations of the complete system by a factor of five for the simulations shown in this paper. In this study only the gravity vector $\vec{g}$ is changed, both in magnitude and direction, while the solidification parameters and other numerical parameters are kept constant. The materials properties for the alloy Ti-45at.%Al are summarized in Table 1, the linear coefficient of density change has been evaluated from numerical calculations by Lopis [7] and the viscosity has been evaluated from experiments by Wunderlich [8]. The temperature dependence of density has been neglected. Directional solidification is modeled by imposing a fixed temperature gradient in z-direction $G_z = 120$ K/cm and a cooling rate r=0.3 K/s, corresponding to a pulling speed of $v_w = 25$ μm/s. The simulations are performed in a rectangular domain of 960μm x 450μm, height and width, respectively, starting with a spherical seed at the bottom of the calculation domain, its width defining the initial spacing. Boundary conditions are periodic in x-direction. For the concentration field the top boundary is treated with a fixed concentration and the bottom boundary is isolated. Flow is constrained to be parallel to top and bottom boundaries and a non-slip condition is used at the solid-liquid interface. This choice of boundary conditions mimics a situation in which the melt in the far field above the mush would be perfectly mixed (by convection). The phase field parameters are adjusted using the solution of the dendrite tip operating point from the KGT model [9] for the case of directional solidification under purely diffusive conditions (zero-$\vec{g}$) and steady state growth. A careful calibration of the interface thickness (δ=1.2 μm) and phase field mobility (μ_{sl}=2$\cdot$10^{-2} cm^4/J/s) were performed to ensure convergence of the steady state solutions of tip radius and tip undercooling to KGT solutions.

Table 1. Physical and numerical parameters used in our simulations

Density (liquid) (ρ)	$3.7 \cdot 10^6$	g/m^3	[7]
Density variation ($\partial\rho/\partial c$)	$1.77 \cdot 10^4$	g/m^3/at%	[7]
Kinematic viscosity (ν)	$1.89 \cdot 10^{-6}$	m^2/s	[8]
Liquidus slope (m_l)	-11.268	K/at%	[10]
Partition coefficient (k_{Al})	0.9	-	[10]
Liquid diffusivity(D_l)	$3 \cdot 10^{-9}$	m^2/s	[11]
Interfacial energy (σ_{sl})	0.1	J/m^2	[12]

3. Simulation results

In Ti-45at.%Al, convective flow is driven by local melt density changes caused by the weak segregation of the lighter element Al. The simulation results are presented below for two distinct cases: (I) $\vec{g}$ is aligned parallel to the growth direction and hence aluminum-lean liquid is transported downwards the dendrite tips and (II) $\vec{g}$ is aligned opposite to the growth direction and hence aluminum-rich liquid is rising from the mush, flowing upwards along the dendrite stem and tip. For each case, the growth velocity (v) and the average melt flow velocity (u) are evaluated and displayed as function of time. The average melt flow velocity was obtained by summation of the z-component of the velocity along each column of the domain divided by the number of liquid cells in each column. The growth velocity refers to the tip velocity of the longest dendrite in the array. Snapshots of the concentration field and the flow field at different selected time are also included. Case III is expanding the scenario (II) by including an additional forced flow in lateral direction.

3.1. Case I : $\vec{g}$ parallel to the growth direction and microgravity

Figure 1 displays the dendrite evolution in 0 $\vec{g}$, 1$\vec{g}$, 5$\vec{g}$ and 15$\vec{g}$. A longer time is simulated for microgravity as to verify that steady state is reached. Indeed one can see that it is reached after a damped oscillation at about 100s, the growth velocity converging to the pulling speed represented with a red dashed line. When increasing gravity the convection rolls feed Al-lean liquid to the dendrite tip region, such that one would expect the central dendrite to grow faster, however this effect is rather weak and the length of the central dendrite or the mushy zone is virtually independent from the magnitude of gravity. More pronounced secondary arms develop, which indicates that the flow amplifies the side branching instability. Furthermore, tip splitting of the central dendrite is observed at 15$\vec{g}$. The flow velocity increases with gravity but the averaged up- and downward flow velocity remains below the growth velocity. The growth velocity reaches the pulling speed at around 100s, being only slightly affected by the gravity magnitude.

3.2. Case II : $\vec{g}$ opposite to the growth direction

The snapshots and velocity profiles in figure 2 display dendrite growth configurations after solidification onset along with the corresponding flow patterns for different magnitudes of $\vec{g}$ alined opposite to the growth direction. For this case, the convective flow and the response of the mushy zone are qualitatively different from case I. One can observe the following effects; first, two convection rolls form on each side of the dominant primary dendrite, now with upward flow on both sides of the tip. With increasing magnitude of the gravity vector $\vec{g}$ the initially dominant single dendrite tip grows less fast, being fed with Al-rich melt by upward flow. In the downward flow regions Al-lean melt is advected and correspondingly tertiary dendrite arms grow faster until they become primary dendrites, eventually leading to a decrease of the dendrite spacing. The average flow velocity in the convection rolls increases with gravity. The flow pattern is symmetric, relative to the central primary dendrite for -1$\vec{g}$. The symmetry of the flow pattern is altered at high $\vec{g}$ because the primary spacing is severely reduced and the convection pattern changes in response to the increased number of primary dendrites. Local tip splitting events are also detected. The solidification dynamics shows the following peculiarities, as can be inferred from the average flow velocity profiles and the growth velocity profiles in figure 2, for the lowest gravity (-1$\vec{g}$), the flow velocity is always lower than the dendrite growth velocity and hence dendrite characteristics are only slightly affected. At 100s the growth velocity reaches the pulling speed, e.g. steady state. For -7$\vec{g}$, between 0 and 10s the central dendrite remains dominant but is slowed down by the upward flow. Between 10 and 18 s the quick growth of the tertiary dendrite arm on the right side of the domain in response to the downward flow in this area leads to an increase of the growth velocity. In fact, one can state that the oscillatory behavior of the growth velocity for t $>$ 10s is due to the competition between the different dendrites and their interaction with flow. Finally a dynamic steady state is reached at about t=60 s with permanent oscillations of the dendritic front kinetics. This behavior is also observed for higher magnitudes or $\vec{g}$ eventually with shorter periods of the oscillations and with more frequent tip splitting events.

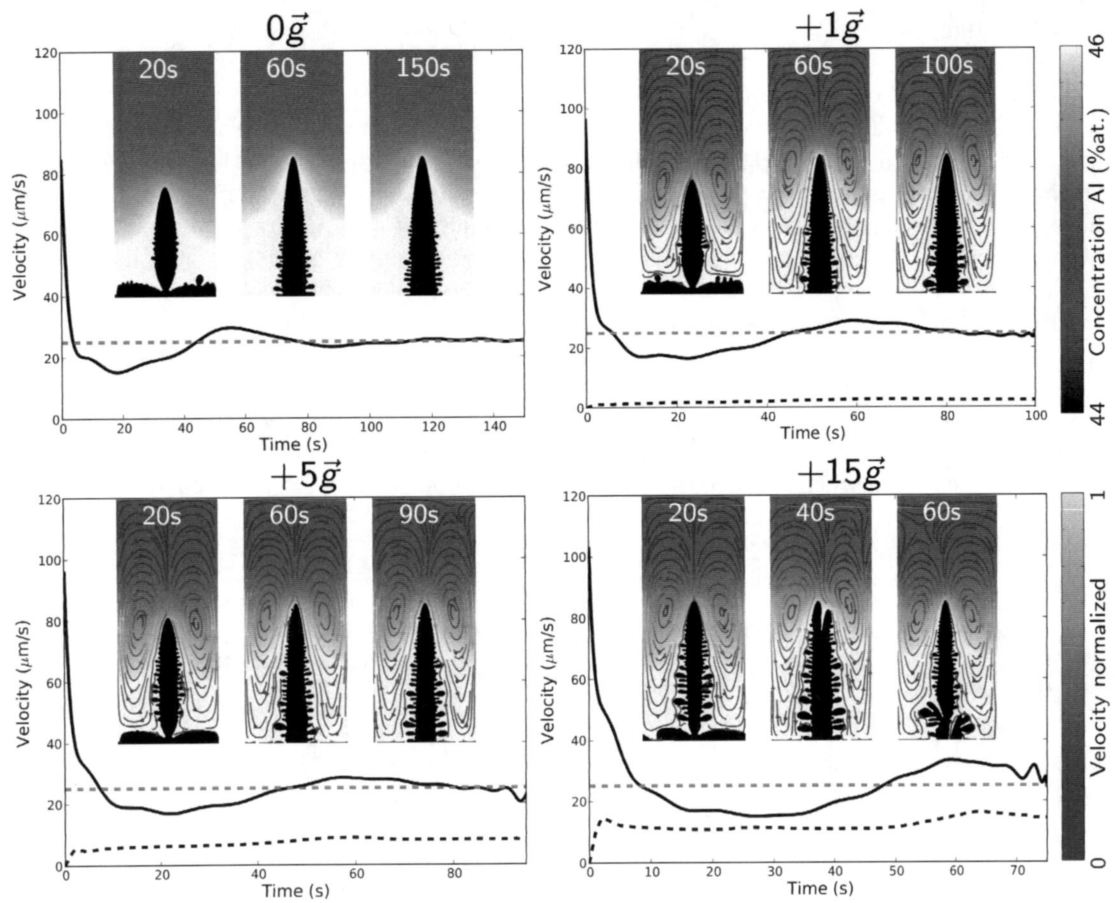

Figure 1. Dendrite growth in Ti-45at.%Al as function of time for case I with the gravity vector oriented parallel to the growth direction. The plots show the growth velocity and the average flow velocity for different magnitudes of $\vec{g}$, black solid line dendrite tip velocity, black dashed line maximal flow z-velocity, red dashed line pulling speed. The inserted snapshots display concentration and flow maps.

3.3. Case III : $\vec{g}$ opposite to the growth direction and forced lateral flow

The snapshots in figure 3 display dendrite growth and flow configurations after 10 seconds of solidification under -17$\vec{g}$ and an additional lateral flow imposed along the x-axis by applying a pressure difference of Δp = 0.1, 1 and 2 mPa, respectively. The simulation parameters are the same as before except for the calculation domain that is 600μm high and 450μm wide. For Δp = 0.1 mPa, one can observe that the lateral flow is weaker than the flow driven by solutal convection and hence little impact on the dendrite growth kinetics and morphology is seen. With increasing Δp the lateral flow becomes gradually more dominant and precludes the effects of the gravity driven solutal convection. Spacing refinement through excessive growth of tertiary arms is no longer operative. Instead one can observe that the secondary arms of the central dendrite are growing preferentially into the flow, a phenomenon that has been observed many times in experiments [13, 14] and simulations [6]. Nevertheless, the lateral flow velocity has to be several times bigger than the convective flow velocity to induce severe changes of the dendritic morphology. For the present case the threshold value of the lateral flow velocity above which convective effects may be neglected ranges at about twice the maximum velocity in the convection rolls at -17$\vec{g}$.

4. Summary and conclusion

Phase field simulations were used to investigate the early growth of β(Ti) dendrites in a directional solidification setup, taking into account the melt convection driven by solutal buoyancy for various hyper-gravity scenarios. The results show distinct effects of gravity on the solidification dynamics and the resulting morphology of the mushy zone, depending on the direction of the gravity vector. For the gravity vector parallel to the growth direction the flow pattern behaves stable and the magnitude of the average flow velocity parallel to the growth direction is always below the pulling speed in the experimentally accessible range (15$\vec{g}$). Consequently, there is only a weak effect on the mushy zone characteristic. The most significant impact of convective flow on the columnar mushy zone is observed if the growth direction is opposite to the $\vec{g}$-vector. In this case, the primary dendrite spacing decrease with increasing magnitude of $\vec{g}$ because flow and solute advection greatly favour the development of tertiary arms into new primary arms. This is due to the fact that the flow pattern becomes unstable, evolves

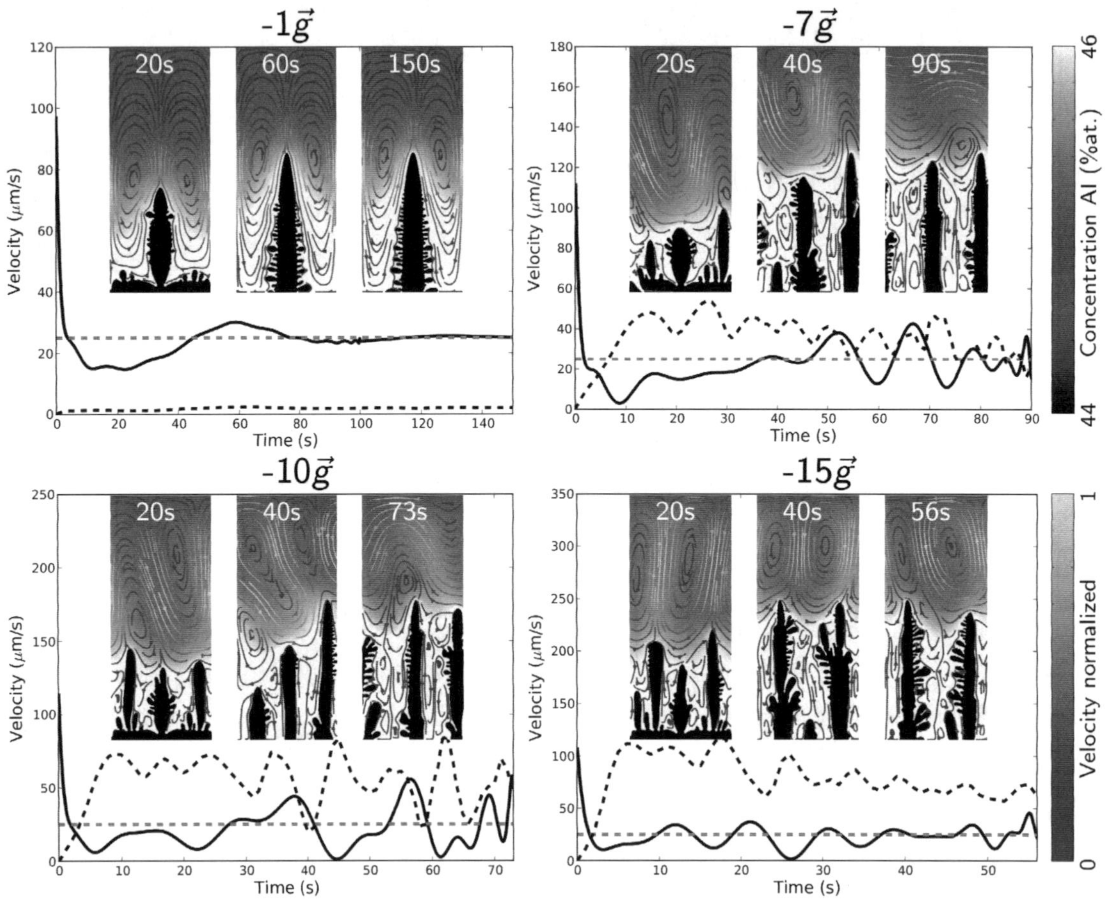

Figure 2. Dendrite growth in Ti-45at.%Al as function of time for case II with the gravity vector opposite to the growth direction. The plots show the growth velocity and the average flow velocity for different magnitudes of $\vec{g}$; black solid line dendrite tip velocity, black dashed line maximal flow z-velocity, red dashed line pulling speed. The inserted snapshots display concentration and flow maps.

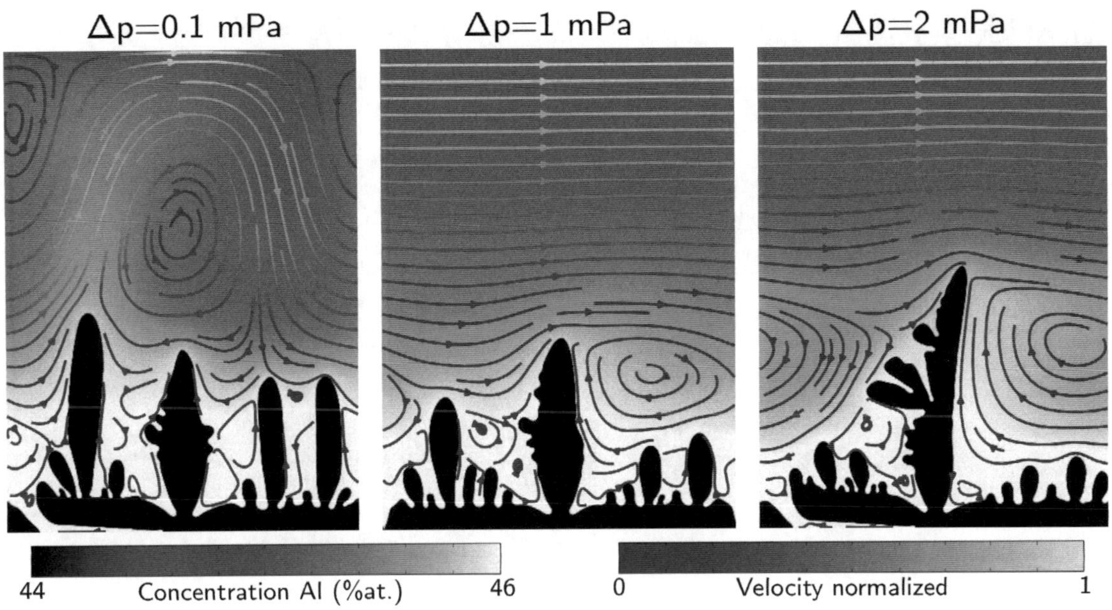

Figure 3. Dendrite growth in Ti-45at.%Al at t=10 seconds under $-17\vec{g}$ with an additional lateral flow imposed by a pressure difference of 0.1 mPa (left), 1mPa (center) and 2 mPa (right).

on larger length scales and thus results in higher flow velocities. These effects are precluded, if lateral flow is strong, e.g. several times stronger than convective flow. For the analysis of centrifugal experiments it is therefore mandatory to verify the transport rates associated with macroscopic flow in the sample and to use the results as an input to meso and microscale models. As discussed in [1] the present simulation results are 2D and not independent from the computational domain size and must be regarded as qualitative results so far. However, they will be used for the calibration of a mesoscale model [15] that will address the 3D problem in future.

Acknowledgements
We gratefully acknowledge the support of this research work by the German Federal Ministry of Research through the German Space Agency DLR under Contract FKZ 50WM1143.

References
[1] Steinbach I 2009 *Acta. Mat.* **57** 2640
[2] Apel M and Diepers 2006 H-J *MCWASP XI Proceedings* 505
[3] Diepers H J and Steinbach I 2006 *Material Science Forum* **508** 145
[4] Bataille C C Grugel R N Hmelo A B and Wang T G 1994 *Metal Trans A* **25** 865
[5] Beckermann C Diepers H-J Steinbach I Karma A Tong X 1999 *J. Comp. Phys.* **154** 468
[6] Lu Y Beckermann C and Ramirez J C 2005 *Journal of crystal growth* **280** 320
[7] Lopis A S Computational simulation of molten titanium-aluminum metal and alloys (available at http://www.mintek.co.za/Pyromet/Files/2010Lopis.pdf).
[8] Wunderlich R 2008 *High Temperature Materials and Processes* **27** 401
[9] Kurz W Giovanola and B Trivedi R 1986 *Acta Mat.* **34** 823
[10] Witusiewicz V T Bondar A A Hecht U Rex S and Ya T 2008 *Alloy Compd.* **465** 64
[11] Binder K and Kargl F 2012 *Gradecet Meeting Budapest 4* Unpublished
[12] Eiken J Apel M Witusiewicz V T Zollinger J and Hecht U 2009 *J. Phys.: Cond.Mat.21* 464104.
[13] Shevchenko N Eckert S Boden S and Gerbeth G 2012 *IOP Conf. Ser.: Mater. Sci. Eng.* **33** 012035
[14] Okamoto T Kishitake K and Bessho I 1975 *J. of Crys. Growth* **29** 131

[15] Založnik M Viardin A Combeau H and Apel M 2014 *submitted to ICASP 4 Proceedings*

IOP Conf. Series: Materials Science and Engineering **117** (2016) 012008 doi:10.1088/1757-899X/117/1/012008

Three-dimensional modeling of a thermal dendrite using the phase field method with automatic anisotropic and unstructured adaptive finite element meshing

C Sarkis[1a], L Silva[1b], Ch-A Gandin[1c] and M Plapp[2d]
[1] MINES ParisTech, CEMEF UMR CNRS 7635, CS10207, 06904 Sophia Antipolis, France
[2] Ecole Polytechnique, LPMC UMR CNRS 7643, 91128 Palaiseau, France

Email: [a]carole.sarkis@mines-paristech.fr, [b]luisa.silva@mines-paristech.fr, [c]charles-andre.gandin@mines-paristech.fr, [d]mathis.plapp@polytechnique.fr

Abstract. Dendritic growth is computed with automatic adaptation of an anisotropic and unstructured finite element mesh. The energy conservation equation is formulated for solid and liquid phases considering an interface balance that includes the Gibbs-Thomson effect. An equation for a diffuse interface is also developed by considering a phase field function with constant negative value in the liquid and constant positive value in the solid. Unknowns are the phase field function and a dimensionless temperature, as proposed by [1]. Linear finite element interpolation is used for both variables, and discretization stabilization techniques ensure convergence towards a correct non-oscillating solution. In order to perform quantitative computations of dendritic growth on a large domain, two additional numerical ingredients are necessary: automatic anisotropic unstructured adaptive meshing [2,[3] and parallel implementations [4], both made available with the numerical platform used (CimLib) based on C++ developments. Mesh adaptation is found to greatly reduce the number of degrees of freedom. Results of phase field simulations for dendritic solidification of a pure material in two and three dimensions are shown and compared with reference work [1]. Discussion on algorithm details and the CPU time will be outlined.

1. Introduction

The phase field approach is a method of choice for simulating interfacial pattern formation phenomena in solidification. The widely recognized appeal of this approach is to avoid the explicit tracking of macroscopically sharp phase boundaries. This makes it better suited than more conventional front tracking methods to simulate time-dependent free boundary problems in three dimensions (3D) or when complex geometries are involved. Tracking is avoided by introducing an order parameter, or phase field ϕ, which varies smoothly from one value in the liquid to another value in the solid across a spatially diffuse interface region related with a thickness W. This field naturally distinguishes the solid and liquid phases and converts the problem of simulating the advance of a sharp boundary to that of solving a stiff system of partial differential equations that govern the evolution of the phase and diffusion fields.

The main difficulty when solving numerically phase field models is due to the very rapid change of the phase field across the diffuse interface, which thickness has to be taken small enough to correctly capture the physics of the phase transformation. A high spatial resolution is therefore needed to describe the smooth transition. In order to reduce the computational time and the number of grid points, adaptive anisotropic and unstructured finite elements have been used. In the past, other authors have performed numerical simulations in 2D and 3D using the finite element method and the phase field model with an adaptive isotropic mesh based on an error estimator [5,[6]. Others used the phase field model with an adaptive structured isotropic mesh using finite difference method [7[**10**] and finite element method [9-[11]. And some of them used the level set model with an adaptive structured isotropic mesh and finite element method [12].

A first numerical goal of this paper is to present the influence of using an adaptive mesh [2, [3] on the computational cost adding the time adaptation and the element adaptation with parallel computations [4]. The second goal of this paper is to demonstrate that accurate quantitative solutions of the free-boundary problem given by the thin-interface limit are recovered by the finite element method. We present the results of simulations of dendritic growth in 2D and 3D. We take advantage of the crystal symmetries to reduce computation time.

2. Modeling

In solidification of a pure system, thermal dendritic growth is generally described by stating conservation of energy and using the Gibbs-Thomson relation to establish the normal velocity of propagation of the solid-liquid interface [13-15], providing a sharp interface formulation of our solidification problem. In a diffuse interface context, instead of solving the equations on each solid/liquid domain with the given interface conditions, we may obtain a set of equations valid in the whole domain by using a free energy approach.

2.1. *Energy diffusion*

We will use the symmetric model (α= const, c_p = const). Let us consider the dimensionless temperature, $q = c_p\left(\frac{T - T_M}{L}\right)$, with T the temperature, c_p the heat capacity, L the latent heat and T_M the melting temperature. We thus solve the energy diffusion equation:

$$\frac{\partial q}{\partial t} - a\,\mathrm{D}q = \frac{1}{2}\cdot\frac{\partial f}{\partial t} \tag{1}$$

with α the dimensionless thermal diffusion. The Gibbs-Thomson equation allows us to establish the relation at the solid/liquid interface $\theta = -\,d_0(\boldsymbol{n})\kappa - \beta(\boldsymbol{n})v_n$ with $d_0 = \Gamma c_p/L$ and $\beta = c_p/\mu_k L$, $\boldsymbol{n}$ being the normal vector, v_n the normal velocity at the interface (v_n = $v.\ \boldsymbol{n}$), $\Gamma = \sigma^{s/l}T_M/\rho L$, the Gibbs-Thomson coefficient, $\sigma^{s/l}$ being the interfacial energy and ρ the density, μ_k the interface mobility. Thus, d_0 is the capillary length and β is the kinetic coefficient.

2.2. *Phase field formulation*

Let us consider the solidification of a pure material. We define ϕ as a function which describes the presence of the liquid and the solid phases in the computational domain, Ω. This function varies between -1 in the liquid and 1 in the solid. Let us suppose that the solution of our phase field problem is $\phi = -\tanh(\eta/\sqrt{2})$. In the expression, η is the signed distance to the solid/liquid interface. The variational derivative of the free energy functional provides the evolution equation for ϕ which, after manipulation and taking into account the solution form to use in our solver, is

$$\frac{1}{M_\phi}\frac{\partial\phi}{\partial t} - W^2(\boldsymbol{n})\Delta\phi - 2\,W(\boldsymbol{n})\,\nabla W(\boldsymbol{n}).\nabla\phi = [\phi - \lambda\theta(1-\phi^2)](1-\phi^2) + \sum_i\frac{\partial}{\partial i}\left(|\nabla\phi|^2\,W(\boldsymbol{n})\frac{\partial W(\boldsymbol{n})}{\partial\,\partial_i\phi}\right) \tag{2}$$

In this expression $\frac{1}{M_\phi} = \tau(\boldsymbol{n}) = \tau_0\,a_s(\boldsymbol{n})^2$, M_ϕ being the molecular mobility, $W(\boldsymbol{n}) = W_0\,a_s(\boldsymbol{n})$ being the interface anisotropy and we write $\lambda = \frac{\alpha\tau_0}{W_0^2\,\mathrm{a}_2}$. $\partial_i\phi = \frac{\partial\phi}{\partial i}$ with $i = x, y$ in 2D and $i = x, y, z$ in 3D. We take $\mathrm{a}_1 = 0.8839$, $\mathrm{a}_2 = 0.6267$, $\tau_0 = 1$ and $W_0 = 1$.

In 2D, several authors have proposed the form [16]:

$$a_s(\boldsymbol{n}) = \left(1 + \varepsilon_4 \cos\left(4 \arctan \frac{\partial \phi_y}{\partial \phi_x}\right)\right) \tag{2}$$

The following derivative terms for the phase field formulation are thus computed

$$\frac{\partial}{\partial x}\left(|\nabla \phi|^2 \, W(\boldsymbol{n}) \frac{\partial W(\boldsymbol{n})}{\partial\, \partial_x \phi}\right) = -\frac{\partial}{\partial x}(W(\boldsymbol{n})\,(W(\boldsymbol{n}))'\, \partial_y \phi)) \text{ and } \frac{\partial}{\partial y}\left(|\nabla \phi|^2 \, W(\boldsymbol{n}) \frac{\partial W(\boldsymbol{n})}{\partial\, \partial_y \phi}\right) = \frac{\partial}{\partial y}(W(\boldsymbol{n})\,(W(\boldsymbol{n}))'\, \partial_x \phi))$$

ε_4 indicates the strength for the anisotropic arms that we have. A more general form in 2D and 3D is given by [17]:

$$W(\boldsymbol{n}) = W_0(1 - 3\varepsilon_4)\left[1 + \frac{4\varepsilon_4}{1 - 3\varepsilon_4} \frac{(\partial_x \phi)^4 + (\partial_y \phi)^4 + (\partial_z \phi)^4}{|\nabla \phi|^4}\right] \tag{3}$$

2.3. *Numerical resolution*

In following section, we will discuss the numerical parameters and the mesh grid, computed from a metric, needed for the simulation, followed by the numerical resolution, using the phase field method, showing the propagation of the solid-liquid interface.

2.3.1. Phase field and energy solver. In what concerns the resolution of these equations, we have used the finite element method with a continuous approximation for both phase field and temperature functions in the mesh and stabilized resolution to take into account very small diffusion coefficients. In fact, numerical solution of convection-diffusion-reaction equations such as the ones we are treating, using a classical Galerkin formulation, normally exhibits global oscillations in convection-dominated problems, especially in the vicinity of sharp gradients. We use the SUPG (Streamline Upwind Petrov Galerkin) stabilization method to solve this problem by adding a perturbation term to the weighting functions with the aim to get an oscillation-free solution. The linear system of equations issuing from the discretization is solved implicitly using the conjugate bi gradient-least squares method (BCGSL). There is also preconditioning to the resolution using the Jacobi method with incomplete factorization LU per block of size 2.

Advantages of solving the phase field equation as a convection-diffusion problem concern mainly the diffusion term, $\Delta\phi$. In fact, this term, treated implicitly, will naturally smooth singularities in the interface shape. However, convergence towards the sharp interface solution is conditioned by the fact that there is a diffusion layer related with W_0, the thickness that is directly involved in such an equation.

2.3.2. Mesh adaptation. The mesh is initially (and throughout time) adapted using a topological mesher [2][3] that is incorporated to our library and that is based on a metric field, given at the nodes of the mesh. M is a unit metric field associated with any unstructured mesh. The metric is built using the affine transformation to a reference element which has to be equilateral of edge length equal to unity. It provides both the size and the stretching of elements. In our case, this field can be computed using the edge vectors of the mesh, $X_{ij} = X_j - X_i$, i and j being the extreme nodes of the edge. Starting from an existing mesh, the new nodal metrics field M_i we provide to the mesher is

$$M_i = \left(\frac{1}{q} \mathrm{S}_{j \in G(i)} \; s_{ij}^{\,2} \, X_{ij} \otimes X_{ij}\right)^{-1} \tag{5}$$

q being the space dimension, Γ (i) being the set of nodes connected to node i. s the stretching factor applied to obtain the new edge size. The edge stretching factor, s_{ij}, is obtained from the *a posteriori* estimated error, and is given by

$$s_{ij} = \left(L/e_{ij}\right)^{1/p} \tag{6}$$

where e represents the edge error and p a stretching exponent, $1 < p < q$, q being the dimension. Starting from a given element, we examine what information we can construct from the set of edges. Since more than only two edges can be encountered for a node, it is necessary then to find an approximation or an averaging process of the information. For this reason, we first state that the length size of the edges sharing a given node is exactly the interpolation of the continuous length distribution function defined in the space at the considered point. In this technique we computed the error along and in the direction of each edge.

$$L = \left(\sum_{i} \sum_{j \in G(i)} e_{ij}^{\frac{p}{p+2}} / A \right)^{\frac{p+2}{p}} \tag{7}$$

The error is computed using the recovered gradient of the solution on which we wish to adapt (phase field, temperature, or both). We construct a solution vector that contains the two fields and we compute its gradient, $U = (\phi, \theta)$.

In fact, $e_{ij} = \max(|\nabla U \,.\, X_{ij}|;\ e_{min}|X_{ij}|^2)$ and $\nabla U = (X_i)^{-1}.U_i$, where $X_i = \frac{1}{|G(i)|}\sum_{j \in G(i)} X_{ij} \otimes X_{ij}$ is the distribution function and $U_i = \sum_{j \in G(i)} U_{ij}\, X_{ij}$. e_{min} is a chosen constant. A is the number of element given.

In what concerns the referred parameters, when we increase the total number of elements, the mesh is enriched around the interface and we decrease the error. If we want a small error, we should decrease the mesh size, but we also need to increase the number of nodes at the same time. If not, we will increase the error because we do not have enough elements to perform the simulation.

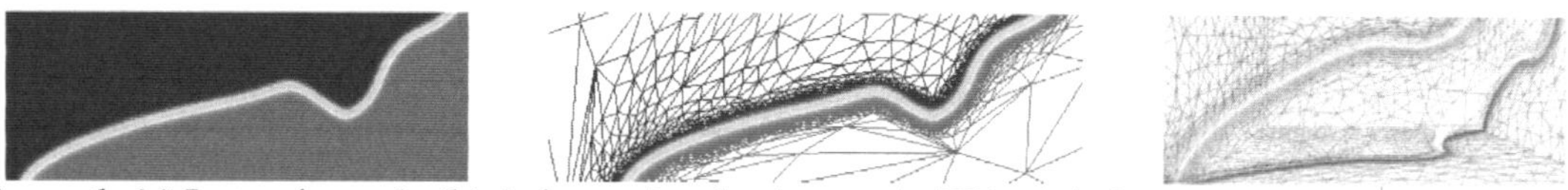

Figure 1. (a) Isotropic mesh, (b) Anisotropic adaptive mesh (2D), c) Anisotropic adaptive mesh (3D).

Figure 1 illustrates the difference between the mesh size inside the interface and outside the interface. We have smaller sizes at the interface and larger outside. The mesh is adapted according to θ and the function ϕ that indicates the interface. The red color represents the solid, the blue one represents the liquid and the green one is the interface.

For a matter of CPU time optimization, we may define a remesh frequency. For that, we compute the norm of the Gibbs-Thomson velocity v and we find the maximum velocity v_{max}, over the entire domain as:

$$v = \frac{-\theta}{\beta} - d_0 K \quad \text{with } \beta = \frac{a_1 \tau}{\lambda\, W(\boldsymbol{n})},\ d_0 = \frac{a_1 W(\boldsymbol{n})}{\lambda},\ \text{the curvature } K = \nabla \cdot \boldsymbol{n}$$

We compute the displacement $v_{max}.\Delta t$ and we sum the displacements for each time step. When the sum of the displacements exceeds $3W_0/2$, remeshing is activated.

2.3.3. Time adaptation. Improvements on CPU time can also be obtained automatically by adapting the time step Δt as:

$$\Delta t = h_{min}/(10 \cdot v_{max}) \tag{8}$$

h_{min} is the minimal mesh size.

2.3.4. Adaptation of the prescribed number of elements. When a mesh adaptation step is performed, the "interface volume", defined as the volume occupied by the interface thickness of the phase field function, $V_{interface}$ (surface in 2D) may be obtained as follows:

$$V_{\text{Interface}} = \int_{-\infty}^{\infty} \delta_\phi \, \mathrm{dV} \text{ with } \delta_\phi = \begin{cases} 0 & \text{for } \phi < -E \\ \dfrac{1}{2E} + \dfrac{1}{2E}\cos\left(\dfrac{\pi\phi}{E}\right) & \text{for } -E < \phi < E \\ 0 & \text{for } \phi > E \end{cases} \quad (9)$$

with $E = 1$. Then we compute the number of elements needed at the interface as

$$N_1 = S_{\text{Interface}}/S_{\text{element}} = 2.S_{\text{interface}}/h_{min}^{\ 2} \quad \text{in 2D}$$

$$N_1 = V_{\text{Interface}}/V_{\text{element}} = 6.V_{\text{interface}}/h_{min}^{\ 3} \quad \text{in 3D}$$

Finally we compute the total number of element needed as $NE = N_1 + N_2$ with N_2 a constant, to add a certain number of elements outside the interface thickness. This adaptation is done because, during dendritic growth, the number of elements should largely increase during the computation because the surface of the interface increases, and so we need more elements to represent it.

3. Numerical Results

3.1. Convergence method.
Let us consider a rectangular domain [0;1000]x[0;300] on which we place an initial seed of size 5. To study the convergence of our method, we compute the tip velocity in the x direction. The anisotropy function for growth is $W(\boldsymbol{n}) = W_0(1 - 3\varepsilon_4)\left[1 + \dfrac{4\varepsilon_4}{1 - 3\varepsilon_4}\dfrac{(\partial_x\phi)^4 + (\partial_y\phi)^4}{|\nabla\phi|^4}\right]$ with $\varepsilon_4 = 0.05$. Other simulation parameters are: $\alpha = 1$, $\tau_0 = 1$, $\Delta t = h_{min}/(10 \cdot V_{max})$. In the following, we compare the solution for different values of the minimal mesh size h_{min}.

In figures 2 and 3, we observe that the tip velocity decreases with time and becomes constant until it decreases instantly to become very small when the tip reaches the border of the domain. On the left, we represent the sensitivity of the solution to the mesh's minimal size; on the right, the steady-state velocity computed value, compared with the value obtained using a green function calculation for an analytical test value. We see that with decreasing the minimal mesh size in the interface the velocity converges to the analytical solution.

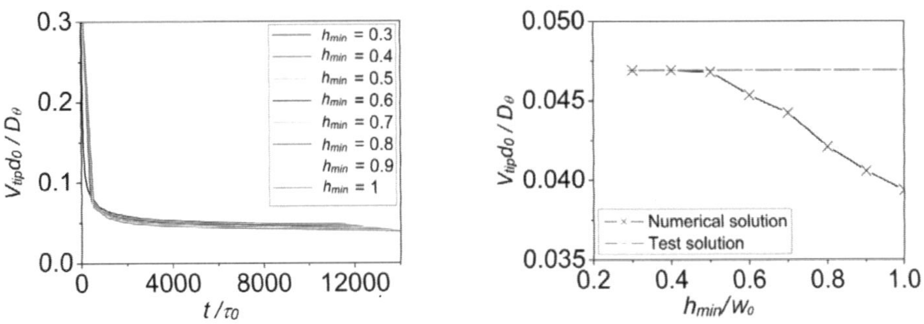

Figure 2. Dimensionless tip velocity as a function of grid spacing for $\Delta = 0.65$, $d_0/W_0 = 0.554$ and $N_2 = 25\,000$. The red line corresponds to the value obtained from the green function calculation [1].

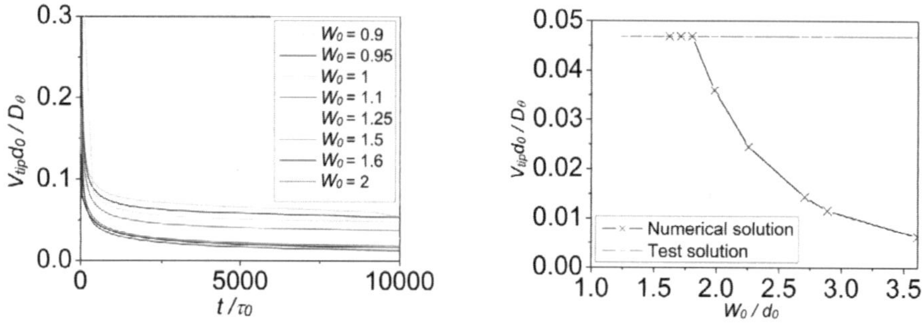

Figure 3. Dimensionless tip velocity as a function of interface thickness for $\theta_0 = -0.65$. The red line corresponds to the value obtained from the green function calculation.

3.2. Computational time

We will show the difference on the computational time with different conditions for a rectangular domain [1000;300] with $\theta_0 = -0.65$, $\varepsilon_4 = 0.05$, $\alpha = 1$ and $e_{min} = 10^{-7}$.

Table 1. Table for different time computation with different conditions

Case	V_{Tip}	*NE*	h_{min}	*Nb_Proc*	*Fr*	**Adapt Time**	**Adapt *NE***	T_{CPU}
1	0.0469	4 638 842	0.4	1	No	No	No	36 000h*
2	0.0469	4 638 842	0.4	4	No	No	No	2 400h*
3	0.0469	4 638 842	0.4	16	No	No	No	1 240h*
4	0.0485	100 000	0.4	16	1	No	No	504h
5	0.051	80 000	0.4	16	Yes	No	No	80h
6	0.05	80 000	0.4	16	Yes	Yes	No	13h
7	0.0469	46 000	0.4	16	Yes	Yes	Yes(N_2 = 25000)	10h

* Predicted time

We observe the influence on CPU time using the parallel computation from the first three cases, using 1, 4 or 16 processors. From case 3 and case 4, we show the decreased CPU time using the mesh adaptation. From case 4 to case 5, we show the influence of the mesh adaptation with the frequency computed above depending on the velocity. In case 6 and 7 we add the time adaptation and the adaption of the number of element. The graphs in figure 4 show the CPU time for the different cases.

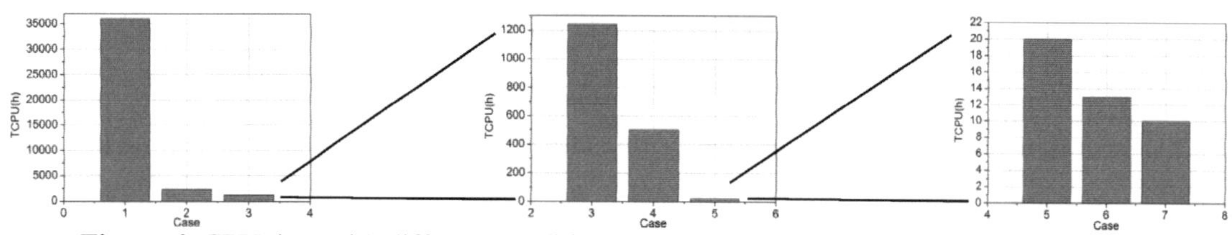

Figure 4. CPU time with different conditions taken to simulate the same dendritic growth.

In Table 2, we show that using our model we can obtain exactly the same tip velocity, for different parameters, computed using the green function supposed as a test solution.

Table 2. Table to compare the tip velocity

Δ	ε_4	α	d_0/W_0	V_{Tip}	V_{Tip}^{GF}	% error	Domain	*NE*	T_{CPU}
0.65	0.05	1	0.554	0.0469	0.0469	0	[1000;300]	46 000	10h
0.55	0.05	2	0.277	0.017	0.017	0	[1000;300]	44 000	6h 30min
0.55	0.05	4	0.139	0.017	0.017	0	[500;150]	34 000	2h 30 min
0.45	0.05	4	0.139	0.00545	0.00545	0	[1000;300]	50 000	5h 50 min
0.55	0.02	2	0.277	0.00685	0.00685	0	[1000;300]	30 000	7h 30 min

4. Conclusion

We have presented a quantitative phase field model that replicates previous published results [1] while taking advantage of an anisotropic adaptive mesh based on an error estimator, with variable number of elements and parallel computations with time adaptation. The computational cost is then shown to be decreased by a factor of 3600 (case 1 to case 7) using 16 CPU.

We have done 3D simulation for this model to show dendritic growth using the adaptive anisotropic and unstructured finite element mesh

Figure **5** and taking advantage of the symmetry. Next step is to simulate a thermal-solute dendritic growth with physical parameters using the kinetic anisotropy for Al-Cu.

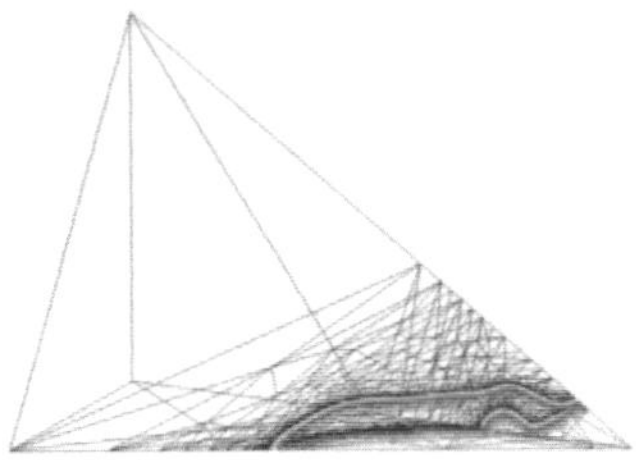

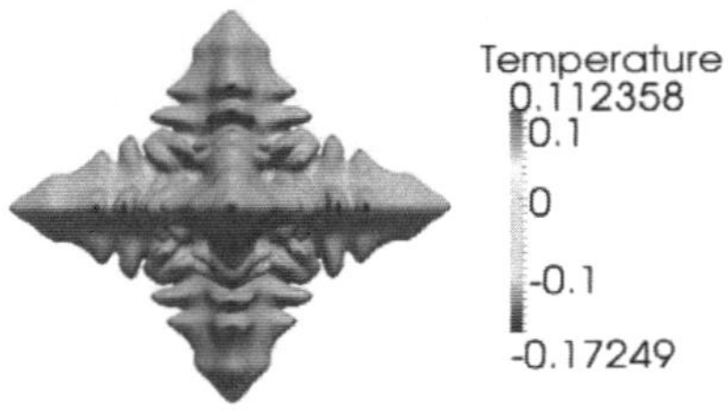

Figure 5. 3D thermal dendritic growth, $\varepsilon_4 = 0.05$, $\alpha = 1$, $\theta_0 = -0.65$, Domain = 1000.

References

[1] Karma A and Rappel W J 1998 Quantitative phase field modeling of dendritic growth in two and three dimensions *Physical review E* **57** 4323-4349

[2] Coupez T 2011 Metric construction by length distribution tensor and edge based error for anisotropic adaptive meshing *Journal of Computational Physics* 230 2391-2405

[3] Coupez T, Silva L and Hachem E (to appear) 2014 Implicit boundary and adaptive anisotropic meshing, *Tetrahedron IV, SEMA SIMAI Springer Series.*

[4] Digonnet H, Silva L and Coupez T 2007 Cimlib: A fully parallel application for numerical simulations based on components assembly *Air Conference Proceedings* **908** 269-274

[5] Narski J and Picasso M 2005 Adaptive finite elements with high aspect ratio for dendritic growth of a binary alloy including fluid flow induced by shrinkage *Elsevier Science* 1-9

[6] Narski J and Picasso M 2006 Adaptive 3D finite elements with high aspect ratio for dendritic growth of a binary alloy including fluid flow induced by shrinkage *FDMP* **1-1** 1-13

[7] Tonhard R and Amberg G 1998 Phase-field simulation of dendritic growth in shear flow *Journal of Crystal Growth* **194** 406-425

[8] Mullis A.M et al. 2008 An adaptive, fully implicit multigrid phase-field model for the quantitative simulation of non-isothermal binary alloy solidification *Acta Materiala* **56** 4559 4569

[9] Mullis A.M et al. 2011 An adaptive, multilevel scheme for the implicit solution of three dimensional phase-field equations *Wiley online library* **27** 106-120

[10] Lan C.W et al. 2011 Adaptive three-dimensional phase-field modeling of dendritic crystal growth with high anisotropy *Journal of Crystal Growth* **318** 51-54

[11] Provatas N et al. 1999 Adaptive Mesh Refinement Computation of Solidification Microstructures Using Dynamic Data Structures *Journal of computational physics* **148** 265-290

[12] Tan L and Zabaras N 2007 A levelset simulation of dendritic solidification of multi-component alloys *Journal of Computational Physics* **221** 9-40

[13] Heringer H, Gandin Ch-A, Lesoult G and Henein H 2006 Atomized droplet solidification as an equiaxed growth model *Acta Materialia* **54** 4427-4440

[14] Rathjen K A and Jiji L M 1971 Heat conduction with melting or freezing in a corner *Journal of Heat Transfer* **54** 101-108

[15] Provatas N, Goldenfeld N and Dantzig J1999 Adaptive mesh refinement computation of solidification microstructures using dynamic data structures *Journal of Computational Physics* **148** 265-290

[16] Plapp M and Karma A 2000 Multiscale Finite-Difference-Diffusion-Monte-Carlo Method for Simulating Dendritic Solidification *Journal of Computational Physics* **165,** 592–619

[17] Friedli J 2011 Interfacial energy anisotropy and growth morphologies in aluminum-zinc alloys *École polytechnique Fédérale de Lausanne* thesis

IOP Conf. Series: Materials Science and Engineering **117** (2016) 012009 doi:10.1088/1757-899X/117/1/012009

CAFE simulation of columnar-to-equiaxed transition in Al-7wt%Si alloys directionally solidified under microgravity

D R Liu[1a], N Mangelinck-Noël[2,3], Ch-A Gandin[4], G Zimmermann[5], L Sturz[5], H Nguyen Thi[2,3] and B Billia[2,3]

[1] School of Materials Science and Engineering, Harbin University of Science and Technology, No. 4 Lin Yuan Road, Xiang Fang District, Harbin 150040, China

[2] Aix-Marseille University and [3] CNRS, IM2NP, UMR CNRS 7334, campus Saint-Jérôme, Case 142, 13397 Marseille Cedex 20, France

[4] MINES ParisTech, CEMEF, CNRS UMR 7635, Sophia Antipolis, France

[5]ACCESS e.V., Aachen, Germany

E-mail: [a] nacylake@126.com

Abstract. A two-dimensional multi-scale cellular automaton - finite element (CAFE) model is used to simulate grain structure evolution and microsegregation formation during solidification of refined Al-7wt%Si alloys under microgravity. The CAFE simulations are first qualitatively compared with the benchmark experimental data under microgravity. Qualitative agreement is obtained for the position of columnar to equiaxed transition (CET) and the CET transition mode (sharp or progressive). Further comparisons of the distributions of grain elongation factor and equivalent diameter are conducted and reveal a fair quantitative agreement.

1. Introduction

Grain morphology has a great influence on the final performance of as-cast products [1]. Depending on the applied processing parameters and the alloy physical properties, solidified grain structures could be purely columnar, or totally equiaxed, or a mixture of columnar and equiaxed. Columnar grains often grow from near the mold surface under a highly imposed thermal gradient [2]. Although, the castings with columnar structure significantly increase the strength in the columnar direction, columnar structures have been found to be more prone to cracking. Castings of equiaxed structure are desirable for many applications, since they exhibit homogeneity in physical and mechanical properties [3]. Therefore, an accurate control of columnar-to-equiaxed transition (CET) and a quantitative understanding of the relationship between CET and casting parameters are essential to industrial application and scientific development.

Within the ESA CETSOL (Columnar to Equiaxed Transition in SOLidification Processing) project, six experiments with different solidification parameters were carried out to study CET transition under microgravity in the Microgravity Science Laboratory (MSL), using the Low Gradient Furnace (LGF) on board the International Space Station (ISS) [4]. Experimental characterizations of the experimental samples were performed by Zimmermann et al. [4], Mirihanage et al. [5] and Liu et al. [6]. Experimental studies are essential to identify mechanisms at the state of CET. However, during solidification, a complex interplay of multi-scale phenomena cannot be observed and then, numerical modelling offers the possibility to study the CET mechanism and the parameter impacts in details.

Direct macroscopic modelling, such as cellular automaton - finite element (CAFE) model, has been developed by Gandin and co-workers with advantages in numerically investigating CET [7,8]: at the size of casting, the topological description of grain structure transition and composition heterogeneity can be realized. Comparisons with benchmark experimental data obtained from microgravity experiments are necessary to guide further improvement of model.

In this publication, CAFE simulations of CET transition during solidification of refined Al-7wt%Si alloys under microgravity are compared with benchmark experimental data. Three aspects are considered: distributions of grain equivalent diameter and elongation factor, CET position and CET transition mode. The influence of pulling-velocity jump on CET is investigated. The reason why simulations cannot quantitatively match some of the experimental results is discussed.

2. Experiment

The schematic setup of the sample cartridge assembly inserted in the low gradient furnace (LGF) available in the Materials Science Laboratory (MSL) on board the International Space Station (ISS) was presented in [4-6]. Rod-like samples of 7.8 mm in diameter and 245 mm in length were placed in a protective Al_2O_3 cylindrical crucible. The main solidification parameters are summarized in table 1[4]. In the present article, we focus on the samples (FM1 / FM5) which are refined Al-7wt%Si alloys with 0.5 wt % Al – 5 wt% Ti – 1 wt% B refiner. Processing conditions are identical except the pulling velocity. In FM1, a pulling-velocity jump is applied from 0.01 mm s^{-1} in stage I to 0.2 mm s^{-1} in stage II. For FM5, the pulling velocity keeps at a low constant value, 0.01 mm s^{-1}, through stages I and II.

Table 1. Processing parameters for samples solidified under microgravity, where V_n is the furnace pulling velocity and S_n the solidification length under V_n.

Sample (μg)	FM1	FM5
Grain refined	Yes	Yes
Homogenization time (s)	600	600
Initial gradient (°C mm^{-1})	0.9	0.9
Stage I		
Time interval (s)	0 ~ 2000	0 ~ 2000
V_1 (mm s^{-1})	0.01	0.01
S_1 (mm)	20	20
Stage II		
Time interval (s)	2000 ~ 2250	2000 ~ 4000
V_2 (mm s^{-1})	**0.2**	**0.01**
S_2 (mm)	50	20
Cooling rate (°C s^{-1})	0.067	0.067
Stage III		
V_3 (mm s^{-1})	3	3

The characterization of the experimental grain structure was performed using electron backscattered diffraction (EBSD). For each identified grain, the grain equivalent diameter and grain elongation factor were calculated based on the numerical analyses of EBSD output data file. The equivalent diameter of a non-spherical object is equal to the diameter of a sphere of equivalent area [6]. The elongation factor is defined as the major axis length to the minor axis length ratio. The two axes intersect orthogonally at the centroid of the grain [6].

The end of columnar growth is determined by visual examination of the polished sample surface. On the one hand, the highest position of dendrites which develop from the beginning of solidification is named as CET_{min}, that is the top boundary of columnar zone. On the other hand, the definition of CET position between transition zone and equiaxed region is dependent on the distribution pattern of averaged elongation factor along the solidification direction. A critical elongation factor equal to 2 [9] is chosen as a criterion. If the profile of the averaged elongation factor varies around 2 within an area, an intermediate zone exists where equiaxed and short elongated grains coexist. If the profile drops below 2 at a position and barely bounces back, an equiaxed region is established and this position is defined as CET_{max}. If there is only CET_{min} identified, a sharp CET from columnar to equiaxed grains occurs. If CET_{min} and CET_{max} coexist, an intermediate region forms and the CET mode is progressive.

The difference in CET transition manner between FM1 and FM5 is explained as: a pulling-velocity jump used in FM1 leads to a higher degree of constitutional undercooling ahead of the solidification front. Under such condition, the nucleation rate is enhanced. With an intensive and frequent nucleation of grains in the liquid, the newly nucleated grains compete over the continued growth of existing grains to form a truly equiaxed zone after columnar zone [9].

3. Modelling

3.1. CAFE modelling

Conservation equations over finite element mesh (FE) are solved using a volume averaging approach. A more detailed presentation of these equations is presented elsewhere [7,8]. For simulation of solidification under microgravity, the computation of momentum equation is switched off. A few assumptions are made. (i) Equal and constant densities in the phases. (ii) No movement of solid phase. (iii) Only heterogeneous nucleation of equiaxed grains is considered.

The FE mesh is divided into a regular lattice of fine square named CA cells. The CA technique is used to track the development of grain envelopes. A Gaussian distribution is adopted to describe nucleation of equiaxed grains [8]. Three input parameters are needed: mean nucleation undercooling ΔT_{mean}, the standard deviation ΔT_{σ} and maximum nucleation density N_{max}. A dendrite tip kinetics defined in [8] is used to calculate dendrite growth.

3.2. Calculation domain, boundary and initial conditions

The experimental system that includes the sample cartridge assembly and the heating system is not directly simulated. The right half of cylindrical geometries, including alloy sample, crucible and plug, are chosen as the computational domains (figure 1). The configuration is two-dimensional and axi-symmetrical. Since the bottom part of the sample ranging from 0 mm to 70 mm remains at solid state through the whole experimental process, the alloy sample is numerically divided into two parts. The top part is simulated using CAFE model and the bottom using only FE model.

To simulate the directional solidification, Dirichlet thermal boundary conditions [10] are used in this study by imposing temperature profiles on the outer surface of three sub-domains "plug", "top crucible" and "bottom crucible".

Twelve input cooling curves (TC1 to TC12) are placed on the outer surface of the top crucible (figure 1) and correspond to experimentally measured profiles. The thermocouple positions range from (TC1) 72.5 mm to (TC12) 182.5 mm defined from the reference plane (0 mm). Another five temperature profiles (SC0: -5.0 mm, SC13: 202.5 mm, SC14: 222.5 mm, SC15: 243.0 mm and SC16: 281.5 mm) are imposed to strictly maintain the unidirectional heat flow through the "Al7Si alloy" domain. These temperature profiles are defined as a function of time by propagating the time evolution of the temperature gradients between thermocouples TC1 and TC2, TC11 and TC12, TC12 and SC13, SC13 and SC14, SC14 and SC15.

At the beginning of the simulation, the whole system has an initial temperature of 800.0 °C. Five nuclei with different growth orientations and with a nucleation undercooling of 0 °C are fixed at the bottom boundary of "Al7Si alloy" domain to reproduce the columnar grain nucleation and competition.

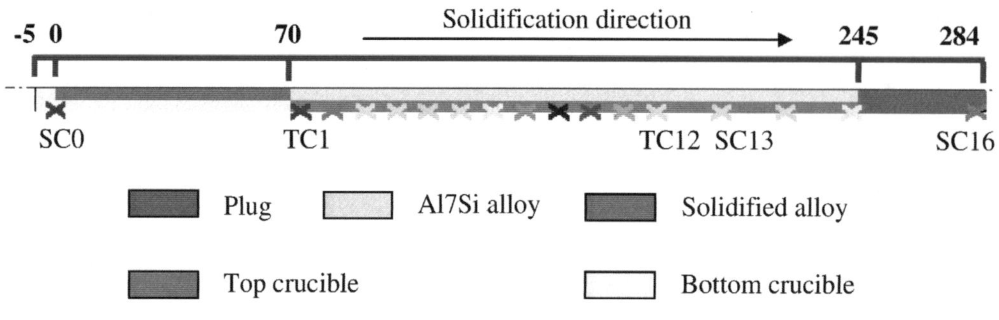

Figure 1. Schematic drawing of casting system with two-dimensional axi-symmetrical coordinates. All dimensions are in mm.

CAFE model is used through the software XR2Sol-CAFE developed by MINES ParisTech CEMEF. The thermo-physical properties of Al-7wt% Si alloy are presented in details in [11]. Here, only critical parameters used in the present CAFE simulations are listed. FE mesh size is 0.3 mm and CA cell size 0.04 mm. The heat transfer coefficients at alloy / crucible, alloy / plug and crucible / plug interfaces are 10^4 W m^{-2} $^oC^{-1}$, 10^3 W m^{-2} $^oC^{-1}$ and 10^3 W m^{-2} $^oC^{-1}$, respectively. The nucleation parameters ΔT_{mean} = 4 °C, ΔT_σ = 0.5 °C and $N_{max} = 5\times10^{10}$ m^{-3} are chosen to get a match with the experimentally observed grain structures in FM1. All these parameters stay the same for different simulation cases. These parameters are kept constant for all the simulations.

4. Results and discussion

4.1. CAFE simulation of grain structure

Figure 2 compares the distributions of averaged grain elongation factor and equivalent diameter along two sections (L4 and L5) cut in the longitudinal direction obtained from EBSD analyses (dashed line) and from CAFE simulations (solid line) [12]. The experimental CET values given here agree with the values obtained from grain size analysis in [13]. According to the definition methodology proposed in section 2, for FM1, the predicted CET_{min} is 130 mm very close to the measured 127 mm (figure 2(a)). For FM5, both simulation and experiment show two CET positions (figure 2(a)). The simulated CET_{min} is 140 mm that is 10 mm higher than the measured 130 mm. The predicted CET_{max} is 165 mm which is 9 mm higher than the measured 156 mm.

A good agreement between simulation and experiment is obtained for the distribution of grain equivalent diameter for FM1 (figure 2(b)). In simulations, 67 % and 78 % grains with a diameter smaller than 1.0 mm on sections L4 and L5 are measured, respectively, which is close to the respective experimental measurements of 65 % and 63%. For FM5 (figure 2(b)), the general tendency that grain size decreases from the intermediate zone to the equiaxed zone is reproduced. Without the application of the pulling-velocity jump, both simulation and experiment present a coarse and non-homogenous grain structure. In simulations, 19% and 31 % grains with a diameter smaller than 1.0 mm on sections L4 and L5 are measured respectively, which are different from the respective experimental measurements of 43% and 53 %. Difference between simulations and experiments is attributed to the fact that the formation and development of the grain structure in the two-dimensional axi-symmetrical simulation do not exactly correspond to the real experiments regarding the number of effective nuclei, the random growing orientations, and the branching in 3D. This discrepancy is less pronounced for the cases where many grains are formed, because statistical reasons minor these effects.

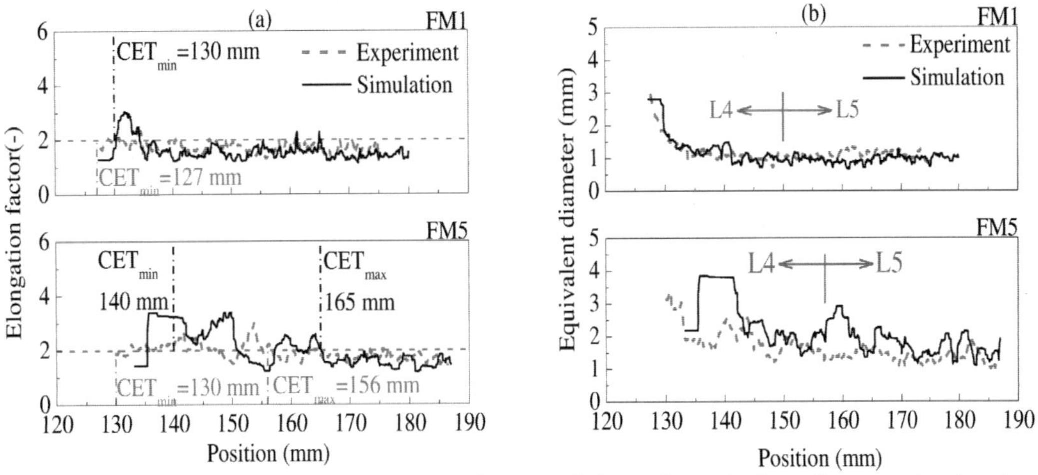

Figure 2. Comparison of (a) grain elongation factor and (b) grain equivalent diameter simulated using CAFE model in black with experimental measurements in red on sections L4 and L5: FM1 (refined Al - 7 wt% Si alloy, $V_2 = 0.2$ mm s^{-1}); FM5 (refined Al - 7 wt% Si alloy, $V_2 = 0.01$ mm s^{-1}). Superimposed red and black vertical dash-dotted lines in (a) correspond to experimentally-measured and predicted CET positions, respectively.

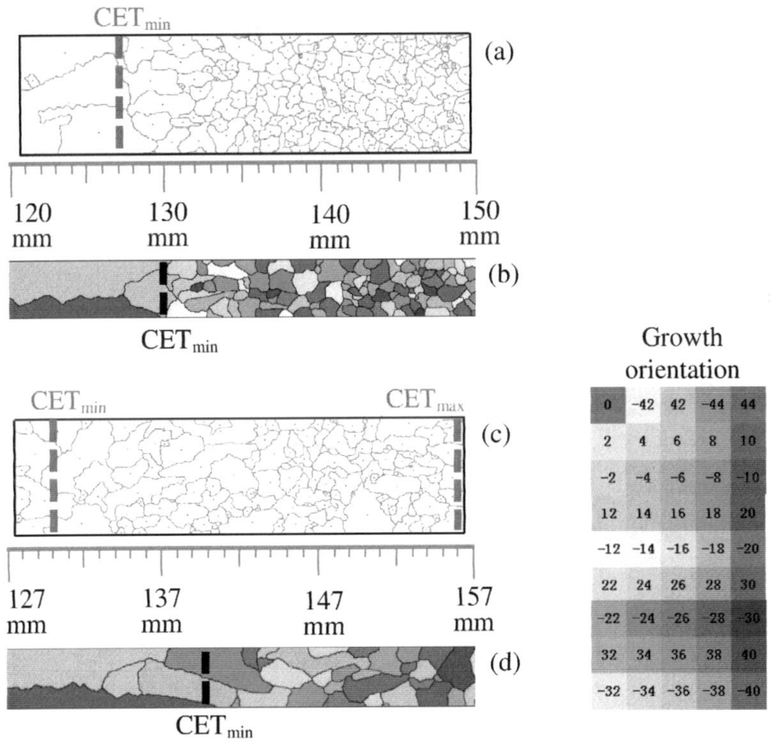

Figure 3. Comparisons of grain structures on section L4 (b) and (d) simulated using CAFE model with (a) and (c) EBSD measurements. (a) and (b) FM1 (refined Al - 7 wt% Si alloy, $V_2 = 0.2$ mm s^{-1}); (c) and (d) FM5 (refined Al - 7 wt% Si alloy, $V_2 = 0.01$ mm s^{-1}).

Figure 3 compares grain structures on section L4 obtained from EBSD analyses (figures 3(a) and (c)) and from CAFE simulations (figures 3(b) and (d)) [12]. From the simulated grain structures, it can be noted that the CET is sharp for FM1 (figure 3(b)) since there is only one CET position (CET_{min}). For FM5 (figure 3(d)), the CET is progressive. The two CET positions (CET_{min} and CET_{max}) are located on sections L4 and L5, respectively. The predicted CET transition mode is consistent with experimental observations.

5. Conclusion

Numerical simulations of two MSL-LGF microgravity experiments (FM1 and FM5) performed using CAFE model are compared qualitatively and quantitatively with the experiments. The influence of the pulling velocity on the grain structure is qualitatively and quantitatively reflected in CAFE simulations. The CET transition mode, be it sharp or progressive, is retrieved. Distributions of grain elongation factor and equivalent diameter are fairly reproduced. The discrepancy between the predicted and measured CET positions is due to the fact that the grain structure development in the two-dimensional axi-symmetrical simulations does not correspond to that in the experiments regarding the nucleation positions and growth orientations. The distributions of grain size and elongation factor are retrieved but cannot be perfectly identical with experimental measurements due to the statistical aspect of nucleation events.

Acknowledgements

This research work is supported by the European Space Agency (Noordwijk, NL) in the frame of the CETSOL project. Authors acknowledge Hydro-Aluminium Gmbh for providing the alloys used in this work. The 2D CAFE model was developed in the software XR2Sol+CAFE by CEMEF. The first author acknowledges financial supports from the National Natural Science Foundation of China (Grant No. 51101045) and Development and Application of Technology Project for young researcher of Harbin (Grant No. 2014RFQXJ026).

References

[1] Spittle JA 2006 *Int. Mater. Rev.* **51** 247
[2] Chen Y, Bogno AA, Xiao NM, Billia B, Kang XH, Nguyen-Thi H, Luo XH, Li DZ 2012 *Acta Mater.***60** 199
[3] Reinhart G, Gandin Ch-A, Mangelinck-Noël N, Nguyen-Thi H, Spinelli JE, Baruchel J, Billia B 2013 *Acta Mater.***61** 4765
[4] Zimmermann G, Sturz L, Billia B, Mangelinck-Noël N, Nguyen-Thi H, Gandin Ch-A 2011 *J. Phys. Conferences Series* **327** 012003
[5] Mirihanage WU, Browne DJ, Zimmermann G, Sturz L 2012 *Acta Mater.* **60** 6362
[6] Liu DR, Mangelinck-Noël N, Gandin Ch-A, Zimmermann G, Sturz L, Nguyen-Thi H, Billia B 2014 *Acta Mater.* **64** 253
[7] Gandin Ch-A 2010 *C. R. Phys.* **11** 216
[8] Guillemot G, Gandin Ch-A, Bellet M 2007 *J. Cryst. Growth* **303** 58
[9] Hunt JD 1984 *Mater. Sci. Eng.* **65** 75
[10] Cruz K SD, Mangelinck-Noël N, Gandin Ch-A, Billia B 2011 *IOP Conf. Ser. Mater. Sci. Eng.* **27** 012017
[11] Gandin Ch-A 2000 *ISIJ Int.* **40** 971
[12] Liu DR, Mangelinck-Noël N, Gandin Ch-A, Zimmermann G, Sturz L, Nguyen-Thi H, Billia B 2015 *Acta Mater.* **In press**
[13] Zimmermann G, Sturz L, Billia B, Mangelinck-Noël N, Liu DR, Nguyen-Thi H, Bergeon N, Gandin Ch-A, Browne DJ, Beckermann Ch, Tourret D, Karma A 2014 *Materials Science Forum* **790-791** 12

IOP Conf. Series: Materials Science and Engineering **117** (2016) 012010 doi:10.1088/1757-899X/117/1/012010

Modelling Al-4wt.%Cu as-cast structure using equiaxed morphological parameters deduced from in-situ synchrotron X-ray radiography

M Ahmadein[1,2,a], M Wu[1,3,b], G Reinhart[4,5], H Nguyen-Thi[4,5] and A Ludwig[1]
[1] Chair for Modeling and Simulation of Metallurgical Processes, University of Leoben, Austria
[2] Department of Production Engineering and Mech. Design, Tanta University, Egypt
[3] Christian-Doppler Laboratory for Advanced Process Simulation of Solidification & Melting, University of Leoben, Austria
[4] Aix- Marseille University & [5] CNRS, IM2NP UMR 7334, Campus Saint-Jérôme, Case 142, 13397 Marseille Cedex 20, France

Email: [a] mahmoud.ahmadein@unileoben.ac.at, [b] menghuai.wu@unileoben.ac.at

Abstract. The as-cast structure of laboratory scale Al-4wt.%Cu was numerically calculated using assumed morphological parameters. Two parameters are identified: The shape factor which correlates the growth velocity of dendrite envelope to that of the tip; and the sphericity of the equiaxed envelope or the circularity of the columnar trunk envelope which is required to calculate the averaged species diffusion flux across the envelope. In the present work, the real-time radiographs of equiaxed solidification experiment carried out on Al-4wt.%Cu at the European Synchrotron Radiation Facility are used to track the development of crystal envelope with time. The growth rate of the equivalent circular envelope was correlated to dendrite tip growth velocity to deduce the shape factor. The sphericity of dendrite envelope is estimated over the time. The average of the deduced morphological parameters is applied to the model to predict the as-cast structure. The results were compared to those obtained by using morphological parameters from literature. The predicted phase quantities, columnar-to-equiaxed transition, and macrosegregation exhibited significant dependence on those parameters. The predicted macrosegregation using the experimentally deduced parameters fits better to the measurements.

1. Introduction

Modelling the formation of the solidification structure is an essential tool to control the mechanical properties of a specific alloy. Numerous models have been proposed over decades. However, few of them can be applied on process scale. Beckermann and co-workers [1,2] used a multiphase, volume-averaged approach to incorporate the microscopic crystal growth kinetics into the macroscopic model for the transport phenomena occurring at the process scale. Rappaz and Thévoz [3,4] proposed a micro–macro model to incorporate the dendritic crystal morphology into the global transport system. The columnar-to-equiaxed transition (CET) was also treated in several works [5-7], but the growth of the columnar dendrites were over-simplified. More sophisticated 3-phase model dealing with mixed columnar-equiaxed solidification was proposed by some of the current authors [8-10]. It was demonstrated that it can account for melt convection and grain sedimentation, macrosegregation, and macrostructure. They extended their model to a 5-phase one that incorporates the dendritic growth of the columnar and equiaxed crystals [11,12]. The 5-phase model was verified to be able to quantitatively predict the as-cast structure of Al-alloy ingots [13,14] and the solidification and melt flow of the ammonium chloride transparent solution [15].

On continued trials to improve the quantitative results of the 5-phase model, the present work is devoted to evaluate the influence of morphological parameters on simulation results. Real-time radiographs of the equiaxed solidification experiment carried out on Al-4wt.%Cu at the European Synchrotron Radiation Facility was used to track the development of crystal envelope with time. The average shape factor and the sphericity of the equiaxed crystals are determined. The predicted results using morphological parameters from literature and those deduced from the current experiment are compared and discussed.

2. Modelling aspects

The 5-phase mixed columnar-equiaxed solidification model with dendritic morphology comprises three hydrodynamic phases: the liquid melt, the equiaxed crystals, and the columnar grains, denoted as e-, c- and l-phases and have corresponding volume fractions; f_l, f_e, and f_c. Globular and dendritic growth of equiaxed crystals and cellular and dendritic growth of columnar grains are considered. In case of dendritic growth, two additional phase regions exist within each of the equiaxed and the columnar crystal envelopes as shown in figure 1: the solid dendrites with corresponding solid fractions α_s^e and α_s^c, and interdendritic melt with corresponding volume fractions α_d^e and α_d^c inside crystal envelopes. It is assumed that the interdendritic melt is transported with the solid dendrites and is generally more solute-enriched than the extradendritic melt surrounding the crystals. In this sense, a fictitious crystal boundary envelope is constructed to separate the interdendritic melt from the extradendritic melt. Consequently, the system encompasses five 'thermodynamic' phases: (1) the solid dendrite and (2) the interdendritic melt within the equiaxed grain, (3) the solid dendrite and (4) interdendritic melt within the columnar crystal envelope, and (5) the extradendritic melt. The corresponding averaged volume fractions are: f_s^e, f_d^e, f_s^c, f_d^c, f_l referring to total volume and they characterize by corresponding solute concentration: c_s^e, c_d^e, c_s^c, c_d^c, c_l.

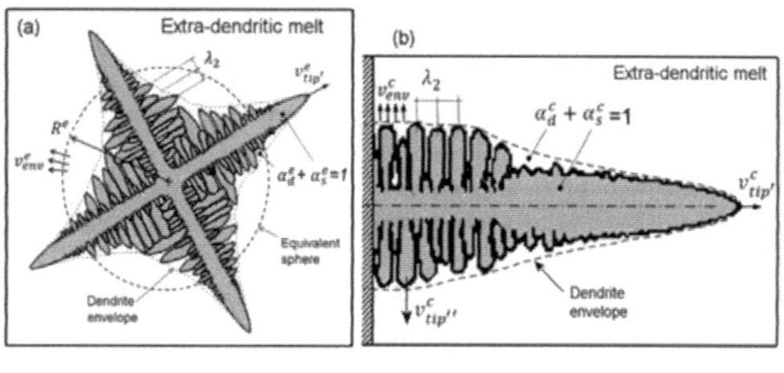

Figure 1. Schematic of the dendrite envelope separating the inter- and extra-dendritic melt for the (a) equiaxed crystal and (b) columnar dendrite.

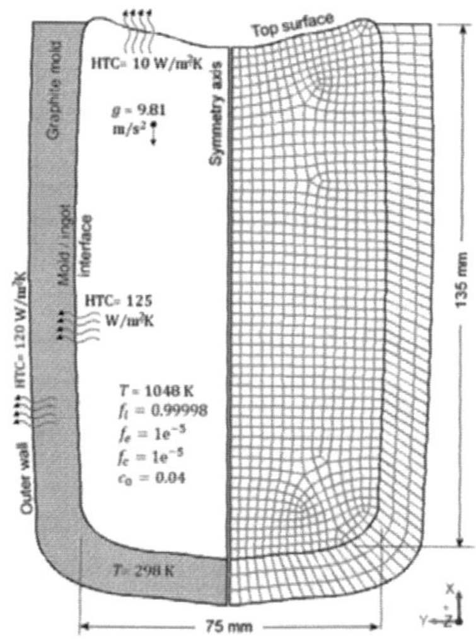

Figure 2. 2D-axisymmetric grid boundary and initial conditions of the ingot casting.

The growth of the grain envelope and the solidification of the interdendritic melt are treated differently. The growth of the envelopes is determined by dendrite growth kinetics. The semi-empirical Kurz-Giovanola-Trivedi (KGT) model [16] is used for the growth of columnar primary dendrite tips with velocity, $v^c_{tip'}$. Lipton-Glicksman-Kurz (LGK) model [17] is applied for the growth of columnar secondary dendrite tips (radial growth of the columnar trunk) and equiaxed primary dendrite tips. The profiles of the equiaxed grain and columnar trunk envelopes are respectively simplified as volume-equivalent sphere with radius R^e and volume-equivalent cylinder with radius R^c. For equiaxed crystals a shape factor, φ^e_{env}, correlates the growth velocity of the equivalent sphere, v^e_{env}, to the primary dendrite tip velocity, $v^e_{tip'}$, as given in equation (1). Similarly, φ^c_{env} for the columnar trunk correlates v^c_{env} to $v^c_{tip''}$ as given in equation (2). To calculate the species diffusive flux across the grain envelope, the real surface area of equiaxed and columnar envelopes, S^e_{env} and S^c_{env}, is respectively correlated to the surface area of equivalent sphere, S^e_{sph}, and the equivalent cylinder, S^c_{cyl}, by a sphericity form factor, φ^e_{sph}, for the equiaxed grain and a circularity shape factor, φ^c_{circ}, for the columnar trunk as given in equations (3) and (4), respectively.

$$v^e_{env} = \varphi^e_{env} \cdot v^e_{tip'} \quad (1)$$
$$v^c_{env} = \varphi^c_{env} \cdot v^c_{tip''} \quad (2)$$
$$S^e_{env} = S^e_{sph}/\varphi^e_{sph} \quad (3)$$
$$S^c_{env} = S^c_{cyl}/\varphi^c_{circ} \quad (4)$$

A heterogeneous nucleation law [18,19] is used to calculate the source term for the transport equation of the equiaxed number density. The advancement of the columnar tip front can be prevented either mechanically (hard blocking) when f_e ahead of the columnar primary tip front exceeds 0.49 [21] or thermodynamically (soft blocking) [6] when liquid melt at the columnar tip is enriched with solute to the extent that the constitutional undercooling vanishes. At this moment columnar-to-equiaxed (CET) transition occurs. Further details of the model are provided in [11,20].

The above model is used to predict the solidification structure of a cylindrical Al-4.0wt.%Cu ingot in a graphite mould using a 2D axisymmetric grid as shown in figure 2. The temperature of mould and melt were initially set to 298 K and 1048 K respectively. The thermophysical properties and the modelling parameters of the alloy are provided in [14]. The conservation equations of mass, enthalpy, species, and number density are solved using CFD software package, ANSYS-Fluent version 14.5.0 based on the control volume method. The time step size to achieve the solution convergence during a solidification time of ~525 s ranged between 1×10-5 and 1×10-3 s.

Five simulation cases were conducted to study the influence of the morphological parameters on the predicted as-cast structure using assumed, literature, and measured morphological parameters as listed in table 1. Real-time radiographs of equiaxed solidification experiment carried out on Al-4wt.%Cu are used to determined φ^e_{env} and φ^e_{sph} as explained in the next section.

Table 1. Morphological parameters of the dendritic envelope from different sources.

Study cases	φ^e_{env}	φ^e_{sph}	φ^c_{env}	φ^c_{circ}	Source
Case#1	0.48[a]	0.4[a]	0.798[a]	0.5[a]	[a] Assumed
Case#2	0.683[b]	0.846[b]	0.798[b]	0.886[b]	[b] Equiaxed: octahedral, columnar: square rod [11]
Case#3	0.478[c]	0.837[c]	0.564[c]	0.56[c]	[c] Equiaxed: 6 orthogonal square pyramids,
Case#4	0.358[d]	0.623[d]	0.798[b]	0.886[b]	columnar: 4 orthogonal square wedges [11]
Case#5	0.358[d]	0.623[d]	0.564[c]	0.56[c]	[d] Current experiment

3. Determination of equiaxed morphological parameters

Directional solidification experiments were carried out on beamline ID19 at the European Synchrotron Radiation Facility (ESRF, Grenoble, France). The sample was made of an Al-4wt.%Cu alloy, 35 mm in length, 6 mm in width and 150-200 µm in thickness. The sample was adjusted into a soft graphite crucible designed to somewhat accommodate the thermal expansion stresses in the region surrounding the solid-liquid interface. The details of the used Bridgman furnace have been described in [23]. The applied temperature gradient G between the two heaters was set to 17.5 K/cm and anti-parallel to gravity. Solidification was triggered by decreasing the temperature of the hot zone at a rate of 1 K/min, while maintaining the same injected power to the bottom heater. To trigger the CET the temperature gradient is continuously decreased to 10 K/cm. Radiographs of the solidifying samples were recorded with an ESRF FReLoN (Fast Read-out Low Noise) camera. Details about the used optics, beam energy and image processing can be found in [24]. The final result is enhanced contrast and almost defect free images with the solidified Al–rich structures appearing in light grey and the Cu–rich liquid regions appearing in dark grey as shown in figure 1.

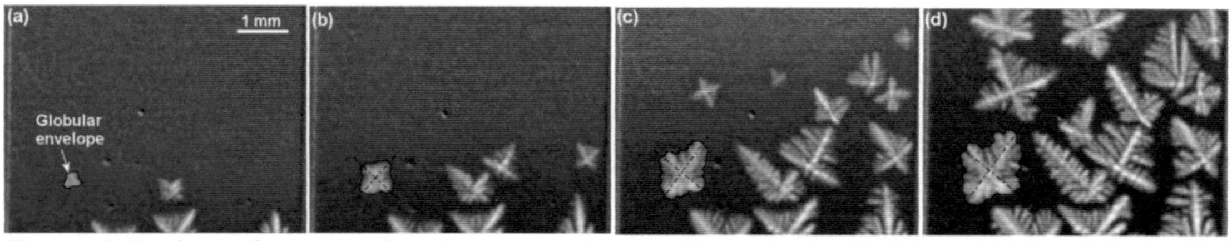

Figure 3. In situ radiographs of equiaxed solidification of Al-4wt.-% Cu at relative times (a) 0 s, (b) 45 s, (c) 126 s, and (d) 225 s, with cooling from the bottom at a rate of 0.5 K/min.

To determine φ_{env}^{e} and φ_{sph}^{e}, the growth of primary dendrite tip and the evolution of dendrite envelope are traced over the solidification time. 2D real-time radiographs of the growing equiaxed crystals from the above experiment are used for this purpose. An example of tracing the envelope development is shown in figure 1. The following procedure is applied to four crystals: (1) the most equiaxially growing crystals are selected, (2) evaluation starts from the onset of crystal growth and ends before grain impingement, (3) two main growth axes and a dendrite envelope are drawn for each crystal, (4) the radius of the equivalent-area circle, R^e, is calculated, (5) from the length development of the each of the dendritic axes, v_{tip}^{e}, is calculated as the average of both, (6) from the development of R^e over time, v_{env}^{e} is calculated, (7) equation (1) is applied at each time to get φ_{env}^{e}, (8) φ_{sph}^{e} is calculated as the ratio of the perimeter of the equivalent-area circle to the perimeter of the actual dendrite envelope.

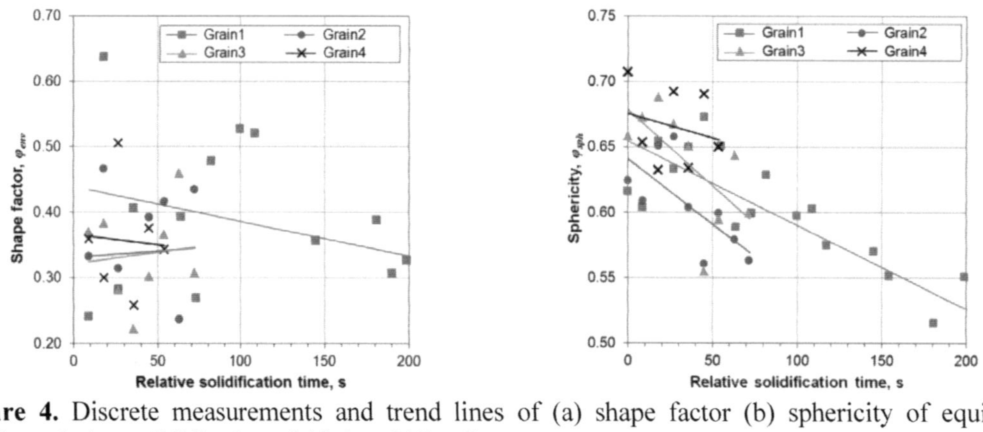

Figure 4. Discrete measurements and trend lines of (a) shape factor (b) sphericity of equiaxed envelope during solidification of Al-4wt.-%Cu alloy.

The development of φ_{env}^{e} and φ_{sph}^{e} is plotted for the 4 crystals versus solidification time as shown in 2a and b, respectively. The measured φ_{env}^{e} and φ_{sph}^{e} are fluctuating with time, in contradiction with the assumptions of the 5-phase model [11]. This can be attributed to melt flow and the unsteady solute distribution in the liquid ahead of the dendrite tips, which affects the tip undercooling and correspondingly the growth rate and envelope morphology. The resulting φ_{env}^{e} is more fluctuating than φ_{sph}^{e} because it is calculated directly from $v_{tip'}^{e}$, which is very sensitive to the tip undercooling. In general, φ_{sph}^{e} decreases with time. High φ_{sph}^{e} is an evidence for the globular growth at the beginning of solidification as shown in 2b. As solidification proceeds the morphology tends to be more dendritic. As a result, φ_{sph}^{e} decreases. For modelling purpose, a constant value of $\varphi_{env}^{e} = 0.358$ and $\varphi_{sph}^{e} = 0.623$ are considered by averaging the measurements of the four crystals over the time.

4. Simulation results

The as-cast structure is calculated using 5 sets of morphological parameter as per table 1. The resulting CET lines are overlaid on the as-cast structure from experiment as shown in fgure 5. The as-cast structure for all cases comprises an equiaxed core, columnar top and ingot peripherals, and a CET surface. Compared to the top CET boundary, the bottom CET boundary is almost unaffected. Nevertheless, the proportion of each phase region differs depending on the morphological parameters.

Some averaged phase quantities in the ingot and inside the CET surface are listed in table 2. By comparing the results (case#2 vs. case#4) and (case#3 vs. case#5), it can be easily concluded that, as φ_{env}^{e} decreases, the formed fraction of equiaxed phase in ingot, f_e decreases. This can be attributed to the reduction of the growth rate of the equiaxed grain envelope v_{env}^{e} according to equation (1).

The sphericity of the equiaxed crystal in the 5-phase model is directly correlated to the envelope diffusion area as given in equation (3). The smaller the sphericity (i.e. more dendritic morphology) the higher the diffusion surface concentration, S_{env}^{e} and correspondingly, the diffusion flux though the grain envelope increases. By increasingS_{env}^{e} a higher species transport rate is expected from the solute-rich interdendritic liquid to the less enriched extradendritic liquid. As a result, inside the dendrite envelope the formed eutectic (residual liquid) fraction, α_{d}^{e}, decreases and the solid fraction, α_{s}^{e}, increases. This is obvious from the results listed in table 2, where φ_{sph}^{e} decreases from 0.846 to 0.623 (case#2 and case#4) the corresponding α_{d}^{e} (eutectic) inside the CET surface decreases from 0.078 to 0.037 in account of solid fraction α_{s}^{e}. Accordingly, the concentration of equiaxed dendrite envelope, c_e, decreases 0.039 to 0.0257. Analogous correlation is achieved between case#3 and case#5. Since φ_{sph}^{e} is almost the same for case#2&3 and for case#4&5, the calculated α_{d}^{e} and c_e are very close for each pair.

Table 2. Calculated average of some phase quantities in ingot and inside the CET surface.

	f_e-ingot	f_c-ingot	α_d^e-core	α_s^e-core	c_e-core
Case#1	0.252	0.711	0.012	0.987	0.0170
Case#2	0.35	0.647	0.078	0.919	0.0390
Case#3	0.311	0.663	0.073	0.925	0.0370
Case#4	0.216	0.778	0.037	0.959	0.0257
Case#5	0.236	0.731	0.033	0.966	0.0252

The calculated solute concentration of phase mixture is used to study the influence of φ_{env}^{e} and φ_{sph}^{e} on macrosegregation. A comparison between the simulation results using theoretical and experimental morphological parameters (case#2 and case#4, respectively) is shown in 4. Both exhibit macrosegregation. However, the latter case is much severe. Compared to the measured macrosegregation map [14], the predicted macrosegregation using morphological parameters from experiment (case#4) exhibit better agreement. Thus the range and distribution of Cu concentration are quite similar. The higher negative macrosegregation of case#4 inside the CET surface can be simply attributed to the reduced eutectic content within equiaxed grain envelope (α_{d}^{e} in table 2) as discussed above. The higher positive segregation in case#4 can also be related to φ_{sph}^{e}. The solute rejected from the interdendritic liquid enriches the extradendritic liquid which accumulates due to melt convection in specific regions and solidifies as eutectic causing positive macrosegregation.

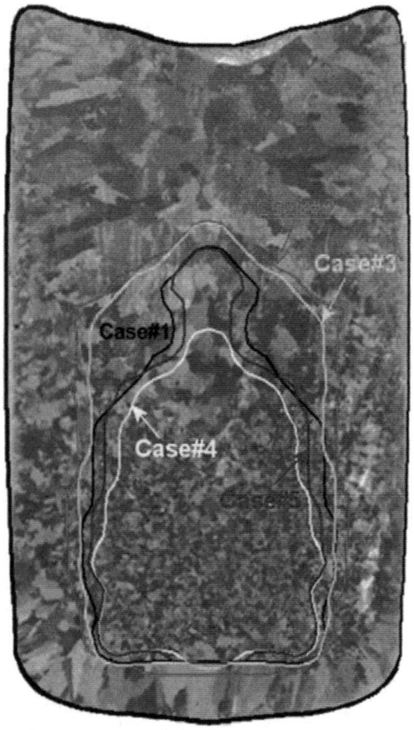

Figure 5. Calculated CET using different morphological parameters as listed in Table 1.

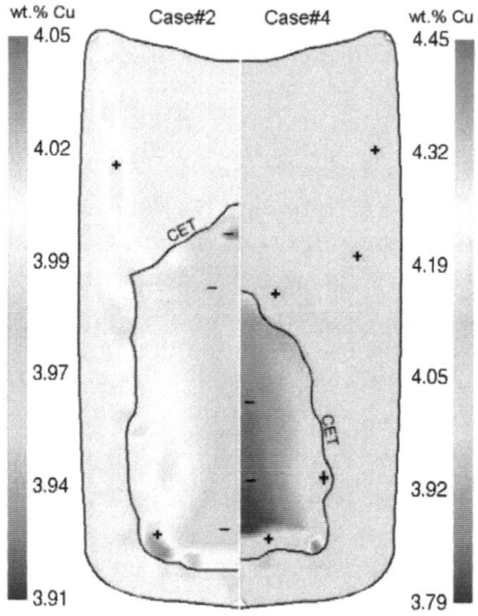

Figure 6. Calculated segregation map with CET for cases #2 and #4.

5. Summary

Shape factor and sphericity of equiaxed dendrite envelope can be extracted from the in situ observation of solidification experiments. In general, φ_{env}^{e} and φ_{sph}^{e} vary with time but φ_{sph}^{e} tends to decrease. The calculated average over time of 4 crystals was $\bar{\varphi}_{env}^{e}$= 0.358 and $\bar{\varphi}_{sph}^{e} = 0.623$. The measured φ_{env}^{e} and φ_{sph}^{e} are relatively lower than the theoretical values from literature. The comparative numerical study shows that CET configuration and the phase quantities are considerably affected by the morphological parameters. However, the variation is not so dramatic. Decreasing φ_{env}^{e} reduces envelope growth rate and correspondingly the formed equiaxed phase fraction. Smaller φ_{sph}^{e} exhibits greater diffusion area and flux thought the fictitious grain envelope. Correspondingly, eutectic fraction and solute concentration in equiaxed crystals decrease. Thus the macrosegregation map is also affected by the morphological parameters. The calculated macrosegregation using the experimental morphological parameters fits better to the measurements. Future improvements can be done by using φ_{env}^{c} and φ_{circ}^{c} estimated from experiments and to study the influence of cooling conditions on the envelope evolution and shape factors. Numerically, variable morphological parameters have to be implemented in the current model.

Acknowledgement
This work is financially supported by the FWF Austrian Science Fund (P23155-N24).

References

[1] Beckermann C and Viskanta R 1993 *Applied Mechanics Reviews* **46** 1
[2] Wang C.Y and Beckermann C 1997 *Metall. Mater.Trans* A **27** 2754
[3] Rappaz M and Thévoz P.h 1987 *Acta Metallurgica* **35** 1487
[4] Rappaz M 1989 *Inter. Mater. Rev.* **34** 93
[5] Wang C.Y and Beckermann C 1994 *Metall. Mater. Trans. A* **25** 1081
[6] Martorano M.A, Beckermann C and Gandin Ch-A 2003 *Metall. Mater. Trans.* A **34** 1657

[7] Ciobanas A.I and Fautrelle Y 2007 *J. Physics* D **40** 3733
[8] Ludwig A and Wu M 2005 *Metall. Mater. Trans.* A **413** 109
[9] Wu M and Ludwig A 2006 *Metall. Mater. Trans.* A **37** 1613
[10] Wu M and Ludwig A 2007 *Metall. Mater. Trans.* A **38** 1465
[11] Wu M and Fjeld A Ludwig 2010 *Comput. Mater. Sci.* **50** 32
[12] Wu M and Ludwig A Fjeld A 2010 *Comput. Mater. Sci.* **50** 43
[13] Wu M, Ahmadein M, Kharicha A, Ludwig A, Li J.H and Schumacher P 2012 *13th MCWASP* (Schladming: Austria) IOP Conf. Ser.: Mater. Sci. Eng. 33
[14] Ahmadein M, Wu M, Li J.H, Schumacher P and Ludwig A 2013 *Metall. Mater. Trans.* A **44A** 2895
[15] Ahmadein M, Wu M, Stefan-Kharicha M, Kharicha A and Ludwig A 2014 *Mater. Sci. Forum* **790-791** 247
[16] Kurz W and Fisher D.J 1998 *Fundamentals of Solidification* 4th ed (Zurich: Trans Tech)
[17] Lipton J, Glicksman M.E and Kurz W 1984 *Mater. Sci. Eng.* **65** 57
[18] Thevoz Ph and Rappaz M 1989 *Metall. Trans.* A **20A** 311
[19] Rappaz M and Gandin Ch.-A 1993 *Acta Metallurgica et Materialia* **41** 345
[20] Wu M and Ludwig A 2009 *Acta Mater.* **57** 5621
[21] Hunt J 1984 *Mater. Sci. Eng.* **65** 75
[22] Bogno A, Nguyen-Thi H, Reinhart G, Billia B and Baruchel J 2013 *Acta Mater.* **61** 1303
[23] Nguyen-Thi H, Jamgotchian H, Gastaldi J, Härtwig J, Schenk T, Klein H, Billia B, Baruchel J and Dabo Y 2003 *J Phys* D*: Appl. Phys.* **36** A83
[24] Buffet A, Nguyen-Thi H, Bogno A, Schenk T, Mangelinck-Noël N, Reinhart G, Bergeon N, Billia B and Baruchel J 2010 *Mater Sci Forum* **649** 331–336

Numerical investigation of solidification and CET of the transparent alloy NPG-37.5 wt.% DC in microgravity "TRACE" experiment

M Ahmadein[1,2,a], M Wu[1,3,b], L Sturz[4,c] G Zimmermann[4,d] and A Ludwig[1]
[1] Chair for Modeling and Simulation of Metallurgical Processes, University of Leoben, Austria
[2] Department of Production Engineering and Mechanical, Tanta University, Egypt
[3] Christian-Doppler Laboratory for Advanced Process Simulation of Solidification & Melting, University of Leoben, Austria
[4] ACCESS e.V., Aachen, Germany

Email: [a] mahmoud.ahmadein@unileoben.ac.at, [b] menghuai.wu@unileoben.ac.at, [c] L.Sturz@access.rwth-aachen.de, [d] G.Zimmermann@access.rwth-aachen.de

Abstract. A solidification experiment "TRACE" of the transparent alloy Neopentylglycol (NPG)-37.5wt.% D-Camphor (DC) was conducted on-board the sounding rocket TEXUS-47 in low-gravity environment to investigate the columnar growth and the columnar-to-equiaxed transition (CET). To improve the fundamental understanding of solidification and CET in microgravity, the current laboratory scale experiment was tried to be numerically reproduced by a recently developed 5-phase volume averaging model. The temperature gradient in the solidification cell is applied to the simulation. In absence of melt flow, the calculated cooling curves, columnar tip position, tip undercooling and velocity, and number density of equiaxed crystals were compared to the results of in-situ real-time observations of the experiment. The CET could be predicted at position close to that of experiment. Simulation reveals the competitive growth between the columnar and equiaxed crystals before CET. Modelling parameters of equiaxed nucleation and columnar tip growth are the key to regulate this competition and to locate the CET. Experimental verification of modelling parameters considering melt flow is intended in the future work.

1. Introduction

Mechanical properties of alloys are directly related to their as-cast structure. Columnar growth is favoured for example in single crystal growth [1] or in some turbine blades [2]. High thermal gradient at solid/liquid interface promotes the columnar growth. Equiaxed grain structure exhibits more isotropic properties and are formed upon nucleation that can be provoked by several mechanisms: (1) heterogeneous nucleation [3]; (2) the 'big band' theory [4]; (3) partial remelting of columnar dendrites [5]; and (4) the showering down of dendrite crystals formed from the casting top surface [6]. A mixed columnar-equiaxed structure with a transition zone (CET) exhibit undesirable anisotropic mechanical properties. CET occurs if the volume faction of equiaxed grains ahead of the columnar tip front is high enough to block the advancement of the columnar tip front (hard blocking) [7]. The advancement of the columnar tip front can be decelerated or even blocked (soft blocking) if the solute concentration in the melt ahead of the front increases to the limit that the constitutional undercooling decreases or vanishes[8].

Understanding the origin of the solidification structure helps to control the structure and correspondingly the mechanical properties. Gravity has a remarkable contribution to the thermal and solutal convection of the melt and the crystal sedimentation. The transport of heat and mass are correspondingly affected. The analysis of solidification phenomenon can be simplified by reducing the aforementioned interactions in low gravity (microgravity) conditions. For this purpose a solidification experiment of the transparent alloy Neopentylglycol (NPG)-37.5wt.% D-Camphor (DC) was conducted on-board the sounding rocket TEXUS-47 in a low gravity environment [9,10]. The current work is dedicated to investigate the columnar/equiaxed competition during the micro-gravity experiment using a 5-phase mixed columnar-equiaxed solidification model [11,12]. A numerical study is conducted to adjust the model parameters for the nucleation law and the semi-empirical model of columnar tip velocity. The numerical predictions were discussed and compared to the experiment.

2. Experimental aspects

The concept of "TRACE" (TRansparent Alloy in Columnar Equiaxed solidification) is to perform directional solidification in a thin (1 mm) sample using the optically transparent model system NPG-37.5 wt.% DC alloy for in-situ and real-time observation of the solidifying dendritic microstructure, see figure 1. The thermal gradient within the container is controlled by superimposing a forced cooling at the bottom (cooler) and the top (heater) as shown in figure 2. The columnar growth is initiated by applying a cooling rate of 1/300 K/s at heater and cooler at a roughly constant gradient. An increase of the cooling rate to 1/30 K/s leads to a transition to equiaxed dendritic growth and thus to CET. Melting, mixing, homogenization of the liquid alloy and columnar growth is carried out prior to the rocket launch. The low-gravity level (below 1mg) is achieved at about 1 min after lift-off of the rocket and the high cooling rate is applied to provoke CET under diffusive conditions. Images are taken in overview and from microstructure details. The thermal profile within the alloy is measured using five type-K thermocouples with diameter 0.25 mm inserted from one side. The thermal data and video data are analysed in terms of the position, velocity and temperature of the solidification interface. The nucleation temperatures and the number density of equiaxed crystals are estimated from the experiment. In addition, the volume fractions of the columnar behind the solidification front and the equiaxed and liquid phases ahead of solidification front are determined. Further details of experimental procedure can be found elsewhere [9,10].

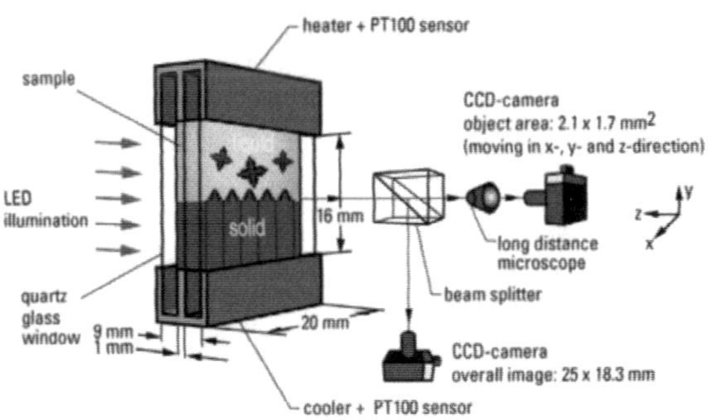

Figure 1. Schematic drawing of the main parts of the experimental setup.

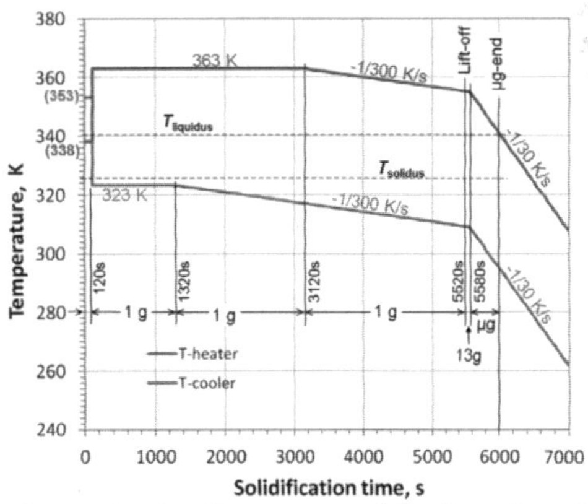

Figure 2. Cooling curves of the forced cooling system at the heater and cooler in µg experiment.

3. Model features

The 5-phase mixed columnar-equiaxed solidification model with dendritic morphology comprises three hydrodynamic phases: the liquid melt, the equiaxed crystals, and the columnar grains, denoted as l-, e- and c-phases and have corresponding volume fractions; f_l, f_e, and f_c. Globular and dendritic growth of equiaxed crystals and cellular and dendritic growth of columnar grains are considered. In case of dendritic growth, two additional phase regions exist within each of the equiaxed and the columnar crystal envelopes: the solid dendrites with corresponding solid fractions α_s^e and α_s^c, and interdendritic melt with corresponding volume fractions α_d^e and α_d^c inside crystal envelopes. Consequently, the system encompasses five 'thermodynamic' phases: (1) the solid equiaxed dendrite and (2) the interdendritic melt within the equiaxed grain envelope, (3) the solid columnar dendrite and (4) interdendritic melt within the columnar crystal envelope, and (5) the extradendritic melt. The corresponding averaged volume fractions are: f_s^e, f_d^e, f_s^c, f_d^c, f_l referring to total volume and they are characterized by corresponding solute concentration: c_s^e, c_d^e, c_s^c, c_d^c, c_l.

The growth of the grain envelope and the solidification of the interdendritic melt are treated differently. The growth of the envelopes is determined by dendrite growth kinetics. The semi-empirical Kurz-Giovanola-Trivedi (KGT) model [12] is fitted by the polynomial given in equation (1) and is used for the growth of columnar primary dendrite tips, where $v_{tip'}^c$: growth velocity of the columnar primary dendrite tip, k_1 and k_2: model parameters. Lipton-Glicksman-Kurz (LGK) model [13] is applied for the growth of columnar secondary dendrite tips (radial growth of the columnar trunk) and equiaxed primary dendrite tips.

A heterogeneous nucleation law [15,16], equation (2), with three fitting parameters; the mean undercooling, ΔT_M, the standard deviation of undercooling, ΔT_σ, and maximum grain number density, n_{max} is used to calculate the source term for the transport equation of the equiaxed number density, n, where ΔT_{con} is the constitutional undercooling. Further details of the model are provided in [11,12].

$$v_{tip'}^c = k_1 \cdot \Delta T_{\text{con}}^2 + k_2 \cdot \Delta T_{\text{con}}^3 \tag{1}$$

$$\frac{dn}{d(\Delta T)} = \frac{n_{max}}{\sqrt{2\pi} \cdot \Delta T_\sigma} \cdot e^{-\frac{1}{2}\left(\frac{\Delta T_{\text{con}} - \Delta T_M}{\Delta T_\sigma}\right)} \tag{2}$$

The advancement of the columnar tip front can be stopped either mechanically (hard blocking) when f_e ahead of the columnar primary tip front exceeds 0.49 [7] or thermodynamically (soft blocking) [8] when liquid melt at the columnar tip is enriched with solute to the extent that the constitutional undercooling vanishes. At this moment columnar-to-equiaxed (CET) transition occurs. The CET position is determined based on the growth competition between the c- and e-phases. The growth rate of c-phase perpendicular to cold bottom wall is calculated from equation (1) and that for e-phase is affected by the growth kinetics and the number density, n, from equation (2).

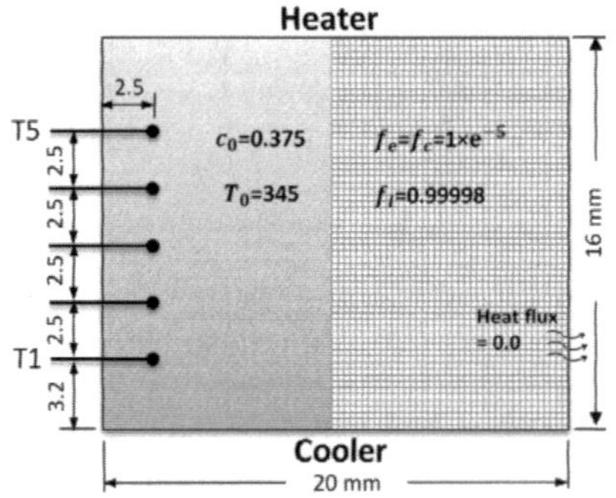

Figure 3. 2D grid of the solidification cell with boundary and initial conditions.

4. Configuration of simulation

The solidification cell of Neopentyl-glycol (NPG)-37.5wt.% D-Camphor (DC) is simplified to a 2D grid with volume element size of 0.25×0.25 mm^2 as shown in figure 3. The grid is provided by 5 measuring points (thermocouples) to record the cooling curves analogous to experiment. Similar to the experiment, a temperature gradient is applied to the simulation grid by superimposing the experiment cooling curves at the cooler and heater (Figure 2) on the corresponding grid boundaries. The conservation equations of mass, enthalpy, species, and number density are solved sequentially at each time using CFD software package, ANSYS-fluent version 14.5.0 based on the control volume method. Time step size of 0.2 s was used to achieve solution convergence over a solidification time of ~7000 s.

The thermophysical properties of NPG-37.5 wt.% DC and the modelling parameters are listed in table 1. Here ΔT_m and ΔT_σ are estimated by a Gaussian fit to experimental data and k_1 is assumed. In microgravity (μg) experiment CET occurs at ~9.5 mm above the cold boundary. This value is used to estimate the fitting n_{max} and k_2 after a numerical parameter study.

Table 1. Thermophysical properties and modelling parameters of NPG-37.5wt.% DC [9,17,18].

Thermo physical properties:	**Nucleation parameters:**
c_0= 37.5 wt.% DC, c_E= 45.3 wt.% DC	n_{max}= 5.0326×10^{10} m^{-3}, ΔT_σ= 4.0 K, ΔT_m = 13.5 K
$T_{liquidus}$ [K]: 340.25, $T_{liquidus}$ [K]: 325.95	**Morphological parameters:**
Liquidus slope: -1.84 K/wt.%, Partition coefficient: 0.085	Equiaxed dendrite:
Diffusion coeff. [m^2/s] liquid: 9.7×10^{-11}, solid: 8.0×10^{-20}	Shape factor: 0.48, Sphericity: 0.4
Thermal capacity [J/kg·K] liquid: 2400, solid: 2650	Columnar dendrite:
Thermal conductivity [W/m·K] liquid: 0.125, solid: 0.265	Shape factor: 0.7979, Trunk circularity: 0.5
Density [kg/m^3]: 1035	λ_1= 400 μm, λ_2= 50 μm
Latent heat of fusion [kJ/kg]: 33.4	**Tip growth parameters for KGT model:**
Liquid viscosity [kg/m·s]: 6.43×10^{-3}	k_1= 1.1x10^{-8}, k_2 =4.55x10^{-8}
Gibbs Thomson coefficient [Km]: 7.8e-8	

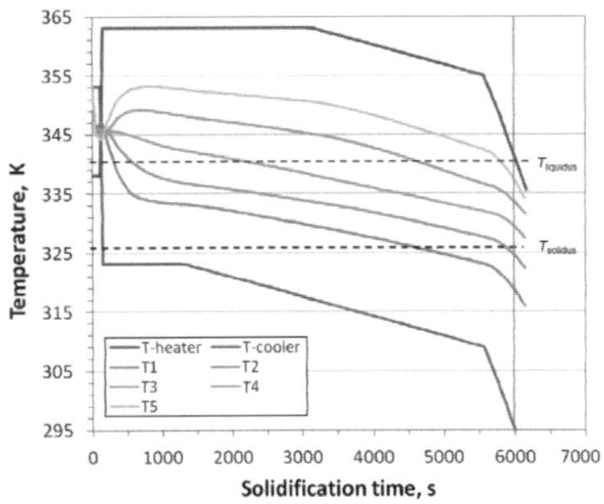

Figure 4. The calculated cooling curves in the solidification cell at positions T1-T5.

5. Results and discussion

The calculated cooling curves of the "thermocouples" 1→5 plotted in figure 4 are very similar to those from experiment [9]. The melt is supercooled at position T1 after ~300 s, whereas position T5 is supercooled at 5740 s. Increasing cooling rate after 5580 s triggers the CET. The low thermal conductivity of the alloy delays the response of the cooling curves to the cooler and heater temperatures. However, a constant temperature gradient is almost maintained.

Solidification starts at the bottom as columnar with an advancing tip front (c-front) towards the top as shown in figure 5a. Later on, equiaxed crystals nucleate and grow ahead of the c-front. The melt below the liquidus isoline is supercooled. Increasing the cooloing rate after ~5500 s expands the supercooled liquid region between c-front and liquidus isoline (figure 5b). Correspondingly, the nucleation rate of e-phase and the growth rates of e- and -phases increase. This competition results in higher f_e ahead of the advancing c-front to surpass the blocking limit (0.49). Thus, the advancement of c-front stops and CET occurs. Thereafter (figure 5c), the solidification of e-phase continues ahead of the c-front, whereas c-phase solidify only behind the c-front. Similar to μg-experiment (figure 5d), the calculated CET occurs between T3 and T4. Nevertheless, in simulation CET occurs relatively earlier than experiment. This can be attributed to the overestimation of k_2 parameter.

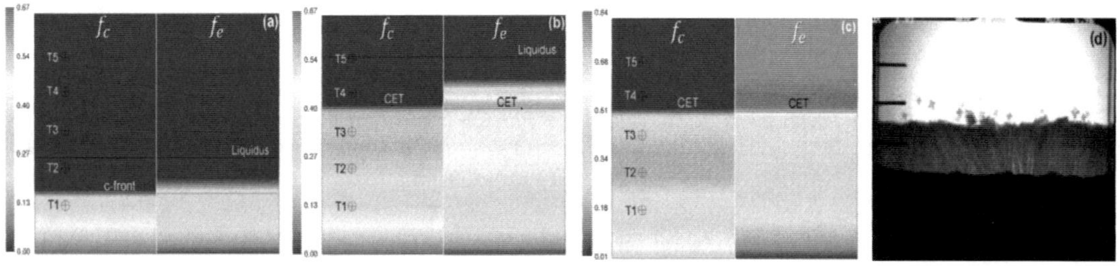

Figure 5. The calculated f_c and f_e at (a) 820 s, (b) 5820 s, and (c) 6400 s and (d) the CCD-image of the solidification cell in μg-experiment at 5955 s.

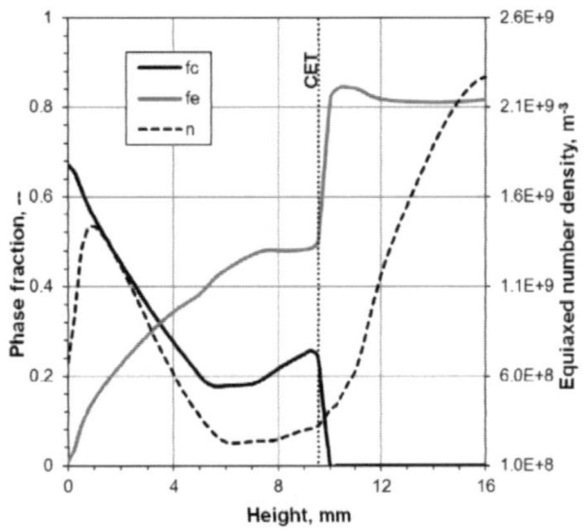

Figure 6. Distribution of f_c, f_e and n versus mould height at 6400 s.

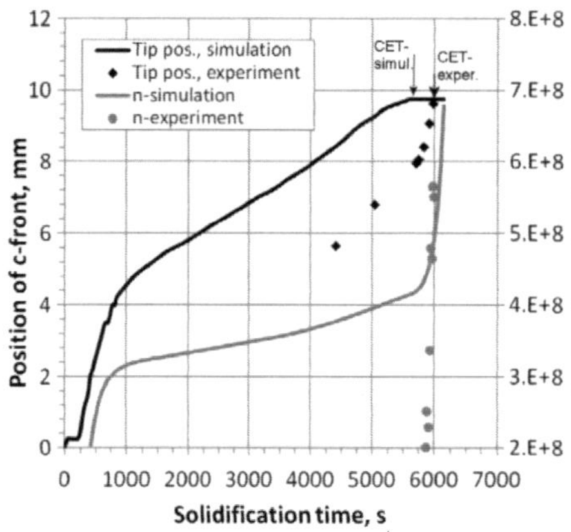

Figure 7. Calculated c-front position and average n *vs* some experimental observations.

The distribution of f_c, f_e and n at 6400 s is plotted versus the mould height as shown in figure 6. Solidification at the bottom starts as columnar. Sudden temperature drop at the bottom after 2 min increases the nucleation rate and correspondingly f_e. The competitive growth between e- and c-phases *is evident below CET (9.5 mm), particularly when the cooling rate increases at 5500 (~8 mm). Thus,* f_c increases but f_e exceeds 0.49 and CET occurs. Ahead of CET f_c= 0 and the nucleation rate increases as the solidification front comes closer to the top.

The development of the position of c-front and the average n in simulation and experiment are plotted during solidification as shown in Figure 7. Initial fast cooling at the bottom causes rapid advancement of c-front which is associated with high nucleation rate. After ~800 s the thermal gradient within the container is stabilized (0). Accordingly, c-front advances linearly and a constant nucleation rate for the equiaxed crystals is obtained. After ~5500 s, the thermal gradient is maintained whereas the cooling rate is increasing. Respectively, the calculated growth rate of c-front is maintained and the nucleation rate of e-phase in the liquid close to c-front increases, leading to blocking of c-front and CET. Calculated and observed CET occur at ~9.5 mm. However, earlier CET is predicted by the model at ~5820 s. A linear growth rate for c-front is observed before 5500 s with relatively faster rate in simulation. After 5500 s the observed c-front advancement rate increases, whereas the calculated one decreases. The calculated n increases at CET, whereas the observed n increases abruptly. According to experiment, almost no equiaxed crystals form before 5500 s. Formation of equiaxed crystals within the columnar mush is also unobservable.

The calculated undercooling using different laws and the measured one are plotted in Figure 8. The behaviour of ΔT_{exp} and ΔT_{th} are similar, but with ~ 4 K deviation. Measuring ΔT_{exp} did not consider melt constitution [9] and is based on the interpolation of thermocouple readings. ΔT_{con} is applied to equation (1) and (2) and is generally smaller than ΔT_{exp} and ΔT_{th}. Increasing cooling rate after 5500 s increases the concentration of rejected solute in melt. Correspondingly, ΔT_{con} decreases and the growth rate of c-front decreases until CET occurs (Figure 7). The calculated n can also be correlated to the development of ΔT_{con}. Additional experimental and numerical investigations to the estimation of ΔT are still demanding. Practically, observation of nucleation is still ambiguous.

The deviation between observations and the calculations can be attributed to using assumed modelling parameters; k_2 and n_{max}. A parameter study is conducted to estimate the best fit of n_{max} and k_2 to the experiment, two parameter sets are used: (I) $k_1 = 1.1\text{x}10^{-8}$, $k_2 = 4.55\text{x}10^{-8}$ with variable n_{max}: 4×10^{10} - 8×10^{10} m^{-3}; (II) $k_1 = 1.1\text{x}10^{-8}$, $n_{max} = 5.0326\times10^{10}$ with variable k_2: 2.66×10^{-8} - 2.8×10^{-8}. The position of CET is plotted versus n_{max} and k_2 for 32 simulation cases as shown in Figure 9. Columnar growth dominates if $n_{max} < 5.0\times10^{10}$ m^{-3} or $k_2 > 2.75\times10^{-8}$ whereas equiaxed solidification dominates if $n_{max} > 5.58\times10^{10}$ m^{-3} or $k_2 < 2.66\times10^{-8}$. It is obvious that CET is very sensitive to the nucleation parameters of e-phase and to the growth parameters of c-phase. The CET position can be numerically be reproduced by several combinations of the parameters. Here the parameter set (I) with $n_{max} = 5.0326\times10^{10}$ m^{-3} and set (II) with $k_2 = 2.7368\times10^{-8}$ achieved CET at 9.5 mm but at different times. Further investigations are recommended to determine the proper modelling parameters.

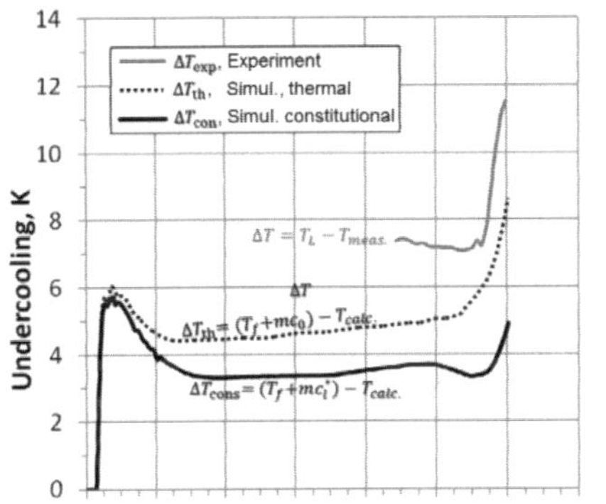

Figure 8. Calculated and measured undercooling at c-front during solidification.

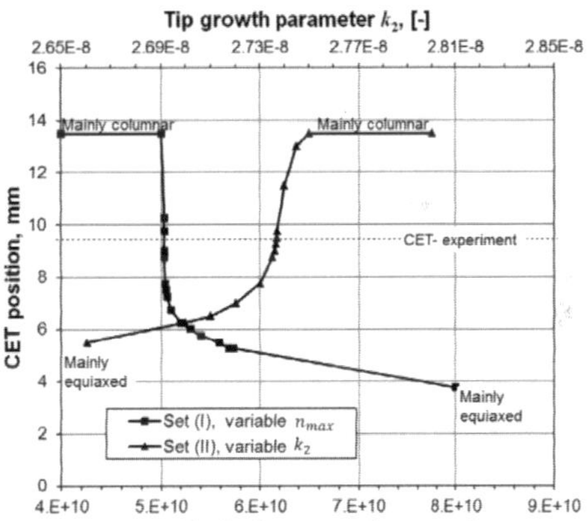

Figure 9. The calculated positions of CET at various n_{max} and k_2 combinations.

6. Conclusion

The 5-phase model might reproduce the solidification structure of µg-experiment including the CET. However, the results are very sensitive to the nucleation parameters of the equiaxed crystals, particularly $\boldsymbol{n_{max}}$, and to the growth parameters of the primary columnar tip, particularly $\boldsymbol{k_2}$. Higher $\boldsymbol{n_{max}}$ promotes the formation of more equiaxed grains and enhances CET. Increasing $\boldsymbol{k_2}$-parameter delays CET. Numerous combinations of $\boldsymbol{n_{max}}$ and $\boldsymbol{k_2}$ can numerically achieve the CET at position which corresponds to experimental one. As future work, the proper set of nucleation and columnar growth parameters has to be experimentally investigated. In addition, the influence of melt flow and crystal sedimentation has to be numerically investigated.

Acknowledgement
This work is financially supported by the FWF Austrian Science Fund (P23155-N24) and the German BMBF (50WM0843).

References

[1] Pötschke M, Gaitzsch U, Roth S, Rellinghaus B and Schultz L 2007 *Magnetism Magnetic Mater.* **316** 383
[2] Wagner A, Shollock B.A and McLean M 2004 *Mat. Sci. Eng. A*, **374** 270
[3] Winegard W and Chalmers B 1954 *Trans. ASM-AIME* **46** 1214
[4] Chalmers B 1963 *J. Austral. Inst. Met.* **8** 255
[5] Jackson K, Hunt J and Uhlmann D Seward III T 1966 *Trans. TMS-AIME* **236** 149
[6] Southin R 1967 *Trans. TMS-AIME* **239** 220
[7] Hunt J 1984 *Mater. Sci. Eng.* **65** 75
[8] Martorano M, Beckermann C and Gandin Ch-A 2003 *Metall. Mater. Trans. A* **34A** 1657
[9] Sturz L and Zimmermann G 2011 *Proc. 20th ESA Symposium on European Rocket and Balloon Programmes and Related Research* vol 700 SP (Hyere: France) p 459
[10] Sturz L and Zimmermann G 2011 J. Physics Conf. Series **327** Article no. 012002
[11] Wu M Ludwig A 2009 *Acta Mater.* **57** 5621
[12] Wu M, Fjeld A and Ludwig A 2010 *Comput. Mater. Sci.* **50** 32
[13] Kurz W and Fisher D.J 1998 *Fundamentals of Solidification* 4th ed (Zurich: Trans Tech)
[14] Lipton J, Glicksman M.E and Kurz W 1984 *Mater. Sci. Eng.* **65** 57
[15] Thevoz Ph and Rappaz M 1989 *Metall. Trans. A* **20A** 311
[16] Rappaz M and Gandin Ch.-A 1993 *Acta Metallurgica et Materialia* **41** 345
[17] Bayrama Ü, Aksöz S and Maraşlı N 2012 *Thermochimica Acta* **531** 12
[18] Witusiewicz V.T, Sturz L, Hecht U and Rex S 2014 *J. Cryst. Growth* **386** 69

Sensitivity analysis of dendritic growth kinetics in a Bridgman furnace front tracking model

R P Mooney[a], S McFadden[b]
Trinity College Dublin, Dublin 2, Ireland.

Email: [a]rpmooney@tcd.ie, [b]shaun.mcfadden@tcd.ie

Abstract. A directional solidification experiment of a Ti–Al–Nb–B–C alloy by power down method is simulated using a Bridgman furnace front tracking model. The effect of varying the dendritic growth parameters; *C*, the columnar dendrite growth coefficient, and *n*, the undercooling exponent, is investigated. A matrix of growth coefficients and undercooling exponents - at three levels each, based around a growth law for Ti–46wt.%Al - is applied in simulations, and the effect on columnar dendrite tip temperature, tip velocity, and tip temperature gradient is observed. The simulation results show that the dendrite tip velocity and temperature gradient at the tip are practically unaffected by the use of different growth parameters. However, the predicted columnar dendrite tip undercooling did vary to give the required dendrite tip velocity. This finding has implications for the analysis of microstructural transitions, such as the Columnar to Equiaxed Transition (CET). In conclusion, it is suggested that, for transient solidification conditions, a CET prediction criterion based on tip undercooling is preferable to one that uses growth velocity.

1. Introduction

Many practical analytical models of dendritic growth use the Ivantsov parabolic model [1] to treat diffusion of heat and solute at the dendrite tip. Notable models include; marginal stability theory models based on the work Langer and Müller-Krumbhaar [2], such as the Lipton–Glicksman–Kurz (LGK) [3] model for equiaxed growth at small undercoolings, the Kurz–Giovanola–Trivedi (KGT) [4] model for columnar growth, and microsolvability theory models, as outlined by Kessler and Levine [5], where surface energy anisotropy is accounted for. In any such models, a velocity–undercooling relationship can be determined and it is possible to fit results to a power law curve so that the growth model can be more readily applied in subsequent numerical simulations. Rebow and Browne [6] demonstrated an example of this where the predicted growth velocity, *V*, was fitted to various levels of dendrite tip undercooling, ΔT, by the power law; $V=C\Delta T^n$, where *C* (the growth coefficient) and *n* (the undercooling exponent) were the fitted variables for directional solidification of Al–4wt.%Cu and Al–2wt.%Si. In a similar way, the KGT model was used by Rebow et al. [7] to model columnar growth of Ti–46wt.%Al, where the stability parameter σ^* was set at 0.0506, i.e. twice the normal value used under the marginal stability criterion.

A Bridgman Furnace Front Tracking Model (BFFTM) [8] was recently applied to a thermal characterisation study [9] to determine the heat transfer coefficients of an experimental vertical Bridgman furnace. This characterisation study was a prerequisite for further experiments, which have since been carried out in the same furnace, investigating columnar to equiaxed transition (CET) in a multicomponent alloy of nominal composition; Ti–45.5Al–4.5Nb–0.2C–0.2B (at.%), using a power down method at four different cooling rates; 15 °C/min, 20 °C/min, 30 °C/min and 50 °C/min. Recently, Lapin et al. [10] reported on preliminary results from the experiment carried out at a cooling rate of 30 °C/min (CET was observed in the sample processed at 30 °C/min only). The work reported on in this article and the work outlined in references [8], [9] and [10] has been carried out as part of the European Space Agency (ESA) backed research project: GRADECET (GRAvity DEpendance of Columnar to Equiaxed Transition in titanium alloys.)

The purpose of this article is to numerically model, using the BFFTM, the conditions at the columnar dendrite tip during the power down experiment carried out by Lapin et al. [10] at a cooling rate of 30 °C/min. The effect of varying the growth coefficient and undercooling index in the dendrite kinetic power law within the model - about central values taken from the Rebow model for TiAl - are simulated, observed and reported on. This *sensitivity analysis* of dendrite growth kinetics is important for this set of experimental data given that the current growth model is based on a binary alloy and therefore may include a margin of error.

There are several approaches to predicting CET given in the literature. A comparison of the alternative CET criteria is presented elsewhere [11]. The CET criterion used in this study is based on modelling columnar mush and comparing conditions at the columnar front on the Hunt [12] G-V diagram.

2. Method

2.1. Experimental method

The experimental method is described elsewhere in detail by Lapin et al. [10]. A schematic of the vertical Bridgman furnace used in the CET power down experiment is given in figure 1. The experimental procedure is summarised here as follows. Samples, 10 mm in diameter and 150 mm in length, were contained in a yttrium oxide crucible with wall thickness of 2.5 mm. The furnace chamber was flushed several times with high purity argon gas and back filled with argon to a pressure of 10 kPa (absolute). The experiment was carried out in three main stages. Firstly, in the heating and stabilization stage ($t < 300$ s), the sample was heated until the fixed thermocouple reached 1720 °C where upon the sample was allowed to settle at this temperature for 300 s. Secondly, in the pulling stage ($300 < t < 372$ s), the sample was drawn 20 mm into the crystalliser by the drive unit at a constant rate of 0.278 mm/s. Finally, in the power down stage ($t > 372$ s), the heater temperature was reduced in a controlled manner so that the reference temperature, given by the fixed thermocouple, reduced at a rate of 30 °C/min.

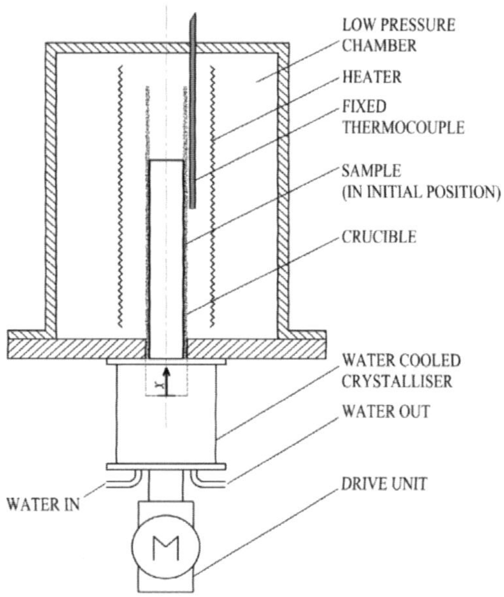

Figure 1. Schematic of the vertical Bridgman furnace apparatus.

2.2. *Numerical method*

The BFFTM [8] was used to model the experimental procedure. The modelled domain consisted of a series of disc-shaped control volumes (CVs) that contained the sample. Heat flow between CVs was permitted in the axial direction only. However, each CV had an individual radial heat flux component to account for heat gained (or lost) radially due to the proximity of the CV to the heater (or crystalliser). The BFFTM solves the following equation for heat flow in a rod (of cross-sectional area A and perimeter p) moving at axial velocity u, using an explicit CV finite difference method,

$$\frac{\partial}{\partial t}(\rho c T)=\frac{\partial}{\partial x}\left(k\frac{\partial T}{\partial x}\right)-u\rho c\frac{\partial T}{\partial x}-\frac{hp}{A}(T-T_\infty)+E \quad , \tag{1}$$

where ρ, c and k are the density, specific heat capacity and thermal conductivity of the alloy respectively, T_∞ and h are the temperature and heat transfer coefficient of the surrounding source (or sink), respectively, and E is the latent heat generated per unit volume. See reference [8] for a complete description of the numerical model. The results from a thermal characterisation study [9] were used to set the boundary conditions at either end of the domain and to set the heat transfer coefficients as a function of axial position in the furnace. Mooney et al. [9] gave a thorough account of the BFFTM algorithm as applied to this Bridgman apparatus using a similar TiAl multicomponent alloy. The BFFTM models columnar growth in the axial direction by tracking a front marker at the dendrite tips, thus providing information on the transient thermal conditions ahead of the columnar front. The information provided can then be used to determine the likelihood of a microstructural transition such as a CET. This method is an example of an indirect CET prediction method where the columnar mushy zone is simulated in the absence of an equiaxed mushy zone. The condition of the bulk undercooled liquid ahead of the columnar mushy zone is analysed with a view to predicting the likelihood of equiaxed nucleation and growth [11].

2.3. *Growth kinetics matrix*

A matrix of values for the growth coefficient C, and undercooling index n, were chosen to give a wide variation in growth rate versus undercooling based about central baseline values taken from the KGT columnar growth model power law fit for TiAl [7]. This range of values represents a ±38% variation in C and a ±35% variation for n, about the central values (C_2 and n_2) in Table 1.

Table 1. Matrix of values for growth coefficient and undercooling index.

Growth coefficient, C [m s^{-1} C^{-n}]	Undercooling index, n		
	$n_1 = 1.79$	$n_2 = 2.79$	$n_3 = 3.79$
$C_1 = 1.63 \times 10^{-6}$	n_1, C_1	n_2, C_1 -----	n_3, C_1 ——
$C_2 = 2.63 \times 10^{-6}$	n_1, C_2	n_2, C_2 -----	n_3, C_2 ——
$C_3 = 3.63 \times 10^{-6}$	n_1, C_3	n_2, C_3 -----	n_3, C_3 ——

Figure 2 shows the resulting set of velocity-undercooling curves when the power law ($V=C\Delta T^n$) is applied using the values from Table 1. Lines of a specific colour (blue, red or green) share the same growth coefficient C, while lines of a specific type (solid, dashed or dotted) share the same the undercooling index n.

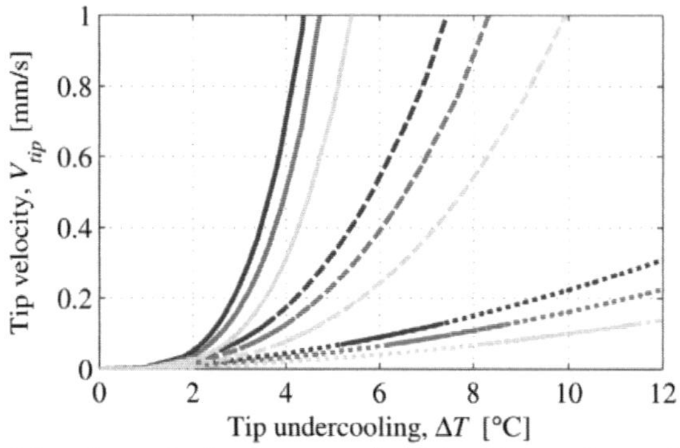

Figure 2. Columnar dendrite tip velocity versus dendrite tip undercooling for values of growth coefficient and undercooling index given in Table 1.

3. Results

Figure 3 shows (a) the simulated front position x, (b) the dendrite tip velocity V, and (c) the temperature gradient at the dendrite tip G, plotted against simulation time t, during the period: 200 < t < 1200 s. Figure 4 shows the simulated columnar dendrite tip velocity and temperature gradient at the dendrite tip plotted against columnar front position. Figure 5 shows the simulated dendrite tip temperature Ttip, and undercooling at the dendrite tip ΔT, plotted against front position x. Figure 6 shows the columnar dendrite tip velocity plotted against the temperature gradient (similar to a Hunt [12] plot). In the analysis carried out by Lapin et al. [10] it was discovered that, for the sample cooled at 30 °C/min, the CET occurred at a position approximately 76 mm from the cold end of the sample. The point at which the front marker reaches this position in simulations is shown in figure 6.

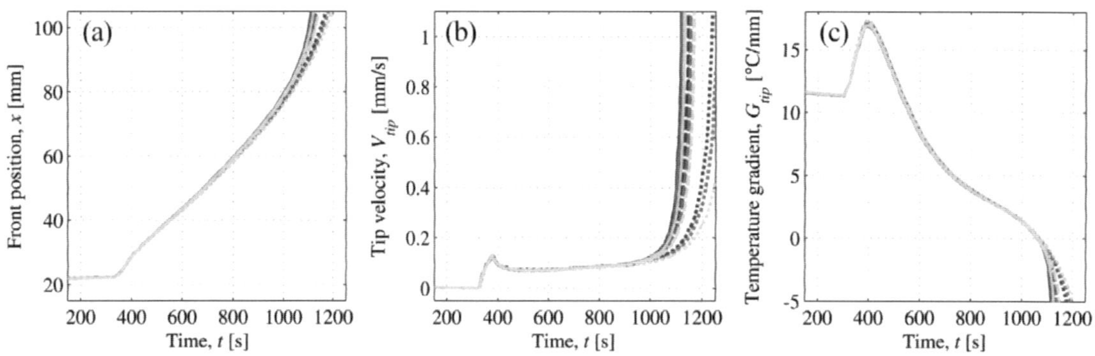

Figure 3. Simulated front position (a), columnar dendrite tip velocity (b) and temperature gradient at the dendrite tip (c) versus time.

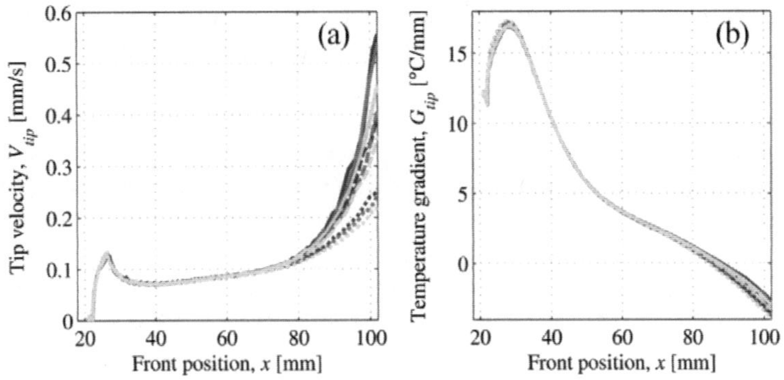

Figure 4. Simulated columnar dendrite tip velocity (a) and the temperature gradient at the dendrite tip (b) versus front position.

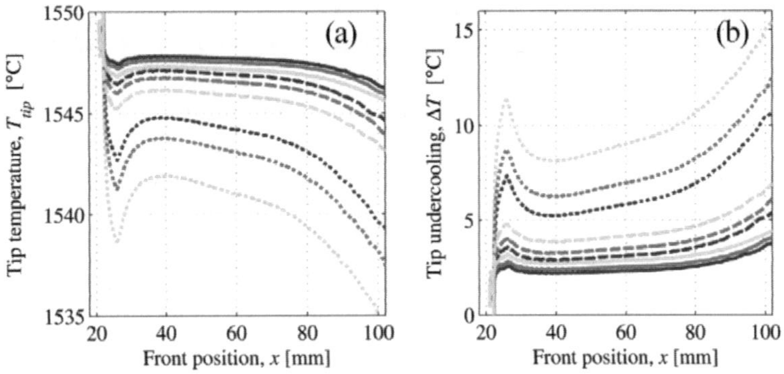

Figure 5. Simulated dendrite tip temperature (a) and undercooling at the dendrite tip (b) versus front position.

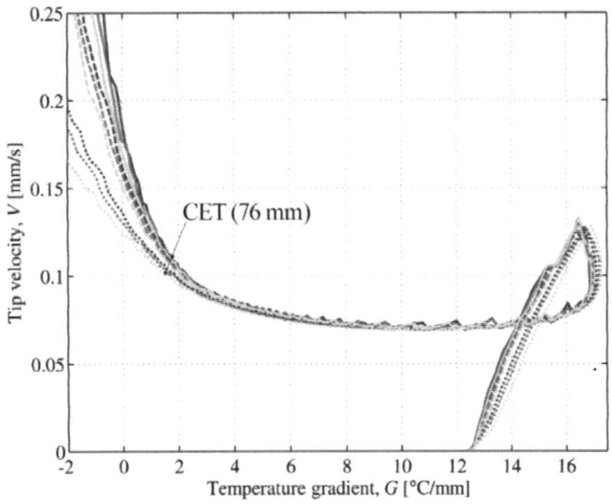

Figure 6. Columnar dendrite tip velocity versus temperature gradient at the dendrite tip.

4. Discussion

In the results it is important to note the timings for the different stages of the experiment/simulation. In the initial heating and stabilization stage of the simulation ($t < 300$ s) no movement or cooling of the sample occurs. This stage is of minimal interest for solidification because the system is stable.

Directional solidification begins when the pulling stage starts at time t = 300 s. During this stage the sample is pulled into the water-cooled crystalliser at a fixed rate. At time t = 372 s the pulling stage ends and the power down stage begins to cool the sample at a rate of 30 °C/min.

In all simulations it is important to note that the CET position measured by post mortem analysis in the experimental sample was 76 mm. Since we are modelling columnar growth only we can say that the simulation is only valid up to this position. According to the model predictions, the equivalent time for the columnar front at 76mm is approximately 975 s regardless of the growth parameters used in the simulation.

In figure 3 (a) we see the temporal progress of the columnar front position for each pair of growth parameters in Table 1. It is clear that, for times up to approximately 1000 s, the simulated progress of the columnar front is practically unaffected by different growth parameters in the model. A similar outcome is observed in the simulated results for velocity curves plotted against time in figure 3 (b) and simulated temperature gradient plotted against time in figure 3 (c).

In figure 4 we see the tip velocity (a) and temperature gradient at the tip (b) plotted against columnar front position. As with results in figure 3 the model's results with different combinations of growth parameters displayed an indifference to tip velocity and temperature gradient up to and including the point where the CET occurred. This point is also well illustrated by the Hunt plot given in figure 6.

In figure 5 the temperature at the dendrite tip (a) and tip undercooling (b) are plotted against front position. In this case the lines are not overlapping and we see the true effect of varying the growth parameters in the dendrite growth law. The temperature at the dendrite tip, and consequently the tip undercooling, varies accordingly so that the power law; $V=C\Delta T^n$, delivers the growth velocity required to satisfy the cooling conditions imposed on the sample by the process. The simulated tip undercooling at CET, for the baseline set of growth parameters (n_2, C_2), was 3.8 °C. The highest simulated tip undercooling at CET, ΔT = 10.1 °C, occurred using the lowest pair of growth parameters (n_1, C_1), and conversely, the lowest simulated tip undercooling at CET, ΔT = 2.5 °C, occurred when using the highest pair of growth parameters (n_3, C_3).

It is clear from references [8] and [9] that the BFFTM is appropriate for situations where the Biot number (the ratio of thermal resistance within the sample to the thermal resistance at its circumference) is less than 0.1. The Biot number did not exceed 0.03 in all of the simulations discussed here. In this respect, an axial heat flow assumption is vindicated. The 1-dimensional heat flow assumption is not appropriate when the axial temperature gradient becomes comparable with the radial temperature gradient. This point is expanded upon in recent results published elsewhere [13].

5. Conclusion

A Front Tracking Model of transient directional solidification in a Bridgman furnace was developed. The model predicted the temperature evolution and growth conditions at the columnar front position, and it was deemed to be an appropriate model of the solidification process up to the point of CET only. A sensitivity study using the model showed that the predicted growth velocity and temperature gradient at the dendrite tip were insensitive to changes in the dendrite kinetics growth parameters. It is postulated that the operating conditions of the furnace (pulling rate and cooling rate) dictate the growth velocity (solidification rate) and temperature gradients. However, the simulated dendrite tip temperature and undercooling at the columnar dendrite tip were shown to be sensitive to the choice of dendrite growth parameters.

It is accepted in literature [11] that the condition of the bulk undercooled liquid ahead of the columnar front is important in predicting the onset of the equiaxed mushy zone and, hence, the CET. The extent of constitutional undercooling ahead of the columnar front is dependent on the tip undercooling and temperature gradient at the columnar front. Hence, accurate knowledge of the thermal conditions at the tip is required.

The results provided here demonstrate a particular shortcoming of the classic Hunt CET analysis [12] when applied to transient directional solidification. The Hunt analysis (developed for the steady-state Bridgman scenario) considers the combination of *G* and *V* as the defining parameter set for predicting CET. We show that the *G–V* plot is insensitive to the selection of dendrite growth parameters. Furthermore, we show that the undercooling at the tip (which is an indicator of conditions for CET), is sensitive to the selected dendrite growth parameters and deserves special attention.

In conclusion it is suggested that, for transient solidification conditions, a CET prediction criterion based on tip undercooling is preferable to one that uses velocity.

Acknowledgments

This work was funded by European Space Agency PRODEX contract 4000107132 via Enterprise Ireland. The authors thank their GRADECET project partners: U. Hecht of ACCESS e.V. and Prof. J. Lapin of the Institute of Materials and Machine Mechanics at the Slovak Academy of Sciences.

References

[1] G P Ivantsov 1947 *Doklady Akademiya Nauk SSSR* **58** 567
[2] J S Langer and H Müller-Krumbhaar 1978 *Acta Metallurgica* **26** 11 1681–1687
[3] J Lipton M E Glicksman and W Kurz 1987 *Metallurgical Transactions A* **18** 2 341–345
[4] W Kurz B Giovanola and R Trivedi 1986 *Acta Metallurgica* **34** 5 823–830
[5] D A Kessler and H Levine 1986 *Physical Review Letters* **57** 24 3069–3072
[6] M Rebow and D J Browne 2007 *Scripta Materialia* **56** 6 481–484
[7] M Rebow D J Browne and Y Fautrelle 2010 *Materials Science Forum* **649** 243–248
[8] R P Mooney S McFadden M Rebow and D J Browne 2012 *Transactions of the Indian Institute of Metals* **65** 6 527–530
[9] R P Mooney S McFadden Z Gabalcová and J Lapin 2014 *Applied Thermal Engineering* **67** 1–2 61–71
[10] J Lapin Z Gabalcová U Hecht R P Mooney and S McFadden 2014 *Materials Science Forum* **790–791** 193–198
[11] S McFadden D J Browne and C-A Gandin 2009 *Metallurgical and Materials Transactions A* **40** 3 662–672
[12] J D Hunt 1984 *Materials Science and Engineering* **65** 1 75–83
[13] R P Mooney U Hecht Z Gabalcová J Lapin and S McFadden 2015 *Kovove Materialy* **53** 3 187

IOP Conf. Series: Materials Science and Engineering **117** (2016) 012013 doi:10.1088/1757-899X/117/1/012013

Mesoscopic modelling of columnar solidification

M Založnik[1], A Viardin[2], Y Souhar[1], H Combeau[1], M Apel[2]

[1] Institut Jean Lamour, CNRS – Université de Lorraine, F-54011 Nancy CEDEX, France
[2] Access e.V., Intzestr. 5, D-52072 Aachen, Germany

E-mail: miha.zaloznik@univ-lorraine.fr

Abstract. We used two complementary modeling approaches for the simulation of columnar growth in directional solidification of organic alloys: a phase-field model and a mesoscopic envelope model of dendritic growth. While the phase-field method captures the details of the dendritic structure and of the growth dynamics, the mesoscopic model approximates the complex dendritic morphology by its envelope. The envelope growth is deduced from the velocities of the dendrite tips, calculated by an analytical LGK-type tip model that is matched to the heat and concentration fields in the stagnant film around the envelope. The computational cost of the mesoscopic model is several orders of magnitude lower and can bridge the gap between phase-field and macroscopic models. We demonstrate the applicability of the mesoscopic model to columnar growth and discuss its possibilities and limitations by comparisons with phase-field simulations for the same conditions.

1. Introduction

The patterns and the dynamics of the nonlinear growth of solidification microstructures are studied with much success using phase-field methods. Phase-field models have been developed into efficient tools giving quantitative results. However, as in any model, its application is subject to certain limitations, mainly due to the necessary computing power. The mesoscopic solidification model of Steinbach, Beckermann et al. [1, 2] offers a complementary tool for the study of dendritic solidification. It can be applied at larger scales, in three dimensions and can include fluid flow at acceptable computing cost. The mesoscopic model is thus more convenient for the study of the dynamics of ensembles with a higher number of dendrites or phenomena that contain larger length and time scales, such as the CET. Computational efficiency is achieved at the price of several simplifications. The details of the dendritic structure are not resolved. Instead, the complex morphology is approximated by its envelope and by a field of phase fractions in the interior of the envelope. Also, the selection of the dendrite-tip operating state is not predicted by the model. The growth of the envelope is deduced from the velocities of the dendrite tips, calculated by an analytical LGK-type tip model that is matched to the concentration fields in the stagnant film around the envelope. Information from the microscopic scale is required to determine the tip selection parameter σ^* and the thickness of the stagnant film.

The mesoscopic model was applied to equiaxed dendritic growth in pure metals [1, 2] and in alloys [3, 4]. It was demonstrated that it can accurately predict envelope shapes of equiaxed dendrites as well as the dynamics of their interactions. The only application of the mesoscopic model to columnar growth was developed by Delaleau et al. [3], who compared the model to in-situ X-ray experiments of solidification of thin samples. To reproduce the experimental conditions, the same disposition of grain nuclei as observed in the experiment was introduced in their simulations. The dynamics of primary spacing selection in columnar growth has been recently successfully modeled by the meso-scale DNN model [5]. Another meso-scale approach was attempted by the so-called tuned cellular-automaton model [6], alas requiring strong unphysical adjustments of the parameters in order to obtain primary spacings of a realistic order of magnitude.

The question whether the mesoscopic model can address the mechanism of spacing selection in columnar growth remained open. Spacing selection is determined by both, the operating state of a single dendrite tip, i.e. its tip radius, and the solutal interaction between neigboring dendrites [7]. While the mesoscopic model should give an accurate representation of solutal interactions, the tip selection, a consequence of the anisotropy of the interface energy, is, clearly, strongly simplified. It is included implicitly (at a larger scale) by its consequences: the selection parameter and the tip growth directions.

In this paper we demonstrate the applicability of the mesoscopic model to spacing selection in columnar growth. As a reference we use phase-field simulations in 2D. We study directional solidification of a transparent organic alloy,. We compare simulations with the two models for the same conditions and we show that the primary arm spacing can be correctly reproduced by the mesoscopic model.

2. The mesoscopic envelope-field model

The core idea of the mesoscopic envelope model [1, 2] is the description of a dendritic grain by its envelope – a virtual smooth surface that links the tips of the actively growing dendrite branches. The velocity of the envelope growth can thus be calculated from the velocities of the dendrite tips. The growth of the dendrite tips is controlled by the solute flux that they reject into their surroundings and is therefore determined by the local conditions in the proximity of the envelope. The branched dendritic structure inside the envelope is only implied and its details are not resolved; the interior of the envelope is rather described in a volume-averaged sense by a phase-fraction field and other volume-averaged quantities. The phase change that determines the evolution of the structure inside the growing envelope is controlled by the solute exchange with the surroundings of the grain.

The model used to describe the kinetics of the dendrite tips is based on the analytical solution of the growth of a single isothermal parabolic tip in an infinite undercooled melt. The classical Ivantsov solution of this problem, recast by Cantor & Vogel [8], gives the supersaturation Ω_δ of the melt at a finite distance δ from the tip with a curvature radius $R_{\rm tip}$ as a function of the tip Péclet number, $\mathrm{Pe}_C = R_{\rm tip}V_{\rm tip}/(2D_{\rm l})$. To determine unique values for the tip velocity and radius, an additional tip selection criterion is required. In the case of a solutally controlled tip (limit of high Lewis numbers) the selection criterion is of the type $R^2_{\rm tip}V_{\rm tip} = \mathrm{const}$. The tip velocity and radius are then given by the system of the Cantor-Vogel function (here for a 2-dimensional tip) and the tip selection criterion, as a function of the local supersaturation Ω_δ:

$$\Omega_\delta = \sqrt{\pi \mathrm{Pe}_C}\, \exp\left(\mathrm{Pe}_C\right) \left[\mathrm{erfc}\left(\sqrt{\mathrm{Pe}_C}\right) - \mathrm{erfc}\left(\sqrt{\mathrm{Pe}_C\left[1+\frac{2\delta}{R_{\rm tip}}\right]}\right)\right] \; ; \quad R^2_{\rm tip}V_{\rm tip} = \frac{d_0 D_{\rm l}}{\sigma^*}. \qquad (1)$$

$d_0 = \Gamma/(m_{\rm L}C^*_{\rm l}(k_{\rm p}-1))$ is the solutal capillary length, σ^* is the selection parameter, Γ is the Gibbs-Thomson coefficient, $m_{\rm L}$ is the liquidus slope, $k_{\rm p}$ is the binary equilibrium partition coefficient, and $C^*_{\rm l}$ is the concentration in the liquid at the tip interface (i.e. the equilibrium liquid concentration at the given temperature). Equations (1) together give the local tip velocity at any point on the grain envelope. The matching distance δ is a model parameter that is best chosen at the order of the diffusion length $\delta \sim D_{\rm l}/V_{\rm tip}$ [4]. In the mesoscopic model the tip growth directions are prescribed, since the key physics for selection of the preferential growth directions is not included. Knowing the local tip velocity and direction, the envelope velocity is calculated: $\vec{v}_n = V_{\rm tip}\vec{n}\cos\theta$, where θ is the angle between the normal to the envelope and the closest of the tip growth directions (i.e. the one forming the smallest angle with the normal).

To track the front that represents the envelope we use the sharp interface tracking method by Sun & Beckermann [9]. This method represents the front by a contour of a continuous indicator function. The function is propagated by a phase-field like propagation equation. The transport at the mesoscopic scale is described by volume-averaged equations, valid in the whole domain. They are equivalent to the classical volume-averaged macroscopic models [10]. The phase change is described at a microscopic scale, where Scheil assumptions are applied. The diffusion of solute in the solid phase is neglected at the mesoscopic scale. This leads to the following conservation equations for solute mass in the liquid and the solid phase:

$$g_{\rm l}\frac{\partial C_{\rm l}}{\partial t} = D_{\rm l}\nabla\cdot(g_{\rm l}\nabla C_{\rm l}) + C_{\rm l}(k_{\rm p}-1)\frac{\partial g_{\rm l}}{\partial t}\,; \qquad \frac{\partial(g_{\rm s}C_{\rm s})}{\partial t} = -k_{\rm p}C_{\rm l}\frac{\partial g_{\rm l}}{\partial t}\,. \tag{2}$$

$g_{\rm l}$ and $g_{\rm s}$ are the liquid and solid phase fraction respectively and $D_{\rm l}$ is the diffusion coefficient in the liquid. This equation is valid inside and outside the envelope. Inside and on the envelope the liquid is assumed to be in thermodynamic equilibrium, such that $C_{\rm l} = (T - T_{\rm f})/m_{\rm L}$, where T is the imposed temperature field. Eq. (2) thus gives the liquid fraction inside the envelope. Outside the envelope $g_{\rm l} = 1$ and eq. (2) is reduced to the diffusion equation.

3. The phase-field model

In the present work, the multi-phase-field formulation proposed by Steinbach et al. [11, 12] is employed to describe the morphological evolution during the solid-liquid phase transformation. The phase-field method relies on a phase-field parameter which varies continuously from liquid "l" (ϕ=0) to solid "s" (ϕ=1) along the interface of width η. The evolution of the phase-field parameter in time is given by:

$$\dot{\phi}_s = M^a_{sl}(\vec{n})\left[\sigma^a_{sl}(\vec{n})K_{sl} + w\Delta G_{sl}\right] \tag{3}$$

$$K_{sl} = \phi_s\nabla^2\phi_l - \phi_l\nabla^2\phi_l + \frac{\pi^2}{\eta^2}(\phi_s - \phi_l)\,; \qquad w = \frac{\pi}{\eta}\sqrt{\phi_s\phi_l}\,; \qquad \Delta G_{sl} = -f_s(\vec{c}_s) + f_l(\vec{c}_l) + \vec{\mu}(\vec{c}_s - \vec{c}_l)$$

where $M^a_{sl}(\vec{n})$ is the anisotropic interfacial mobility between liquid and solid as a function of the interface orientation (described by the normal $\vec{n}$) as $\sigma^a_{sl}(\vec{n})$ is the effective anisotropic surface stiffness. K_{sl} is related to the local curvature of the interface. The thermodynamic driving force ΔG_{sl} depends on the respective bulk contributions $f_s(\vec{c}_s)$ and $f_l(\vec{c}_l)$, $\vec{\mu}$ is the generalized chemical potential introduced to conserve local concentration due to the constraint of quasi-equilibrium. This constraint postulates that all the reduced chemical potentials are the same for the coexisting phase i.e $\vec{\mu} = \vec{\mu}_s(\vec{c}_s(\vec{x})) = \vec{\mu}_l(\vec{c}_l(\vec{x}))$. The evolution of the concentration field is given by:

$$\dot{\vec{c}} = \nabla(\phi_s D_s \nabla\vec{c}_s) + \nabla(\phi_l D_l \nabla\vec{c}_l), \tag{4}$$

where $D_{\rm s}$ and $D_{\rm l}$ are respectively the diffusion coefficients in the solid and liquid phase.

4. Results and discussion

We compared the results of the phase-field and the mesoscopic model for several different configurations. The system consisted of a succinonitrile-2 at%acetone alloy solidifying in a fixed gradient of $G_T = 2\cdot10^4$ K/m with a cooling rate of $\dot{T} = -1.16$ K/s ($V_{\rm pull} = 58\,\mu$m/s). The growth was initiated with regularly spaced nuclei injected at the bottom of the domain at an initial supersaturation of 0.20. Several different spacings of the initial nuclei λ_0 and several lateral confinements were used. The computational domain is a moving frame, travelling with the dendrite tips, leaving a minimum distance ahead of the most advanced tip. The frame height was 500 μm and the tip distance was 100 μm. At the vertical sides a zero flux boundary condition was used in the mesoscopic model and a periodic boundary condition in the phase-field model. A fixed concentration C_0 was imposed at the top boundary and a zero derivative at the bottom boundary. The thermophysical properties of the alloys are identical as used in [13]. The tip selection parameter σ^* for the mesoscopic model was determined from the steady-state tip radius obtained by the phase-field model. For that, we used the tip selection criterion $R^2_{\rm tip}V_{\rm tip} = (d_0 D_{\rm l})/\sigma^*$, where the tip velocity is known and the liquid concentration at the tip used in the capillary length was determined from the supersaturation $\Omega_0 = (C^*_{\rm l} - C_0)/((1-k_{\rm p})C^*_{\rm l})$ given by the Ivantsov model (Eq. (1) for $\delta \to \infty$): $\Omega_0 = {\rm Iv}(Pe_{\rm tip})$. We obtained $\sigma^*_{\rm SCN-ace} = 0.0328$. The grid spacing was 0.5 μm in the phase-field simulations and 2.5 μm in the mesoscopic simulations. For a domain width of 250 μm (shown in Figs. 1 and 2) the computation time for the phase-field simulations was 36 h on 12 CPU cores and for the mesoscopic simulations it was 5 h on a single CPU core. The computations were done on different computers, therefore these values do not give a precise comparison but rather an order of magnitude.

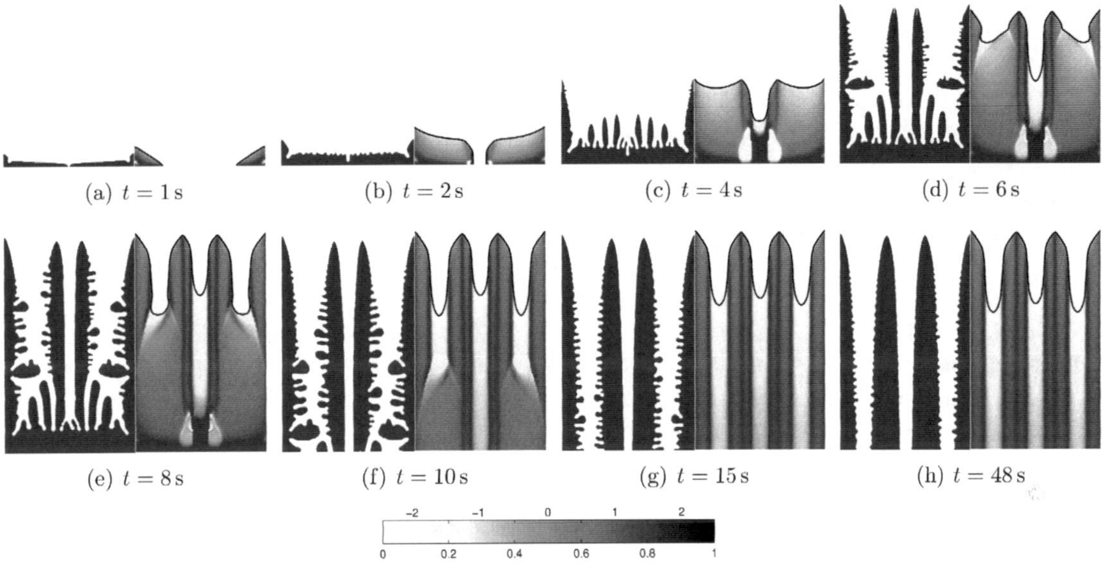

Figure 1. The evolution of the columnar grains from two nuclei placed in the bottom corners. Solid phase calculated by phase field (left) and solid fraction field calculated by the mesoscopic model (right). The domain width is 250 μm.

The columnar growth undergoes an initial transient, during which the tip velocities, the tip undercoolings and the dendrite spacing evolve until a steady state is reached. The transient depends strongly on the spacing of the initial nuclei and on the confinement. To illustrate the transient we show the evolution in the case of SCN-acetone in a confinement of $W = 250\,\mu$m, growing from two nuclei placed in the bottom corners of the domain, i.e. with an initial spacing of $\lambda_0 = 250\,\mu$m, in figures 1 and 2. We can see that after the seeding dendritic branches develop and the horizontal lateral arms first grow very fast along the bottom of the domain, where the supersaturation is the highest. Simultaneously, but initially at a slower rate, vertical primary branches develop in the direction of the temperature gradient. When the tips of the lateral branches meet in the middle of the domain (figure 1(b)), the diffusion fields emanating from the two vertical primary branches have not yet entirely blended (the diffusion time is $\sim (W/2)^2/D_{\rm l} = 12$ s). The supersaturation is thus the highest at mid-distance between the

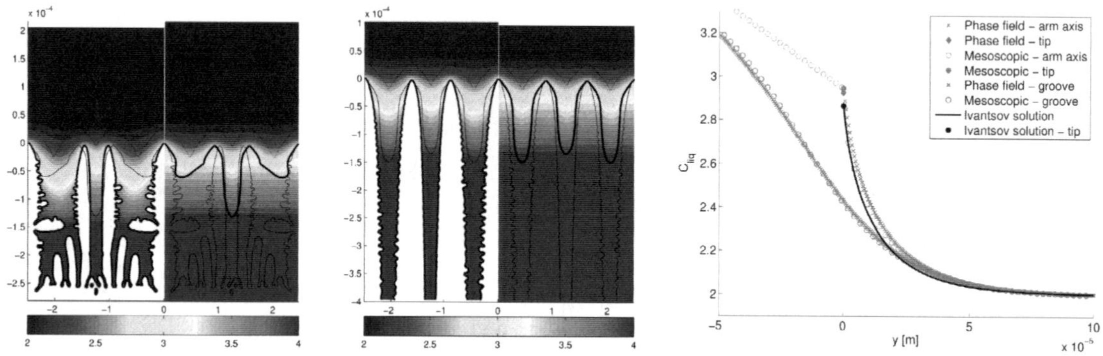

(a) $t = 6\,\mathrm{s}$ (b) $t = 48\,\mathrm{s}$ (c) $t = 48\,\mathrm{s}$

Figure 2. Comparison of concentration fields in the liquid given by the phase-field and the mesoscopic models. (a–b) Field contours (left: phase field, right: mesoscopic). (c) Steady-state concentration profiles along the axis of a primary arm (at $x = 0$) and along a groove between two primary arms (at $x = 125\,\mu\mathrm{m}$), and comparison with the analytical Ivantsov solution.

grains. This drives a fast growth of vertical secondary branches in the middle of the domain (figure 1(c)). In the mesoscopic model this is represented by a vertical expansion of the envelope (at a low solid fraction). Finally, two of the secondary branches prevail and stabilize. In the mesoscopic model this is manifested by the growth of the cusps at the end of the lateral wings of the envelope. In the steady state, attained at approximately $t = 30\,\mathrm{s}$, the tips grow at the rate imposed by the pulling speed of the thermal field ($58\,\mu\mathrm{m/s}$), with a constant tip undercooling and a stable spacing. We can note that the spacing is not uniform but is symmetric.

We performed simulations for several domain widths (lateral confinements) and initial nucleus spacings in order to check the reliability of the spacing selection mechanism reproduced by the mesoscopic model. On the one hand, configurations with an initial spacing larger than the final primary arm spacing ($\lambda_0 > \lambda_1$) were used. In this case the spacing selection proceeds by the development of secondary branches growing vertically from the horizontal lateral arms. The spacing of these branches is small. In a second stage a competition of the secondary branches results first in an increase of the spacing and finally in a stabilisation of a primary spacing. In the mesoscopic model a similar competition of protuberances of the envelope occurs and leads to a stable primary spacing. On the other hand, configurations with ($\lambda_0 < \lambda_1$) were used. In this case, the lateral primary branches are quickly blocked and no significant secondary branching occurs. It is blocked by the diffusion fields emanating from the vertical primary branches, which quickly meet. The spacing selection proceeds by a competition between the vertical primary branches. The phenomenon is reproduced by the mesoscopic model. Simulations initialised with a planar envelope were also performed with the mesoscopic model. After an initial destabilisation of the planar envelope, protuberances with a small spacing develop. They then develop into branches and undergo a competition until a stable primary spacing develops.

The primary spacings obtained in the different configurations are summarized in figure 3. We can see that the final spacing depends not only on the confinement, which can be expected, but also on the initial spacing. The predicted spacings can be compared to classic geometrical models, which give a unique average primary spacing for steady growth as a function of the temperature gradient, the pulling velocity and the solute concentration. They can be resumed by $\overline{\lambda}_1 = a\sqrt{\ell_{\mathrm{sl}} R_{\mathrm{tip}}}$, where the constant a can be alloy-specific [14]. For our configuration different models of this type predict spacings between $32\,\mu\mathrm{m}$ and $150\,\mu\mathrm{m}$; a range across almost an order of magnitude. The predictions of the phase-field and of the mesoscopic model fall

IOP Conf. Series: Materials Science and Engineering **117** (2016) 012013 doi:10.1088/1757-899X/117/1/012013

easily within these limits, however their range of variation is much tighter. In contrast to the simple geometrical models, theoretical analyses [15] and experiments [16] showed that a selection mechanism that would uniquely determine the primary spacing does not exist. A given spacing is stable down to a critical minimum velocity.

If we wish to simulate grain interactions, for example interactions of a columnar front with equiaxed grains and the columnar-to-equiaxed transition, an important aspect is a faithful representation of the undercooling field ahead of the columnar front. In figure 2(a–b) we show the solute concentration fields in the liquid around and ahead of the columnar dendrites during the transient and in the steady state. In the left part of each image the concentration field from phase-field and in the right part from the mesoscopic model is shown. The differences in the liquid are relatively small. A more detailed representation is given by a plot of the concentration profiles in the liquid ahead of a primary arm and in the groove along two arms for the steady state in figure 2(c). We can see that the differences are barely perceptible along both profiles. A theoretical Ivantsov solution for an isolated tip is shown for comparison. It shows that the solutal undercooling of the tips of the columnar front is slightly higher than that of an isolated tip. This indicates a slight solutal interaction between the tips. This interaction is captured with good accuracy by both models.

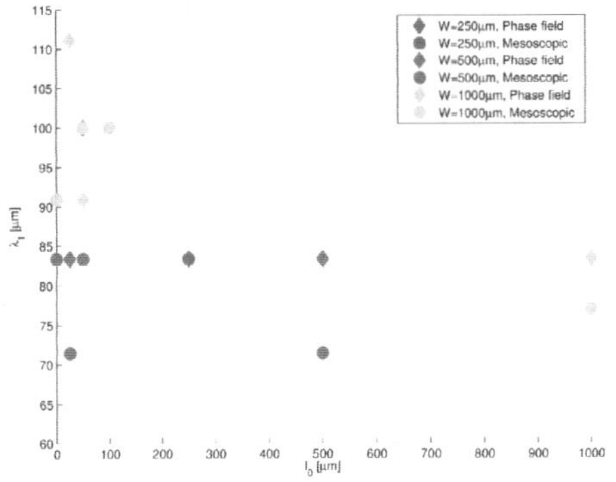

Figure 3. Dependence of the final dendrite spacing on the spacing of the initial nuclei λ_0 and on the width of the confinement W. Mesoscopic simulations with an initially plane front are characterized by an initial spacing $\lambda_0 = 0$.

5. Conclusions and perspectives

We demonstrated the applicability of the mesoscopic envelope model to columnar growth in directional solidification. By comparisons to phase-field computations, used as a reference, we showed that the mesoscopic model: (i) provides a good approximation of the dynamics of the columnar front; (ii) captures the solute diffusion at the scale of the primary spacing and gives an accurate prediction of the solutal undercooling ahead of the columnar tips; (iii) correctly reproduces the phenomena leading to a stable primary spacing by branching and by elimination of branches. The computational cost is substantially smaller than for phase-field computations. The principal reason is the fact that the required grid spacing is around one order of magnitude larger than for the phase-field method.

Further analysis is needed to fully understand the mechanisms of the rebranching of the envelope and of growth competition in the framework of the mesoscopic model. Obvious extensions of this work are 3D computations of columnar growth and the inclusion of fluid flow. Both can be achieved with reasonable computational load. The mesoscopic model can provide a bridge between macroscopic process-scale models and microstructure models, especially in situations involving a strong coupling between the microstructure evolution and the macroscale melt flow.

Acknowledgments

We gratefully acknowledge the support to this research by the German Federal Ministry of Research through the German Space Agency DLR under Contract FKZ 50WM1143. This work was supported by the French State through the program "Investment in the future" operated by the National Research Agency (ANR) and referenced by ANR-11 LABX-0008-01 (LabEx DAMAS).

References

[1] Steinbach I, Beckermann C, Kauerauf B, Li Q and Guo J 1999 *Acta Mater.* **47** 971–982
[2] Steinbach I, Diepers H J and Beckermann C 2005 *J. Cryst. Growth* **275** 624–638
[3] Delaleau P, Beckermann C, Mathiesen R H and Arnberg L 2010 *ISIJ Int.* **50** 1886–1894
[4] Souhar Y, De Felice V F, Založnik M, Combeau H and Beckermann C 2015 *IOP Conf. Series: Mater. Sci. Eng.* (in press)
[5] Tourret D and Karma A 2013 *Acta Mater.* **61** 6474–6491
[6] Kharicha A, Stefan-Kharicha M, Wu M and Ludwig A 2012 *IOP Conf. Series: Mater. Sci. Eng.* **33** 012115
[7] Steinbach I 2008 *Acta Mater.* **56** 4965–4971
[8] Cantor B and Vogel A 1977 *J. Cryst. Growth* **41** 109–123
[9] Sun Y and Beckermann C 2007 *J. Comput. Phys.* **220** 626–653
[10] Beckermann C and Viskanta R 1993 *Appl. Mech. Rev.* **46** 1–27
[11] Steinbach I, Pezzolla F, Nestler B, Seeßelberg M, Prieler R, Schmitz G and Rezende J 1996 *Physica* D **94** 135–147
[12] Eiken J, Böttger B and Steinbach I 2006 *Phys. Rev.* E **73** 066122
[13] Melendez A J and Beckermann C 2012 *J. Cryst. Growth* **340** 175–189
[14] Steinbach S 2005 *Einfluss von Strömungen auf die Entwicklung des Mikrogefüges bei der gerichteten Erstarrung von Al-Si und Al-Si-Mg Legierungen* Ph.D. thesis RWTH Aachen
[15] Warren J A and Langer J S 1993 *Phys. Rev.* E **47** 2702–2713
[16] Losert W, Shi B Q, Cummins H Z and Warren J A 1996 *Phys. Rev. Lett.* **77** 889–891

IOP Conf. Series: Materials Science and Engineering **117** (2016) 012014 doi:10.1088/1757-899X/117/1/012014

The role of the stagnant-film thickness in mesoscopic modeling of equiaxed grain envelopes

Youssef Souhar[1], **Valerio F. De Felice**[1], **Miha Založnik**[1], **Hervé Combeau**[1], **Christoph Beckermann**[2]

[1] Institut Jean Lamour, CNRS – Université de Lorraine, F-54011 Nancy CEDEX, France
[2] Department of Mechanical and Industrial Engineering, University of Iowa, Iowa City, Iowa 52242, USA

E-mail: `youssef.souhar@univ-lorraine.fr`

Abstract. The mesoscopic envelope model overcomes the limitations of phase-field methods. It can be applied at larger scales and can include fluid flow at reasonable computing cost. It consists of the description of a dendritic grain by an envelope that links the active dendrite branches. The grain is modelled as an evolving porous medium and the liquid-solid phase change and solute transport are modelled by volume-averaged equations. The velocities of the dendrite tips are determined by the local solute-concentration field in the proximity of the envelope through an analytical stagnant-film model. In this publication, we present our implementation of the model for a binary alloy and we discuss the influence of the stagnant-film thickness, the principal model parameter, on the predicted 3D equiaxed grains by comparisons with the scaling laws for binary-alloy dendrites obtained in recent experiments by Melendez and Beckermann.

1. Introduction

The growth of dendritic (treelike) crystal grains is governed by an intricate interplay between the transport of heat and solute as well as capillary and surface effects. Furthermore, it is influenced by adjacent grains due to the overlap of thermal and chemical species fields emanating from each growing grain. The patterns and the dynamics of the strongly nonlinear growth of solidification microstructures are successfuly studied with phase-field methods. These have been developed into efficient tools giving quantitative results. However, as in any model, their practicable scale of application is limited. To bridge the gap between purely microscopic and macroscopic approaches, several meso-scale models have been proposed [1–4], based on various approximations. The mesoscopic envelope model [5–7] offers an accurate representation of group interactions in dendritic solidification. It can be applied in three dimensions at the scale of several grains at acceptable computing cost. The details of the dendritic structure are not resolved. Instead, the complex morphology is approximated by its envelope and by a field of phase fractions in the interior of the envelope. Also, the selection of the dendrite-tip operating state is not predicted by the model. The growth of the envelope is deduced from the velocities of the dendrite tips, calculated by an analytical LGK-type tip model that is matched to the concentration fields in the stagnant film around the envelope. In the model, the tip selection parameter σ^* is an input parameter coming either from measurements or from phase-field simulations.

The mesoscopic model was previously applied to equiaxed dendritic growth in pure materials [5, 6] and in alloys [7]. It was demonstrated that it can accurately predict envelope shapes and the growth dynamics of pure-material equiaxed dendrites. In this paper, we aim at validating the model for equiaxed growth of alloys by comparing 3D simulations to the scaling laws for binary-alloy dendrites obtained experimentally by Melendez and Beckermann [8]. We discuss the influence of the main parameter of the model, the stagnant-film thickness, on the accuracy of the prediction of the primary-tip velocity, of the envelope shape and of the internal solid fraction of the grains. We give a recommendation for the stagnant-film thickness that leads to a good approximation of the scaling laws.

2. The mesoscopic envelope-field model

The core idea of the mesoscopic envelope model [5–7] is the description of a dendritic grain by its envelope – a virtual smooth surface that links the tips of the actively growing dendrite branches (see figure 1). The velocity of the envelope growth can thus be calculated from the velocities of the dendrite tips. The growth of the dendrite tips is controlled by the solute flux that they eject into their surroundings and is therefore determined by the local conditions in the proximity of the envelope. The branched dendritic structure inside the envelope is only implied and its details are not resolved; the interior of the envelope is rather described in a volume-averaged sense by a phase-fraction field and other volume-averaged quantities. The phase change that determines the evolution of the structure inside the growing envelope is controlled by the solute exchange with the surroundings of the grain.

The model used to describe the kinetics of the dendrite tips is based on the analytical solution of the growth of a single isothermal parabolic tip in an infinite undercooled melt. The classical Ivantsov solution of this problem, recast by Cantor and Vogel [9] gives the supersaturation Ω_δ of the melt at a finite distance δ from the tip with a curvature radius R_{tip} as a function of the tip Péclet number, $\mathrm{Pe}_C = R_{\mathrm{tip}} V_{\mathrm{tip}}/(2D_{\mathrm{l}})$. In three dimensions the Cantor-Vogel function is

$$\Omega_\delta = \mathrm{Pe}_C \exp\left(\mathrm{Pe}_C\right)\left[\mathrm{E}_1\left(\mathrm{Pe}_C\right) - \mathrm{E}_1\left(\mathrm{Pe}_C\left[1+\frac{2\delta}{R_{\mathrm{tip}}}\right]\right)\right]. \qquad (1)$$

To determine unique values for the tip velocity and radius, a tip selection criterion is required. In the case of a solutally controlled tip (limit of high Lewis numbers) the selection criterion is

$$R_{\mathrm{tip}}^2 V_{\mathrm{tip}} = \frac{d_0 D_{\mathrm{l}}}{\sigma^*}, \qquad (2)$$

where $d_0 = \Gamma/(m_{\mathrm{L}} C_{\mathrm{l}}^* (k_{\mathrm{p}} - 1))$ is the solutal capillary length, σ^* is the selection parameter, Γ is the Gibbs-Thomson coefficient, m_{L} is the liquidus slope, k_{p} is the binary equilibrium partition coefficient, and C_{l}^* is the concentration in the liquid at the tip interface (i.e. the equilibrium liquid concentration at the given temperature). Equations (1) and (2) together give the local tip velocity at any point on the grain envelope. The stagnant-film thickness δ is a model parameter that is discussed in the next section. In the mesoscopic model, the tip growth directions are prescribed since the key physics for selection of the preferential growth directions is not included. Knowing the local branch-tip velocity and direction, the envelope velocity is calculated: $\vec{v}_n = V_{\mathrm{tip}} \vec{n} \cos\theta$, where θ is the angle between the normal to the envelope and the closest of the tip growth directions (i.e. the one forming the smallest angle with the normal).

To track the front that represents the envelope we use the sharp interface tracking method by Sun and Beckermann [10]. In this method the tracked front is given by the iso-surface of the smooth indicator field ϕ at $\phi = \phi_{\mathrm{env}}$. The transition of the field ϕ follows a hyperbolic tangent profile, $\phi(n) = \frac{1}{2}\left[1 - \tanh\left(\frac{n}{2W}\right)\right]$, in the vicinity of the front. The phase-field equation that is used to propagate the field contains a stabilization term (right hand side of Eq. (3)), which ensures that the transition of the field ϕ retains its shape and width W.

$$\frac{\partial \phi}{\partial t} + |\vec{v}_n|\, \vec{n} \cdot \nabla\phi = -b\left[\nabla^2\phi - \frac{\phi\,(1-\phi)\,(1-2\phi)}{W^2} - |\nabla\phi|\, \nabla\cdot\left(-\frac{\nabla\phi}{|\nabla\phi|}\right)\right], \qquad (3)$$

The transport at the mesoscopic scale is described by volume-averaged equations, valid in the whole domain. They are equivalent to the classical volume-averaged macroscopic models [11]. The phase change is described at a microscopic scale, where Scheil assumptions are applied. The diffusion of solute in the solid phase is neglected at the mesoscopic scale. This leads to the following conservation equations for solute mass in the liquid and the solid phase:

$$g_l \frac{\partial C_l}{\partial t} = D_l \nabla \cdot (g_l \nabla C_l) + C_l (k_p - 1) \frac{\partial g_l}{\partial t}; \qquad \frac{\partial (g_s C_s)}{\partial t} = -k_p C_l \frac{\partial g_l}{\partial t}. \tag{4}$$

g_l and g_s are the liquid and solid phase fraction respectively and D_l is the diffusion coefficient in the liquid. This equation is valid inside and outside the envelope. Inside and on the envelope the liquid is assumed to be in thermodynamic equilibrium, such that $C_l = (T - T_f)/m_L$, where T is the imposed temperature of the isothermal system. Equation (4) thus gives the liquid fraction inside the envelope. Outside the envelope $g_l = 1$ and Eq. (4) is reduced to the diffusion equation.

3. Results and discussion

Steinbach et al. [5, 6] demonstrated the validity of the mesoscopic model for a pure substance. In this paper, we study the validity of the model for the growth of an isolated equiaxed grain into an infinite and uniformly supercooled melt for a binary alloy. This is done through comparisons with the recent experiments of Melendez and Beckermann [8]. They observed the same generalised envelope shape across a wide range of concentrations and undercoolings. The correlations were close to those observed for pure materials in microgravity. We can therefore assume that the same generalised envelope shape also holds for a purely solutal dendrite with negligible thermal undercooling. Our comparisons are based on this premise and our validation is done for an isothermal, purely solutally driven growth. The computations were performed for a similar range of solutal undercoolings as the experiments. The supersaturation range found in the experiments was $\Omega_0 = 0.05$–0.45. We performed simulations for $\Omega_0 = 0.05$–0.25. Moreover, this validation is done for purely diffusive conditions, although the experiments of Melendez and Beckermann [8] showed that the influence of natural convection on the primary tip velocity was not negligible. However, the small difference between the scaling laws obtained in these experiments and the scaling laws from micro-gravity experiments for pure materials indicates

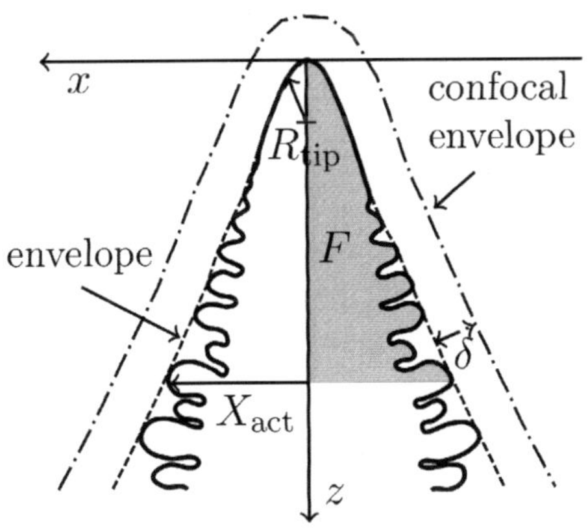

Figure 1. Schematic illustration of the envelope, the confocal envelope, the envelope width X_{act} and the solid projection area F of an equiaxed dendrite.

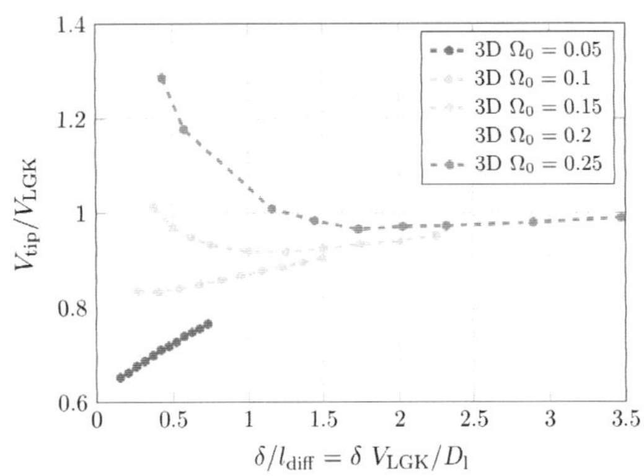

Figure 2. Dependence of the predicted normalised primary-tip velocity on the normalised stagnant-film thickness.

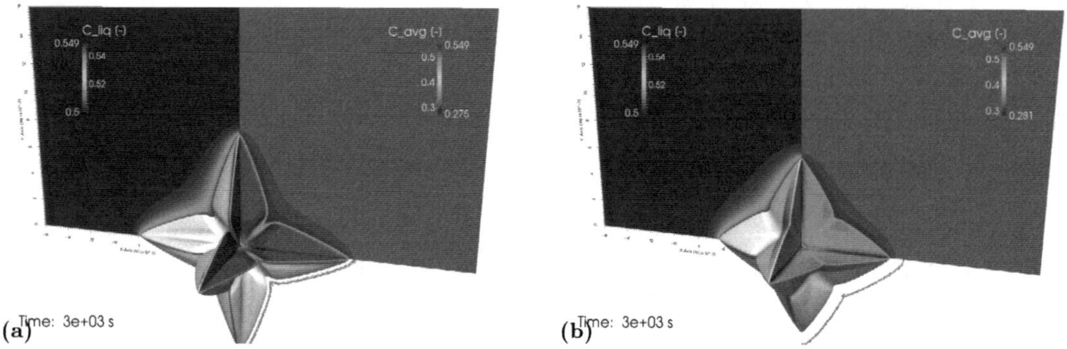

Figure 3. Comparisons of equiaxed grains growing at a supersaturation of $\Omega_0 = 0.1$ for different values of the stagnant-film thickness: (a) $\delta/R_{\mathrm{LGK}} = 5$ and (b) $\delta/R_{\mathrm{LGK}} = 20$.

that the natural convection did not significantly affect the envelope shape in the experiments of Melendez and Beckermann.

We are interested in the tip velocity, the envelope shape, and the solid fraction in the steady state of the dendrite envelope. We can consider that the steady state is reached when both the primary-tip velocity and the envelope shape no longer vary. The theoretical tip velocity and radius, V_{LGK} and R_{LGK}, are calculated with the LGK model, using the same tip selection parameter $\sigma^* = 0.02$ as in the mesoscopic simulations. For purely solutal growth this corresponds to using Eqs. (1–2) with an infinite stagnant-film thickness δ.

The dependence of the primary tip velocity on the stagnant-film thickness for different supersaturations is shown in figure 2. We observed that the best way to scale the stagnant-film thickness is by the diffusion length $l_{\mathrm{diff}} = D_{\mathrm{l}}/V_{\mathrm{LGK}}$. We can see that the steady-state primary tip velocity depends on δ/l_{diff} due to the matching between the analytical solution and the numerical solute diffusion field. Within a large range of supersaturations the error is less than 20 % when $\delta/l_{\mathrm{diff}} > 1$ and less than 10 % when $\delta/l_{\mathrm{diff}} > 1.5$.

Similarly, if we consider the predicted shape of the equiaxed grains, we can notice in figure 3 that the dimensionless stagnant-film thickness δ/l_{diff} significantly affects the shape of the grain envelope. This influence can be studied by comparing our results with the scaling law determined by Melendez and Beckermann [8], which links the dimensionless sidebranch envelope widths $X_{\mathrm{act}}/R_{\mathrm{tip}}$ with the dimensionless longitudinal distance from the tip z/R_{tip} :

$$\frac{X_{\mathrm{act}}}{R_{\mathrm{tip}}} = 0.84 \left(\frac{z}{R_{\mathrm{tip}}} \right)^{0.85} \tag{5}$$

Except near the tip, where the mesoscopic envelope has no physical meaning, the computed envelopes are in relatively good agreement with the experiments. As can be seen in figure 4, the agreement is better when $\delta/l_{\mathrm{diff}} \leq 1$. However, with a too small stagnant-film thickness, the computed primary-tip velocity becomes less accurate and thus a compromise has to be made.

Melendez and Beckermann [8] determined another interesting scaling law for the projected area F of the solid dendrite structure:

$$\frac{F}{R_{\mathrm{tip}}^2} = 0.58 \left(\frac{z}{R_{\mathrm{tip}}} \right)^{1.77} \tag{6}$$

In our case, this projected area is obtained by first extracting the plane of the ridge of the sidebranches, (i.e. the plane at $y = 0$) and then calculating the area integral of the solid fraction

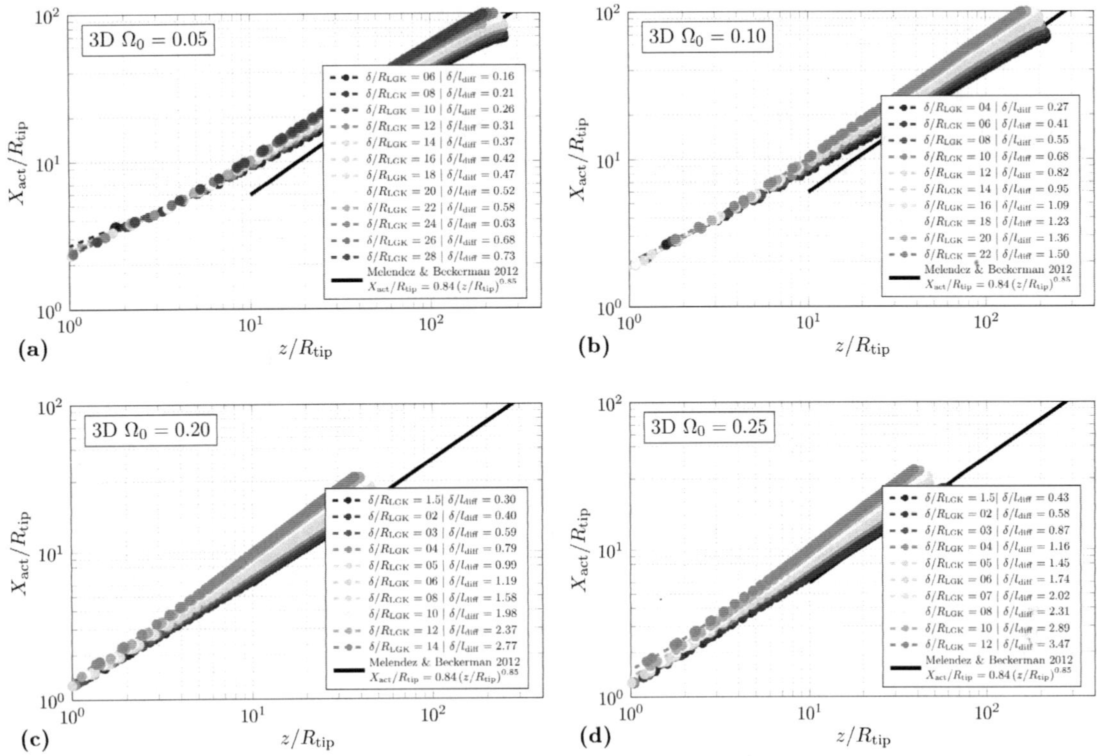

Figure 4. Dimensionless sidebranch envelope width X_{act}/R_{tip} as a function of the dimensionless longitudinal distance from the tip z/R_{tip} and its sensitivity to the stagnant-film thickness δ for four different supersaturations: (a) $\Omega_0 = 0.05$, (b) $\Omega_0 = 0.10$, (c) $\Omega_0 = 0.20$ and (d) $\Omega_0 = 0.25$.

as a function of the distance from the primary tip in this plane. With the notations of figure 1, the projected area is calculated using the formula:

$$F(z) = \int_0^z \left(\frac{1}{X_{act}(z)} \int_0^{X_{act}(z)} g_s(x,z)\,\mathrm{d}x \right) X_{act}(z)\,\mathrm{d}z = \int_0^z \int_0^{X_{act}(z)} g_s(x,z)\,\mathrm{d}x\,\mathrm{d}z \qquad (7)$$

In figure 5, we can see that the internal solid fraction is in very good agreement with the experiments. The exponent factor of the scaling law is very close to the experimentally determined value. For high supersaturations, namely $\Omega_0 = 0.2$ and $\Omega_0 = 0.25$, the pre-factor is also similar. Moreover, we can note that the stagnant-film thickness has little influence on the projected area and finally we can recommend $\delta/l_{diff} = 1$ as a good compromise to obtain realistic simulations with the mesoscopic envelope-field model in terms of envelope shape, internal solid fraction and primary-tip velocity.

4. Conclusions and perspectives

The mesoscopic envelope-field model of solidification gives physically realistic results for equiaxed dendritic growth in binary alloys. We have shown this by comparisons of three-dimensional simulations to experimental results for a wide range of undercoolings. The model can correctly approximate the velocity of both primary and secondary dendrite arms (i.e. the envelope shape) as well as the internal solid fraction in the envelope. To obtain accurate predictions, a proper choice of the stagnant-film thickness, a model parameter, is required. We have shown that a

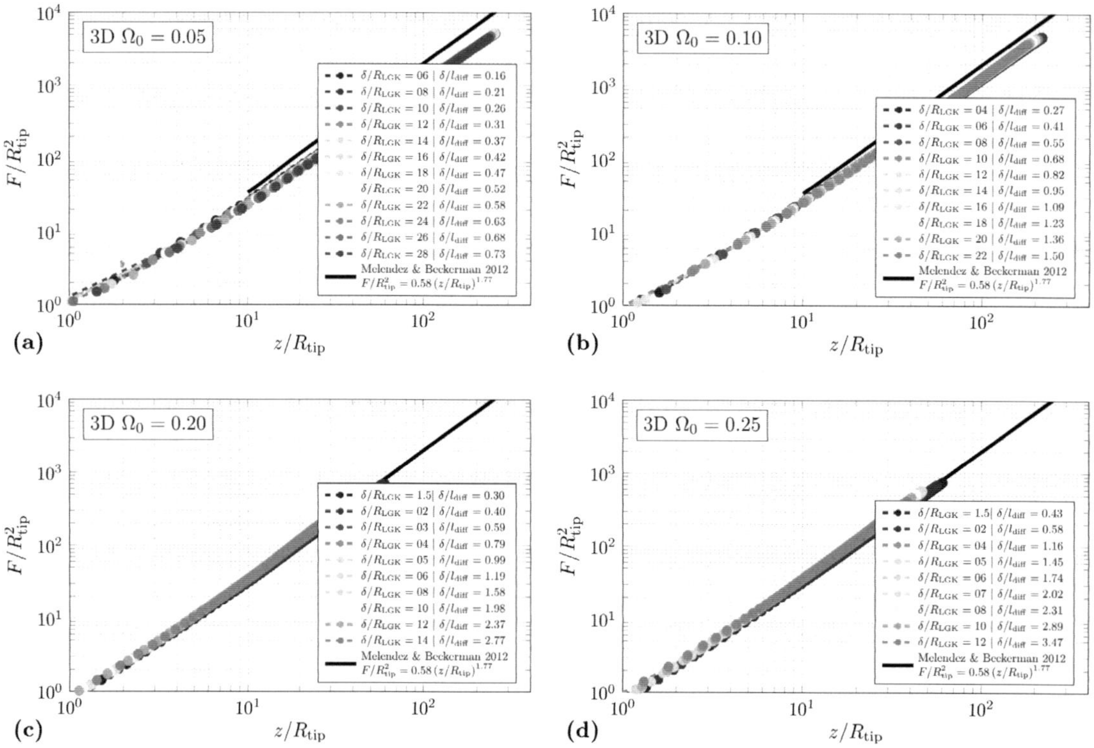

Figure 5. Normalised projection area F/R_{tip}^2 as a function of the dimensionless longitudinal distance from the tip z/R_{tip} and its sensitivity to the stagnant-film thickness δ for four different supersaturations: (a) $\Omega_0 = 0.05$, (b) $\Omega_0 = 0.10$, (c) $\Omega_0 = 0.20$ and (d) $\Omega_0 = 0.25$

stagnant-film thickness of the order of the primary-tip diffusion length D_l/V_{tip} leads to accurate predictions of the dynamics, of the shape and of the internal solid fraction of equiaxed grains.

Upcoming work will consist of studying interactions between grains with multiple orientations, and of modeling of natural and forced convection, as well as of convection due to grain settling.

Acknowledgement

This work was supported by the French State through the program "Investment in the future" operated by the National Research Agency (ANR), referenced by ANR-11 LABX-0008-01 (LabEx DAMAS) and by NASA under grant number NNX14AD69G.

References

[1] Rappaz M and Gandin C A 1993 *Acta Metall. Mater.* **41** 345–360
[2] Zhu M F and Hong C P 2001 *ISIJ Int.* **41** 436–445
[3] Kharicha A, Stefan-Kharicha M, Wu M and Ludwig A 2012 *IOP Conf. Ser. Mater. Sci. Eng.* **33** 12115
[4] Tourret D and Karma A 2013 *Acta Mater.* **61** 6474–6491
[5] Steinbach I, Beckermann C, Kauerauf B, Li Q and Guo J 1999 *Acta Mater.* **47** 971–982
[6] Steinbach I, Diepers H J and Beckermann C 2005 *J. Cryst. Growth* **275** 624–638
[7] Delaleau P, Beckermann C, Mathiesen R H and Arnberg L 2010 *ISIJ Int.* **50** 1886–1894
[8] Melendez A J and Beckermann C 2012 *J. Cryst. Growth* **340** 175–189
[9] Cantor B and Vogel A 1977 *J. Cryst. Growth* **41** 109–123
[10] Sun Y and Beckermann C 2007 *J. Comput. Phys.* **220** 626–653
[11] Beckermann C and Viskanta R 1993 *Appl. Mech. Rev.* **46** 1–27

Bent dendrite growth in undercooled Fe-B alloy melts

C Karrasch[1,2], T Volkmann[1], J Valloton[1,3], M Kolbe[1] and DM Herlach[1,2]
[1]Institute für Materialphysik im Weltraum, DLR, 51170 Cologne, Germany
[2]Institut für Experimentalphysik IV, Ruhr-Universität Bochum, 44780 Bochum, Germany
[3]Advanced Materials and Processing Laboratory, University of Alberta, Edmonton, Canada

E-mail: Christian.Karrasch@dlr.de

Abstract. Dendritic growth is the main solidification mode in alloy casting. In order to control dendrite growth for materials design from the melt it is important to fully understand the influence of process conditions. This study stands as an experimental note observing bent dendrite growth in Fe-B alloys and suggesting possible explanations as induced by fluid flow, thermal, and concentrational diffusion or impurities. Electromagnetic levitation technique (EML) is used for containerless processing of undercooled melts under 1g and reduced gravity conditions in parabolic flight. Further investigations are needed to find a suitable explanation for the observed bent dendrite growth behaviour.

1. Introduction

The dendritic structure formed during the solidification process determines to some extent the final microstructure and therefore can have significant influence on material properties. In alloy casting dendritic solidification is the main growth mode, for example in steel production. To design materials from the melt it is important to understand the influence of fluid flow, convectional effects, and impurities on the dendrite growth process.

Electromagnetic levitation technique (EML) is applied to undercool droplets of metallic melt accessible for *in-situ* diagnostics of the solidification processes. In order to verify models for dendritic growth in undercooled melts we measured growth velocities of pure Fe and Fe-B (1, 5, and 10 at.% B) alloys. In the case of cubic metals such as Fe, dendrite tips grow along the preferred <100> direction [1]. During our investigations on the Fe-B binary system, our measurements showed unexpected bent growing dendrites. This work shows empirically that at least the path of the dendrite tip can change during growth and deviates from a straight line.

2. Experimental

To achieve large undercoolings the use of high-purity raw materials is essential. Samples of Fe-1, 5, and 10 at.% B were prepared from commercial high-purity elements 6N Fe and 6N B purchased from ALFA AESAR. Components were alloyed in an arc melting furnace previously evacuated (10^{-6} mbar) and subsequently filled with high purity Argon (6N) atmosphere (1 bar).

For *in-situ* studies of deeply undercooled Fe-B melts the electromagnetic levitation technique (EML) [3] is used as schematically shown in Figure 1. This containerless technique provides large undercoolings prior to solidification and allows direct observation of the solidification process. The transformation of the undercooled liquid phase into the solid phase leads to a visible contrast due to the release of latent heat during rapid solidification, which is recorded by a high-speed video camera.

The temperature is measured contactlessly by an infrared-pyrometer. High purity 6N He-gas is used to cool down the sample.

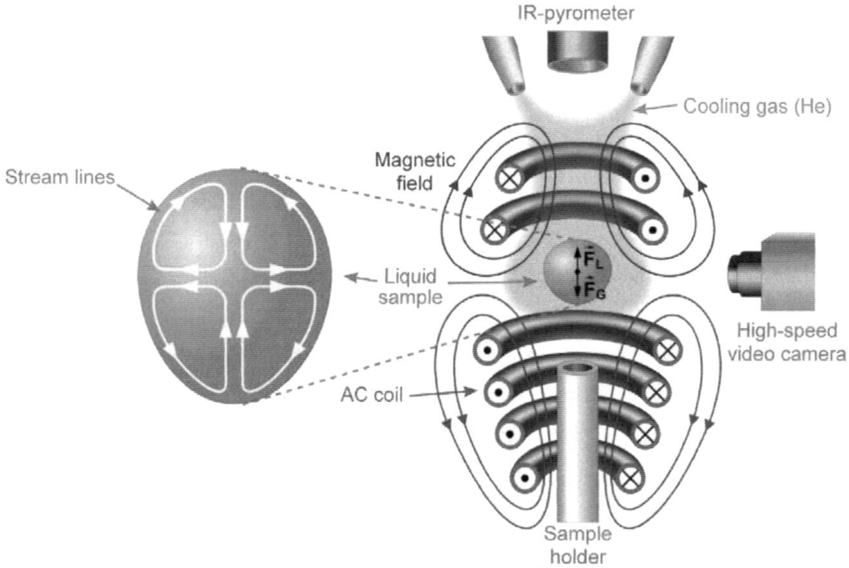

Figure 1. Schematic view of an electromagnetic levitation facility and levitated liquid sample with stream lines of fluid flow as indicated inside the melt [2].

3. Results

Investigations of Fe with 1, 5, and 10 at.% B alloys show bent dendrite growth up to certain undercoolings in ground-based 1g-EML. Figure 2 shows typical results for the trajectories of the growing dendrites for each composition. The dendrite growth velocities are in the order of 10^{-1} m/s, therefore in the same order of magnitude as the fluid flow velocities in 1g-EML (0.3 m/s) according to R.W. Hyers [4]. In Table 1 the undercooling regimes and velocity results are listed. Different types of bent patterns are observed for the three compositions. The trajectories of the dendrite tips show spiralling (Fe-1 at.% B), zigzagging (Fe-5 at.% B) and U-turn (Fe-10 at.% B) growing dendrites. This behaviour is reproducible and characteristic for each composition.

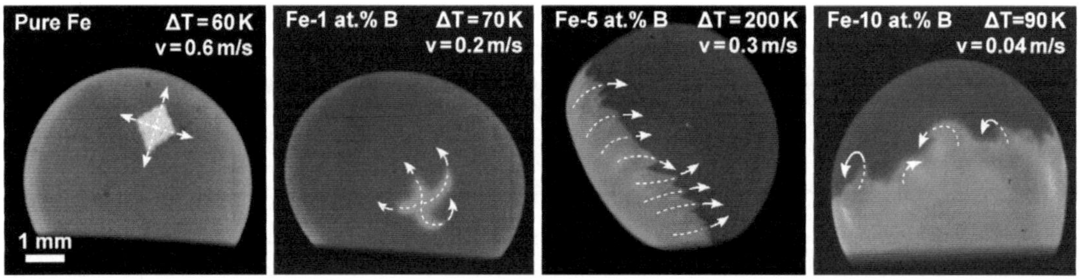

Figure 2. High-speed video images of electromagnetic levitated samples. The dark grey area is the undercooled liquid. The light grey region corresponds to the growing solid, which appears brighter due to the release of latent heat during rapid solidification. Arrows indicate the directions and trajectories of the growing dendrites showing straight (pure Fe), spiral like (Fe-1 at.% B), zigzagging (Fe-5 at.% B), and U-turn (Fe-10 at.% B) patterns in 1g-EML.

Table 1. Highest undercoolings ΔT and corresponding dendrite growth velocities v for Fe-1, 5 and 10 at.% B where bent dendrite growth has been observed so far. Data for pure Fe is added for comparison.

		Undercooling ΔT [K]	Dendrite growth velocity v [m/s]	Observed pattern
1g EML	Pure Fe	60	0.6	Straight, no bending
1g EML	Fe-1 at.% B	70	0.2	Spiral growth
1g EML	Fe-5 at.% B	200	0.3	Zigzagging
1g EML	Fe-10 at.% B	110	0.05	U-turn
TEMPUS	Fe-1 at.% B	100	0.14	Zigzagging

3.1. Bent growing dendrites under reduced gravity conditions

In order to verify the influence of convection on the growth morphology described in this study, parabolic flight experiments were performed under reduced gravity conditions using the TEMPUS facility [5]. In µg-EML the fluid flow is about 0.05 m/s [4], one order of magnitude lower than in 1g-EML but still in the order of the slowest observed growth velocities. Figure 3 shows a bent growing dendrite under microgravity during parabolic flight. The undercooling ΔT is 98 K and the growth velocity v is about 0.14 m/s.

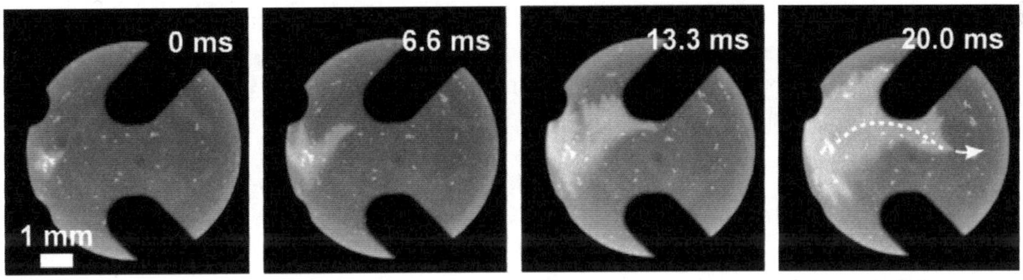

Figure 3. Snapshots of an undercooled solidifying Fe-1 at.% B liquid sample (dark grey) under reduced gravity conditions during parabolic flight showing bent dendrite growth (light grey).

Figure 4. SEM and EBSD images of a cross section of Fe-10 at.% B solidified at an undercooling ΔT of 110K in 1g-EML showing a bent dendrite. The bending occurs mainly in the light blue region and at its transitions to the dark blue and green region, respectively.

3.2. Microstructure
Figure 4 shows the cross section microstructure of a bent dendrite with SEM and EBSD of an Fe-10 at.% B sample. The sample was processed in 1g-EML and solidified at an undercooling ΔT of 110 K.

4. Discussion
The trajectories of the dendrite tips show spiralling, zigzagging, and U-turn patterns for growing dendrites. So far we did not observe bent growing dendrites in any other metallic materials we investigated except multicomponent steel alloy. Our findings appear to be relevant only for Fe-based alloys.

In general, the dendritic solidification is a competition between heat/solute diffusion and surface energy [1]. B is poorly soluble in Fe [6]. According to growth models, a solute pile up exists at the solid-liquid interface of the dendrite tip, which slows down the dendrite growth velocity. Additionally, fluid flow influences the thermal and solute gradients ahead of the solid-liquid interface. In particular, the fluid flow velocity inside the melt in 1g-EML is in the same order of magnitude as the growth velocity itself. Figure 1 shows schematically the calculated fluid flow loops inside a laminar liquid sample according to P. Galenko et al. [2]. There seems to be a weak correlation between the fluid flow loops inside the liquid sample and the trajectories of the observed bent growing dendrites. More advanced investigations are needed to confirm this. Even under reduced gravity conditions with weaker fluid flow velocity the bent dendrite growth occurs for Fe-1 at.% B.

The microstructure of solidified samples showed that the effect of bent dendrites seems to be mainly sample-surface dominated. Up to now we were unable to find significant bent dendrites in the bulk microstructure of the samples. This could be explained due to coarsening and fragmentation. The cross section microstructure of solidified sample (Figure 4) indicates a stepwise breaking of the dendrite. Different crystal orientations of the grains are represented in different colours. A bending occurs mainly in the middle part (light blue region) and at its transitions to the dark blue and green region.

Regarding bent growing dendrites, one possible explanation was given by A. M. Mullis in 1999 [7]. He simulated dendritic bending and rosette formation during solidification under forced fluid flow perpendicular to the growth direction in a shearing flow. So far it is unclear whether our experiments show an actual bent growth or deformation during growth. Phase-field modelling could reproduce the measured bending/deformation during solidification as reported by M. Yamaguchi and C. Beckermann [8]. They computed elasto-viscoplastic deformation of growing solid under a linear shear velocity field using the material point method. According to them, a phase-field model for simultaneous solid deformation and liquid flow is still not available.

Another possible explanation for the observed bent dendrite growth is given by L. Gránásy et al. stating that impurities perturb the solidification process by deflecting the dendrite tip during growth [9]. These foreign particles act as orientation pinning centres. In the melt a random field of impurities may exist leading to zigzagging or spiralling dendrites.

Our study has several limitations for the analysis of the dendrite growth morphologies. First, the strong oscillation and rotation of the liquid droplet has to be taken into account. In particular, the shape of the sample changes during solidification. Second, the nucleation point appears statistically at a random point on the sample surface, so it is not possible to reproduce the same experimental conditions. Third, the fluid flow inside the melt is a nonlinear chaotic process and non-predictable. Fourth, the influence of impurities cannot be neglected.

5. Conclusion
Our research shows bent growing dendrites under different fluid flow conditions. This phenomenon has been observed *in-situ*, as far as we know, for the first time in solidifying metals during levitation. In the case of Fe-10 at.% B in 1g-EML the effect was stronger than for Fe-1 and 5 at% B. Bent dendrite growth could also be observed for Fe-1 at.% B under micro-gravity conditions during

parabolic flight with weaker fluid flow compared to 1g-EML. Microstructure analyses of this sample are still in progress.

The cause of the observed bent dendrite growth is unclear. Impurities cannot be neglected and should be considered as a possible explanation to the growth behaviour. Unless it seems to be the influence of fluid flow on the thermal and/or concentration gradients at the growing dendrite tip. A more systematic approach with defined fluid flow conditions is necessary. It is well known that a growing dendrite follows the largest temperature gradient. For instance, future research using an infrared camera could help to clarify and visualise the thermal field around the growing dendrite.

Finally the effect of bent growing dendrites may be used to manipulate the microstructure development during solidification. For example, dendrites could be bent during growth to follow the curving of a turbine blade, or guided growing dendrites of semiconductor on substrates could lead to new technologies.

Acknowledgments

The authors are grateful to DLR Space Administration Bonn for the parabolic flight opportunity and to the TEMPUS-Crew performing the experiment. Special thanks are extended to G.J. Ehlen, J. Gegner, R. Kobold, and S. Klein for support and fruitful discussions. This research work was financially supported by the German Research Foundation DFG within contract HE1601/18, the European Space Agency ESA under contract numbers 4200020277 and 4200014980, and the Canadian Space Agency CSA FAST program.

References

[1] Dantzig J A and Rappaz M, Solidifcation 2009 *Engineering Sciences, Materials* (EPFL Press) **289**
[2] Galenko P K, Binder S and Ehlen G J, 2012 *Solidification of Containerless Undercooled Melts*, ed. D.M. Herlach and D.M. Matson Wiley-VCH, 349-362
[3] Herlach D M, 1991 *Annual Review of Material Science* **21** 23-44
[4] Hyers R W, *2005 Meas. Sci. Technol* 16 394-401
[5] Lohöfer G and Piller J, 2002 Proceedings 40th AIAA 2002-0764
[6] Cameron T B and Morral J E, 1986 *Metallurgical Transactions* **17A** 1481-1483
[7] Mullis A M, 1999 *Acta Materialia* **47** 1783-1789
[8] Yamaguchi M and Beckermann C, 2013 *Acta Materialia* **61** 4053-4065
[9] Gránásy L, Pusztai T, Warren J A, Douglas J F, Börzsönyi T and Ferreiro V, 2003 *Nature Materials* **2** 92-96

A simple model for spherical growth in alloy solidification

Z Fan[a] and S Z Lu
BCAST, Brunel University London, Uxbridge, Middlesex, UB8 3PH, UK.

Email: [a]Zhongyun.fan@brunel.ac.uk

Abstract. During alloy solidification, spherical growth is the initial growth stage immediately after heterogeneous nucleation but before the morphological instability for dendritic growth. A number of analytical solutions are available for spherical growth and Zener's approximation has been used in nearly all such models. These models, however, are only applicable to solidification with very small supersaturation. In this paper, based on the theoretical analysis of precipitation growth in solid state we present a simplified analytical expression for spherical growth applicable to solidification with an entire range of supersaturation.

1. Introduction

Crystal growth during alloy solidification has two distinctive stages, spherical growth and dendritic growth [1]. When a liquid alloy is cooled down to a temperature below its liquidus to reach the required undercooling for heterogeneous nucleation tiny crystals are nucleated. The initial growth after nucleation is spherical and this is followed by dendritic growth after morphological instabilities take place.

Spherical growth keeps a minimum ratio of interface area to the sphere volume so it is stable at the beginning. The Gibbs-Thomson effect due to curvature is the main driving force for growth when the sphere is very small. The solid crystal grows in spherical shape while continuously rejecting solute to the liquid ahead of the solidification front. As a result, the growing sphere is gradually surrounded by a solute rich layer [1]. With the increase of the sphere size, more and more solute is accumulated at the growth front and the curvature effect is getting less while solute will have more and more effect on growth. The solute rich layer in the liquid restricts the growth and becomes a destabilizing factor for the growth. The sphere particle grows in a balance between the stabilizing factor from the curvature and the destabilizing factor from the solute accumulated at the solid-liquid interface until the solute effect dominates the growth and interface instability occurs [2]. A transition from spherical growth to dendritic growth will then occur to accommodate the interface instability and surface energy anisotropy [3].

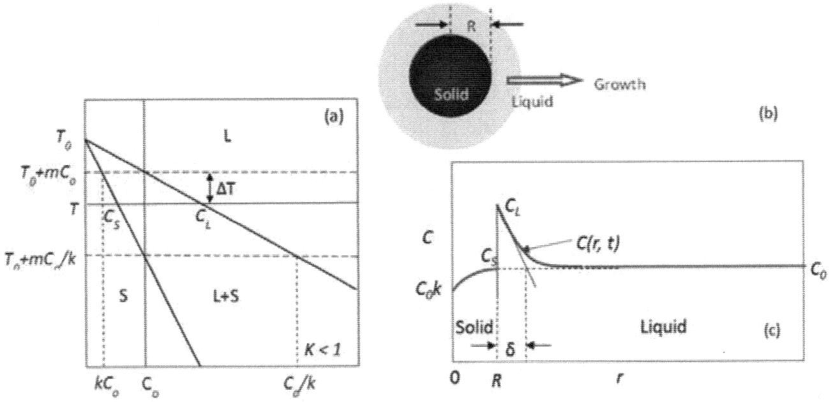

Figure 1. (a) A typical phase diagram with single phase solidification for an alloy of composition, C0, with the equilibrium solute partition coefficient, k, freezing range, ΔT_0, and the undercooling, ΔT. (b) Schematic representation of spherical growth model with the radius, R; a solute-rich liquid layer is around the solid sphere; (c) composition profile at the growth front. δ is defined as the characteristic diffusion length.

Figure 1 (a) shows a part of a typical phase diagram for single phase solidification of an alloy with composition C_0 for the case of equilibrium partition coefficient, k < 1. The solidification with spherical growth occurs in a spatially isothermal melt at a temperature T with an undercooling, ΔT, as indicated in figure 1 (a). Assume that there is no convection in the liquid and there is no interaction between solid spheres during solidification, the growth is completely controlled by diffusion and local equilibrium is then always maintained at the solid-liquid interface. As illustrated in figure 1 (b), r is the radial distance and R is the radius of the solid sphere. The solute profile in the liquid at the growth front is schematically represented in figure 1 (c). The compositions at the interface for the solid and the liquid are C_S and C_L, respectively. The growth rate depends on how fast the solute diffuses away from the interface to the bulk liquid.

Because the spherical growth is the very early stage of crystal growth with a short time duration, there has been so far very little research effort on spherical growth and the majority of the effort has been focused on dendritic growth [1, 4-9]. Zener, in his classical paper [10], provided an analysis for diffusion controlled solid state transformation in spherical form. This analysis is referred as "Zener Approximation" and has been used to treat the spherical growth in solidification [1, 10-13]. In Zener Approximation, the characteristic diffusion length is taken the same as the particle radius, R, and a linear concentration profile is assumed. This resulted in a simple expression for the velocity of spherical growth (V)

$$V = \alpha \frac{D_L}{R} \tag{1}$$

and

$$\alpha = \frac{C_L - C_0}{C_L - C_S} \tag{2}$$

where α is the solute supersaturation defined by Zener [10, 11]. Equation (1) has been used in models for grain size prediction [12, 13] because of its simplicity. In equation (1) the interface velocity is a sole linear function of α, solute supersaturation, for a given alloy. However, this expression can only be used when $\alpha \ll 1$ because the growth velocity calculated by equation (1) is contradictory to the classical theory of diffusion controlled growth, in which interface velocity becomes infinity as α goes to unity [11]. This is also equivalent to the steady state growth as $\Delta T \rightarrow \Delta T_0$ where ΔT_0 is the freezing range as shown in figure 1 (a).

In general, the interface velocity, V, is defined as

$$V = \frac{dR}{dt} = \frac{1}{2}\lambda\sqrt{\frac{D_L}{t}} = \frac{1}{2}\lambda^2\frac{D_L}{R} \tag{3}$$

where λ is a dimensionless interface parameter, $\lambda = \frac{R}{\sqrt{D_L t}}$, R is the radius of the solid sphere, D_L is the solute diffusion coefficient in liquid and is assumed to be constant, and t is the time. $\lambda^2/2$ is treated as the growth coefficient for a given size of sphere.

Maxwell and Hellawell [14] have used a simple spherical growth model to study grain refinement and revealed that not all the potent inoculants would trigger heterogeneous nucleation because of the recalescence. They followed the treatment by Aaron et. al. [15] for the diffusion controlled growth in solid state and developed an expression for spherical growth in their solidification model [14].

$$\lambda = \frac{\alpha}{\sqrt{\pi}} + \sqrt{\frac{\alpha^2}{\pi} + 2\alpha} \tag{4}$$

Figure 2 is a plot of $\lambda^2/2$ against the supersaturation, α, for both Zener approximation ($\lambda^2/2=\alpha$ in equation 1) and the Maxwell and Hellawell's model (equation 4). It can be seen that neither of the two equations above satisfies the classical theory of diffusion limited growth, in which interface velocity should reach infinity as $\alpha\rightarrow 1$. As will be discussed later both Zener approximation and Maxwell and Hellawell's model are only accurate when $\alpha\rightarrow 0$, they will not be applicable to solidification with large supersaturation.

In this paper we present an exact analytical solution and a simplified analytical expression for spherical growth applicable to solidification with entire range of supersaturation.

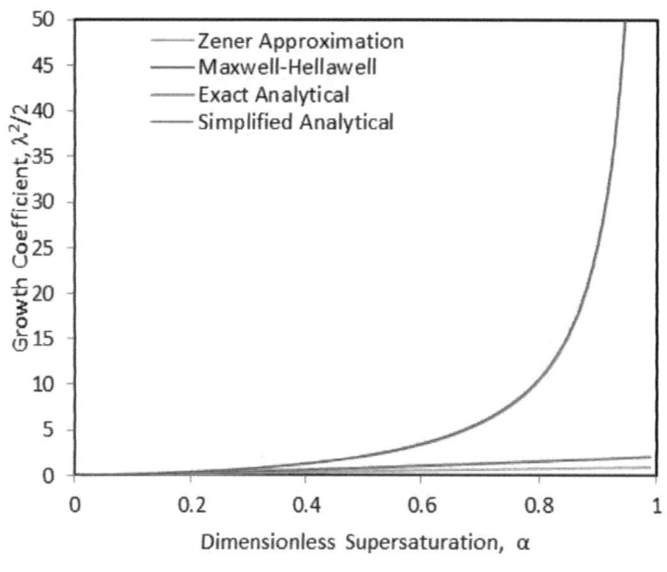

Figure 2. Plot of the Growth coefficient $\lambda^2/2$, against Dimensionless Supersaturation, α, for equation (1) of the approximate expression with Zener Approximation, where $\lambda^2/2 = \alpha$, equation (4) of the Maxwell-Hellawell expression, equation (14) of the analytical solution and equation (17) of the simplified expression.

2. A simple model

Consider the growth of a single solid sphere which does not interact with other spheres in an isothermal melt at temperature T (see figure 1), the sphere size is time dependent, $R=R(t)$ and $r=R$ is the solid-liquid interface position. The concentration profile at the S/L interface is a function of the radial distance, *r*, and time, *t*, $C=C(r,t)$. The governing equation for the diffusion controlled spherical growth is the Fick's second law:

$$\frac{\partial C}{\partial t} = D_L[\frac{\partial^2 C}{\partial r^2} + \frac{2}{r}\left(\frac{\partial C}{\partial r}\right)] \tag{5}$$

Equation (5) has the following initial and boundary conditions:

$$C(r \geq R, t = 0) = C_0 \quad C(r < R, t \geq 0) = C_s$$

$$C(r = R, t > 0) = C_L \quad C(r \to \infty, t > 0) = C_0$$

It has also to satisfy the mass balance at the solid-liquid interface described by the following equation:

$$V(C_L - C_S) = -D_L\left(\frac{\partial C}{\partial r}\right)_{r=R} \tag{6}$$

The analytical solution is actually the same as given for solid state precipitation in [16] and [17] because it is the same in terms of diffusion theory:

$$\frac{\lambda^2}{2}\left[1 - \frac{1}{2}\sqrt{\pi}\lambda \exp\left(\frac{\lambda^2}{4}\right) \operatorname{erfc}\left(\frac{\lambda}{2}\right)\right] = \alpha \tag{7}$$

Equation (7) shows that the growth coefficient, $\lambda^2/2$, is an implicit function of solute supersaturation, α. For a given supersaturation, a unique value of growth coefficient can be numerically obtained and, consequently, the growth velocity can be calculated using equation (3). The numerically calculated growth coefficient from equation (7) is plotted in figure 2 as a function of solute supersaturation in the same way as the Zener approximation and the Maxwell-Hellawell expression above. It is clear in figure 2 that based on present analytical solution (equation 14) the growth coefficient $\lambda^2/2$ becomes very large when the supersaturation is approaching 1, which is consistent with the prediction by the classical diffusion controlled theory [11]. Neither the Zener approximation nor the Maxwell and Hellawell solution offer correct prediction when $\alpha \to 1$. Therefore, equations (1) and (4) can only be applied when $\alpha \to 0$. This is equivalent to the condition with a very small undercooling. When the solute undercooling is very large approaching the freezing range the results from equations (1) and (4) will deviate further from the exact analytical solution. It is also the case for very dilute alloys in which the freezing range is very small so that any small undercooling may result in a large supersaturation.

It should be pointed out that although the current analytical solution is derived for the case of $k < 1$, it is equally applicable to the cases of $k > 1$.

During solidification the growing solid sphere is surrounded by a solute rich liquid layer that is characterized by the characteristic diffusion lengths, δ (see figure 1 (c)). In diffusion controlled growth the growth velocity of the solid particle depends on how fast the solute accumulated in the layer can diffuse away from the solid/liquid interface to the bulk liquid. In other words, the solute accumulated at the interface will restrict crystal growth. It is expected that the thinner the solute boundary layer, the faster the growth. It is shown in figure 1 that the characteristic diffusion length is defined as:

$$\delta = \frac{C_0 - C_L}{\left(\frac{\partial C}{\partial r}\right)_{r=R}} \tag{8}$$

where $\left(\frac{\partial C}{\partial r}\right)_{r=R}$ is the concentration gradient in the liquid at the solid/liquid interface.

Following the same treatment as the diffusion controlled solid-solid transformation [17] the ratio of the characteristic diffusion length to the sphere size, δ/R is thus

$$\frac{\delta}{R} = \frac{2\alpha}{\lambda^2} \tag{9}$$

The δ/R ratio is a sole function of supersaturation since λ is also α dependent. A plot of the δ/R ratio against the supersaturation, α, is presented in figure 3. It is learned from figure 3 that the maximum ratio is when α approaches zero when the diffusion characteristic length, δ, is close to the sphere radius, R. This is the same as the Zener approximation for solid state transformation. It is then evident that the Zener approximation can only be accurate when α is close to zero. As the value of α is getting larger and larger the δ/R ratio is getting smaller and smaller until it goes to zero when α approaches one. This also suggests that the growth restriction at the growth front due to solute pile-up becomes smaller and smaller as the supersaturation is getting larger and larger and there is no growth restriction as the supersaturation reaches unity.

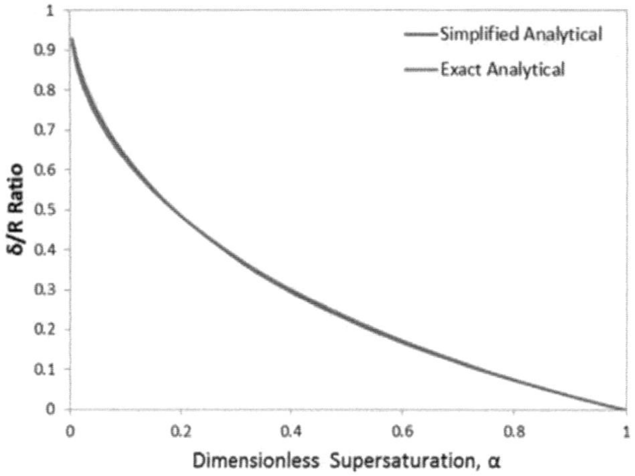

Figure 3. Plot of δ/R ratio against solute supersaturation, α, for both analytical solution and simplified analytical expression in which δ is the characteristic diffusion length and R is the radius of growing solid sphere.

The present model provides an accurate solution for diffusion controlled spherical growth during alloy solidification. As described in equation (3) and equation (7), the growth velocity is solely α dependent. However, the equation (7) is implicit and inconvenient to use. Therefore, there is a need to find a function of α that is a good fit of the exact solution over the entire range of α. Zener used asymptotic expansion to show that $\lambda^2/2 = \alpha$ as α goes to zero and $\frac{\lambda^2}{2} = \frac{3}{1-\alpha}$ approaches infinity as α goes to unity [10]. Through curve fitting for equation (7) and taking account of $\frac{\alpha}{1-\alpha} = \frac{C_0 - C_L}{C_S - C_0}$, a simplified analytical expression which satisfies asymptotic limits is obtained.

$$\frac{\lambda^2}{2} = \frac{\alpha}{1-\alpha}(1 + \sqrt{\alpha} + \alpha) \tag{10}$$

This simplified analytical expression, equation (10), is plotted together with the exact solution, i.e., equation (7), in figure 2. The two curves are almost overlapping over the whole range of the plot. Error analysis shows that the deviation of the simplified analytical expression from the exact analytical solution is less than 2% at the most from α=0.01 to 0.97. Therefore equation (10) can be a good replacement for equation (7) and it is much more convenient for practical application. Using the same replacement, the equation (9) for the normalized diffusion characteristic length can thus be presented as:

$$\frac{\delta}{R} = \frac{1-\alpha}{1+\sqrt{\alpha}+\alpha} \quad (11)$$

Equation (11) is also plotted in figure 3 in comparison with the analytical solution of equations (9). The two curves are again almost perfectly overlapping in whole range of α.

3. Discussion

As mentioned earlier, the Zener approximation can be used only when the supersaturation, α, is very small and this is the case when undercooling is small compared with the freezing range. For a very dilute alloy, the freezing range can be very small, it will be very easy to reach high supersaturation, and therefore, the approximate expressions, equations (1) and (3), can no longer be used. However, the present analytical model can provide an exact solution to such problems. Consider that two alloy melts are cooled below their respective liquidus but with the same undercooling (e.g., $\Delta T=0.25K$) which allows heterogeneous nucleation to occur. One alloy has a relatively large freezing range, Al-3 wt. % Cu (all compositions are in wt.% unless stated otherwise), for example; the other is a dilute alloy, e.g., Al-0.01 Ti. Due to the large difference in freezing range, the supersaturation will be different: α=0.04 for Al-3 wt. % Cu alloy and $\alpha=0.45$ for Al-0.01 Ti alloy. It is not a problem to use the approximate solutions for Al-3 Cu alloy, but there will be a large error for Al-0.01 Ti alloy. These data in comparison are listed in table 1. The growth velocities calculated with the approximate solutions and with present analytical solution are very close for Al-3 Cu alloy (a factor of 1.3) but will be very different for Al-0.01 Ti alloy (a factor of 4, see table 1).

Table 1. Data comparison for two alloy systems of Al – 3 wt.% Cu and Al – 0.01 wt.% Ti including Equilibrium Solute Partition Coefficient, k; Liquidus Slope, m; Freezing Range, $\Delta T_0=mC_0(k-1)$; Dimensionless Undercooling, $\theta=\Delta T/\Delta T_0$; Dimensionless Supersaturation, α; ratio of Characteristic Diffusion Length to Sphere Radius, δ/R; Growth coefficient, $\lambda^2/2$; and the ratio of the Interface Growth Velocities calculated by Analytical Solution to that calculated by the Approximate Solution. $\Delta T = 0.25\ K$ is assumed for both alloy systems.

					$\Delta T = 0.25\ K$				
	C_0 (wt.%)	k	m (K/wt.%)	ΔT_0 (K)	θ	α	δ/R	$\lambda^2/2$	Growth Velocity Ratio (Anal./ Appr.)
Al-Cu	3	0.14 [7]	-2.6 [7]	48	0.005	0.04 (α→0)	0.774	0.052	1.3
Al-Ti	0.01	7.67 [18]	33.3 [18]	0.3	0.83	0.45 (α>>0)	0.259	1.735	3.85

On the other hand, in the above cases with the same value of undercooling but different supersaturation in the two alloy systems, the characteristic diffusion lengths of the solute rich boundary layers are different for the same size of the particle since the ratios of δ/R is different. The calculated results using expression (11) shows that δ/R is 0.774 for Al-3 Cu alloy and 0.259 for Al-0.01 Ti alloy. This means that with the same undercooling ($\Delta T=0.25K$) the characteristic diffusion length for Al-3 Cu alloy is about three times thicker than that for the dilute Al-0.01 Ti alloy that has much higher supersaturation. It is conceivable that the growth will be more restricted with a thicker solute boundary layer at the solid/liquid interface, and consequently the growth rate will be slower. The calculated results of growth coefficient, $\lambda^2/2$, using equation (10) are indicators of the growth velocities that are consistent with the above understanding. These data are also listed in table 1 and they indicate that for the same solid particle size and undercooling, the growth velocity of dilute Al-0.01 Ti alloy is much higher than that of the Al-3 Cu alloy.

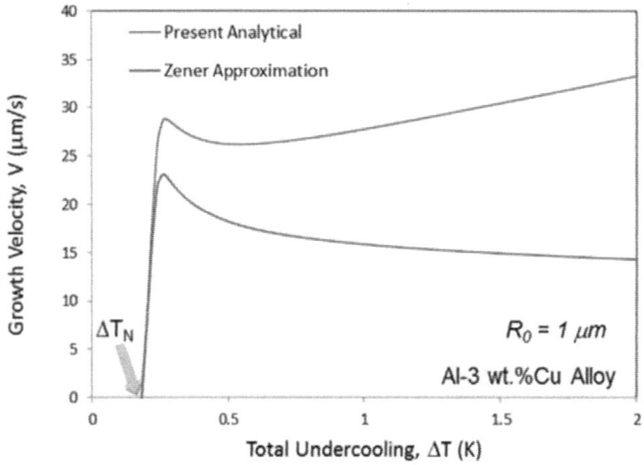

Figure 4. Plot of interface velocity against undercooling for Al- 3 wt. % Cu alloy calculated with the simplified analytical expression and with the approximate solution. ΔT_N is the initial undercooling after nucleation of the assumed solid sphere with radius $R_0 = 1\mu m$.

As described by equation (3), the velocity of spherical growth is proportional to the growth coefficient, $\lambda^2/2$, where $\lambda^2/2$ is a sole function of solute supersaturation, α, which in turn is a function of alloy composition, C_0, liquidus slope, m, and the solute partition coefficient, k. Considering the increasing radii during the spherical growth a plot of interface velocity against undercooling for Al- 3 Cu alloy is shown in figure 4. The undercooling is taken as the sum of the curvature term, $\Delta T_r = \frac{2\Gamma}{R}$ where $\Gamma = \frac{\sigma_{SL}}{\Delta S}$ is the Gibbs-Thomson coefficient, Γ=0.9 x 10^{-7} mK [7], and the solute term, $\Delta T_S = m(C_0 - C_L)$. The initial size of the sphere, R_0, after nucleation is assumed to be 1 μm in radius and the diffusion coefficient of copper in liquid aluminium, D_L = 3 x 10^{-9} m^2/s is taken in calculation [7]. The interface velocity calculated with the Zener approximation (equation 1) is also plotted in figure 4 for comparison. It is obvious that the Zener approximation does not give the correct trend that interface velocity goes up as the undercooling/supersaturation increases. The present model, on the other hand does give the right trend that velocity increases with the undercooling or supersaturation. As demonstrated above, the simplified expression is applicable to dilute alloys and can be used to examine the growth restriction effect of solute element during solidification.

4. Summary

Following the theoretical analysis of precipitation growth in solid state we have derived an exact solution for the growth of spheroids in liquid alloys, which describes the growth coefficient as an implicit function of the solute supersaturation. For convenience we have also converted the exact solution into a simplified analytical expression where the growth coefficient is an explicit function of the solute supersaturation. The present analytical solutions are applicable to spherical growth with the entire range of supersaturation ($0<\alpha<1$), and is a significant improvement over both the Zener approximation and the Maxwell and Hellawell's model in variant size approximation.

Acknowledgement

Professor J. D. Hunt had participated in this work before he passed away in 2012. The financial support from EPSRC (UK) under the grant for the EPSRC Centre-LiME is gratefully acknowledged.

References

[1] Dantzig J A and Rappaz M 2009 Solidification EPFL Press
[2] Mullins W W and Sekerka R F 1963 *Journal of Applied Physics* **34** 323
[3] Kessler D A Koplik J and Levine H 1988 *Advances in Physics* **37** 255
[4] Ivantsov G P Dokl 1947 *Akad. Nauk SSSR* **58** 567.
[5] Horvay G and Cahn J W 1961 *Acta Metall.* **9** 695
[6] Lipton J Glicksman M E and Kurz W 1984 *Materials Science and Engineering* **65** 57

[7] Kurz W and Fisher D J 1989 Fundamentals of Solidification, Lausanne: Trans. Tech.
[8] Trivedi R and Kurz W 1994 *International Materials Reviews* **39** 49
[9] Asta Beckermann M C Karma A Kurz W Napolitano R Plapp M Purdy G Rappaz M Trivedi R 2009 *Acta Materialia* **57** 941
[10] Zener C 1949 *Journal of Applied Physics* **20** 950
[11] Christian J W 1975 *The Theory of the Transformation of Metals*, Pergamon Press
[12] Wu M Ludwig A 2009 *Acta Materialia* **57** 5621
[13] Hunt J D and Fan Z 2011 Solidification Science and Technology, Proceeding of the John Hunt International Symposium, 12-14 December 2011, Brunel University Press 59
[14] Maxwell I and Hellawell A 1975 *Acta Metalurgica* **23** 229
[15] Aaron H B Fainstein D and Kotler G R 1970 *Journal of Applied Physics* **41** 4404
[16] Frank F C 1950 Proc. Roy. Soc. London **A201** 856
[17] Glicksman M E 2000 *Diffusion in Solids*, John Wiley & Sons, Inc.
[18] Desnain P Fautrelle Y Meyer J L Riquet J P Durand F 1990 *Acta Metall. Mater.* **38** 1513

IOP Conf. Series: Materials Science and Engineering **117** (2016) 012017 doi:10.1088/1757-899X/117/1/012017

The influence of cooling rate and Fe/Cr content on the evolution of Fe-rich compounds in a secondary Al-Si-Cu diecasting alloy

A Fabrizi[a] and G Timelli
University of Padua, Department of Management and Engineering, Stradella S. Nicola, 3 - 36100 Vicenza, Italy

E-mail: [a]fabrizi@gest.unipd.it

Abstract. This study investigates the morphological evolution of primary α-Al(Fe,Mn,Cr)Si phase in a secondary Al-Si-Cu alloy with respect to the initial Fe and Cr contents as well as to the cooling rate. The solidification experiments have been designed in order to cover a wide range of cooling rates, and the Fe and Cr contents have been varied over two levels. Metallographic and image analysis techniques have been used to quantitatively examine the microstructural changes occurring at different experimental conditions. The morphological evolution of the α-Fe phase has been also analysed by observing deep etched samples. By changing the cooling rate, α-$Al_{15}(Fe,Mn,Cr)_3Si_2$ dodecahedron crystals, as well as Chinese-script, branched structures and dendrites form, while primary coarse β-$Al_5(Fe,Mn)Si$ needles appear in the alloy with the highest Fe content at low cooling rates.

1. Introduction

From several decades, secondary Al-Si casting alloys have been widely used for the production of automotive components. A key-factor to be considered in the recycling of Al alloys consists in controlling the presence of impurities, such as the Fe level. The presence of Fe induces the precipitation of brittle Fe-rich compounds deleterious for the mechanical properties as well as the castability and the machinability of castings. In high-pressure diecast (HPDC) Al alloys, Fe-rich compounds crystallize with three distinct morphologies, such as needle-like, Chinese-script and star-like or polyhedral morphology [1]. According to the crystal structure and chemistry, the needle-like Fe-rich compounds are commonly referred as the monoclinic β-Al_5FeSi phase, while Chinese-script and compact particles are generally indicated as α-$Al_{15}(Fe,Mn)_3Si_2$, or α-$Al_{12}(Fe,Mn)_3Si$, phase with a body centred cubic (bcc) crystal structure [2]. However, morphology, size and volume fraction of each specific Fe-rich compound depend largely on the alloy composition and the solidification condition [3]. The Fe, Mn and Cr contents in the alloy, as well as their reciprocal ratio, significantly affect the morphology of the Fe-rich compounds [1,4]. It was found that a greater amount of needle-like particles is achieved for high Fe:(Mn+Cr) ratio, while more Chinese-script and/or polyhedral, star-like and blocky particles precipitate as the ratio decreases. As aforementioned, the morphology as well as the formation of Fe-rich compounds is also strongly affected by the cooling rate because of the diffusion and solubility variation of Fe, Mn and Cr during the solidification [5]. In the present work, the morphological evolution and the growth mechanism of primary α-Fe phase in a secondary Al-Si-Cu alloy was studied as function of the cooling rate. 2D microstructural investigations coupled to 3D characterizations were performed on HPDC specimens as well as on specimens solidified at lower cooling rates. Furthermore, the chemical composition of the base Al-Si-Cu alloy was varied by modifying the initial Cr and Fe contents.

2. Experimental procedure

Three secondary AlSi9Cu3 alloys with different Cr and Fe levels were used in this study. The melting procedure and the production of the HPDC specimens are described elsewhere [6]. The chemical compositions of the experimental alloys are given in table 1. About 80 g-piece was drawn from each HPDC alloy and remelted inside a cylindrical ceramic crucible with an inner diameter of 20 mm and a height of 80 mm by using an electric-resistance muffle furnace. The melting temperature was set at 800±5°C with 1 hour holding time in order to ensure homogeneity and, potentially, dissolution of the present intermetallics. After manually skimming, the molten metal was solidified according to two different cooling rates. For the first batch of specimens, the crucible was cooled in still air outside the furnace (hereafter called *medium cooling rate*); while the second batch was cooled more slowly to room temperature leaving the material inside the furnace at controlled cooling (hereafter called *low cooling rate*). The cylindrical as-cast samples were longitudinally sectioned and prepared for microstructural investigations by standard metallographic procedure. The microstructural characterization was performed by using optical microscope, scanning electron microscope (SEM), energy-dispersive spectroscopy (EDS) and electron back-scattered diffraction (EBSD). The microstructural scale was estimated by measuring the secondary dendrite arm spacing (SDAS) at equivalent positions in the as-cast specimens. The cooling rate, R, for each solidification condition was then determined from the SDAS values by using the equation: SDAS = $28.13 \cdot R^{-0.4}$ [7]. The 3D morphology of primary Fe-rich compounds was revealed by complete dissolution of Al matrix with a 20% NaOH water solution at 65°C for 15 minutes. Finally, the EDS analyses were carried out to study the chemical composition of the Fe-rich compounds with respect to the morphology and the cooling condition, while the combination of EDS and EBSD was performed to identify the phase and the crystallographic orientation of the compounds.

Table 1. Chemical compositions of the experimental alloys (wt.%); *SF* = Sludge Factor.

Alloy	Al	Si	Cu	Fe	Mn	Cr	Mg	Zn	Ni	Ti	Fe:Mn	*SF*
1	bal.	8.26	2.447	0.718	0.588	0.058	0.193	1.041	0.049	0.040	1.22	2.07
2	bal.	8.14	2.407	0.716	0.556	0.104	0.189	1.033	0.047	0.041	1.29	2.14
3	bal.	8.20	2.388	1.415	0.59	0.103	0.207	1.019	0.053	0.037	2.40	2.90

3. Results and discussion

The microstructures of the HPDC alloys are reported in figure 1. Primary polyhedral and star-like Fe-rich compounds with a mean diameter of ~20 μm, usually termed *sludge*, are uniformly distributed in all the alloys. The tendency of an Al alloy to form sludge particles strictly depends on the Fe, Mn and Cr contents according to the following empirical formula [8]:

$$\text{Sludge Factor } (SF) = (1 \times \text{wt.}\%Fe) + (2 \times \text{wt.}\%Mn) + (3 \times \text{wt.}\%Cr) \quad (1)$$

By comparing the different microstructures, the alloy 3 with the highest *SF* value shows a larger amount of primary Fe-rich compounds. Due to the high cooling rate in HPDC, all the alloys show a refined microstructure with very fine eutectic Si; the mean SDAS value results ~10 μm and no significant variation is noted among the alloys; moreover, the estimated cooling rate is about 800 °C/min. Beside blocky particles, few primary Fe-rich compounds with a Chinese-script structure appear in all the experimental alloys (figure 2a). Generally, particles having a size comparable to the SDAS are considered primary (or pre-dendritic) phases. The EBSD investigations revealed that the blocky particles and the Chinese-script structures well match a cubic structure (space group, *s.p.*, Pm-3) with the lattice parameter a ~ 1.26 nm, associated to the α-Fe phase (figure 2b). Further, from the orientation map of the α-Fe phase in figure 2c, it results that all the distinct branched of the Chinese-script structure have identical crystallography orientation. The 3D characterization of deep-etched samples revealed the Chinese-script phase as a whole interconnected and branched structure [7,9].

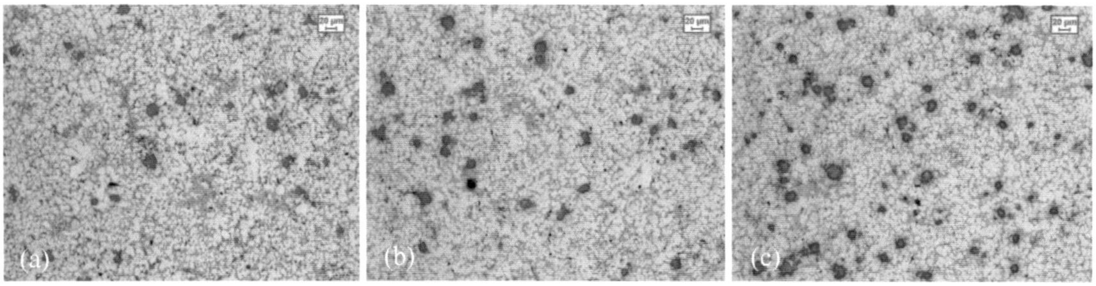

Figure 1. Microstructures of the HPDC alloys: (a) alloy 1, (b) alloy 2 and (c) alloy 3.

The chemical composition of the α-Fe phase was studied as function of the morphology and the alloy composition through EDS, and the results are reported in figure 3. The stoichiometry of both Chinese-script and compact compounds can be considered consistent with both α-$Al_{12}(Fe,Mn,Cr)_3Si_2$ and $Al_{15}(Fe,Mn)_3Si_2$ phases [10,11]. However, the Chinese-script morphology contains greater Si content and higher Fe:(Mn+Cr) ratio than the blocky particles. The EDS analyses also revealed that the chemical composition of α-Fe phase depends on the alloy chemistry: higher the Cr and Fe levels in the alloys (alloys 2 and 3, respectively), higher the Cr and Fe contents in the Fe-rich compounds, regardless to their morphology.

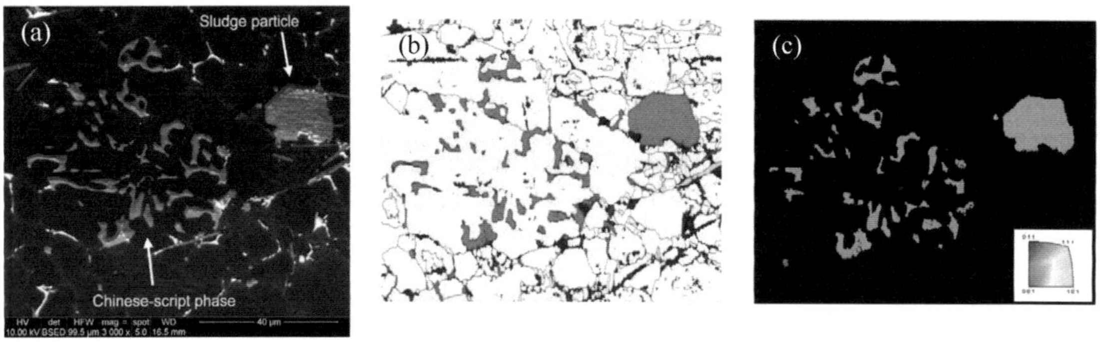

Figure 2. (a) SEM micrograph showing Fe-rich compounds with compact shape and Chinese-script morphology; corresponding (b) EBSD phase map (yellow = α-Al (cubic, *s.p.* Fm-3m); red = α-Fe (cubic, *s.p.* Pm-3); green = β-Fe (monoclinic, *s.p.* 2/m); blue = Al_2Cu (tetragonal; *s.p.* I4/mcm); violet = Si) and (c) orientation map of α-Fe phases.

Nevertheless, due to reciprocal substitution of Fe, Mn and Cr atoms in the cubic lattice of the α-Fe phase, the total Fe+Mn+Cr content remains approximately constant for all the distinct morphologies. At high magnification, it is observed how many eutectic Si particles lie along the blocky Fe-rich compounds, suggesting blocky compounds constitute suitable site for nucleation of Si crystals. On the other hand, the Chinese-script compounds are separated from the eutectic Si through the α-Al phase suggesting a peritectic reaction occurred. Concerning the blocky morphology of Fe-rich compounds, it has been recently found that most of the polyhedral and star-like 2D-morphologies derive from regular or hollowed rhombic dodecahedron with a cavity on all the 12 rhombic facets, respectively [12]. The rhombic dodecahedron has totally 14 vertices, six of them are located on the three orthogonal axes and they correspond to the intersection common point of the four facets belonging to {011} crystal planes. After dissolution of α-Al matrix by deep-etching, some hollowed dodecahedrons seem to exhibit an initial (figure 4a) or more advanced growth of the tips on the six different vertices, extending on the orthogonal, cube-edge <100> directions (figure 4b). A 3D reconstruction of degenerated hollowed dodecahedron was carried out and insert in figure 4b. The relative cross-section of the 3D reconstruction seems to properly correspond to the observed 2D morphology (figure 4c).

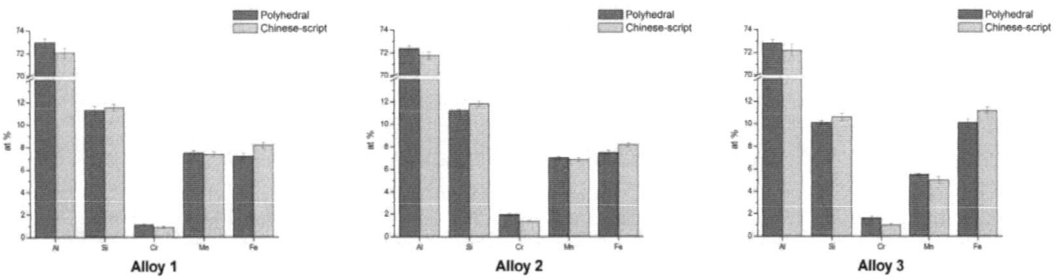

Figure 3. Chemical compositions (at.%) of α-Fe phase as function of the different morphologies in the experimental alloys.

Figure 4. 3D microstructures of hollowed dodecahedron with (a) initial and (b) advanced growth of tips along orthogonal directions; (c) 2D morphology of a degenerated star-like structure. In the insets, the 3D CAD model and the associated cross-section are shown.

Figure 5 illustrates the typical microstructures of as-cast alloys solidified at *medium cooling rate*. In general, it can be clearly seen that the microstructural scale increased with respect to HPDC alloys; the mean SDAS value raises to ~40 μm for all the analysed alloys and the estimated cooling rate decreases to 25 °C/min. The Fe-rich compounds show mainly Chinese-script morphology in alloys 1 and 2 but occasionally they assume a skeleton- or dendrite-like morphology. The fraction of coarse dendrites increases with the Cr content and even more with the Fe level. Considering the 3D morphology of the α-Fe dendrites, a morphological evolution from the hollowed dodecahedron model to coarse dendrite is proposed (figure 6). By continuous growing of the tips on the six vertices during crystal growth, the primary dendrite arms can formed along preferential <100> direction and, further, secondary and ternary branches can be created from the primary arms growing along the orthogonal directions (figure 6c). The EDS results show that the chemical composition of the two types of α-Fe compounds slightly differs. In general, the Chinese-script compounds show higher Fe content than the dendrites; further, the chemical composition of the α-Fe phase is sensitive to the initial composition of the melt, as previously observed in the HPDC alloys. At the *low cooling rate*, the microstructural scale of the alloys further increases (figure 7): in fact, the mean SDAS value is ~110 μm and the eutectic Si particles appear as coarse platelets while the cooling rate reaches ~ 2°C/min. The previous coarse dendrites disappear and large blocky Fe-rich particles with an average size of ~ 350 μm are revealed at the bottom of the specimens. As reported by Shabestri et al. [11], primary α-Fe crystals forming at temperatures above the liquidus cause gravity segregation due to their higher density. Furthermore, the Chinese-script structures become larger in the alloys 1 and 2. Comparing the microstructures in figure 5 with those resulting at low cooling rate (figure 7), it is evident that the secondary arm spacing of the Chinese-script phase increases as the cooling rate decreases.

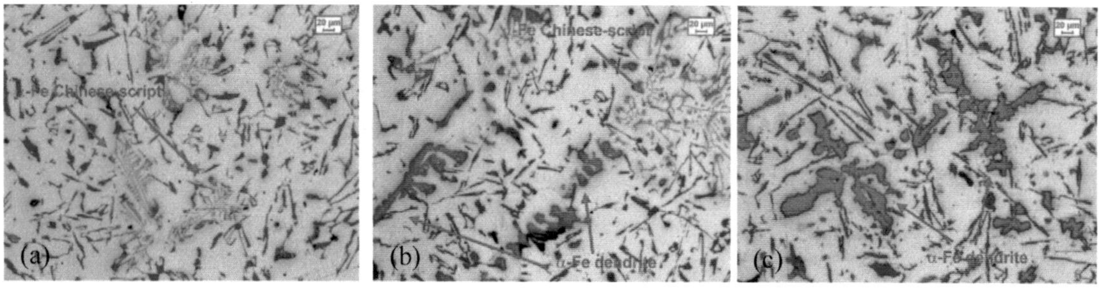

Figure 5. Microstructures of as-cast (a) alloy 1, (b) alloy 2 and (c) alloy 3 solidified at medium cooling rate; α-Fe Chinese-script compounds and coarse α-Fe dendrites are indicated.

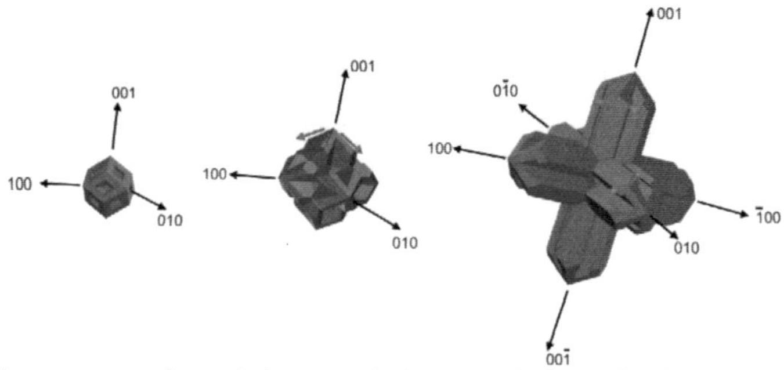

Figure 6. Schematic representation of the morphology evolution of primary α-Fe compounds from hollowed dodecahedron to coarse dendrite. The growth direction of secondary and ternary arms are indicated by red and green arrows, respectively.

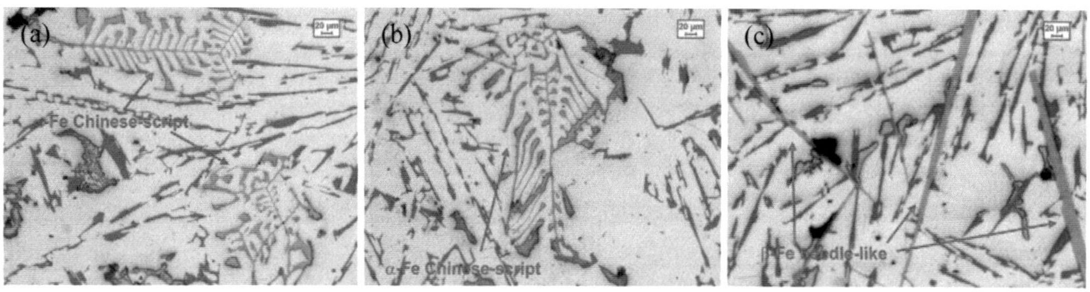

Figure 7. Microstructures of as-cast (a) alloy 1, (b) alloy 2 and (c) alloy 3 at low cooling rate: α-Fe Chinese-script compounds and β-Fe needle-like particles precipitate.

As shown in figure 7c, coarse primary needle-like phases appear in the alloy 3 with the highest Fe:Mn ratio. According to the EDS analyses, the stoichiometry of the needle-like structures is $Al_5(Fe,Mn)Si_{1.4}$ and then identified as β-Fe phase. By considering the 2D and 3D microstructural investigations, an explanation of the growth mechanism and morphology evolution of primary α-Fe phase is proposed. Generally, at high cooling rates, the great undercooling promotes the formation of numerous and fine α-Fe nuclei. The initial growth of these nuclei is diffusion controlled and, at this stage, the radial growth rate of such isolated seeds is fully isotropic. This leads to the formation of spherical primary α-Fe phase. As the solidification proceeds, the nuclei grow and the cubic unit cells of α-Fe phase can arrange spontaneously by growing in a faceted manner and by forming a rhombic dodecahedral structure. However, during the growth process, the α-Fe crystal might exceed a critical size losing its stability. The growth velocity of the edges can prevail on the growth rate of the rhombic facets and, consequently, the formation of cavities at the facets' centres might occur (*skeletal growth* [13]), thus, hollowed rhombic dodecahedrons can form. Under such solidification conditions, the advancing growth of primary α-Fe crystals is interrupted as the α-Al dendrites rapidly begin to form. On the contrary, at slower cooling rates but still far away from a steady-state condition, the further growth of the primary α-Fe crystals can continue before Al-dendrite formation and, at a more advanced stage, the six orthogonal vertices of the hollowed dodecahedron begin to rapidly protrude leading to the development of primary dendrite arms. Subsequently, secondary arms can be generated from the primary ones in the four orthogonal planes, forming branching facets. A similar growth process was also observed for Si and Mg_2Si crystals [14,15]. However, further investigations need to be performed in order to verify the growth mechanism of α-Fe dendrites.

4. Conclusions

The effects of cooling rate and Fe/Cr content on the formation and morphology evolution of primary Fe-bearing particles have been studied in a secondary Al-Si-Cu alloy. In the HPDC alloys, blocky compounds and small Chinese-script particles are detected. By reducing the cooling rate, large α-Fe Chinese-script structures are identified to be the major intermetallic phase in the lower Fe content alloys, while coarse α-Fe dendrites form by increasing the Cr and Fe level. At the slowest solidification regime, α-Fe Chinese-script structures enlarge and coarse α-Fe crystals grow and sediment at the bottom of the molten bath. Primary needle-like β-Fe particles result the dominant phase as the Fe:Mn ratio increases above 2. By studying the 3D morphologies of α-Fe phase, a growth mechanism is presented to support the morphological transition of initial dodecahedral particles to coarse dendrites or to coarse crystals by varying the solidification conditions.

References

[1] T Narayanan L A, Samuel F H and Gruzlesky J E 1994 *Metall. Mat. Trans.* A **25** 1761
[2] Sweet L, Zhu S M, Gao S X, Taylor J A and Easton M A 2011 *Metall. Mater. Trans.* A **42** 1737
[3] Makhlouf M and Apelian D 2002, *Casting Characteristics of Aluminium Die Casting Alloys*, (Work Performed Under Contract No. DEFC07-99ID13716) pp 1-46
[4] Gowri S and Samuel F H 1994 *Metall. Mater. Trans.* A **25** 437
[5] Narayanan L A 1994, *Crystallization and dissolution studies of iron intermetallics in Al-Si alloys*, (McGill University, Montreal, Canada)
[6] Timelli G, Ferraro S, Grosselle F, Bonollo F, Voltazza F and Capra L 2011 *Metall. Ital.* **103** 5
[7] Sivarupan T, Caceres C H and Taylor J A 2013 *Metall. Mater. Trans.* A **44** 4071
[8] Jorstad J L 1986 *Die Casting Engineer* **30** 30
[9] Warmuzek M, Mrowka G and Sieniawski J 2004 *J. Mater Process. Technol.* **157-158** 624
[10] Shabestari S.G. 2004 *Mater. Sci. Eng.* A **383** 289
[11] Shabestari S G and Gruzleski J E 1995 *Metall. Mater. Trans.* A **26** 995
[12] Fabrizi A, Ferraro S and Timelli G 2014 *Shape Casting: 5th Int. Symp. 2014 - TMS 2014*, eds. M Tiryakioğlu, J Campbell, G Byczynski (Hoboken: John Wiley and Sons) pp 277-84
[13] Li C, Wu Y Y, Li H and Liu X F 2011 *Acta Mater.* **59** 1058
[14] Xu C L, Wang H Y, Liu C and Jiang Q C 2006 *J. Cryst. Growth* **291** 540
[15] Qin Q D, Zhao Y G, Zhou W and Cong P J 2007 *Mater. Sci. Eng.* A **447** 186

3D characterization by tomography of beta $Al_9Fe_2Si_2$ phase precipitation in a Al6.5Si1Fe alloy

D Ferdian[1,2a], L Salvo[3], J Lacaze[1], C Tenailleau[4], B Duployer[4] and B Malard[1]
[1] CIRIMAT, Université de Toulouse, UMR CNRS 5085, ENSIACET, BP 44362, 31030 Toulouse Cedex 4, France
[2] Dept. Metallurgy and Materials, Universitas Indonesia, Kampus Baru UI Depok 16424, Indonesia
[3] Univ. Grenoble Alpes, SIMAP, F-38000 Grenoble, France
[4] CIRIMAT, UMR CNRS 5085, Université Paul Sabatier, 31062 Toulouse Cedex 9, France

Email: [a] deni@metal.ui.ac.id

Abstract. The microstructure evolution of beta phase during solidification of a synthetic Al6.5Si1Fe (wt.%) alloy has been investigated by in-situ synchrotron micro-tomography and post-mortem tomography. In-situ solidification was observed at a constant cooling rate of 10°C min^{-1}, from above the alloy's liquidus with the melt at 618°C down to 575°C which is just above the (Al)-Si-beta invariant eutectic reaction. Primary (Al) dendrites nucleated at 608°C, followed by the formation of beta-$Al_9Fe_2Si_2$ phase starting at 593°C. After a rapid growth stage until 587°C as thin plates, beta phase continued to grow at a paced rate. Thickening of the plates was also evaluated and it was observed that the decrease in the lateral growth rate of the plates did not lead to an increase of their thickening rate. It was noted that the interconnectivity between beta precipitates increased as the solidification progressed. While nucleation of beta phase has previously been reported to occur on the alumina scale formed at the outer surface of the material, it is shown from post mortem tomography that bulk nucleation can occur as well.

1. Introduction

Tomography in material science is used nowadays as a non-destructive method to visualize internal structure of objects and acquire information on their three dimensional geometries and properties [1, 2]. Within the past 10 years, the development of in-situ tomography for the study of solidification phenomena has increased significantly. This is due to the development of fast cameras to record the images which facilitate the observation of in situ solidification process [3]. The use of in-situ method with X-ray can be seen in the work by Mathiesen and co-workers [4, 5] who studied dendritic growth and the dynamics of columnar to equiaxed transition. Kim et al. [6] also studied the effect of iron on an Al-Si-Cu alloy with in-situ radiography. Furthermore, the development of in-situ experiments was achieved by allowing 3D imaging by rotating the sample during in situ solidification [7, 8].

The present work considers the solidification of a high purity Al6.5Si1Fe (in mass.%) alloy, and more particularly the precipitation of beta-$Al_9Fe_2Si_2$ phase that occurs after primary (Al) and before the invariant (Al)-Si-beta eutectic. Early studies on beta phase precipitation and growth morphology using in situ X-ray tomography and in situ synchrotron can be seen in the works by Wang et al. [9], Terzi et al. [10] and Puncreobutr et al. [11] which were all performed on Al-Si-Cu-Fe alloys. Observation by Wang et al. [9] using in situ synchrotron radiography suggested several stages in beta phase formation in a Al7.5Si3.5Cu0.8Fe alloy. They observed a rapid growth in early stage which was later constrained by primary aluminum dendrites, followed by attachment and diffusion growth where the plate thickened. Terzi et al. [10] in a Al8Si4Cu0.8Fe alloy showed the beta phase nucleated at or near the surface or the outer skin surface after the primary aluminum nucleation, where the outer surface was most probably γ-Al_2O_3. Recent work by Puncreobutr et al. [11] quantified the nucleation of intermetalic as a function of undercooling in Al7.5Si3.5Cu0.6Fe alloy. In this case also, the plates displayed a very rapid lateral growth and slow thickening. They also found that some eutectic silicon precipitates attached to a beta plate or outer oxide when they first appear. Further review regarding the beta phase precipitation can be seen in the work by Ferdian et al. [12] The objective of this study is to characterize the morphology and the growth of beta phase in high purity Al-Fe-Si ternary system using in-situ as well as post-mortem tomography.

2. Experimental method

Two types of tomography experiments, namely post mortem and in-situ, were conducted for studying the growth of beta phase. The in-situ synchrotron tomography was conducted at European Synchrotron Radiation Facility (ESRF) ID-19, with the holding furnace developed by SIMAP, Grenoble. The Al-6.5Si-1Fe sample was prepared as a cylinder with 1 mm in diameter and 3 mm in height. At the bottom, a 0.55 mm hole was drilled for inserting a thermocouple. A pink beam of 17.6 KeV with 700 projections was used with a scan time of 0.75 sec resulting in a total of 36 tomography data images (1.1 $\mu m.voxel^{-1}$). The experiment was recorded at a constant cooling rate of 10°C min^{-1} from the fully melt condition at 618°C to 575°C prior to the (Al)-Si-beta eutectic reaction.

Post–mortem tomography has been applied to differential thermal analysis (DTA) samples cooled at various cooling rates, in the range 0.2 to 5°C min^{-1}. DTA samples were 3.9 mm in diameter and 5mm in height. Post mortem tomography was carried out by using GE Phoenix Nanotom equipped with a 180 kV/15W X-ray tube. A monochromatic beam was transmitted and produced 1440 images as the specimen was rotated by 360° (2.5 $\mu m.voxel^{-1}$). An image stack of volume was constructed using DatosX (GE Sensing and Inspection GmbH) and VG Studio Max (Volume Graphic GmbH, Germany).

The tomography data was then reconstructed and post-treated with image analysis by extracting the beta precipitates through series of operation using imageJ (National Institute of Health, USA) and Aphelion (ADCIS, France). Due to the similar contrast of beta phase and the oxide skin, an additional masking operation consisting of Gaussian-blur filtering, thresholding, iteration of erosion and dilation [7], was added to remove the outer oxide layer.

3. Result and discussion

Analysis to the reconstructed volume of in-situ experiment showed the precipitation of primary (Al) dendrites occured at 608°C, while nucleation of beta phase started at 593°C in several locations as the one marked by a circle in figure 1. The beta phase grew in interdendritic regions along with the continuing growth and coarsening of the primary (Al) dendrites. This precipitation of beta phase in the alloy agrees with DTA experiments and calculations according to scheil model or lever rule [13]. Further analysis on the nucleation of beta phase was performed by defining a region of interest of 1008x1008x500 voxels taken from the middle region. The 3D volume reconstructed rendering shows the nucleation of beta phase with each separate particle coloured differently, as illustrated in figure 2. The first observed nucleation of beta phase appeared in five sites located on the outer oxide skin; however due to oxide skin removal operation and weak phase contrast, some of the initiation sites may have been wiped off during image analysis.

As the solidification progressed, a rapid increase in the number of beta precipitates was detected and this continued until saturation. Observation showed the rapid growth of beta phase and notice was made that no new nucleation occurred below 587°C. Analysis upon the region of interest indicated that approximately 32 nucleation events of beta particles in total were observed during the solidification process.

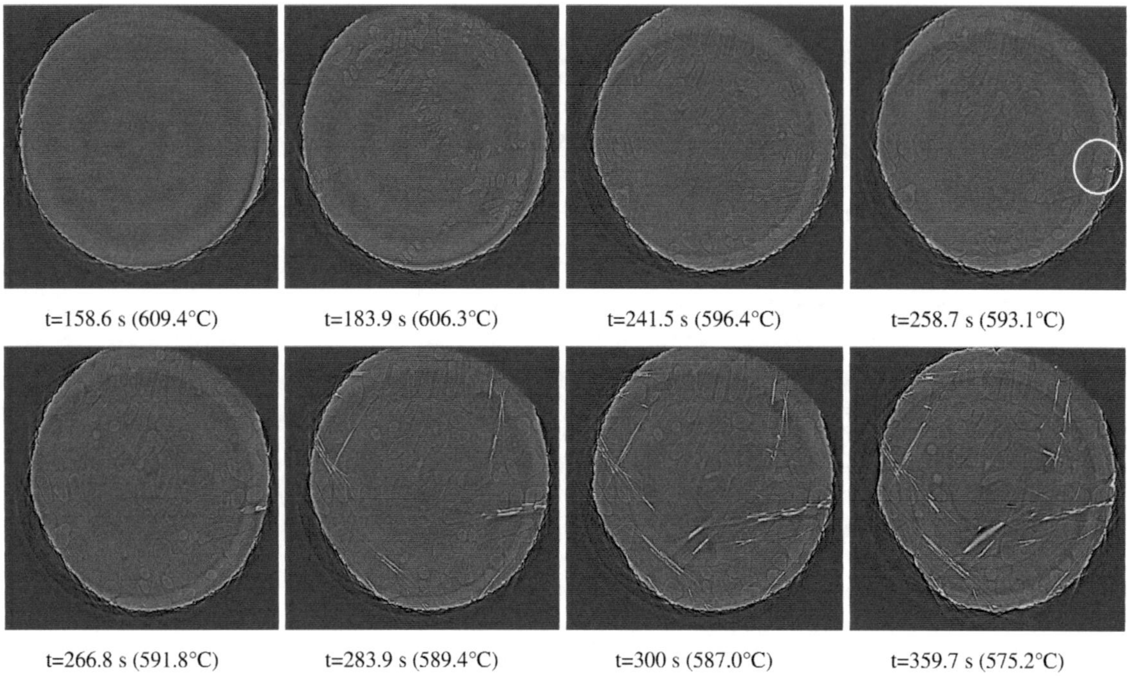

Figure 1. Image slices extracted horizontally from the middle of the reconstructed volume obtained during in situ solidification of Al6.5Si1Fe.

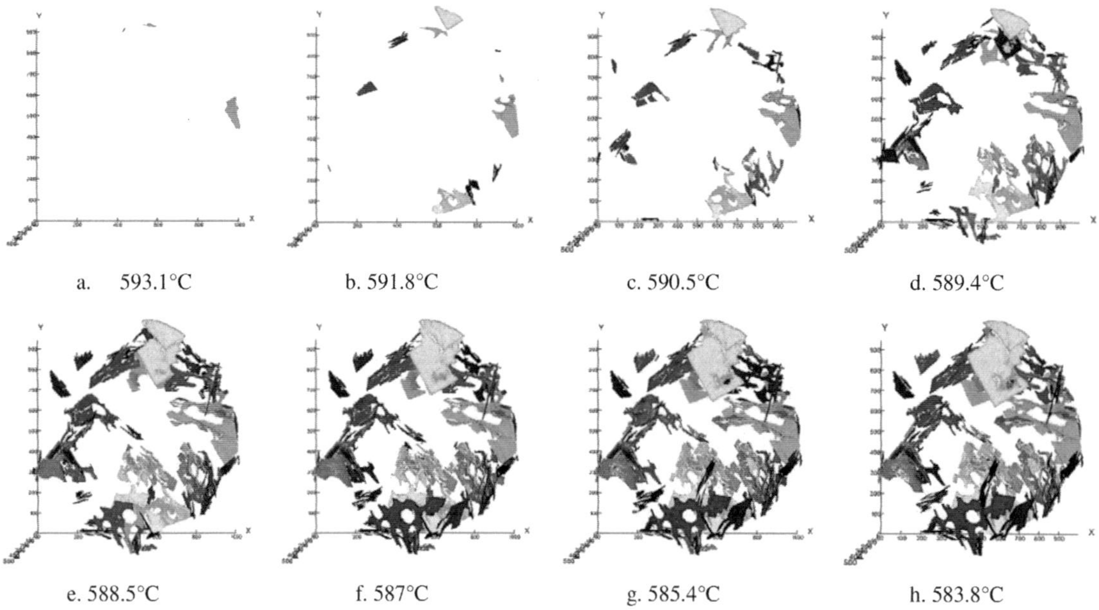

Figure 2. 3D visualization of beta phase nucleation where each colour represents single beta particle. Note that some particles impinged and connected in a later stage.

The tomography images showed that sometimes dendrite arms of primary (Al) blocked the growth path of the beta precipitates. However, in such a case, it could be seen that the beta phase growth continued by surrounding the dendrite and some of the precipitates reconnected after passing the obstacle. During beta growth some branching and bending were also observed. Branching was noticed in an intermediate stage growth and the interconnectivity was increased in a later stage as the temperature dropped and the connection between precipitates created clusters of large beta phase.

A series of growth steps of a single beta phase is shown in figure 3. Examination showed a robust growth of beta phase occurred in an early growth stage, where the phase length could be doubled in one tomography step. As well known, the beta phase growth is characterized by a high lateral growth rate that leads to the development of elongated 2D plates. The growth began to slow down as the temperature reached 585°C and this was accompanied by the increase of interconnectivity of beta precipitates.

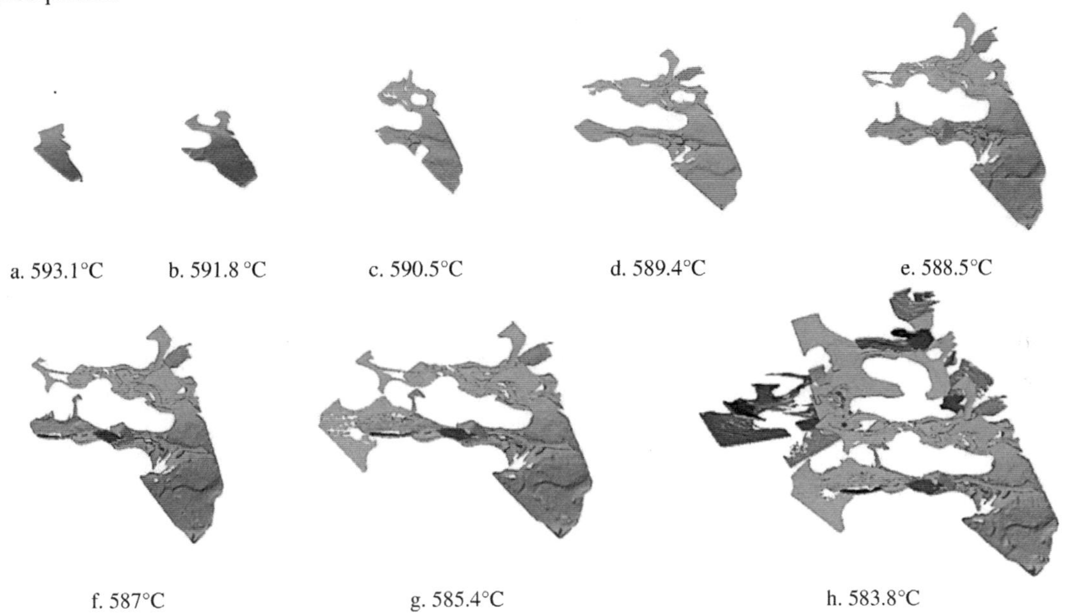

Figure 3. A series of reconstructed 3D images from a single nucleated beta precipitate taken at several steps during growth. As the solidification continues other precipitates got connected to the selected one (voxel size 1.1 μm).

Lateral length measurements were performed on several beta precipitates at each tomograph step. From these length measurements, five beta phase particles which are labelled 1 to 5 were selected for further analysis of their growth kinetics. In general, we could classify the beta phase growth into three stages based on the length curve shape as: initial stage (I), peak stage (II) and decline or saturation stage (III) as seen in figure 4a. Individual analysis of beta phase particles showed that the lateral growth rate varied among them and also between each step. The highest detected rate in stage I was at 34.5 $\mu m.s^{-1}$ and then fluctuated in stage II within the range of 5 - 25 $\mu m.s^{-1}$. The variation between steps from a single phase could be caused by the blockage on (Al) dendrites and other obstacles in the interdendritic region that slowed down the growth rate as already mentioned.

Analysis of the thickness of beta precipitates showed it increased during solidification especially in the region close to the surface where the nucleation initiated. Thickness rate was measured from the highest plate thickness from each tomography step. The results showed that the beta precipitates thickened at a rate of 0.76 $\mu m.s^{-1}$ during the first stage and this slowed down to 0.21 $\mu m.s^{-1}$ during the next stage. The thickening rate in the later stage further decreased to 0.1 $\mu m.s^{-1}$.

The thickening mechanism was largely overtaken by the high lateral beta growth which is 20 times faster. As seen from figure 4b which compares both rates, the lateral growth rate was dominant in the early stage, marked with very high values and slowed down as the solidification progressed. However, the thickening rate did not increase during the later stage, but rather seemed to decrease along with decreasing lateral growth rate.

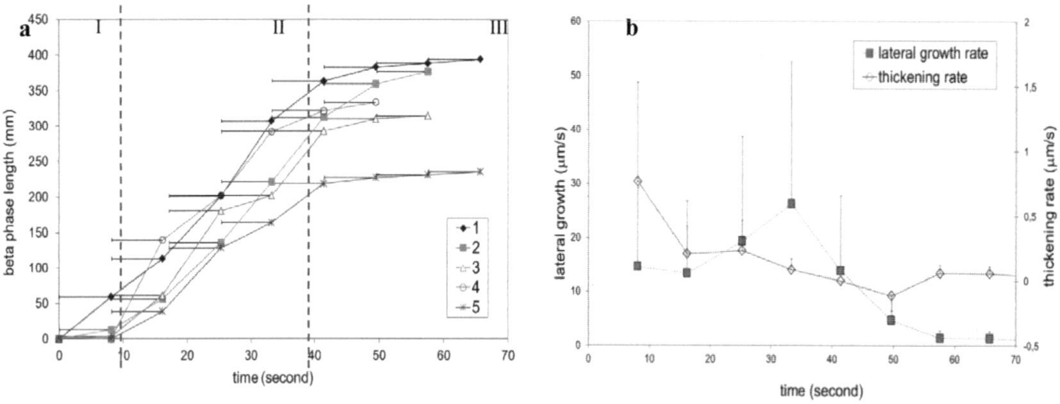

Figure 4. Evolution with time of the maximum length of beta precipitates (a), comparison of lateral and thickening growth rates of beta precipitates (b).

The volume fraction of beta phase taken from in-situ tomography data increases rapidly in the beginning and saturates as the solidification continues as seen in figure 5a. The final value at the end of the experiment, at a temperature just above the ternary (Al)-Si-beta eutectic, agrees with the prediction made following the Scheil's model using Thermocalc and the TCAL2 database [13]. Comparison of the experimental and predicted evolution shows that the beta phase appeared with an undercooling of 10°C. Such a high undercooling explains the large growth rate experienced by the beta precipitates at the beginning of their growth.

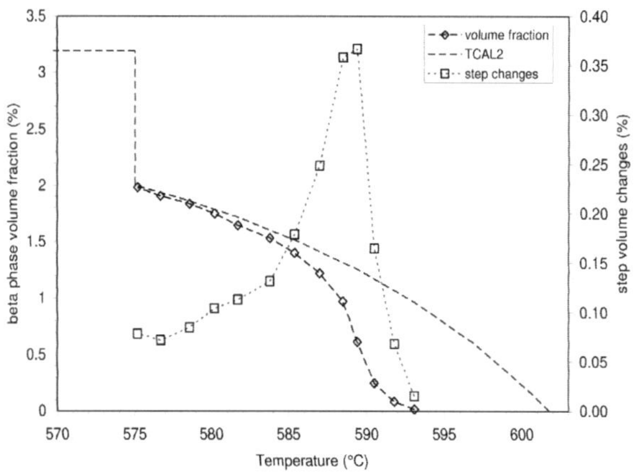

Figure 5. Measured volume fraction of beta phase compared with predicted evolution carried out with Thermocalc and the TCAl2 database.

Analysis to the post-mortem tomographs showed the nucleation sites for the beta phase are mostly located again on the outer oxide surface (skin). The growth followed by the beta precipitates expanding in radial direction toward the sample centre, as observed in sample cooled at 0.2°C.min^{-1} in figure 6. However, observation at low cooling rate also showed small fraction of beta precipitates not associated with the large plates that nucleated on the outer skin. At a higher cooling rate, the initiation of beta precipitates seemed dispersed within the sample, between the (Al) dendrites. Unfortunately, the image quality generated by tomography is inadequate for detailed analysis, as clearly seen at a higher cooling rate where very thin plates could hardly be detected due to a too low spatial resolution.

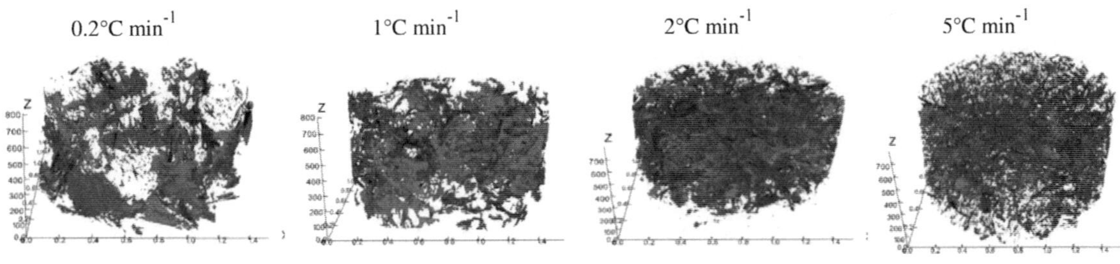

Figure 6. Rendering of post-mortem tomograph with various cooling rates (voxel size: 2.5 μm)

4. Conclusion

In situ tomography experiment showed that beta phase growth is dominated by its lateral extension. Initial growth showed a high growth rate which slowed down after saturation level was reached. This evolution could be related with the high undercooling necessary for beta phase nucleation. The slowing down of the lateral growth did not cause the thickening rate to increase. Instead, it appears that the thickening rate also decreased along with lateral growth rate decreasing. Observation on samples solidified at low cooling rates by post-mortem tomography seems showing nucleation of beta precipitates occurs both on the outer oxide skin of the samples and in the bulk of the material.

Acknowledgment

Financial support from the European Space Agency (contract #4200014347) through the MICAST program and the Education Ministry - Republic of Indonesia are greatly appreciated. Synchrotron beam time from ESRF also highly acknowledge.

References

[1] E. Maire, J.Y. Buffiere, L. Salvo, J.J. Blandin, W. Ludwig, J.M. Letang 2001 *Adv. Eng.Mater* **3** 539
[2] J. Kastner, B. Harrer, H.P. Degischer 2011 *Mater. Charact* **2** 99
[3] O. Ludwig, M. Dimichiel, L. Salvo, M. Suéry, P. Falus, 2005, *Metall. Mater. Trans. A* **36A** 1515
[4] R.H. Mathiesen, L. Arnberg, H. Nguyen-Thi, B. Billia 2012 *JOM* **64** 76
[5] R.H. Mathiesen, L. Arnberg 2007 *JOM* **59** 20
[6] B. Kim, S. Lee, S. Lee, H. Yasuda, 2012, *Mater. Trans.* **53** 374
[7] S. Terzi, L. Salvo, M. Suéry, A.K. Dahle, E. Boller 2010 *Acta Mater* **58** 20
[8] D. Tolnai, P. Townsend, G. Requena, L. Salvo, J. Lendvai, H.P. Degischer 2012 *Acta Mater* **60** 2568
[9] J. Wang, P.D Lee, R.W Hamilton, M. Li and J. Allison 2009 *Scripta Mater* **60** 516
[10] S. Terzi, J.A. Taylor, Y.H. Cho, L. Salvo, M. Suéry, E. Boller, A.K. Dahle 2010 *Acta Mater* **58** 5370
[11] C. Puncreobutr, A.B. Phillion, J.L Fife, P. Rockett, A.P. Horsfield, P.D. Lee 2014 *Acta Mater* **79** 292
[12] D. Ferdian, C. Josse, P. Nguyen, N. Gey, N. Ratel-Ramond, P. de Parseval, Y. Thebault, B. Malard, J. Lacaze, L. Salvo, *Metall. Mater. Trans. A, in press*
[13] D. Ferdian, B. Suharno, B. Duployer, C. Tenailleau, L. Salvo, J. Lacaze 2012 *Trans. Ind. Inst. Met* **65** 821

IOP Conf. Series: Materials Science and Engineering **117** (2016) 012019 doi:10.1088/1757-899X/117/1/012019

Formation of intermetallic phases in AlSi7Fe1 alloy processed under microgravity and forced fluid flow conditions and their influence on the permeability

S Steinbach[1,a], L Ratke[2], G Zimmermann[3] and O Budenkova[4]
[1] Institut für Materialphysik im Weltraum, Deutsches Zentrum für Luft- und Raumfahrt (DLR), 51170 Köln, Germany
[2] Institut für Werkstoff-Forschung, Deutsches Zentrum für Luft- und Raumfahrt (DLR), 51170 Köln, Germany
[3] ACCESS e.V., Intzestrasse 5, 52072 Aachen, Germany
[4] SIMAP/EPM, 1130 rue de la Piscine - BP 15, 38402 St. Martin D'Heres, France

E-mail: [a] sonja.steinbach@dlr.de

Abstract. Ternary Al-6.5wt.%Si-0.93wt.%Fe alloy samples were directionally solidified on-board of the International Space Station ISS in the ESA payload Materials Science Laboratory (MSL) equipped with Low Gradient Furnace (LGF) under both purely diffusive and stimulated convective conditions induced by a rotating magnetic field. Using different analysis techniques the shape and distribution of the intermetallic phase β-Al_5SiFe in the dendritic microstructure was investigated, to study the influence of solidification velocity and fluid flow on the size and spatial arrangement of intermetallics. Deep etching as well as 3-dimensional computer tomography measurements characterized the size and the shape of β-Al_5SiFe platelets: Diffusive growth results in a rather homogeneous distribution of intermetallic phases, whereas forced flow promotes an increase in the amount and the size of β–Al_5SiFe platelets in the centre region of the samples. The β-Al_5SiFe intermetallics can form not only simple platelets, but also be curved, branched, crossed, interacting with dendrites and porosity located. This leads to formation of large and complex groups of Fe-rich intermetallics, which reduce the melt flow between dendrites leading to lower permeability of the mushy zone and might significantly decrease feeding ability in castings.

1. Introduction

Aluminium–silicon cast alloys are widely used in industry due to their low densities, low cost, and acceptable mechanical properties. The mechanical properties of aluminium alloys are, however, adversely affected by even a small amount of iron, which leads to the formation of intermetallic phases. In Al-Si-Fe alloys the β-Al_5FeSi phase is the most harmful one (detrimental to the mechanical properties of the alloy), which solidifies as long platelets [1-4]. Insight in the three-dimensional structure of the β-Al_5FeSi phase was provided by serial sectioning [5] or x-ray computed tomography (CT) [6]. Recent investigations demonstrate that the formation of the β-Al_5FeSi phase in the interdendritic region may significantly reduce the melt flow through the mush and therefore also modify the secondary dendrite arm spacing [7].

One parameter used to characterize the fluid flow in the mushy zone is permeability, i.e. a tensor measuring the ease of fluid flow through the solid network [8]. Permeability is a critical parameter for casting models and is related to the geometry of the liquid channels and the grain morphology [9]. For columnar dendritic structures, the permeability is anisotropic due to directional nature of the dendrite arms [10]. To investigate the anisotropy of the permeability, Poirier [9] used a multilinear regression analysis of experimental data to propose an empirical relationship based on the Blake-Kozeny equation. Puncreobutr et al. [11] continued work in this area and developed an analytical expression that takes into account the effects of intermetallic particles on permeability.

To understand the effect of convective flow on the microstructure of an alloy with intermetallic phases precipitated in the mush and to understand the effect of intermetallic phase formation on permeability, directional solidification experiments with a hypoeutectic Al-Si alloy enriched in Fe under well-defined boundary conditions in microgravity environment were performed in the frame of the ESA-MAP project MICAST (MIcrostructure Formation in CASTing of Technical Al-Alloys under Diffusive and Magnetically Controlled Convective Conditions) on-board the International Space Station (ISS). In the ESA payload MSL LGF [12] AlSi7Fe1 alloy samples were directionally solidified under both purely diffusive and stimulated convective conditions induced by a Rotating Magnetic Field (RMF). This allows for systematic studies of the impact of the solidification velocity and fluid flow on the formation of intermetallic β-Al_5FeSi phase (IMP).

2. Experimental set-up and sample analysis

In the microgravity environment provided by the ISS, four rod-like samples of diameter 7.8 mm and of length 245 mm from unmodified Al-6.5wt%Si-0.93wt%Fe alloy [13] were directionally solidified in the Bridgman-type furnace insert MSL LGF (moving furnace, fixed sample) [12] in the same temperature gradient of 2 K/mm, but with different values of the furnace movement velocity: $v = 0.02\ mms^{-1}$ (MICAST1#4), $v = 0.04\ mms^{-1}$ (MICAST1#3), $v = 0.06\ mms^{-1}$ (MICAST1#2) and $v = 0.08\ mms^{-1}$ (MICAST1#1). After a certain solidified length (50 mm on average), which may allow for a steady-state columnar growth under diffusive conditions, the rotating magnetic field (RMF) was switched on. The operating frequency of 57 Hz and the magnetic field strength of 5.7 mT are sufficient to induce forced melt flow in the melt.

Axial cross-sections of the samples were investigated to determine the distribution of the intermetallic β-Al_5FeSi phase in 2D. To allow for an unambiguous identification of the β-Al_5FeSi phases a special preparation technique with a basic final polishing step was developed. As a result the β-Al_5FeSi phases appear in an intense dark colour with respect to the primary α-Al phase and the Al-Si eutectic. Digital image analysis was used for binarization of the images. The β-Al_5FeSi phases could be identified automatically and a detailed analysis of the size and distributions of the phases was performed. Figure 1 shows as an example the β-Al_5FeSi phases for MICAST1#4 sample classified in groups of different lengths, determined with image analysis using the Feret-maximum value.

For the 3D CT analysis of the intermetallic phases samples of dimension 1x1x8mm normal to the growth direction of the sample rod (8 mm in diameter) were prepared. The scans were carried out in the x-ray tomography system Phoenix nanotom® (Phoenix |X-Ray GmbH) available at DLR. The nanofocus x-ray tube coupled with a 2304×2304 pixel detector produces a volume of 2304 × 2304 × 2304 voxels. The magnification was chosen such that an effective pixel size of 1.0 μm is obtained. A total of 800 radiographs during 360 degree of sample rotation were acquired at a tension of 100 kV and a current of 135 μA. Image corrections and ring artifact corrections were applied before the final reconstruction. After image acquisition, significant post-processing was necessary to segment the three phases: primary α-Al, intermetallics, and eutectic. From these images, tomographic 3D-volumes of 1000 x 1000 x 1000 voxels with a voxel size of 1.0 μm were generated (datos|x-reconstruction software, GE Sensing and Inspection GmbH) for each prepared sample segment. One reconstruction cube was chosen in the sample center and one was chosen at the outer part of the sample. Further image analysis was performed using AvizoFire 8.1. An example of the results is shown in figure 2: The β-Al_5FeSi phases form a complex and interconnected network.

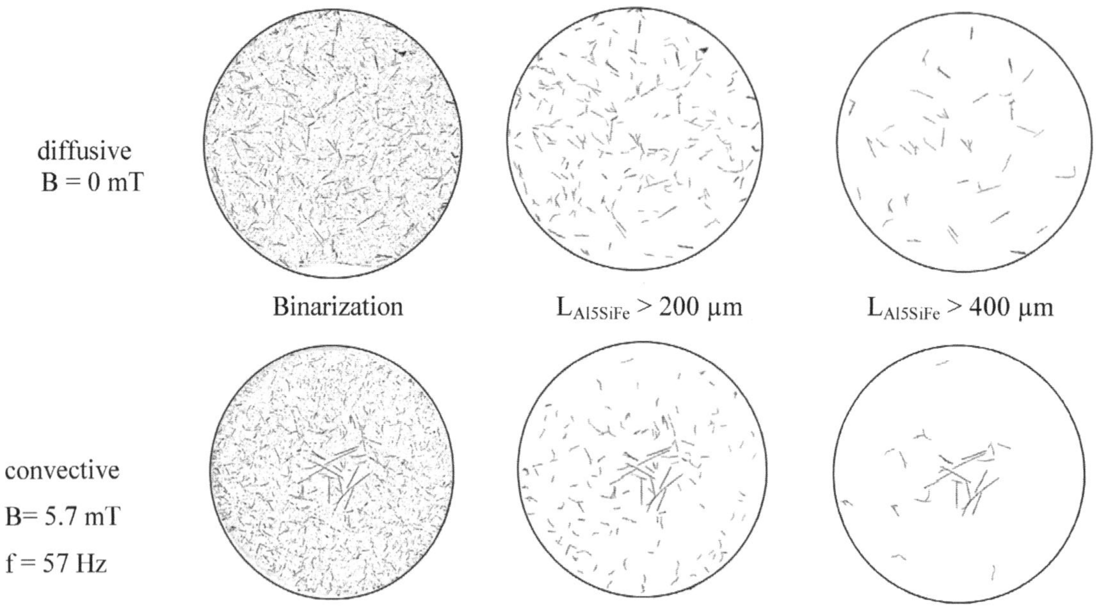

Figure 1. Binarized image (left), all β-Al_5FeSi phases longer than 200µm (middle) and all β-Al_5FeSi phases longer than 400 µm (right) of sample MICAST1#4. The top images represent diffusive growth conditions and bottom images are results of solidification with forced melt flow. The areas represent the entire rod diameter.

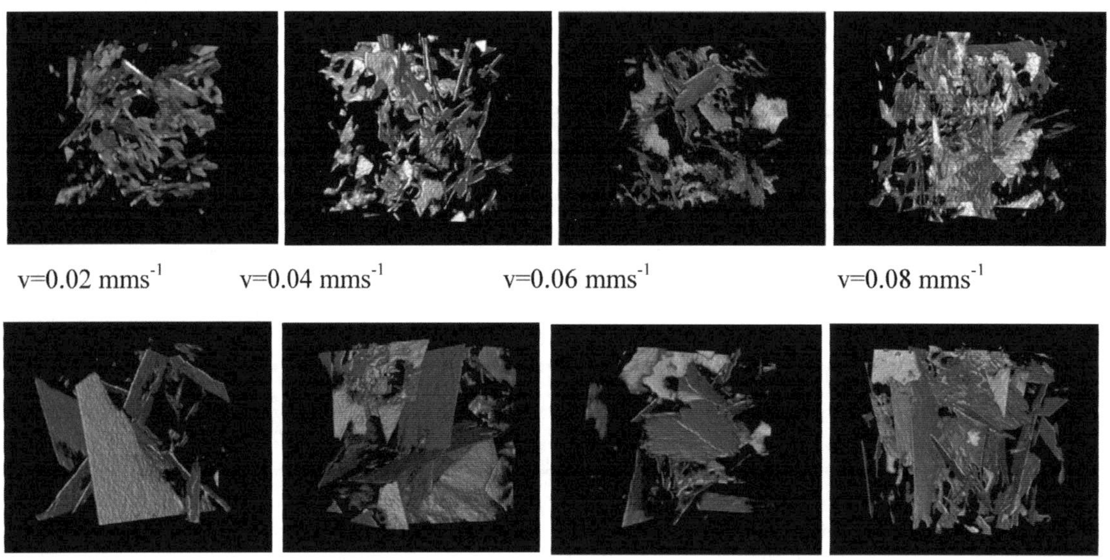

Figure 2. Segmented reconstruction cubes of 650 x 650 x 650 μm^3 of samples processed with controlled melt flow and different furnace movement velocities. Top images were taken from the outer parts of the sample and bottom images were taken from the sample center.

In order to determine the influence of the intermetallic phases on the permeability of the dendritic network, the flow problem based on the Stokes equations was simulated using the reconstructed cubes (outer part of samples solidified with RMF). The commercial computational flow dynamics software Avizo XLab Hydro (VSG, France), which utilizes a finite volume method, was applied directly to the tomographic datasets for flow predictions. The flow studies were performed in each of the three directions along the x, y and z axes of the 3D-volume. Inlet flow along the z-axis gives the permeability for liquid flow parallel to the primary dendrite arms, while permeability in the normal direction was taken as the value calculated with inlet flows along the x- and y-axes. The boundary conditions were: free pressure outlet, no-slip conditions, liquid viscosity was given a value of 0.00125Pa s.

3. Results and Discussion

3.1. 2D-analysis

Based on axial 2D cross-sections obtained from regions of pure diffusive growth and also from regions solidified with forced melt flow, the sizes and distribution of β-Al_5FeSi phases were determined for different solidification velocities (figure 1). In the diffusive case [14] the phases are larger and more homogeneously distributed than in the case of induced melt flow. In general, the size of the β-Al_5FeSi phases growing in the interdendritic region is limited by the spacing of the dendritic area and the total amount of Fe available for formation of the intermetallic phase. The RMF induced a secondary melt flow [15], which results in an enrichment of Si and Fe in the centre part of the samples due to strong macrosegregation. Therefore, especially for lower solidification velocities (figure 2, bottom left), the enrichment of Si and Fe in the sample centre part results in a few but rather large β-Al_5FeSi particles.

3.2. 3D-analysis

As seen in figure 2 the intermetallics are large and plate-like. They form a complex and interconnected network. It is immediately clear that they are able to block flow through the interdendritic network of channels ('barrier effect'). The volume fraction of the β-Al_5FeSi phase measured for the different samples (solidified with RMF, outer part of the sample) compared to the calculated one (dashed line) is shown in figure 3. The decrease of the volume fraction at 0.02 mms^{-1} and the increase at 0.08 mms^{-1} can be explained by the local decrease and increase of the Fe content as measured by SEM EBSD microprobe analysis, respectively.

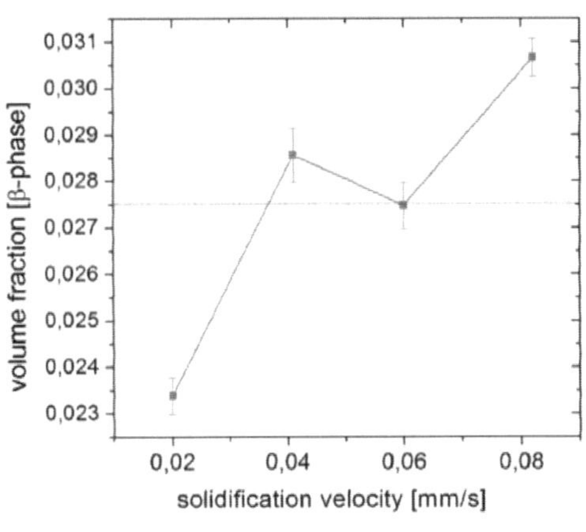

Figure 3. Volume fraction of the β-Al_5FeSi phase measured on the out of centre part of samples solidified with RMF.

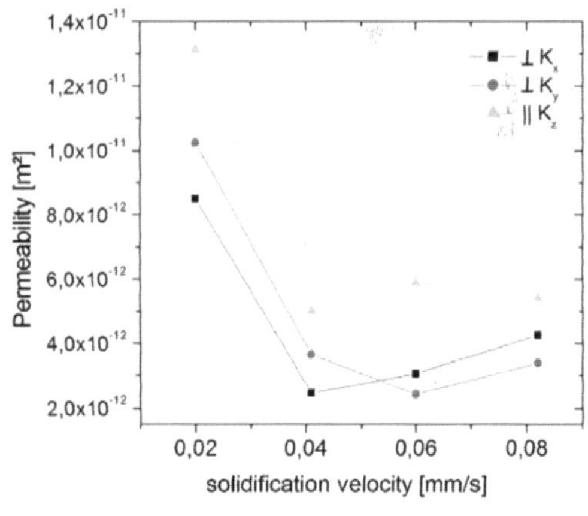

Figure 4. Permeability for liquid flow parallel (z) and perpendicular (x, y) to the primary dendrites simulated with Avizo XLab Hydro.

The permeability values extracted from the flow simulations with Avizo XLab Hydro are shown in figure 4 for flow parallel (z) and perpendicular (x, y) to the primary columnar dendrites. With decreasing solidification velocity (larger primary dendrite spacing) and decreasing volume fraction of the β-Al_5FeSi phase, the permeability increases.

The calculated permeability value (z-direction; MICAST1#2) has been fitted (figure 5) to an empirical relationship based on the Blake-Kozeny equation following the procedure of Poirier [9] and Puncreobutr [11] with an additional term, $1-\beta f_{IMP}$, where β represents the impact factor of the intermetallic phase:

$$K_p = (1 - \beta f_{IMP})\, K_p^{orig} \quad (1)$$

$$K_p^{orig} = C_1\, (f_l^3\, \lambda_1^2)/(1 - f_l) \quad (2)$$

where K_p is the permeability in the direction parallel to the primary dendrites, K_p^{orig} is the original expression of Poirier [9], λ_1 is the primary dendrite spacing, f_l is the liquid volume fraction and C_1 is a fitting constant. The values for f_l and f_{IMP} were obtained from thermodynamic calculations with a linearized phase diagram [16]. It can be seen that in the first solidification phase the permeability decreases monotonically with temperature as expected because of the thickening of the primary dendrites as solidification progresses. The permeability is significantly lower for the case with fluid flow as compared to the one without, because of the reduced primary dendrite spacing. During the solidification phase, after the beginning of intermetallic phase precipitation, the loss of permeability that occurs due to the intermetallic phases can be directly attributed to the barrier effect of the β-Al_5FeSi phases. A further analysis is ongoing, trying to answer the question, why the effect is not bigger (grid size of the Avizo model etc.).

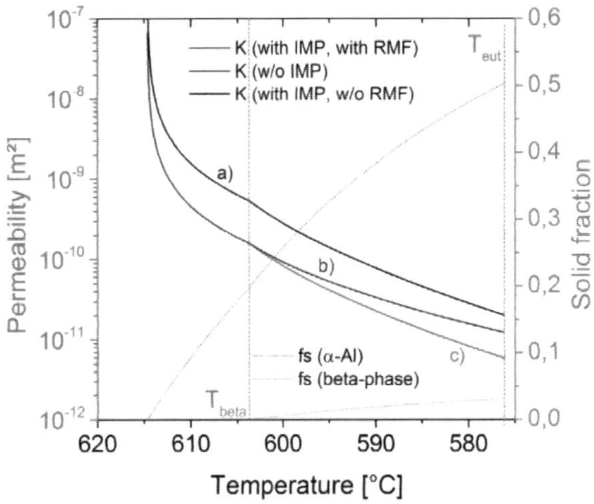

Figure 5. The evolution of permeability with temperature for flow parallel to the primary columnar dendrites a) without fluid flow but with IMP, b) with fluid flow but without IMP and c) with IMP and fluid flow (MICAST1#2). For comparison the evolution of the fraction solid of the α-Al dendrites and the β-Al_5FeSi phase with temperature is also provided. The dotted line indicates the temperature where the nucleation of the intermetallic phases started.

4. Conclusions

This paper provides an analysis of the formation of intermetallic phases in AlSi7Fe1 alloy in samples processed onboard the ISS. Based on the axial 2D cross-sections obtained from regions of purely diffusive growth and also from regions solidified under forced melt flow, the sizes and distribution of β-Al_5FeSi phases were determined for different solidification velocities. In the diffusive case the phases are larger and more homogeneously distributed than in the case of induced melt flow. Especially for lower solidification velocities, the enrichment of Si and Fe in the centre part of the sample by fluid flow results in only a few but rather large β-Al_5FeSi platelets.

The coupling of the 3D CT analysis of the structure of the β-Al_5FeSi particles, with numerical simulations demonstrates qualitatively the blocking effect of intermetallics on fluid flow and the loss in permeability in the direction parallel to the primary dendrites. This can explain the decrease in the feeding abilities in castings and could therefore explain the increased susceptibility of such alloys for casting defects as hot tearing and microporosity.

Acknowledgments
This work was conducted within the ESA-MAP programme MICAST', contract 14347/01/NL/SH, and also funded by the German Space Agency DLR. The sample material was kindly provided by Hydro Aluminium Deutschland GmbH, Bonn, Germany. The authors would like to thank the MICAST team for fruitful discussions.

References
[1] Murali S, Raman K S and Murty K S S 1994 *Materials Characterization* **33** 99
[2] Tang S K and Sritharan T 1998 *Materials Science and Technology* **14** 738
[3] Steinbach S and Ratke L 2007 *Transactions of the Indian Institute of Metals* **60** 137
[4] Mikolajczak P and Ratke L 2011 *Materials Science and Engineering* **27** 012024
[5] Dinnis C M, Taylor J A and Dahle A K 2005 *Scripta Materialia* **53** 955
[6] Terzi S, Taylor J A, Cho Y H, Salvo L, Suery M, Boller E and Dahle A K 2010 *Acta Materialia* **58** 5370
[7] Steinbach S und Ratke L 2009 *International Journal of Cast Metals Research* **22** 1
[8] Flemings M C 1974 *Solidification processing* New York, McGraw-Hill
[9] Poirier D R 1987 *Metallurgical Transactions* **18** 245
[10] Murakami K, Shiraishi A, Okamoto T 1983 *Acta Metallurgica* **31** 1417
[11] Puncreobutr C, Phillion A B, Fife J L and Lee P D 2014 *Acta Materialia* **64** 316
[12] Enz T, Steinbach S, Simicic D, Kasperovich G and Ratke L 2011 *Microgravity Science and Technology* **23** 345
[13] Ferdian D and Lacaze J 2012 *Transactions of the Indian Institute of Metals* **65** 821
[14] Zimmermann G, Schaberger-Zimmermann E, Steinbach S and Ratke L 2014 *Materials Science Forum* **790** 40
[15] Noeppel A, Ciobanas A, Wang X D, Zaidat K, Mangelinck N, Moreau R, Weiss A, Zimmerman G and Fautrelle Y 2010 *Metallurgical and Materials Transactions* **B41** 193
[16] Budenkova O et al. 2014 *Materials Science Forum* **790** 46

IOP Conf. Series: Materials Science and Engineering **117** (2016) 012020 doi:10.1088/1757-899X/117/1/012020

Impurity effects on the solidification of primary $Al_3(Sc,Zr)$ phase in Al alloys

J H Li[1a] and P Schumacher[1,2]
[1] Chair of Casting Research, Montanuniversität Leoben, A8700 Leoben, Austria
[2] Austrian Foundry Research Institute, A8700 Leoben, Austria

Email: [a]jie-hua.li@hotmail.com

Abstract. The nucleation and growth of primary Al_3Sc phase has been investigated in high purity Al alloys and commercial purity Al alloys, with a special focus on the impurity effects. In the case of high purity Al alloys, most primary Al_3Sc phases were pushed to grain boundaries during moving solidification front. Such type of primary Al_3Sc phase does not contribute to the heterogeneous nucleation and thereby no significant grain refinement of Al alloys was observed, although some Al_3Sc particles remain the same orientation with the Al matrix. In the case of commercial purity Al alloy, the presence of impurities, e.g. Ti, Fe and Si, enhances the heterogeneous nucleation of primary Al_3Sc phase. Most primary Al_3Sc phases are located within the Al matrix, and keep the same orientation with the Al matrix. Furthermore, the presence of impurities also changes the growth mode of primary Al_3Sc phase. In the case of commercial purity Al alloy, a layer by layer growth was observed. This investigation demonstrates that impurities have important effects on the nucleation and growth of primary Al_3Sc phases in Al based alloys.

1. Introduction

In the early 1960s, the potential of Sc in controlling microstructure of Al based alloys has been recognized [1, 2]. Significant improvements in properties of commercial 5×××, 7××× have been achieved through Sc additions. Most previous research focused on the decomposition of supersaturated Al-Sc solid solutions, which leads to the formation of the fully coherent, secondary $L1_2Al_3Sc$ phase. However, much less attention has been paid to the primary Al_3Sc phase solidified directly from the liquid. Although Sc only exhibits a small growth restriction in Al alloys, even smaller grain sizes than with Al-Ti-B grain refiner addition can be obtained with hypoeutectic addition of Sc into Al based alloy. The grain refinement effect may be mainly dominant by heterogeneous nucleation. Furthermore, grain refinement by Sc addition was found to be accompanied by a change in growth morphology from unrefined dendritic, to fine spherical grains with a divorced eutectic appearing on the grain boundaries in the refined castings [3].

In order to further decrease the expensive Sc addition, a combined addition of Zr and Sc was found to enable a considerable grain refinement in Al alloys, which is much stronger than that of Zr or Sc alone. Furthermore, an important impurity effect has been reported in [4-6]. The presence of such types of the impurities (i.e. Fe, Si) affects the nucleation and growth of primary $Al_3(Sc,Zr)$ phase, and thereby enhance the heterogeneous nucleation of α-Al during solidification.

In this paper, the nucleation and growth of primary $Al_3(Sc,Zr)$ phase has been investigated in high purity Al alloys and commercial purity Al alloys, with a special focus on the impurity effects. The nucleation and growth modes are also discussed.

2. Experimental methods

High purity Al-0.25Sc-0.25Zr alloy was prepared using high purity Al ingots (5N, 99.998), Al-10 wt. % Zr and an Al-2.2 wt. % Sc master alloys (wt. %, used throughout the paper unless noted otherwise), respectively. For comparison, commercial purity Al-0.25Sc-0.25Zr alloy was also prepared using commercial purity Al ingots (2N8, 99.8), Al-10Zr and an Al-2.2Sc master alloy, respectively. It should be noted that in high purity Al, Si (7 ppm), Fe (6 ppm), Cu (8 ppm), Mg (4 ppm), Ti (< 1 ppm) were measured in high purity Al. While, in commercial purity Al, Si (0.0675), Fe (0.805), Cu (0.0011), Mg (0.0113), Ti (0.0117) were measured in commercial purity Al. Clearly, significant impurities (Si, Fe, Mg, Ti) are present in commercial purity Al alloys.

Each batch, weighting about 6 kg, was melted in a resistance furnace at 720 °C. A reference sample was taken from the melt in order to identify the grain size before the additions of Al-10Zr and Al-2.2Sc master alloys. The melt was stirred with a graphite rod for 20 s after inoculation. The samples were taken from the melt after the addition of Al-10Zr and Al-2.2Sc master alloys (about 5 min) and tested using standard TP-1 test.

According to the TP-1 test, the samples were sectioned 38 mm from the bottom surface. Standard metallographic procedures were performed to prepare these sections for grain size measurements. The samples were etched using a mixture of 13 g boric acid, 35 g HF, 800 ml H_2O at a voltage of 20 V for 45 seconds. All images used for grain size measurement were taken from the centre of the samples using optical microscopy in a polarized mode. The reported grain sizes were measured from at least 20 images using the line-intersect method.

The samples for scanning electron microscopy (SEM) and electron backscatter diffraction (EBSD) investigation were mechanically ground and electron-polished in a solution consisted of 5% perchloric acid and 95% methanol at -30 °C. EBSD investigation was performed using a Zeiss 1525 scanning electron microscope equipped with an EDAX EBSD system. Evaluation of scans was made with orientation imaging microscopy (OIM) software. The EBSD investigation can be used to determine the orientation relationship between the primary $Al_3(Sc,Zr)$ particles and the α-Al matrix. In order to further elucidate the enrichment of the impurities, electron probe micro analyzer (EPMA, JEOL JXA-8500F) was also employed.

3. Results

Figure 1 shows one primary $Al_3(Sc,Zr)$ particle in high purity Al alloy with the addition of 0.25Sc and 0.25Zr. The primary $Al_3(Sc,Zr)$ particle is located at the grain boundaries, as marked in figure 1a. However, the local region of the primary $Al_3(Sc,Zr)$ particle is located within the α-Al matrix, as enlarged in figure 1b. Such kind of primary $Al_3(Sc,Zr)$ phases is believed to be engulfed during solidification. No low angle orientation relationship (e.g. cube to cube) was observed between the primary $Al_3(Sc,Zr)$ particle and the α-Al matrix, indicating that this pro-eutectic $Al_3(Sc,Zr)$ particle did not contribute to the heterogeneous nucleation and thereby grain refinement. A coarse grain size was observed, as marked in figure 1c.

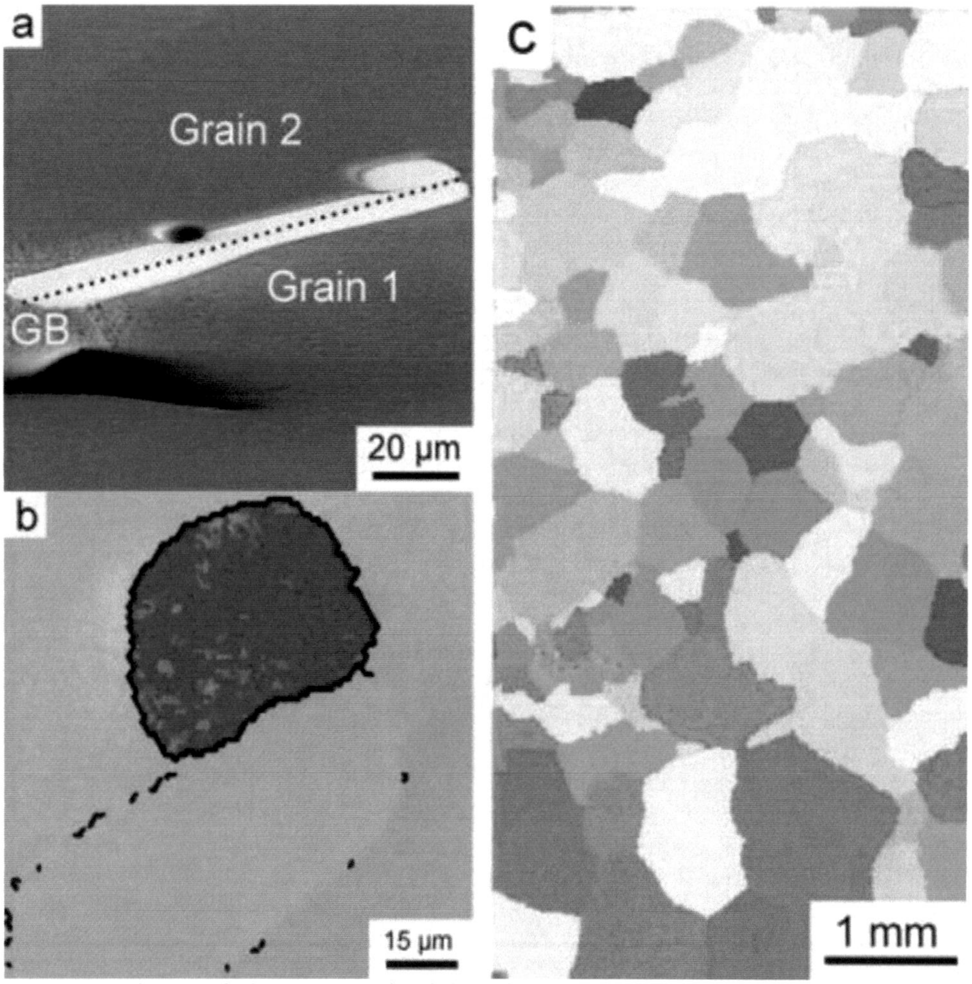

Figure 1. Primary $Al_3(Sc,Zr)$ phase in high purity Al alloy.

Figure 2. Primary Al_3Sc phase with a layer by layer growth in commercial purity Al alloy.

Figure 2 shows one primary $Al_3(Sc,Zr)$ particle in commercial purity Al alloy with the addition of 0.25Sc and 0.25Zr. The "star-like" primary $Al_3(Sc,Zr)$ particle was found to be located within the α-Al matrix. The orientation of the primary $Al_3(Sc,Zr)$ particle is identical to the orientation of the α-Al matrix [7]. Such type of primary $Al_3(Sc,Zr)$ phase is believed to contribute to the heterogeneous nucleation during solidification. A fine grain size was observed, as shown in figure 2b. Furthermore, a very interesting observation is that the primary $Al_3(Sc,Zr)$ particle appears to grow layer by layer, as shown in figure 2a. Some layers with a weaker contrast were observed. These layers are believed to be enriched in impurities (e.g. Si, Fe, Cu, Mg, Ti), which are always present in commercial purity Al alloys. In fact, the layered structure is typical for commercial purity Al alloys with Sc and Zr addition.

In order to further elucidate the enrichment of the impurities within the layered structure, figure 3 shows a series of EPMA mapping. Clearly, Al is enriched in the α-Al matrix between the layers of the $Al_3(Sc,Zr)$ phase. While, Sc, Ti and Zr are enriched in the $Al_3(Sc,Zr)$ phase. It should be noted that, similar to Al, other impurities (i.e. Cu, Si, Fe) may be also present in the α-Al matrix between the layers of the $Al_3(Sc,Zr)$ phase [8].

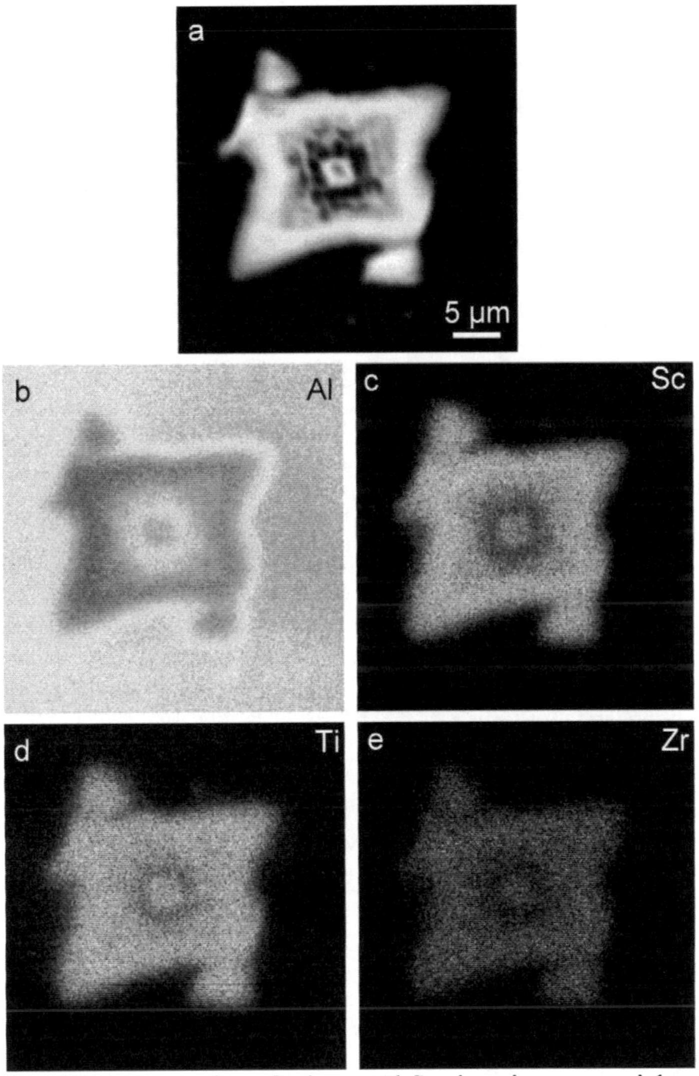

Figure 3. A series of EPMA mapping of primary Al_3Sc phase in commercial purity Al alloy.

4. Discussions

The final grain size of Al alloy is dependent on the number density, size and size distribution of primary $Al_3(Sc,Zr)$ phases and the growth restriction of Al. The growth restriction of Al alloy is dependent on the cooling rate, liquid flow field and growth restriction factor of the alloying elements. In this study, the cooling rate, liquid flow field can be regarded to be identical because the standard TP-1 test was performed. As there is also little influence of growth restriction on Al caused by Zr, Sc addition and only low addition of other elements, the factors affecting the final grain size is mainly by the number density, size and size distribution of primary $Al_3(Sc,Zr)$ phases. The number density, size and size distribution of primary $Al_3(Sc,Zr)$ phases itself is dependent on the impurity and the growth restriction on $Al_3(Sc,Zr)$ phase. The presence of impurities (i.e. Si (0.0675), Fe (0.805), Cu (0.0011), Mg (0.0113), Ti (0.0117) in commercial purity Al-0.25Sc-0.25Zr alloy results into significant effects on the number density size and size distribution of primary $Al_3(Sc,Zr)$ particles [7].

The observed layered structure of the $Al_3(Sc,Zr)$ particle can be discussed on the basis of the solidification sequence [8]. Firstly, by a peritectic reaction, the faceted $Al_3(Sc,Zr)$ phase forms. Once the peritectic particle has formed further solidification is by the peritectic reaction:
Al_3Zr + liquid $\rightarrow$ Al, which leads to the formation of the first Al layer. This Al layer can act as nucleation surface for the initial eutectic reaction: liquid $\rightarrow$ α-Al + $Al_3Sc(X)$ where the second phase (here Al_3Sc) has to be formed. The very close orientation relationship and small mismatch between Al and Al_3Sc permits readily nucleation of Al_3Sc on Al. The subsequent eutectic reaction occurs within the diffusion field of rejected elements (i.e. Al) not participating easily in the peritectic or eutectic reaction. This subsequent eutectic transformation cannot occur readily as the Al-Al_3Sc is a heavily divorced eutectic system so that no cooperative growth is seen but there are subsequent eutectic reactions of Al_3Sc on Al and Al on Al_3Sc. However, this growth by subsequent eutectic reactions is hindered by other elements present in the diffusion field surrounding the growing particle. In order to avoid long range diffusion other elements are incorporated into a ternary cell structure with the $Al_3Sc(X)$ layer. While the Al layer can easily accommodate the solute elements present. This peritectic growth mechanism can terminate until the eutectic transformation ends and the last Al layer has reached a critical size at which free growth for a given undercooling occurs [9-11].

5. Conclusions

In the case of high purity Al alloys, primary $Al_3(Sc,Zr)$ particles were pushed to grain boundaries by the moving solidification front. Such type of primary $Al_3(Sc,Zr)$ phase did not contribute to the heterogeneous nucleation, and thereby grain refinement of Al alloys. In the case of commercial purity Al alloy, the presence of impurities not only enhances the heterogeneous nucleation of primary $Al_3(Sc,Zr)$ phase, but also induced a transition on primary $Al_3(Sc,Zr)$ phase from a peritectic to eutectic reaction. This investigation demonstrates that impurities have important effects on the nucleation and growth of primary $Al_3(Sc,Zr)$ phases and subsequently on Al in Al alloys. Controlling impurities and their concentrations can enhance the grain refinement of Al alloys.

References

[1] RØyset J and Ryum N 2005 Scandium in aluminium alloys *Int. Mater. Rev.* **50** 19-44.
[2] Toropova L S, Eskin D G, Kharakterova M L and Dobatkina T V 1998 Advanced Aluminium Alloys Containing Scandium: Structure and Properties. (Gordon and Breach Science Publisher, Amsterdam, The Netherlands).
[3] Norman A F, Prangnell P B and Mcewen R S 1998 The solidification behaviour of dilute aluminium–scandium alloys *Acta Mater.* **46** 5715-5732.
[4] Knipling K E, Seidman D N and Dunand D C 2011 Ambient- and high-temperature mechanical properties of isochronally aged Al–0.06Sc, Al–0.06Zr and Al–0.06Sc–0.06Zr (at.%) alloys *Acta Mater.* **59** 943-954.
[5] Morrison C B, Seidman D N and Dunand D C 2011 Coarsening resistance at 400 °C of precipitation-strengthened Al–Zr–Sc–Er alloys *Acta Mater.* **59** 7029-7042.

[6] Morrison C B, Seidman D N and Dunand D C 2012 Effect of Er additions on ambient and high-temperature strength of precipitation-strengthened Al–Zr–Sc–Si alloys *Acta Mater.* **60** 3643-3654.

[7] Li J H, Oberdorfer B, Wurster S and Schumacher P 2014 Impurities effects on the nucleation and grwoth of primary Al2(Sc,Zr) phase in Al alloys *J. Mater. Sci.* **49** 5961-5977.

[8] Li J H, Wießner M, Albu M, Wurster S, Sartory B, Hofer F and Schumacher P 2014 *Mater. Character.* unpublished.

[9] Quested T E and Greer A L 2004 The effect of the size distribution of inoculant particles on as-cast grain size in aluminium alloys *Acta Mater.* **52** 3859-3868.

[10] Quested T E, Dinsdale A T and Greer A L 2005 Thermodynamic modelling of growth-restriction effects in aluminium alloys *Acta Mater.* **53** 1323-1334.

[11] Quested T E and Greer A L 2005 Athermal heterogeneous nucleation of solidification *Acta Mater.* **53** 2683-2692.

Modification of Ohnaka back diffusion equation

A Turkeli
Department of Metallurgical and Materials Engineering, Engineering Faculty, University of Marmara, Kuyubasi, 34722 Kadıkoy Istanbul, Turkey

Email: aturkelister@gmail.com

Abstract. The first and simplest analytical model for a binary alloy came from Gulliver-Scheil, considering a constant partition coefficient at the interface during solidification without back diffusion in solid. Later, Brody-Flemings proposed an approximate analytical model, taking into account one dimensional diffusion of solute in solid during solidification in plate-like dendrite arms for parabolic growth. However, this model does not work well at high back diffusion values. Ohnaka modified this back diffusion parameter. The Ohnaka model works well at the low and high back diffusion values but it underestimates between them. In this paper, Voller-Ohnaka back diffusion model is modified further to predict concentration profiles more accurately than all others analytical models for the parabolic growth rate.

1. Introduction

During the solidification of an alloy, the solute rejected at the solid-liquid interface is redistributed in the dendritic structure, which is known as microsegregation. This controls the amount of interdendritic eutectic, the amount of inclusions, solidification temperature range, and hot tear, and also decreases mechanical and corrosion properties of final product.

The first attempts to predict microsegregation quantitatively were derived by Gulliver and Scheil and the equation is known as Gulliver-Scheil equation. It is assumed that there is complete mixing of solute in the liquid but no diffusion of solute in the solid can occur and local thermodynamic equilibrium exists at the solid-liquid interface, described by a constant equilibrium distribution coefficient (k) and negligible undercooling occurs at the tip. This equation is [1-2]:

$$C_s = kC_o(1-f)^{k-1} \tag{1}$$

where:
Cs = the solid composition (wt%),
Co = the average composition of an alloy (wt%),
f = fraction of solid.

It predicts the worst segregation values. Since the introduction of the probe microanalysis technique in the late 1950s, it is shown that the assumption of no solid state diffusion is unjustified. Brody and Flemings recognised this and presented an analysis which quantifies the effect of solid state diffusion occuring between the Gulliver-Scheil and the lever rule cases. Their analytical solution for parabolic growth rate is [3-4]:

$$C_s = C_o k(1-(1-2\alpha k)f)^{\left(\frac{k-1}{1-2\alpha k}\right)} \tag{2}$$

where: $\alpha=(D_s t_f/L^2)$ back diffusion parameter at the solid-liquid interface,
Ds the diffusion coefficient in solid (m^2/s),
t_f the solidification time (s),
L the half of the solidified distance (m).

In this equation the dimensionless parameter, α, determines the extent of diffusion in the solid. However, this treatment does not conserve solute, especially for fast diffusion elements, i.e. the applicability of this equation is limited to $k\alpha < 0.1$.

2. Back diffusion models

Since Brody-Flemings introduced this back diffusion parameter into the microsegregation equation, several researchers have attempted to modify back difffusion parameter to predict concentration profiles more accuraletely than the Brody-Flemings model.

Clyne and Kurz examined the influence of rapid solid state diffusion in solidification. They derived a relationship for large α values. In their model the Ω replaced by α is [5]:

$$\Omega = \alpha\left(1 - exp\left(-\frac{1}{\alpha}\right)\right) - 0.5exp\left(-\frac{1}{2\alpha}\right) \quad (3)$$

This equation works well at low and high α values but not between them. Ohnaka has criticised their model and solved back diffusion equation approximately for the planar solidification [6]. He assumed that solute profiles in the solid can be expressed with a quadratic equation, then he obtained the following back diffusion equation.

$$\Gamma = \frac{2\alpha}{1+2\alpha} \quad (4)$$

Comparing with the Kurz-Fisher equation he replaced 2Ω with Γ . His results showed that his equation can estimate concentration profiles better than the Brody-Flemings's solution but it gives similar results to the Clyne-Kurz solution.

Voller suggested two different approaches for the back diffusion at the solid-liquid interface which are called the profile model and the parameter model. In the parameter model he replaced β with Γ in the Ohnaka's model. His model for parabolic growth rate is [7]:

$$\beta = \frac{2\gamma\alpha}{2\alpha+\gamma} \quad (5)$$

where γ is an integration factor, and

$$\gamma = \frac{A\alpha k}{A\alpha k+1} \quad (6)$$

where A is a constant.

He calculated A=4 in this parameter by comparing Kobayashi's exact results with his suggested equation. These two equations can be simplified as follows:

$$\beta = \frac{2\alpha}{1+2\alpha+cor} \quad (7)$$

Comparing to Ohnaka's back diffusion model he only added cor (cor=(1/2k)) to the denominator part of the Ohnaka's equation. Voller showed that his model can predict very closely the maximum concentration of Kobayashi's model. However, the prediction of this model for an amount of eutectic and therefore, for concentration profiles is not very close to the exact model as shown in this paper. In order to improve his back diffusion model, a fitting function is introduced to his equation as given below:

$$cor = \frac{f^{10+20f^{12}}}{2k} \tag{8}$$

As you can see in equations 7 and 8, back diffusion parameter is not constant but is a function of k, α, and f. cor is not an integration factor anymore as Voller suggested but is a correction variable for back difffusion model.

On the other hand, Ganesan and Poirier wrote a numerical programs for the planar dendritic solidification with the parabolic growth rate. Then comparing this numerical model with Brody-Flemings analytical model for large number of α and k values and using a statistical package they developed an alternative back diffusion equation for parabolic growth rate. In their model the μ replaced by α is [8]:

$$\mu = \frac{exp\left(-\frac{Af}{(\alpha k)^B}\right)}{2+\frac{Cf}{\alpha^D}} \tag{9}$$

where A, B, C, and D are constants and A=0.1523, B=0.4331, C=1.1093, and D=0.9367. Similarly, their results showed that back diffusion is not constant during solidification but is a function of α, k, and f. It is a good agreement with the present suggestion.

3. Results and comparisons

Ohnaka, Voller, Ganesan-Poirier, and present modified back diffusion equations are plotted as a function of fraction of solid for two different α values (α=0.1 and α=1) in the figure 1. It can be easily recognized that modified back diffusion model decreases gradually from Ohnaka back diffusion model to Voller back difffusion model after f=0.6 and it is equal to the Voller back diffusion model at f=1, therefore it does not affect its prediction of maximum concentration. Ganesan-Poirier model shows similar behavior but its prediction is slightly less than the present curve.

The prediction of amount of eutectic structure of several analytical models are compared to each other in figure 2-4 for C_{eu}/C_o=3, 6.5 and 12 values, respectively. As you can see in the figure 2, the prediction of amount of eutectic of the present analytical model (with modified back diffusion parameter) is very close to the Kobayashi's results [9] whereas, the prediction of Ohnaka's model underestimates them. On the other hand, Voller model overestimates the amount of the eutectic stucture. The estimations of Wang-Beckermann model [10] is also very close to the exact solution of Kobayashi at the low C_{eu}/C_o values.

For C_{eu}/C_o=6.5, all available analytical models in literature are compared with the exact analytical model of Kobayashi in the figures 3a and 3b. The behaviour of all analytical models are almost the same as the previous figure. Kurz-Fisher and Ohnaka models underestimate eutectic values, whereas, Voller's model overestimates them. The prediction of Wang-Beckermann model at low partition coefficients are very close to the exact model, however, increasing k values it overpredicts the amount of eutectic because this model becomes infinite when the fraction of solid approaches to one. Ganesan-Poirier estimates very closely to the exact model of Kobayashi. The prediction of

Nastac-Stefanescu model [11] has its own unique profiles because no growth rate is dictated in the assumption of this model.

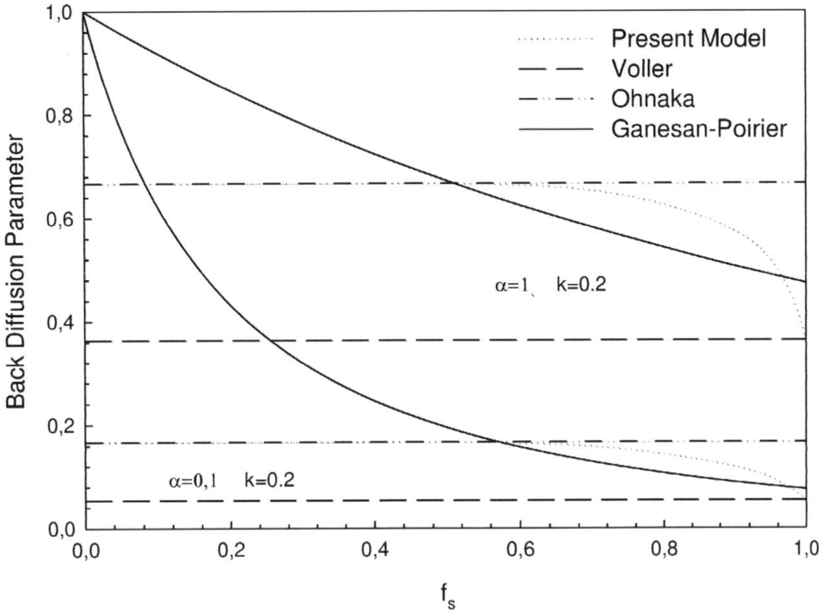

Figure 1. Comparisons of back diffusion models as a function of fraction of solid for two different α values (k=0.2, α=0.1 and α=1).

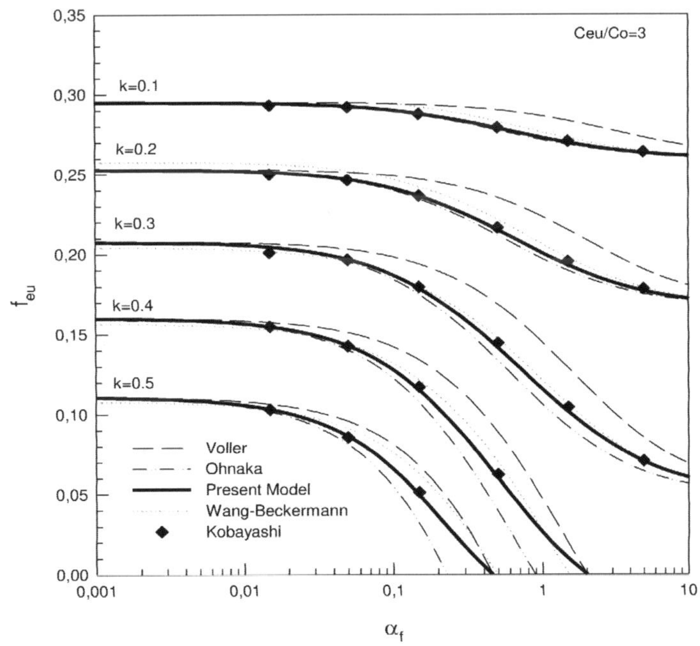

Figure 2. Prediction of eutectic structure as a function of back diffusion parameter for several models (for $C_{eu}/C_o=3$).

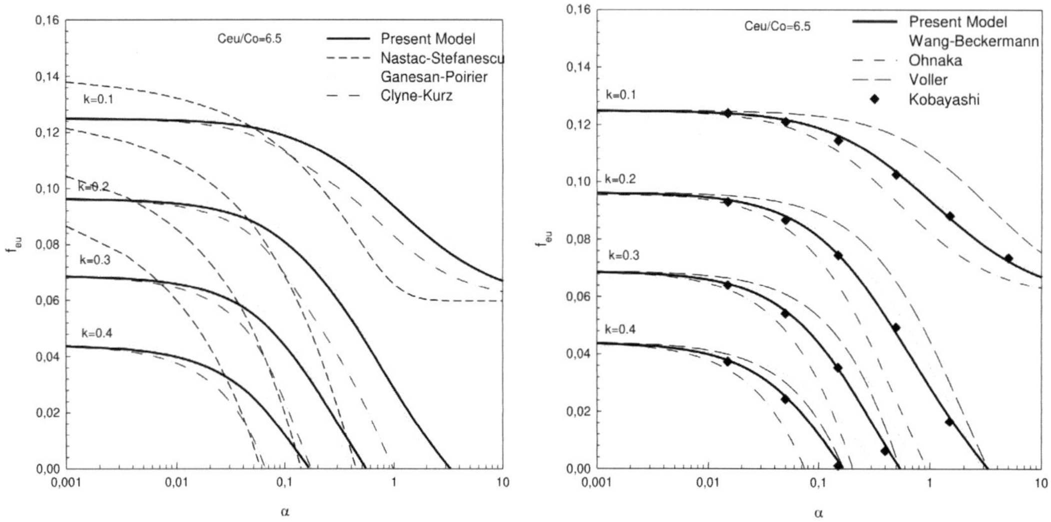

Figure 3. a and b. Prediction of eutectic structure as a function of back diffusion parameter for several models (for C_{eu}/C_o=6.5).

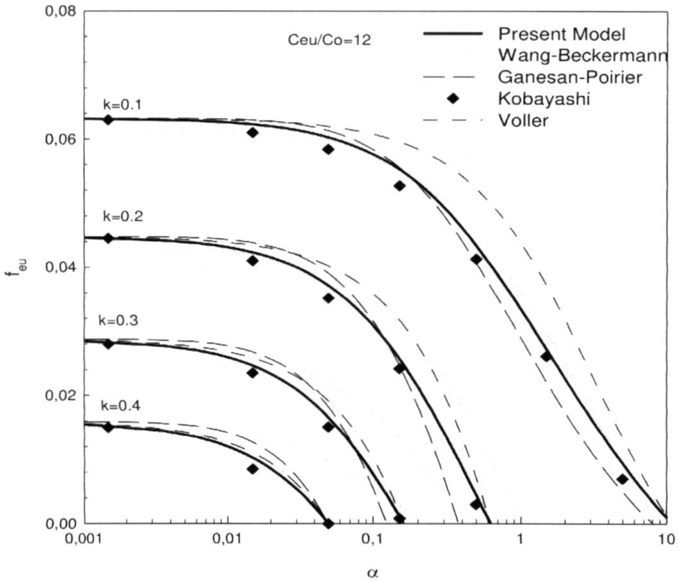

Figure 4. Prediction of eutectic structure as a function of back diffusion parameter for several models (for C_{eu}/C_o=12).

Some models which are very close to the exact solution are compared to each other only for C_{eu}/C_o=12. As shown in the figure 4, the prediction of present analytical and Ganesan-Poirier models are very close to the the prediction of exact solution of Kobayashi whearas the Wang-Beckermann model overestimates it with increasing k values.

4. Conclusion

Voller-Ohnaka back diffusion equation is modified and compared with other analytical models. It is shown that the prediction of the amount of eutectic structure using the modified back diffusion equation is very close to the exact solution of Kobayashi model. The calculation of the present model is much easier than the exact Kobayashi equation.

Acknowledgement

This work was supported by a grant from the University of Marmara, FEN-D-120514-0187.

References

[1] Gulliver G.H. 1922 *Metallic Alloys,(Griffin, London)*
[2] Scheil T.E. 1942 *Zeitschrift für Metallkunde*, **34** 70
[3] Brody H.D. and Flemings M.C.1966 *Trans. Of Metal. Society, of AIME* **236** 615
[4] Flemings M.C. 1974 *Solidification Processing* (McGraw-Hill, New York)
[5] Clyne T.W. and Kurz W. 1981 *Metall. Trans.* **12A** 965
[6] Ohnaka I. 1986 *Trans. ISI of Japan* **26** 1045
[7] Voller V.R. 2001 *Journal of Crystal Growth* **226** 562
[8] Ganesan S. and Poirier D.R. 1989 *Journal of Crystal Growth* **97** 851
[9] Kobayashi S. 1988 *Trans. ISI of Japan*, **28** 728
[10] Wang C.Y. and Beckermann C. 1993 *Materials Science and Eng*, **A171** 199
[11] Nastac L. and Stefanescu D.M. 1993 *Metall. Trans.* **24A** 2107

IOP Conf. Series: Materials Science and Engineering **117** (2016) 012022 doi:10.1088/1757-899X/117/1/012022

The influence of Cu, Mg and Ni on the solidification and microstructure of Al-Si alloys

A Darlapudi[a], S D McDonald and D H StJohn
Centre of Advanced Materials Processing and Manufacturing, The University of Queensland, St Lucia QLD 4072 Australia

E-mail: [a] a.darlapudi@uq.edu.au

Abstract. The influence of alloying elements (Cu, Mg, and Ni) on eutectic nucleation, eutectic grain morphology and the final microstructure of an Al-10Si commercial purity alloy in unmodified and Sr-modified conditions was investigated. It was found that the nucleation and eutectic grain growth morphologies of both the unmodified and Sr-modified Al-Si eutectic were significantly influenced by the addition of ternary alloying elements to a degree dependent on when the intermetallic phase formed during the solidification of the alloy with respect to the Al-Si eutectic. In cases where an intermetallic phase nucleated prior to the onset of the Al-Si eutectic reaction, the eutectic nucleation frequency was affected by changes to the available nuclei population. In cases where the intermetallic nucleated after the Al-Si eutectic, segregation of the ternary solutes in front of the Al-Si eutectic interface changed the nucleation and macroscopic growth dynamics. The changes in nucleation and growth dynamics of the Al-Si eutectic due to the presence of solute altered the morphology of the eutectic silicon considerably. This study has revealed a number of insights into the mechanisms of nucleation and growth of the Al-Si eutectic.

1. Introduction

Many commercial hypo-eutectic Al-Si alloys contain 5-12wt.%Si [1]. Cu and Mg are often added to achieve strengthening [1, 2]. Recent research has shown that changes in the nucleation frequency and macroscopic evolution of the Al-Si eutectic influence the final microstructure and formation of casting defects [3-6]. Impurity modification, alloy purity, the presence of solute and the solidification conditions are some of the factors that can influence eutectic nucleation, macroscopic growth patterns and the scale and morphology of the microstructure [7-9]. However, with the exception perhaps of the modifying elements Sr, Na and Sb, the influence of common alloying elements on eutectic nucleation and the macroscopic evolution of the eutectic phases in Al-Si alloys are relatively unexplored.

The solidification of a binary hypo-eutectic Al-Si alloy starts with the nucleation and growth of Al dendrites followed by the nucleation and growth of the Al-Si eutectic phases. It is reasonable to assume that the addition of a ternary alloying element will influence the number and distribution of nuclei present in the melt. In addition, solute pile-up ahead of both the primary Al phase and the eutectic growth interface can create constitutionally supercooled (CS) zones [10] which will influence both nucleation and growth. The purpose of this work is to investigate the influence of alloying elements (Cu, Mg and Ni) on eutectic nucleation, grain morphology and the final microstructure.

2. Materials and experimental procedures

Al-10Si-X (X = Cu, Mg, or Ni) unmodified and Sr-modified alloys were made from commercial purity aluminium, high purity silicon and master alloys (as indicated in Table 1).

Table 1. Nominal alloy compositions and master alloys used in the present work.

Alloy addition	Nominal alloy composition	Master alloy used
Cu	Al-10Si-(1, 2,4,6) wt.%Cu	Al-33 wt.%Cu
	Al-10Si-(1,2,4,6,8,10,15) wt.%Cu -350ppmSr	Al-33 wt.%Cu, Al-10 wt.%Sr
Mg	Al-10Si-(0.5,1.5,2.5) wt.%Mg	Pure Mg (Commercial grade)
	Al-10Si-(0.5,1.5,2.5) wt.%Mg-350ppmSr	Pure Mg, Al-10 wt.%Sr
Ni	Al-10Si-(0.5,2.5,4.5) wt.%Ni	Al-20 wt.%Ni
	Al-10Si-(0.5,2.5,4.5) wt.%Ni-350ppmSr	Al-20 wt.%Ni, Al-10 wt.%Sr

The alloys were melted at 760°C in an electric resistance furnace. Twenty minutes before casting, Sr if required, was added to the melt by an Al-10wt.%Sr master alloy. Samples were taken by dipping pre heated stainless steel cups into the melt. A type N thermocouple covered in a stainless steel sleeve was lowered quickly into one sample to record the temperature. The other samples were quenched into a water bath at room temperature at different times during eutectic solidification. After sectioning parallel to the vertical axis, the samples were polished using standard metallographic techniques.

Thermal analysis data was used in combination with the optical micrographs to observe the evolution of different phases during solidification. Quantitave metallographic studies were performed to understand the influence of increasing additions of each alloying element on the Al-Si eutectic nucleation frequency. The Sr-modified eutectic grain growth morphology was studied using the micrographs obtained from quenched samples using a stereo microscope. In unmodified samples the total number of Si-particles/m^2 of sample was used to approximate the nucleation frequency of the eutectic grains. In the Sr-modified alloys, eutectic grains were counted from the macrographs.

3. Results and discussion

3.1 Eutectic nucleation

The area density of eutectic grains in unmodified and Sr-modified alloy quenched samples are plotted against the solute content in figure 1 in which a considerable amount of scatter can be observed. In the case of Al-10Si-1.5Mg, Al-10Si-1.5, 2.5Ni and Al-10Si-8, 10, 15Cu samples counting the eutectic grain size was not possible due to the complex interface morphology of the eutectic grains as shown in figure 2(b) for example.

From figure 1, it can be observed that in the case of Cu additions the eutectic nucleation frequency steadily increased with an increase in the solute content with the exception of some minor variations. In the case of Mg additions the eutectic nucleation frequency increased up to 0.5wt.% in the case of unmodified alloys and 1.5wt.% in the case of Sr-modified alloys and then decreased as the solute addition increased further. In the case of Ni additions to the unmodified Al-10Si alloy, the eutectic nucleation frequency increased. Where as in the case of Sr-modified alloy the eutectic nucleation frequency slightly increased at 0.5wt.% Ni addition level and then decreased.

The different effects of the various elements at different concentrations can be rationalised by considering the solidification path of each alloy. As we observe the solidification path of the Al-10Si-X (X = Cu, Mg, or Ni) alloys using the thermal analysis curves and optical micrographs, it is observed that the addition of a ternary alloying element resulted in the formation of an intermetallic phase for all compositions. When the intermetallic nucleated before the Al-Si eutectic reaction (circled data points in figure 1), the Al-Si eutectic nucleation frequency decreased in the unmodified and Sr-modified conditions, relative to samples with lower additions of the same elements. If the formation of the intermetallic phase occurred after the Al-Si eutectic, the ternary alloy addition resulted in an increase in the eutectic nucleation frequency in unmodified and Sr-modified alloys.

In commercial purity unmodified alloys there is an abundance of AlP particles [11-14]. These particles are considered highly potent with respect to their ability to nucleate polyhedral silicon particles, which then initiate the formation of the eutectic grains. In cases where the ternary alloying elements result in the formation of intermetallic phases before the Al-Si eutectic forms, they can interact with the available nuclei population and hence change the eutectic nucleation frequency. In this case, it is important to consider the nucleation mechanism of these intermetallics and their ability to nucleate the Al-Si eutectic. It is worth noting that for the Mg and Ni additions, when intermetallic formation occurred prior to the Al-Si eutectic solidifying nucleation frequency decreased in both the unmodified or Sr-modified conditions. This may indicate that AlP is the most potent nucleant for the Al-Si eutectic but that none of these intermetallics (Mg_2Si and Al_3Ni) were able to nucleate the Al-Si eutectic efficiently.

When an alloying element added to the Al-Si alloy is not involved in any solidification reactions prior to formation of the Al-Si eutectic, the rejected solute from both the eutectic constituents accumulates ahead of the interface as the eutectic grain is growing. Due to this solute segregation ahead of the eutectic interface the equilibrium melting point of the liquid changes locally. The actual temperature of the melt is less than the equilibrium liquidus temperature and, hence, the melt in this boundary layer is constitutionally supercooled [10]. In addition to the changes in the equilibrium liquidus temperature other changes might be expected. These include variations in the surface tension which can influence the wetting of nuclei present in the boundary layer and the melt, and changes to the chemical driving force for nucleation [15]. As such, significant changes can be expected in the nucleation behaviour due to solute segregation.

Easton et al. [16] proposed that the presence of a CS zone ahead of a primary phase can promote a wave of nucleation events ahead of the interface. In the case of unmodified Al-Si eutectic in the presence of highly potent nuclei, nucleation events readily occur ahead of the growing interface of existing eutectic grains leading to a high density of eutectic grains throughout the casting. Hence, the presence of ternary segregating elements has a relatively minor effect on the eutectic grain density in the unmodified alloys.

The addition of eutectic modifiers to Al-Si alloys significantly decreases the density and potency of the nuclei of the eutectic grains. The addition of Sr deactivates the AlP particles probably due to the nucleation of Sr--intermetallics on AlP [11, 17], which significantly reduces the potency of the nucleant particles. Thus, more CS is needed to activate these nuclei. The presence of a ternary segregating solute can provide CS ahead of the eutectic grain-liquid interface and, therefore, increase the number of eutectic grains. Despite this increase in the number of grains with increasing ternary element concentration the number of grains is significantly less than for the unmodified alloys.

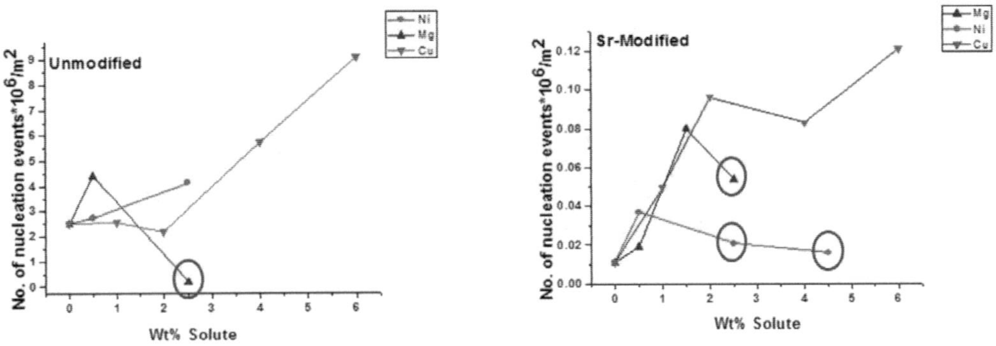

Figure 1. The number of nucleation events in unmodified and Sr-modified alloys with different alloying additions. The circled points indicate that an intermetallic containing the ternary alloy nucleated before the Al-Si eutectic reaction commenced. Note the difference in scale on these plots.

3.2 Macroscopic growth

The abundant nucleation in unmodified alloys influenced the eutectic growth in two ways. For a given rate of heat extraction the interface growth rate will be low due to the presence of more solid-liquid interfacial area. In addition, as the eutectic grains are much closer to each other their solute diffusion fields quickly overlap decreasing the available driving force for growth. For Al-Si binary alloys this solidification condition leads to many small discrete eutectic grains with an irregular interface structure and it is very difficult to observe the influence of ternary alloying elements on their macroscopic growth morphology. This remains an area that needs more research.

In the case of the Sr-modified eutectic, it was observed that the effect of ternary alloying elements on eutectic macroscopic growth was dependent on when intermetallic evolution occurs. When the intermetallic is formed before the Al-Si eutectic nucleation commenced, the intermetallic influenced the eutectic interface growth by acting as a physical barrier for eutectic growth. In this case, the intermetallic phase morphology was found to be vital.

In cases where the Al-Si eutectic nucleated before the intermetallic, it was observed that at low solute addition levels the spherical morphology of the eutectic interface slightly destabilised late in solidification. As the solute addition level increased the interface destabilised sooner. Further increases in solute concentration result in eutectic grains with a dendritic morphology as shown in figure 2 which shows the macrostructure of quenched samples of Al-10Si-0.5Mg-350Sr (figure 2a) and Al-10Si-1.5Mg-350Sr (figure 2b) alloys. During eutectic solidification a diffusion layer forms due to the rejection of ternary elements at the eutectic grain-liquid interface. A CS zone can then form immediately ahead of the advancing interface and perturbations eventually grow into the CS zone. With the increase in addition of a ternary segregating element the increase in the amount of CS on one hand destabilised the eutectic interface. On the other hand, it increases the eutectic nucleation frequency which in turn decreases the interface velocity contributing towards stability of the eutectic-liquid interface. However, the results from this research that the interface destabilises earlier with an increase in the concentration of ternary alloy additions, suggests that the increase in the amount of CS ahead of the interface and its contribution towards interface destabilisation is dominant.

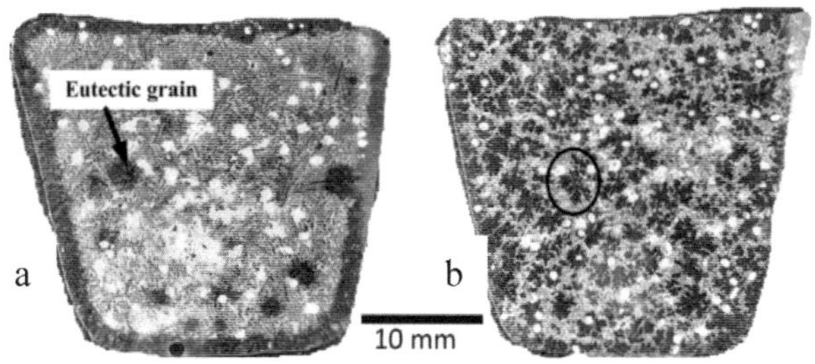

Figure 2. Macrographs of quenched Sr-modified samples of Al-10Si (a) 0.5Mg and (b) 1.5Mg alloys. The circled region shows a eutectic grain with a dendritic morphology.

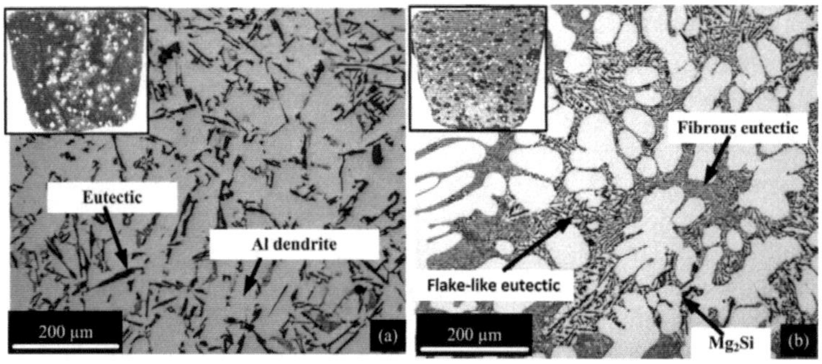

Figure 3. Micrographs of quenched unmodified samples of Al-10Si (a) 0.5Mg and (b) 2.5Mg alloys. The insets show the macrographs of the alloys

3.3 Eutectic microstructure

The changes in nucleation frequency and the growth morphology of the eutectic grains affected the morphology of eutectic Si. In the cases where the Al-Si eutectic nucleation frequency decreased due to the addition of solute (Mg with concentration high enough to form pre-eutectic intermetallics), the final eutectic silicon morphology changed from flake-like to a mixture of flake-like and fibrous silicon as shown in figure 3. Figure 3 shows the micrographs and macrographs of unmodified quenched samples of Al-10Si-0.5Mg (figure 3a) and Al-10Si-2.5Mg (figure 3b) alloys.

With an increase in the Al-Si eutectic nucleation frequency, for example the case of Cu additions to Sr-modified alloys, the silicon morphology changed from completely fibrous to a mixture of flake-like and fibrous morphologies as shown in figure 4. Figure 4 shows the micrographs and macrographs of the Al-10Si-2Cu-350Sr (figure 4a) and Al-10Si-8Cu-350Sr (figure 4 b) alloy samples.

The origin of the morphological change from flake-like to fibrous Si with the addition of eutectic modifiers is not fully understood. Studies have generally focused either on the effect of modifiers on the nuclei [11, 18-22] or on the effect of modifiers on the eutectic Si growth [23-26]. However, it is likely that a modifier element has a combined effect on both the nucleation and the growth of the eutectic Si.

Most faceted phases, including Si, can undergo a roughening transition (faceted to non-faceted) when the driving force for growth or the interface growth velocity increases [27]. Apart from twinning [23, 28] it is believed that the driving force ahead of the solidifying interface and the growth velocity of the interface controls the Si morphology [20]. These two parameters change during the solidification of a particular alloy and they are also influenced by the changes in the nucleation and macroscopic growth modes that are caused by the addition of a ternary alloying element. The relative impact of these two factors is difficult to separate due to the complex interactions between them.

This work has shown that CS plays an important role in each step of the solidification of Al-Si alloys from nucleation and eutectic grain growth to the degree of modification of the Si phase. First, CS promotes the nucleation of eutectic grains throughout the casting. Second, increasing the amount of CS causes destabilisation of the eutectic grain interface from spherical to dendritic. Third, the growth rate of the eutectic grain interface is controlled by the amount of CS developed between the eutectic grains. The growth rate is initially high when the grain spacing is large as a higher amount of CS can be sustained over longer distances before the diffusion fields overlap decreasing the amount of CS and thus the growth velocity, causing a transition from fibrous to flake-like Si (dependent on the concentration of modifier). The interdependence between eutectic nucleation and growth influences the final eutectic microstructure. It is important to take factors such as the solidification conditions, nuclei population, macroscopic growth mode and intermetallic evolution into consideration when predicting the final eutectic microstructure of a commercial foundry alloy.

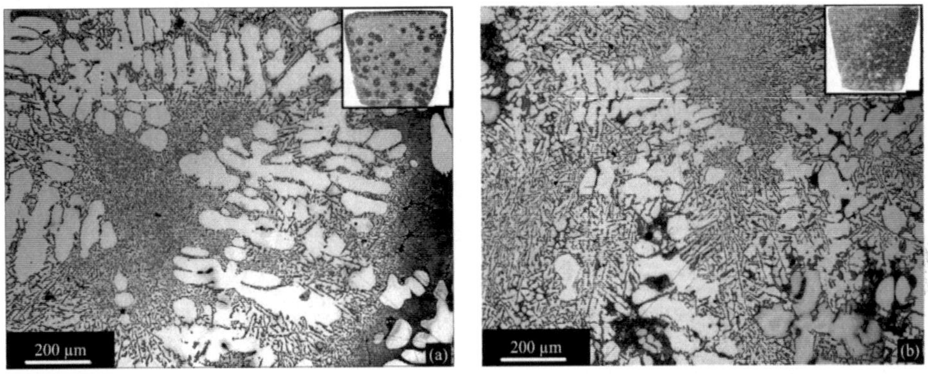

Figure 4. Influence of the copper concentration on the morphological transition of the eutectic Si in eutectic cells of samples with (a) 2 wt.%Cu and (b) 8 wt.%Cu. The insets show the macrographs.

4. Conclusions

- The addition of a ternary alloying element (Cu, Mg, and Ni) to an Al-10Si alloy influences the eutectic nucleation frequency, growth morphology and the final alloy microstructure considerably.
- When the addition of a ternary element resulted in the nucleation of an intermetallic phase as the primary phase or as a binary eutectic phase prior to Si nucleation, there is a decrease in the eutectic nucleation frequency both in unmodified and Sr-modified conditions probably due to a change in the nuclei population.
- Presence of a constitutionally undercooled zone ahead of the eutectic interface does not result in major changes in the unmodified eutectic nucleation frequency due to the presence of a high number of potent nuclei (probably AlP). In Sr-modified Al-Si alloys where the eutectic nucleation frequency is very low compared to the unmodified alloy, the addition of solute (even 1 wt.% of Cu) results in increased nucleation. Further increases in the solute content results in an increase in the number of eutectic grains and a decrease in their size. This important result indicates that the basic principles relating to the effect of solute on grain refining can be extended to eutectic growth as well.
- In Sr-modified alloys solute pile-up ahead of the eutectic interface (due to the presence of ternary alloying element) influences the eutectic macroscopic growth morphology significantly. After a critical value of solute addition a dendritic growth pattern emerged in the eutectic (i.e. the eutectic grains developed primary arms).
- It follows that the eutectic silicon morphology depends both on eutectic nucleation and growth dynamics and the influence of solute must be considered.

Acknowledgements
The authors acknowledge support from ARC project DP120101672 and the CAST Cooperative Research Centre.

References
[1] Davis J R, Properties and selection: nonferrous alloys and special-purpose materials, 1990.
[2] Mondolfo L F, Aluminium Alloys; Structure and Properties, London-Boston:Butterworths, 1976.
[3] McDonald S, Dahle A, Taylor J, StJohn D, *Metallurgical and Materials Transactions* B, **35** (2004) 1097-1106.
[4] Knuutinen A, Nogita K, McDonald S D, Dahle A K, *Journal of Light Metals*, **1** (2001) 241-249.
[5] Iwahori H, *Trans. Am. Foundrymen's Soc.* (AFS Trans.), **98** (1990) 167-173.
[6] Laslaz G, Fourth International Conference on Molten Aluminium Processing, (1995) 459-480.
[7] Mathiesen R, Arnberg L, Li Y, Meier V, Schaffer P, Snigireva I, Snigirev A, Dahle A, *Metallurgical and Materials Transactions* A, **42** (2011) 170-180.
[8] Heiberg G, Nogita K, Dahle A K, Arnberg L, *Acta Materialia*, **50** (2002) 2537-2546.
[9] Mc Donald S in, University of Queensland, 2002.
[10] Kurz W, Fisher D J, *Fundamentals of Solidification*, 4th ed., Trans Tech Publications 1998.
[11] Cho Y H, Lee H C, Oh K H, Dahle A K, *Metallurgical and Materials Transactions* A, **39** (2008) 2435-2448.
[12] Crosley P B, Mondolfo L F, AFS Transactions, **74** (1966) 53-64.
[13] Dahle A K, Nogita K, Zindel J W, McDonald S D, Hogan L M, *Metallurgical and Materials Transactions A: Physical Metallurgy and Materials Science*, **32** (2001) 949-960.
[14] Nogita K, McDonald S D, Tsujimoto K, Yasuda K, Dahle A K, *Journal of Electron Microscopy*, **53** (2004) 361-369.
[15] Yao X, Dahle A K, Davidson C J, StJohn D H, *Journal of Materials Research*, **21** (2006) 2470-2479.
[16] Easton M, StJohn D, *Metallurgical and Materials Transactions A*, **30** (1999) 1625-1633.
[17] Dinnis C M, Taylor J A, Dahle A K, in, The University of Queensland, 2004.
[18] McDonald S D, Nogita K, Dahle A K, *Acta Materialia*, **52** (2004) 4273-4280.
[19] Nogita K, Dahle A K, Materials Characterization, **46** (2001) 305-310.
[20] Flood S C, Hunt J D, *Metal science*, **15** (1981) 287-294.
[21] Hanna M, Lu S-Z, Hellawell A, *Metallurgical and Materials Transactions A*, **15** (1984) 459-469.
[22] Kim C B, Heine R W, *Journal Institute of Metals*, **92** (1963-1964) 367.
[23] Lu S-Z, Hellawell A, *Journal of Crystal Growth*, **73** (1985) 316-328.
[24] Lu S-Z, Hellawell A, *Metallurgical and Materials Transactions A*, **18** (1987) 1721-1733.
[25] Jenkinson D C, Hogan L M, *Journal of Crystal Growth*, **28** (1975) 171-187.
[26] Kobayashi K F, Hogan L M, *Journal of Materials Science*, **20** (1985) 1961-1975.
[27] Sunagawa I, Morphology of crystals, Terra Sci, Tokyo, 1987.
[28] Timpel M, Wanderka N, Schlesiger R, Yamamoto T, Lazarev N, Isheim D, Schmitz G, Matsumura S, Banhart J, *Acta Materialia*, **60** (2012) 3920-3928.

IOP Conf. Series: Materials Science and Engineering **117** (2016) 012023 doi:10.1088/1757-899X/117/1/012023

In situ synchrotron study of liquid phase separation process in Al-10 wt.% Bi immiscible alloys by radiography and small angle X-ray scattering

W Q Lu[a], S G Zhang[b] and J G Li[c]
School of Materials Science and Engineering, Shanghai Jiao Tong University, 800 Dongchuan Rd., Shanghai 200240, China

E-mail: [a] flytsing1003@sjtu.edu.cn, [b] sgzhang@sjtu.edu.cn, [c] lijg@sjtu.edu.cn

Abstract. Liquid phase separation process of immiscible alloys has been repeatedly tuned to create special structure for developing materials with unique properties. However, the fundamental understanding of the liquid phase separation process is still under debate due to the characteristics of immiscible alloys in opacity and high temperature environment of alloy melt. Here, the liquid phase separation process in solidifying Al-Bi immiscible alloys was investigated by synchrotron radiography and small angle X-ray scattering. We provide the first direct evidence of surface segregation prior to liquid decomposition and present that the time dependence on the number of Bi droplets follows Logistic curve. The liquid decomposition results from a nucleation and growth process rather than spinodal decomposition mechanism because of the positive deviation from Porod's law. We also found that the nanometer-sized Bi-rich droplets in Al matrix melt present mass fractal characteristics.

1. Introduction

Aluminium-based immiscible alloys have shown promising applications on bearing if a unique microstructure consisting of homogenous dispersed soft phase within an Al matrix is obtained [1]. However, liquid decomposition occurs mainly due to the density difference between two liquid and temperature gradient because of the Marangoni motion in solidifying immiscible alloys. Many studies have been carried out to further study the liquid phase separation process (LPSP) [2-8]. It was shown that it occurs as the formation of minor droplets by the spinodal decomposition [3, 5] or the nucleation and growth mechanism [9-11]. However, as the LPSP either occurs at high temperature or is hidden by solidification, it is the major obstacle and challenge for the experimental investigation.

Recently, synchrotron radiography experiments have been performed to investigate the LPSP of the Al-Bi immiscible alloys [2, 3, 12]. Schaffer et al. [2, 3] reported that the interaction between the Bi droplets and solidification front is decisive for final solidification structure. Wang et al. [12] found that the Bi droplets seriously aggregated with the presentation of the electric current pulse. However, they did not show the changes in Bi droplets number with time during the LPSP, and they also did not discuss the exact mechanisms of the LPSP.

Small angle X-ray scattering (SAXS) has always been used to determine whether a phase transformation results from spinodal decomposition mechanism or nucleation and growth process [13-15]. It is also a powerful means to investigate the liquid structure in nanometer scale [16]. According to the Al-Bi phase diagram, the mole fraction of Bi in the Bi-rich minor droplets changes with temperature during the LPSP [17]. Thus, the structure of the Bi-rich minor droplets is different at different temperatures. However, there is no experimental investigation in the LPSP by SAXS. Here, besides direct imaging of the behaviour of the Bi-rich droplets and its effect on the morphology evolution in solidifying Al-10 wt.% Bi immiscible alloy, *in situ* synchrotron SAXS technique is used to investigate the structure of the Bi-rich minor droplets in nanometer scale.

2. Experimental procedures

The Al-10 wt.% Bi alloy spheres were prepared from pure Al(99.99%) and pure Bi(99.99%) by one-step fabrication method, the detailed preparation process can be found elsewhere [18]. The imaging and SAXS experiments were carried out at beam lines BL13W1 and BL16B1 of Shanghai Synchrotron Radiation Facility (SSRF), respectively. Figure 1 shows the sketch of the experimental device configuration at SSRF. An alloy sphere was placed on a Cu sheet for imaging or a MgO sheet for SAXS experiment in furnace body, the furnace body was evacuated to 10^{-4} Pa and backfilled argon to 0.1 MPa. The sample was heated and cooled by a 100 W CO_2 laser with the focused beam hitting the specimen from the top. The temperature was measured by infrared pyrometer. The imaging and SAXS experiments were carried out at the same cooling rate. However, in order to guarantee the uniformity and stability of the melts temperature, the specimen used in SAXS experiment was placed on the MgO Sheet.

At imaging beam line station, the energy of the synchrotron X-ray beam was 30 keV. The exposure time per frame was 1 s. The effective pixel size was limited to 0.74 μm. Image Pro Plus 6.0 image analyzer software was used to measure the number of the Bi-rich minor droplets.

At SAXS beam line station, the energy and wavelength of the incident beam are 17 keV and 0.73 Å, respectively. The distance between specimen and CCD detector was 5.396 m. A Mar CCD 165 detector was used to record the scattered X-ray. The range of scattering vector (q) and scattering angle (2θ) are 0.086~0.637 nm^{-1} and 0.058~0.424°, respectively. The scattering intensities were obtained from 2-D SAXS pattern after background subtracting by Fit-2D software.

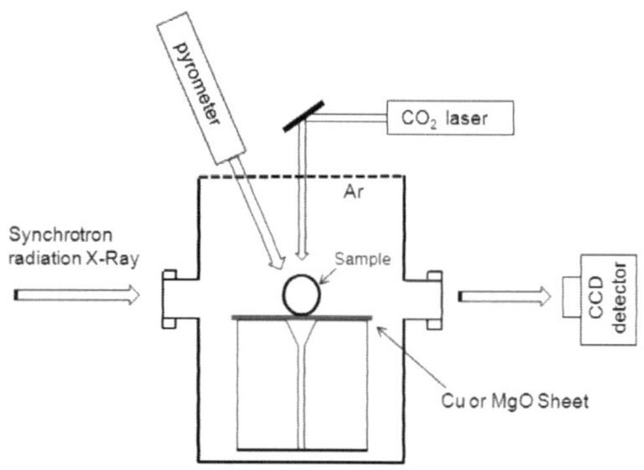

Figure 1. The schematic of experimental device configuration at SSRF.

3. Results and discussions

3.1 Segregation morphology evolution and larger sized droplets behaviour

Figure 2 shows the segregation evolution in solidifying Al-10 wt.% Bi immiscible alloy. The light grey phase is Al-rich phase, while the dark phase is Bi-rich phase because of the different absorption coefficient of X-ray between Al and Bi. The bright region in the bottom of the alloy melt is a hydrogen bubble which nucleates on the Cu sheet and increases its size with time (figure 2a). From figure 2b, Bi-rich droplets appear in the bubble area prior to other parts of the alloy melt.

There is a temperature gradient across the alloy sphere from bottom up because the sample was placed on a Cu sheet and heated by a CO_2 laser from the top. Furthermore, the energy change is about −1.4 ev for exchanging a Bi atom in the bulk with an Al atom on the surface [19]. Thus, Bi atom approaches to bottom and surface to decrease the system free energy (figure 2b). The amount of the precipitation Bi-rich phase increases with decreasing temperature (figure 2b-c). After surface segregation, some Bi-rich droplets separate out from matrix melt outside of the bubble area, meaning that bulk phase separation occurred. This experiment directly confirms that surface segregation takes place before the LPSP, yielding consistent results with the numerical simulations [20].

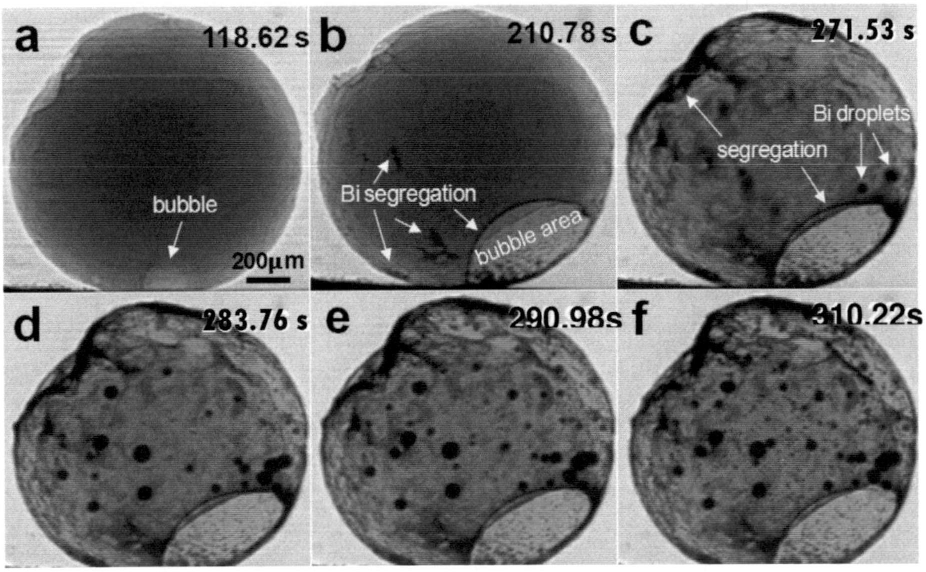

Figure 2. The morphology evolution during the solidification of the Al-10 wt.% Bi immiscible alloy with the cooling rate of 1.6 K/s. The temperature of the sample at time zero is 1417 K.

Once the Bi-rich droplets nucleated, they were instantaneously exposed to a force field [3, 21]. Besides Marangoni motions [1], they were also acted on by the Stokes motion [3]. As seen in figure 2, the position of the Bi-rich droplets did not change because the Marangoni motion can't overcome the Stokes motion.

From figure 2, the number of Bi-rich droplets inside the bubble area is unchanged after 210.78 s. However, the number of Bi droplets number of the bubble area changes with the extension of time. Figure 3 shows the change in Bi droplets number over time during LPSP. It can be seen that, the droplets number increases with time at the initial stage of the LPSP, while it reaches a constant value at the later stage. On cooling a homogenous melt of Al-Bi immiscible alloy, the continuous decrease of temperature increases the supersaturation of Bi atoms in Al matrix [22]. Its increase leads to a rapid increase of the nucleation rate, thus the droplets number increase rapidly. Meanwhile, as the surface segregation and the formation of Bi-rich droplets consume a large amount of Bi atoms in Al matrix, the size of newly formed Bi-rich droplets decreases (figure 2e-f). At the final stage of the LPSP, the Bi solute is almost all deposited from the Al matrix melt, leading to saturation of the number of the Bi-rich droplets. Therefore, characterized by a saturation level characteristic [23], the Logistic population model should be applied satisfactorily to the change of the number of Bi-rich droplets during the liquid phase separation period, and it can be described as:

$$N = 300.3 + \frac{32 - 300.3}{1 + \left(\frac{t}{293.7}\right)^{78.7}} \tag{1}$$

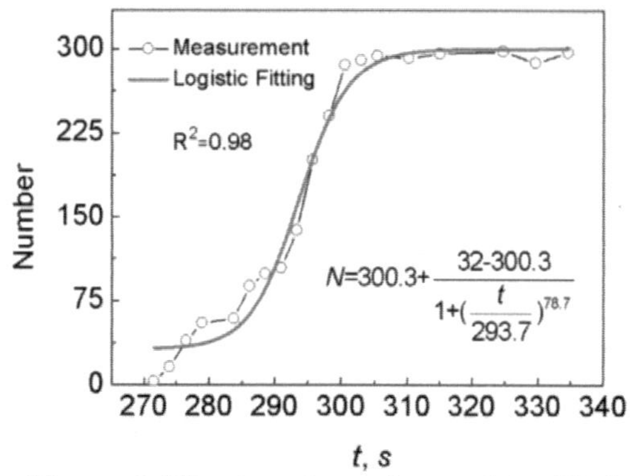

Figure 3. The time dependence of the Bi-rich minor droplets number.

3.2 Occurrence mechanism of the liquid phase separation

The relative scattering intensity curves I(q) vs. q as a function of temperature obtained from Al-10 wt.% Bi immiscible alloy are presented in figure 4a. It can be seen that the scattering intensity significantly increases at 863 K due to crystallization. The scattering intensities are different at low q region but reach a constant value at high q region.

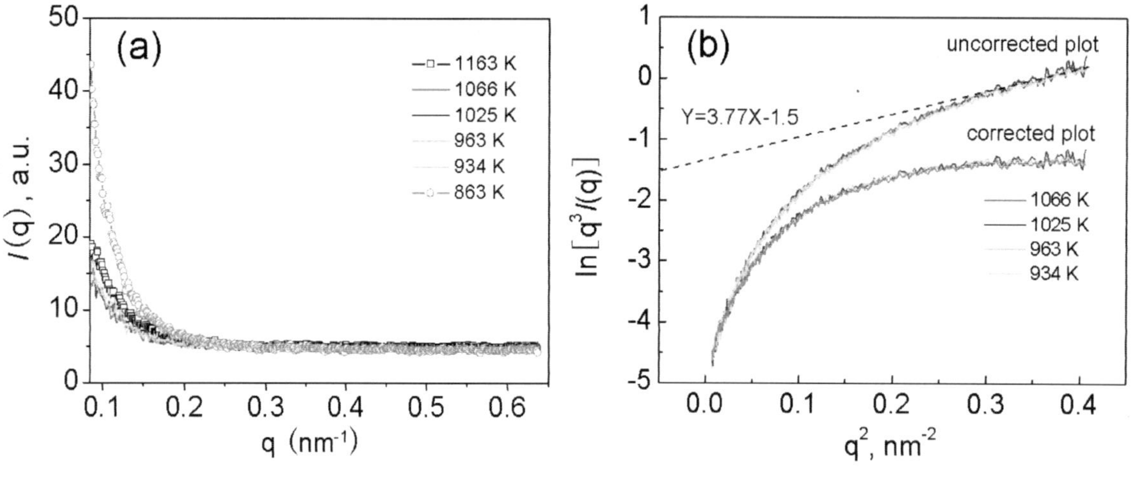

Figure 4. (a) I(q) vs. q as a function of temperature obtained from Al-10 wt.% Bi immiscible alloy with the cooling rate is 1.2 K/s, the error limit of measure temperature is ±2.4 K. (b) ln[q^3I(q)] vs. q^2 at different temperatures, the positive deviation from Porod's law was corrected.

As an ideal two phase system has a certain constant electron density difference and sharp boundary, the scattering intensity plot obeys Porod's law, i.e. $\lim_{q\to\infty}[q^3I(q)] = K'$, where K' is the Porod constant [24]. However, there is positive or negative deviation from Porod's law for a non-ideal system when electron density fluctuation or a diffuse interface layer exists in any phase [16]. Therefore, Porod's law can be adopted to determine whether the LPSP of Al-Bi immiscible alloys results from a nucleation and growth process or spinodal decomposition mechanism. Figure 4b shows the plot of ln [q^3I(q)] vs. q^2 calculated from figure 4a during the LPSP of Al-10 wt.% Bi immiscible alloy. It can be seen that there is a positive deviation from Porod's law. According to Li et al. [16] and Li [25], this positive deviation means that the electron density changes abruptly in the Al-matrix melt and there is a sharp boundary rather than a diffuse interface layer between Bi-rich droplets and Al-matrix melt. So we think that the LPSP occurs as the nucleation of Bi-rich droplets directly from the supersaturated Al-matrix melt rather than a homogeneous spinodal reaction.

From figure 4b, the plot of ln[q^3I(q)] vs. q^2 was fitted by $\ln[q^3 I(q)] = \ln K' + \sigma^2 q^2$ at high q region. Suggested by Li [26], the positive deviation was corrected to calculate K'. And the width (L) of electron density fluctuation can be calculated by $L = \sqrt{2\pi}\sigma$ with σ^2 representing the slope of the ln [q^3I(q)] vs. q^2 at high q region [27]. Based on the above equations, L=4.86 nm, which was independent of the temperature, yielding consistent results with the electron density fluctuation in Al-15 wt.% Cu melt [16].

3.3 Liquid structure of the nanometer-sized droplets

The size of the scattering objects (here is the Bi-rich minor droplets) can be analyzed by the Guinier approximation, which can be described by $I(q) \propto \exp\left[-(R_G q)^2/3\right]$ with R_G representing gyration radius. R_G can be calculated by $R_G = (-3k)^{1/2}$ with k is the initial slope of the plot (ln I(q) vs. q^2) [16, 28]. For spherical particle, $R_G = \sqrt{\frac{3}{5}}R$ [29]. Therefore, the radius of the newly formed Bi-rich droplets is about 14.2 nm.

According to the phase diagram of Al-Bi immiscible alloy [17], the temperature dependence of the mole fraction of Bi in the Bi-rich droplets can be described as [28]:

$$C(Bi) = -0.1737 + (2.71\times10^{-3} - 1.74\times10^{-6}T)T \tag{2}$$

From the equation (2), the relationship between the mole fraction of Bi in the Bi-rich droplets and temperature is plotted in figure 5. It can be seen that the mole fraction of Bi in the Bi-rich droplets increases with decreasing temperature during the liquid decomposition period. Thus, the liquid structure of the Bi-rich droplets under different temperature is different. Similar to the clusters of metallic melts [29], the nanometer-sized Bi-rich droplets can present fractal characteristic. The scattering intensity on a fractal structure is often described by $I(q) \propto q^{-\alpha}$, where α is related to fractal dimension and it can be calculated by the slope ($-\alpha$) of ln I(q) vs. ln (q). For mass dimension, $D_m = \alpha + 1$ [16].

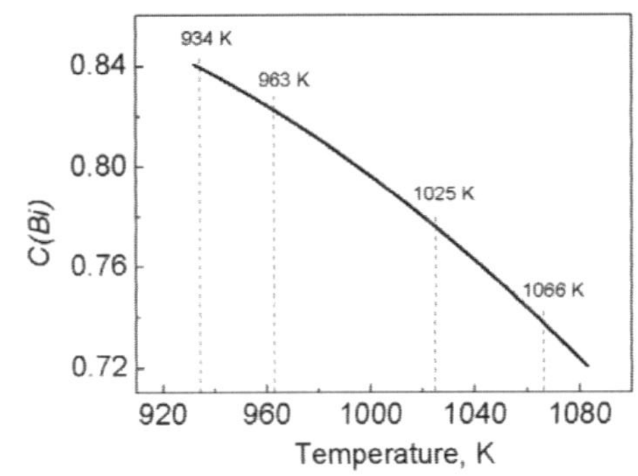

Figure 5. Bi concentration in the Bi-rich droplets as a function of temperature.

Figure 6 shows the changes of fractal dimension with the temperature during liquid phase separation stage. It can be seen that the maximum value of the fractal dimension appears at 934 K close to the monotectic transition temperature. This means that the newly formed Bi-rich droplets at 934 K are most compact. Suggested by Sonntag and Russel [30], if the fractal dimension of the aggregate is about 2.48, its structure is 3D flocs. As the dimensions of the Bi-rich droplets is in the range of 2.39~2.56, the newly formed Bi-rich droplets during liquid phase system. Thus, the fractal dimension of the Bi-rich droplets is different from the diffusion limited cluster aggregation (DLCA) model [31]. Therefore, the Bi-rich droplets formed by nucleation and growth have compact structure during the initial stage of growth.

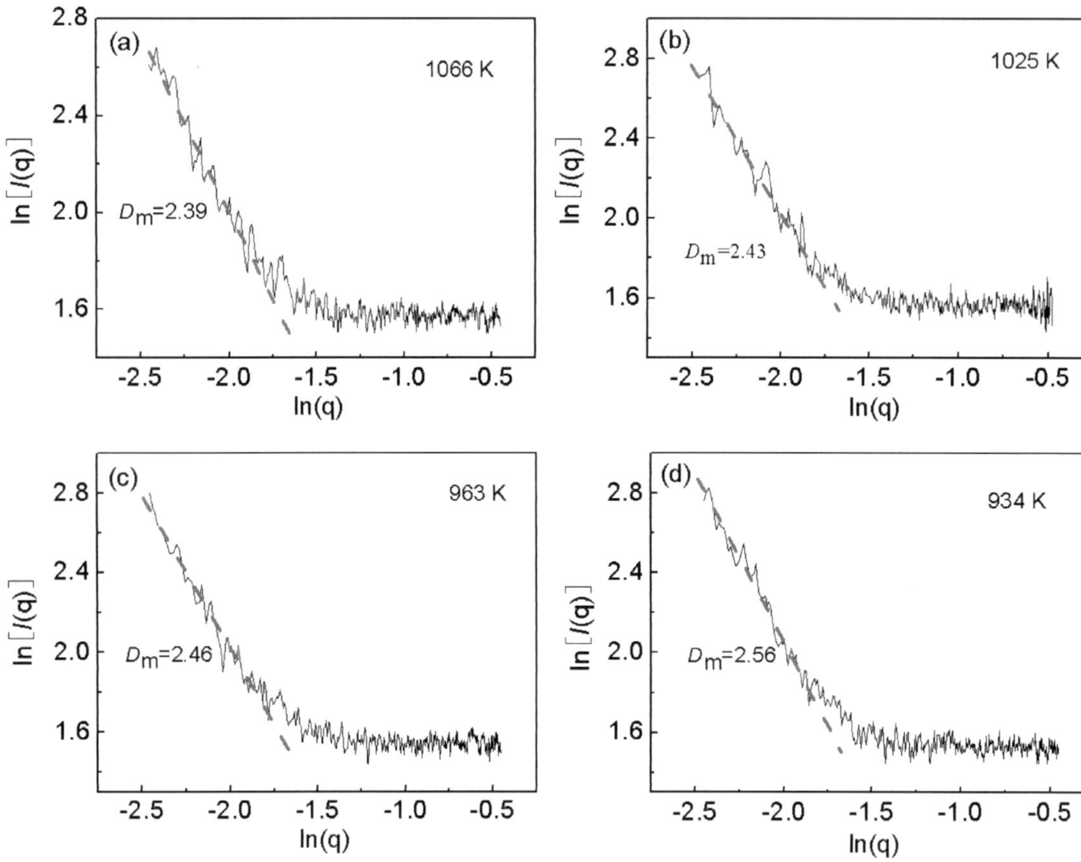

Figure 6. ln [I(q)] vs. ln (q) as a function of temperature separation process are 3D flocs. It should be noted that the nature of metallic melts is significantly different from the flocculation of colloidal.

4. Summary

Synchrotron radiography and small angle X-ray scattering have been used to investigate the LPSP of the Al-Bi immiscible alloys. We found that the LPSP occurs by the nucleation of Bi-rich droplets, which present mass fractal characteristics. We also found that the time dependence on the number of Bi droplets follows Logistic curve and provide the first experimental evidence that surface segregation occurs before liquid decomposition.

Acknowledgments

We thank the staff members of the BL13W1 and BL16B1 beam lines of SSRF and Z.Y. Dai for helping to prepare for the synchrotron experiments. This work is supported by the National Natural Science Foundation of China (Nos. 51027005 and 51271119), the Shanghai Science & Technology Committee (No. 11JC1405900).

References

[1] Ratke L and Diefenbach S 1995 *Mater. Sci. Eng. R* **15** 263
[2] Schaffer P L, Mathiesen R H and Arnberg L 2009 *Acta Mater.* **57** 2887
[3] Schaffer P L, Mathiesen R H, Arnberg L, Sabatino M D and Snigirev A 2008 *New J. Phys.* **10** 1
[4] Wang W L, Li Z Q and Wei B 2011 *Acta Mater.* **59** 5482
[5] Wang C P, Liu X J, Ohnuma I, Kainuma R and Ishida K 2002 *Science* **297** 990
[6] Shi R P, Wang C P, Wheeler D, Liu X J and Wang Y 2013 *Acta Mater.* **61** 1229

[7] Luo B C, Liu X R and Wei B 2009 *J. Appl. Phys.* **106** 053523
[8] Ratke L 1995 *Mater. Sci. Eng. A* **203** 399
[9] Wu M, Ludwig A and Ratke L 2003 *Metall. Mater. Trans. A* **34** 3009
[10] Tegze G, Pusztai T and Granasy L 2005 *Mater. Sci. Eng. A* **413-414** 418
[11] Wu M H, Ludwig A, and Ratke L 2003 *Simul. Mater. Sci. Eng.* **11** 755
[12] Zhu J, Wang T M, Cao F, Fu H W, Fu Y N, Xie H L and Xiao T Q 2013 *J. Mater. Eng. Perform.* **22** 1319
[13] Chai Z G and Meng C F 1998 *J. Appl. Cryst.* **31** 7
[14] Meng F L, Chai Z G and Wang Y M 2000 *J. Mater. Sci. Lett.* **19** 1837
[15] Wu Z D, Lu X H, Wu Z H, Kui H W 2014 *J. Non-Cryst. Solids* **385** 40
[16] Li F G, Zhang J, Dai Y B, Bian F G, Han Y F, Sun B D 2014 *Mater. Chem. Phys.* **143** 471
[17] 1986 Binary Alloy Phase Diagrams vol 1, ed T B Massalski (Ohio: American Society of Metals)
[18] Dai R, Zhang S G, Li Y B, Guo X and Li J G 2011 *J. Alloys Compd.* **509,** 2289
[19] Zhang X M, Liu J C, Tang J G, Li L, Chen M G, Liu S D, Zhu B 2010 *Appl. Surf. Sci.* **256** 7300
[20] Qin T, Wang H P and Wei B 2007 *Sci. Chin. G* **37** 409
[21] Lu W Q, Zhang S G and Li J G 2013 *Mater. Lett.* **107** 340
[22] Guo J J, Liu Y, Jia J, Su Y Q, Ding H S, Zhao J Z and Xu X 2001 *Scr. Mater.* **45** 1197
[23] Wu H, Chakraborty A, Li BL and Kenerley C M 2009 *Ecol. Complex.* **6** 403
[24] Porod G 1951 *Kolloid-Z* **124** 83
[25] Li Z H 2013 *Chin. Phys. C* **37** 108002
[26] Li Z H, Zhao J P, Wu D, Sun Y H, Wang J, Liu Y, Sheng W J and Dong B Z 2000 *Acta Chim. Sin.* **58** 1147
[27] Hashimoto T, Fujimura M and Kawai H 1980 *Macromolecules* **13** 1237
[28] Zhao J Z, Li H L, Li H Q, Xing C Y, Zhang X F, Wang Q L and He J 2010 *Comput. Mater. Sci.* **49** 121
[29] Ding G H, Zu F Q, Li X F, Xi Y and Shen R R 2006 *Acta Phys. Sin.* **55** 4188
[30] Sonntag R C and Russel W B 1987 *J. Colloid Interface Sci.* **115** 378
[31] Liu J, Shih W Y, Sarykaya M and Aksay I A 1990 *Phys. Rev. A* **41** 3206

Misorientations in spheroidal graphite: some new insights about spheroidal graphite growth in cast irons

J Lacaze[1a], K Theuwissen[1], L Laffont[1] and M Véron[2]

[1] CIRIMAT, Université de Toulouse, ENSIACET, CS 44362, 31030 Toulouse, France
[2] SIMAP, CNRS-Grenoble INP, BP 46, 38402 Saint-Martin d'Hères, France

Email: [a]Jacques.lacaze@ensiacet.fr

Abstract. Local diffraction patterning, orientation mapping and high resolution transmission electron microscopy imaging have been used to characterize misorientations in graphite spheroids of cast irons. Emphasis is put here on bulk graphite, away from the nucleus as well as from the outer surface of the spheroids in order to get information on their growth during solidification. The results show that spheroidal graphite consists in conical sectors made of elementary blocks piled up on each other. These blocks are elongated along the prismatic *a* direction of graphite with the *c* axes roughly parallel to the radius of the spheroids. This implies that the orientation of the blocks rotates around the spheroid centre giving low angle tilting misorientations along tangential direction within each sector. Misorientations between neighbouring sectors are of higher values and their interfaces show rippled layers which are characteristic of defects in graphene. Along a radius of the spheroid, clockwise and anti-clockwise twisting between blocks is observed. These observations help challenging some of the models proposed to explain spheroidal growth in cast ions.

1. Introduction

Graphite in cast iron may be of lamellar or spheroidal form, or of various intermediate or degenerate morphologies. Thanks to modern means of investigation, it appears quite clearly established that all of these forms consist of elementary blocks that are elongated along the prismatic direction of graphite. In this paper, some recent observations are reported and discussed. Use of the classification proposed by Austerman et al. [1] will be made, according to which twist disorder denotes cases where basal planes are rotated while remaining parallel and tilt disorder is for cases when basal planes do not remain parallel to one another.

2. Observations

Micrographs in figure 1 show images of lamellar graphite obtained by transmission electron microscopy (TEM). The bright field image in figure 1a shows at rather low magnification the piling up of four long blocks which are 1-2 µm in thickness. Blocks are generally of high crystallinity while misorientations may show up between blocks as tilts or twists. In some rare cases, high misorientations may be observed between neighboring blocks in lamellar graphite. In such a case, the 0002 lattice fringe image in figure 1b shows that graphite layers (graphene) can bend so as to develop a continuous transition between the two blocks, and this transition takes place on a thickness of about 10 nm.

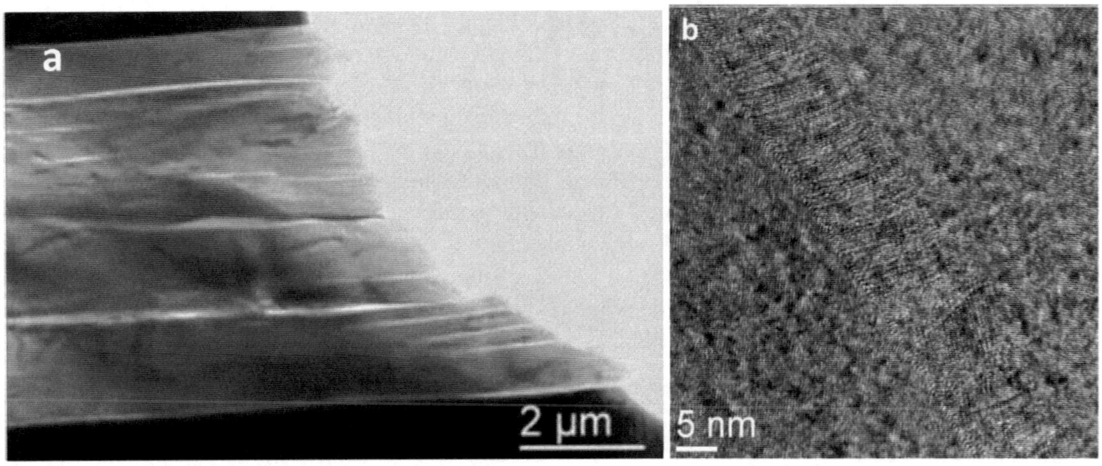

Figure 1. TEM bright field image showing parallel blocks in lamellar graphite (a) and high resolution image of a transition zone between two highly misoriented blocks.

Figure 2 shows TEM micrographs of so-called chunky graphite that is a degenerate form of spheroidal graphite. Figure 2a shows that this form of graphite consists also of clustered blocks. Figure 2b illustrates again the capability of graphite layers to bend, here a 180° turn defining one outer boundary of a block.

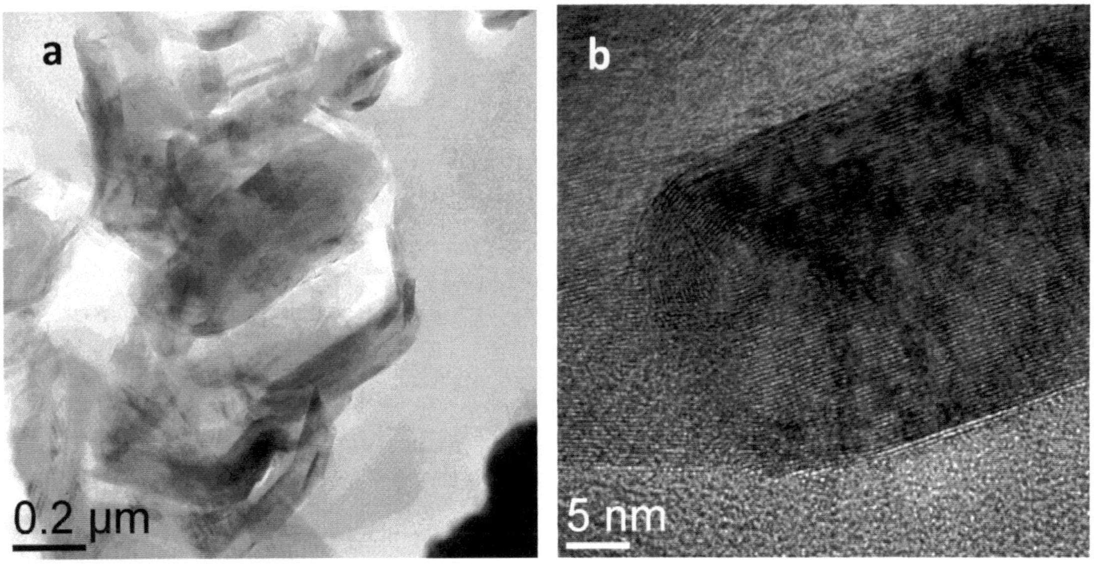

Figure 2. TEM bright field micrograph of a piece of chunky graphite (a) and HRTEM micrograph of a particular graphite block (b).

Spheroidal graphite morphology is characterized by the stacking of the blocks with the graphite *c* direction being radially oriented. This has been known for long thanks to optical observations as well as scanning and transmission electron microscopy. The blocks are generally a few microns in length, i.e. much shorter than in most cases of lamellar graphite, with thicknesses ranging between about 10 and 200 nm. One well known feature obtained long ago from optical microscopy with polarized light is that spheroids consist of sectors in which graphite crystals are similarly oriented, with high misorientations showing up between these sectors. Figure 3a shows the diametrical section of a spheroid that was obtained by focused ion beam (FIB) technique and submitted to local diffraction patterning with the ACOM system in a JEOL 3010 TEM operated at 300 kV. Graphite orientation was determined in points located along a grid with a 17 nm step size. Figure 3b shows the projection of the measured orientations parallel to the Z direction (perpendicular to the image plane) and figure 3c the triangle giving the color code for the orientations. Figure 3b shows first that all (0002) planes of the studied spheroid section are effectively contained in the section plane as only blue and green show up and no red (the small red and pink areas in the upper right of the image are matrix inclusions). Figure 3b further demonstrates that graphite is organized in sectors which appear to be strongly misoriented between each other as the transition between blue and green areas is quite sharp. Both observations are in full agreement with optical observations.

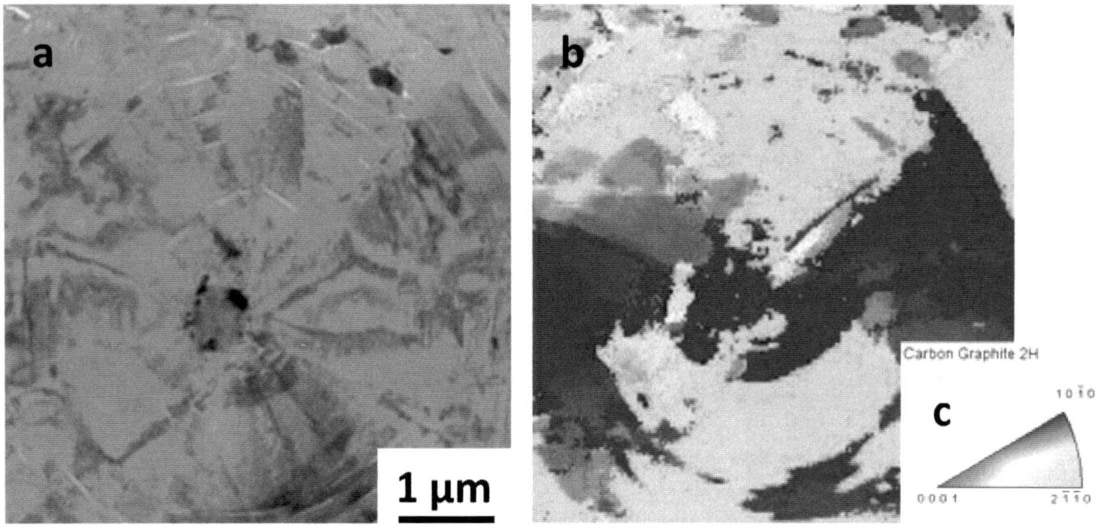

Figure 3. TEM bright field micrograph of a diametrical section of a spheroid (a), orientation map projected along the Z direction (perpendicular to the image) (b) and color code (c).

Figure 4a shows the projection along the Y direction (in plane vertical direction) of graphite orientations from the same mapping as in figure 3, with the color code being the same as in figure 3c. It is clearly seen that the sectors observed in figure 3b are divided in sub-areas in which graphite orientations are similar but twisted between each other. The increase of the spread angle of the sectors leads to some rotation of the graphene layers in the tangential direction. This results in low angle misorientations to progressively produce a conical object (a sector).

If the development of the spheroid was the result of continuous growth of bending layers, the expected orientation map would correspond to the one presented in figure 4b, where continuous bending would lead to continuous colour changes in the orientation map. Such features are only observed within widespread sectors (e.g. lower right part of figure 4a) while high angle misorientations are observed between neighbouring sectors.

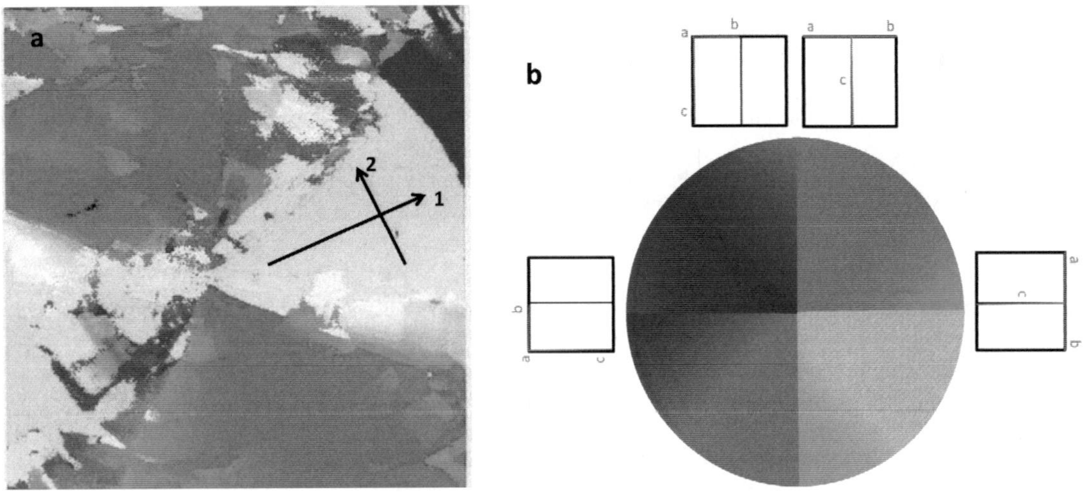

Figure 4. Projection along the Y direction (vertical in-plane direction) (a) and expected orientation map if the rotation of the (0002) axis were continuous (b).

Figure 5a shows the misorientation between two successive points and the cumulative misorientation along line 1 in figure 4a. The orientation change between successive records can go up to 13° but is most often lower than 5° and corresponds to a twist. The fact that the cumulative plot shows both decreases and increases demonstrates that this twist may be in either direction (clockwise or counterclockwise) around the (0002) axis. Figure 5b shows the change in graphite direction along line 2 in figure 4a. Combining the information given by the map in figure 4a with the data in figure 5b, it is clearly seen that within this sector the graphite blocks undergo a continuous twist of their (0002) orientation in the tangential direction. The same observation could be repeated at various positions of the line along the spheroid radius within the same sector, as well as in the sector to the bottom of the spheroid section. In other areas that show many colour changes between blue and green in figure 3b, such line scanning of the orientations showed mainly large misorientations between neighboring sub-sectors.

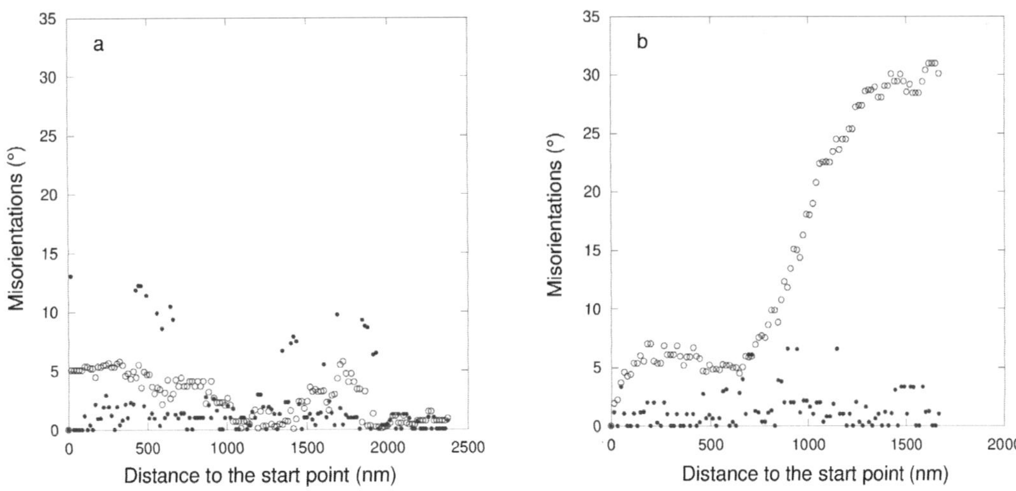

Figure 5. Misorientation between successive points (dots) and cumulative misorientation (open circles) along the lines 1 (a) and 2 (b) in figure 4b respectively.

3. Discussion

There have been several models proposed to explain spheroidal growth of graphite in cast iron, amongst which three are worth of consideration. The first one originates from the work of Hillert and Lindblom [2] and suggests that spheroidal growth proceeds by helicoidal growth around screw dislocations emanating from the center of the spheroid, generated by foreign (Mg, Ce) atoms entrapped in the graphite lattice. If this were so, the misorientations recorded along a spheroid radius should be zero or constant and not changing positively or negatively as seen in figure 5a. The second model due to Sadocha and Gruzleski [3] proposed a continuous growth of a graphite layer around the outer surface of the spheroid. Such a model is in contradiction with the observation of well-defined sectors at the boundary of which large misorientations are observed as illustrated in figure 6. Finally, Double and Hellawell [4] suggested that nucleation of new layers at the surface of the spheroids may be easy, particularly at the edges between sectors. These layers should then extend laterally along the spheroid's surface. It is this model that seems the most probable, and it has been suggested that these are not steps but elementary blocks which nucleate and extend along the spheroid's surface [5]. Simple calculation showed that the critical nucleation radius is of the same order of magnitude as the reported thickness of the elementary blocks, i.e. 10-200 nm [6].

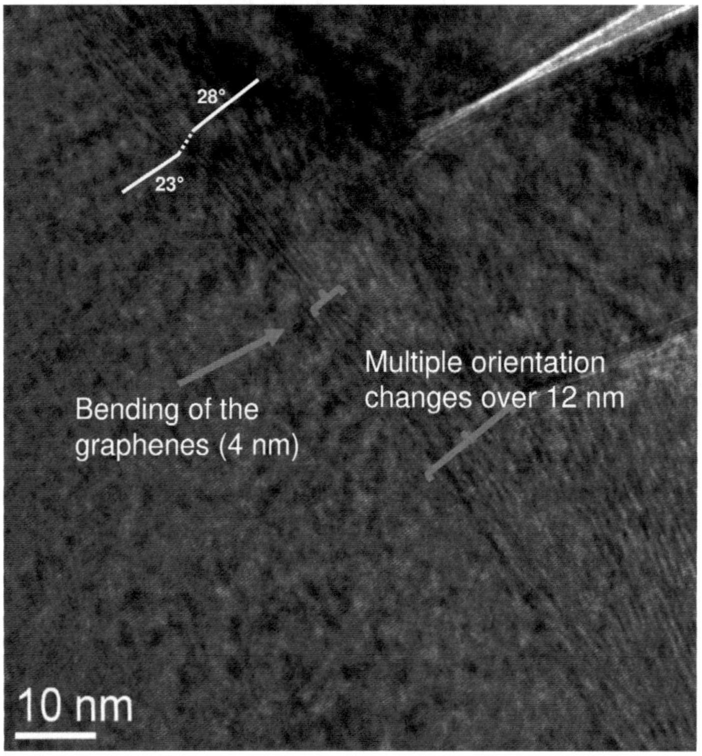

Figure 6. HRTEM micrograph showing bending of the graphite layers at the boundary between two neighboring sectors.

Apart for deoxidizing and desulfuring the melt, the role of spheroidizing elements such as Mg and Ce is certainly to limit the lateral extension of the blocks when extending along the spheroid's surface. Certain authors have detected the presence of adsorbates in graphite and graphene using STM and HR-TEM [7-9]. These were mostly found in domain boundaries and the authors related this to the existence of dangling bonds or exposed edges which favour binding of carbon atoms or foreign element at these locations. Furthermore, recent research on carbon materials has led to the understanding of defects and grain boundaries in graphite and graphene. Grain boundaries in graphite contain a number of defects and the misorientation between the two domains is related to the defect configuration. Moreover, these defects often produce an inflexion of the graphitic layers, i.e. out of plane wrinkles of the layers [10-12] which can be compared to those presented in figure 6. It is interesting to note that according to these authors, the occurrence of such defects does not depend on the misorientation angle but rather on the formation energy of such defects. However, the most energetically favoured angles correspond to misorientations predicted by the coincidence site lattice theory [13] with angles that were commonly observed during the present work and previously reported [6].

4. Conclusion

The results presented in this study bring to light new information on the crystallography of graphite spheroids in cast iron. Graphite spheroids present conical sectors within which growth blocks are twisted along the radial direction and tilted with low angle misorientations along the tangential direction. The high misorientations between neighbouring sectors are accommodated by local bending of the graphite layers through the formation of defects. These certainly play an important role in the growth mechanisms by providing sites for adsorption of foreign elements, as well as favouring nucleation of new blocks. It seems likely that spheroidal growth results from a compromise between nucleation and growth of new elementary blocks at the spheroid's surface.

References

[1] Austerman SB, Myron SM and Wagner JW 1967 *Carbon* **5** 549
[2] Hillert M and Lindblom Y 1954 *J. Iron Steel Inst.* **148** 388
[3] Sadocha JP and Gruzleski JE 1975 Proc. "The metallurgy of cast iron" (Saint-Saphorin: Georgi) p. 443
[4] Double DD and Hellawell A 1995 *Acta Metallurgica et Materialia* **43** 2435
[5] Ghergu R, Magnusson Åberg L and Lacaze J 2014 *Mater. Sci. Forum* **790-791** 435
[6] Theuwissen K, Lafont MC, Laffont L, Viguier B and Lacaze J 2012 *Trans. Indian Inst. Met.* **65** 627
[7] An J, Voelkl E, Suk JW, Li X, Magnuson CW, Fu L, Tienmeijer P, Bischoff M, Freitag B, Popova E and Ruoff RS 2011 *ACS Nano* **5** 2433
[8] Gao L, Guest JR and Guisinger NP 2010 *Nano letters* **10** 3512
[9] Akhukov MA, Fasolino A, Gornostyrev YN and Katsnelson MI 2012 *Physical Review B* **85** 115407 1
[10] Zhang J and Wang C 2014 *Computational materials science* **87** 26
[11] Zhang J, Zhao J and Lu J 2012 *ACS Nano* 7 2704
[12] Amini S and Abbaschian R 2012 *Materials Letters* **88** 129
[13] Minkoff I and Myron S 1969 *Philosophical magazine* **19** 379

Microstructures in a ternary eutectic alloy: devising metrics based on neighbourhood relationships

A Dennstedt[1], A Choudhury[2], L Ratke[3], and B Nestler[4]

[1]Institut für Materialphysik im Weltraum, Deutsches Zentrum für Luft- und Raumfahrt (DLR), 51170 Köln, Germany
[2]Indian Institute of Science, Department of Materials Engineering, 560012 Bangalore, India
[3]Institut für Werkstoffforschung, Deutsches Zentrum für Luft- und Raumfahrt (DLR), 51170 Köln, Germany
[4]Institut für Angewandte Materialien, Karlsruhe Institut für Technologie, 76131 Karlsruhe, Germany

E-mail: [1]anne.dennstedt@gmx.com, [2]abhiknc@gmail.com, [3]lorenz.ratke@dlr.de, [4]britta.nestler@kit.edu

Abstract. Ternary eutectics, where three phases form simultaneously from the melt, present an opportunity to study the fundamental science of microstructural pattern formation during the process of solidification. In this paper we investigate these phenomena, both experimentally and by phase-field simulations. The aim is to develop necessary characterisation tools which can be applied to both experimentally determined and simulated microstructures for a quantitative comparison between simulations and experiments. In SEM images of experimental cross sections of directionally solidified Ag-Al-Cu ternary eutectic alloy at least six different types of microstructures are observed. Corresponding 3D phase-field simulations for different solidification conditions and compositions allow us to span and isolate the material parameters which influence the formation of three-phase patterns.

Both experimental and simulated microstructures were analysed regarding interface lengths, triple points and number of neighbours. As a result of this integrated experimental and computational effort we conclude that neighbourhood relationships as described herein, turn out to be an appropriate basis to characterise order in patterns.

1. Introduction

During solidification of a eutectic melt several solid phases appear at the same time while no melt will be left. In the case of binary eutectics two phases are created. They can be arranged as lamellae or as fibres in a matrix. Which pattern will be created depends on the eutectic composition. In the case of ternary eutectics three phases are created. As a result many more arrangements are possible. Lewis et al. published a sketch showing five types of ternary eutectics [1] that were before verbally listed by Ruggiero and Rutter [2]. However, in reality ternary eutectics are much more complex as shown in this sketch ([3, 4, 5, 6, 7] and refs. therein) and their structures generally fail to be describable by the simple geometrical approaches [8].

The system Ag-Al-Cu provides one ternary eutectic at a composition of 18.1 at% Ag, 69.1 at% Al and 12.8 at% Cu [9]. After solidification of such a melt the phase fractions change due to solid state reactions to almost equal volume fractions of the three phases α-Al, Ag_2Al and Al_2Cu [4]. Here, we describe different microstructures observed in directionally solidified ternary eutectic Ag-Al-Cu alloy and determined neighbourhood relationships both in these experimental images as well as in images of simulated microstructures.

2. Experimental procedure and simulation technique

Ag-Al-Cu alloys of eutectic composition were cast into stainless steel moulds. Subsequently, rods were machined to fit the ARTEMIS facility. The ARTEMIS facility [10] was used to perform directional solidification with velocities of 0.08 to 3.8 μm/s and applied gradients of 1.5 to 6.2 K/mm. Experiments with a single solidification velocity as well as experiments with two or three different solidification velocities were performed in one run with one sample. Cross sections of the directionally solidified rods were prepared using standard metallographic procedures. Scanning electron microscope (SEM) images were collected using the backscattered mode to show microstructures of this ternary eutectic. The contrast between the three phases is excellent: white represents the Ag_2Al phase, black the solid solution α-Al, and gray the Al_2Cu phase. Comparing all images different microstructures are differentiable. Six images showing characteristics of these microstructures were chosen to discuss characteristics of these microstructures in ternary Ag-Al-Cu alloy.

The first phase-field simulations of this alloy were performed by Apel et al. (Chapter 29 in [11]). In the present analysis our aim is to investigate the material parameters and conditions which lead to the selection of the different patterns in a ternary eutectic alloy. Phase-field simulations were performed on the basis of the quantitative phase-field model presented in [12]. However, the thermodynamic description of the bulk free-energy densities is modified wherein information from the CALPHAD databases [9] was incorporated by forming simplisitic parabolic free energies, with local thermodynamic extrapolation of the equilibrium chemical potential and the equilibrium solidus and liquidus compositions along the tie-lines given by the eutectic compositions. The modification ensures an easier and efficient coupling with the databases than what was previously used in the ideal energy solution model in [12]. The entire description ensures that the equilibrium compositions of phases and the slopes of the liquidus are exactly mapped, correspondingly as in the information from the thermodynamics databases, near the ternary eutectic point. Additionally the variation of the composition as a function of the chemical potentials is derived accurately until second order in $\mu - \mu_{eq}$, (μ being the chemical potential and μ_{eq} the equilibrium value) which is a consequence of the actual free-energy densities from the databases being approximated by second-order polynomials. This ensures that the Gibbs-Thomson coefficient which at leading order comprises of the ratio of the surface energies and the second derivatives of the free-energy w.r.t composition is accurately represented. A detailed overview can be found in a recent publication [15].

For the comparison with the present experimental results, wherein the volume fractions differ from those predicted from the CALPHAD databases [9] because of kinetic reasons, we shift the ternary eutectic point corresponding to the volume fractions achieved in experiments, while keeping all the liquidus and solidus slopes as derived from the thermodynamic databases. This is essential since volume fractions are a key parameter influencing pattern formation. The simulated microstructures are thereafter analysed in a comparable way like the experimental microstructures.

3. Results and discussion

Comparing all cross sections at least six different types of microstructures were observed. Figures 1a to 1f show examples for these structures.

An aligned structure (figure 1a) was detected that is called ladder or brick-like structure and which is expected for slowly solidified ternary Ag-Al-Cu alloy [3, 4, 13]. Therein two intermetallic phases Ag_2Al and Al_2Cu alternate in a row and thus one could imagine them as one lamella consisting of two sub-lamellae. In between the α-Al phase separates as parallel lamellae the

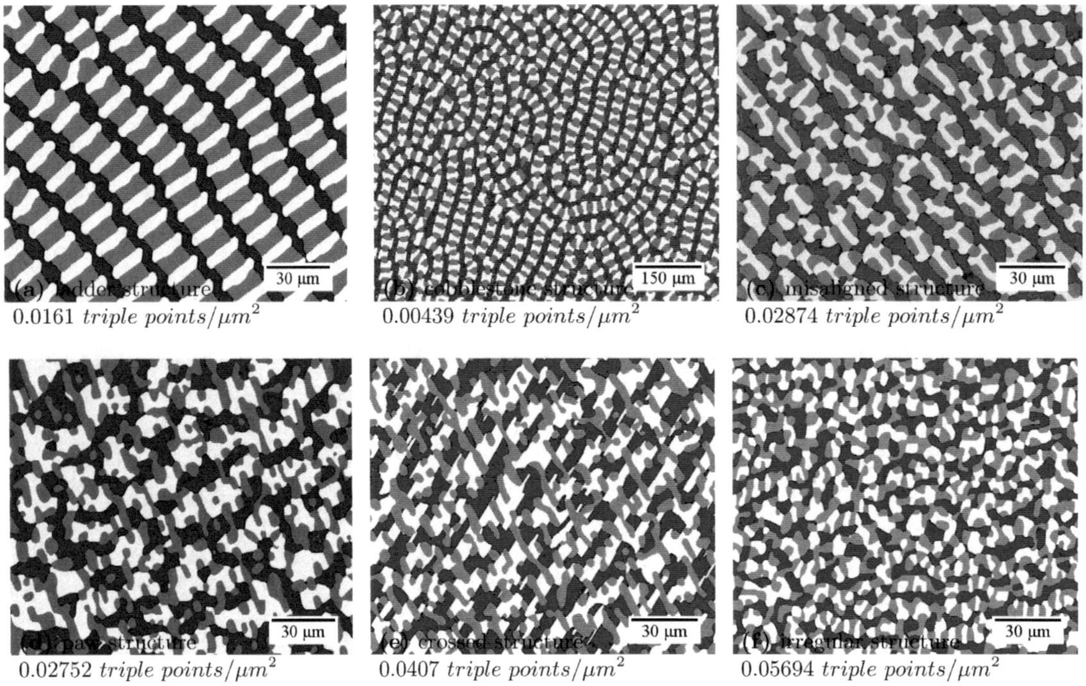

Figure 1. SEM images of Ag-Al-Cu alloy solidified with (a) 0.39 μm/s and 6.2 K/mm; (b) 0.08 μm/s and 2.2 K/mm; (c) 0.72 μm/s and 3.3 K/mm; and (d) to (f) 1.9 μm/s and 3.2 K/mm. White = Ag_2Al, gray = Al_2Cu, black = Al.

intermetallic super-lamellae. In contrast thereto a structure was observed showing bended α-Al lamellae or particles (figure 1b). The intermetallic superlamella still shows alternating Ag_2Al and Al_2Cu phases. We called this structure a cobblestone (pavement) structure. Alternatively, in a third microstructure (figure 1c), the α-Al lamellae still develop elongated particles which are arranged in parallel and can be identified as lamellae. However, the intermetallic lamellae are internally unaligned. On the basis of the brick-like structure we called this microstructure a misaligned structure. The microstructures shown in figure 1d to 1f were detected in a single cross section in different regions, and corresponding regions were also detected in cross sections later in the solidification. The reasons for these different microstructures are until now not known. In the microstructure in figure1d the Al_2Cu particles can be thought as threaded phase particles. The Ag_2Al phase is arranged in between the Al_2Cu particles in such a way that the α-Al phase still is in lamellae and thus the different Ag_2Al shapes sometimes mimic an H or a doubled H. If there are many more than four neighboured Al_2Cu particles attached to a single Ag_2Al particle the shape of this arrangement resembles an animal climbing on a tree. Thus, this microstructure is called a paw structure. Figure 1e shows not only elongated Al_2Cu particles but also elongated Ag_2Al particles. These elongations are oriented in different directions, and we called the microstructure a crossed structure. After all, a further microstructure was observed where all three phases are placed in an unaligned and irregular way (figure 1f). The intermetallic phases Ag_2Al and Al_2Cu are irregularly connected. Thus a higher number of intermetallic neighbours is possible and the particles of the α-Al phase are smaller and curved in comparison to the more ordered structures.

We analysed each image concerning triple point density, number of particles per image area

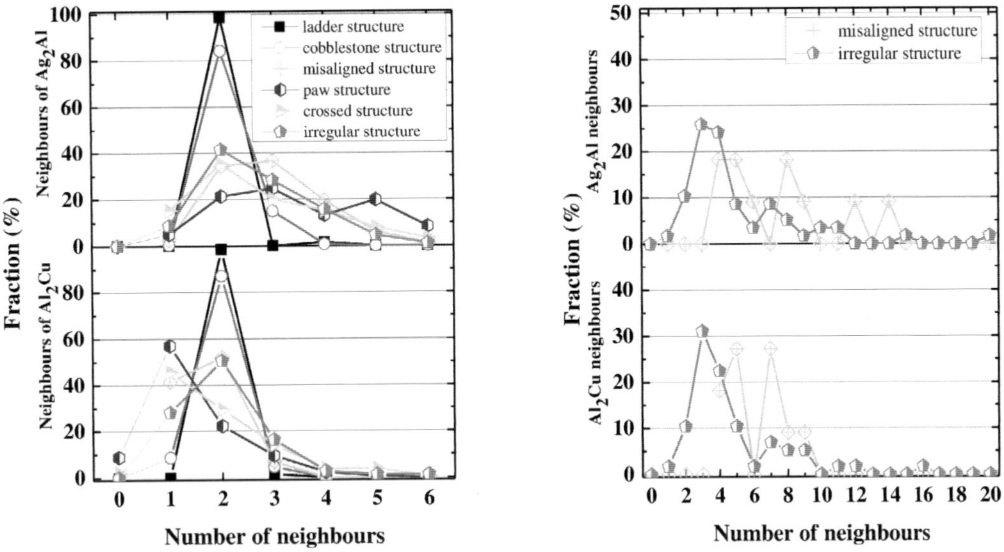

(a) Around intermetallic particles (b) Around Al particles

Figure 2. Results from experimental microstructures: fraction of particles with (a) up to 6 intermetallic neighbours around the intermetallic particles, and (b) up to 20 neighbours around the Al particles in misaligned and irregular structure.

as well as size and perimeter of the particles. These parameters strongly dependent on the solidification conditions. As known already [14], larger particles are observed for solidifications with lower solidification velocities. This means that not only size but also perimeter of the particles increases with decreasing solidification velocity as well as the number of particles and triple point density decrease. Because microstructures shown here were not obtained with same solidification velocity a comparison of all microstructures using these four parameters is not useful. However, triple point densities determined on the images are written below the single images 1a to 1f. Different values are observed for paw, crossed and irregular structure even so they were formed under the same solidification conditions. Thus, the triple point density depends not only on solidification parameters but also on the microstructure created.

A parameter not depending on solidification velocity is the number of neighbours of the intermetallic particles [4]. However, we detect clear differences between the different microstructures. Figure 2a shows in the upper part how many of the Ag_2Al particles feature 0, 1, 2 or more Al_2Cu neighbours. In the lower part of this figure the Al_2Cu particles were analysed. In the ladder structure nearly all intermetallic particles have two intermetallic neighbours (figure 2a). In the cobblestone structure the fraction of intermetallic particles with two neighbours decreased. The other microstructures show an even further decreased fraction of intermetallic particles with two neighbours. As a consequence of this more Ag_2Al particles show more than 2 neighbours whereas more Al_2Cu particles show less than 2 neighbours.

In figure 3 the fraction of Ag_2Al particles with two Al_2Cu neighbours is plotted against the fraction of Al_2Cu particles with two Ag_2Al neighbours. The fraction of Ag_2Al particles is mostly similar to the fraction of Al_2Cu particles thus all points are plotted near a line x=y. The most ordered structure is the ladder structure and it's point is plotted most far away from the point of

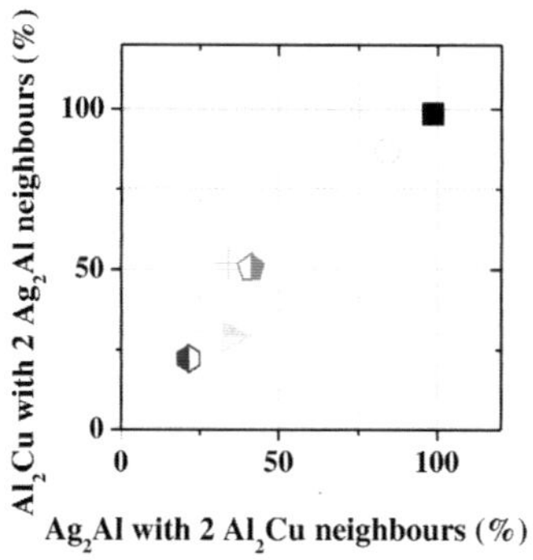

Figure 3. A fraction-fraction plot for the experimental microstructures.

origin. The second in this row is the cobblestone structure showing fractions of more than 80% with two neighbours. The points for misaligned and irregular structure (figure 1c and 1f) are plotted near each other, but a comparison of the images shows that these two microstructures are not the same. The distributions of intermetallic neighbours around the intermetallic particles are very similar (figure 2a). However, the distributions of intermetallic neighbours around the α-Al particles differ strongly (fig.2b). In the case of the misaligned structure the Al particles feature more neighbours compared to the α-Al particles in the irregular structure.

Figures 4 shows results of a 3D phase-field simulation resulting in a microstructure similar to a brick-like structure. The figures from left to right highlight the various stages during the early stage of solidification until a uniform microstructure is achieved.

We tiled the simulated images to analyse also the larger particles touching the image borders. Similar to neighbour relationships in the experimental microstructures the fraction of Ag_2Al particles as well as of Al_2Cu particles with two intermetallic neighbours is changing, resulting to high fractions in the case of more order (figure 5). The analysis tools developed enable the quantification of the increase in order of the simulated microstructure with increase in solidification time. Additionally, the apparent similarity between the ladder-structure and the simulation images is adequately captured by the analysis tools. While there are certain small differences between the neighbour fractions of 1 and 3 neighbours between the simulations and the experiments, it is possible that with increase in the simulation box-size, that an improved statistics are possible with reduced differences.

While, we do not present a more thorough analysis into the actual mechanisms of pattern selection and the influence of the several material and process parameters since it is outside the scope of the present paper, we will be presenting a more detailed analysis in a forthcoming publication [16], where we characterize the influence of solid-liquid interfacial energies and volume fractions of the solids on pattern formation during three-phase growth.

4. Conclusions

In this paper 6 different microstructres observed in directionally solidified ternary eutectic Ag-Al-Cu alloy experiments are described. We describe neighbourhood relationships that allow us to characterise order and thereby quantitatively distinguish each of the observed microstructures. The developed analysis tools are generally applicable even for characterizing simulated microstructures. Phase-field simulations are utilised to understand early stage pattern formation resulting in the development of one of the experimental microstructures which is described as the ladder-structure in the paper. Quantitative investigation with the help of analysis tools show the increasing order in the simulated microstructure with increasing

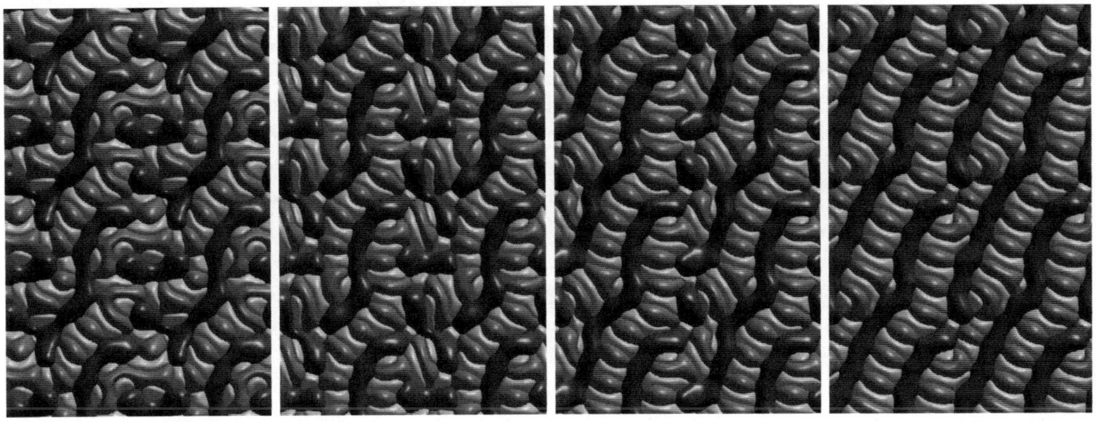

Figure 4. Simulation images for ternary eutectic alloy showing microstructure at different time steps (2*3 tiles)(increasing time, from left to right). Green = Ag_2Al, red = Al_2Cu, blue = Al.

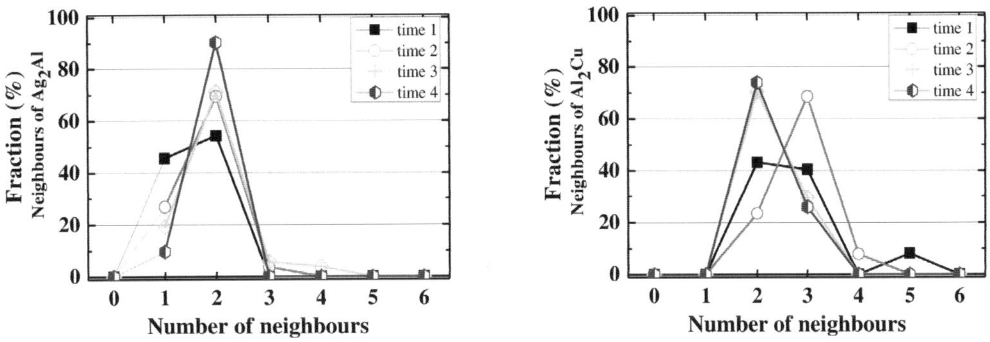

Figure 5. Results from simulated microstructures: fraction of particles with up to 6 intermetallic neighbours around the intermetallic particles.

solidification time, and additionally confirms the resemblance between the simulated and experimental microstructure to a good deal of accuracy.

Acknowledgment

The authors gratefully acknowledge the financial support by the Deutsche Forschungsgemeinschaft under contract numbers RA537/14-2 and NE822/14-2.

References

[1] Lewis D, Allen S, Notis M, Scotch A 2002 J. Electron. Mater. **31** 161
[2] Ruggiero M A, Rutter J W 1997 Mater. Sci. Technol. **13** 5
[3] Genau A and Ratke L 2011 IOP conf. Ser.: Mater. Sci. Eng **27** 012032
[4] Genau A and Ratke L 2012 Int. J. Mat. Res. **103** 469
[5] Dennstedt A and Ratke L 2012 Trans Indian Inst Met **65** 777
[6] Dennstedt A, Ratke L 2013 Metallogr. Microstruct. Anal. **2** 140
[7] Hawksworth A, Rainforth W M, Jones H 1998 Scr. Mater. **39** 1371
[8] Himemiya T and Umeda T 1999 Mater. Trans. JIM **40** 665
[9] Witusiewicz V T, Hecht U, Fries S G and Rex S 2005 J. Alloys Compd. **387** 217
[10] Alkemper J, Sous S, Stöcker C, Ratke L 1998 J Cryst. Growth **191** 252

[11] Apel M, Böttger B, Witusiewicz V T, Steinbach I 2004 *Solidification and Crystallization, WILEY-VCH Verlag GmbH & Co. KGaA, Weinheim ISBN: 3-527-31011-8*
[12] Choudhury A, Nestler B Phys. Rev. E **85** 021602
[13] McCartney D G, Hunt J D, Jordan R M 1980 Metall. Mater. Trans. A **11** 1243
[14] Kurz W, Fisher D J 1989 *Fundamentals of Solidification* (Trans Tech Publications Ltd, Switzerland)
[15] Choudhury A, Kellner M, Nestler B Current Opinion in Solid-State and Materials Science *Accepted article*
[16] Choudhury A, Dennstedt A, Ratke L, Nestler B, Physical Review E *Article in review*

Use of a phenomenological chemical scale for the identification of high distribution coefficient impurities within the ITS-90

D H Lowe
National Physical Laboratory, Teddington, UK
E-mail: dave.lowe@npl.co.uk

Abstract. The International Temperature Scale of 1990 standardises practical temperature measurement based on the reproducibility of the solidification phase transition temperature of highly pure metals. The effect of impurities can be the biggest block to establishing international equivalence of temperature metrology and so it is important to have knowledge of the distribution coefficients for all potential impurities. In particular impurities with a high distribution coefficient cause problems with applying corrections, and so it is helpful to identify which impurities are likely to be an issue. By plotting the published measured distribution coefficients of impurities in the metals used as references in the International Temperature Scale of 1990, as a function of a quantum mechanical based scale intended to provide separation of binary alloy crystal structure, an apparent resonance peak is found. This could allow a simple fit equation to be applied to determine whether a given impurity is likely to cause an error to any correction strategy.

1. Introduction

The effect of impurities on the measured freezing temperature of temperature fixed-points is often the single largest component in the uncertainty of realisation of the International Temperature Scale of 1990 (ITS-90) [1] and can cause problems when it comes to confirming international equivalence of scales [2]. The solidification behaviour of Ga, In, Sn, Zn, Al, Ag, Au and Cu is of interest. The recommended method [1] to assess the effect of impurities on the measured solid-liquid phase transition of these metals requires a full chemical analysis and knowledge of the liquidus slope, m, and distribution coefficient, k, for each element. The liquidus slope and distribution coefficient are related to each other by [3]

$$m = (k - 1)/A \quad (1)$$

where $A = L/(RT^2)$ is the first cryoscopic constant, with L the latent heat of fusion of the solvent (kJmol^{-1}), T its freezing temperature in kelvin and R is the gas constant (kJmol). Since A is known for the metals listed, knowledge of k is sufficient to assess the effect of an impurity. At the uncertainty levels required the detection limits of the analysis of the impurities have to be considered and so <u>all elements</u> need to be characterised, not just those detected as being present. Ideally then, until there is reliable knowledge of the distribution coefficients for all the binary systems involving ITS-90 fixed points some predictor of k is desirable [4].

In fact, if it can be established that impurities with k greater than some threshold are not present in significant quantities then corrections based on analysis of the freezing curve shape may be possible. If the threshold is $k \approx 0.1$ a single parameter (the unknown concentration, c) fit to the freezing curve (temperature against fraction melted) gives the reduction in temperature from the pure system [1] as

$$\Delta T = \frac{c}{FA} \tag{2}$$

where F is the solid fraction. If the threshold is at $k \approx 1.3$ then it is possible to correct with an uncertainty of about 30% of the correction by fitting the Scheil-Gulliver model to the freezing curve [5].

It can therefore be useful just to have an indication that a metal fixed-point cell is likely to show an erroneous freezing temperature if a particular impurity could be present in substantial amounts (and by substantial, meaning at the level of 1 ppm). When trying to predict k it has been known for some time that there is a periodicity with atomic number [6]. Recently a review of all available published k values looked for patterns in k based on the Hume-Rothery rules [4]. However, there was no clear correlation between either electronegativity, atomic size or valency with k. To see if improvement could be made to predicting which impurities would have high k the experimental data given in that publication were analysed with respect to an arbitrary chemical scale rather than atomic number or any other single atomic property.

2. Chemical Scale, χ

Structure maps were introduced to classify and aid the identification of new binary compounds with specific crystal structure type and perhaps with particular desired property [7]. Their use involves defining a suitable co-ordinate system on two perpendicular axes. Often these are based on some aspect of the Hume-Rothery rules for solid solutions. Recognising that the choice of axes could be arbitrary, in [8][9] the periodic table was reduced to a continuous one dimensional array with the Mendeleev number M giving the position in the array and a value χ assigned to each M. The reasoning behind the concept was fully explained in [10]. As it is not a well known concept (at least within temperature metrology), some of the reasoning of that publication is paraphrased here:

In order to construct a structure map one or more monotonic scales are needed. The Hume-Rothery rules often used for this purpose are classical in nature. Classical concepts are inadequate to describe the behaviour of valence electrons and the bonding between atoms. The periodic table, though, does reflect the fundamental quantum properties of the valence electrons. However its two dimensional nature means it is not suitable to be used in structure maps. To this end the two dimensional arrangement is reduced to one dimension, in effect by running a string through each element and then "pulling" the table into a chain. The elements are then numbered by their position on this chain, the "Mendeleyev number", M. Since the way different atoms chemically combine into different structure types depends on the strength and direction of the bonds, which depend on a fundamentally quantum property – the shape of the electron orbitals – a "correct" string (more useful string) will have patterns and trends in a structure map of crystal structure, just as the periodic table has patterns and trends in chemical properties. It is then a matter of arranging the stringing of the periodic table such that it groups binary compounds according to their structure type [10]. Subsequently this scale was refined by assigning a monotonically increasing value to each M such that the separation was further improved, while requiring that there was no overlap between groups. It is this scale, χ that is used here.

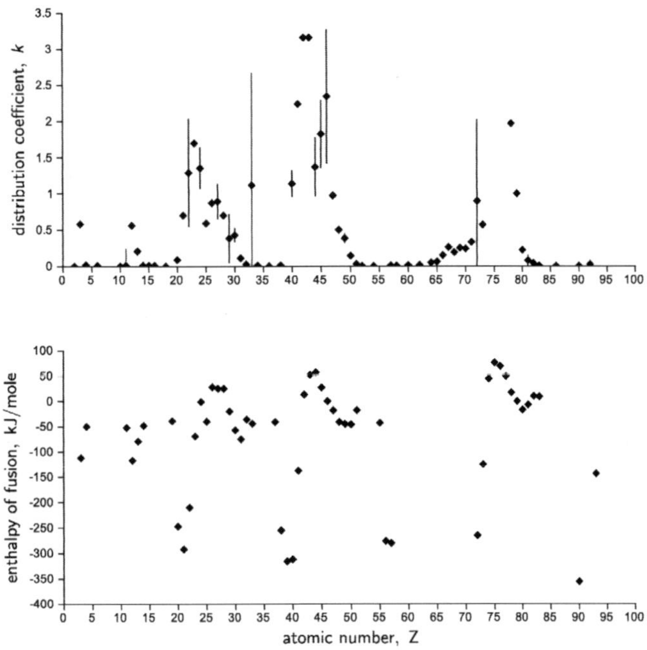

Figure 1. Distribution coefficient k and calculated heat of fusion ΔH_{sol} plotted as a function of solute atomic number for gold as a solvent

3. Results

3.1. Comparison of chemical scale to atomic number

How an impurity behaves in a solidifying melt is closely tied up with compound formation. This in turn is related to the heat of formation of the two atomic species, with bare patches in structure maps of stable compounds corresponding to regions of positive heat of formation [9]. If ΔH is negative a compound can form, if positive it is preferable for the impurity to be dissolved as an alloy with no particular crystal structure. The heat of formation of an impurity in a metal can be estimated using the heat of solution (ΔH_{sol}) of alloys in the liquid phase, which can in turn be approximately calculated using known quantities (electronic work function, bulk modulus and molar volume) [11]. In figure 1 the measured values for k in gold taken from [4] and calculated ΔH_{sol} taken from [11] are plotted against atomic number Z. The error bars (both here and elsewhere) are the standard deviation of reported measured k values, not the actual measurement uncertainties. The first thing to note is that where more than one measurement is reported there is often a big difference in the value for k. From this figure it appears there is some correlation between the two properties, and that there is a periodicity related to the atomic number.

In figure 2 the same quantities are plotted against chemical scale χ. The measured k and calculated ΔH_{sol} values are now more ordered. Also shown are the magnitude (3) and phase (4) of a resonance based on the Lorentz equations for classical bound electrons [12] using a least squares fit of (3) to k. This was then used to calculate a phase using (4), visually normalised to H. There is a discrepancy between the central value when comparing k to H, but it should be noted that the former is measured while the latter is a calculated approximation. It was found that the other ITS-90 fixed point materials had similar behaviour for heat of solution, and so

(3) was fitted to the measured k of the other ITS-90 metals

$$k(\chi) = \frac{\chi}{C_1(\chi_0^2 - \chi^2)^2 + C_2\chi^2} \tag{3}$$

$$\Delta H(\chi) = 1 + \frac{C_3(\chi_0^2 - \chi^2)}{C_1(\chi_0^2 - \chi^2)^2 + C_2\chi^2} \tag{4}$$

No sensible results were had for In 3(f) or Sn 3(g). In the other cases there are more or less well defined peaks at around $\chi = 1$.

4. Discussion

Like the periodic table, the chemical scale is phenomenological in its construction. It turns the two dimensional periodic table into a one dimensional scale that groups together binary compounds depending on their particular crystallographic type when used as the basis of a structure map. However this mapping of different crystal structures into different domains is reflecting the underlying quantum nature of inter-atomic bonds. The χ-scale is in this sense more fundamental than the widely used Hume-Rothery rules which use classical concepts such as the size of an atom [13]. Many of the ITS-90 relevant metals show a peak in k when plotted as a function of χ. It appears that as χ increases the likelihood of compounds forming switches from high to low (with H_{sol} behaving as the phase), and at this switchover k increases rapidly - the likelihood of an impurity being preferentially taken up into the solid phase increases. This increase in k only occurs over a fairly narrow range of χ. In the application of a resonance function to fit the peak in k there is the suggestion that the solid is behaving as the Einstein model with a collective single value oscillation into which suitable χ valued impurities are readily absorbed. Outside this value the impurity is likely to be insoluble ($k \approx 0$) or to form inclusions of

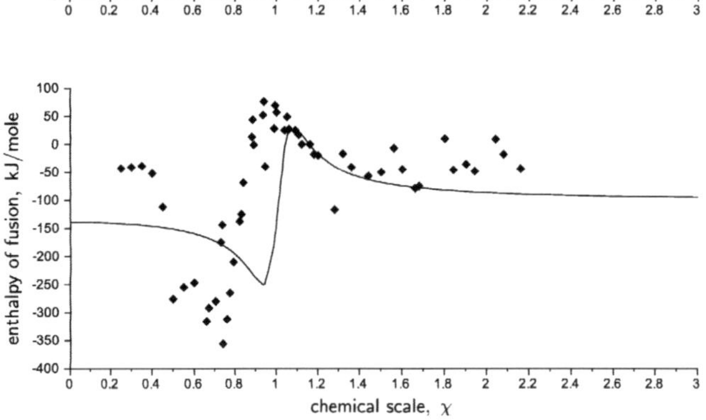

Figure 2. Slice through the structure map for k and H at gold, $\chi = 1.16$ together with fits based on a central value χ_0 derived from equation 3

compounds as clusters of atoms with a definite crystal structure and not affecting the solid-liquid interface temperature.

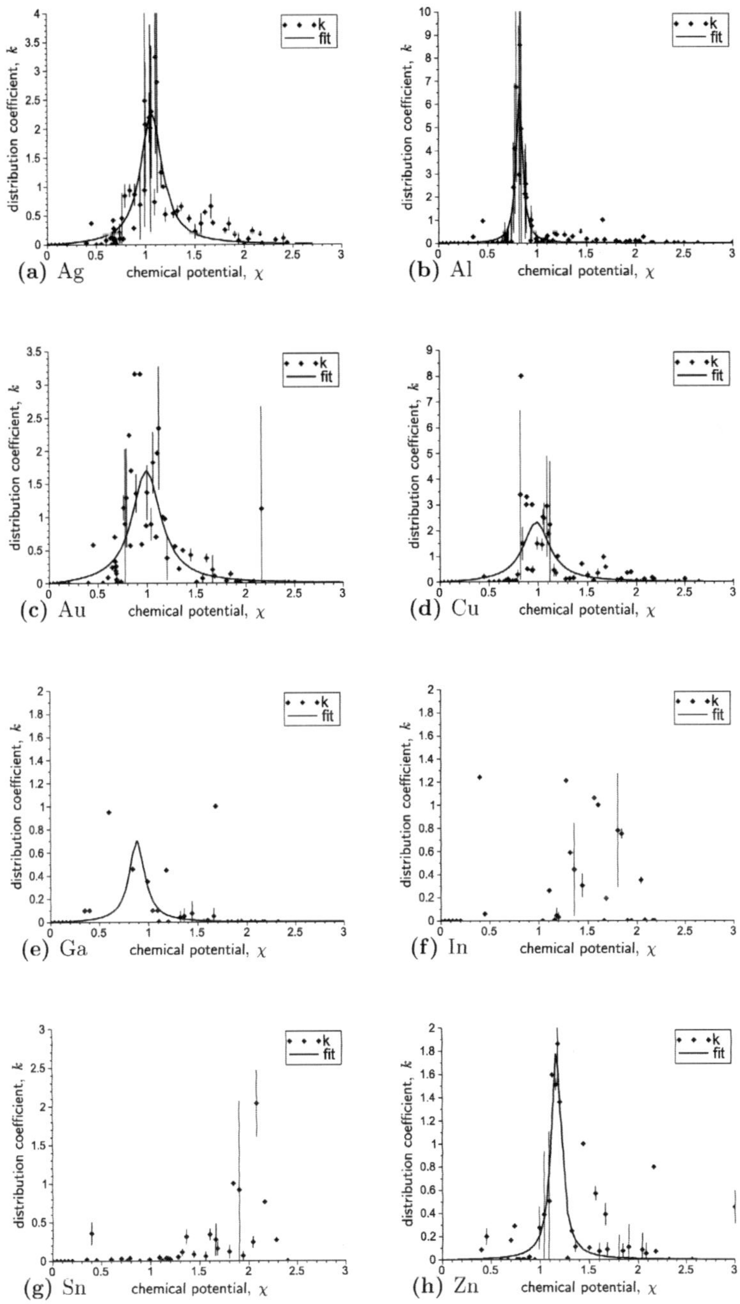

Figure 3. Least squares fit to the k value of impurities in metals used as ITS-90 fixed points. No fit could be obtained for In and Sn. Error bars are the standard deviation of different measurements where more than one published value is available.

5. Conclusion

It is apparent that there is a periodicity in k relating to atomic number (figure 1). It is known that this periodicity can be related to the periodic table [6]. Applying the classical Hume-Rothery rules fails to improve the predictive possibilities of the periodicity [4]. Use of the quantum mechanical chemical scale, χ, which is connected to the variation in bonding due to the arrangement of an elements valence electron orbitals leads to a single peak for high k at $\chi \approx 0$. An assumed association between compound formation and solid solubility suggested considering heat of fusion of elements in the liquid state as something that may be related but is readily calculable. The results look like a resonance in χ with k as the amplitude and ΔH_{sol} as the phase. The impurities where $\Delta H_{sol} < 0$ should be considered as likely to form compounds and have no effect on the measured freezing temperature. There is a potential benefit to being able to predict the magnitude of k, first to assess unknown binary systems and secondly to give some element of corroboration to data that might have widely varying published values (consider the error bars in figures 2 and 3, which are often as large as the determined values). Using the chemical scale χ to order k in place of Z gives clearer separation of high and low k impurities. The patterns suggest only elements with a χ between 0.8 and 1.2 need to be considered when deciding if $k \approx 0$ can reasonably be assumed so as to allow a correction based on freezing curve shape. This corresponds to transition metals from about group 5 to group 11. The application of the chemical scale χ might therefore aid in getting better values for k and the effect of impurities and so help demonstrate international equivalence between temperature scales.

Acknowledgements

Thanks to Professor Peter Lee of Manchester University who suggested applying the chemical scale.

References

[1] "Methodologies for the estimation of uncertainties and the correction of fixed-point temperatures attributable to the influence of chemical impurities" Fellmuth B., Hill K., Bloembergen P., de Groot M., Hermier Y., Matveyev M., Pokhodun A., Ripple D. and Steur P. CCT/05-08 `http://www.bipm.org/cc/CCT/Allowed/23/CCT_05_08_rev.pdf`
[2] Renaot E., Valin M. H. and Elgourdou M. Int. J. Thermophys **29** 852–860 (2008)
[3] Fellmuth B. and Hill K. Metrologia **43** 1 (2006)
[4] Pearce. J. V. Int. J. Thermophys DOI 10.1007/s10765-014-1585-5
[5] Malik Z., Hunt J. D., Davies H., Lee P. D., Lowe D. and Quested P. N. Int. J. Thermophys **32** 1589–1601 (2011)
[6] Drapala J., Lunacek J., Kuchar L. and Kuchar L. Jr. Materials Science and Engineering, A173 (1993) 73-77
[7] Pettifor D. G., Physica B **149** (1988) 3–10
[8] Pettifor D. G., Solid State Communications Vol 51 pp 31–34 (1984)
[9] Pettifor D. G., J. Phys. C: Solid State Phys. 19 285 (1986)
[10] Pettifor D. G., New Scientist 110 48 (1986)
[11] Miedema A. R., de Boer F. R. and Boom R. CALPHAD **1** (1977) 341–359
[12] Electronic properties of materials p157–160 Hummel R. E. Springer-Verlag (1985)
[13] Interatomic forces in condensed matter Finnis M. Oxford University Press (2003) ISBN:9780198509776

IOP Conf. Series: Materials Science and Engineering **117** (2016) 012027 doi:10.1088/1757-899X/117/1/012027

Effects of TiB_2 particles and Ag on the activation energy of Ω phase in Al alloys

F Melotti[1,a], T Hirst[2,b], A Dustan[3,c] and W D Griffiths[1,d]
[1]University of Birmingham, Edgbaston, Birmingham, B152SE, UK
[2]Controls and Data Services, Shaftmoor Lane, Hall Green, Birmingham, B28 8SW, UK
[3]Aeromet International PLC, Cosgrove Close, Worcester, WR3 8WA, UK

Email: [a]FXM295@bham.ac.uk, [b]terry.hirst@aeroenginecontrols.com, [c]alan.dustan@aeromet.co.uk, [d]w.d.griffiths@bham.ac.uk

Abstract. This work analyses the effects of TiB_2 reinforcement particles and silver additions on the activation energy of the GP zones and the Ω phase in an aluminium matrix composite (AMC). Several additions of TiB_2 and Ag were made to a 2xxx series alloy. Differential scanning calorimetry (DSC) was used to identify the temperature peaks and the Kissinger approach used to calculate activation energies. Results showed that the activation energy for the Ω phase was greatly reduced by the addition of both elements; however, the TiB_2 particles were more effective. In addition, experimental results show that the lowest value of the activation energy, 68 $kJmol^{-1}$, was not affected by the silver content. This value is lower than any value found in literature and suggests that the formation of the Ω phase is related to the pipe diffusion of Cu in Al.

1. Introduction

Metal matrix composites (MMCs) are used in aerospace industries to satisfy the demand for a light weight, high performance material. Al-based MMCs play a fundamental role with their great specific strength and improved wear resistance. If a heat treatable aluminium alloy is used as a matrix, the total strength of the aluminium matrix composite (AMC) is the combination of all the strengthening mechanisms, including the ceramic particles and the precipitates.

One of the main effects caused by the addition of ceramic particles is an increase in hardness and a shortening in time to reach the peak-aged condition. However, in a 2xxx series alloy, the addition of silver has shown extremely positive results improving the hardness and reducing the time to peak aging [1, 2]. The reason for this behaviour is the increased number of sites for the heterogeneous nucleation of the Ω phase, which is the main strengthening phase in the alloy [3].

Differential scanning calorimetry (DSC) provides information on a thermally active phase, such as a precipitate, providing a means of analysis of the precipitates present in the alloy, and their nucleation energy and temperature, for a macroscopic sample.

Several reports of the activation energy for the formation of the Ω phase in a non-reinforced Al alloy have been made and most of these experiments found it to be around 150 $kJmol^{-1}$ [4, 5]. This value indicates that the process for the formation of the Ω phase is controlled by the diffusion of Cu in Al (E_d ~ 140 $kJmol^{-1}$). However, no studies have been carried out on understanding the effects of ceramic particles (specifically TiB_2) on the formation activation energy of the Ω phase.

The major drawback of using DSC is the identification of the precipitates, and in particular being able to link a nucleation or dissolution event to its corresponding phase. For this reason, DSC is used in conjunction with microscopy to correctly identify the precipitates. However, for the 2xxx series, an extensive literature exists on the identification and correlation between DSC peaks and precipitates.

The present paper describes the use of DSC to study and understand the effects of TiB_2 particles and Ag content on the activation energy of the GP zones and the Ω phase in an Al-Cu alloy. In particular, the Kissinger-Akahira-Sunose method [6-8], which is known as the generalized Kissinger approach, will be used to study the effects of these two additions on the activation energy.

2. Experimental procedure

Five different alloys were produced by varying the composition of the A205 alloy. This alloy, denoted alloy A, is an Al-Cu-Mg with additions of TiB_2 ceramic particles and silver. Alloys B and C were cast with varying amounts of Ag, resulting in an amount of Ag of 0.5 and 1.5, with respect to that of alloy A. Alloy D was prepared by diluting the amount of TiB_2 in alloy to half that in alloy A. Alloy E was a TiB_2 free alloy with half the amount of Ag of the initial alloy A. The composition variations of the five alloys were analysed by XRF and are summarized in table 1.

The alloys were melted in an induction furnace and poured into sand moulds to obtain test bars of 10 mm diameter and 100 mm length. All alloys used in this work are cast alloys. A solution heat treatment was carried out, and then the alloys were quenched into cold water to achieve a super saturated solid solution. In order to avoid natural aging, samples were stored in a refrigerator until used.

The samples used were machined to 5 mm diameter and 0.5 mm height, with masses of around 28 mg, and then solution heat treated. A Perkin-Elmer DSC 7 was used to measure over a temperature range from 350 K to 650 K, with different scanning rates ranging from 10 $Kmin^{-1}$, 20 $Kmin^{-1}$ and 40 $Kmin^{-1}$.

Table 1. Chemical compositions, given as ratios, of an initial Al-Cu alloy, to which were added varying Ag and TiB_2 additions.

Alloy	Ag	TiB_2	Al
A	1	1	rem
B	0.5	1	rem
C	1.5	1	rem
D	1	0.5	rem
E	0.5	0	rem

All DSC samples were run twice; the first run activated the precipitates and the second run was recorded as a baseline. The baseline was then subtracted from the first signal. This experimental procedure was carried out in order to enhance the signal from the DSC and eliminate instrument noise. The Kissinger approach was then used to calculate the activation energy for the formation of the two phases present in the alloy. This approach is based on the theory that the temperature at which the given peak occurs is shifted with increasing scan rate.

3. Results

In a DSC plot of an Al-Cu-Mg alloy, after solution heat treatment, two endothermic and two exothermic peaks were found. These peaks were identified as related to the nucleation and dissolution of the GP zones and the Ω phase, as suggested in the literature [9]. Because each event is connected, the endothermic and exothermic peaks are adjacent to each other, meaning that a phase is first nucleated and then dissolved.

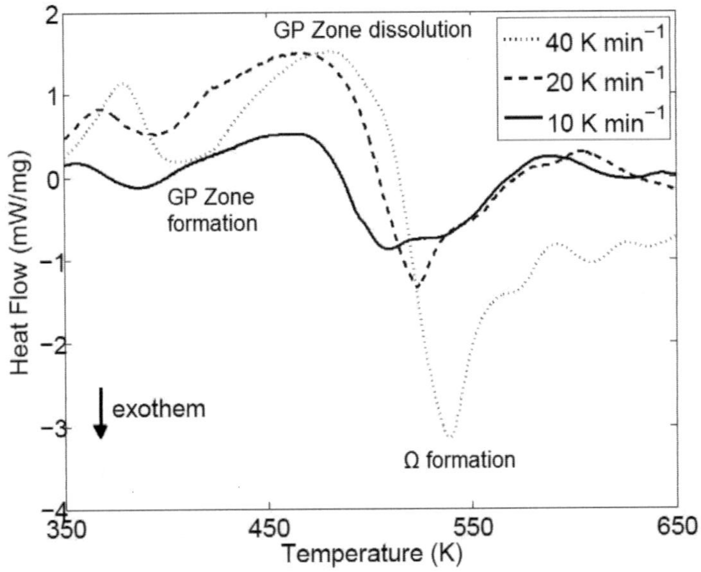

Figure 1. Plot of the heat flow versus temperature for alloy D. Scanning rates ranging from 10 Kmin^{-1} to 40 Kmin^{-1}.

Figure 1 shows DSC curves for alloy D after the solution heat treatment process for three different scanning rates. In an Al-Cu-Mg alloy, the GP zones are the first precipitate to form, at around 370 K. The dissolution of this phase starts at about 430 K reaching a peak at around 470 K. This peak is followed by an exothermic peak at around 540 K, which is correlated to the formation of the Ω phase. The Ω phase is the main strengthening mechanism in 2xxx series alloys, and it nucleates from an Ag-Mg cluster. The Ω phase contributes to the strength of the 2xxx series alloy because it nucleates on the {1,1,1} planes which are the slip plane of the aluminium matrix.

The dissolution of the Ω phase and the nucleation of the θ^{i} phase are not clearly distinguishable in the plot, because they partially overlap [10]. The S phase was not observed in this alloy due to the high Cu:Mg ratio.

Activation energies are derived through the Kissinger methods which correlate the scanning rate (ϕ) to the peak temperature (T_p) of the process. Equation (1) shows the relationship between these variables. The activation energy (E_a) for the nucleation of the GP zones and Ω phase, can be derived from an Arrhenius equation by plotting $ln(T_p^2/\phi)$ vs $1/T_p$, as seen in figure 2. The resulting activation energies are summarized in Table 2. In carrying out the Kissinger analysis, the heating rate (ϕ) must be converted to Ks^{-1}.

In each case, equation (1) fits the data points with an R^2 value of > 0.98.

$$ln\left(\frac{T_p^2}{\phi}\right) = \left(\frac{E_a}{RT_p}\right) + C \quad (1)$$

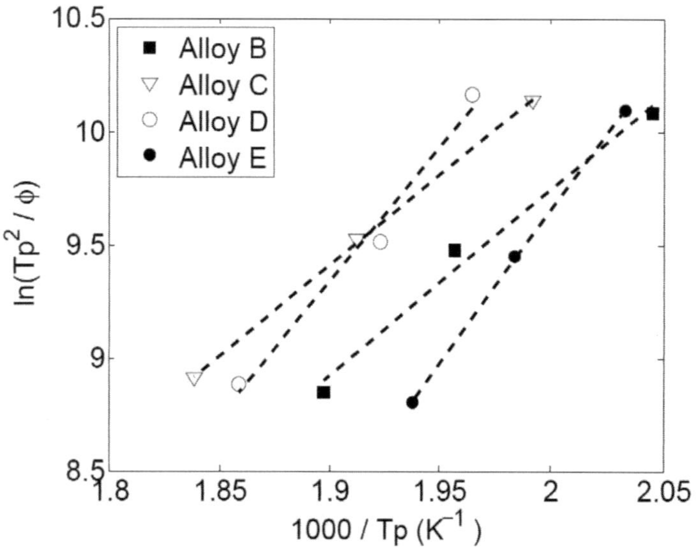

Figure 2. Comparison of Arrhenius plots for the Kissinger analysis of the exothermic peaks due to the Ω phase nucleation in alloys B, C, D and E. The lines are the fits of Equation (1).

Figure shows that by varying the scanning rate, a peak can be enhanced; in fact, at faster scanning rates (e.g. 40 Kmin^{-1}), the area underneath the peaks is greater. Because of this and because of the partially overlapping peaks, it was not possible to detect an exothermic peak linked to the nucleation of the Ω phase in alloy A.

Table 2. Results for the activation energies of the GP zones and the Ω phase calculated by the Kissinger approach.

	E_a (kJ mol^{-1})	
Alloy	GP	Ω
A	63	ND
B	75	69
C	76	68
D	87	98
E	59	113

ND: not detectable

From the results reported in table 2, it is possible to suggest that the TiB_2 particles are more effective in reducing the activation energy of the Ω phase than the silver. In fact, increasing the TiB_2 content twofold (alloys D and B) lowered the activation energy by almost 30% for the Ω phase, while the addition of Ag, from alloy B to C, did not have any effect on the activation energy for the Ω phase. None of the additions had a significant effect on the activation energy of the GP zones. Alloy E was taken as a reference for the activation energy for non-reinforced alloys.

4. Discussion

The identification of phases by DSC can be difficult, and for this reason most research involves DSC and microscopy to interpret their results. However, in a 2xxx series alloy, with a high Cu/Mg mass ratio, the S phase is suppressed. Therefore, the precipitation sequence in the temperature range studied in this work was: GP zones, Ω phase and θ^{i} phase.

Dutta *et al.* [11] analysed a 2xxx series alloy with addition of Al_2O_3 particles and observed the θ^{i} phase nucleating between 550 K and 560 K, at a scanning rate of 20 Kmin^{-1}, which is in the same range as in this work. In addition, Oguocha *et al.* [12], studying AA2618 with Al_2O_3 particles, found the nucleation of S and θ^i phase at around 570 K. As a consequence, it is possible to correlate the exothermic peak found in Figure at around 540-550 K with the nucleation of the θ^i phase. The exothermic peak found at 500-520 K, must therefore be associated with the nucleation of the **Ω** phase.

In addition, if a non-reinforced alloy is taken into consideration, the dissolution of GP zones is found at around 470 K [13] and so the nucleation of the Ω phase occurs at 490-500 K [9, 14]. These results are comparable with this work. However, the first endothermic peak at around 370 K was not linked to the dissolution of any phases. In agreement with the literature [12, 15], this work showed that the addition of TiB_2 particles does not affect the nucleation temperature of any phases. In fact, Chang et al. [14] has demonstrated that the shift in the exothermic peak is related to the Cu:Mg ratio.

The value of the activation energy for the Ω phase reported for a non-reinforced alloy is close to that for lattice diffusion of Cu in Al, which is around 140 kJ mol^{-1} [4, 16]. However, the value reported in this study for a reinforced alloy is well below the lattice diffusion of Cu in Al, suggesting another process occurs, as reported by Dutta *et al.* [17]. In a reinforced alloy the difference in the coefficients of thermal expansion generates a high density of dislocation lines during quenching. Therefore, the diffusivity of Cu atoms in the alloy is improved because the Cu atoms will migrate through the dislocation lines, in a process called pipe diffusion [18]. The activation energy for pipe diffusion of Cu atoms in Al is 66 kJ mol^{-1}[15] which, as a consequence, will be the lowest activation energy value possible to nucleate the **Ω** phase.

The combined effect of both silver and TiB_2 particles in alloy B shows a value for the activation energy lower than any value reported so far for the Ω phase. This reduction due to different alloying additions is also reported in the literature. In fact, Borrego et al. [15] was able to reduce the activation energy of the phases studied by 25 % by increasing the volume fraction of the SiC reinforced particles in a 6xxx series alloy. Nieh at al [19] reported decreasing the activation energy by 37% when comparing a reinforced 6xxx series composite with a non-reinforced alloy. The lower activation energy seems therefore to be due to the higher density of dislocations with a higher amount of reinforced particles. This will facilitate the diffusion of Cu atoms and the formation of the Cu bearing precipitates such as Ω and θ^i phase.

The activation energy for the formation of the GP zones reported in this work was around 68 $kJmol^{-1}$, and it remained unchanged with alloying additions. The results of this work are comparable to values reported in literature [13, 20].

In summary, this work offers a first examination of the activation energies for the GP zones and the Ω phases in metal matrix composite. In particular, the effects of the reinforcing particles (TiB_2) and the silver content have been analysed with the Kissinger approach. The results suggest that in a reinforced alloy, the activation energy for the Ω phase can be reduced. This value for the activation energy in a reinforced alloy suggests that the formation of the Ω phase is due to pipe diffusion of Cu in Al.

5. Conclusions

1. The activation energy for the formation of the Ω phase is greatly reduced by the addition of TiB_2 particles and silver, while the activation energy for the formation of the GP zones is unchanged.
2. In MMC alloys, the most effective mechanism to reduce the activation energy of the formation of the Ω phase is to increase the volume fraction of the reinforcing particles.
3. The silver additions did not show any effect on the activation energy of the Ω phase.
4. The lowest value for the activation energy was found to be 68 $kJmol^{-1}$ phase. This value was found when both silver and TiB_2 were added.

Acknowledgments
The authors are grateful for financial support from the Technology Strategy Board. The authors also wish to acknowledge the technical assistance of by A J Caden and F Biddlestone.

References

[1] Zhou, X.W., et al., The influence of various Ag additions on the nucleation and thermal stability

of Omega phase in Al-Cu-Mg alloys. Materials Science and Engineering A-Structural Materials Properties Microstructure and Processing, 2013. **564**: p. 186-191.

[2] Liu, X.Y., et al., *Microstructure and mechanical properties of Al–Cu–Mg–Mn–Zr alloy with trace amounts of Ag.* Materials Science and Engineering: A, 2009. **525**(1): p. 128-132.

[3] Chang, Y. and J. Howe, *Composition and stability of Ω phase in an Al-Cu-Mg-Ag alloy.* Metallurgical and Materials Transactions A, 1993. **24**(7): p. 1461-1470.

[4] Afify, N., et al., *Precipitation Kinetics in Supersaturated Al-2.0 at% Cu-1.0 at% Mg Alloy.* Materials Transactions, 2010. **51**(2): p. 317-320.

[5] Ovono Ovono, D., I. Guillot, and D. Massinon, *Determination of the activation energy in a cast aluminium alloy by TEM and DSC.* Journal of Alloys and Compounds, 2007. **432**(1–2): p. 241-246.

[6] Kissinger, H.E., *Reaction Kinetics in Differential Thermal Analysis.* Analytical Chemistry, 1957. **29**(11): p. 1702-1706.

[7] Kissinger, H.E., *Variation of peak temperature with heating rate in differential thermal analysis.* Journal of Research of the National Bureau of Standards, 1956. **57**(4): p. 217-221.

[8] Mittemeijer, E., *Analysis of the kinetics of phase transformations.* Journal of Materials Science, 1992. **27**(15): p. 3977-3987.

[9] Zhang, J.-b., et al., *Identification of thermal effects involved in DSC experiment on Al-Cu-Mg-Ag alloys with high Cu:Mg ratio.* International Journal of Minerals, Metallurgy, and Materials, 2011. **18**(6): p. 671-675.

[10] Li, Q. and R.N. Shenoy, *DSC and TEM characterizations of thermal stability of an Al–Cu–Mg–Ag alloy.* Journal of Materials Science, 1997. **32**(13): p. 3401-3406.

[11] Dutta, I., C.P. Harper, and G. Dutta, *Role of Al2o3 Particulate Reinforcements on Precipitation in 2014 Al-Matrix Composites.* Metallurgical and Materials Transactions A-Physical Metallurgy and Materials Science, 1994. **25**(8): p. 1591-1602.

[12] Oguocha, I.N.A. and S. Yannacopoulos, *A calorimetric study of S ' and theta ' precipitation in Al2O3 particle-reinforced AA2618.* Journal of Materials Science, 1999. **34**(14): p. 3335-3340.

[13] Smith, G.W., W.J. Baxter, and R.K. Mishra, *Precipitation in 339 and 2124 aluminum: A caveat for calorimetry.* Journal of Materials Science, 2000. **35**(15): p. 3871-3880.

[14] Chang, C.-H., et al., *Impact of Cu/Mg ratio on thermal stability of hot extrusion of Al-4.6 pct Cu-Mg-Ag alloys.* Metallurgical and Materials Transactions A, 2007. **38**(11): p. 2832-2842.

[15] Borrego, A. and G. González-Doncel, *Calorimetric study of 6061-Al–15 vol.% SiCw PM composites extruded at different temperatures.* Materials Science and Engineering: A, 1998. **245**(1): p. 10-18.

[16] Rylands, L.M., et al., *Coarsening of Precipitates and Dispersoids in Aluminum-Alloy Matrices - a Consolidation of the Available Experimental-Data.* Journal of Materials Science, 1994. **29**(7): p. 1895-1900.

[17] Dutta, I. and D.L. Bourell, *A theoretical and experimental study of aluminum alloy 6061-SiC metal matrix composite to identify the operative mechanism for accelerated aging.* Materials Science and Engineering: A, 1989. **112**(0): p. 67-77.

[18] Love, G.R., *Dislocation pipe diffusion.* Acta Metallurgica, 1964. **12**(6): p. 731-737.

[19] Nieh, T.G. and R.F. Karlak, *Aging Characteristics of B4c-Reinforced 6061-Aluminum.* Scripta Metallurgica, 1984. **18**(1): p. 25-28.

[20] Smith, G.W., *Precipitation kinetics in solutionized aluminum alloy 2124: II. Effect of prior Guinier–Preston zone formation.* Thermochimica Acta, 1998. **323**(1–2): p. 123-130.

IOP Conf. Series: Materials Science and Engineering **117** (2016) 012028 doi:10.1088/1757-899X/117/1/012028

Correlation between microstructure and hardness of a Bi-1.5wt%Ag lead-free solder alloy

J E Spinelli[1,a], R A Macedo[1], B L Silva[1] and A Garcia[2]
[1] Department of Materials Engineering, Federal University of São Carlos – UFSCar, 13565-905, São Carlos, São Paulo, Brazil.
[2] Department of Materials Engineering, University of Campinas, UNICAMP, PO Box 6122, 13083-970 Campinas, SP, Brazil.

E-mail: [a] spinelli@ufscar.br

Abstract. In the present study a hypoeutectic Bi-1.5wt%Ag alloy was directionally solidified under transient heat flow conditions and the microstructure was analysed. Bi-Ag alloys are considered as potential alternatives to replace Pb-based alloys as solder materials for metallic connections under high temperatures. However, a lack of understanding regarding the effects of solidification thermal parameters (growth rate - V_L, the cooling rate - $\dot{T}$) on microstructural aspects is reported in literature. Another challenge is to improve properties and reliability. The results of the present study include the determination of the tip growth rate and the cooling rate by cooling curves recorded by thermocouples positioned along the casting length, metallography, X-ray fluorescence (XRF) and Vickers hardness. The entire directionally solidified Bi-1.5Ag microstructure was arranged by faceted Bi-rich dendrites surrounded by a eutectic mixture (Bi+Ag). The primary and secondary dendrite arm spacing (λ_1 and λ_2), the interphase spacing (λ) and the diameter of Ag-rich particles were also measured along the casting length; and experimental growth laws. Relating these microstructural features to the experimental thermal parameters are proposed.

1. Introduction

During the last decades lead-based eutectic or near-eutectic alloys have been massively used as solders in electronic packaging and assembly. Due to the inherent negative impacts to human health and environment, nowadays lead (Pb) is intended to be banned from such applications. As a consequence, alternative solder alloys have to be selected and tested in order to replace the traditional lead-based material. In the case of high-temperature applications the usage of the Pb-5wt.%Sn solder alloy is still considered since lead-free alternative alloys do not achieve enough performance considering characteristics such as wettability, reliability and cost. As a consequence, further research is necessary since the use of high lead content alloys is growing considerably [1]. In this context, near eutectic Bi-Ag alloys become promising solder alternatives once Bi is the least toxic of the heavy metals, allowing its application as a substitute for Pb. This would become a step toward an environmentally friendly soldering technique. Further, the temperature criterion is achieved since the melting temperature of the eutectic Bi-2.5wt%Ag is quite close to that of the Pb-5wt.%Sn alloy. This criterion is based on the fact that high-temperature solders usually have melting temperature in the range between 260°C and 400°C. It is obviously realized that both cited characteristics are not enough to guarantee soundness for a candidate alloy in order to substitute high Pb-alloy solders.

The Bi-2.5Ag alloy is considered an interesting option due to compatible melting point and hardness as compared with Pb-based traditional solder alloys, with an actual possibility of having ductility increased. As a consequence, a Bi-Ag alloy has already been developed as a die attach solder for power devices and light-emitting diodes (LEDs) [2].

In the case of Bi-Ag alloys, the development of non-equilibrium solidification microstructural features have been barely identified by some few studies [3,4]. It has been noted a lack of studies emphasizing the effects of the cooling rate variation on the final solder fillet microstructure. In this context, unidirectional solidification systems can be very useful for investigating the evolution of microstructures in solder alloys. Solder fillets develop an as-cast microstructure as a consequence of the solidification process. In this case, arrangements formed by dendrites and cells are well known to have significant influence on the final properties like mechanical strength and corrosion resistance [5,6]. The parameters characterizing such microstructures are the primary/secondary dendrite arm spacings ($\lambda 1$, $\lambda 2$) and the cell spacing (λc).

The results reported by Masayoshi and co-authors [7] demonstrated a gradual increase in the tensile strength with the increase in the alloy Ag content. In this study, the dispersed Ag phase with average size of 18μm is reported to act as an effective arrester against propagating cracks. Recently, the examination of the Bi-1.5, 2.5 and 3.5wt.%Ag alloys [2] revealed that the wetting angle decreases as the silver content increases. A 11.17° wetting angle is reported for the Bi-3.5wt.%Ag alloy, which means that it could easily spread on a Cu substrate surface.

The aim of this study is to correlate the microstructural parameters such as primary/secondary dendrite arm spacing (λ_1/λ_2) and eutectic spacing (λ) with the solidification thermal parameters during directional solidification of the Bi-1.5wt%Ag alloy. This will allow realizing which experimental growth laws may be fitted for this alloy. Examination of the combined effects of macrosegregation and microstructure parameters on the microhardness profile along Bi-1.5wt.%Ag alloy is also planned.

2. Experimental procedure

A directional solidification setup was used so that heat was directionally extracted only through a water-cooled bottom made of low carbon steel. Further details about the solidification system used in the present research have been described in a previous study [8]. The Bi-1.5wt%Ag, alloy was melted in situ by controlling the power of the radial electrical wiring. The electric heaters were disabled so that solidification could be initiated, and at the same time the controlled water flow was opened. Continuous temperature measurements in the casting were performed during solidification by fine type J thermocouples (0.2mm diameter wire sheathed in 1.0mm outside diameter stainless steel tubes) placed along the casting length.

The data acquisition system employed allows accurate determination of the slope of the experimental cooling curves. Hence, the cooling rates associated with the liquidus isotherm have been determined by considering the thermal data recorded immediately after the passage of this isotherm by each thermocouple.

An etching solution of 96% (vol.) C_2H_5OH and 4% (vol.) HNO_3 applied during 30s was used to reveal the microstructures. The primary dendrite arm spacing ($\lambda 1$) was measured on transverse sections of the casting. The triangle method was used in order to perform such measurements [9]. Further, the intercept method was adopted (also on transverse specimens) in order to determine the interphase spacing (λ), which is in brief the distance between two adjacent Ag spheroids positioned side by side inside the eutectic mixture. The ImageJ software for processing and analysis was used to measure the cited spacings and their distribution ranges. At least 30 measurements were performed for each selected position of each examined Bi-Ag alloy. Microstructural characterization was completed using a Field Emission Gun (FEG) - Scanning Electron Microscope (SEM-EDS) Philips (XL30 FEG). Vickers microhardness tests were performed on the cross sections of the samples using a test load of 500 g and a dwell time of 15s. The adopted hardness value of a representative position was the average of at least 10 measurements on each sample. Samples for segregation analysis underwent a fluorescence spectrometer (FRX), model Shimadzu EDX-720 to estimate local average concentration through an area of 100 mm^2 probe.

3. Results and discussion

The temperature-time profiles registered for the hypoeutectic Bi-1.5wt%Ag alloy are depicted in figure 1. The experimental cooling curves refer to thermocouples located at specific distances from the cooled surface of the directionally solidified casting, i.e., 3mm, 8.5mm, 13.5mm, 18.5mm, 23mm and 42.5mm. Such thermal readings have been treated to provide plots of position from the metal/mold interface and the corresponding time of the *liquidus* isotherm passing by each thermocouple. The derivative of a generated function with respect to time gave values for the tip growth rate, which varied from 0.07 to 0.4mm/s. A large cooling rate spectrum was allowed to be obtained due to the employed directional solidification technique. The cooling rates varied from 0.03 K/s to 9.7K/s which is a range that encompasses the typical values applied during reflow procedures in industrial practice. Preceding studies give complete information about the experimental set-up used in the directional solidification trials [5] [8].

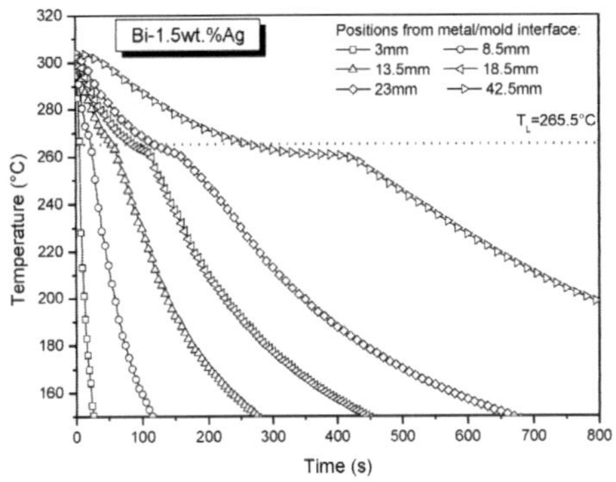

Figure 1. Cooling curves acquired for the Bi-1.5wt%Ag alloy at different positions from the cooled bottom of the casting. T_L means liquidus temperature.

Some representative transverse and longitudinal microstructures are depicted in figure 2a and figure 2b, respectively. The solidification thermal parameters like growth rate and cooling rate were included inside each microstructure. It can be noted that the size of the microstructure is connected with these parameters. Thus, as the cooling rate is high close to the bottom of the casting and decreases toward the top, the microstructural length scale increases as a result. Bi-rich β phase dendrites prevail for the entire range of experimental growth rates (0.4mm/s – 0.07 mm/s) determined during solidification of the hypoeutectic Bi-1.5wt%Ag alloy. The light regions are constituted by the primary solid formed (β phase) and the eutectic mixture is represented by the dark interdendritic regions.

The dendrites appear as anomalous broken microstructures (see figure 2). This kind of feature was reported earlier by Song and Chuang [10], which considered such appearance as a consequence of non-equilibrium solidification conditions. The average primary dendritic spacing determined for the hypoeutectic Bi-1.5wt%Ag was found to vary between 121μm and 1024μm while the average eutectic interphase spacing measured based on the presence of Ag-rich spheroids in the interdendritic regions was in the range 0.3μm to 2.2μm.

It can be seen that the Bi-rich phase tends to grow anisotropically as faceted crystals. A SEM image is depicted in figure 3 so that the contour formed between a secondary dendrite arm (Bi-rich dark region) and a eutectic portion could be appreciated. Nodules at the interface between the phases (indicated by black arrows) are homogeneously spaced between each other. This is in disagreement with previous studies [11]. The referred results on the literature were conducted under stationary heat flow conditions, permitting long periods of temperature maintenance on levels above the eutectic point. Transient conditions enabled in the present experimental conditions together with the high cooling rates may allow Bi to grow in a faceted manner.

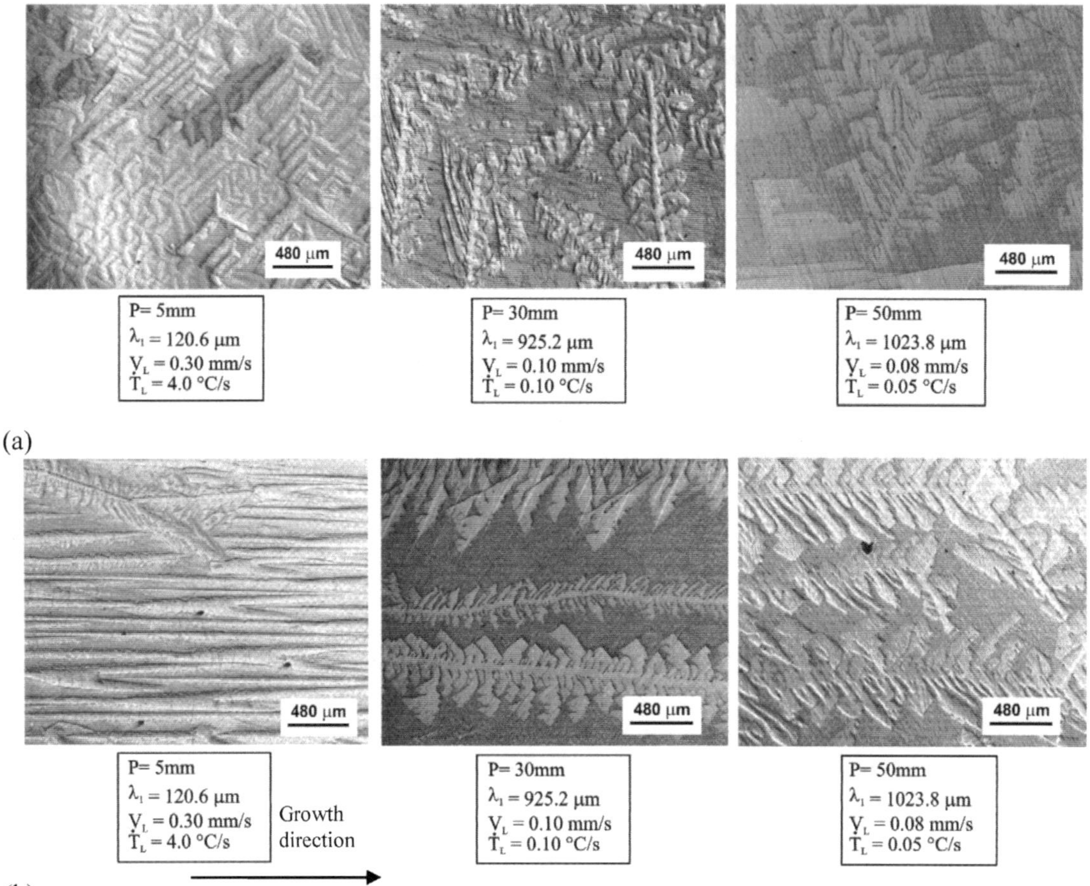

Figure 2. (a) Transverse and (b) longitudinal microstructures of the directionally solidified Bi-1.5wt%Ag alloy.

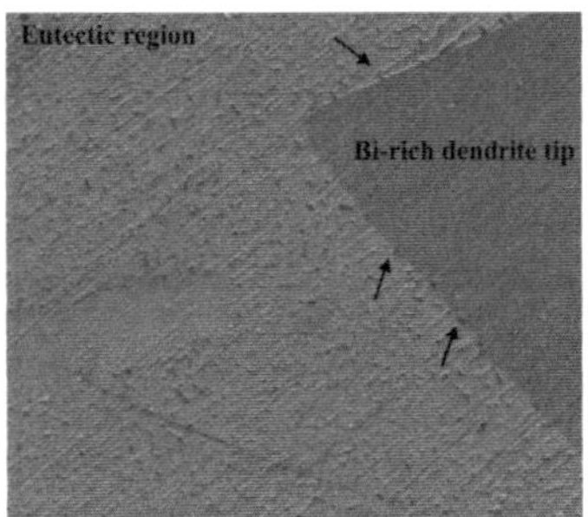

Figure 3. SEM image detailing the faceted contour between a Bi-rich secondary dendrite branch (dark gray region) and the eutectic region found in the Bi-1.5wt%Ag alloy.

Figures 4a and 4b show average microstructural spacing values (λ_1 and λ_2) and their standard deviation for the Bi-Ag alloy examined, plotted as a function of cooling rate and growth rate, respectively. The exponents usually reported for the growths of single-phase dendrites [12] were considered representative for the near eutectic Bi-Ag composition. These exponents are -0.55 and -2/3 and they have been found to apply for experimental power growth laws relating the spacings to the growth rate (λ_2=A $v^{-2/3}$) and the cooling rate (λ_1=B $\dot{T}^{-0.55}$), respectively, for the examined alloy.

The effects of growth rate and temperature gradient on the eutectic interphase spacing have been determined as can be seen in figure 4c. Points are experimental results and line represents an empirical fit to the experimental points, with spacing being expressed as a power function of G x v. The main theoretical models for cellular/dendritic growth, which were developed for steady-state conditions, propose specific exponents to be used in this kind of correlation (−1/2 for G and −1/4 for v) [13,14]. In the present experimental investigation, which was conducted under conditions of unsteady-state growth, the exponents −1/2 for G and −1/4 for V_L were found to be appropriate for the eutectic growth of the Bi–1.5 wt%Ag alloy. The same exponents have been adopted by Goulart and co-authors [15] for eutectic growth of the directionally solidified Al-1.5wt%Fe alloy. As can be seen through the SEM images in figure 4c finer Ag spheroids are associated with regions close to the cooled surface of the casting. Figure 5a shows the experimental macrosegregation profile of Ag along the length of the directionally solidified Bi-1.5wt%Ag alloy casting. This segregation profile has the typical trend of a normal segregation distribution for a solute redistribution coefficient (k_0) less than unity. Hardness is affected by both the dendritic spacing evolution (figure 5b) and the silver (Ag) distribution along the casting length. Lower dendritic spacings allow a more extensive distribution of the Ag-rich spheroids. Higher hardness values are associated with regions where such distribution is optimized together with the lower silver content, i.e., regions close to the bottom of the casting.

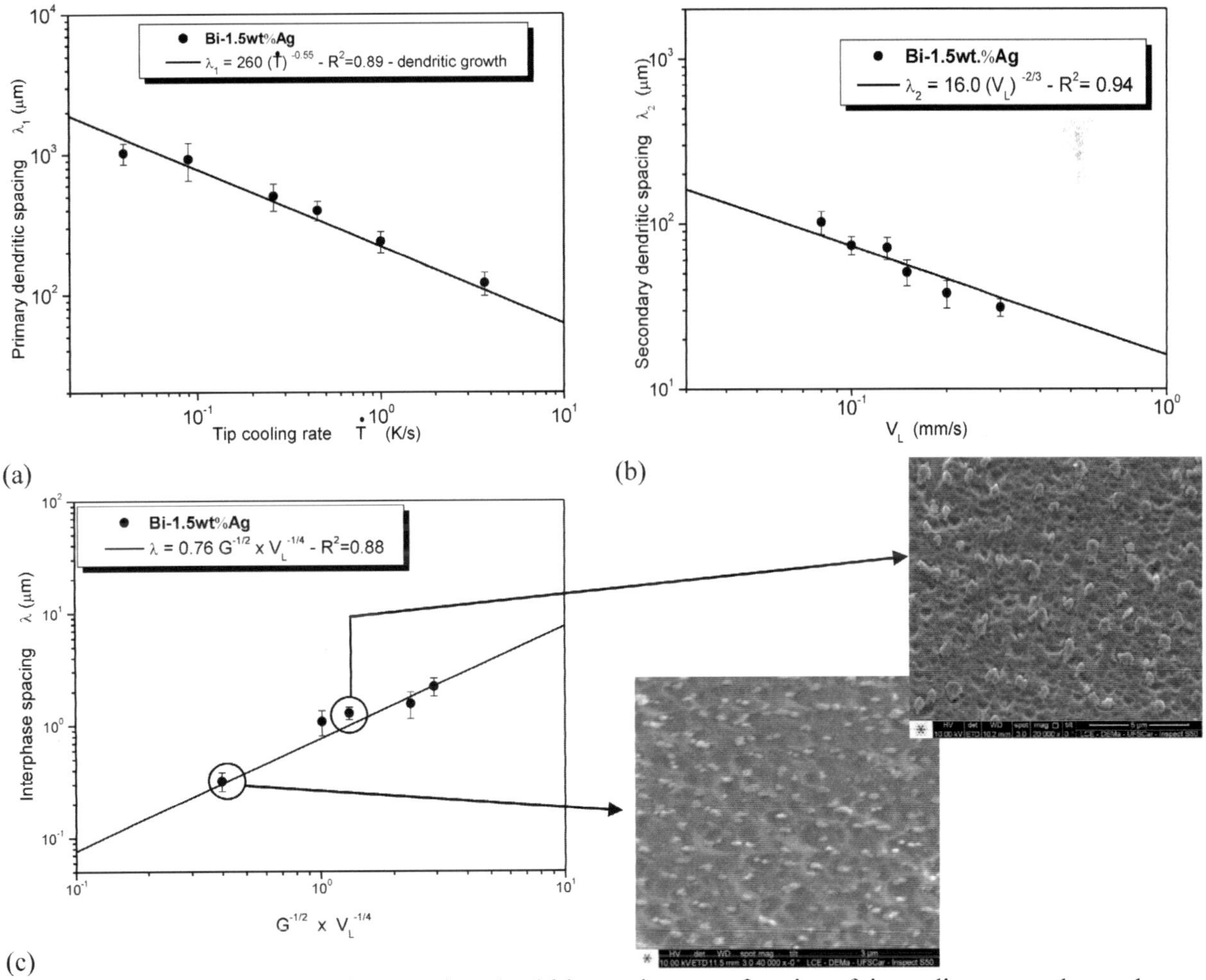

Figure 4. (a) Primary and (b) secondary dendritic spacings as a function of tip cooling rate and growth rate for the Bi-1.5wt%Ag alloy. (c) Interphase spacing as a function of the $G^{-1/2}.v^{-1/4}$. R^2 is the coefficient of determination.

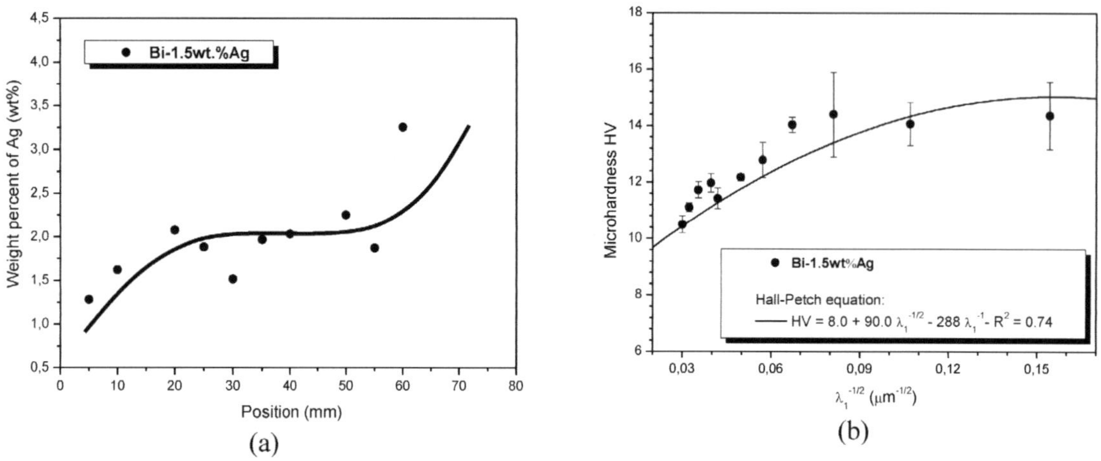

Figure 5. (a) Experimental Ag segregation profile along the length of the directionally solidified Bi-1.5wt%Ag alloy; (b) Vickers microhardness as a function of $\lambda_1^{-1/2}$.

4. Conclusions

Experimental power growth laws relating λ_1 and λ_2 to the cooling rate and the growth rate, respectively, are proposed. A modified Hall-Petch equation was able to represent the experimental hardness scatter of the Bi-1.5wt%Ag alloy. This has been attributed to both the primary dendritic spacing evolution and the silver distribution along the casting length.

Acknowledgements

The authors acknowledge the financial support provided by FAPESP (grants 2013/08259-3 and 2013/13030-5) and CNPq.

References

[1] Kim B, Lee C, Lee D and Kang N 2014 Effect of Sb addition on Bi–2.6Ag–0.1Cu solders for high-temperature applications *J Alloys Compd* **592**: 207-12.

[2] Zeng G, McDonald S and Nogita K 2012 Development of high-temperature solders: Review. *Microelectron Reliab* **52**: 1306–22.

[3] Song J, Chuang H and Wu Z 2006 Interfacial reactions between Bi-Ag high-temperature solders and metallic substrates *J Electron Mater* **55**: 1041-49.

[4] Song J, Chuang H and Wen T 2007 Thermal and tensile properties of Bi-Ag alloys. *Metall Mater Trans A* 2007; **38**: 1371-75.

[5] Spinelli J E and Garcia A 2013 Microstructural development and mechanical properties of hypereutectic Sn-Cu solder alloys. *Mater. Sci. Eng. A* **2013**; 586: 195-201.

[6] El-Daly A, Hammad A E, Al-Ganainy G A and Ibrahiem A A, 2013 Enhancing mechanical response of hypoeutectic Sn–6.5Zn solder alloy using Ni and Sb additions. *Mater Des* **52**: 966-73.

[7] Marayoshi S, Tomohiro Y, Kunio S, Hiroshi N and Tadashi, T, Effects of Ag content on the mechanical properties of Bi-Ag alloys substitutable for Pb based solder. Trans JWRI 2012; **41**: 51-4.

[8] Silva A P, Spinelli J E, Mangelinck-Noel N and Garcia A 2010 Microstructural development during transient directional solidification of hypermonotectic Al-Bi alloys. *Mater Des* **31**: 4584-91.

[9] Gunduz, M, Çardili, E.. Directional solidification of aluminium-copper alloys. 2002 *Mater Sci Eng A* **327**: 167-185.

[10] Song J, Chuang H and Wen T. 2009 Faceting behavior of primary Ag in Bi-Ag alloys for high temperature solder applications. *Mater Trans* **50**: 1902-04.
[11] Crocker M N, Fidler R S and Smith R W 1973 The characterization of eutectic structure. *Proc Roy Soc. London A* **335**: 15-37.
[12] Kurz W and Fisher DJ Fundamentals of Solidification. Switzerland: Trans. Tech. Public 1992.
[13] Goulart P R, Spinelli J E, Cheung N, Mangelinck-Nöel N amd Garcia A. 2010 Al-Fe hypoeutectic alloys directionally solidified under steady-state and unsteady-state conditions. *J Alloys Compd* **504**: 205-10.
[14] Trivedi, R 1984 Interdendritic Spacing: Part II. A Comparison of Theory and Experiment. *Metall. Mater. Trans. A* **15**: 977-82.
[15] Trivedi R and Kurz W 1994 Dendritic growth. *Int. Mater. Rev.* **39**: 49-74.

Applications of the directional solidification in magnetic shape memory alloys

Y J Huang[1], J Liu[2], Q D Hu[1], Q H Liu[1], I Karaman[3,4] and J G Li[1,a]
[1] School of Materials Science and Engineering, Shanghai Jiao Tong University, Shanghai 200240, China
[2] Key laboratory of Magnetic Materials and Devices, Ningbo Institute of Material Technology and Engineering, CAS, Ningbo 315201, China
[3] Department of Mechanical Engineering, Texas A&M University, College Station, TX 77843, USA
[4] Department of Material Science and Engineering, Texas A&M University, College Station, TX 77843, USA

E-mail: [a]lijg@sjtu.edu.cn

Abstract. A zone melting liquid metal cooling (ZMLMC) method of directional solidification was applied to prepare highly-oriented $Ni_{52}Fe_{17}Ga_{27}Co_4$ magnetic shape memory alloys. At high temperature gradient and low growth velocity, the well-developed preferred orientation for coarse columnar crystals was obtained. Such a structure leads to a large complete pseudoelastic recovery of 5% at 348 K. Moreover, the pseudoelastic behaviours and the kinetics of the martensitic transformation (MT) are significantly affected by the intersection angle between the loading direction and the grain boundaries.

1. Introduction

Magnetic shape memory alloys (MSMAs) have been considered as potential substitutes for giant magnetostrictive and piezoelectric materials in actuator application because they exhibit a large magnetic field induced strain (MFIS) of up to 10% [1]. Moreover, in comparison with other conventional shape memory alloys (SMAs), MSMAs show a better combination of high strain and high actuation frequencies, which has led to the search for new MSMAs including Ni-Mn-X (X=Ga, In, Sn, Sb) [2, 3], Ni-Fe-Ga-(Co) [4] and Co-Ni-Al [5] alloys. Among them, Ni-Fe-Ga-(Co) MSMAs with an significantly improved ductility [4, 6] are regarded as a promising candidate owing to low field requirement for triggering MFIS [7], and high magneto-crystalline anisotropy energy [8]. Interestingly, the MT temperatures and Curie temperature of Ni-Fe-Ga-(Co) FSMAs can be tailored over a wide range via compositional modifications and heat treatments [9, 10]. More importantly, a relatively large MFIS was obtained in Ni-Fe-Ga-(Co) alloys at room temperature [7]. Experimental and theoretical transformation strains for Ni-Fe-Ga-(Co) alloys exceed that in most other commercial SMAs [11]. Additionally, Ni-Fe-Ga single crystals can exhibit complete pseudoelastic recovery over a wide temperature range (>300 K) as compared to other SMAs [12]. Large, fully recoverable tensile strains greater than 10% have been measured in selected single crystal orientations [13].

However, their excellent functional properties are usually achieved in the single crystals [1, 14], which is difficult and time consuming. Fortunately, recent documents show that a large reversible magnetic field-induced strain could be achieved in highly textured polycrystals [15, 16], as the anisotropy of such a structure is analogous to that of single crystals, which can be prepared by the directional solidification [17]. Therefore, one aim of this work is to obtain such highly textured polycrystals of $Ni_{52}Fe_{17}Ga_{27}Co_4$ MSMA by using directional solidification. Meanwhile, the role of grain boundary on the MT in this kind of MSMAs is still unclear because previous investigations mainly focus on single crystals. Therefore, another purpose of this work is to study the role of grain boundary on pseudoelastic behaviours in highly-oriented polycrystals.

2. Experimental

Homogeneous ingots with nominal composition of $Ni_{52}Fe_{17}Ga_{27}Co_4$ (in at. %) were firstly prepared from Ni, Fe, Ga, Co (all 99.99% purity) by arc melting four times and then cast into a graphite cylinder mold. Subsequently, directional solidification was carried out in the self-made ZMLMC furnace. The process of the directional solidification has been described in details elsewhere [17]. In this work, the directionally solidified $Ni_{52}Fe_{17}Ga_{27}Co_4$ samples were conducted at three different parameters: (i) G_L=250 K/cm, V=100 μm/s; (ii) G_L=1200 K/cm, V=15 μm/s; (iii) G_L=1200 K/cm, V=5 μm/s.

The crystal structure and preferred orientation was identified by X-ray diffraction analysis (XRD; Rigaku, Ultima IV) using Cu K_α radiation. Microstructure was investigated by optical microscopy (OM; Axio Imager A1m).

To test the role of grain boundary on the MT, compression samples were cut from the directionally solidified rods into rectangular prisms with nominal dimensions of 3mm×3mm×6mm, and with one of the three different intersection angles between the compression axis and the crystal growth direction (CGD). It should be noted that, for better understanding the role of grain boundaries, all specimens were used in as-cast form without any heat treatments to avoid precipitation of γ phase [9] and cut from the same rod that was directionally solidified at the condition (iii). Figure 1 shows the loading and crystal growth directions for samples cut at 0°, 45° and 90°. The compression experiments were conducted using an MTS screw driven test frame with a heating/cooling rate of 10 K/min. A high temperature extensometer was used to measure the strain from the tips of the compression platens. After compression, the surface topography was observed by OM.

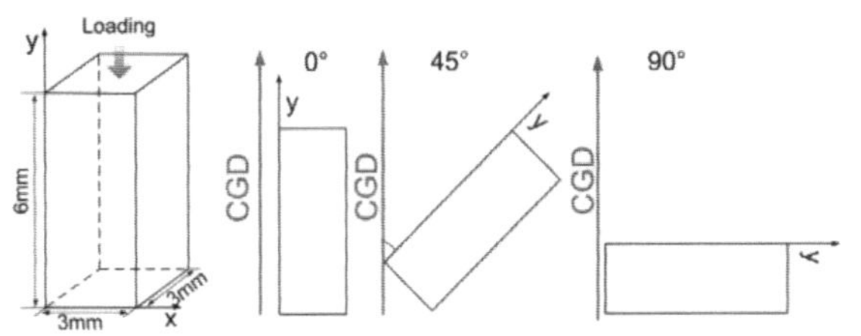

Figure 1. Schematics of the compressive samples and their crystal growth direction (CGD).

3. Results and discussions

3.1 Microstructure of the directionally solidified samples

Figure 2a shows the XRD patterns a transverse section of the arc-melting master alloys and the directionally solidified samples. The master alloys show the random orientations of $L1_0$ martensite [9], which is in accordance with the microstructural observation indicating the typical equiaxial grains (figure 2b). As for the directionally solidified samples, at condition (i), two peaks of (004) and (440) $L1_0$ are detected at 56.6° and 69.5°, respectively. Additionally, there are small signals of (220) and (222) $L1_0$. However, at condition (ii) and (iii), only one sharp peak of (222) $L1_0$ at 43.5° is observed. It is suggested that the preferred orientation of the (222) $L1_0$ is well developed at high G_L and low V. On the other hand, the microstructure of steady growth zones in the longitudinal section shows that six columnar grains grow along the rod axis at condition (i) (figure 2c). In contrast, at condition (ii), only three columnar grains containing some large martensitic blocks can be observed (figure 2d); at condition (iii), only two columnar grains can be observed in the longitudinal section (figure 2e).

Moreover, some differently shaped martensitic lamellae are exhibited in different blocks (figure 2d). Micro-twinning substructures are found between the parallel martensitic lamellae, which have been studied by the electron backscatter diffraction (EBSD) [18].

Generally, highly crystal anisotropy leads to the high energy of magnetocrystalline anisotropy [19], which is a key to obtain large MFIS [1, 8]. However, magnetic anisotropy is usually dissipated in multi-variant state, which is induced by the self-accommodating manner when the MT occurs. Therefore, it is very meaningful that the whole columnar grain (figure 2e) is occupied by one large martensitic block, so called the giant single-variant set, in this case where uniformly oriented martensitic lamellae are well aligned. Subsequently, we will present the superelasticity behaviours in such a highly-oriented structure.

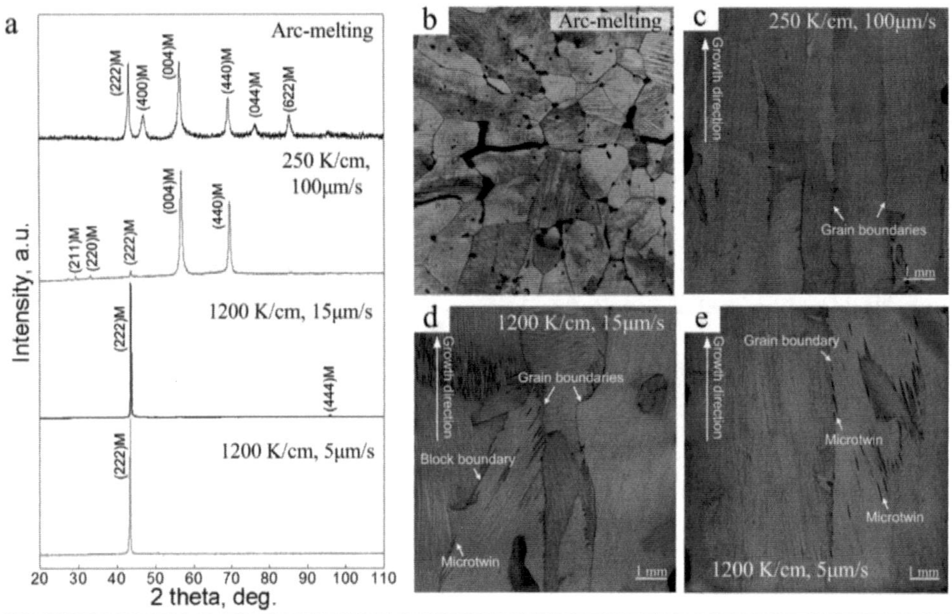

Figure 2. (a) XRD patterns of the transverse section of the master alloy and the directionally solidified samples, (b-e) the corresponding microstructure for the arc-melting (b) and longitudinal section morphology of the steady growth zone in directionally solidified samples under various conditions: (c) G_L=250 K/cm, V=100 μm/s, (d) G_L=1200 K/cm, V=15 μm/s, (e) G_L=1200 K/cm, V=5 μm/s.

3.2 Pseudoelastic behaviours

Before the compression tests, the martensitic start (M_s= 336 K) and finish (M_f= 326 K) temperatures and the austenite start (A_s = 337 K) and finish (A_f = 348 K) temperatures were determined using differential scanning calorimetry (DSC, NETZSCH) with a heating/cooling rate of 10 K/min.

The compressive stress-strain responses for three orientations at 348K (A_f) are shown in figure 3. A complete pseudoelastic recovery is found for the 0° samples when loaded to 5% strain (figure 3a). This strain magnitude is comparable to that of other single crystalline MSMAs (6.5% for $Ni_{49}Fe_{18}Ga_{27}Co_6$ [20], 4.2% for $Ni_{54}Fe_{19}Ga_{27}$ [21], 5% for $Ni_{53.1}Mn_{26.6}Ga_{20.3}$ [22], etc.). The onset stress (σ_s) for the inelastic deformation stage is steady at 56 MPa until the fifth cycle. On the fifth cycle, σ_s drops to 40 MPa, and the stress hysteresis ($\Delta\sigma$) increases by 32 MPa. This implies that no cyclic hardening is produced as observed in NiFeGa [12], and a good compatibility of strain at the grain boundaries is satisfied at the angle of 0°.

For the other two samples (45° and 90°), partial pseudoelastic recovery is observed (figure 3b and c). On the first loading cycle for the 45° sample, a residual strain of 0.61% is recorded after a recovery of 0.39% upon unloading. Of the recovered strain, 0.12% is ascribed to the elastic recovery (ε_{el}) using an apparent elastic modulus of 29 GPa for the martensite, and the remaining 0.27% (ε_M) is apparently associated with the reverse transformation to austenite. On subsequent cycles, the ε_M decreases, which is attributed to the microstrain accumulated in the vicinity of grain boundaries during the cyclic loading. This suggests that the strain compatibility at the grain boundaries is reduced. A similar phenomenon is also evident in the 90° samples (figure 3c). It is noted that the compressive stress level of the 45° sample is close to half that in the 0° and 90° samples, which may be attributed to the greater Schmid factor and more active correspondent variant pairs (CVPs) [12].

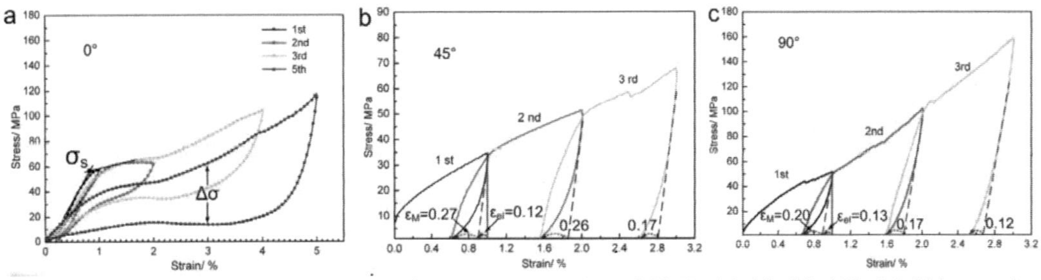

Figure 3. The compressive stress-strain responses at A_f = 348 K: (a) 0°, (b) 45°, (c) 90° samples.

To further investigate the effects of grain boundaries and angles with loading direction, the surface topography for the compressed samples after testing are presented in figure 4. From the enlarged images around the grain boundaries (figure 4d, e and f), for the 0° samples, needle- or plate-like martensite variants are symmetrically formed and then grow into the parent phase (the white arrows in figure 4d). Such morphology minimizes the elastic strain energy generated during the MT and provides a good strain compatibility of the grain boundaries. In addition, for the 45° samples, more fine multi-variant lamellae are resulted from the self-accommodation mechanism [23] (figure 4e). As the intersection angle increases to 90°, some cracks are produced at triple junctions (figure 4c and f). Also, coarse martensite lamellae are formed in the grains at two ends of the samples (A in figure 4f), while the austenite phase is retained in the central grains (B and C in figure 4f).

It is clear that from the aforementioned results, the angle between grain boundaries and the loading direction has a critical role on the MT. Meanwhile, those phenomena suggest that there is a stress field surrounding the grain boundaries, which leads to a certain incompatibility across grain boundaries. Based on the geometric mechanics, the applied stress on grain boundary plane can be divided into two parts of the shear stress and the compressive stress (figure 5a). Moreover, the resolved compressive stress increases as the intersection angle (θ) increases. The resolved shear stress promoting the growth of the martensite is hardly constrained by the grain boundaries [24]. In contrast, the transport of the resolved compressive stress across the grain boundaries is easier to be hindered.

For the intersection angle of 0°, it is reasonable to deduce that the compressive stress at grain boundaries is lower and the strain compatibility in neighbouring grains is better. In other words, when each single grain is transformed to various variants, the same displacement in each grain is generated on the grain boundary plane (figure 5b). Thus, the symmetrical morphology of martensite is formed in the neighbouring grains near the grain boundaries (figure 4d), and the complete pseudoelastic recovery of 5% is obtained at the angle of 0° (figure 3a).

The resolved compressive stress on the grain boundary plane for the angle of 90° is the larger. Furthermore, the buffering effect of the grain boundaries for transporting the applied stress leads to stress concentrations at grain boundaries. Thus, during the cyclic loading, the applied stress in the grains at two ends of the 90° samples increases more quickly than that of the central grains. This suggests that the martensitic phase firstly nucleates at grain boundaries and preferably grows into the grains at two ends (figure 4f). At the beginning of the MT, in order to maintain the strain continuity at the boundary plane, the central grains are required to accommodate plastically because their free energy for the MT is not satisfied, as schematized by figure 5c showing the strain incompatibility on the grain boundary plane. On subsequent loading, the plastic deformation may be produced in the vicinity of grain boundaries, because the potency of the boundaries as nucleation sites for the martensite is always reduced by the prior deformation, even if it has been small. It should be noted that a large stress concentration (indicated by the white dot circles in figure 4c) would be formed by the plastic deformation at the crystal defects, especially at the triple junctions, and then lead to the cracks (figure 4f). As a result, a high fraction of retained austenite phase is observed in the central grains (figure 4c and f) while the coarse martensite lamellae in the grains at two ends, and only partial pseudoelastic recovery is presented (figure 3c). It is reasonable to infer that a similar effect is also presented in the 45° samples, reducing the strain compatibility across grain boundaries because the resolved compressive stress is between that of the 90° and 0° samples (figure 5a).

On the other hand, two kinetic roles of the stress concentrations on the MT should be considered. One is to activate additional slip, leading to different displacement in neighbouring grains (figure 5c) and asymmetric morphology (figure 4f). Another is to decrease the non-chemical free energy required for the MT, promoting the nucleation of the martensite and resulting in an apparent increase of the MT temperatures [24]. Thus, it is reasonable to conclude that the M_s of the 90° samples is shifted to a higher temperature. In other words, the compression temperature (348 K) for the 90° samples may be below its M_s, resulting in the partial pseudoelastic recovery. The compressive strain – temperature curves at a constant stress of 50 MPa (figure 6) show that the MT temperatures for the 90° samples are higher by 17 K than that for the 0° samples. Note that their MT temperatures without external stress field are very close to each other (Inset in figure 6). A lower transformation strain for the 90° samples is also presented due to the strain incompatibility at the grain boundaries.

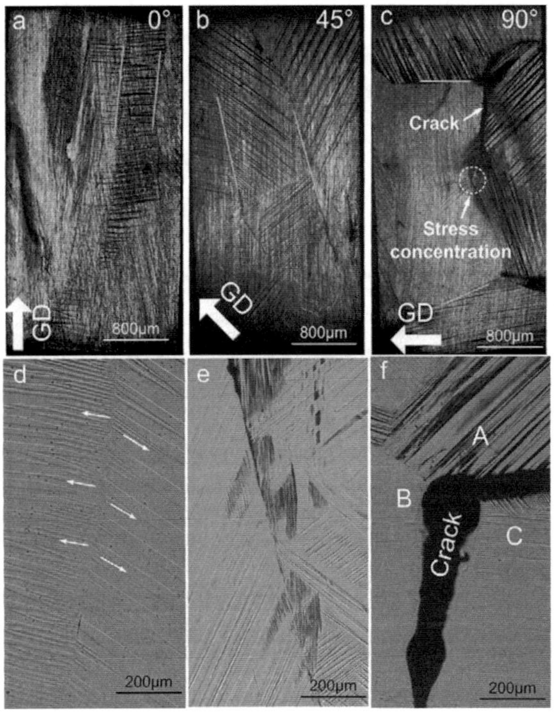

Figure 4. Optical images of surface topography for the compressed samples: (a, d) 0°, (b, e) 45°, (c, f) 90°. The enlarged images around grain boundaries (d, e and f) correspond to (a, b and c), respectively. The green lines indicate the grain boundaries.

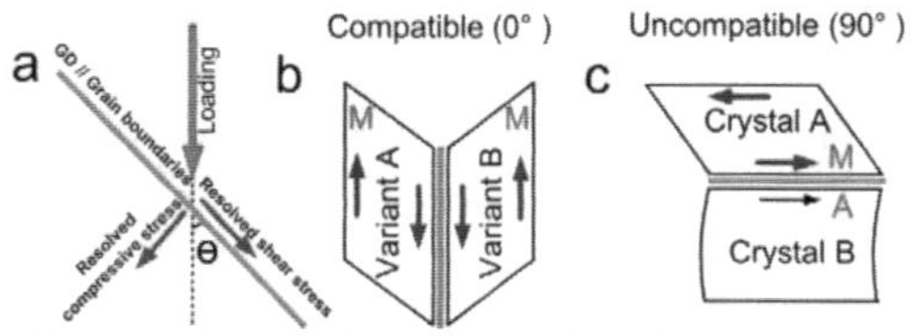

Figure 5. (a) A schematic of the stress distributions in the stress field surrounding the grain boundaries (green lines); (b) and (c) schematics of the relationship between the strain compatibility at grain boundaries, corresponding to 0° and 90°, respectively. The red letters of "M" stand for the martensitic phase.

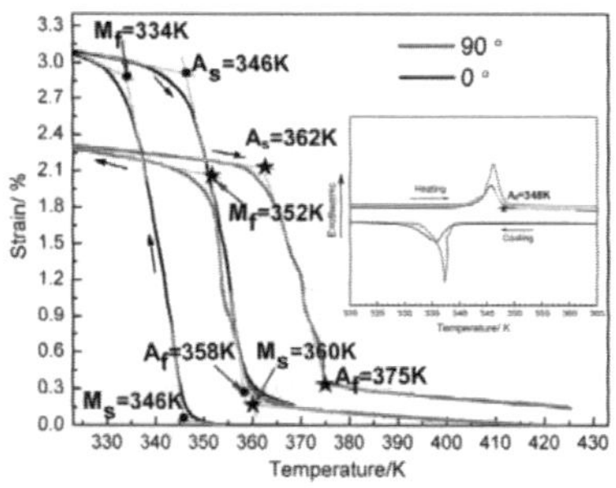

Figure 6. The compressive strain–temperature curves at a constant stress of 50 MPa. Inset shows that their MT temperatures determined by DSC, which are very close to each other

4. Conclusion

The ZMLMC directional solidification was successfully applied to obtain highly oriented polycrystals of $Ni_{52}Fe_{17}Ga_{27}Co_4$ MSMAs. The solidification conditions have a great influence on the microstructure and crystal orientation of alloys. At G_L=1200 K/cm and V=5 μm/s, columnar tricrystal with giant martensitic single-variant sets is obtained. Such a highly-oriented structure results in a complete pseudoelastic recovery of 5% at 348 K when loaded to the crystal growth direction (0°). At higher intersection angles (45° and 90°) between the loading direction and the grain boundaries (the crystal growth direction), only partial pseudoelastic recovery is observed. In comparison with the 0° samples, the MT temperatures under stress for the cases with the intersection angle of 90° are higher. The microstructural observation shows that different martensite morphologies are formed for different intersection angles.

Acknowledgements
This work was financially supported by the National Natural Science Foundation of China (Grant nos. 51010425 and 51027005). The authors are grateful to Instrumental Analysis Center of Shanghai Jiao Tong University for technical support. I.K. acknowledges the financial support from U.S. National Science Foundation, Division of Materials Research, Metals and Metallic Nanostructures Program, Grant No. 1108396, under the umbrella of Materials World Network Initiative.

References

[1] Sozinov A, Likhachev A A, Lanska N and Ullakko K, 2002 *Appl. Phys. Lett.* **80** 1746
[2] Sutou Y, Imano Y, Koeda N, Omori T and Kainuma R, et al, 2004, *Appl. Phys. Lett.* 85 4358
[3] Huang Y J, Hu Q D and Li J G, 2012 *Appl. Phys. Lett.* **101** 222403
[4] Font J, Muntasell J, Santamarta R and Masdeu F, et al, 2006 *Mater. Sci. Eng. A* **438-440** 919.
[5] Liu J and Li J G, 2006 *Scr. Mater.* **55** 755
[6] Sui J, Gao Z, Yu H, Zhang Z and Cai W, 2008 *Scr. Mater.* **59** 874
[7] Morito H, Oikawa K, Fujita A, Fukamichi K and Kainuma R, et al, 2005 *Scr. Mater.* **53** 1237
[8] Morito H, Fujita A, Oikawa K, Fukamichi K, Kainuma R, Kanomata T and Ishida K, 2009 *J. Phys.: Condens. Matter.* **21** 076001
[9] Liu J, Scheerbaum N, Hinz D, Gutfleisch O, 2008 *Acta Mater.* **56** 3177
[10] Liu J, Scheerbaum N, Gutfleisch O, 2008 *IEEE Trans. Magn.* **44** 3025
[11] Hamilton R F, Sehitoglu H, Chumlyakov Y and Maier H J, 2004 *Acta Mater.* **52** 3383
[12] Efstathiou C, Sehitoglu H, Kurath P, Foletti S and Davoli P, 2007 *Scr. Mater.* **57** 409
[13] Hamilton R F, Sehitoglu H, Efstathiou C and Maier H J, 2007 *Acta Mater.* 55 4867
[14] Pötschke M, Weiss S, Gaitzsch U, Cong D, Hürrich C and Roth S, et al 2010 *Scr. Mater.* **63** 383
[15] Liu J, Aksoy S, Scheerbaum N, Acet M and Gutfleisch O, 2009 *Appl Phys Lett* **95** 232515
[16] Gaitzsch U, Potschke M, Roth S, Rellinghaus B and Schultz L, 2009 *Acta Mater.* **57** 365
[17] Huang Y J, Hu Q D, Liu J, Zeng L, Zhang D F, Li J G, 2013 *Acta Mater.* **61** 5702
[18] Liu Q H, Liu J, Huang Y J, Hu Q D and Li J G, 2013 *J. Alloys Compd.* **572** 186
[19] Morito H, Fujita A, Fukamichi K and Kainuma R, et al, 2002 *Appl. Phys. Lett.* **81** 1657
[20] Panchenko E, Chumlyakov Y, Maier H J and Timofeeva E, et al, 2010 *Intermetallics* **18** 2458
[21] Chumlyakov Y, Panchenko E, Kireeva I, Karaman I, Sehitoglu H, Maier H J, Tverdokhlebova A and Ovsyannikov A, 2008 *Mater. Sci. Eng., A* **481–482** 95
[22] Chernenko V A, L'vov V, Pons J and Cesari E, 2003 *J. Appl. Phys.* **93** 2394
[23] Yamanaka A, Takaki T and Tomita Y, 2010 *Int. J. Mech. Sci.*, **52** 245
[24] Ueda M, Yasuda H Y and Umakoshi Y, 2001 *Acta Mater.* **49** 3421

IOP Conf. Series: Materials Science and Engineering **117** (2016) 012030 doi:10.1088/1757-899X/117/1/012030

Crystallographic investigation of grain selection during initial solidification

H Esaka[1a], Y Kataoka[2] and K Shinozuka[1]
[1]National Defense Academy, Dept. of MSE, Yokosuka, Japan
[2]National Defense Academy, Graduated student, Yokosuka, Japan

E-mail: [a]hesaka@nda.ac.jp

Abstract. Normally, macroscopic solidified structure consists of chill, columnar and equiaxed zones. In a chill zone, many fine grains nucleate on the mold surface and grow their own preferred growth direction. Only a few of them continue to grow because of grain selection. In order to understand the grain selection process, crystallographic investigation has been carried out in the zone of initial solidification in this study. 10 g of Al-6 wt%Si alloy was melted at 850 °C and poured on the thick copper plate. Longitudinal cross section of the solidified shell was observed by a SEM and analyzed by EBSD. The result of EBSD mapping reveals that crystallographic orientation was random in the range of initial solidification. Further, some grains are elongated along their <100> direction. Columnar grains, whose growth directions are almost parallel to the heat flow direction, develop via grain selection. Here, a dendrite whose growth direction is close to the heat flow direction overgrows the other dendrite whose growth direction is far from the heat flow direction. However, sometimes we observed that dendrite, whose zenith angle is large, overgrew the other dendrite. It can be deduced that the time of nucleation on the mold surface is not constant.

1. Introduction

When molten alloy is poured into a mold cavity, many small and random grains nucleate on the mold surface [1, 2]. These small grains are called chill grains. Some of them continue to grow inward but some stop growing, because of the grain selection [2]. A grain is composed by some columnar dendrite arms, which have the same crystallographic orientation and may be grown from one nucleus. If the growth direction of these dendrites is not favoured, their growth is interfered with by the growth of dendrites which form another grain. Finally this former grain disappears during growth. In the same way, favoured grains continue growing and well-developed columnar zones would form. Understanding of the solidified structure of the as-cast product is very important to evaluate the mechanical properties of products. Since the process of chill to columnar zone transition is continuous, understanding of formation of chill zone and chill to columnar transition is very important. Therefore, this study has been performed to understand the process of grain selection, which takes place in the initial solidification region using EBSD.

2. Experimental procedure

The substrate used in this study was thick copper plate and its surface was polished unidirectionally by emery paper of #80. Al-6wt%Si alloy was prepared from pure aluminum and silicon. The purities of raw materials were 4N and 6N, respectively. 10g of Al-6 wt%Si alloy was melted at 850ºC and cast on the substrate. Solidified shell was approximately 25mm in diameter and 7 to 8 mm in height. Longitudinal cross section of the solidified shell, which is perpendicular to the direction of polishing, was observed metallographically. The surface of the specimen was carefully prepared using colloidal silica and treated with Ar-milling. Then crystallographic orientation was analyzed using EBSD. In order to visualize the growth of dendrites in this region, the model dendrites are constructed by 3D-CAD.

3. Experimental results and discussions

SEM image of analyzed region is shown in figure 1. The growth direction is from bottom to top. Eutectic grains can be observed in the interdendritic region and were aligned in a certain direction. This indicates that dendrites, which were in a solidified grain, grew in the same direction.

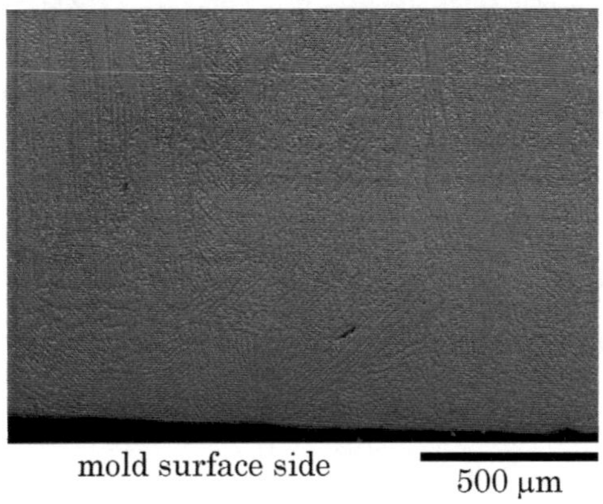

Figure 1. SEM image of longitudinal section of Al-6 wt%Si alloy.

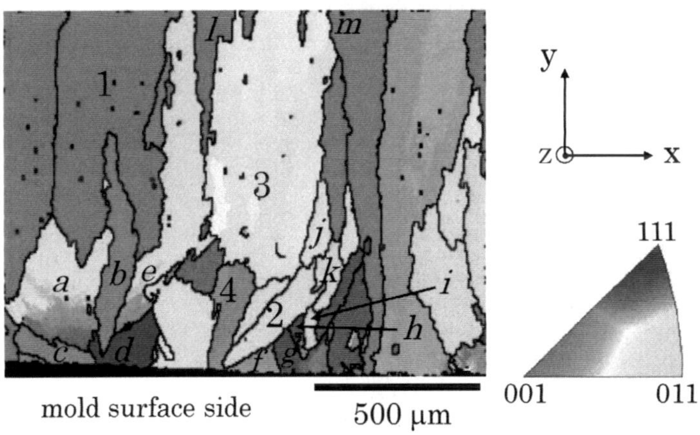

Figure 2. EBSD mapping result in y-direction of longitudinal cross section.

Figure 2 shows the EBSD mapping result in y-direction with rainbow colour, which is the same area as shown in figure 1. Crystallographic misorientations exceeding 10° are indicated by thick black lines. At the bottom of this mapping, the grains were small and the crystallographic orientation was random. Therefore, it is safe to judge that the surface area of the solidified shell was chill zone. On the contrary, approximately 1 mm from the mold surface, which is the top of this mapping, the columnar zone developed and the grains were elongated vertically. Furthermore, the width of these grains was large. These grains were reddish and this means that the crystallographic orientations in y-direction were close to <001>. Therefore, chill-to-columnar transition took place in this area indicated in this mapping. Here, typical grains are named 1, 2, 3 and 4 as shown in figure 2. EBSD mapping results in x-direction and in z-direction are also shown in figures 3 and 4, respectively.

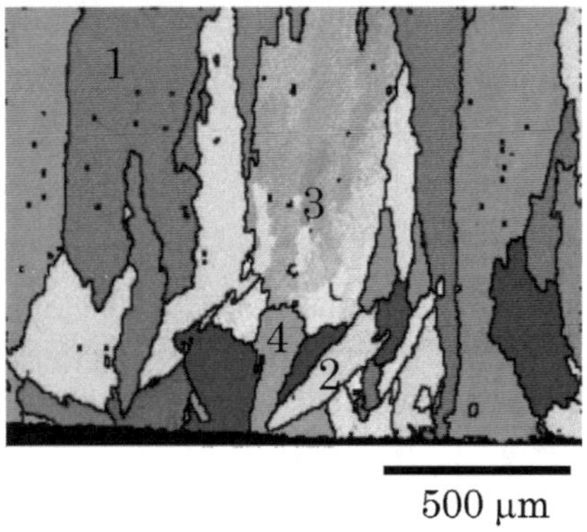

Figure 3. EBSD mapping result in x-direction of Al-6wt%Si alloy.

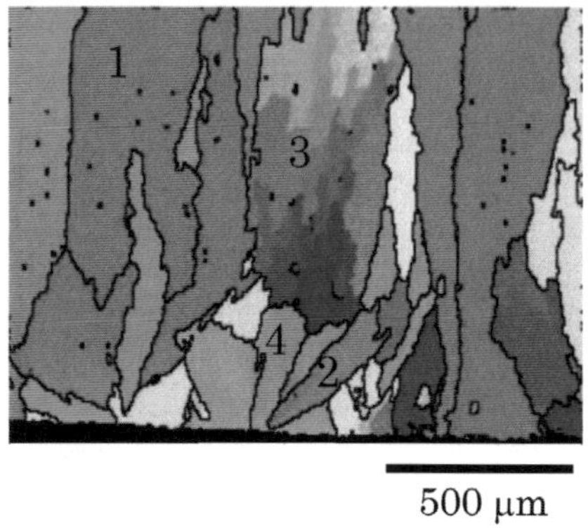

Figure 4. EBSD mapping result in z-direction of Al-6wt%Si alloy.

3.1 Normal grain selection

The growth of grain #1 is analyzed in detail. The grain #1 was a rather long columnar grain. All <001> directions of this grain were parallel to the direction of x, y and z. Because all colours indicated in figures 2-4 were almost red. Therefore, the grain #1 was perfectly aligned. The Euler angles of the grain were obtained from EBSD analysis. Using these angles, the zenith angle, which is the angle between dendrite trunk and heat flow direction, was calculated. The zenith angle of the grain #1 was 7.2 degrees.

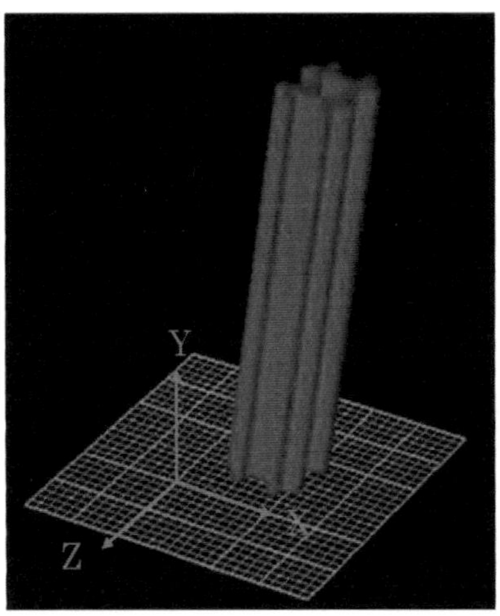

Figure 5. Model dendrites of grain #1 constructed by 3D-CAD.

A model dendrite in grain #1 was constructed with 3D-CAD and is shown in figure 5. Here, the array of dendrite is thought to be square and secondary dendrite arms sintered together and form a plate with uniform thickness [3-5]. The secondary dendrite arm plates were almost parallel to the x- and z- direction. This description agrees well with the EBSD mapping result described above. Furthermore, the dendrites in grain #1 were almost perpendicular to the mold surface. The grain #1 overgrew other small grains, *a*, *b*, *c*, *d* and *e* as shown in figure 2. The zenith angles of these grains are 44.4°, 21.7°, 21.9°, 49.4° and 43.7°, respectively. In this case, the dendrites with low zenith angle (#1) were selected. After that it became a well-developed columnar grain. Unfortunately, since the grain #1 did not touch the bottom surface, the nucleation point of the grain #1 was not observed on this cross section. It must be a certain distance far from this cross section in z-direction.

As shown in figure 2, the grain #2 was small but elongated. It inclined approximately 45°. The colours of this grain in x- and y-directions were almost the same and green. Thus, the orientation in x-direction and y-direction were approximately <011>. Therefore, the growth direction of grain #2 was <001>, or primary dendrite trunk inclined in this direction. Thus, a dendrite can grow along its preferred growth direction, regardless of heat flow direction as pointed out by Kurz and Fisher [2].

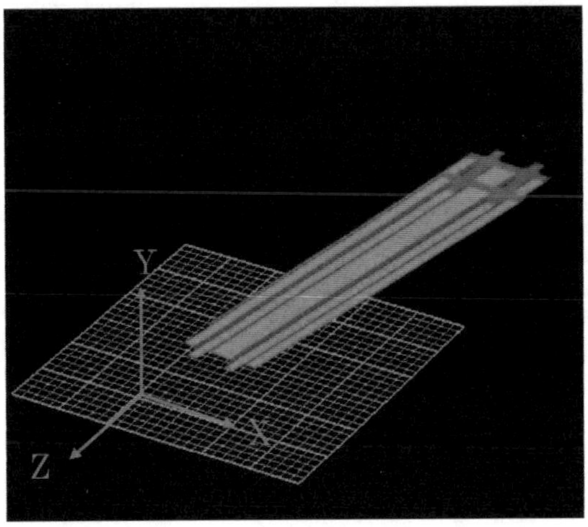

Figure 6. Model dendrites of grain #2 constructed by 3D-CAD.

A model dendrite of this grain is shown in figure 6. The direction of primary dendrite can be again confirmed. The zenith angle of this dendrite is 45.1°. This angle is quite large and this growth direction is not favourable for the growth of columnar grain. Therefore, it indeed overgrew some small grains (*f-i* as shown in figure 2) but was stopped growing by other favorable grains, *j* and *k* as shown in figure 2.

Almost all grain selections analyzed are the same as indicated above, namely favourable grains are selected and unfavourable grains disappear.

3.2 Abnormal grain selection

Occasionally, unfavourable grain overgrew favourable grain. Competitive growth between grains #3 and #4 was analyzed. Here, the growth of grain #4 was stopped by the grain #3 as shown in figure 2. The zenith angles of the grain #3 and #4 were calculated from the Euler angles and to be 38.4° and 25.4°, respectively. A bird's eye view of model dendrites of grain #3 and #4 is shown in figure 7. It is easy to understand that the grain #3 overgrew the grain #4 and both dendrites inclined in the same direction. If the normal grain selection takes place, the grain #4 would overgrow the grain #3, because the zenith angle of grain #4 is smaller than that of grain #3. However, the experimental result was different: the grain #3 overgrew grain #4, which did not agree with the description in the textbook [2].

A probable hypothesis is that the nucleation of grain #3 should be earlier. The grain #3 might nucleate and grow earlier, it could overgrow the grain #4. However, at approximately 1 mm from the mold surface, the local non-uniformity of initial solidification might disappear. Therefore, the grain #3 was stopped growing by other favourable grains (*l* and *m* as shown in figure 2). Because the grain #3 is not favourable for columnar dendrites.

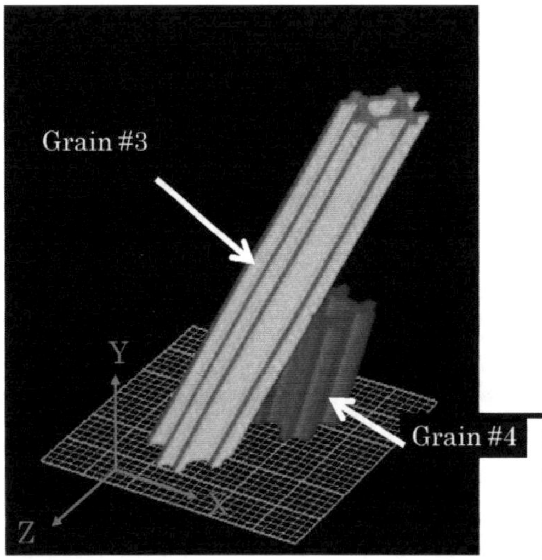

Figure 7. Model dendrites of grain #3 and #4 indicating the competitive growth of two grains constructed by 3D-CAD.

The condition of initial solidification may be complex. For example, flow of molten alloy may not be uniform. Furthermore, the contact between molten alloy and mold surface may not be uniform. Therefore, some unsteady growth may locally take place in this region. One example is that the nucleation does not take place simultaneously. Surface defects of cast products, such as surface cracking, formation of stray crystals, may be originated in the non-uniformity of initial solidification.

4. Conclusions

Crystallographic investigation of solidified grains in initial solidification region has been carried out using Al-6wt%Si alloy by EBSD. The abnormal grain selection as well as the normal grain selection has been observed. In abnormal grain selection, the unfavourable grain overgrew the other favourable grains. This may be due to the fact that the nucleation does not take place simultaneously.

References

[1] Flemings M C 1974 *Solidification Processing* (New York: McGraw-Hill) pp 134-176
[2] Kurz W and Fisher D J 1992 *Fundamentals of Solidification* (Aedermannsdorf: Trans Tech Publication) pp 63-92
[3] Esaka H, Shirakawa M, Shinozuka K and Tamura M 2008 *ISIJ Int.* **48** 264
[4] Esaka H, Yoshimoto A, Sirakawa M, Shinozuka K and Tamura M 2009 *Tetsu-to-Hagané* **95** 844
[5] Esaka H, Yoshimoto A, Mizuno H, Shinozuka K and Tamura M 2009 *Tetsu-to-Hagané* **95** 851

Aging characteristics of the Al-Si-Cu-Mg cast alloy modified with transition metals Zr, V and Ti

F Czerwinski[1,a], S K Shaha[2], W Kasprzak[1], J Friedman[2] and D L Chen[2]
[1] CanmetMATERIALS, Natural Resources Canada, 183 Longwood Road South, Hamilton, Ontario, L8P 0A5 Canada
[2] Department of Mechanical and Industrial Engineering, Ryerson University, 350 Victoria Street, Toronto, Ontario, M5B 2K3 Canada

Email: [a] Frank.Czerwinski@nrcan.gc.ca

Abstract. The hypoeutectic Al-7Si-1Cu-0.5Mg base alloy was modified with different contents of Zr, V and Ti. The wedge-shape samples with varying solidification rates during casting were subjected to isochronal aging at temperatures up to 500 °C. Moreover, as-cast and solution treated alloys were subjected to long-term isothermal aging at 150 °C. As a reference, the A380 alloy, seen as commercial standard for the automotive application target, was used. The modified alloys exerted different aging characteristics than the A380 grade with higher peak hardness and lower temperature of alloy softening. Besides, the influence of the applied solidification rates on hardness after aging was less pronounced in modified alloys than in the A380 grade. For three combinations of Zr, V and Ti tested with contents of individual elements ranging from 0.14 to 0.47 %, no essential differences in aging characteristics were recorded. The results are discussed in terms of the role of chemistry and heat treatment in generating precipitates contributing to the thermal stability of Al based alloys.

1. Introduction

The lightweighting features of Al-based alloys make them very attractive structural materials for automotive applications, including strategic powertrain components [1, 2]. It is well established that an improvement in effectiveness of the combustion engine requires increasing its service temperature and internal pressure. Thus, there is a search for new Al alloys with sufficient thermal stability at continuously increasing temperatures [3, 4, 5, 6].

The Al-Si based grades, i.e., Al-Si-Cu and Al-Si-Mg, have limited thermal stability and lose their strength above approximately 150 °C [6]. To improve the mechanical properties at elevated temperatures, the base Al-Si-Cu-Mg chemistry has to be modified and additions of transitional metals were found to be effective in this respect. According to recent findings, such elements as V, Zr, or Ti, combined with the Al-Si matrix, can improve high temperature stability [4, 6, 7, 8]. As documented in the literature, transition metals develop stable precipitates, which do not coarsen at temperatures as high as 300 °C [9, 10].

In our previous papers many aspects of the transition metals Zr, V and Ti in the Al-Si-Mg-Cu matrix were reported including work hardening and texture [11], tensile and compressive deformation behaviour [9, 12], hot deformation behaviour [13], role of dislocations during deformation at increased temperatures [14], fatigue behaviour [8, 9, 15] and thermal stability of $(AlSi)_x(ZrVTi)$ intermetallic phases formed with transition metals [10]. However, the aging characteristics of the Al-Si-Cu-Mg modified with addition of Zr-V-Ti are still unknown. Thus, the objective of this study was to

IOP Conf. Series: Materials Science and Engineering **117** (2016) 012031 doi:10.1088/1757-899X/117/1/012031

determine the alloy aging characteristics as a function of the alloy chemistry (Zr, V, Ti), the aging temperature and solidification rate.

2. Experimental details

Several alloy melts with experimental chemistries were prepared (table 1). The 30 kg batches were melted in an induction furnace by mixing commercial master alloys of Al–50%Cu, Al–35%Si, Al–6.1%Ti, Al–85%V, Al–10%Zr, Al–10%Sr and pure metals Mg and Al. Wedge-shaped castings were prepared in a copper mold cooled with flowing water at a temperature of 60 °C (figure 1). As a reference, the A380 alloy, seen as the commercial standard for the automotive application target, was used. To ensure melting of all ingredients, the alloy was overheated to 860 °C for approximately 20 minutes. The alloy temperature directly before pouring into the wedge mold was kept at 740 °C. The shape of the wedge castings after sectioning is shown in figure 2. The cooling rate of the wedge samples during solidification was measured using 3 k-type thermocouples positioned along the centerline at 25, 50 and 100 mm from wedge base as it was located in the casting mold, named as bottom, center and top. The typical cooling curves are shown in figure 3. The cooling rate varied from 9 °C/s at the top to 70 °C/s at the bottom of the mold.

Table 1. Chemistry of Al-Si-Cu-Mg alloys modified with addition of Zr-V-Ti and a reference alloy A380.

Alloy ID	Alloying elements in wt.%								
	Si	Cu	Mg	Fe	Sr	Mn	Zr	Ti	V	Al
A	6.77	0.93	0.51	0.095	0.010	0.010	0.14	0.14	0.22	Bal.
B	7.06	0.96	0.51	0.094	0.047	0.007	0.42	0.18	0.17	Bal.
C	7.02	0.95	0.48	0.090	0.012	0.005	0.47	0.21	0.30	Bal.
A380	9.5	3.4	0.07	0.90	<0.001	0.19	0.0073	<0.001	0.0087	Bal.

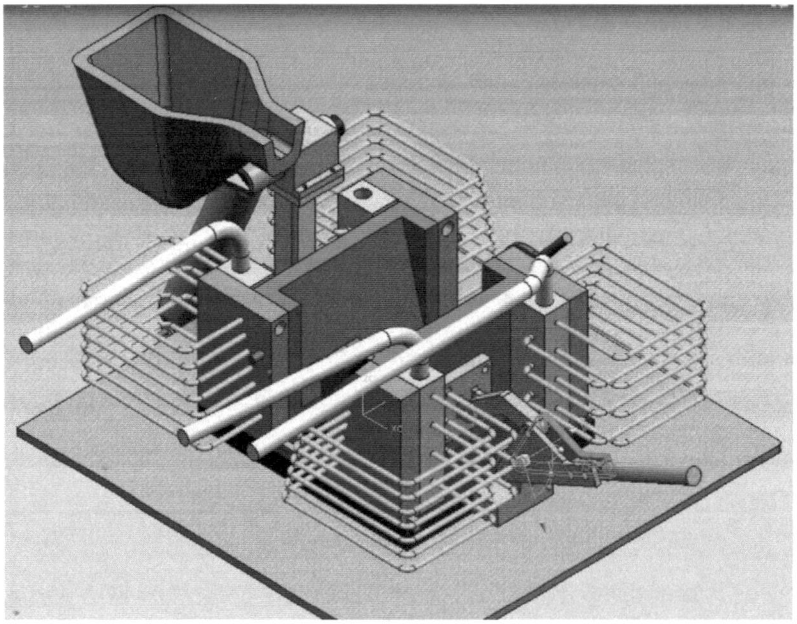

Figure 1. Copper wedge mold with automatic pouring and cooling control with circulating water.

Figure 2. Wedge casting and slices (base 60 x 10 mm and height 120 mm) used for aging experiment and hardness measurements.

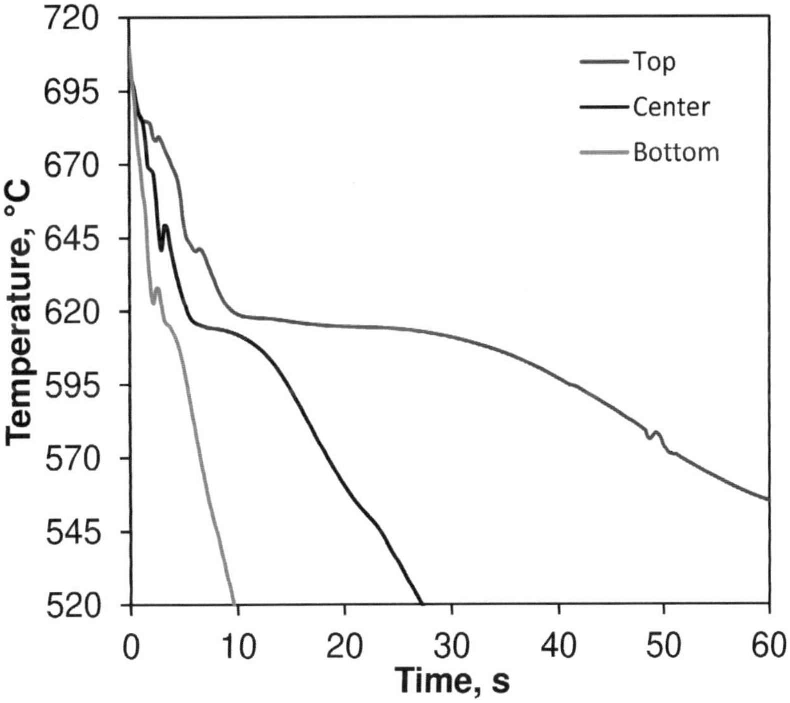

Figure 3. Examples of cooling curves at different locations indicated within the wedge casting.

After casting, the wedges were sliced and hardness was measured as a function of the distance from the wedge base using a Rockwell hardness tester and F scale with a load of 60 kg and a 1/16´´ steel ball indenter. To examine the role of heat treatment the alloy was isochronally aged in 50°C increments each lasting 15 min, beginning at 50 °C and terminating at 500 °C. After each aging step, specimens were water quenched and subjected to hardness testing. The solution treatment consisted of a two-step heating at 510 °C for 0.5 h and at 525 °C for 4.5 h, followed by water cooling. The isothermal aging of as-cast or solution treated alloys was conducted at 150 °C for time intervals up to 128 h.

For microstructural observations, samples were collected from several positions in the wedge. The surfaces were ground using SiC papers up to a grit number of 1200, polished using first an Al_2O_3 slurry and then colloidal silica. The microstructure was examined in the unetched condition using a light optical microscope equipped with a quantitative image analyser and SEM/EDX techniques. The average secondary dendrite arm spacing (SDAS)) values were determined using the linear intercept technique, which involves measuring the distances between secondary dendrite arms along a line normal to the dendrite arms. For the cooling rate applied the approximate change in SDAS was between 10 and 35 μm.

3. Results and discussion

3.1. Alloy microstructure

The characteristic features of the experimental and reference A380 alloys are Al dendrites, surrounded by Al-Si eutectic structures with intermetallics distributed within the interdendritic region. It is well established that at relatively fast cooling rates, such as at the bottom of the wedge casting, and chemically modified alloy, the Al-Si eutectics are much finer and the silicon is seen as having a fibrous/spheroidal morphology (Figure 4). The SEM/EDX spectra across two intermetallics revealed the presence of Ti-, V- and Zr-containing phases with needle like shape, generally referred to as Al(TiVZr)Si [10, 16].

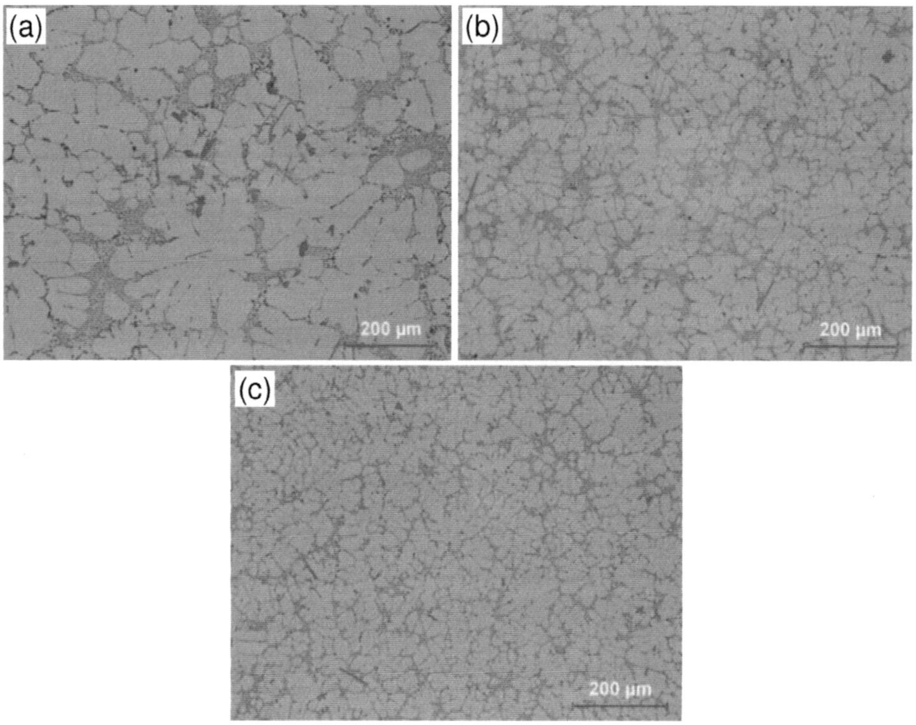

Figure 4. Microstructure of the experimental alloy C from different locations corresponding to figure 3; top (a), center (b) and bottom (c) of the wedge mould in as-cast state.

At the same time, intermetallics observed in A380 are Fe-rich phases $Al_{15}(FeMnCr)_3Si_2$, mostly with a shape of a skeleton. In this alloy, the eutectic Si particles and Fe-phases can act as nucleation sites for Cu-rich phases. The Mg-containing Q-phase and Fe-containing π-phase are also present in both alloys. Such micro-size phases are believed to form as primary phases during initial alloy solidification. According to EDX analysis of element distribution, in the experimental alloys the particles contained mainly (Si, Zr, V, Ti) and (Mg, Cu). On the other hand, in the A380 alloy, particles with (Si, Fe, Mn, Cr) and (Mg, Cu) dominated. It is believed that these micro-size phases in the experimental alloy would, most likely, contribute to the strength retention at elevated temperatures since they were more thermally stable than Cu- and Mg-based phases.

As portrayed in Figure 5, heat treatment changed the alloy microstructure. The metallographic analysis of the T6 heat treated alloy with the optical microscopy and the SEM/EDS techniques confirmed primarily dissolution of Cu-containing phases. Some other phases including eutectic silicon experienced only partial dissolution. In addition to this, some phases were not affected by the solutionizing temperature of 510/525 °C, which indicates their high thermal stability. Details are published in our previous studies [10].

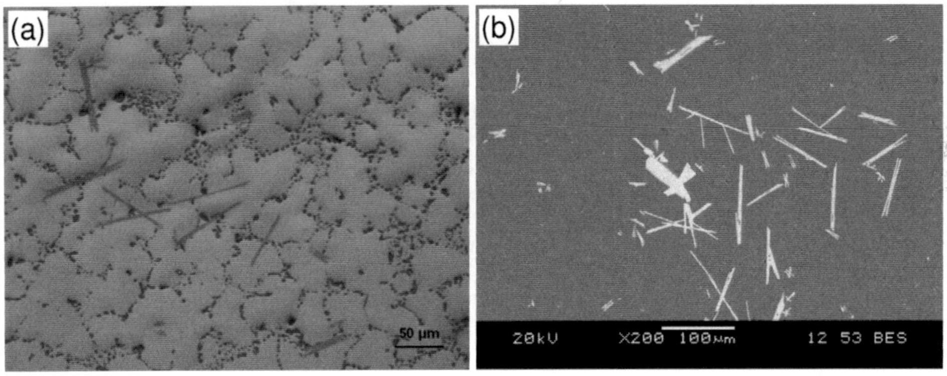

Figure 5. Typical LOM (a) and SEM (b) microstructures of the experimental alloy C in T6 heat treated condition at the center of the casting wedge.

3.2. Hardness after isochronal aging

The results of isochronal aging behaviour of the tested alloys are summarized in figure 6. For all alloys the peak aging conditions are attained at approximately 200 °C. There is a hardness difference at the peak between the experimental alloys and A380; 93HRF versus 87 HRF, respectively. It is of interest that for all three alloys A, B, C the peak hardness is almost identical. For other aging temperatures the hardness difference between A, B and C chemistries was rather small. The initial hardness of the A380 alloy of 83 HRF is over 10 units higher than that of the experimental A, B, and C alloys. For temperatures up to 150 °C, the A380 grade experienced hardness reduction instead of the increase observed for A, B, C alloys.

3.3. Hardness after isothermal aging

The results of isothermal aging of as-cast and solution treated alloys are plotted in figure 7. For as cast experimental alloys, the first hours of aging led to a reduction in hardness from 80-84 HRF to just below 80 HRF. Then, hardness increased attaining a maximum. It was revealed that as-cast A380 showed very similar hardness change, reaching a maximum of 92 HRF after approximately 50 h.

The solution treatment of experimental alloys at 510/525 °C reduced their hardness from 80-82 HRF to around 70 HRF. The subsequent aging led to an increase in hardness with a maximum close to 100 HRF after 50 h exposure and then a slight reduction for up to aging time of 100 h. The solution treatment of the A380 alloy increased its hardness slightly, which after 10 h led to maximum of over 100 HRF, the highest value of all alloys tested.

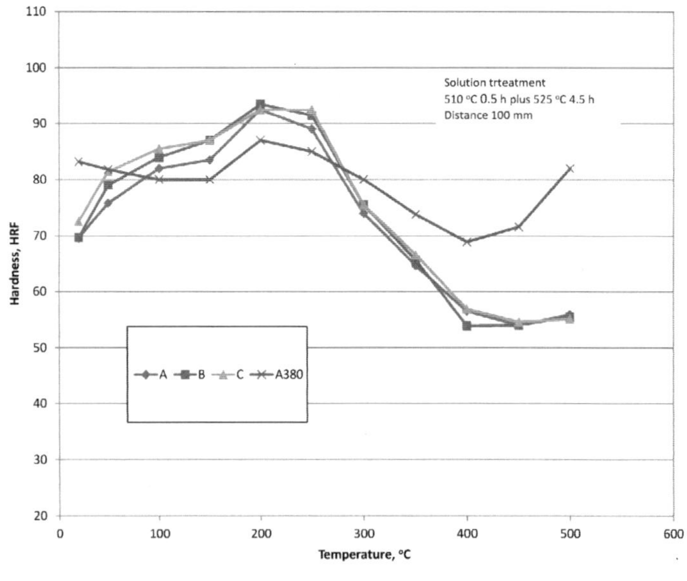

Figure 6. Hardness of the experimental and reference alloys after isochronal aging.

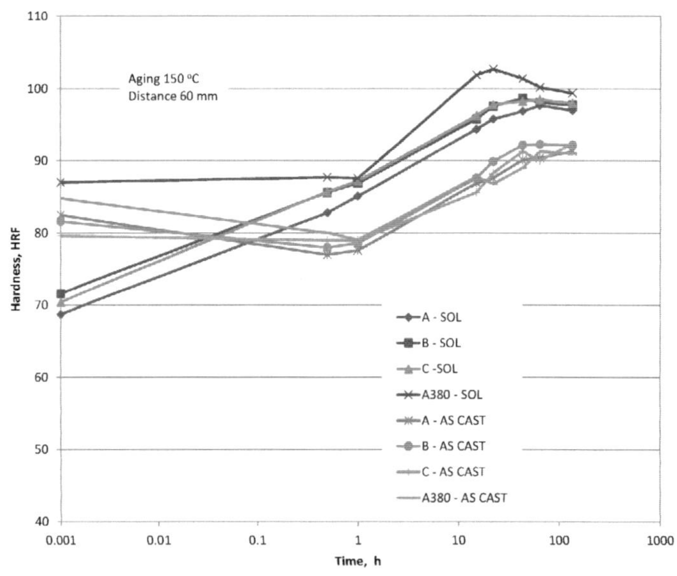

Figure 7. Effect of time on hardness of as cast and solution treated alloys after aging at 150 °C.

3.4. General observations on aging behaviour

It should be stated that no substantial effect of cooling rate, especially in experimental alloys, on hardness measured directly after casting or after heat treatment was detected. Moreover, there was no distinct effect of the amounts of Zr, V and Ti added to experimental alloys on hardness measured. While comparing the experimental alloys with the A380 grade, the latter shows lower response to post casting heat treatments, at the same time preserving higher hardness to higher temperatures, especially above approximately 250 °C. The experimental alloys, however, showed during isochronal aging slightly higher hardness. It is obvious that observed differences in hardness after casting and heat treatments are caused by differences in alloy chemistry, in particular, by the presence of transition metals Ti, Zr and V. The precipitates they form, as identified and described earlier [10], are responsible for the observed hardness differences. This study revealed rather low sensitivity of alloy hardness to contents of transition metals but this requires additional verification. This verification is essential to find the optimum content of transition metals to generate the maximum effect in improving thermal stability of the Al-Si alloys.

4. Conclusions

During isochronal aging, the modified alloys exerted different aging characteristics than the A380 grade with a higher peak hardness and a lower temperature of alloy softening. Besides, the influence of the applied solidification rates on hardness after aging was less pronounced in modified alloys than in the A380 grade.

For three combinations of Zr, V and Ti tested with contents of individual elements ranging from 0.14 to 0.47 %, no essential differences in aging characteristics were observed.

Acknowledgements

The authors would like to acknowledge the financial support of the ecoEII Program of Natural Resources Canada and the Natural Sciences and Engineering Research Council of Canada (NSERC). One of the authors (D.L. Chen) is grateful for the financial support by NSERC-DAS Award, Premier's Research Excellence Award, Canada Foundation for Innovation, and Ryerson Research Chair program. The authors also thank the Innovative Casting team from CanmetMATERIALS for casting experiments.

References

[1] Chaudhury S K and Apelian D, Effects of rapid heating on solutionizing characteristics of Al-Si-Mg alloys using a fluidized bed, *Metallurgical and Materials Transactions A,* vol. 37A, pp. 763-778, 2006.

[2] Ran G, Zhou J E and Wang Q G, Precipitates and tensile fracture mechanism in a sand cast A356 aluminum alloy, *Journal of Materials Processing Technology,* vol. 207, pp. 46-52, 2008.

[3] Lee J A, Automotive Alloys 2003, in *TMS*, San Diego, 2003.

[4] Kasprzak W, Deng Z, Powell J and Niewczas M, Aging characteristics, dimensional stability and assesment of high temperature performance of cast Al-Si alloy for power train applications, *12th International Conference on Aluminum Alloys*, Yokohama, Japan, 2010.

[5] Voorhees P W, Alloys: Scandium overtakes zirconium, *Nature Materials,* vol. 5, pp. 435-436, 2006.

[6] Kasprzak W, Emadi D, Sahoo M and Aniolek M, Development of aluminium alloys for high temperature applications in diesel engines, *Materials Science Forum,* Vols. 618-619, pp. 595- 600, 2009.

[7] Kasprzak W, Chen D L and Shaha S K, Heat treatment development for a rapidly solidified heat resistant cast Al-Si alloy, *Journal of Materials Engineering and Performance,* vol. 22, pp. 1838-1847, 2013.

[8] Kasprzak W, Amirkhiz B S and Niewczas M, Structure and properties of cast Al-Si based alloy with Zr-V-Ti additions and its evaluation of high temperature performance, *Journal of Alloys and Compounds,* vol. 595, pp. 67-79, 2014.

[9] Elhadari H A, Patel H A, Chen D L and Kasprzak W, Tensile and fatigue properties of a cast aluminum alloy with Ti, Zr and V additions, *Materials Science and Engineering A,* vol. 528, pp. 8128-8138, 2011.

[10] Shaha S K, Czerwinski F, Kasprzak W, Friedman J and Chen D L, Thermal stability of (AlSi)x(ZrVTi) intermetallic phases in the Al–Si–Cu–Mg cast alloy with additions of Ti, V, and Zr, *Thermochimica Acta,* vol. 595, pp. 11-16, 2014.
[11] Shaha S K, Czerwinski F, Kasprzak W and Chen D L, Work hardening and texture during compression deformation of the Al-Si-Cu-Mg alloy modified with V, Zr and Ti, *Journal of Alloys and Compounds,* vol. 593, pp. 290-299, 2014.
[12] Shaha S K, Czerwinski F, Kasprzak W and Chen D L, Tensile and compressive deformation behavior of the Al–Si–Cu–Mg cast alloy with additions of Zr, V and Ti, *Materials and Design,* vol. 59, pp. 352-358, 2014.
[13] Shaha S K, Czerwinski F, Kasprzak W, Friedman J and Chen D L, Effect of Zr, V and Ti on hot compression behavior of the Al–Si cast alloy for powertrain applications, *Journal of Alloys and Compounds,* vol. 615, p. 1019–1031, 2014.
[14] Shaha S K, Czerwinski F, Chen D L and Kasprzak W, Dislocation slip distance during compression of Al–Si–Cu–Mg alloy with additions of Ti–Zr–V, *Materials Science and Technology,* vol. 31, no. 1, pp. 63-72, 2015.
[15] Shaha S K, Czerwinski F, Kasprzak W, Friedman J and Chen D L, "Monotonic and cyclic deformation behavior of the Al-Si-Cu-Mg cast alloy with micro-additions of Ti, V and Zr," *International Journal of Fatigue,* vol. 70, pp. 383-394, 2015.
[16] Meng Y, Cui J, Zhao Z and Zuo Y, Effect of vanadium on the microstructures and mechanical properties of an Al–Mg–Si–Cu–Cr–Ti alloy of 6XXX series, *Journal of Alloys and Compounds,* vol. 573, pp. 102-111, 2013.

A meshless approach to thermomechanics of DC casting of aluminium billets

B Mavrič[1,a] and B Šarler[1,2,b]
[1] Institute of Metals and Technology, Lepi pot 11, SI-1000 Ljubljana, Slovenia
[2] University of Nova Gorica, Vipavska 13, SI-5000 Nova Gorica, Slovenia

Email: [a] bostjan.mavric@imt.si, [b] bozidar.sarler@ung.si

Abstract. The ability to model thermomechanics in DC casting is important due to the technological challenges caused by physical phenomena such as different ingot distortions, cracking, hot tearing and residual stress. Many thermomechanical models already exist and usually take into account three contributions: elastic, thermal expansion, and viscoplastic to model the mushy zone. These models are, in a vast majority, solved by the finite element method. In the present work the elastic model that accounts for linear thermal expansion is considered. The method used for solving the model is of a novel meshless type and extends our previous meshless attempts in solving fluid mechanics problems. The solution to the problem is constructed using collocation on the overlapping subdomains, which are composed of computational nodes. Multiquadric radial basis functions, augmented by monomials, are used for the displacement interpolation. The interpolation is constructed in such a manner that it readily satisfies the boundary conditions. The discretization results in construction of a global square sparse matrix representing the system of linear equations for the displacement field. The developed method has many advantages. The system of equations can be easily constructed and efficiently solved. There is no need to perform expensive meshing of the domain and the formulation of the method is similar in two and three dimensions. Since no meshing is required, the nodes can easily be added or removed, which allows for efficient adaption of the node arrangement density. The order of convergence, estimated through an analytically solvable test, can be adjusted through the number of interpolation nodes in the subdomain, with 6 nodes being enough for the second order convergence. Simulations of axisymmetric mechanical problems, associated with low frequency electromagnetic DC casting are presented.

1. Introduction

The thermomechanical phenomena that occurs during DC casting of aluminium alloys plays an important role in designing the casting device and determining the process parameters. Therefore, there exists a great interest in developing numerical models that are able to describe this kind of phenomena. The effort reflects not only in improvement of casting process itself but also in a substantial amount of related publications (see for example [1, 2] and references therein).

The majority of existing numerical models are based on the established finite element method (FEM) [3], which allows for efficient solutions of involved physical models which take into account the thermal expansion, linear elasticity and viscoplastic phenomena on a transient growing domain of the DC cast ingot [4, 5]. By the best knowledge of the present authors, the meshless numerical methods have not been used for the related purpose until now. It is the objective of the present work to fill this gap.

The meshless methods, as compared to the established mesh-based methods, have several advantages such as: the meshless formulation requires no additional topological structure to be established (the spatial positions of nodes suffice), allowing for efficient addition or removal of the computational nodes. The method allows for unstructured node arrangements to be used and for the solution in complicated geometries to be easily obtained. Last but not least, the method can be generalized to higher dimensional problems needing virtually no modifications in formulation [3].

2. Physical model

The primary focus of the present paper is in applying a new numerical method to the considered engineering problem. Therefore, the physical model is kept quite simple. A linear thermoelastic model, resulting from the stress equilibrium, Hooke's law for inhomogeneous isotropic body and linear isotropic thermal expansion is considered. The following governing equation, written in terms of the deformation field $\mathbf{u}$, is obtained [3]

$$G\nabla^2\mathbf{u}+(G+\lambda)\nabla\nabla\cdot\mathbf{u}+\nabla\lambda\nabla\cdot\mathbf{u}+\nabla G\left(\nabla\mathbf{u}+(\nabla\mathbf{u})^T\right)=\nabla\left(\beta\left(T-T_{ref}\right)\right)-\mathbf{f}. \quad (1)$$

Here G stands for shear modulus, λ for Lamé parameter and $\mathbf{f}$ for the body force. The coupling with the temperature field is described by the coefficient β defined as $\beta=(3\lambda+2G)\alpha$, where α is the coefficient of linear thermal expansion. T_{ref} stands for reference temperature at which the thermal expansion is considered to be zero.

The displacement, traction and symmetry boundary conditions are considered. The boundary condition that limits the displacement in the contact with the rigid mold is solved by iteration.

Since the metal undergoes a phase change, the elastic properties substantially vary with temperature. The related dependencies for the considered alloy were obtained from the JMatPro database [3].

3. Solution procedure

The present thermomechanics model is used in a post-processing mode after the calculations with in-house developed model [4] of heat and mass transfer are performed. The stationary state temperature field, obtained from the model, is used as a known input.

3.1. Local radial basis function collocation method

The local radial basis function collocation method (LRBFCM) is used to solve the boundary value problem, given by equation (1). A simplified method formulation is given in the following subsections, with a more elaborate formulation stated in [6].

N_Ω nodes are distributed over the computational domain Ω and N_Γ nodes on its boundary Γ, totalling in $N=N_\Omega+N_\Gamma$ nodes. Every node is assigned a subdomain consisting of ${}_lN$ nodes. For each subdomain the node selection induces a mapping ${}_ls$ from the local to the global node numbering. The solution $\mathbf{u}$ can be constructed by radial basis function interpolant on each subdomain. Multiquadrics Φ_i augmented by constant and first order monomials $p_i(\mathbf{p})$, with $\mathbf{p}$ standing for position vector, are employed in the present work. The multiquadrics, centred in point $\mathbf{p}_i$, are given by

$$\Phi_i(\mathbf{p})=\sqrt{\sum_\xi\left(p_\xi-p_{i,\xi}\right)^2/s_\xi^2+c^2}, \quad (2)$$

where c is the shape parameter and s_ξ the maximum distance between subdomain points in ξ coordinate. The purpose of adding the monomials to the set of basis functions is to allow for exact reproduction of constant and linear deformation fields. The interpolant is given by

$$u_\xi(\mathbf{p})\approx\sum_{i=1}^{{}_lN}{}_l\alpha_{i,\xi}\Phi_{s(i)}(\mathbf{p})+\sum_{i={}_lN+1}^{{}_lN+m}{}_l\alpha_{i,\xi}p_i(\mathbf{p})=\sum_{i=1}^{{}_lN+m}{}_l\alpha_{i,\xi}\psi_i(\mathbf{p}), \quad (3)$$

with the expansion coefficients ${}_l\alpha_{i,\xi}$ determined by collocation and m denoting the number of augmentation monomials. One of the key points of the method is the way the boundary conditions are treated. Let us introduce the linear boundary conditions operator at point j specified by the equation $B^j_{\xi\zeta}u_\zeta(\mathbf{p}_j)=b^j_\xi$. In case the subdomain of a certain point includes a point that lies on the boundary, the corresponding collocation equation is replaced by the equation specifying the boundary condition. To obtain a non-degenerated system, the augmentation equations are added. The system of equations can be written in a matrix form as

$$\sum_{i,\zeta} {}_lA_{ji,\xi\zeta}\,{}_l\alpha_{i,\zeta} = {}_l\gamma_{j,\xi}\,. \tag{4}$$

where indices l, j count discretization nodes, i basis functions and ξ,χ,ζ the spatial directions. The matrix elements of interpolation matrix A and the components of the right hand side vector γ are given by

$${}_lA_{ji,\xi\chi} = \begin{cases} \Psi_i(\mathbf{p}_{{}_ls(j)})\delta_{\xi\chi} & \text{if } \mathbf{p}_{{}_ls(j)} \in \Omega \\ B_{\xi\chi}(\mathbf{p}_{{}_ls(j)})\Psi_i(\vec{r}_{{}_ls(j)}) & \text{if } \mathbf{p}_{{}_ls(j)} \in \Gamma \\ \mathrm{p}_j(\mathbf{p}_{{}_ls(i)}) & \text{if } j >{}_lN \end{cases}, \qquad {}_l\gamma_{j,\xi} = \begin{cases} \mathrm{u}_\xi(\mathbf{p}_{{}_ls(j)}) & \text{if } \mathbf{p}_{{}_ls(j)} \in \Omega \\ b_\chi(\mathbf{p}_{{}_ls(j)}) & \text{if } \mathbf{p}_{{}_ls(j)} \in \Gamma \\ 0 & \text{if } j >{}_lN \end{cases}.$$

The differential operators can be evaluated, by using the described interpolation, in an arbitrary point of the computational domain. By inverting the collocation matrix A, the interpolant can be written in terms of unknown values in collocation nodes. This fact can be used to elegantly discretize the governing equation.

The governing equation is written in a short form as $D_{\xi\zeta}u_\zeta = g_\xi$. To allow for a reasonably compact notation of the discretized system of equations, the domain $\Upsilon^\Omega_{j,\xi}$ and boundary $\Upsilon^\Gamma_{j,\xi}$ indicators are introduced. The indicators evaluate to one if j-th point is a member of the corresponding subset and to zero otherwise. The system of equations is given by

$$\sum_{k,\kappa}\Upsilon^\Omega_{{}_ls(k),\chi}u_{{}_ls(k),\chi}\sum_{i,\zeta}{}_lA^{-1}{}_{ik,\zeta\chi}D_{\xi\zeta}\psi_i(\mathbf{p}_l) = g_{l,\chi} - \sum_{k,\chi}\Upsilon^\Gamma_{{}_ls(k),\chi}b_{{}_ls(k),\chi}\sum_{i,\zeta}{}_lA^{-1}{}_{ik,\zeta\chi}D_{\xi\zeta}\psi_i(\mathbf{p}_l)\,. \tag{5}$$

In this manner a system of equations with dimensions $2N\times 2N$ is obtained. Since the interpolation is local, the obtained system of equations is sparse. This allows the use of highly efficient specialized solvers [5].

3.2. *Treatment of the mold contact*

An axisymmetric billet DC casting is considered in the present work, described in a related coordinate system $\mathbf{p}=p_r\mathbf{i}_r+p_z\mathbf{i}_z$ with p_r and p_z standing for axisymmetric coordinates and $\mathbf{i}_r$ and $\mathbf{i}_z$ for the related base vectors.

The deformation of the top portion of the billet is limited by the presence of a rigid mold. The boundary conditions in this area are highly nonlinear. They are given by inequalities $u_r \le 0$ and $t_r \le 0$ where t_r stands for traction in radial direction. Since the deformations are sensitive to the boundary conditions, the boundary conditions should be changed gradually. In order to achieve this, an error estimator function w is constructed. At the beginning, all the values of w are set to zero and the boundary conditions in the mold region to $u_r = 0$. After each iteration, the function value is updated by using the following heuristic formula $w_{n+1}=w_n+\tau\bar{w}$, where the value of $\bar{w}$ is determined in the following way: if $u_r>0$ then $\bar{w}=u_r/\eta_{prec}a_x$ and if $t_r>0$ then $\bar{w}=-t_r/\eta_{prec}E_{max}$. In these equations the symbol a_r stands for the billet radius and E_{max} for the maximum value of Young's modulus. Parameter η_{prec} specifies the precision. The precision is increasing with decrease of parameter η_{prec}. In this work the value of 0.01 is used for η_{prec}. The parameter τ represents the damping factor and is set to 0.2. The displacement boundary condition $u_r = 0$ is applied at a point if $w > 1$ and the traction boundary condition $t_r = 0$ if $w < -1$. The cases not covered by the formulas are not physical as $u_r > 0$ can occur only when boundary condition $t_r = 0$ is used and $t_r > 0$ can occur only when $u_r = 0$.

3.3. Convergence test

The performance of the method is tested on a two-dimensional domain specified by a unit square $[0,1]\times[0,1]$. The temperature field is assumed to be $T(\mathbf{p}) = \sin 2\pi p_x$. Assuming the boundary conditions given in figure 1 the problem has analytical solution of the form

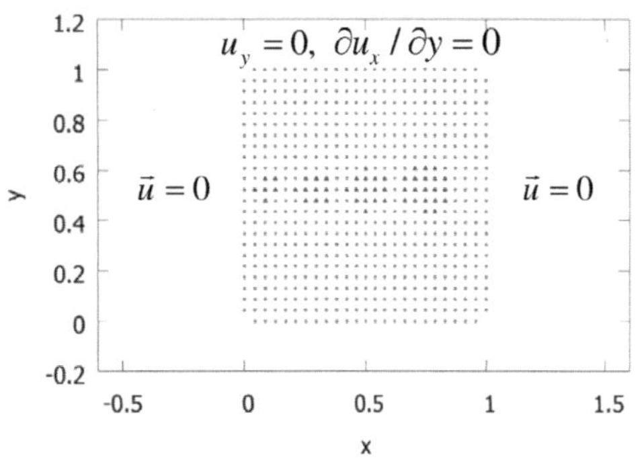

Figure 1. Node arrangement consisting of 1000 nodes with indicated boundary conditions. The triangles show examples of 6, 10, 15 and 21 noded subdomains.

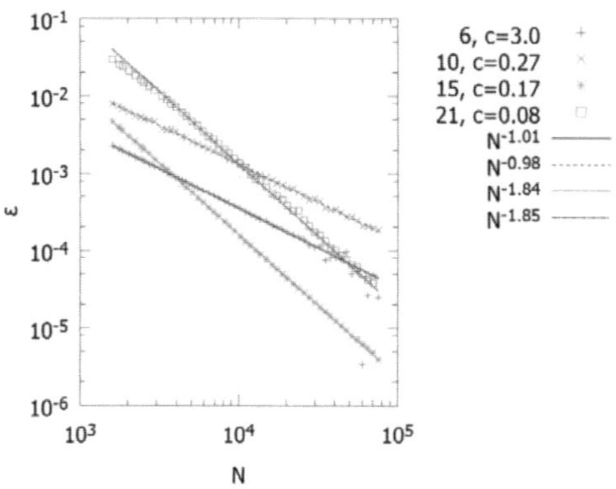

Figure 2. Error dependence on the number of discretization points for different numbers of subdomain nodes. The second order convergence can be observed for 6 and 10 nodes in each subdomain and fourth order for 16 and 21.

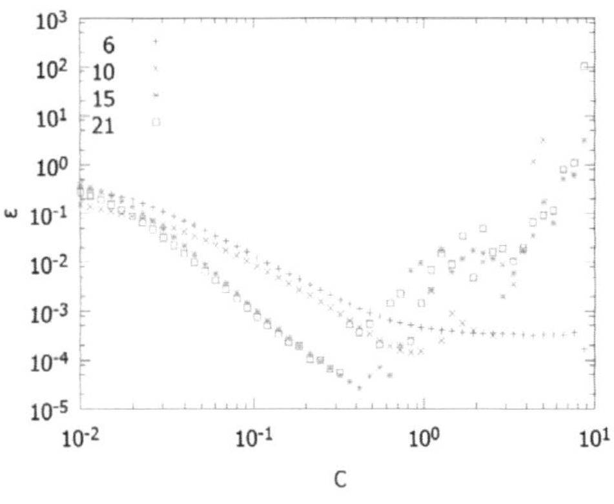

Figure 3. Error dependence on the shape parameter for node arrangement with 10 000 nodes.

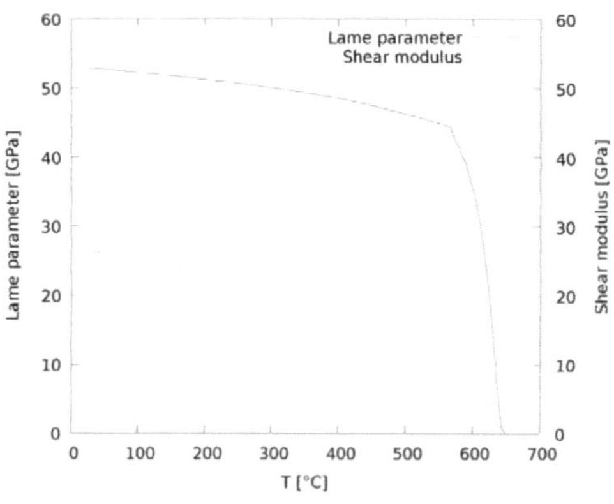

Figure 4. Temperature dependance of Lame parameter λ and shear modulus G for alloy AA6082 as obtanied from JMatPro [4].

$\hat{u}_x(\mathbf{p}) = -\alpha(1+\nu)(\cos 2\pi p_x - 1)/(2\pi(1-\nu))$ with $u_y = 0$. The error estimate is calculated by

$$\epsilon = \left(\sum_l (\mathbf{u}_l - \hat{\mathbf{u}}(\mathbf{p}_l))^2 / \sum_l \hat{\mathbf{u}}(\mathbf{p}_l)^2 \right)^{1/2} \tag{6}$$

The convergence order is shown in figure 2. We can see that the order of convergence depends on the number of the nodes used in the subdomain. The method is of the second order if 6 or 10 nodes are
used and of the fourth order if 15 or 21 nodes are used in each of the subdomains.

Another important issue in meshless methods based on radial basis functions is that they employ a shape parameter [6]. The dependence of the error on the shape parameter is shown in figure 3. We can see that for each number of subdomain points, there exists a certain optimal shape parameter. The optimal value gets smaller as the number of subdomain points is increased.

3.4. *The DC casting example*

The temperature field is assumed from our compatible meshless heat and mass transport model (HMM) for low frequency electro-magnetic DC casting, elaborated in [4]. The computational domain extends from 0 to 1 m height with the diameter of 0.15 m. The mechanics domain is smaller than the one used in the HMM, since only the area below the highest level of liquidus isotherm is considered, resulting in top 0.03 m of the domain being cut off. The mold extends from 0.905m to 1m. The procedure described in subsection 3.2 is used to treat the boundary conditions. The rest of boundary conditions are illustrated in figure 5. The material properties used in the model correspond to AA6082 alloy and are displayed in figure 4.

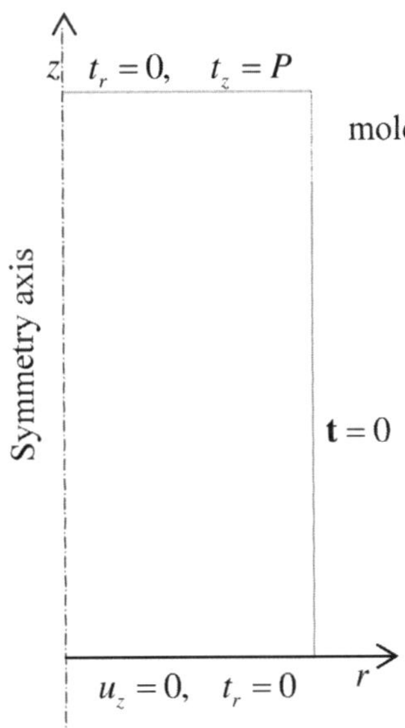

Figure 5. Boundary conditions used in the solid mechanics model.

In figure 6 the deformation in radial direction is shown, along with all three diagonal components of the stress tensor. The shrinkage at the bottom of the computational domain agrees with the expected shrinkage due to the difference in density between the solid and the liquid phase the temperature field for this case is shown in the leftmost plot in figure 7.

The fact that the stress state is very sensitive to the changes in temperature field is demonstrated in figure 7. In the figure we see the comparison between stress fields obtained for two different amplitudes of the driving current. The changes in temperature field are quite hard to notice. The sump depth is slightly larger and the mushy zone is a bit thinner for the larger driving current. The change in the stress state is however more dramatic. In the centre of the billet the nature of the stress changes from compression to tension, which may reduce the quality of the produced billet due to hot-tearing.

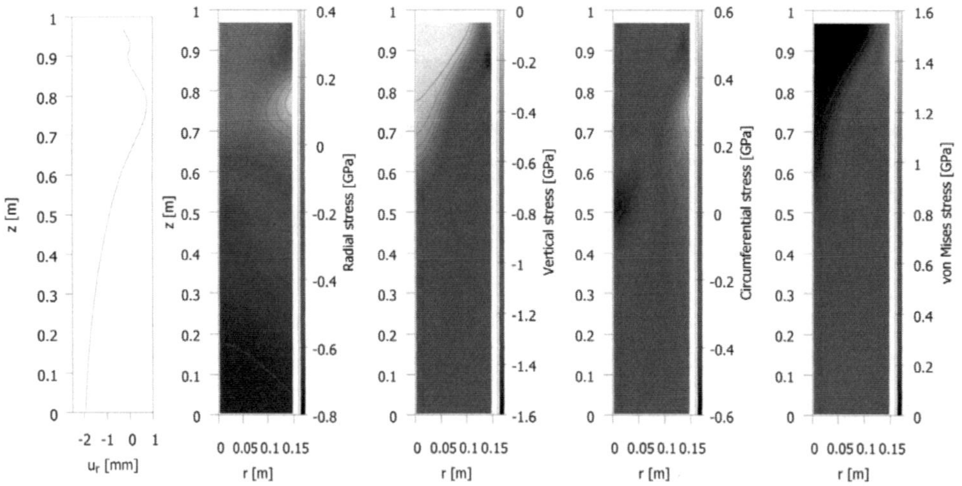

Figure 6. The radial deformation and stress results obtained using the LRBFCM model.

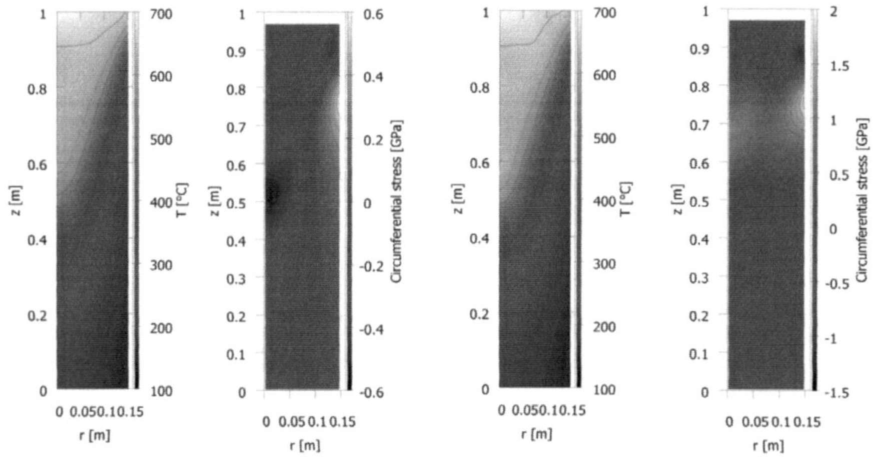

Figure 7. The effect of the amplitude of the driving current on the stresses in the billet. The temperature fields were obtained using in-house HMM [4].

4. Conclusions

It is evident that the developed LRBFCM is a viable novel method for modelling the linear thermoelastic problems as used in a basic thermomechanics model of DC casting. As is well known from the literature, the order of the convergence depends on the number of the nodes in each subdomain and the overall convergence strongly depends on the value of the shape parameter. Solutions obtained by the method for DC casting in the presence of low frequency electro-magnetic field are shown. The method performs well for the simple material model. The described developments thus successfully complement our previous work on meshless methods for fluid mechanics problems [8, 9]. In future we plan to add a viscoplastic constitutive model to the framework in order to more realistically describe the alloy behaviour at elevated temperatures.

Acknowledgements

Support from the Slovenian Grant Agency in the framework of the project L2-6775 and young researchers grants, Slovenian Ministry for Economy in the framework of the Voucher projects and IMPOL Aluminium Industry is kindly acknowledged.

References

[1] Kumar P P, Nallathambi A K, Specht E and Bertram A 2012 Mechanical behavior of mushy zone in DC casting using a viscoplastic material model Technische Mechanik 32 342-357
[2] Fjaer H G and Mo A 1990 ALSPEN-A mathematical model for thermal stresses in direct chill casting of aluminum billets Metallurgical Transactions B 21 1049-1061
[3] Drezet J M and Rappaz M 1996 Modeling of Ingot Distorsions During Direct Chill Casting of Aluminium Alloys Metallurgical and Materials Transactions A 27 3214 - 3225
[4] Lalpoor M, Eskin D G, Ruvalcaba D, Fjær H G, Ten Cate A, Ontijt N and Katgerman L 2011 Cold cracking in DC-cast high strength aluminum alloy ingots: An intrinsic problem intensified by casting process parameters Materials Science and Engineering: A 6 2831-2842
[5] Drezet J M and Phillion A 2010 As-Cast Residual Stresses in an Aluminum Alloy AA6063 Billet: Neutron Diffraction Measurements and Finite Element Modeling Metallurgical transactions A, 41 3396-3404
[6] Gerace A, Erhart K, Kassab A and Divo E 2013 A model-integrated localized collocation meshless method (MIMS) Computer Assisted Methods in Engineering and Science 20 207-225
[7] Fung Y C and Tong P 2008 Classical and computational solid mechanics (Singapore: World Scientific)
[8] Saunders N, Guo U K Z, Li X, Miodownik A P in Schillé J P 2003 Using JMatPro to model materials properties and behavior JOM: the journal of the Minerals, Metals & Materials Society, 55, 60-65
[9] Košnik N, Vertnik R and Šarler B 2014 Simulation of low frequency electromagnetic DC casting Materials Science Forum 790 390-395
[10] Mavrič B and Šarler B 2015 Local radial basis function collocation method for linear thermoelasticity in two dimensions International Journal of Numerical Methods for Heat & Fluid Flow To appear.
[11] Schenk O and Gartner K 2004 Solving Unsymmetric Sparse Systems of Linear Equations with PARDISO Journal of Future Generation Computer Systems, 20 475-487
[12] Liu G R 2003 Mesh Free Methods (Boca Raton: CRC Press)
[13] Vertnik R and Šarler B 2014 Solution of a continuous casting of steel benchmark test by a meshless method Engineering analysis with boundary elements 45 45-61
[14] Kosec G and Šarler B 2014 Simulation of macrosegregation with mesosegregates in binary metallic casts by a meshless method Engineering analysis with boundary elements, 45 36-44

$BaZrO_3$ refractory applied to the directional solidification of TiAl alloys

J He[1,a], C Wei[1,b], S Wang[1,c], D Meng[1,d], X Lu[1,2,e], H Wang[1,2,f], C Li[1,2,g]
[1] Shanghai Key Laboratory of Modern Metallurgy & Materials Processing, Shanghai University, Shanghai 200072, China
[2] Shanghai Special Casting Engineering Technology Research Center, Shanghai 201605, China

Email: [a] zhedudiao@163.com, [b] weichao1989@shu.edu.cn, [c] zmddahu@shu.edu.cn, [d] mengdz@sdju.edu.cn, [e] luxg@shu.edu.cn, [f] whb@shu.edu.cn, [g] chli@staff.shu.edu.cn

Abstract. Recently, much attention has been paid to the refractory used for the directional solidification process of TiAl intermetallics, the Y_2O_3 crucible/Y_2O_3 coated moulds seem to be the suitable candidate. However, the use of Y_2O_3 is limited by its low thermal shock resistance and high cost. In this work, a novel $BaZrO_3$ refractory was introduced to the directional solidification of TiAl intermetallics. The melt of this alloy contained in $BaZrO_3$ crucibles were heated for 30 minutes at 1600, 1650, 1700 °C, respectively, then cooled within the crucible for the investigation of the interface between the melt and the refractory. The scanning electron microscopy (SEM) was used to evaluate the surface topography and microstructure of the samples, and the energy dispersive spectroscopy (EDS) was used to analyse the chemical composition of the samples. The results indicated that no interfacial interaction layer and no obvious element diffusion were observed between the crucible and the metal, which may imply that the $BaZrO_3$ is a promising candidate of refractory for the directional solidification of TiAl alloys. This work can provide a basis for the further study of directional solidification of TiAl alloys by using $BaZrO_3$ shell mould.

1. Introduction

As a potential candidate of light-weight high-temperature structural materials, TiAl alloys have attracted much attention in recent years. However, the widespread use of TiAl alloys is limited by its intrinsic brittleness and poor ductility at room temperature [1-3]. It is demonstrated that TiAl alloys with a fully or nearly lamellar microstructure consisting of γ(TiAl) phase and α_2(Ti_3Al) phase have a good combination of strength and ductility, especially when the lamellar orientation is along the growth direction. However, it is difficult to control the lamellar orientation by simple casting process, thus, the directional solidification (DS) is proposed [4-6]. Recently, two typical processes were applied to discuss the DS of TiAl alloys, namely, the Bridgman process with a ceramic crucible, and the container less process like an optical floating zone method [7]. Compared with the latter, the Bridgman process enables higher superheating temperature, improved microstructure/composition homogeneity, as well as more complex geometry. However, this process was subjected to the high activity of titanium alloys and the inevitable reaction between the melts and the crucible at elevated temperatures, and until now, no refractory materials were found to be absolutely inert against the melt of titanium. In fact, much attention has been paid to the effect of refractory materials such as Al_2O_3,

CaO, and Y_2O_3 on the contamination of TiAl intermetallics during the DS process, and it seems that the Y_2O_3 crucible/Y_2O_3 coated moulds may be the suitable one. However, the use of Y_2O_3 was limited by its low thermal shock resistance and high cost. Therefore, it is necessary to develop a new generation ceramic refractory for DS of TiAl alloys [8-13].

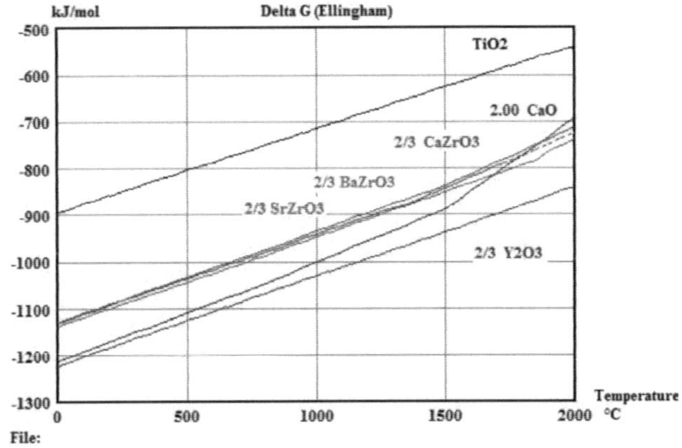

Figure 1. Variation of standard free energies of formation of some relative oxides with temperature.

Figure 1 shows the variation of the standard free energies of formation of some relative oxides with temperature. Generally, the Gibbs free energy of oxides below TiO_2 may be regarded as a candidate ceramic refractory for the DS of TiAl alloys, and other than the common oxides CaO and Y_2O_3, the zirconate family ($CaZrO_3$, $SrZrO_3$ and $BaZrO_3$) also show potential for the application. The zirconate family belongs to typical cubic perovskite, possess good mechanical strength, high thermal and chemical stability, and low coefficiency of thermal expansion. However, they are usually reported as high temperature protonic conductors rather than as the refractory used in melting titanium alloys. In our previous work, the zirconate family were made into ceramic crucibles and introduced to the vacuum induction melting of several titanium alloys, TiAl, TiNi, TiFe and Ti6Al4V [14-17], and it was found that $BaZrO_3$ exhibits a good chemical stability against the titanium melt. However, the long-term metal-crucible contact during the DS process required a higher chemical stability of the refractory than the refractory used in the induction melting. In this study, the melt of TiAl alloy contained in $BaZrO_3$ crucibles were heated for 30 minutes at 1600, 1650 and 1700 °C, respectively, then cooled within the crucible to investigate the long-term metal-crucible interaction. At the same time, the thermal shock resistance of home-made $BaZrO_3$ crucible was evaluated. The purpose is to provide a basic research for the further study of directional solidification of TiAl alloys by using $BaZrO_3$ shell mould.

2. Experiment procedure

The nominal composition of the raw alloy is Ti-48Al, with a sample size of Ø20mm×40 mm, and its chemical compositions are listed in table 1. Before melting, the alloy bars were polished, cleaned, and dried. The $BaZrO_3$ crucible was shaped by the isostatic pressing moulding technology with 2 wt% additional TiO_2 as a flux added into the $BaZrO_3$ powder, the green body was pre-sintered at 900 °C for 3 hours to remove organic compounds and achieve bisques with much higher mechanical strength. Subsequently, the final sintering was carried out at 1750 °C for 3 hours. Samples for thermal shock resistance test were shaped into disks of Ø20mm×1.8mm by isostatic pressing molding technology using 120MPa, and the test process followed the YB/T376.2-1995, which included four procedures: a) the disk samples were dried for 3 hours at 110°C to remove water; b) then quickly moved into a furnace which was preheated to 1100°C; c) after 20 minutes, the samples were moved out to cool in the air to room temperature; d) repeat processes b and c until cracks or distortions were observed on the disk surface. The final number of repetition was recorded to measure the thermal shock resistance.

Table 1. Composition of experiment used TiAl alloys.

Element	Al	Fe	O	N	Other	Ti
wt%	32.93	0.072	0.0473	0.0087	<0.01	Bal.

To well stimulate the long-term metal-crucible interaction during the DS process, the TiAl melt contained in the $BaZrO_3$ crucibles were heated for 30 minutes at 1600, 1650, and 1700 °C, respectively, in a carbon resistance furnace monitored by a Marathon type double colour infrared thermometer, and then cooled within the crucible. In order to reduce the oxygen content to a minimum level, the chamber was evacuated up to less than 10^{-2} Pa and backfilled with argon gas 5 times before heating, and the melting experiment was performed under a dry argon gas protective atmosphere at 0.06MPa.

The melted ingots were sectioned perpendicular to the cross section by electrical discharge machining (EDM) to samples with appropriate sizes and then mechanically polished following the standard metallographic procedures. The microstructure and interface morphology of the samples were examined using Scanning Electron Microscopy (JSM6700F), their elemental compositions were analysed by Energy Dispersive Spectrometer (JSM6700F), and the overall oxygen increase was measured by Nitrogen Oxygen Analyzer (TC-436) with a diameter of 2mm. Samples of length 10mm were cut from the alloys with the surface layer removed by grinding, and the samples were carefully drilled to achieve fine chips for oxygen measurements.

3. Results and discussion

3.1. Thermal shock resistance

The results of thermal shock resistance tests indicate that after 40 times heating-cooling cycle (heated to 1100°C in the furnace and then quickly moved out to the air and cooled to room temperature), no cracks or distortions are observed on the disk surface, which may imply that $BaZrO_3$ possesses a good thermal shock resistance. In fact, $BaZrO_3$ was successfully used as the crucible refractory in preparing the YBCO single crystal superconductor [18], and its thermal shock resistance can meet the requirement of DS process.

3.2. Interaction analyses

Due to the high activity of TiAl melt at elevated temperatures, a severe reaction layer might form between the melt and the crucible, and this layer may directly affect the solidification process. It was reported that when the Al_2O_3 crucible was used in the DS of TiAl alloys, the Al_2O_3 particles might dissolve and /or involve into the alloy matrix and form a reaction layer, while this reaction layer can be avoided when the Y_2O_3 crucible was applied [7]. However, due to its poor thermal shock resistance, Y_2O_3 was usually used as a coating or crucible with a properly designed porous structure, and this will bring the pollution of Y_2O_3 particles into the alloy [7,12,13,19].

Figure 2 shows the photographs of the crucible before and after melting (heated at 1650°C for 30 minutes). In order to analyse the metal-crucible interface, the whole crucibles containing the alloy were embedded with bakelite. It is observed that the alloy can be easily separated from the crucible when samples are sectioned perpendicular to the cross section by a water-cooled diamond wheel; the metal-crucible surface is still remained smooth, and no interaction layer is observed between the metal and the crucible, as seen in figure 2(b). This result is consistent with the prediction of thermodynamics, that the stability of $BaZrO_3$ tends to be higher than TiO_2 and no interaction exists between the TiAl metal and $BaZrO_3$.

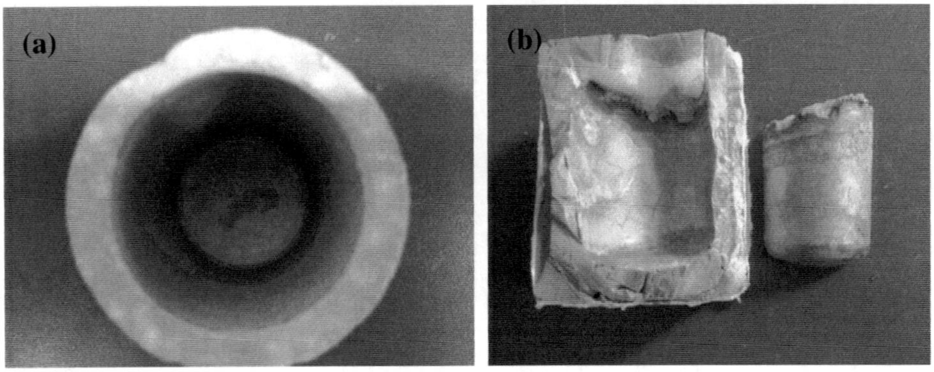

Figure 2. Photographs of crucible before and after melting (heated at 1650°C for 30 min).

Figure 3 shows the typical metal-crucible interfacial microstructures using the $BaZrO_3$ crucibles. It is reported that the Al_2O_3 particles may dissolve and /or became included into the alloy matrix and form a reaction layer between the metal-crucible surface, and the thickness of interaction layer will increase with the increasing of heating temperature and reaction time [7]. Compared with the Al_2O_3 crucible, the interface between the $BaZrO_3$ crucible and the TiAl melt remains quite clear at each temperature, as shown in figure 3 (a), (b), and (c). A layer with average thickness of about 15-60 um is attached to the alloy surface at 1600°C and 1700°C, and this layer is identified to be $BaZrO_3$ using the EDS analyses, as seen in table 2 (Point A, B and C). The formation of this layer may be affected by the density of crucible, but the maximum density of the crucible can only reach 97% in our lab. In addition, the exiting porous may become the channel of the penetration of metal into the refractory during the period of melting and solidification, and after solidification, the alloy finally stripped from the crucible surface to form a layer. By comparison, Y_2O_3 particles can be observed in the alloy matrix when it was used in the DS of TiAl alloys, as seen in figure 4(a). Meanwhile, the $BaZrO_3$ is just attached to the alloy surface, and no $BaZrO_3$ particles were found in the alloy matrix, as seen in figure 4(b). There exist some light-grey areas (point G and H), but their composition is still metal and not oxide, and the EDS results of alloy at different areas in table 2 also indicates that no obvious element diffusion is observed between the crucible and the metal.

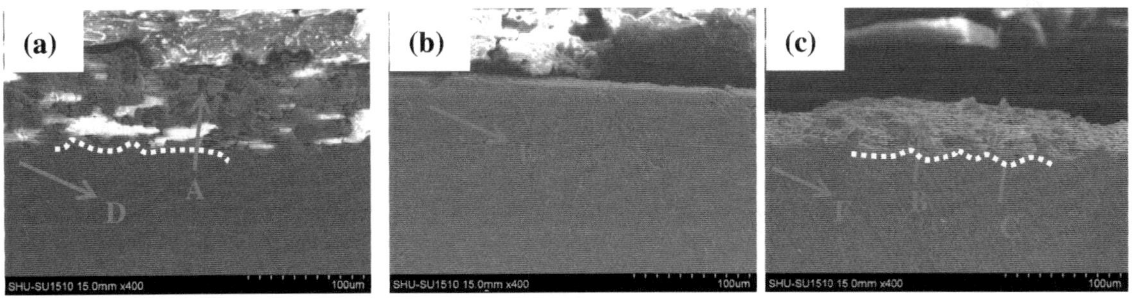

Figure 3. Metal–crucible interfacial microstructures of the ingots using $BaZrO_3$ crucibles, after 30 min at (a) 1600°C, (b) 1650 °C and (c) 1700 °C.

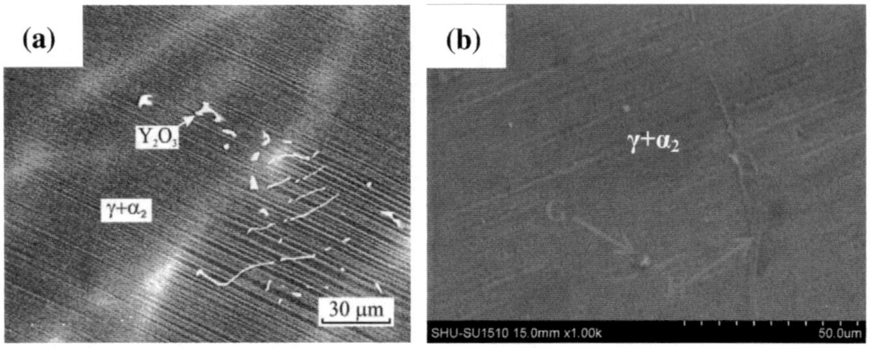

Figure 4. Microstructure of TiAl alloys melted by (a)Y_2O_3 crucible[19] and (b)$BaZrO_3$ crucible.

Table 2. The EDS results of alloy at different areas (at%).

	A	B	C	D	E	F	G	H
Ba	16.54	12.90	16.38	-	-	-	-	-
Zr	18.59	14.46	15.09	-	-	-	-	2.79
O	64.87	68.08	64.40	-	-	-	-	-
Ti	-	1.98	-	45.71	47.43	48.04	35.60	79.95
Al	-	2.59	4.13	54.29	52.57	51.96	64.40	17.26

3.3. Increase of oxygen in ingots

TiAl alloys are sensitive to interstitial atoms such as nitrogen, hydrogen, and oxygen. Among these, the oxygen concentration possesses a great effect on the properties and solidification process. During the DS of TiAl alloys in a ceramic crucible, there is unavoidable contact between the TiAl melt and the ceramic material and this might cause the increase of oxygen in ingots.

Figure 5 is the oxygen contents in ingots at heating temperatures of 1600, 1650, and 1700°C, and they are 2870, 3110, and 3730 ppm, respectively. Generally, it is accepted that the oxygen content in cast TiAl-based alloys should not exceed the limit of about 1200 ppm [19]. Although the recent results surpasses the limit by about 2000 ppm, the maximum oxygen content of 3730 ppm at 1700°C is still below a critical value of about 4800 ppm, which would lead to a change of primary solidification β phase to the α phase[20]. The reason for the oxygen increase may be from two main sources [12,19]; one is from the original master alloy, and the other is from the atmosphere since the vacuum of the present experiment is doubtful. This may contribute to the high oxygen content, but the truth is still unclear, and further study is necessary before the $BaZrO_3$ refractory can be commercially applied to the DS of TiAl alloys.

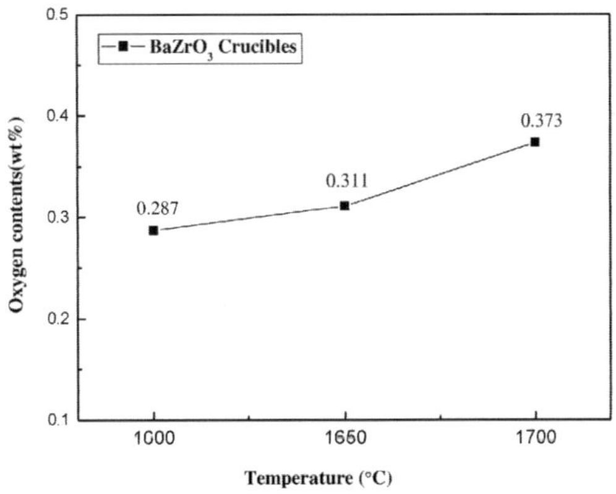

Figure 5. Oxygen contents in ingots at heating temperatures 1600, 1650 and 1700°C.

4. Conclusion

To well stimulate the long-term metal-crucible interaction during the DS process, the TiAl melt contained in $BaZrO_3$ crucibles were heated to 1600, 1650, and 1700 °C, respectively, in a vacuum furnace, and the thermal shock resistance of the crucible was measured, and the metal-crucible interaction and increase of oxygen in the alloy were also investigated. The main conclusions are as follows:

1. The thermal shock resistance test indicates that $BaZrO_3$ possesses a good thermal shock resistance and can meet the requirement of DS process.
2. At 1600, 1650 and 1700 °C, no interaction layer forms between melt of TiAl and $BaZrO_3$ crucible, no obvious element diffusion is observed between melt of TiAl and $BaZrO_3$.
3. No $BaZrO_3$ particles are found in the alloy matrix, the maximum oxygen content (3730 ppm for the samples of 1700°C) is below the critical value of about 4800 ppm.

Acknowledgments

The authors thank the National Basic Research Program of China (973 Program No. 2014CB643403), the National Nature Science Foundation of China (Grant No. 51225401 and No. 51374142) and the Science and Technology Fund of Scientific Committee of Shanghai (No.11520500100 and No. 14JC1491400).

References

[1] Clemens H, 2000 Heinrich Kestler. Processing and applications of intermetallic γ-TiAl-based alloys [J]. *Advanced engineering materials,* **2** 551-569.

[2] Saari H, Beddoes J, Seo D Y, etal. 2005 Development of directionally solidified γ-TiAl structures [J]. *Intermetallics,* **13** 937-943.

[3] Yamaguchi M, Inui H, Ito K. 2000 High temperature structure intermetallics [J]. *Acta mater,* **48** 307-322.

[4] Kishida K, Johnson D R, Masuda Y, etal. 1998 Deformation and fracture of PST crystals and directionally solidified ingots of TiAl-based alloys [J]. *Intermetallics,* **6** 679-683.

[5] Kim S E, Lee Y T, Oh M H, etal. 2000 Directional solidification of TiAl-Si alloys using a polycrystalline seed [J]. *Intermetallics,* **8** 399-405.

[6] Zhang H, Tang X, Zhou C, etal. 2013 Comparison of directional solidification of γ-TiAl alloys in conventional Al2O3 and novel Y2O3-coated Al2O3 crucibles [J]. *Journal of the European ceramic society,* **33** 925-934.

[7] Lapin J, OndrúŠ L, Nazmy M. 2002 Directional solidification of intermetallic Ti-46Al-2W-0.5Si alloy in alumina moulds [J]. *Intermetallics,* **10** 1019-1031.
[8] Lapin J, Gabalcová Z. 2011 Solidification behavior of TiAl-based alloys studied by directional solidification technique [J]. *Intermetallics,* **19** 797-804.
[9] Muto S, Yamanaka T, Johnson D R, etal. 2002 Effects of refractory metals on microstructure and mechanical properties of directionally solidified TiAl alloys [J]. *Materials science and engineering A,* **329-331** 424-429.
[10] Kuang J P, Harding R A, Campbell J. 2000 Investigation into refractories as crucible and mould materials for melting and casting γ-TiAl alloys [J]. *Materials science and technology,* **16** 1007-1016.
[11] Renjie C, Huarui Z, Xiaoxia T, etal. 2011 Interactions between γ-TiAl melt and Y2O3 ceramic material during directional solidification process [J]. *Trans. Nonferrous met. Soc.China,* **21** 2415-2420.
[12] Tetsui T, Kobayashi T, Kishimoto A, etal. 2012 Structural optimization of an yttria crucible for melting TiAl alloy [J]. *Intermetallics,* **20** 16-23.
[13] Zhao Z, Kailiang Z, Lanjie L, etal. 2013 Preparation of BaZrO3 Crucible and Its Interfacial Reaction with Molten Titanium Alloys [J]. *Journal of the Chinese ceramic society,* **41** 1278-1283.
[14] Zhao Z, Fangyuan X, Ming Z, etal. 2013 Vacuum induction melting of TiNi alloys using BaZrO3 crucibles[J]. *Materials science forum,* **765** 316-320.
[15] Lanjie L, Guangyao C, Zhao Z, etal. 2013 Interface Reaction Between Refractory BaZrO3 and TiM(M=Fe, Ni) Melt[J]. *Hot working technology,* **42** 49-54.
[16] Bo Y, Kailiang Zhu, Xionggang Lu, etal. 2012 Preparation of TiFe based alloy melted by CaZrO3 crucible and its hydrogen storage properties [J]. *The Chinese tournal of process engineering,* **12** 849-856.
[17] Erb A, Walker E, Fliikiger R. 1995 BaZr03: the solution for the crucible corrosion problem during the single crystal growth of high-TC superconductors REBa2Cu3O7-δ; RE = Y, Pr[J]. *Physica C,* **245** 245-251.
[18] Lapin J, Gabalcová Z, Pelachová T 2011. Effect of Y2O3 crucible on contamination of directionally solidified intermetallic Ti-46Al-8Nb alloy [J]. *Intermetallics,* **19** 396-403.
[19] Zollinger J, Lapin J, Daloz D, etal. 2007 Influence of oxygen on solidification behavior of cast TiAl-based alloys[J]. *Intermetallics,* **15** 1343-1350.

Effect of cerium addition on casting/chill interfacial heat flux and casting surface profile during solidification of Al-14%Si alloy

V Vijeesh and K N Prabhu[a]
Department of Metallurgical and Materials Engineering, National Institute of Technology Karnataka Surathkal, Mangalore 575025, India

E-mail: [a]prabhukn_2002@yahoo.co.in

Abstract. In the present investigation, Al-14 wt. % Si alloy was solidified against copper, brass and cast iron chills, to study the effect of Ce melt treatment on casting/chill interfacial heat flux transients and casting surface profile. The heat flux across the casting/chill interface was estimated using inverse modelling technique. On addition of 1.5% Ce, the peak heat flux increased by about 38%, 42% and 43% for copper, brass and cast iron chills respectively. The effect of Ce addition on casting surface texture was analyzed using a surface profilometer. The surface profile of the casting and the chill surfaces clearly indicated the formation of an air gap at the periphery of the casting. The arithmetic average value of the profile departure from the mean line (R_a) and arithmetical mean of the absolute departures of the waviness profile from the centre line (W_a) were found to decrease on Ce addition. The interfacial gap width formed for the unmodified and Ce treated casting surfaces at the periphery were found to be about 35μm and 13μm respectively. The enhancement in heat transfer on addition of Ce addition was attributed to the lowering of the surface tension of the liquid melt. The gap width at the interface was used to determine the variation of heat transfer coefficient (HTC) across the chill surface after the formation of stable solid shell. It was found that the HTC decreased along the radial direction for copper and brass chills and increased along radial direction for cast iron chills.

1. Introduction

The simulation of casting solidification is useful for determination of temperature distribution inside the casting, fraction solid, location of defects like shrinkage porosity, hot tear and also for prediction of the microstructure and mechanical properties of the casting. The accuracy and reliability of a simulation model mainly depends on the input parameters like thermo-physical properties of the mold/ casting and the boundary interfacial heat transfer coefficient [1,2]. The casting/chill interfacial heat transfer is dependent on factors like alloy, super heat, latent heat of fusion and chill variables like surface roughness, chill size and thermophysical properties of the chill [3,4]. However, the influence of Ce melt treatment on interfacial heat transfer is not yet investigated.

Al-Si alloys with Si content greater than 13 wt. % are categorized as hypereutectic Al-Si alloys. Recently, Ce addition to hypereutectic Al-Si alloys has gained attention due to its ability in refining and modifying primary and eutectic silicon simultaneously [5]. In hypoeutectic Al-Si alloys (Si<12%), the coarse acicular eutectic silicon is transformed into a fine fibrous form by the addition of elements like sodium, salts of sodium and master alloys of strontium. The process is known as 'modification'. The modification process is known to increase the mechanical and physical properties of the alloy [6].

Prabhu and Ravishankar [7] studied the effect of Na modification on Al-13%Si alloy and found that the modifying of the melt improved the heat transfer from the solidifying casting to the chill material. This was attributed to the increased ability of the liquid metal to wet the chill surface and decreased surface tension of liquid due to the addition of sodium.

The magnitude of interfacial heat transfer depends on the nature of surfaces in contact and the interfacial gap formed during solidification. The effect of varying surface roughness on interfacial heat transfer coefficient was studied by several researchers. Muojekwu et al. [8] studied the effect of chill surface roughness on Al-Si alloys and found that the heat transfer decreased marginally by increasing the roughness of the chill by 500 times. Griffiths and Kayikci [9] found that variation in interfacial heat transfer coefficient with surface roughness was negligible. They found that the sum roughness parameter ($R_z = (R^2_{z(chill)} + R^2_{z(casting)})^{1/2}$) remained constant for all chill surface roughnesses and the negligible variation in interfacial heat transfer coefficient was attributed to unchanged sum roughness parameter. They also reported a curvature formation on the surface of the casting. The casting surfaces were seen to deform convexly soon after pouring, indicating that the transfer of heat is preferentially at the centre of the casting where the two surfaces are in contact.

In the present investigation, the influence of Ce addition on heat transfer at the casting/chill interface was assessed during solidification of Al-14%Si alloy against copper, brass and cast iron chills. The study also aimed to measure the variation of casting surface roughness, surface profile and correlate it with the interfacial heat transfer.

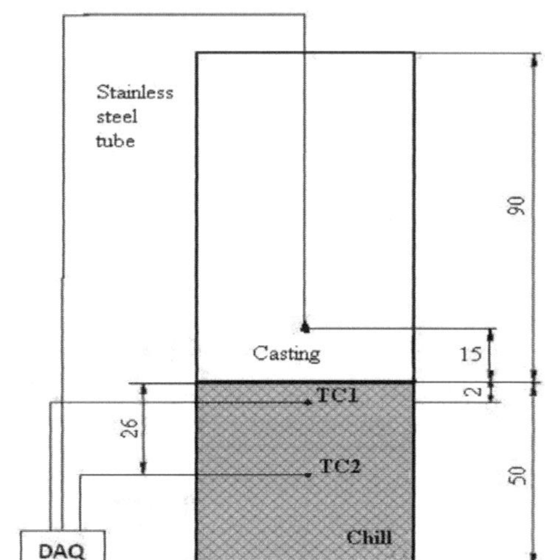

Figure 1. Schematic sketch of the solidification set up.

2. Experimental

In the present investigation, heat transfer studies were carried out by using copper, brass and cast iron chills. Chills having varying thermal conductivities were selected to obtain different cooling rates. Surface roughnesses of chills before and after casting were measured using a surface profilometer (Form Talysurf Series 50mm Intra). The chill roughness before experiment was set at 0.5±0.05μm. A schematic sketch of the experimental setup is shown in Figure 1. A stainless steel tube of 50 mm outer diameter with a wall thickness of 1 mm was attached at the top of the chill. Stainless steel was selected as it has a low thermal conductivity of 16 W/mK. Two holes of 1 mm diameter and 25 mm depth were drilled into the cylindrical surface of the chill at distances of 2 mm (TC1) and 26 mm (TC2) respectively. Two mineral insulated thermocouples (K-type) were inserted into the drilled holes and were connected to a PC through a data acquisition system (NI USB 9162). A preheated crucible containing about 400g of Al-14%Si alloy was heated to about 750°C in an electric furnace. The melt was then maintained at that temperature for about 30 minutes. The molten alloy was degassed by introducing about 5g of hexachloroethane tablet into the melt. Cerium was added to the melt in varying quantities (0.5wt. %, 1wt. %, 1.5wt. % and 2wt. %), thirty minutes prior to the start of thermal analysis experiments. The liquid metal was quickly poured into the stainless steel tube with a chill at the bottom. The thermal history at thermocouple locations was recorded using a computerized data acquisition system. The measured temperatures at TC1 and TC2 were used as input to inverse heat conduction model TmmFE Inverse solver (Thermet solutions Private Ltd, Bangalore) to determine the heat flux across the casting/chill interface and the chill.

3. Results and discussion

Figures 2 (a) and (b) show the measured temperature inside the copper chill at two locations (TC1 and TC2) without Ce and doped with 1.5% Ce respectively. TC1 and TC2 show temperature at locations 5 and 26 mm from the casting/chill interface, respectively. It is evident from the figures that the Ce addition results in increases in temperatures inside the chill. The temperatures at TC1 showed a peak value of 265 °C at about 4-5s after pouring the alloy without any Ce. With Ce addition, the peak at TC1 increased to 300°C. The chill surface temperatures and the heat flux across the chill/casting interface were estimated by solving the inverse heat conduction problem. The surface temperature of the chill at the interface corresponding to the peak temperature at TC1 for the alloy without addition was 270.4 °C. Similar temperature profiles were obtained for brass and cast iron chills during solidification. The peak surface temperature for brass and cast iron chills for solidification of alloy without Ce were found to be 388.8 °C and 445 °C respectively.

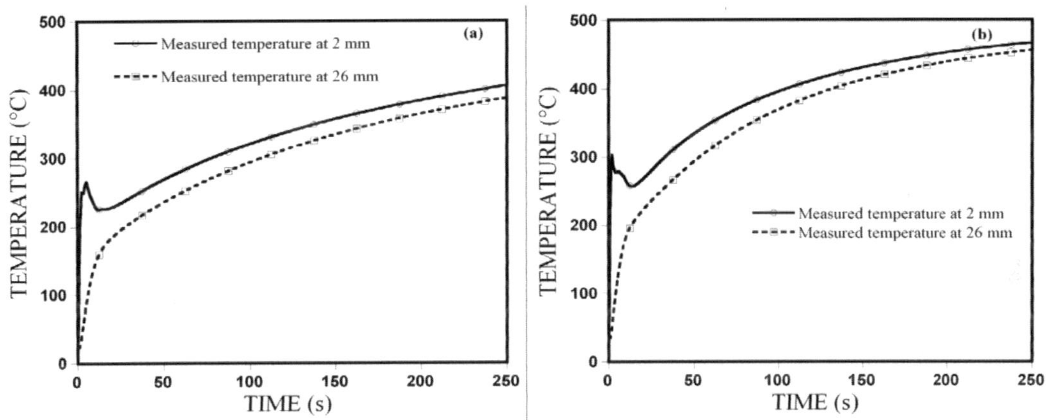

Figure 2. Temperature-time curves at different locations from interface in the copper chill (a) without addition (b) with 1.5 wt. % of Ce.

The heat flux transients estimated using the inverse solver is shown in figure 3(a). The interfacial heat flux shows a steep increase and reaches a peak value. The actual contact between chill surface asperities and the casting is an important factor that determines the rate of interfacial heat transfer. As the solidification proceeds, the interfacial heat flux decreases to a lower value due to the formation of a gap/ non-conforming contact at the interface. The peak heat flux for copper, brass and cast iron chilled alloys without Ce were found to be 6909, 4790 and 3348 kW/m^2 respectively. The addition of Ce to the alloy increased the heat transfer to the chill in all cases. The increase in peak heat flux with Ce addition is shown in figure 3 (b). For example, the peak heat flux during solidification of 1.5%Ce alloy against the copper chill was 9541 kW/m^2, showing 38 percent increase in the heat flux. The corresponding increases in peak heat flux for brass and cast iron chills were 42% and 43%, respectively.

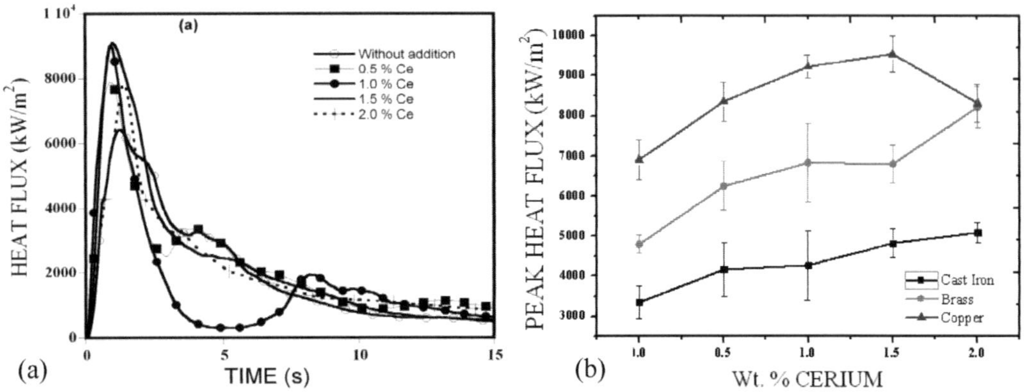

Figure 3. Effect of Ce melt treatment on (a) heat flux of copper chill (b) peak heat flux of various chills.

The improvement in heat transfer due to Ce addition was mainly due to two factors: (i) improved contact between melt and the chill surfaces due to low surface tension of the melt, and (ii) higher thermal conductance of the solid shell formed due to the transformation of silicon morphology from acicular to fibrous [7]. Since the atomic radius ratio of Ce to Al alloy is greater than 15%, the solid solution of Ce in aluminum is difficult. Moreover, Ce is a surface active element which tends to get concentrated at the surface. According to the Gibbs Adsorption equation,

$$\tau = -\frac{C}{RT}\frac{d\sigma}{dc} \tag{1}$$

where, C is solute concentration, R is Boltzmann constant, T is temperature, τ is the mass fraction of the enriched solutes per unit area on the surface of the molten metal, σ is the surface tension of the molten alloy. τ is positive for a surface active element and for $dc>0$, and the change in surface tension ($d\sigma$) becomes negative [10]. This implies that the surface active element decreases the surface tension of molten alloy.

Ce is known to simultaneously modify and refine both eutectic and primary silicon in hypereutectic Al-Si alloys [11]. Previous studies show that Si modification in Al-Si alloys improved the mechanical properties as well as the physical properties [12]. It has been reported that the addition of Ce to Al-0.3Si-Mg alloy significantly increased in the electrical conductivity of the alloy. The presence of 0.3%Ce in the alloy increased the %IACS from 52.1 to 56.3 respectively [13]. Chang et al. suggested that the Ce addition modifies the solid-liquid interfacial energy for silicon modification [14]. Yet in another study it was reported that the Ce addition decreased the surface tension of molten Mg-9Al alloy [15]. Hence, the increase in heat flux is due to the combined effect of decreased surface tension and improved conductivity of the alloy.

To analyze the influence of Ce addition on contacting surfaces, the surface roughness values (R_a) and waviness (W_a) values of castings were measured. Dimensionless parameters based on surface roughness and waviness were defined as ($R_{a(casting)}/R_{a(chill)}$) and ($W_{a(casting)}/W_{a(chill)}$) respectively. The variation of these dimensionless parameters with Ce addition is shown in figure 4(a) and 4 (b). The results indicate that the roughness ratio ($R_{a(casting)}/R_{a(chill)}$) decreased with Ce addition for casting solidified against copper and brass chills and it increased with Ce additions for castings solidified against the cast iron chill. The waviness ratio ($W_{a(casting)}/W_{a(chill)}$) remained more or less constant for copper and brass chilled castings, but it increased significantly for alloys solidified on cast iron. The surfaces of cast iron chilled castings showed higher waviness (W_a) values up to 38 μm for 1.5 wt. % Ce compared with copper and brass chills. Due to the lower thermal conductivity of the cast iron chill, during solidification of Ce added alloys the surface temperature attained at the casting/chill interface was well above the eutectic temperature. For example, the surface temperature of the cast iron chill was found to be 590 °C for 1.5 wt. % Ce additions. The exposure of the casting surface to such a high temperature would have resulted in re-melting of the solid shell formed, thus affecting the casting surface roughness and profile.

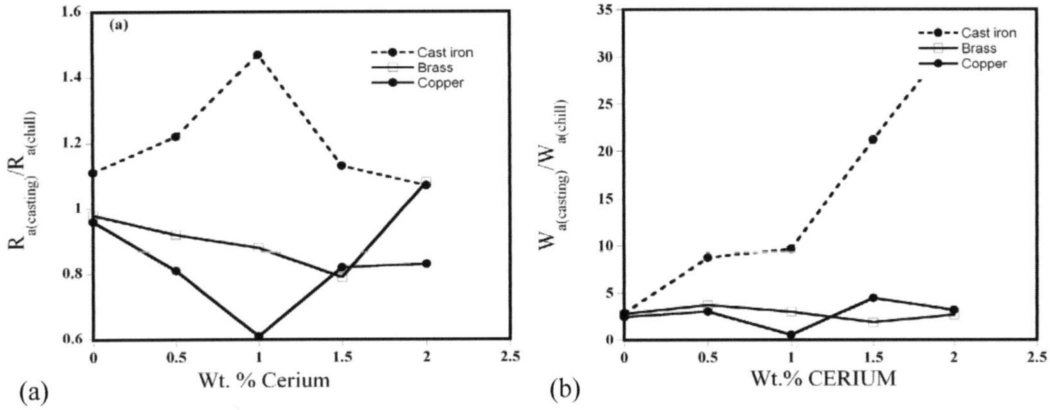

Figure 4. Effect of varying cerium content on (a) roughness factor (b) waviness factor.

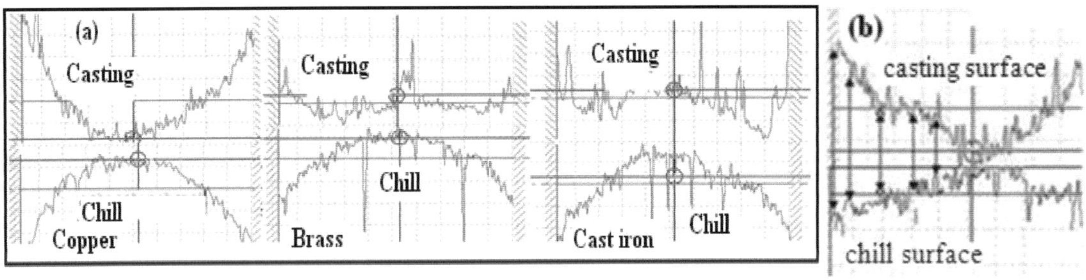

Figure 5. (a) Selected surface profiles across the chill and casting for different chill solidification (b) Schematic sketch of the gap width measurement using waviness of casting and chill surfaces.

The waviness profiles taken across the casting surfaces showed a curvature and were found to vary significantly with chill material. The profile was convex towards copper and brass chill, whereas, the profile was concave towards cast iron chill. The examples of these profiles are shown in figure 5(a). The waviness profile of copper and brass chills indicates that the heat transfer would have occurred at the centre of the profile interface. The gap between the casting and the chill increased in the radial direction. Assuming that the heat transfer across the casting/chill interface would be maximum at central contact point and the gap formed in the radial direction will be filled by atmospheric air, the gap width across the radial direction was determined. Figure 5(b) shows a schematic sketch showing the measurement of gap width using surface profiles for copper and brass castings. Since the cast iron castings are concave towards the chill, it is probable that casting and chill are in perfect contact in circumferential region and a gas gap was formed at the centre.

The gap width data was used to determine the heat transfer coefficient (HTC) using the equation

$$h = \frac{k_a}{\delta} \tag{2}$$

where, k_a is the thermal conductivity of the air at the casting/chill interface, δ is the gap width at the chill and casting interface. Figure 6 shows the spatial distribution of heat transfer coefficient estimated by measuring the gap width along the radial direction. The results show that for copper and brass chilled castings the HTC values decrease with increase in radial distance from the centre due to the increase in gap width.

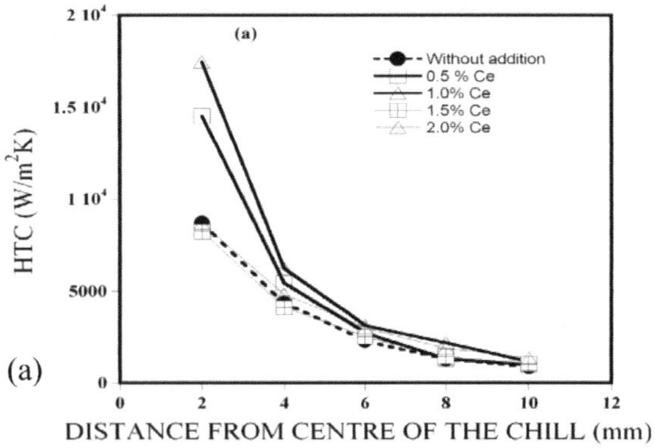

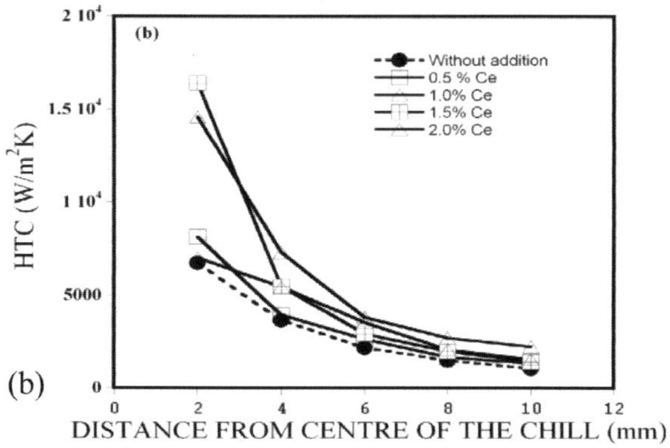

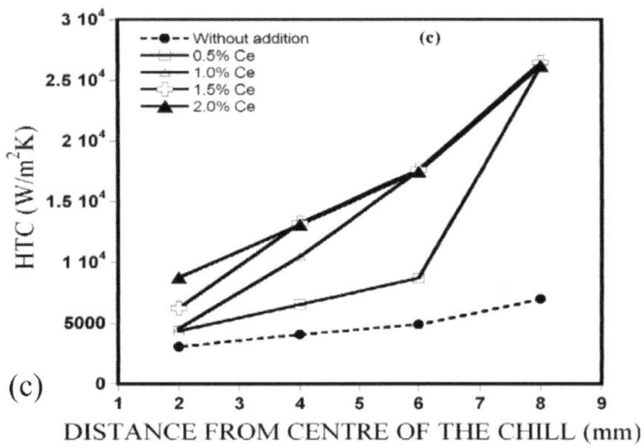

Figure 6. Spatially dependent heat transfer coefficient values along the radial direction for (a) copper (b) brass (c) cast iron.

The HTC increased in radial distance from centre for cast iron chilled castings. The minimum HTC was at the centre of the contacting surfaces. The Ce added alloys showed higher HTC values than the alloys without any addition. For casting surfaces solidified on the copper chill, the widths of the gap at the circumference region for alloys without Ce and with 1.0 wt. % Ce were found to be 35 μm and 13 μm respectively. On Ce treatment the gap width was reduced by about 62%. The reduction in the gap width between casting and chill surfaces results in the increase in heat flux across the interface. The casting solidified against the cast iron chill also showed an increase in the spatially dependent HTC with Ce addition. This was due to the improved contact of the chill and casting surfaces in the circumferential region.

4. Conclusions

- The melt treatment of Al-14%Si alloy using Ce resulted in higher interfacial heat flux for all chilling conditions. On 1.5% Ce addition, the peak heat flux increased by 38%, 42% and 43% for solidification against copper, brass and cast iron chills, respectively.
- The surface profile analysis revealed that the Ce addition resulted in smoother casting surfaces in the case of copper and brass chills and a rougher surface in the case of the cast iron chilled alloys. Dimensionless ratios for roughness ($R_{a(casting)}/R_{a(chill)}$) and waviness ($W_{a(casting)}/W_{a(chill)}$) decreased with Ce addition for copper and brass chills and increased with cast iron chilled alloys.
- The casting surface profile was convex towards the copper and brass chill surfaces and concave towards the cast iron chill surface. The opposite trend observed in cast iron chilled alloys was due to the formation of rough surface caused by re-melting of the solidified shell formed.

- The spatially dependent gap width was measured and was used to calculate the HTC along the radial direction. The HTC values were found to be lower in the circumferential region and higher at the centre for copper and brass chills. An opposite trend was observed in the cast iron chills with HTC at the centre. The spatially dependent HTC increased with Ce addition for all castings due to the reduction in the gap formed between the contacting surfaces.

Acknowledgment
One of the authors (VV) thanks National Institute of Technology Karnataka, (NITK) Surathkal, and Mangalore, India for the Research Scholarship.

References
[1] Kumar T S P, and Prabhu K N 1991 *Metall. Trans. B* **22 B** 717.
[2] Abis S 1986 *Metall. Trans. B*, **17 B** 209.
[3] Velasco E, Talamantes J, Cano S, Valterra S, Mojica J E, and Colas R 1999 *Mater. Trans. B* **30B** 773.
[4] Gafur M A, Haque M N, and Prabhu K N 2003 *J. Mater. Process. Tech.* **133** 257.
[5] Kores S, Voncina M, Kosec B, Mrvar P, and Medved J 2010 *Mater. Tech.* **44 (3)** 137.
[6] Hegde S, and Prabhu K N, J. 2008 *Mater. Sci.* **43** 3009.
[7] Prabhu K N, and Ravishankar B N, 2003 *Mater. Sci. Engg A*, **360** 293.
[8] Muojekwu C A, Samarashekara I V, and Brimacombe J K 1995 *Met. Mat. Trans B* **26B** 361.
[9] Griffiths W D, and Kayikci R J 2007 *Mater. Sci.* **42** 4036.
[10] Zhenghua D, Huaji L, Wanjun Z, and Weibin L 2013 *China Foundry* **10(2)** 108.
[11] Li Q, Xia T, Lan Y, Zhao W, Fan L, and Li P 2013 *J. Alloys. Comp.* **562** 25.
[12] Vijeesh V, Ravi M, and Prabhu K N 2013 *Mater. Perform. Character.* **2(1)** 296.
[13] Liao H C, Liu Y, Lü C L., and Wang Q G, 2015 *Int. J. Cast Met. Res.*, DOI 10.1179/1743133615Y.0000000002.
[14] Chang J Y, Kim G H., Moon I G and Choi C S, 1998 *Scripta Mater.* **39(3)** 307.
[15] Zhenghua D, Huaji L, Wanjun Z, and Weibin L 2013 *Res. Dev.* **10(2)** 108.

Effect of Ce melt treatment on solidification path of ZA8 alloy

R Sudheer, V Vijeesh and K N Prabhu[a]
Department of Metallurgical & Materials Engineering, National Institute of Technology Karnataka, Surathkal, Mangalore 575 025, India

E-mail: [a]prabhukn_2002@yahoo.com

Abstract. The solidification path of ZA8 alloy with Ce addition was characterized using Newtonian technique of thermal analysis. The solidification events were determined using cooling curve and its first derivative curve. The microstructure and chemical composition of various phases in the alloy were studied using EDS, SEM and XRD techniques. It was found that the addition of Ce did not cause formation of new phases. However, it hinders the nucleation of stable β dendrites in the alloy. The presence of Ce promotes the eutectoid phase transformation and increases the hardness of the alloy. Latent heat of solidification and heat of eutectoid transformation were found to increase on Ce addition. The upward solidification of the alloy against Cu chill was analysed. Chilling had significant influence on solidification parameters, and caused refinement of the microstructure. The addition of Ce to the melt had no effect during chill casting of the alloy.

1. Introduction

Zinc–Aluminium base alloys are used for many industrial applications by virtue of their excellent physical and mechanical properties such as good machinability, high as-cast strength, excellent bearing properties, corrosion resistance, fine surface finish, ability to cast thin sections, as well as low energy requirements (for melting). Among ZA alloys only ZA8 can be hot chamber die cast due to its low melting point. This reduces the production cost of ZA8 for industrial purpose compared to that of other ZA alloys [1,2]. The mechanical properties of ZA alloys mainly depend on the presence and amount of various impurities, and the final microstructure of casting [2].

The effect of melt treatment on the cooling curve, microstructure and various mechanical properties of the alloy is discussed by various researchers. Turk et al. studied the effect of addition of Mn on the microstructure and mechanical properties of ZA8 alloy. The addition of Mn modifies the as cast microstructure by refining the β dendrites and increasing the amount of eutectic phase, and also forms a fine and irregularly shaped intermetallic compound (MnAl6). The hardness and creep strength of the ZA8 alloy increased with manganese content, whereas 0.2% yield, tensile and impact strengths of the alloy improved with increasing Mn up to 0.045 wt-% and then reduced with a further increase in manganese. [1]. Turk et al. reported that ZA-8 alloy when modified with Pb, Sn and Cd resulted in higher wear resistance. Metallographic studies showed that the addition of Pb, Sn and Cd resulted in modification of the microstructure of ZA8 alloy [3].

Ramesh et al. studied the effect of Mn addition to ZA8 alloy on thermal analysis parameters, heat transfer and microstructure. The addition of Mn to ZA8 alloy increased the contact angle, indicating decreased wettability of the modified alloy on the chill surface. The microstructure of ZA8 with Mn showed an increased amount of β phase and a decreased amount of eutectic [2]. Ramesh and Prabhu found that the chill material and coating had a significant effect on the cooling curve of ZA8 alloy. It was found that chilling during solidification causes the morphology of dendrites transform to nearly rounded shape with refinement of lamellar eutectic [4,5].

Zhu et al. observed that the Cu modified ZA8 alloy consists of a tree stem shaped eutectic structure together with flower-like cores of solidification, which were surrounded by a large amount of eutectoid structure. Further ageing at 150°C results in decomposition of zinc rich η phase and four-phase transformation, $\alpha+\varepsilon\rightarrow T'+\eta$ [6].

Krupinska et al. reported that the increase in cooling rate results in refinement of the microstructure as well as increase in the alloy hardness. They found that the addition of La and Ce into ZnAl8Cu1 alloy causes only refinement of grains and sub grains in the microstructure but no occurrence of new phases or eutectics during solidification of the alloy [7].

Thermal analysis is a useful tool for the non-destructive assessment of melt quality. The Newtonian and Fourier techniques have been successfully used to calculate latent heat and solid fraction distribution using cooling curve analysis [8]. Haq et al. have shown that to obtain the baseline for calculating the latent heat, polynomial and exponential fitting methods show inaccuracy and inconsistency in the obtained results. On the other hand, the results obtained by fitting the first derivative curve (FDC) linearly are reproducible [9].

Among ZA alloys which are known for superior hardness, wear strength and other mechanical properties, ZA8 alloy has found limited industrial applications. Melt treatment of ZA8 alloy with Ce to modify microstructure and improve mechanical properties can prove to be beneficial. Though thermal analysis of various Al alloys have been done in the past, the solidification path of ZA8 alloy has not been characterized. The data obtained from thermal analysis would be useful for melt quality control in foundries.

2. Experimental

The nominal composition of ZA8 alloy is given in Table 1. ZA8 samples of about 800g were heated in a clay-graphite crucible up to 480°C, and Ce ingots (Alfa-Aesar, USA) were then added to prepared samples having varying Ce compositions of 0.25%, 0.5% and, 1.0% by weight. The furnace holding time was 30 minutes.

Table 1. Composition of ZA8 alloy.

Elements	Al	Cu	Mg	Fe	Sn	Cd	Pb	Ni	Zn
Wt-%	8.2-8.8	0.8-1.3	0.02-0.03	0.065	0.005	0.005	0.005	...	Balance

Calibrated K-type thermocouples (0.45ϕ) enclosed in twin bore ceramic beads were used as temperature sensors. A scanning frequency of 10Hz was selected during the temperature data acquisition. A Data Acquisition System (NI USB 9162) was used for this purpose. The temperature data acquired from thermal analysis was later processed and analyzed to obtain cooling curves and first derivatives of cooling curves (FDC).

Samples were tested in X-ray diffractometer (JEOL JDX-8P-XRD) to obtain XRD patterns. Polished samples were etched with 10% HF. JEOL Analytical SEM (JSM 6380LA) and ZEISS optical microscope (AXIOIMAGER) were used for microstructure examination. Energy Dispersive Spectroscopy (EDS) analysis was also carried out. For hardness testing, samples of ZA8 alloy having 0.05m thickness and 0.05m diameter were prepared and tested as per the ASTM E10 standards. The Brinell Hardness Number (BHN) was measured at 500kg load with a steel ball indenter of 10mm diameter. Cu chills of length 0.05m and diameter of 0.05m were prepared to study the influence of high cooling rates on upward solidification of ZA8 alloy.

3. Results and discussion

Figure 1 shows the cooling curve of ZA8 alloy. Crystals of Zn rich β phase nucleate as the alloy cools below the liquidus temperature. The occurrence of Zn rich eutectic η is observed at the eutectic arrest temperature and, the simultaneous growth of these two phases arrest the cooling curve at the eutectic temperature until the whole liquid alloy gets solidified. Later, at eutectoid temperature, the β phase transforms into eutectic η and Al rich α phase.

From the cooling curve analysis the liquidus, solidus and eutectoid transformation temperatures were found to be 406.7°C, 375.5°C and 252.5°C respectively for ZA8 alloy. The addition of Ce resulted in a drop in the liquidus temperature with undercooling while, the other parameters remained unchanged. The magnified region around the liquidus temperature is shown in figure 2. Table 2 gives the effect of Ce addition on liquidus temperature during solidification of ZA8.

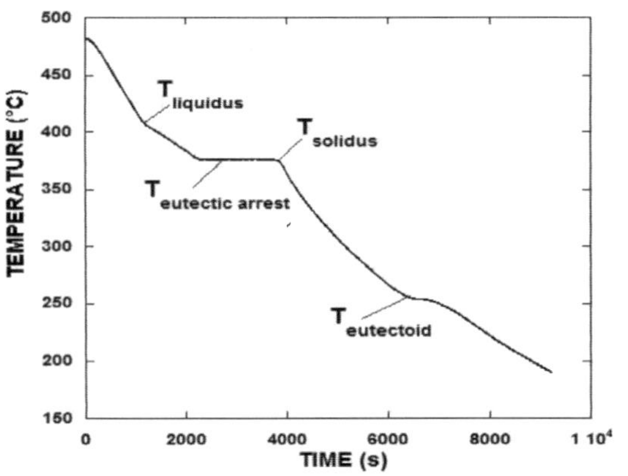

Figure 1. Cooling curve of ZA8 alloy.

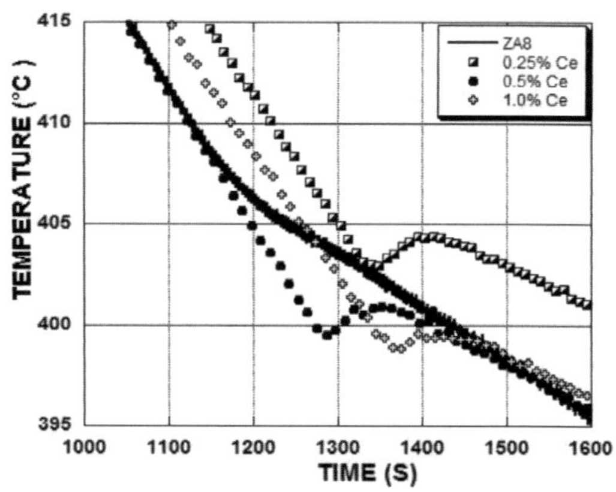

Figure 2. Effect of Ce on liquidus temperature.

Table 2. Effect of Ce addition on Liquidus Temperature during solidification of alloy.

	Liquidus Temperature (°C)					Undercooling (°C)
	Trial 1	Trial 2	Trial 3	Mean	Std. Deviation	
ZA8	406.6	406.7	406.9	406.7	0.152	-
ZA8+ 0.25% Ce	404.7	403.8	403.7	404.0	0.550	2.0
ZA8+ 0.50% Ce	401.5	399.6	400.9	400.6	0.971	2.0
ZA8+ 1.0% Ce	398.6	400.3	399.7	399.5	0.862	1.0

In figure 3, the cooling rate curve of ZA8 alloy shows a small valley between point **a** and point **b**. The presence of peaks **a'** and **a''** are attributed to the melting and solidification of unstable β phase dendrites. On addition of Ce, the peak **a'** became more pronounced in the same region. It clearly confirms that Ce hinders the nucleation of stable β dendrites. The peak at the liquidus point **a**, was attributed to the liquidus undercooling on addition of Ce.

The binary phase diagram of Al-Ce and Zn-Ce system show that Ce has almost zero solubility in Al and Zn at the melt treatment temperature of 480°C. Ce has a higher atomic radius compared to Zn and Al and its accumulation around the nucleation sites must hinder the diffusion of Zn atoms in the melt which in turn decreased the nucleation rate of β dendrites. As solidification progressed, the increase in concentration of Ce in the melt further decreases the rate of formation of stable β phase. This was observed as a more pronounced peak in the cooling rate curves.

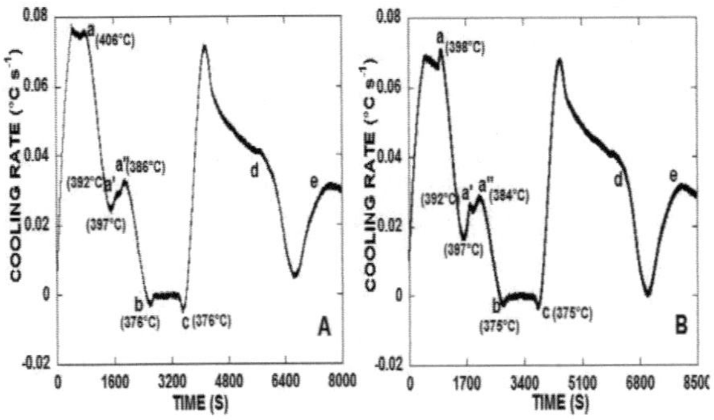

Figure 3. Cooling rate curve, (A) ZA8, (B) ZA8+1.0%Ce.

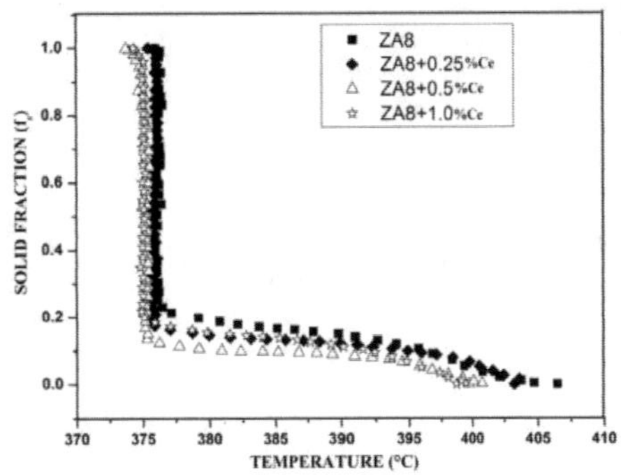

Figure 4. Effect of Ce on Solid Fraction curves.

The Newtonian technique of computer aided cooling curve analysis was used to quantify the effect of Ce addition on latent heat of solidification and the heat of eutectoid transformation. In the first derivative curve (FDC), the start and end points of solidification and eutectoid transformation were fitted linearly to obtain the baseline. The difference between the area under the FDC and the baseline, when multiplied with specific heat capacity (C_p) of the alloy yields the latent heat evolved and the heat of eutectoid transformation. The C_p value of alloy was taken as 435J/kg.K [10]. Table 3 shows that the addition of Ce increases the latent heat evolved and the heat of eutectoid transformation. Increase in latent heat is reflected in the cooling curve as a slight increase in the solidification time on addition of Ce. The significant increase in the heat of eutectoid transformation on addition of Ce shows that Ce promotes eutectoid transformation in the alloy.

Table 3. Effect of Ce addition of heat of phase transformation.

Specimen composition	ZA8	ZA8+0.25%Ce	ZA8+0.5%Ce	ZA8+1.0%Ce
Latent heat (kJ/kg)	55.68	59.85	60.55	60.90
Heat of eutectoid transformation (kJ/kg)	11.20	12.30	12.43	13.04

In figure 4, solid fraction (f_s) curves corresponding to alloys with Ce, lags behind that of ZA8 alloy during nucleation and growth of β phase. This confirms the fact that Ce hinders the formation of stable β dendrites. The f_s values at the eutectic arrest temperature decreased on addition of Ce. Thus, it is evident that the η to β ratio at the end of fusion should be high on addition of Ce. The increase in latent heat of fusion can be attributed to this effect. The XRD pattern of casting samples showed no additional peaks. It indicates that melt treatment did not cause occurrence of any new phase or eutectics during solidification of the alloy.

Figure 5 shows the effect of Ce addition on microstructure of ZA8 alloy. Addition of Ce influenced the morphology of retained β dendrites. A significant decrease in the size and distribution of β, and an increase in the volume fraction of eutectic η were observed on addition of Ce. Ce promotes the disintegration of β dendrites into eutectic η and the Al rich α phase. The mechanism of eutectoid transformation is not well understood.

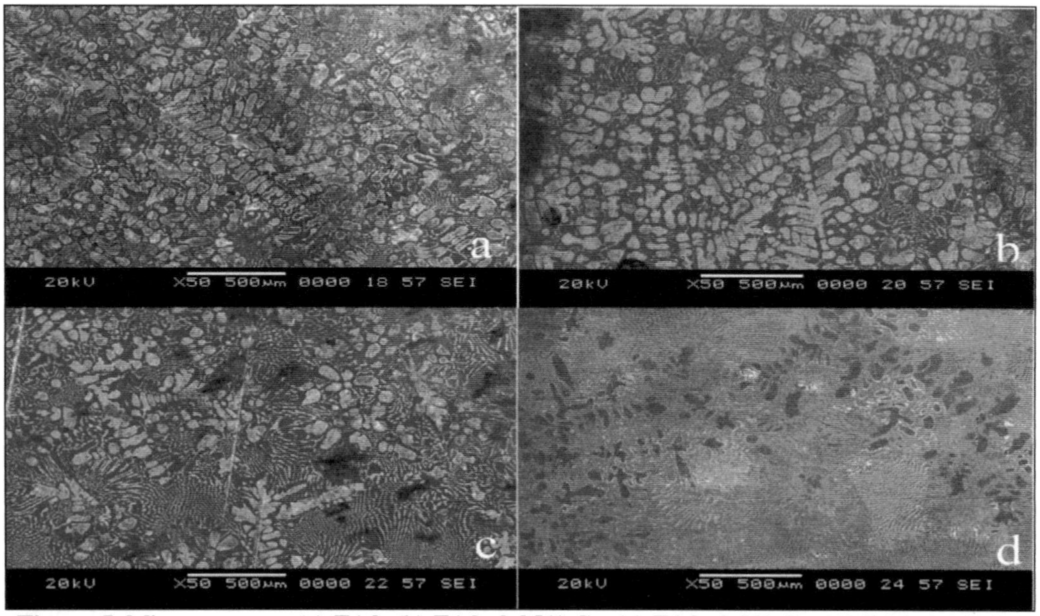

Figure 5. Microstructure (a) ZA8, (b) ZA8+0.25% Ce, (c) ZA8+0.5% Ce, (d) ZA8+1.0% Ce.

EDS analysis of the casting samples were carried out. The presence of Ce was detected only in β phase and eutectic η while, it remained absent in α phase, as shown in figure 6. Further, Ce was detected only in the outer shell of the β dendrite due to intergranular segregation of Ce. After eutectoid transformation, the Ce atoms, dispersed in the newly formed α phase, diffuses into the retained β phase and newly formed eutectic η.

The effect of chilling on solidification parameters and microstructure was investigated using Cu chills. The solidification range of ZA8 alloy reduced to 9°C between 410°C and 401°C on chilling, as shown in figure 7. The eutectoid transformation was observed at around 343°C. The nucleation rate being very high on chilling, the role of Ce in hindering the formation of stable β phase was found to be insignificant.

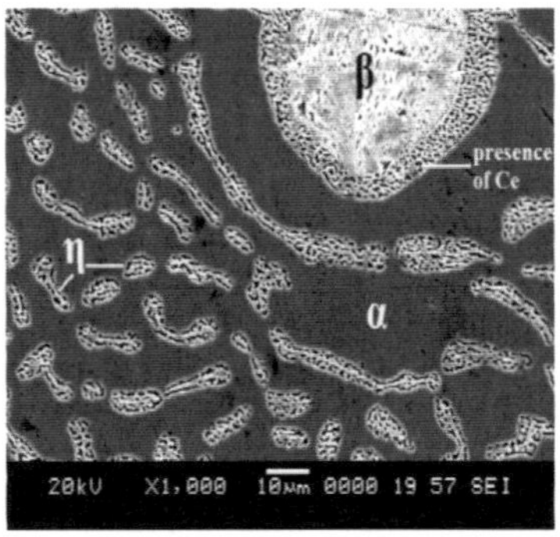

Figure 6. SEM image showing the accumulation of Ce at outer shell of β dendrite

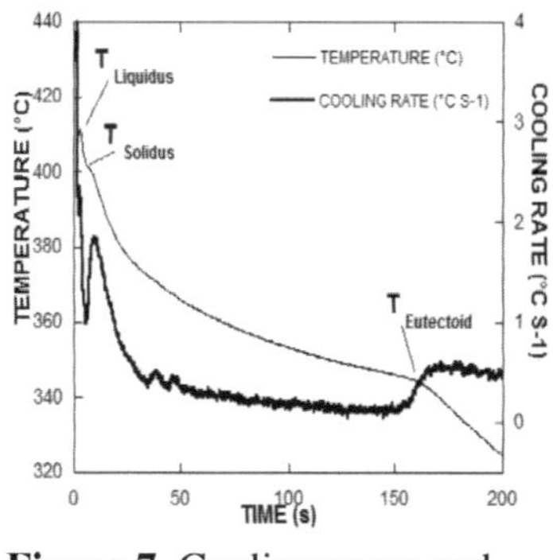

Figure 7. Cooling curve and Cooling rate curve on chilling

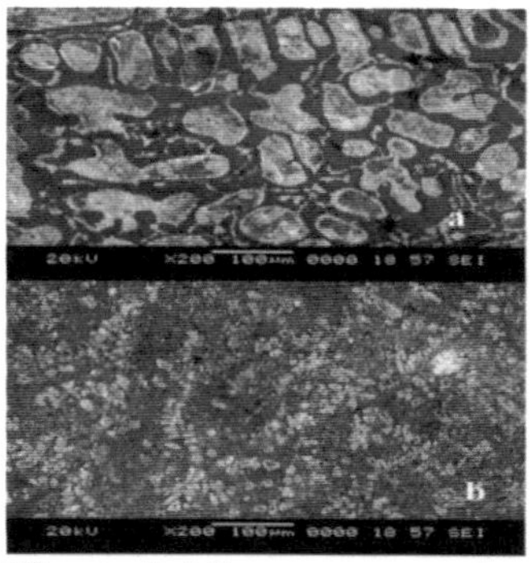

Figure 8. Microstructure of ZA8 (a) crucible cooled (b) chill cast

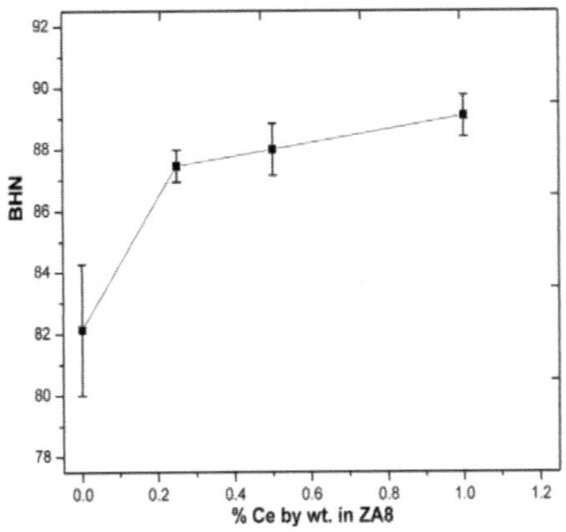

Figure 9. Effect of Ce on BHN of ZA8

The microstructure study of the chilled ZA8 sample showed very fine structure of β phase and eutectic η, as shown in figure 8. Addition of Ce had no effect on the microstructural evolution of the chilled samples. Unlike furnace cooled cast samples, the chilled alloy samples show the presence of Ce in the α phase. Complete diffusion of Ce atoms from the α phase back into the retained β and into the newly formed eutectic η could not occur due to high cooling rates. The addition of Ce had an effect on hardness of ZA8 alloy. Brinell hardness of graphite crucible cooled ZA8 sample was observed to increase with increase in Ce content as shown in figure 9. This change was due to the increase in volume fraction of fine eutectic η particles.

From the phase diagram of Zn-Al system, we can predict that on equilibrium cooling rates, complete transformation of β phase into α phase and very fine eutectic η can be obtained without addition of Ce. The refined microstructure can also be attained at high cooling rates. Though the microstructure obtained is refined in both the cases, the constituent phases are different. The former offers well dispersed η phase grains in the α phase matrix while, the latter contains retained β phase dendrites as well. Addition of Ce offers well dispersed β phase and eutectic η in the α phase matrix even at moderate cooling rates.

4. Conclusions

- The melt treatment of ZA8 alloy with Ce hinders the nucleation of stable β phase. A drop in liquidus temperature and additional peaks in the cooling rate curve indicate the presence of unstable β phase.
- The addition of Ce promotes eutectoid phase transformation in ZA8 alloy. The refined microstructure of ZA8 alloy with Ce addition showed an increase in the volume fraction of η-phase.
- Latent heat of solidification and the heat of eutectoid transformation increased on addition of Ce.
- The addition of Ce increases the hardness of the as cast ZA8 alloy.
- Chilling had a significant effect on the solidification parameters, and caused the refinement of microstructure of ZA8 alloy. Addition of Ce had no significant influence on solidification and microstructure of chill cast ZA8 alloy.

References

[1] Turk A, Durman M, and Kayali E S 2007 *J. Mater. Sci.* **42** 8298.
[2] Ramesh G, Vishwanath H M, and Prabhu K N 2012 Mater. Sci. Technol. 28(11) 1301.
[3] Turk A, Kurnaz C, and Sevik H 2006 Mater. Design. 28 1889.
[4] Ramesh G, and Prabhu K N 2012 *Inter. J. Cast Met. Res.* **25**(3) 160.
[5] Ramesh G, and Prabhu K N 2012 *Trans. Indian. Inst. Met.* **65**(6) 719.

[6] Zhu Y H, Lee W B, To S 2003 *J. Mater. Sci.* **38** 1945.
[7] Krupińska B, Rdzawski Z, and Labisz K 2011 *J. Achievements Mater. Manuf. Eng.* **46** (2) 154.
[8] Djurdjevic M B, Odanovic Z, and Talijan N 2011 *J. Minerals Met. Mater. Soc.* **30**(11) 51.
[9] Haq I, Shin J, and Lee Z 2004 *Met. Mater. Int.* **10**(1) 89.
[10] Alloy data 2006 NADCA Product Specification Standards for Die Castings A-3-14-06.

Numerical simulation on the solidification structure of Ø600mm continuous casting round bloom

Q Fang[1], H W Ni[1a], S J Wang[2] and H Zang[1]
[1] Key Laboratory for Ferrous Metallurgy and Resources Utilization of Ministry of Education, Wuhan University of Science and Technology, Wuhan 430081, China
[2] Chengde Jianlong Special Steel Co Ltd, Chengde 067201 China

E-mail: [a]nihongwei@wust.edu.cn

Abstract. A FE (Finite Element)—CA (Cellular Automation) coupling model was developed for the simulation of solidification structure formation during the Φ600mm round bloom continuous casting process of Q345E steel. The simulation result of the temperature field was consistent with the nail-shooting experimental result, and the simulated solidification structure of the bloom was in great agreement with corrosion testing under the same casting condition. The simulation results showed that the centre equiaxed crystal ratio increased slightly with the increase of secondary cooling water rate and decreased with the increase of casting temperature and casting speed. When the secondary cooling water rate was over 0.08L/kg, it had less effect on the solidification structure. As the casting temperature increased by 1°C or the casting speed increased by 0.01m/min, the centre equiaxed crystal ratio would decrease by 0.4%~1.2% and 0.3%~0.8% respectively. According to the simulation results, the optimized continuous casting process of Ø600mm round bloom should be the secondary cooling water rate of 0.08L/kg, the casting temperature of 1532°C~1539°C and the casting speed of 0.20m/min~0.22m/min. It was found that the solidification structure of Ø600mm Q345E steel round bloom was much improved after the optimized continuous casting process was adopted in practical production.

1. Introduction

The solidification structure of continuous casting bloom has an important effect on the quality of final steel products. A columnar structure will lead to the deterioration of mechanical properties and the centre equiaxed grains are helpful for improving the mechanical properties of the bloom. So it is of great significance to study the proper continuous casting technology to enhance the centre equiaxed grains in the bloom. In recent years, there is an increasing demand for large section continuous casting bloom. However, it is difficult to get the optimum process with the help of practical industrial trials, particularly for large section continuous casting. A dozen years ago, Rappaz and Gandin [1-3] and Zhu et al [4, 5] successfully predicted the transition from columnar to equiaxed zone by coupling the CA method with the finite element method (FEM). Then, the CAFÉ model was widely cited by the world's metallurgists [6-9] to improve the solidification structure of their research objects. But the method is rarely used on the continuous casting process and research on blooms with a diameter over 500mm is rarely reported. This makes the present study valuable to the continuous casting process of large-size blooms.

In this paper, the CAFÉ method was developed to simulate the solidification structure of Ø600mm continuous casting round bloom of Q345E steel. The simulated results of the temperature field and solidification structure were verified by a nail shooting experiment and corrosion testing respectively. The effects of casting parameters on the solidification structure were discussed based on production. Then the proper casting condition to enhance the internal quality of the bloom could be found to guide the production of the specific steel.

2. Model description

2.1. Heat transfer model of FE

2.1.1. Solidification model of heat transfer. The heat transfer characteristics and industrial conditions of continuous casting round bloom are taken into consideration to simplify the calculation process of numerical simulation. Based on the assumptions in Ref. [10, 11], the heat transfer during the continuous casting process of Ø600mm round bloom can be simplified to a two-dimensional unsteady heat transfer model, thus a sheet shifting method can be developed to build a geometric model for the round bloom during continuous casting which is subsequently calculated by ProCAST. Suppose that a circular slice is formed at the meniscus, and the heat of the slice passes from the centre to the surface during the solidification process. Then a solidified shell gradually grows from the surface to the centre until the bloom is completely solidified.

2.1.2. Initial and boundary conditions. The temperature of liquid steel in the slice is supposed to be equally distributed and equal to the temperature of liquid steel in the tundish. A continuous caster's cooling area consists of the mould, secondary cooling zone and air cooling zone, which have different heat transfer conditions. The detailed boundary conditions of heat transfer in each part can be handled as follows:

In the mould, the boundary condition of heat transfer is described by average heat flux, which is calculated by the following equation [12]:

$$q = \frac{\rho_w \times c_w \times W \times \Delta T}{S_{eff}} \tag{1}$$

where q is the average heat flux of the slice in the mould, W/m^2; ρ_w is the density of cooling water, kg/m^3; c_w is specific heat, J/(kg×K); W is the flow rate of cooling water, m^3/s; ΔT is the temperature difference between the water at inlet and outlet, K; S_{eff} is the active area of mould, m^2.

In the secondary cooling zone, the heat transfer coefficient in the water spray region is calculated by equation (2); in the air-mist spray region, the heat transfer coefficients are calculated by equation (3).

$$h = 581 w^{0.541}(1 - 0.0075 t_w) \tag{2}$$

$$h = 350w + 130 \tag{3}$$

where h is the heat transfer coefficient at the bloom surface, W/(m^2×°C); w is the spraying water density, L/(m^2×s).

In the air cooling zone, the heat flux density can be calculated by the following equation:

$$q = \varepsilon\sigma(T^4 - T_w^4) \tag{4}$$

where σ is the Steven—Boltzmann constant, 5.67×10^{-8} W×m^{-2}×K^{-4}; ε is blackness; T and T_w are the surface temperature of the bloom and the ambient temperature respectively, K.

2.2. CA model

2.2.1. Heterogeneous nucleation. A continuous nucleation distribution function $dn/d(\Delta T)$ is used to describe the grain density change which is induced by an increase in the under-cooling. The function is described by following equation [2].

$$\frac{\mathrm{d}n}{\mathrm{d}(\Delta T)}=\frac{n_{max}}{\sqrt{2\pi}\Delta T_{\delta}}\exp\left(-\frac{1}{2}\left(\frac{\Delta T-\overline{\Delta T}}{\Delta T_{\sigma}}\right)^{2}\right) \quad (5)$$

where n_{max} is the maximum nucleation density, which integrates from 0 to ∞ according to a normal distribution when all the nucleation sites are activated; ΔT is the mean nucleation undercooling temperature (°C); ΔT_{σ} is the nucleation undercooling temperature for the standard deviation (°C). In this case, $\Delta T_{s,max}$ =2°C , $\Delta T_{s,\sigma}$=1°C , $n_{s,max}=10^8$ m^{-2}; $\Delta T_{v,max}$=5°C , $\Delta T_{v,\sigma}$ =2°C , $n_{v,max}=10^9$ m^{-3}("s" is the index of surface nucleation, "v" is the index of volume nucleation).

2.2.2. Kinetics of dendritic tip growth. In the calculation of the solidification process, the growth kinetics of both columnar and equiaxed morphologies can be calculated with the aid of the KGT model [13]. In the actual simulation process, in order to accelerate the computation, the KGT model is fitted and the following equation is gained.

$$V(\Delta T)=a_2\Delta T^2+a_3\Delta T^3 \quad (6)$$

where ΔT is the total undercooling of dendrite tip (°C); *a*2 and *a*3 are the coefficients of the multinomial of dendrite tip growth velocity, which are calculated to be a_2=0 and a_3 = 8.712× 10^{-6} m×s^{-1}×K^{-3} based on the data of partition coefficient, liquidus slope, diffusivity in liquid and Gibbs-Thomson coefficient of Fe-X (X = C, Si, Mn, P, S) alloys.

2.3. Casting parameters

The main chemical composition of Q345E steel is showed in table 1, and most of the thermo-physical parameters of the steel can be calculated by inputting the chemical composition into ProCast. There, the liquidus and solidus temperature of Q345E steel is 1517°C and 1472°C respectively.

Table 1. Main chemical composition of Q345E steel.

Main Elements	C	Si	Mn	P	S	Al	V	Nb	Ti
Mass%	0.15	0.25	1.34	0.007	0.003	0.025	0.035	0.032	0.006

3. Results and discussions

3.1. Examination of model

The nail-shooting experiment is an effective and widely used method to measure the solidified shell thickness of continuous casting billets in many steelmaking plants. In this paper, the nail-shooting experiment was used to examine the simulation result of temperature field, the results are shown in figure 1 and table 2; the simulated solidification structure was compared with a corrosion test, which can be seen in figure 2 (owing to the centre symmetry and area of the bloom on the same cross-sectional, a one-fourth cross-sectional was taken to simulate the solidification structure).

As shown in figure 1 and table 2, the error of simulation results for solidified shell thickness was under 4% compared with the results of the nail-shooting experiment. Figure 2 shows the comparison between the corrosion test and the simulated solidification structure, in which about 22.5% center equiaxed grain ratio could be observed in the simulation result and about 23.2% centre equiaxed grain ratio could be counted in the corrosion test. The simulation results for error of solidification structure was about 3% compared with the corrosion test.

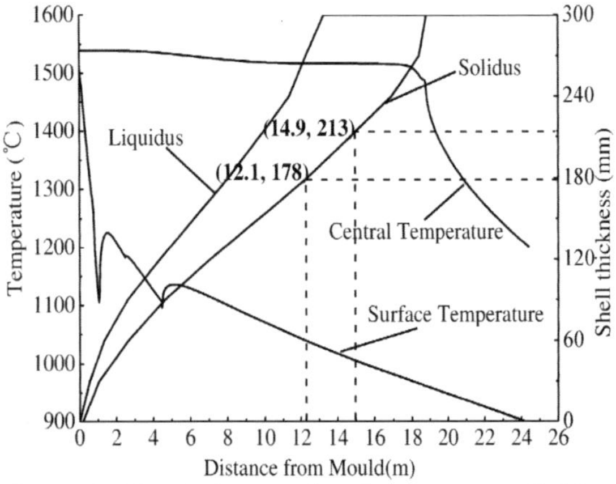

Figure 1. Simulation results of temperature field.

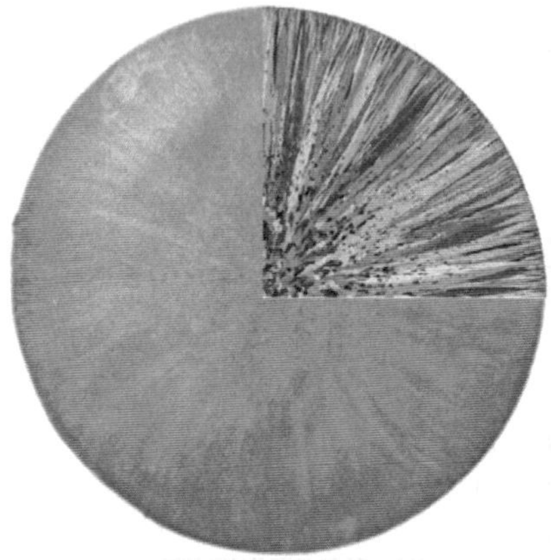

Figure 2. Comparison between corrosion test and simulation result.

Table 2. Simulation and measured results of solidified shell thickness of the bloom.

Cross section	Distance from mould (m)	Experimental results (mm)	Simulation results (mm)	Relative error (%)
600mm	12.1	185	178	3.78
	14.9	215	213	0.93

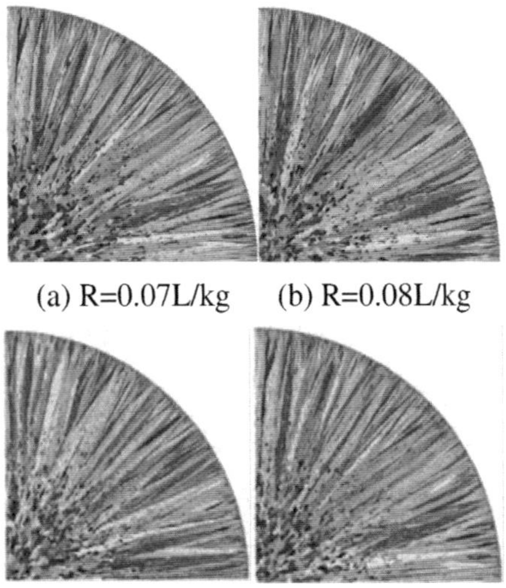

(a) R=0.07L/kg (b) R=0.08L/kg

(c) R=0.09L/kg (d)R=0.10L/kg

Figure 3. The simulated solidification structure under different water rates.

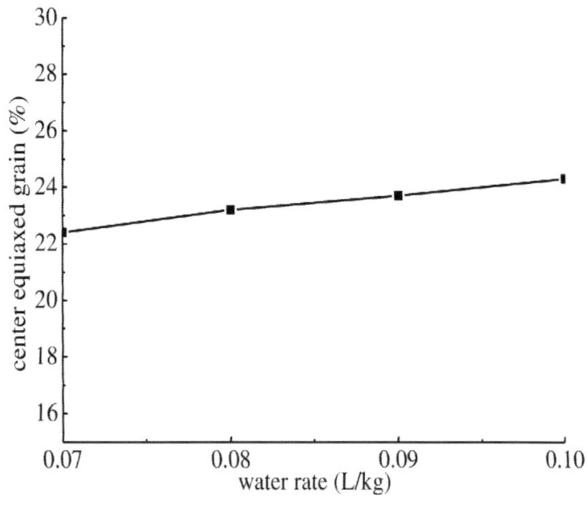

Figure 4. The proportion of equiaxed grain at different water rates.

3.2. Secondary cooling water rate

The solidification structures under different water rate (0.07L/kg~0.10L/kg) were investigated to find a proper value of water rate for the secondary cooling zone. The simulation results are shown in figure 3, and the centre equiaxed grain variation is presented in figure 4.

As can be seen in figure 3 and figure 4: (1) with the increase of secondary cooling water rate, the proportion of centre equiaxed grains increased slowly; (2) as water rate increases by 0.01L/kg, the proportion of equiaxed grains increased by about 0.5%, and when the secondary cooling water rate was over 0.08L/kg, it had less effect on the solidification structure; (3) by comprehensive consideration of the impact of water rate on the solidification structure and production costs, the water rate should be settled at a specific value of 0.08L/kg.

3.3. Casting temperature and casting speed

The influences of casting temperature and casting speed on the solidification structure of the bloom were discussed to determine a group of reasonable casting parameters for the continuous casting bloom. In this research, four different casting speeds and six groups of casting temperature were discussed to find the specific effects of casting speed and casting temperature on the solidification structure of Ø600mm round bloom. The variables of casting speed and temperature are shown in table 3.

Table 3. Variables of casting temperature and temperature speed.

Casting temperature (°C)	1527	1532	1539	1547	1552	1557
Casting speed (m/min)	0.20	0.22	0.24	0.25		

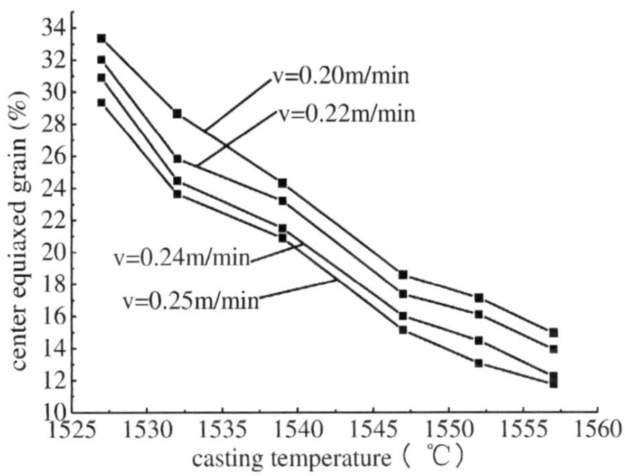

Figure 5. The influence of casting speed and casting temperature on the solidification structure

The simulation results of centre equiaxed grain variation under different casting speeds and casting temperatures are presented in figure 5. With the increase of casting speed and casting temperature, the proportion of centre equiaxed grains apparently decreased. As the casting temperature increased by 1°C or the casting speed increased by 0.01m/min, the centre equiaxed grain ratio in the bloom decreased by about 0.4%~1.2% and 0.3%~0.8% respectively.

The casting temperature should be controlled within a certain range to ensure stable direct motion and the internal quality of the bloom, and the casting speed is also one of the key parameters that affect the surface quality of the bloom and the whole efficiency of the casting process; the casting temperature of Ø600mm round bloom Q345E steel should be properly controlled within 1532°C ~1539°C, and the casting speed should be set within the range of 0.20m/min~0.22m/min. The casting parameters mentioned above met the plant conditions and the centre equiaxed grain ratio was enhanced to a higher degree.

5. Conclusions

1. The simulation results of Ø600mm Q345E steel during the continuous casting process was consistent with the experimental results, as the relative error of temperature field was under 4% compared with the nail-shooting experiment, and the simulated solidification structure of the bloom was in great agreement with the corrosion test.
2. The proper continuous casting process of Ø600mm Q345E steel should be water flow rate at 0.08L/kg, casting temperature within 1532°C~1539°C and casting speed in the range of 0.20m/min~0.22m/min.
3. After the optimized continuous casting process was put into practice, it was found that the proportion of centre equiaxed grains of Ø600mm round bloom Q345E steel apparently increased and the quality of the practical product improved greatly.

References

[1] Gandin C A, Desbiolles J L and Rappaz M 1999 *Metall. Mater. Trans. A* **30** 2187
[2] Rappaz M and Gandin C A 1993 *Acta Metall. Mater.* **41** 345
[3] Gandin C A and Rappaz M 1994 *Acta Metall. Mater.* **42** 2233
[4] Zhu M F, Kim J M and Hong C P 2001 *ISIJ Int.* **41** 436
[5] Zhu M F and Hong C P 2001 *ISIJ Int.* **41** 992
[6] Wang J L, Qiao A Y, Shi X X and Zhang X G 2012 *Foundry Technology* **33** 294
[7] Min Y, Liu C J, Wang D Y and Jing M F 2011 *Journal of Iron and Steel Research* **23** 38
[8] Luo Y Z, Zhang X, Xiao C, Hu Z F, Yuan Y Y and Chen S D 2012 *Ironmaking & Steelmaking* **39** 26
[9] Cheng G Y, Ni H W, Zhang H, Zhan W T, Cheng R J and Liu J G 2009 *Foundry Technology* **30** 48
[10] Lally B, Biegle L and Henein H 1990 *Metal. Trans.* B **21** 761
[11] Tieu A K and Kim I S 1997 *International Journal of Mechanical Sciences* **39** 185
[12] Mostata A, Ahmad J J and Omid A 2008 *ISIJ Int.* **48** 161
[13] Kurz W, Giovanola B and Trivedi R 1986 *Acta Metall.* **34** 823

IOP Conf. Series: Materials Science and Engineering **117** (2016) 012037 doi:10.1088/1757-899X/117/1/012037

Numerical simulation of casting process to assist in defects reduction in complex steel tidal power component

E J Guo, S C Zhao[a], L P Wang, T Wu, B P Xin, J J Tan and H L Jia
School of Materials Science and Engineering, Harbin University of Science and Technology, No.4 Linyuan Road, Xiangfang District, Harbin150040, Heilongjiang Province, People's Republic of China

E-mail: [a] zscwr@163.com

Abstract. In order to reduce defects and improve casting quality, ProCAST software is performed to study the solidification process of discharge bowl. Simulated results of original casting process show that the hot tearing is serious at the intersection of blades and outer or inner rings. The shrinkage porosity appears at the bottom of discharge bowl and the transition area of wall thickness. Based on the formation mechanisms of the defects, the structure of chills attached on the outer surface of discharge bowl casting is optimized. The thickness of chills ranges from 25mm to 35mm. The positions of chills corresponded to the outer surface of the T-shaped parts. Compared to the original casting design (without chills), the hot tearing and shrinkage porosity of the discharge bowl are greatly improved with addition of chills.

1. Introduction

The tidal power devices are demanded higher performance with the rapid development of tidal power technologies. The discharge bowl is a key component of tidal power devices. The material of the discharge bowl is CD4MCu duplex stainless steel [1]. Although this material is resistant to corrosion and good mechanical properties, casting performance is poor. However, only the sand casting method is suitable for producing the discharge bowl, because the structure of the discharge bowl is complex. The diameter of the discharge bowl is 1700mm. The thickness of the blades are about 20mm. The hot tearing is serious at the intersection of blades and outer or inner rings. The shrinkage porosity appears at the bottom of the discharge bowl and the transition area of wall thickness. These defects seriously degrade the mechanical properties and quality of castings. Therefore, it is necessary to research the formation mechanisms of hot tearing and shrinkage porosity and find ways to reduce these defects of discharge bowl casting. Although physical experiment is a possible way, it is still difficult to use those experiments to explain the results in the real castings because of the opaqueness of the alloy melt [2, 3]. Therefore, the ProCAST software is used to study the solidification of discharge bowl casting.

2. Model and boundary conditions

The 3D model of discharge bowl casting should be meshed, and the boundary and initial conditions should be determined before calculation. The 3D model of casting is meshed by Hypermesh. The total number of grid is about 2 million. The material of discharge bowl is CD4MCu, which thermophysical properties come from the Magmasoft's built-in database. The temperature of chills and ambient is 25℃. Thermal, flow and stress modules are activated in run parameters.

3. Numerical simulation of the original casting process

The solidification process and temperature field of the original casting process are shown in figure 1. Red indicates the highest temperature of 1580 ℃. Purple indicates the solidus temperature of 1370℃. Due to the thickness of blades, outer ring, inner ring and bottom ring is smallest, the cooling rate of them are fastest. However, because the center of the feeder is the location of hot spot, the cooling rate and time of feeders are slowly and long.

The locations of three key points are shown in figure 2. The point 1, point 2 and point 3 locate in the intersection of the blade and the bottom ring, the blade and the inner ring and the blade and the outer ring, respectively.

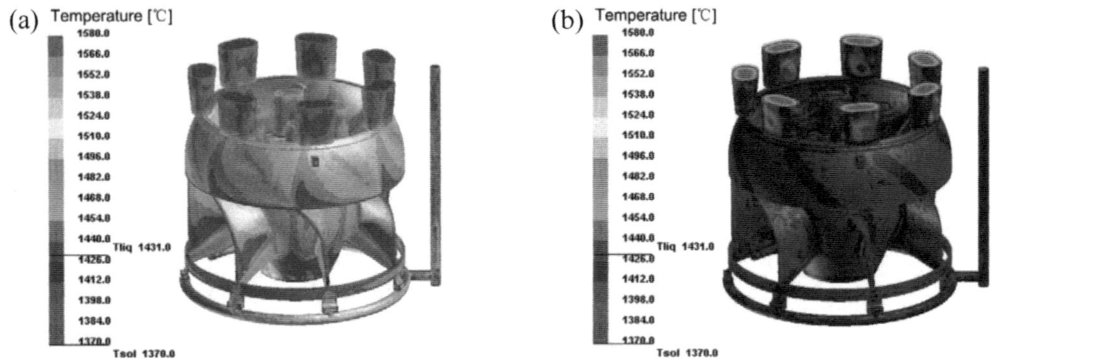

Figure 1. The solidification process and temperature field of the original casting process. (a) 7s after filling, (b) 90s after filling.

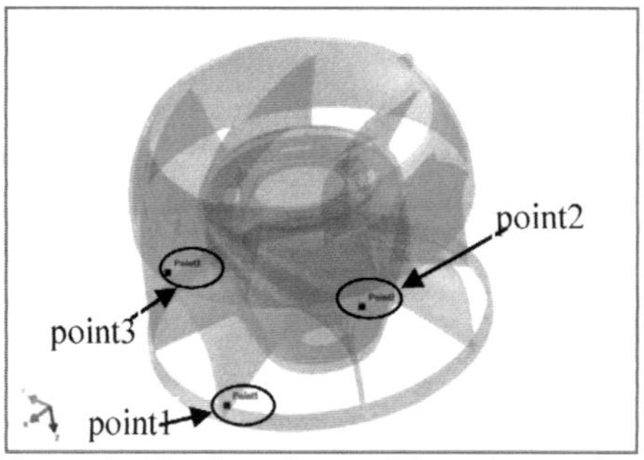

Figure 2. The locations of three key points.

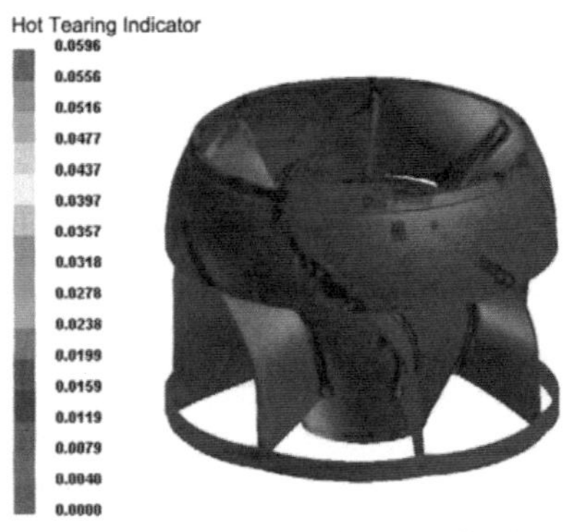

Figure 3. Hot tearing indicator of original casting process.

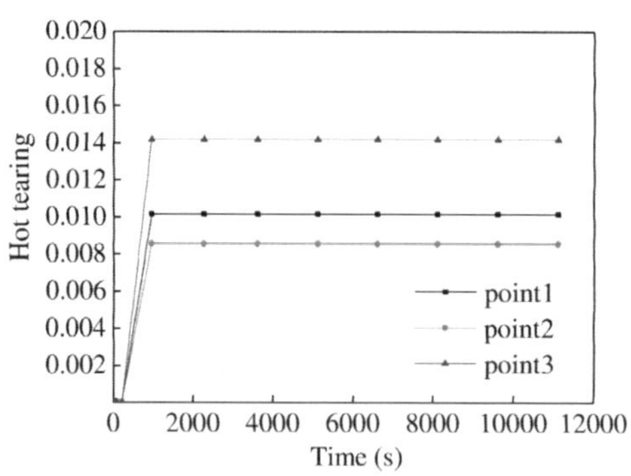

Figure 4. Hot tearing indicator curves of original casting process.

The hot tearing indicator of the original casting process is shown in figure 3. The hot tearing indicator is based on the physics of solidification and deformation. Hot tearing forms when the mushy zone is starved of liquid feeding and deformed in tension. The unfed tensile deformation causes a small additional porosity. This additional porosity or porosity due to solid deformation is a locator for initiation sites for hot tears in the casting [4]. Simulated results show that the hot tearing is serious at the intersection of blades and outer or inner rings (T-shaped parts). The T-shaped parts are the transition zones of wall thickness. The cooling rate of T-shaped parts is uniformity, so the large stress is formed. Hot tearing indicator curves of original casting process are shown in figure 4. The value of hot tearing tendency of point3 is higher than point 2 or point 1. Effective stress and effective plastic strain of original casting process are shown in figure 5. If the value of effective stress is higher than the strength of the alloy, the hot tearing would be generated in T-shaped parts.

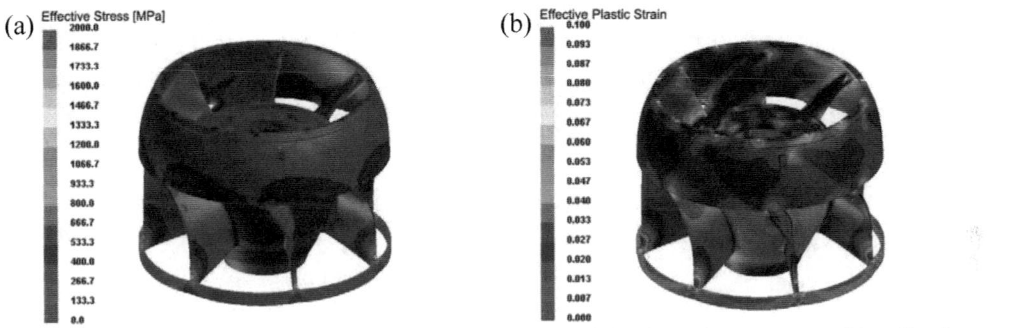

Figure 5. Effective stress and effective plastic strain of original casting process. (a) Effective stress, (b) Effective plastic strain.

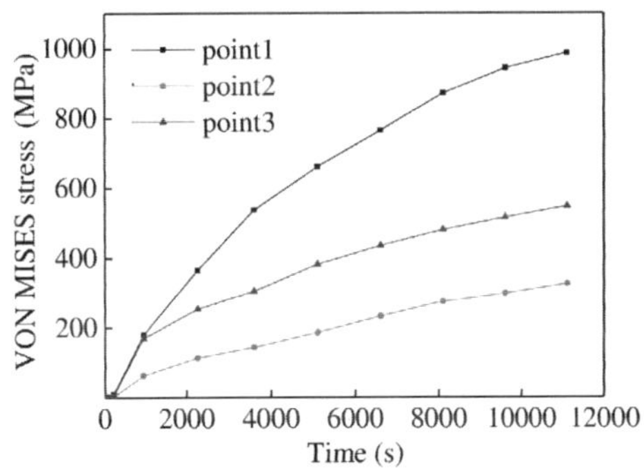

Figure 6. The Von Mises stress curves of original casting process.

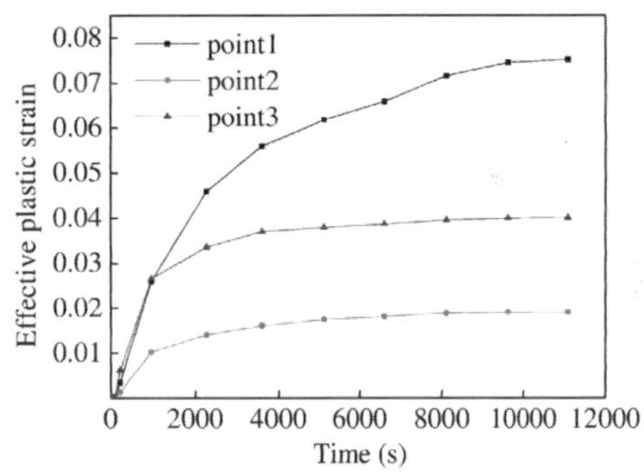

Figure 7. Effective plastic strain curves of original casting process.

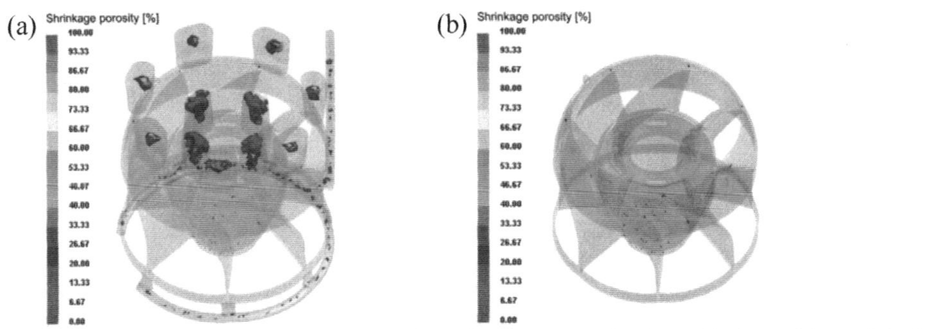

Figure 8. Shrinkage porosity predicted of original casting process. (a) Overall view, (b) Partial view.

The Von Mises stress and effective plastic strain curves of original casting process are shown in figure 6 and figure 7, respectively. The Von Mises stress and effective plastic strain increase over time. Shrinkage porosity predicted of original casting process is shown in figure 8. Shrinkage porosity is quantitatively predictable by Magmasoft's built-in "shrinkage porosity criterion" function. The criterion function automatically calculates the degree of solidification shrinkage and feeding for each of the control volumes [5]. The shrinkage porosity appears at feeders, runners, bottom of discharge bowl and the T-shaped parts.

4. The optimized parameters of chills and simulation results

4.1. The basic principle of reducing hot tearing by chills

Schematic diagram of reducing hot tearing is shown in figure 9. The temperature of T-shaped parts is higher than the adjacent regions. This indicates that the solidification time of T-shaped parts is longer than blades. The T-shaped parts are easy to generate hot tearing, the reasons may be the solidified regions (blades) generate larger linear shrinkage than mushy regions (T-shaped parts) in solidification process, and the tensile strength of mushy regions is smaller than solidified regions. If the chills are well attached on the outer surface of transition regions, the solidification sequence of T-shaped parts would be as same as blades. Therefore, the hot tearing will be effectively reduced by chills.

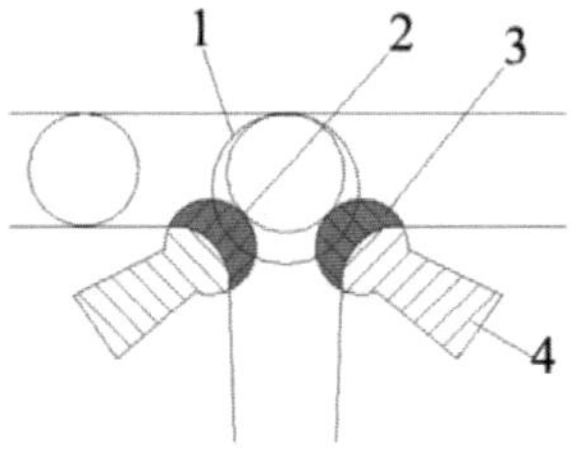

Figure 9. Schematic diagram of reducing hot tearing. 1- Inscribed circle is not chilled, 2- Inscribed circle is chilled, 3- The chilled regions, 4- Chills.

4.2. The optimized parameters of chills

Based on the formation mechanisms the defects, the structure and position of chills attached on the outer surface of discharge bowl casting are optimized. The thickness of chills ranges from 25mm to 35mm. The positions of chills correspond to the outer surface of the T-shaped parts. The material of chills is steel. The optimized structure of chills is shown in figure 10.

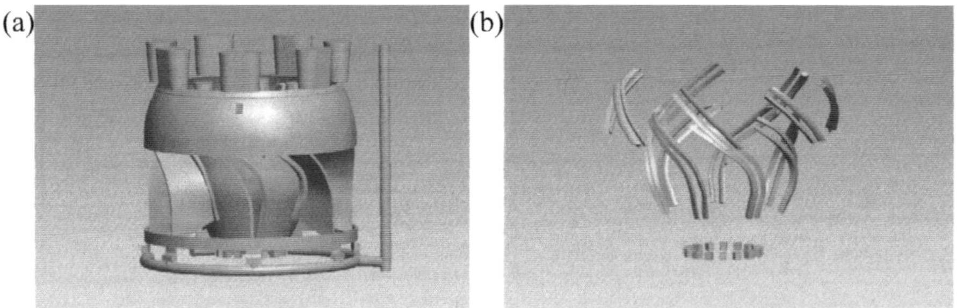

Figure 10. The optimized structure of chills. (a) Foundry methods drawing, (b) Structure drawing of chills.

4.3. Simulation results after optimization

Pouring temperature is 1580℃, and filling time is 30s. The solidification process and temperature field after optimization is shown in figure 11. T-shaped parts are chilled in the solidification process. The hot tearing indicator after optimization is shown in figure 12. The hot tearing of T-shape parts is significantly reduced. The hot tearing indicator curves after optimization is shown in figure 13. Compared with the simulation results without chills, the hot tearing value of point 1, point 2 and point 3 decrease from 0.010 to 0.004, 0.008 to 0.0005 and 0.014 to 0.005, respectively. The change in hot tearing value of point2 is largest. The value of hot tearing of three key points is reduced by at least an order of magnitude.

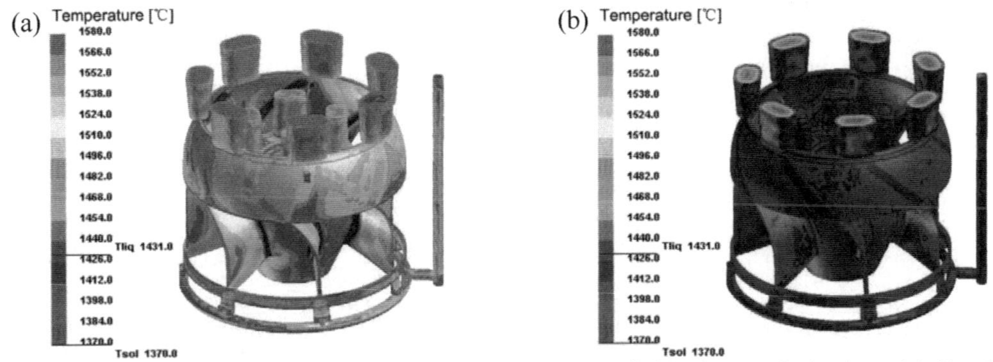

Figure 11. The solidification process and temperature field after optimization. (a) 7s after filling, (b) 90s after filling.

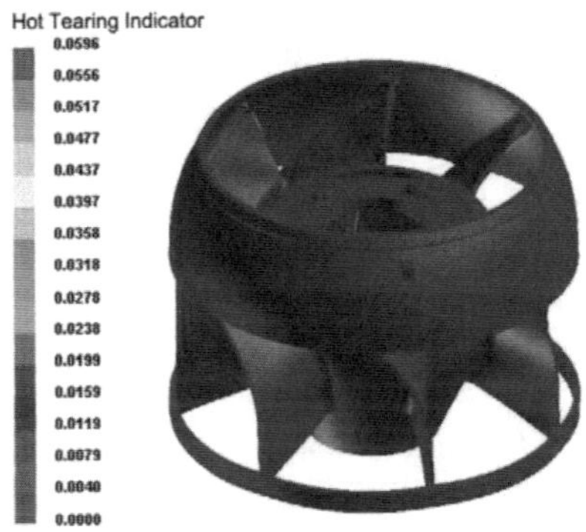

Figure 12. Hot tearing indicator after optimization.

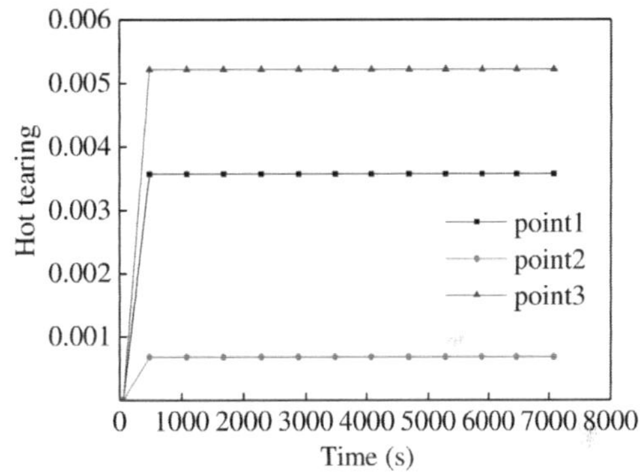

Figure 13. Hot tearing indicator curves after optimization.

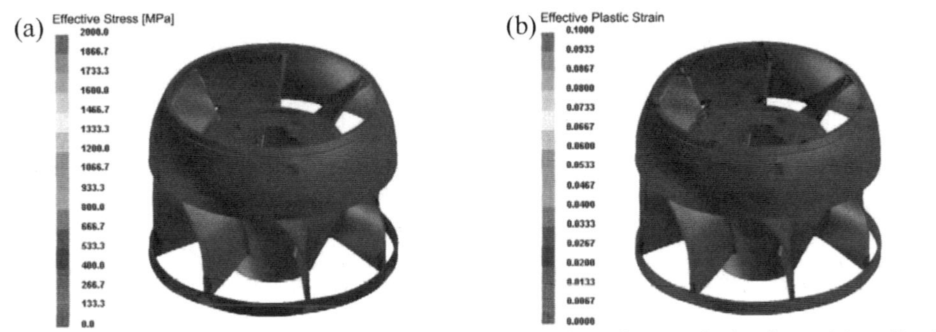

Figure 14. Effective stress and effective plastic strain after optimization. (a) Effective stress, (b) Effective plastic strain.

The effective stress and effective plastic strain after optimization are shown in figure 14. The effective stress and effective plastic strain of T-shape parts are obviously reduced by chills. The Von Mises stress curves after optimization are shown in figure 15. The Von Mises stress of point 1, point 2 and point 3 decrease to 400MPa, 200MPa and 100MPa at 7000s. The effective plastic strain curves after optimization is shown in figure 16. The Von Mises stress of point 1, point 2 and point 3 decrease to 0.020, 0.010 and 0.0006 at time 7000s.

The shrinkage porosity predicted is shown in figure 17. The solidification sequence of T-shaped part is uniform, when the chills are attached on the outer surface of transition regions. The shrinkage porosity is less than without chills.

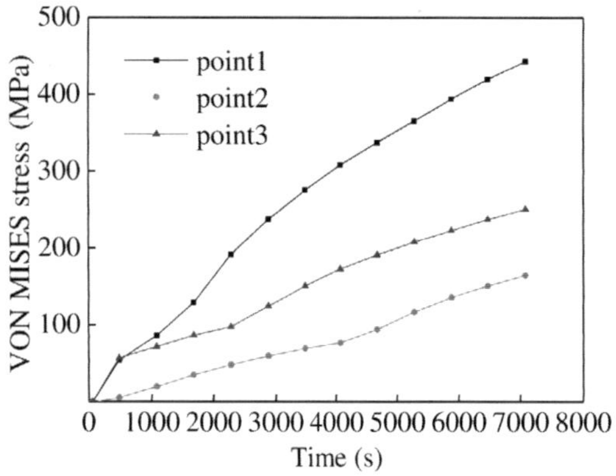

Figure 15. The Von Mises stress curves after optimization.

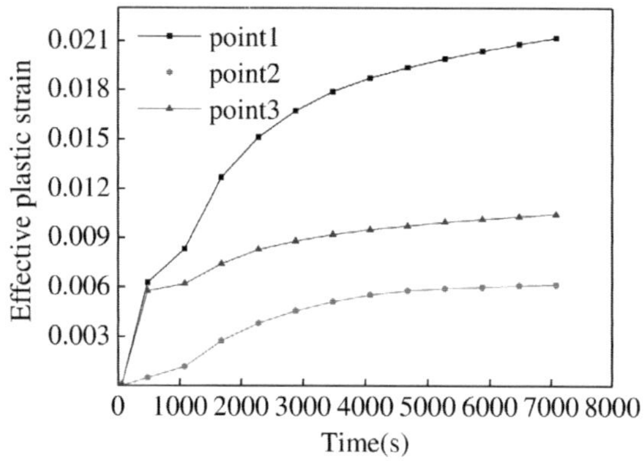

Figure 16. Effective plastic strain curves after optimization.

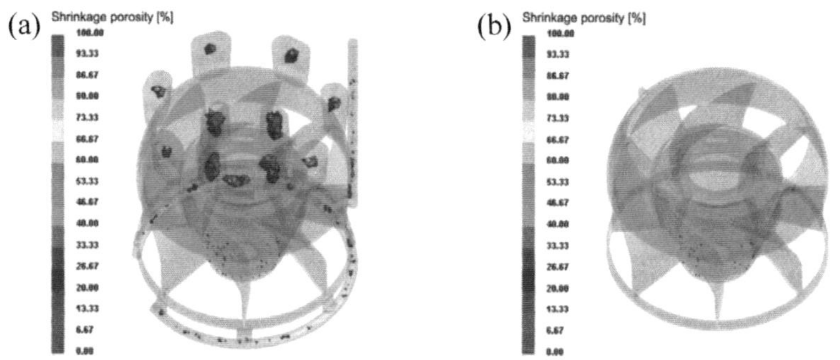

Figure 17. Shrinkage porosity predicted after optimization. (a) Overall view, (b) Partial view.

5. Conclusions

The simulation results of original casting process show that hot tearing tendency is serious at the intersection of blades and outer or inner rings. The shrinkage porosity appears at the bottom of discharge bowl and the transition area of wall thickness. The optimized size parameters of chills range from 25mm to 35mm, and the positions of chills correspond to the outer surface of the T-shaped parts. The hot tearing and shrinkage porosity of the discharge bowl are greatly improved with addition of chills compared with the original casting design (without chills).

Acknowledgments

The authors gratefully acknowledge the financial support from the National Natural Science Foundation of China (Grant No. 51174068 and 51374086).

References

[1] Elin M W, Bengt B and Staffan H 2008 Weldability aspects of a newly developed duplex stainless steel LDX2101 *Mater. Technol* **79** 473-481

[2] Milkowska-Piszczek K and Falkus J 2014 Calculation of the boundary conditions in the continuous casting of steel process *Metalurgija* **53** 571-873

[3] Abdullin A D 2013 Detecting microporosity defects in steel castings by computer modeling of the casting operation in ProCAST *Metallurgis* **57** 167-171

[4] Monroe C, Beckermann C 2005 Development of a hot tear indicator for steel castings *Mat sci eng A-struct* **413** 30-36

[5] Wu M, Sahm P R 1999 Numerical study of porosity in titanium dental castings *J mater sci mater m* **10** 519-525

Surface formation in direct chill (DC) casting of 6082 aluminium alloys

N Bayat[a] and T Carlberg[b]
Department of Natural Science, Mid Sweden University, 851 70 Sundsvall, Sweden

E-mails: [a] nazlin.bayat@miun.se, [b] torbjorn.carlberg@miun.se

Abstract. Surface defects in aluminium billet production are a real problem for the subsequent extrusion procedure. Extrusion productivity can be influenced by the surface properties, which is defined as surface appearance, surface segregation zone depth and large Mg_2Si and β-particles (Al_5FeSi). In this research the surface formation during DC casting of 6082 aluminium billets produced by the air slip technology is studied. The surface microstructures of 6082 aluminium alloys with smooth and wavy surface appearances were investigated, including segregation zone depths and phase formation. The results were discussed based on the exudation of liquid metal through the mushy zone. The specific appearance of the wavy surface of 6082 alloys was correlated to how the oxide skin adheres to the underlying mushy zone and coupled to the dendritic coherency and surface tension of the skin. The occurrence of different phases at the very surface and in the layer just below was explained by variations in solidification directions and subsequent segregation patterns.

1. Introduction

The 6xxx series Al alloys are basically produced for extrusion. Among 6xxx alloys, the 6082 with high content of magnesium and silicon is the strongest alloy with widespread applications in building, aircraft and automotive industry [1]. The surface zone of the mentioned alloy, when produced by direct chill (DC) casting, has an important role on the quality of the extruded materials. Although the introduction of the air–slip technique in DC casting has improved the surface quality to some extent, strong variations in surface appearance and surface defects can be found in this alloy type [2-4]. The depth of the surface segregation should also be considered as an important factor in surface quality, and can be defined as a layer in which the alloy contents deviate from the bulk [5, 6].

In the recent paper [7] the surface formation in low-alloyed 6000 series alloys was investigated. In the present paper the study is extended to the more highly alloyed 6082 with a more severe surface segregation, and the different types of surface appearances and phase formations are discussed.

2. Experimental procedure

Microstructures and surface segregation from 6082 Al alloy billets, 178 mm diameter, with banded or wavy–defect surfaces were compared to defect free or smooth surfaces. The compositions of mentioned alloy are given in table 1, and grain refiner in shape of Al-5Ti-1B master alloy at a level of 1 Kg/tonne had been added. Metallographic samples were cut at the surface and along the growth direction to show the structure of the surface zone. The surface segregation and occurring phases were studied by light optic microscope (LOM). A scanning electron microscope (SEM) was used to observe the transverse cuts, but also to study unprepared surface structures directly on the billet surfaces. The secondary phases were analysed by energy–dispersive X-ray spectroscopy (EDX). A part of the surface with a wavy appearance was deeply etched in a solution of methanol, iodine and tartaric acid to partially dissolve the aluminium matrix to give a better view of the secondary phase distribution. Stereo light microscope (SLM) was also used to study the surface appearances.

Table 1. Chemical composition of 6082 aluminium alloy in wt%.

Alloy	Si	Fe	Cu	Mn	Mg	Cr	Zn	Ti
6082	0.96	0.19	0.001	0.46	0.63	0.0006	0.002	0.011

3. Results

In figure 1, the SLM images of smooth surface of 6082 alloys and wavy samples surfaces of 6082 and 6060 alloys are shown. It is clear that the wavy 6082 surface has an irregular pattern and the distances between waves are not constant, as could be observed in low-alloyed of 6xxx series [7].

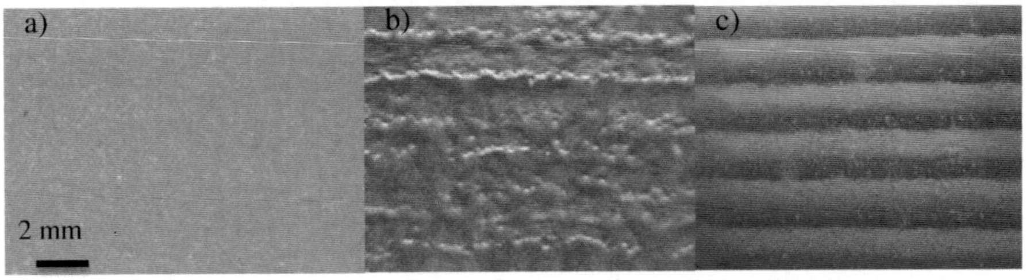

Figure 1. The stereo light microscope images: a) Smooth surface; and b) Wavy surface for 6082 alloy and c) Wavy surface for 6060 alloy [9].

On many places the surface is bulged out between points and patches in the transverse direction, i.e. in a band-like pattern without forming bands.

Figure 2 illustrates the surface segregations are defect free and wavy-defect surfaces. At the defect free surface the segregation thickness is more uniform and about 50-80 μm, while the surface segregation depth in the wavy sample is 50-150 μm. It is clear from the figure that the surface segregation depth at the wavy surface is clearly not uniform, and also the phases occurring in this zone are different.

Figure 3 illustrates the surface microstructures viewed by SEM directly on the outer surfaces. In figure 3a the microstructure of a defect free billet can be observed. Light phases in plate-like (fan shaped) morphology are dominating at the smooth surface all over the surface and only a few other phases can be seen randomly. The corresponding surface in the transverse cut is shown in figure 2a and b, where the plate–like structure can be seen growing out towards the surface in the cell boundaries. Figures 3b shows a wavy surface structure and it is obvious that different phases dominate at different positions along the waves.

In figure 4 results from EDX analysis of different intermetallic phases are shown. The plate-like structure (figure 4a), which dominates the smooth surface, but occur on all surfaces, has a Fe:Si ratio (atomic %) of about 1:2. The phases in figure 4b and 4c can also be found in all surfaces but are more frequently seen in the wavy surface, the ratios are (Fe,Mn):Si (atomic %) 1.5:1 and Mg:Si 2.3:1 in 4b and 4c, respectively. In figure 4d, where a secondary phase is filling out the cell boundaries, the (Fe,Mn): Si ratio (atomic %) is about 1:1.

Figure 5, shows the same sample after deep etching where it can be seen how a secondary phase in a dendritic like morphology, has been grown in the intercellular areas and out on the surface.

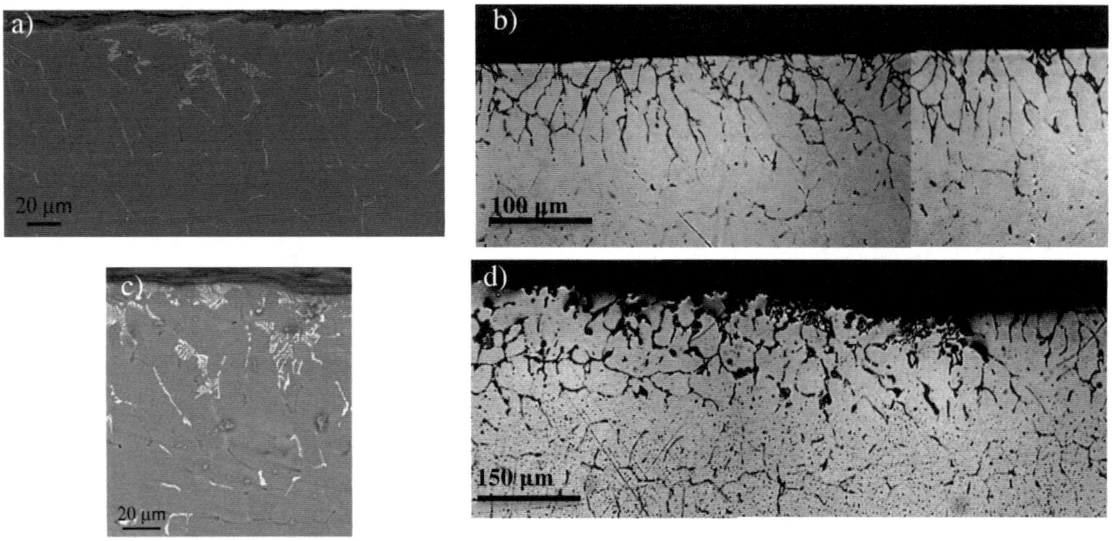

Figure 2. Transverse cross section of 6082 Al alloys a, b) Defect free surface along growth direction; c, d) wavy surface along growth direction.

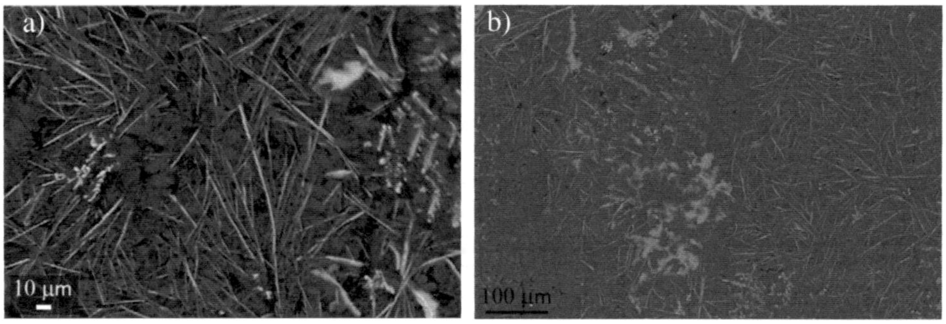

Figure 3. SEM untreated surface images of a) smooth surface; and b) wavy – defect surface.

4. Discussion

At smooth surfaces the segregation zone has a relatively constant thickness, Figures 2a and b, indicating that this type of surface is obtained when the pressure fluctuations in the melt, and thus the meniscus movements, are minimized. At such conditions, the melt uniformly exudates from the mushy zone to the layer between the oxide skin and the coherent solid [7].

In a previous study [7] it was shown that the segregation zone thickness of 6060 and 6005 alloys are 50-70 μm and 120-150 μm, respectively. The mentioned alloys have lower alloy contents than 6082, and it can therefore be expected that 6082 should have a thicker segregation zone. However, the results here show a zone thickness for 6082 of the same level as for 6060 but thinner than for 6005, which have an intermediate alloy level. In a comparison of these alloys the amount of secondary phases, eutectic structures and alloy contents [5] in the surface zone increase with alloy content, but the thickness of the layer does not. This might be explained by the fact that all alloys end with eutectic solidification at about 557 °C, and although the amount of eutectic increases, the solidification interval decreases for the higher alloyed 6082, figure 6. A factor influencing the thickness of segregation zone is also the metal head height [8], which might have varied for the studied billets.

When an unsmooth surface like the one discussed above is formed, a so-called wavy surface, as in Figure 1b can occur. The surface has irregular grooves and bulged zones, which occur in a semi-periodic pattern. Due to metal head pressure changes the meniscus movements will form a wavy - defect surface [7, 9]. Surface tension forces and pressure differences will define the curvature of the meniscus, and increasing surface tension coefficient will decrease the surface curvature [10]. In [7] the formation of surface zone during casting of a billet of 6005 and 6063 Al alloys were discussed, and exudation of alloying elements through the mushy zone was described. Exudation of liquid with high alloying content will affect the surface tension and the plasticity of the meniscus at mushy zone. For instance, it is reported that more than 0.2% of Mg can lead to the formation of MgO instead of Al_2O_3, and MgO seems to be weaker than Al_2O_3 [3]. While the billet is moving downwards the surface tension changes and gravitational forces over the meniscus will no longer be balanced, and at some point liquid will flow over the solid shell. This type of irregular wavy defect can occur in 6082 Al alloys, while the formations of waves on 6060 and 6005 alloys are more regular [7]. This irregular appearance might look as if exudated liquid is pressed out on the oxidized surface and solidified in an irregular way. However, studying the cross sections along many surfaces it is clear that liquid never exudates out through the oxide skin, see e.g. figure 2d. This type of surface only occurs in more highly alloyed 6xxx alloys in contrast to e.g. 6060, which show a very regular banding or lapping pattern, as in figure 1c, [7, 9, 10]. The difference can be explained by differences in coherency points for these alloys. When the meniscus fluctuates, and a wavy surface is formed the oxide skin moves inwards over the mushy zone during low liquid pressure [7]. At this movement, in a low-alloyed billet, the oxide skin will effectively adhere to a rigid coherent semi-solid and form a ring contact. If the solid fraction is under the coherency point, as in a high alloy billet, the oxide skin will adhere more randomly, i.e. in patches, points or shorter lines as in figure 1b. When the pressure increases the oxide skin is bulged out, and in the low-alloyed case a perfect ring droplet will form, while in the high alloy case the oxide skin is pressed out irregularly between the contact points. The differences in the coherency point can be seen in figure 6, where the solid fractions as a function of temperature are shown from thermodynamic calculations, (JMatPro). The coherency point for these class of alloys are in the solid fraction of about 0.5 [11], which means that it will occur at 650 °C for 6060 but at 640 °C for 6082.

In figure 5, eutectic α-phase, growing from the intercellular areas up to the surface, can be seen, but there is no connection to surface phases. It can therefore be concluded that the surface phases rather loosely has grown in the surface layer. The phases shown in figure 3, a and b, and in figure 4, a, b, c and d, are found on the surface of the billets when using a SEM. Similar observations have been done on other 6xxx alloys in [7, 8] but the formation of these top surface structures have not been satisfactory explained. In [7] the plate-like phase was identified as β-phase, Al_5FeSi, and the dendritic phase as α-phase, $Al_{15}(Fe,Mn)_3Si_2$. This can also be confirmed in this study as the Fe:Si ratio is 1:2 in the plates in figure 4a and the (Fe,Mn):Si ratio is 1.5:1 in the dendrites in figure 4b. The third phase, also in a dendritic pattern but with a weaker contrast, figure 4c, has the Mg:Si ratio 2.3:1, which indicates Mg_2Si phase. The phase occurs rather frequently and on many places together or below the other phases, as can be seen in figure 4b. In figure 5 an attempt was made to show how the surface phases are connected to the structure in the intercellular spacing's. In [7] the outwards solidification of the exudated melt to a top most layer of about a few microns was discussed, and strongly increasing solute concentrations were concluded.

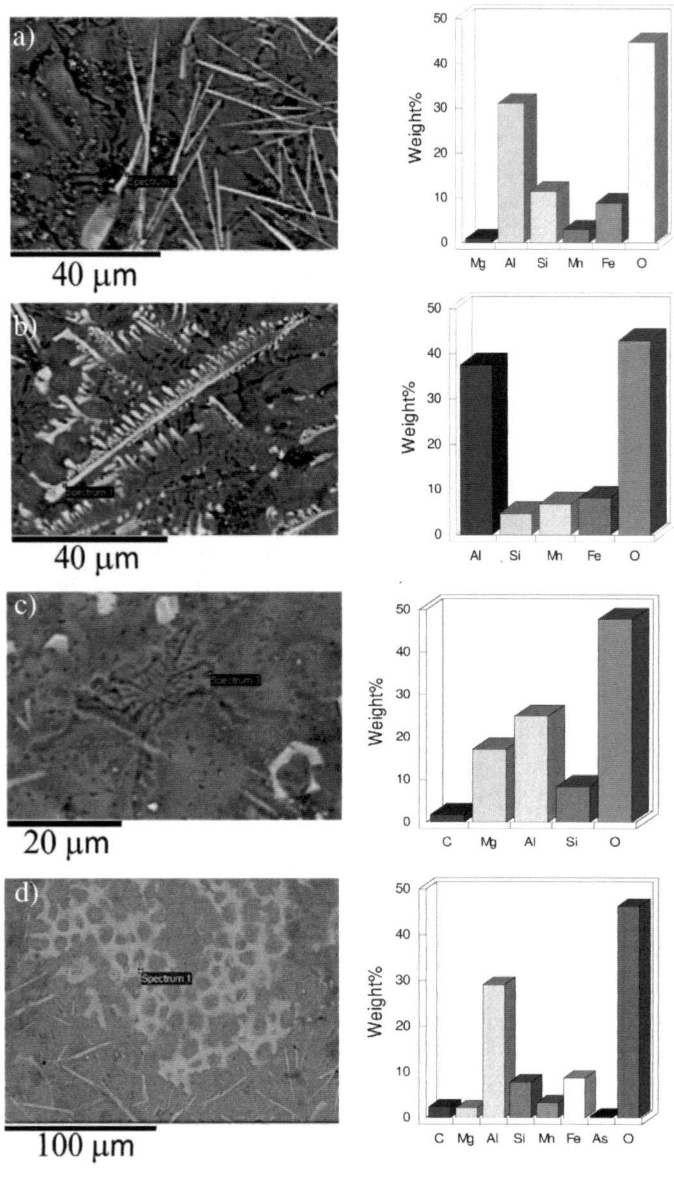

Figure 4. EDX analysis of different secondary phases at a, b) defect free sample; c, d) wavy surface.

The shapes of the phases, i.e. plates or dendrites that have grown in that layer indicate that they grow as primary precipitations but the growth is not followed by a eutectic growth to finalize the solidification. Instead, the thin liquid layer with the last melt will be sucked down between the cells due to solidification shrinkage during the final stages of the solidification. As it is illustrated in figure 3b, different phases can be observed on the wavy surfaces. Also the structures in the cross sections, figure 2d, show the variations on a place where the surface starts to bulge out. In a short section to the right a structure similar to a smooth surface can be seen with a cellular structure growing outwards and with mostly β-phase in the cell-boundaries. Then from the right to the left the surface is pushed out and there eutectic areas with α-phase dominate. In such regions, occurring around patches and point contacts, the structure indicates a more equiaxed growth, as there is no clear outward directed cell structure. It may be so that during the outwards cellular growth the segregation gives a higher Si content, and thus more β-phase is formed in cell-boundaries and on the surface, while the bulged out areas solidifies more equiaxed with larger eutectic areas between grains in which α-phase dominates.

Figure 5. Secondary phases of wavy surface after deep etching.

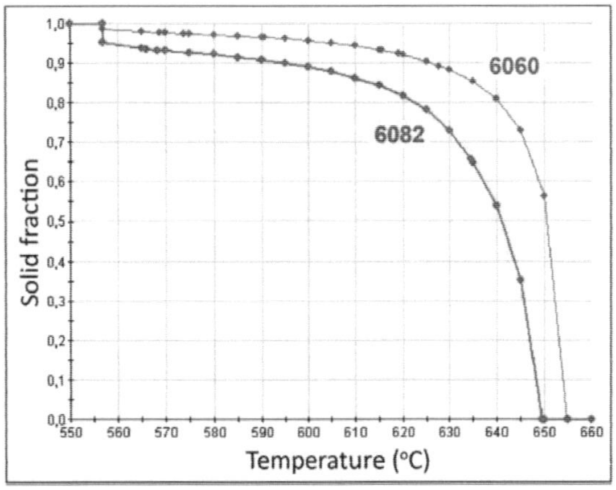

Figure 6. Solid fractions calculated with JMatPro for 6082 and 6060 alloys.

5. Conclusions

Smooth surfaces and surface segregation zones of uniform thicknesses are obtained when the pressure fluctuations in the melt, and thus the meniscus movements, are minimized.

The surface segregation zone thickness in 6082 alloys, for both smooth and wavy surfaces, is not wider than for 6060 and 6005 alloys but the amount of solute and secondary phases in this zone is larger.

The formation of the specific surface appearance on wavy 6082 surfaces can be correlated to how the oxide skin adheres to the underlying mushy zone.

In 6082 alloys the phases Mg_2Si, alpha and beta were observed in the surface zones of all studied surfaces but the extent of the different phases vary with surface appearance. At smooth surfaces beta is dominating, while at wavy surfaces, especially at the rim of bulged areas, alpha and Mg_2Si are more frequently appearing.

References

[1] Marioara C D et al 2003 *Acta Mater.* **51** 789-796.
[2] Erdegren M and Carlberg T 2011 *Light Metal TMS* 675-680.
[3] Bainbridge I F et al 2004 *Light Metal TMS* 693-698.
[4] Carlberg T and Jarfors A E W 2014 *Metall. Mater. Trans. B* **45B** 175-181.
[5] Buxman K 1978 *Light Metal TMS* 768-782.
[6] Nadella R et al 2008 *Prog. Mater. Sci.* 421-480.
[7] Bayat N and Carlberg T 2014 *JOM* **66 No. 5** 700-710.
[8] Håkonsen A et al 2014 *Light Metal TMS* 873-878.
[9] Benum S et al 1999 *Light Metal TMS* 737-742.
[10] Iversen F 2002 *PhD Thesis* NTNU Trondheim.
[11] Velman N J M et al 2001 *Metall. Mater. Trans. A* **32A** 147-155.

The role of TiB_2 in strengthening TiB_2 reinforced aluminium casting composites

Z Chen, H Kang, Y Zhao, Y Zheng and T Wang[a]

School of Materials Science and Engineering, Dalian University of Technology. No.2 Linggong Road, Ganjingzi District, Dalian City, P. R. China.

Email: [a]tmwang@dlut.edu.cn

Abstract. With an aim of developing high quality in situ TiB_2 reinforced aluminium foundry alloy based composites, the conventional direct synthesis method was modified into a two-step route. In step one we optimized the halide salt route to fabricate in situ TiB_2 particulate reinforced aluminium matrix composites and in step two we investigated the effects of the Al−5wt.% TiB_2 composite, as a "master composite", on strengthening the practical foundry alloys. The in situ formed TiB_2 particles play two roles while strengthening the composites: (1) The grain refinement effect that improves the quality of the alloy matrix; and (2) The interactions between the hard particulates and the matrix add extra increment to the material strength. In different alloy systems, TiB_2 may play distinct roles in these two aspects (figure 1). Further analysis of the strengthening mechanisms shows that particle agglomeration behaviour during solidification is responsible for the latter one. The present work details the role of TiB_2 in strengthening TiB_2 reinforced aluminium casting composites.

1. Introduction

The use of K_2TiF_6 and KBF_4 salts, in the halide salt route to produce aluminum matrix composites (AMCs), suffers from two predominant drawbacks. One is that a substantial portion of the charge is of no use, giving rise to detrimental fluoride-bearing emissions and troublesome decanting operation. The other, more intractable, is that the unclear reaction mechanism adds extra difficulties to eliminating unwanted salt compounds, oxides, metastable intermetallic compounds, etc. and composition adjustment owing to the loss of raw materials. These drawbacks can be resolved well if we adopt a remelting and diluting (RD) process using master composites that have been pre-fabricated.

In the literature regarding in situ aluminum alloy composites, efforts were made either on composites with pure Al matrix [1-7] or on those with alloy matrix [4, 5, 8-13] through direct synthesis with reactants. In Part I and II of our previous work we described an improved halide salt route to produce in situ TiB_2 reinforced aluminum foundry alloy based composite with enhanced mechanical properties [14,15].

While it seems the RD approach is generally suitable to prepare high quality aluminum foundry alloy based composites, the reinforcing efficiencies may vary because of different agglomeration behaviors of TiB_2 in different alloy systems. Yet the key question that needs to be answered is: What roles does the TiB_2 play in reinforcing the matrix?

This work was undertaken to discuss this question.

2. Experimental

99.85 wt.% Al, 99.7 wt.% Si, 99.8 wt.% Mg, 99.97 wt.% Cu and 99.8 wt.% Zn were used as the starting materials. Al−5wt.% TiB_2 master composite was prepared via the improved halide salt route described in Ref. [14]. The Ti and B recoveries in the as-received composites were measured using an inductive coupled plasma (ICP) emission spectrometer calibrated for Ti and B. The materials were cut, weighted and melted in an electric resistance furnace to produce a series of AlSi7Mg0.3, AlCu4.5Si1.1 and AlZn6Mg0.5 alloy matrix composites with 0, 1, 2, 3 and 4 wt.% TiB_2.

Mechanical stirring was made 2 min before casting to bring back the settled particles from the lower part of the melt. The pouring temperature was 720 °C. X-ray diffraction (XRD) analysis was conducted to confirm the presence of TiB_2.

The etched specimens were observed using field emission secondary electron microscope (FESEM). Tensile tests (0.2% offset yield strength ($\sigma_{0.2}$), ultimate tensile strength (UTS) and elongation (δ)) were conducted on a computerized universal testing machine.

3. Results and discussion

A 98.6% Ti recovery and 90.8% B recovery were obtained in the final composite using the improved halide salt method. Figure 1 shows the microstructures of the in situ composite. It can be seen that the TiB_2 particles in sample 1 and 2 were formed in agglomerates, most of which segregated at the α-Al grain boundaries (figure 1a). In the inset of figure 1a we can see that the TiB_2 existed in near-equiaxed shape with two shrunk (0001) planes.

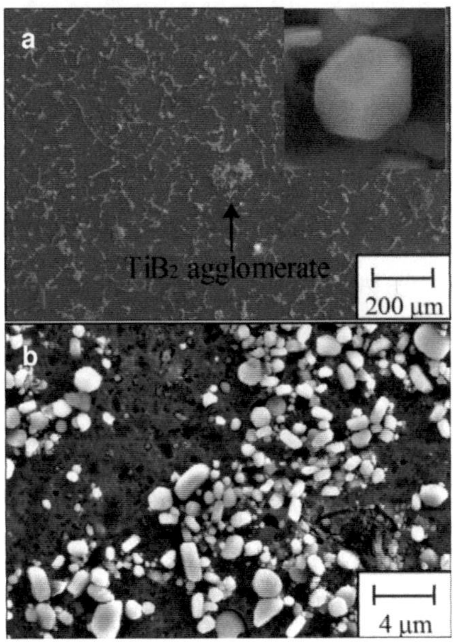

Figure 1. SEM micrographs of the in situ Al−5TiB_2 master composite at a (a) lower magnification and (b) higher magnification. Shown in the inset is the morphology of in situ TiB_2 particles.

Equiaxed particles, profiting from desired symmetry and shape, are most desirable. The composite recorded a 140% improvement in UTS compared to the matrix (from 43.2 to 103 MPa) without compromising its elongation (from 37.1% to 36.2%).

The XRD spectra for the alloys and experimental composites, as given in figure 2, reveals only TiB_2 reflections besides those of the matrix alloys. This is taken to indicate that TiB_2, once formed, is not to react with the solute elements, including Si, Mg, Cu, and Zn, in Al. The increasing intensity of TiB_2 signals suggests a higher weight percentage of TiB_2 in the composite.

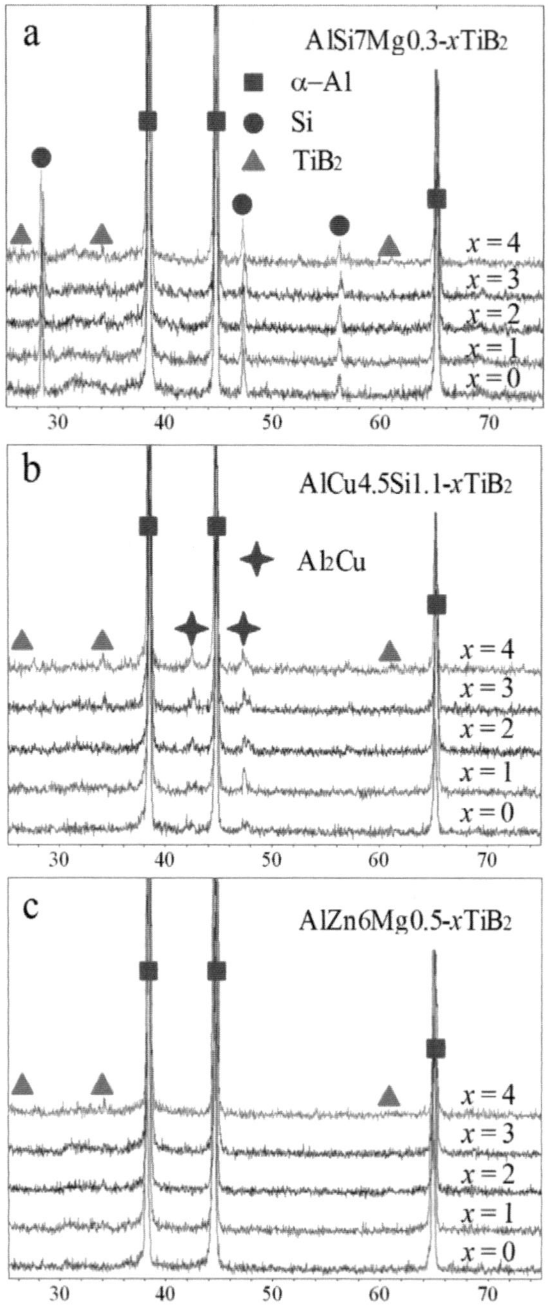

Figure 2. XRD spectra of the experimental (a) AlSi7Mg0.3−xTiB_2, (b) AlCu4.5Si1.1−xTiB_2 and AlZn6Mg0.5−xTiB_2 (x = 0, 1, 2, 3, 4) composites prepared by remelting and diluting approach.

Figure 3 shows the microstructure changes of the alloys incorporated with TiB_2 particles in different addition levels. It is expected that different reinforcing efficiencies may be owing to different roles the TiB_2 plays in two aspects upon solidification: (1) The grain refinement effect that improves the quality of the alloy matrix; and (2) The interactions between the hard particulates and the matrix add extra increment to the material strength.

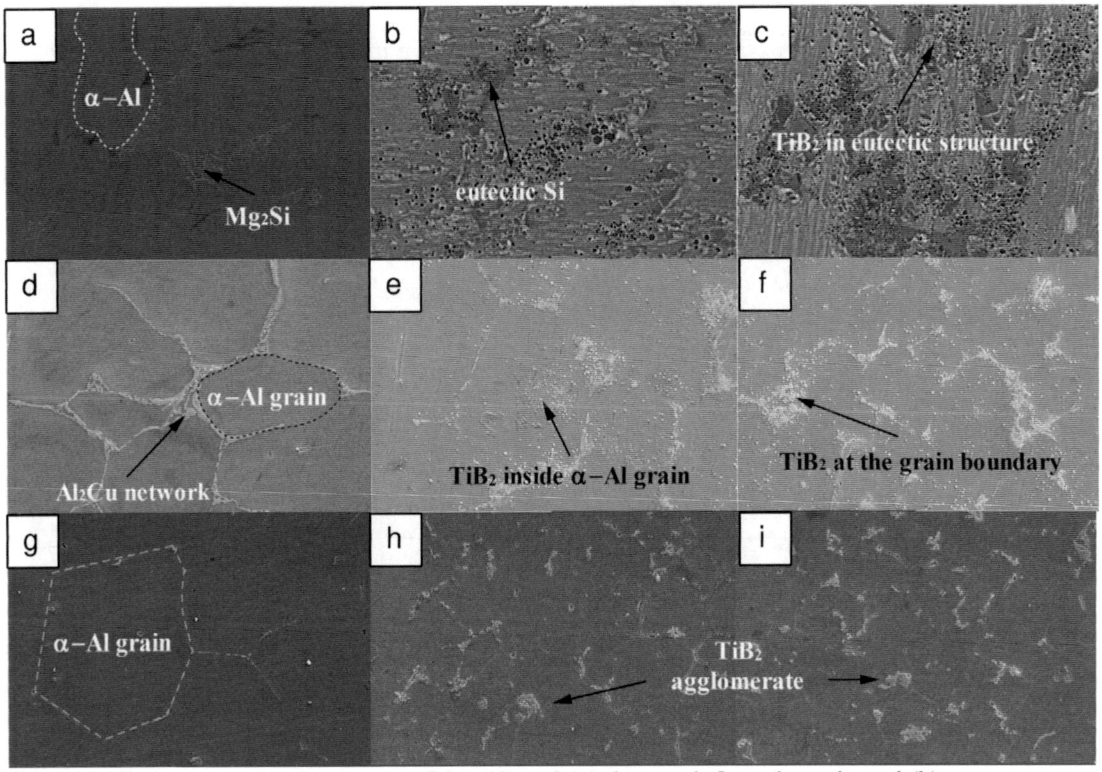

Figure 3. The as-cast microstructures of (a), (d) and (g) the unreinforced matrix and (b) AlSi7Mg0.3−2TiB_2, (c) AlSi7Mg0.3−4TiB_2, (e) AlCu4.5Si1.1−2TiB_2, (f) AlCu4.5Si1.1−3TiB_2, (h) AlZn6Mg0.5−1TiB_2, (i) AlZn6Mg0.5−2TiB_2, respectively.

Figure 4 shows the grain boundary spacing, including grain size and SDAS, versus *Vp* for the three systems. TiB_2 is but less efficient in reducing the SDAS of AlSi7Mg0.3, despite its evident efficiency in refining the primary grain structure confirmed by microstructural analysis.

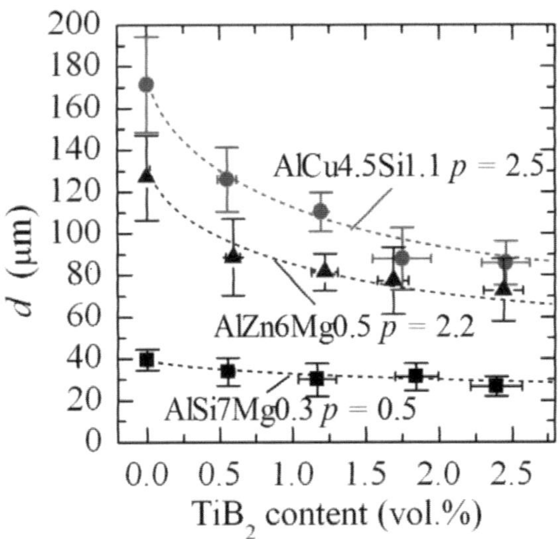

Figure 4. Average grain size versus TiB_2 content (vol.%) for the three systems. p was obtained by fitting the experimental data with $d = d_0 \cdot (1 + pV_p)^{-1/3}$.

The tensile properties, i.e. $\sigma_{0.2}$, UTS and δ of the composites with different alloy systems in as-cast state are shown in figure 5. From figure 5a–c we can see that all three systems can be successfully strengthened, but the reinforcing efficiency was rather discrepant for each alloy system.

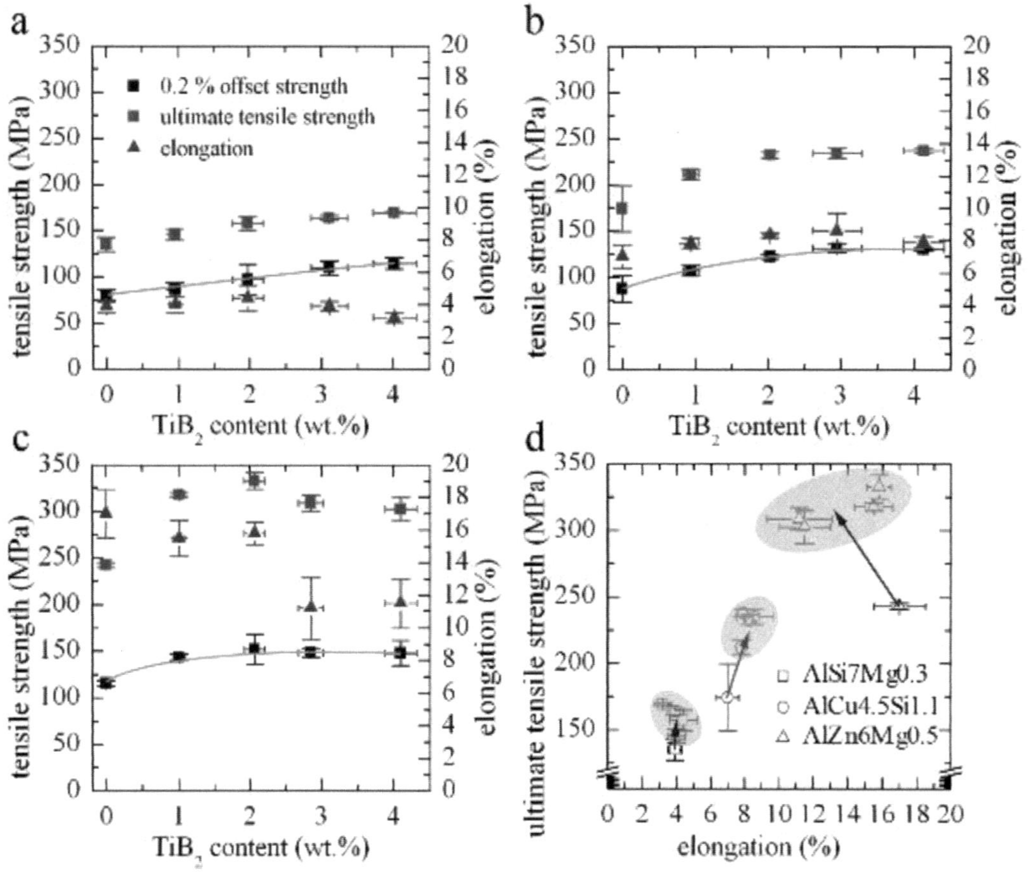

Figure 5. As-cast tensile properties of the (a) AlSi7Mg0.3, (b) AlCu4.5Si1.1 and (c) AlZn6Mg0.5 samples as a function of the TiB_2 content. Dashed lines represent the UTS and δ of the unreinforced matrix and shaded areas depict those of the diluted composites. (d) is an overview of the reinforcements by taking δ as the x axis and UTS as the y axis.

In figure 6, the predicted yield strength of the composites, are higher than the measured values. This is not surprising since practically the TiB_2 particles are distributed unevenly in the matrix. Considering clustering of TiB_2, we can replace Vp by $Vp^* = \varepsilon Vp$ and then adjust ε to fit the curve with the experimental data using the following equation [15].

$$\Delta\sigma_{total} = \Delta\sigma_{gf} + ((\Delta\sigma_{Orowan})^2 + (\Delta\sigma_{CTE})^2)^{1/2} \quad (1)$$

where $\Delta\sigma_{total}$ is the total increment of yield strength owing to the grain refinement, Orowan and CTE strengthening mechanisms. ε were estimated to be 0.14, 0.56, and 0.35 for the AlSi7Mg0.3, AlCu4.5Si1.1 and AlZn6Mg0.5, respectively, indicating that ~35%, 56%, and 14% of the added TiB_2 particles can be uniformly distributed to participate in strengthening. Compared with the microstructural features shown in figure 3, these values were found quite reasonable.

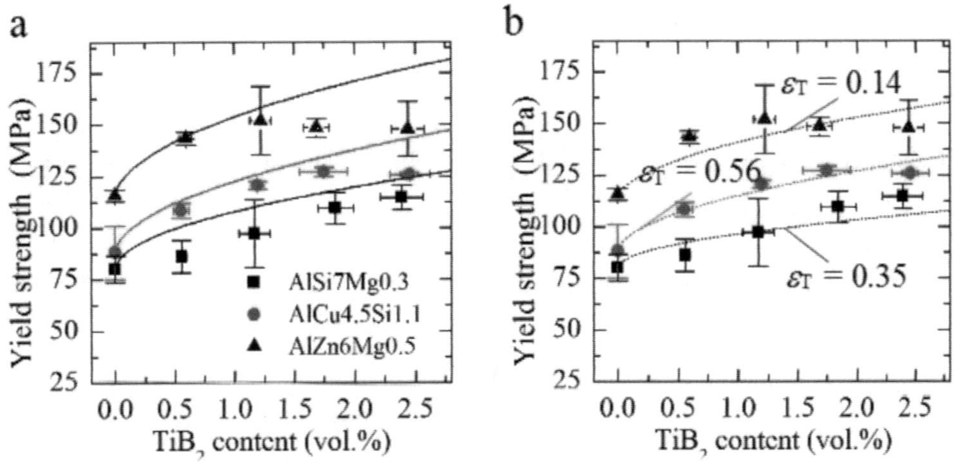

Figure 6. Comparison of the theoretically predicted yield strength (a) before and after adjusting *Vp* with the experimental data from tensile tests.

We now compare the estimated ε values with the measured percentage of dispersed TiB_2 particles analyzed from the microstructural features in figure 7. It can be seen that the measured TiB_2 percentages dispersed near the estimated ε values well, indicating that these values were quite reasonable.

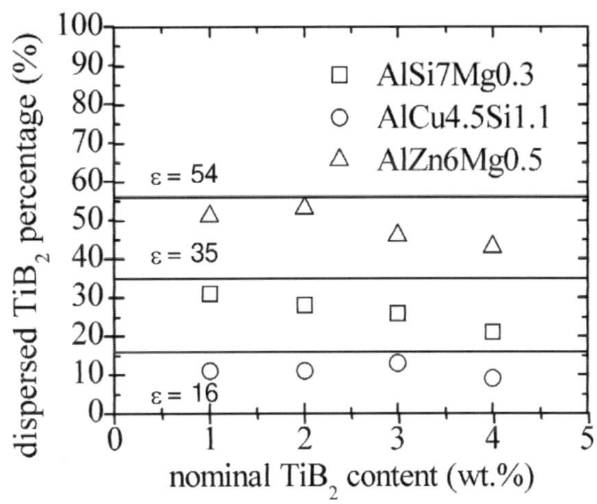

Figure 7. Comparison of the theoretically fitted ε values with the measured percentage of dispersed TiB_2 particles analyzed from the microstructural features.

4. Conclusions

The homogeneity of TiB_2 is strongly dependent on the alloy system. By way of incorporating pre-fabricated Al-TiB_2 master composite in different alloy systems, improved mechanical properties can be obtained in all three series of composites. But for different alloy systems, the strengthening efficiency may vary, owing to the different agglomeration behaviour of TiB_2 particles. In AlSi7Mg0.3−$x$$TiB_2$, the predominant mechanisms are Orowan and CTE strengthening. In AlCu4.5Si1.1−$x$$TiB_2$, all the three mechanisms contribute to the final yield strength. In AlZn6Mg0.5−$x$$TiB_2$, the main mechanism is grain refinement and the contribution of Orowan and CTE strengthening is very little.

References

[1] Davies P, Kellie J L F, Wood J V. UK Patent No 2 257985A; 1992.

[2] Wood J V, Davies P and Kellie J L F 1993 *Key Engineering Materials* 77-78:357-9.

[3] Lakshmi S, Lu L and Gupta M. In situ preparation of TiB_2 reinforced Al based composites. 1998 *J Mater Process Tech* **73**:160-6.

[4] Tee K L, Lu L, Lai M O. 1999 Synthesis of in situ Al–TiB_2 composites using stir cast route. *Compos Struct* **47**:589-93.

[5] Feng C and Froyen L. 2000 Microstructures of in situ Al/TiB_2 MMCs prepared by a casting route. *J Mater Sci*. **35:**837-50.

[6] Ravi K R, Saravanan M, Pillai R M, Mandal A, Murty B S, Chakraborty M, et al. 2008 Equal channel angular pressing of Al–5wt% TiB_2 in situ composite. *J Alloy Compd*. **459**:239-43.

[7] Xue J, Wang J, Han Y, Li P, Sun B. 2011 Effects of CeO_2 additive on the microstructure and mechanical properties of in situ TiB_2/Al composite. J Alloy Compd. **509**:1573-8.

[8] Lu L, Lai M O and Chen F L. 1997 Al-4 wt% Cu Composite reinforced with in-situ TiB_2 particles. Acta Mater. **45**:4297-309.

[9] Lü L, Lai M O, Su Y, Teo H L and Feng C F. 2001 In situ TiB_2 reinforced Al alloy composites. Scripta Mater. **45**:1017-23.

[10] Ramesh C S, Ahamed A, Channabasappa B H and Keshavamurthy R. 2010 Development of Al 6063–TiB_2 in situ composites. Mater Design. **31**:2230-6.

[11] Li P, Li Y, Wu Y, Ma G, Liu X 2012 Distribution of TiB_2 particles and its effect on the mechanical properties of A390 alloy. Mater Sci Eng A. **546**:146-52.

[12] Michael Rajan H B, Ramabalan S, Dinaharan I and Vijay S J 2013 Synthesis and characterization of in situ formed titanium diboride particulate reinforced AA7075 aluminum alloy cast composites. Mater Design. **44**:438-45.

[13] Zhao D G, Liu X F, Pan Y C, Liu Y X and Bian X F. 2006 Microstructure and mechanical behavior of AlSiCuMgNi piston alloys reinforced with TiB_2 particles. J Mater Sci. **41**:4227-32.

[14] Chen Z, Wang T, Zheng Y, Zhao Y, Kang H and Gao L. 2014 Development of TiB_2 reinforced aluminum foundry alloy based in situ composites–Part I: An improved halide salt route to fabricate Al–5 wt.%TiB_2 master composite. Mater Sci Eng A. **605**:301-9.

[15] Wang T, Chen Z, Zheng Y, Zhao Y, Kang H and Gao L. 2014 Development of TiB_2 reinforced aluminum foundry alloy based in situ composites–Part II: Enhancing the practical aluminum foundry alloys using the improved Al–5wt.%TiB_2 master composite upon dilution. *Mater Sci Eng A*. **605**:22-32.

Study on rheo-diecasting process of 7075R alloys by SA-EMS melt homogenized treatment

G Zhihua[a], X Jun[b], Z Zhifeng[c], L Guojun[d] and T Mengou[e]
General Research Institute for Non-Ferrous Metals, Beijing, China

Email: [a]gzhzsm@msn.com, [b]xujun@grinm.com, [c]zhangzf@grinm.com, [d]liuguojunx@163.com, [e]menciustang@126.com

Abstract. An advanced melt processing technology, spiral annular electromagnetic stirring (SA-EMS) based on the annular electromagnetic stirring (A-EMS) process was developed for manufacturing Al-alloy components with high integrity. The SA-EMS process innovatively combines non-contact electromagnetic stirring and a spiral annular chamber with specially designed profiles to in situ make high quality melt slurry, and intensive forced shearing can be achieved under high shear rate and high intensity of turbulence inside the spiral annular chamber. In this paper, the solidification microstructure and hardness of 7075R alloy die-casting connecting rod conditioned by the SA-EMS melt processing technology were investigated. The results indicate that, the SA-EMS melt processing technology exhibited superior grain refinement and remarkable structure homogeneity. In addition, it can evidently enhance the mechanical performance and reduce the crack tendency.

1. Introduction

The current shape cast components usually contain a coarse and non-uniform microstructure and various casting defects, offering poor mechanical performance. A grand challenge is to develop solidification processing technologies which can ensure a fine and uniform as-cast microstructure free from cast defects, so that the cast products can be either directly used in the as cast state, or only require minimal thermo-mechanical processing [1]. Characterized by non-pollution, non-contact, low cost, precisely control and easy combination with industry, electromagnetic stirring (EMS), which is considered as a main melt conditioning method for gaining fine casting products, has been successfully applied in industry [2-5]. However, inhomogeneous microstructure still occurs in the casting products especially large-sized ones due to the skin effect resulting from electromagnetic induction. To solve the problem, an advanced metal melt conditioning process, namely, the annular electromagnetic stirring (A-EMS) was developed, and an intensively forced shearing could be achieved at a higher shear rate inside the annular chamber at the commercial frequency. The A-EMS process exhibits better grain refinement and microstructure homogeneity. Moreover because of the intrinsic characteristics of electromagnetic stirring, electromagnetic force in the melt showed the circumferential distribution. The circumferential flow was strong, but the axial flow and radial flow were weak. It was inevitable that temperature gradient and composition gradient distributed along the radial and axial directions [6-9]. To solve the problem, the spiral annular electromagnetic stirring (SA-EMS), an optimized A-EMS, was developed. Compared with A-EMS, the new electromagnetic stirring melt processing technology (SA-EMS) could not only avoid the skin effect efficiently, impose a high shear rate and a high intensity of turbulence to the liquid metal, but also could cause alloy melt flow to turbulent flow eventually, leading to temperature field and composition field of alloy melt in the whole melt volume distribute uniformly by means of specially designed spiral profiles. In this paper, the solidification microstructure, mechanical properties and crack tendency of 7075R alloy die casting connecting rod conditioned by the SA-EMS melt processing technology were investigated.

2. Experiment

The nominal chemical composition of the experimental 7075R alloy is shown in table 1. The liquidus and solidus temperature measured by DSC of the test alloy are 641° C and 477° C.

Table 1. Nominal chemical composition of the test alloy (*wt. %*)

Element	Zn	Mg	Cu	Cr	Fe	Si	Ti	Mn	Al
Content	5.69	2.28	1.63	0.18	0.24	0.36	0.12	0.27	Bal

The experimental procedures were as followed: The 7075R alloy melt was prepared in a clay graphite crucible in an electric resistance furnace. The metal melt was held at 750°C for ~2 h to be homogenized. By using a graphite rod, the dross was removed from the surface of the melt, and the metal melt was stirred manually for purposes of composition uniformity. Predetermined melt was transferred to two small stainless steel crucibles which are preheated to 400°C for conducting experiments (see figure 1). One was directly poured into the die-casting chamber without electromagnetic stirring and the cooling core rod (N-EMS) when the melt was cooled to 650 °C. The other one was conditioned by the SA-EMS from 750°C to 650°C, then the melt was poured into the die-casting chamber. Stirring current was 30A and stirring frequency was 50HZ. The mould temperature was 350 °C. The injection speed was 2m/s. All of them were die-casting formation and cooled to the room temperature. The die-casting connecting rod was cut along the longitudinal section.

Microstructures observation and hardness analysis will be carried out at six different positions, as shown in figure 2. The samples were cut, ground and polished using a standard metallographic technique. The samples were anodized with Barker's reagent (4%HBF4 in distilled water) and then viewed under polarized light using a Zeiss optical microscope. A Zeiss optical imaging system was utilized for the optical microscopy observations and the quantitative measurements of the microstructure information. HBE-3000A electronic Brinell hardness tester was carried out to test the hardness of the samples.

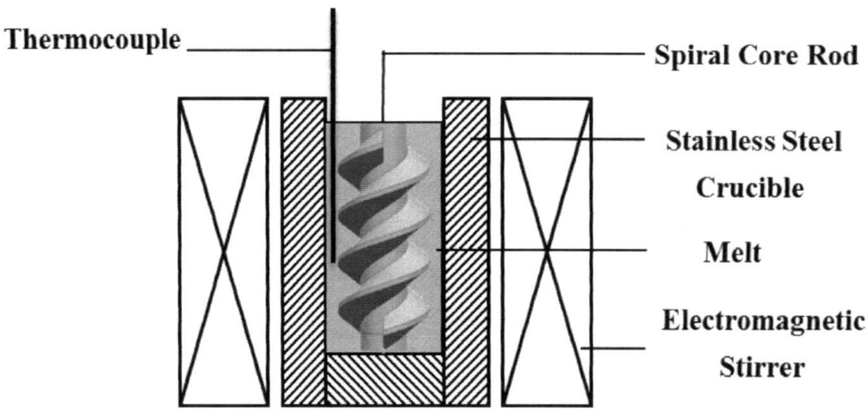

Figure 1. Schematic diagram of SA-EMS melt conditioning experimental device.

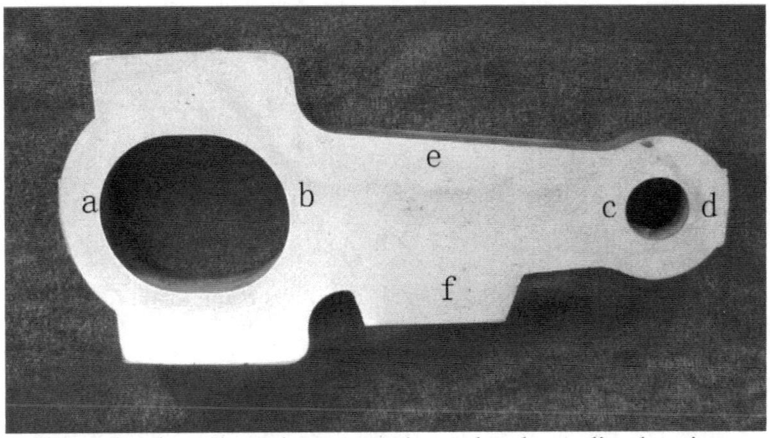

Figure 2. Rheo-diecasting connecting rod and sampling locations.

3. Results

Figure 3 shows the microstructures distribution of conventional die-casting part in six different positions. As can be seen, the size and morphology of microstructures in different positions is very non-uniform, and it is also very non-uniform in the same position, including coarse dendrite grains and small equiaxed grains. But for the SA-EMS rheological die-casting part, as shown in figure 4, the microstructures in different positions are very fine and uniform, and the microstructures are composed of fine equiaxed grains at the same position. Figure 5 shows the grain size and hardness distribution of the parts at different positions under the N-EMS and SA-EMS conditions. As can be seen, compared with common die-casting forming (N-EMS), under the condition of SA-EMS, the microstructures of rheo-diecasting part distributed uniformly fine and showed little difference in different position. Brinell hardness of rheo-diecasting parts increased significantly, and also showed little difference in different positions.

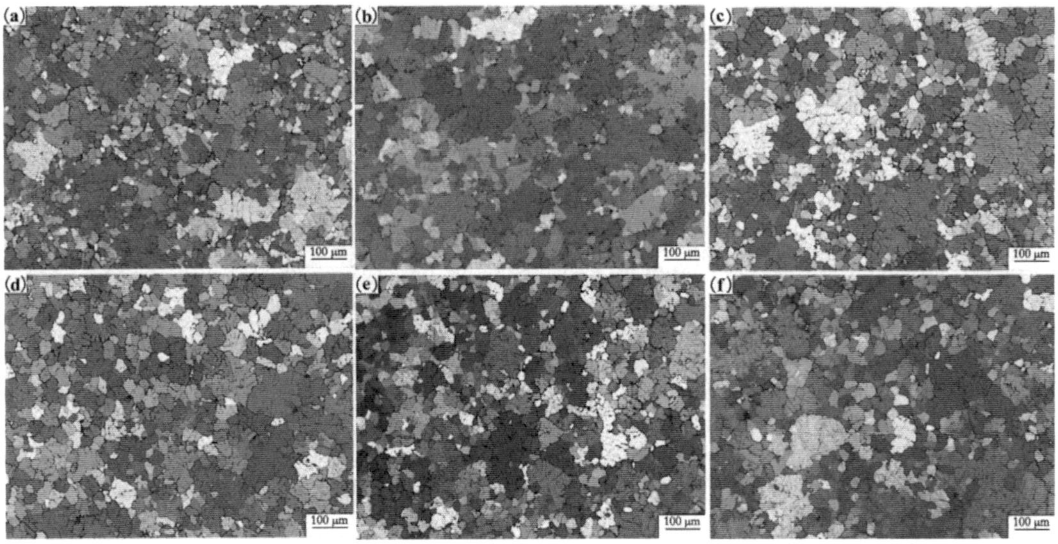

Figure 3. Microstructures of die-casting part (N-EMS) in six different positions.

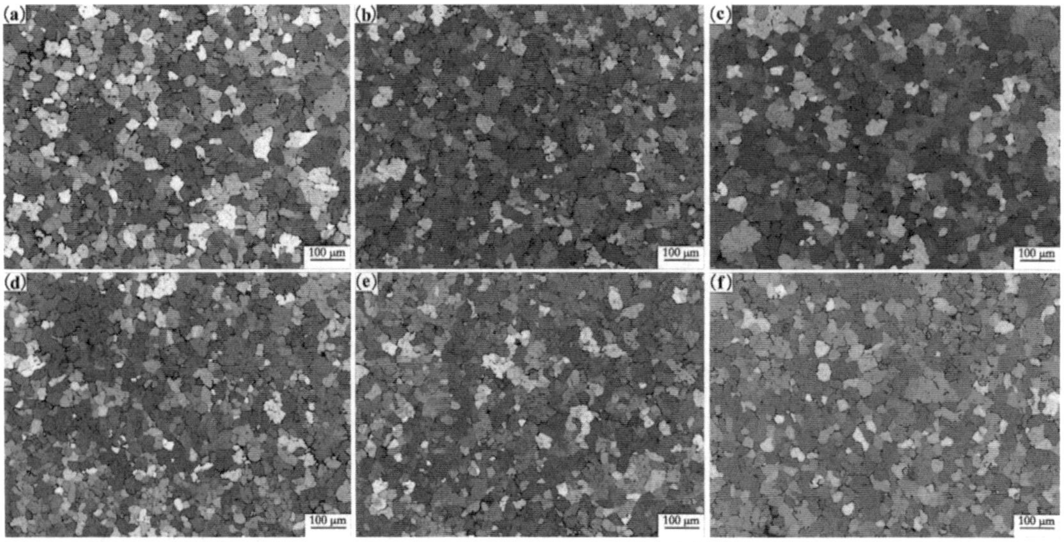

Figure 4. Microstructures of rheo-diecasting part (SA-EMS) in different positions.

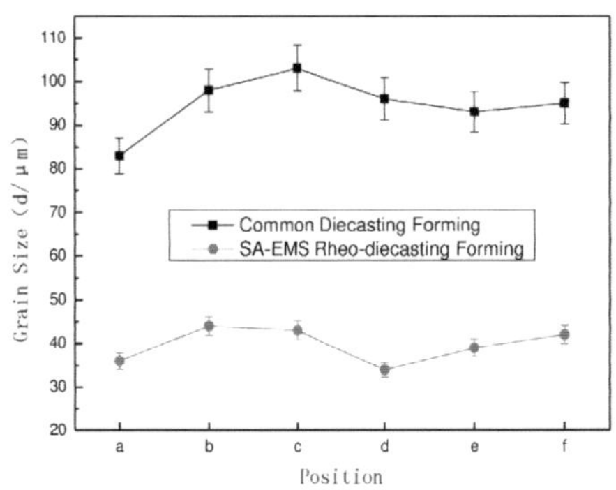

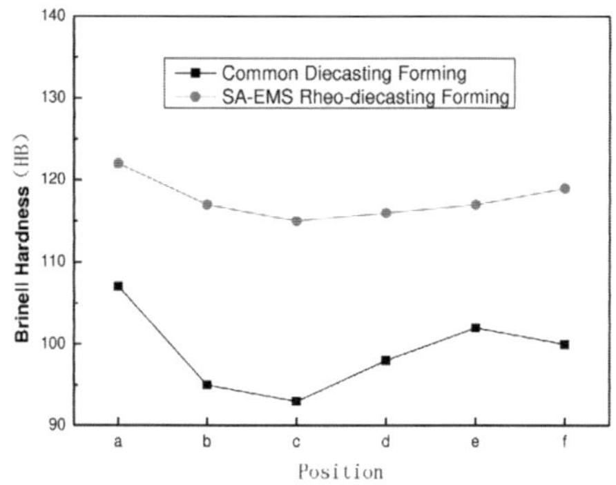

(a) (b)

Figure 5. Distribution of grain size and hardness in different positions of die-casting parts under conditions of N-EMS and SA-EMS: (a) grain size, (b) hardness.

Figure 6 shows the crack tendency of die-casting parts under the N-EMS and SA-EMS conditions. On the basis of the above research, we can know that the thick-thin wall transition zone of connecting rod part is at the b and c positions, and the change of microstructures is very big, easily leading to cracking (seeing figure 6a). But for the SA-EMS rheological die-casting part, as can be seen from figure 6b, there are no obvious cracks in b and c positions, and the crack tendency significantly decreased. This is because the solidification microstructures of the connecting rod part after the SA-EMS processing become very fine and uniform and show little differences in different position, and the crack is not easy to produce.

Figure 6. The crack tendency of die-casting parts under different melt treatment (a) N-EMS, (b) SA-EMS.

4. Discussion

For the N-EMS sample, according to the classical nucleation theory, nucleation behaviour is triggered by thermal undercooling through the mould wall. Once the superheated liquid metal is poured and in touch with the relatively cold mould surface, "Big Bang" nucleation takes place, and numerous nuclei are created. However, the majority of those nuclei are dissolved into the oncoming superheated liquid, and only a small proportion of nuclei in the vicinity of the mould surface can survive, resulting in columnar and coarse dendritic structures [10]. However, the fluid flow inside the SA-EMS unit is characterized by a high shear rate and high intensity turbulence where large interfacial area provides improved heat transfer and strong dispersive mixture. This intensive melt shearing ensures the temperature field, chemical composition field and tiny heterogeneous particles of the melt extremely uniformly distribute throughout the entire volume of the melt. Under such conditions, heterogeneous nucleation will take place throughout the whole volume of the liquid metal, and all the nuclei created will have a chance to survive and eventually contribute to the microstructural refinement. After nucleation, numerous nuclei will compete for growth by consuming the liquid around them. And the initial spherical growth will continue until the instability of the solid/liquid interface occurs [11, 12]. Due to the effective nucleation, solidification may have completed before reaching this instability. Therefore, the effective volume nucleation of the melt after intensive shearing usually results in fine and equiaxed grains throughout the entire casting.

5. Conclusions

The SA-EMS melt processing technology could avoid the skin effect efficiently, impose a high shear rate and a high intensity of turbulence to the liquid metal and increase the area of heat dissipation of unit volume melt by means of innovatively combining noncontact electromagnetic stirring and an spiral annular chamber with specially designed profiles, so that the conditioned liquid metal has uniform distribution of temperature field, chemical composition field and tiny heterogeneous particles. The SA-EMS process exhibits superior grain refinement and remarkable structure homogeneity. In addition, it can evidently enhance the mechanical performance and reduce the crack tendency.

Acknowledgements
This work was financially supported by the National Basic Research Program of China (973 Program) (No.2011CB606302) and International Science and Technology Cooperation Program of China (2013DFA51370).

References

[1] Fan Z, Xia M, Zhang H, Liu G, et.al. 2009 Melt conditioning by advanced shear technology (MCAST) for refining solidification microstructures. *International Journal of Cast Metals Research*, **22**, 221(4): 103-107.

[2] Kang C G, Bae J W, and Kim B M 2007 The grain size control of A356 aluminum alloy by horizontal electromagnetic stirring for rheology forging. *J. Mater. Process. Technol.*, 187-188: 344.

[3] Zoqui E J, Paes M, and Es-Sadiqi E 2002 Macro-and microstructure analysis of SSM A356 produced by electromagnetic stirring. *J. Mater. Process. Technol.*, 120: 365.

[4] Steinbach S and Ratke L The effect of rotating magnetic fields on the microstructure of directionally solidified Al-Si-Mg alloy. *Mater. Sci. Eng. A*, 2005, 413-414: 200.

[5] Jia Q M, Zheng Y L, Chen J Y 2001 *Electromagnetics*. Beijing: Higher Education Press, 262. (in Chinese)

[6] Xu J, Zhang Z F, Bai Y L, et a1. 2008 A Device of Preparing Semi-Solid Alloy Slurry or Billet：China Patent，2008101 16181. 9.. (in Chinese)

[7] Bai Y L, Xu J, Zhang Z F, Shi L K 2009 Annulus electromagnetic stirring for preparing semisolid A357 aluminum alloy slurry. *Trans. Nonferrous Met. Soc.* China., 19(5): 1104

[8] Zhu G L, Xu J, Zhang Z F, Bai Y L 2009 Annular electromagnetic stirring--a new method for the production of semi-solid A357 aluminum alloy slurry. *Acta Metallurgica Sinica* (English Letters), 22(6): 408

[9] Tang M O, Xu J, Zhang Z F, Bai Y L 2011 Effects of annulus gap on flow and temperature field in electromagnetic direct chill casting process. *Trans. Nonferrous Met. Soc.* China., 21(5):1123

[10] H T Li, M Xia, Ph Jarry, G M Scamans, and Z Fan 2011 Grain refinement in a AlZnMgCuTi alloy by intensive melt shearing: A multi-step nucleation mechanism. *Journal of Crystal Growth*, 214: 285-292.

[11] Fan Z, Liu G, 2005 Hitchcock M Solidification behaviour under intensive forced convection. *Materials Science and Engineering*: A, 413-414:229-235.

[12] Fan Z, Liu G, Hitchcock M, Das A 2007 Solidification behaviour under intensive forced convection. *Materials Processing Under the Influence of External Fields*, 201-212.

On the role of solidification modelling in Integrated Computational Materials Engineering "ICME"

G J Schmitz[a], B Böttger and M Apel
ACCESS e.V, Intzestr. 5, D-52072 Aachen, Germany

E-mail: [a]g.j.schmitz@access.rwth-aachen.de

Abstract. Solidification during casting processes marks the starting point of the history of almost any component or product. Integrated Computational Materials Engineering (ICME) [1-4] recognizes the importance of further tracking the history of microstructure evolution along the subsequent process chain. Solidification during joining processes in general happens quite late during production, where the parts to be joined already have experienced a number of processing steps which affected their microstructure. Reliable modelling of melting and dissolution of these microstructures represents a key issue before eventually modelling 're'-solidification e.g. during welding or soldering. Some instructive examples of microstructure evolution during a joining process obtained on the basis of synthetic and simulated initial microstructures of an Al-Cu binary model system are discussed.

1. Introduction

Segregation of alloy elements in a component is not only important for the properties of this component (e.g. solid solution hardening or formation of secondary phases etc.) but also for its further processing. Especially welding processes - typically happening at later stages of the production sequence of a component – strongly depend on the initial conditions for the materials to be joined. Scopes of the present paper are (i) to demonstrate effects of initial segregation patterns on microstructure formation during a typical laser welding process and (ii) to assess effects of simplifying assumptions for the segregation in the initial microstructure on the simulation of melting and solidification.

2. Simulation settings and materials data

Microstructure simulations were performed using MICRESS (- the MICRostructure Evolution Simulation Software) [5] version 6.154. MICRESS is based on the multi-phase-field concept [6, 7] and allows for online-coupling to thermodynamic and mobility databases [8].

The simulation domains are two dimensional using a cubic grid with 2000x500 grid points. The grid spacing is 0.5 μm. Boundary conditions for temperature were either defined as isothermal holding when generating the initial microstructures or using the 1-D thermal field functionality of MICRESS [13] for modelling the solidification and laser-welding conditions.

2.1. Materials data

Numerical simulations require data on both the materials and the process conditions. For the materials, especially the phase-diagram data, the interfacial energies between the phases involved, the diffusion constants of the components in the individual phases, and thermal conductivities are required. A binary Al-Cu alloy has been selected as a simple example in order to demonstrate the different effects on melting, grain growth in the liquid and eventually nucleation during re-solidification.

Thermodynamic data for the system Al-Cu have been taken from a binary subset of the COST507 database [9]. The material being considered for this basic study is a binary Al-Cu alloy with a nominal composition of 3 at.% Cu. The system reveals the liquid, the FCC and the θ phase (Al_2Cu). Equilibrium calculations using Thermo-Calc Software [10] predict a fraction of approx. 0.8% of the θ phase in equilibrium at room temperature. The diffusion coefficient for Cu was assumed to be constant at $D= 2*10^{-4}$ cm²/s in the liquid phase and to be $D= 1*10^{-8}$ cm²/s in FCC. No diffusion was assumed in the θ-phase. Thermal conductivities were estimated as λ_{liquid} =1.2 W/cmK and λ_{solid_phases} = 1.3 W/cmK. Data for specific heat and latent heat were determined from the thermodynamic database.

The properties of the interfaces relevant for the simulation are listed in table 1. They were either identified from literature or have been estimated. Details on the exact specification of these different entities in MICRESS can be found in [11, 12].

Table 1. Physical properties of the interfaces.

phase-pair	interfacial energy (J/cm²)	interfacial mobility (cm⁴/(Js))	anisotropy
Liquid/FCC	10^{-5}	0.2	kinetic: 0.2 static:0.2
Liquid/ θ	$1.2*10^{-4}$	0.05	isotropic
FCC/θ	$1.2*10^{-4}$	10^{-8}	isotropic
FCC/FCC*	10^{-5}	$5*10^{-3}$	isotropic

*to allow also description of grain growth phenomena

2.2. Generation of initial microstructures

Different initial microstructures were generated as input for the subsequent simulations of a typical laser welding process:

- an as solidified microstructure
- a solidified and heat treated microstructure
- a synthetic microstructure
- a recrystallized microstructure

Nucleation of the FCC and θ phases in all simulations was allowed in the bulk liquid using the seed density model being implemented in MICRESS [14] and a typical but arbitrary seed density distribution. Further, nucleation was also allowed at interfaces with a critical nucleation undercooling of ΔT_{crit}= 4 K for both the FCC phase and the θ- phase.

The as solidified microstructure (#1) was generated by simulating an equiaxed solidification process starting from the homogeneous and isotropic melt at T = 950 K. Heat has been extracted at 1.5 W/cm²K from the simulation domain leading to full solidification after approx. 0.5 s. An iterative homoenthalpic approach [13] has been used to obtain a realistic equiaxed microstructure with an average grain size of approx. 150 μm and to assure self-consistency between heat extraction and release of latent heat.

The heat treated microstructure (#2) was generated by using the as solidified microstructure (#1) and subjecting it to an isothermal holding for 5h at 800K.

A synthetic microstructure (#3) was generated from microstructure #1 by a heat treatment at a higher temperature (5h at 850K). At this temperature the θ-phase dissolves. The resulting microstructure thus only contains grains of the FCC phase. Due to the treatment at higher temperatures and accordingly faster diffusion no residual segregation can be observed in this microstructure. Note that the secondary dendrite arms disappear during this process. This type of microstructure – i.e. grains with constant concentration – could also have been generated by a kind of Voronoi tessellation and this microstructure thus has been called "synthetic". The procedure used here to generate this synthetic microstructure additionally provides the same topology of the grains as the microstructures #1 and #2 and thus allows for a better comparison of the results.

A recrystallized microstructure (#4) was simulated on the basis of the solidified microstructure (#1). The recrystallization functionality of MICRESS® was used to generate a recrystallized fine grained microstructure. Values for a stored elastic energy originating e.g. from a forming process and revealing typical values of 0-10 J/cm^3 were assigned to the different FCC grains in this microstructure for this purpose. During the re-crystallization process new grains nucleate and the grain growth is driven by stored elastic energy. It should be noted that grain boundaries are pinned by the θ-phase particles during the simultaneous recrystallization/homogenization process, and that the recrystallization process does not affect the segregation pattern. The homogenization heat treatment parameters were identical with those used for microstructure #2

A summarizing overview of the initial microstructures is depicted in figure 1. Note the almost similar microstructures of #1 and #2 (two phase structures differing essentially in their Cu concentration fields) and the strong differences of #3 (FCC phase only, no Cu segregation) and #4 (small grains, Cu-concentration field same as #2)

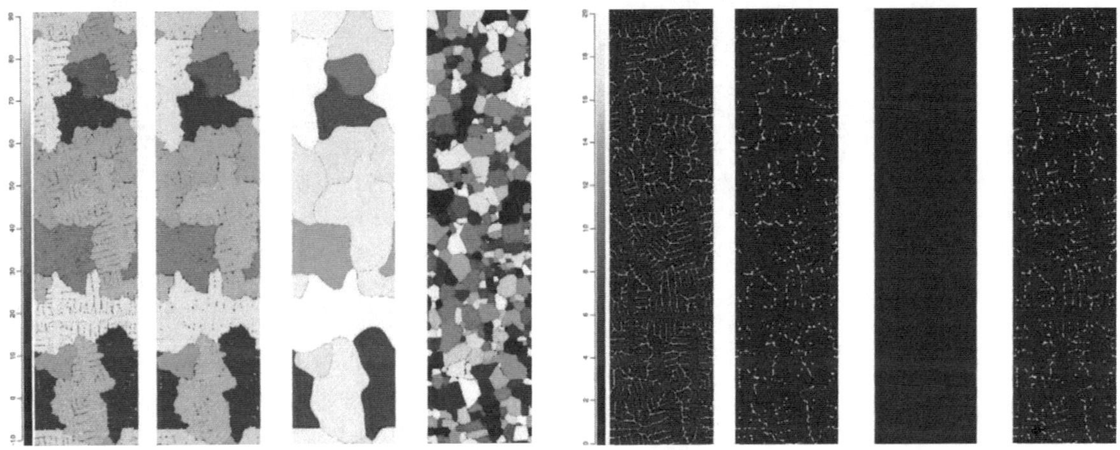

Figure 1. Overview of the four different initial microstructures (#1- #4 ; from left to right) for the grain orientations resp. grains (left) and for the Cu-concentration fields (right).

2.3. Modeling laser welding

The thermal boundary conditions for the subsequent simulation of the joining process - being identical for each of the initial microstructures - qualitatively correspond to a laser welding situation and were realized using the 1D-thermal field functionality of MICRESS®. The laser process was modelled as a heat flux boundary condition to the 1D temperature field as depicted in figure 2. This simulation setting leads to a partial melting and re-solidification of the microstructures as shown in figure 3 for the example of microstructure #1. Nucleation of the liquid phase inside the bulk solid phases and at grain/phase boundaries was allowed where a local overheating of 2 K was reached.

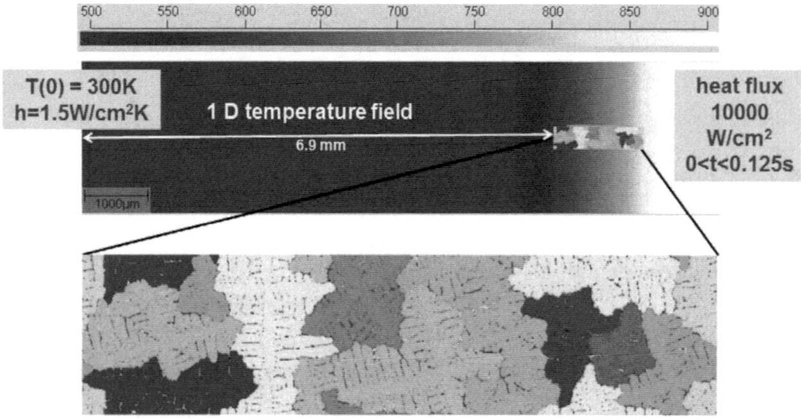

Figure 2. Schematic sketch of the 1D temperature field at t=0.13 s, its initial and boundary conditions and its relative position with respect to the 2D microstructure domain.

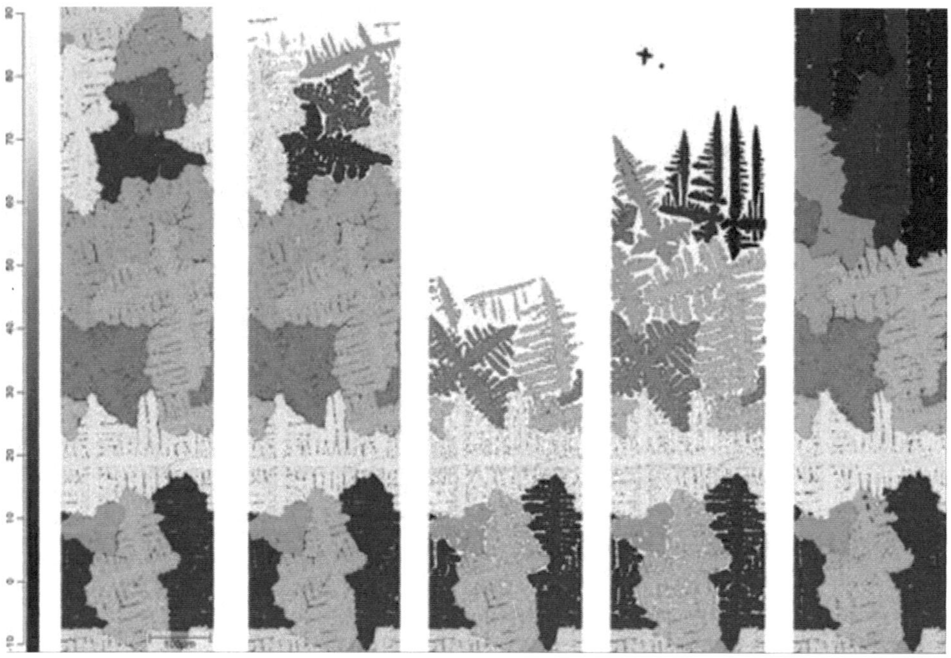

Figure 3. Melting and re-solidification of microstructure #1 as a time sequence (after 0, 0.124, 0.19, 0.26 and 2 seconds). Shown is the grain orientation. Note the directional dendritic structure in the re-solidified upper region.

3. Results and discussion

An overview of the phenomena occurring in the different initial microstructures, which will be discussed with respect to melting, grain growth in the mushy region, nucleation and subsequent re-growth in the following, is depicted in figure 4.

In the as solidified microstructure #1 melting as to be expected proceeds along the interdendritic regions, which solidified last during the generation of this microstructure and thus reveal the lowest melting point. The result is a continuous liquid film extending quite far into the bulk of the sample.

For the heat treated sample #2 in contrast no coherent liquid film is created, but different liquid regions are formed which also extend into the bulk of the sample.

Melting of the synthetic microstructure #3 significantly differs from samples #1 and #2. The homogeneous concentration field of Cu and its single phase FCC structure lead to a sharply localized melting front with no extension into the bulk of the sample. Upon melting a large number of small fragments of the initial grains are created.

The recrystallized microstructure #4 does not melt along its grain boundaries. Its melting behaviour is essentially determined by the original Cu-segregation which was not altered during the recrystallization process. Molten areas are also here found in the bulk phases of the sample.

Re-solidification for all cases starts from grains, which have not fully dissolved in the melt. Additionally the formation of some new nuclei in the bulk liquid can be observed in all cases. Major differences are identified for the recrystallized microstructure – revealing more and smaller grains as starting points for the re-solidification of dendrites. In case of the synthetic microstructure a large number of small fragments of the initial microstructure can be observed during melting. Each of these fragments should be assigned to become a new grain. In this case thus a very fine polycrystalline array of dendrites would have to be expected in contrast to what is depicted in figure 4.

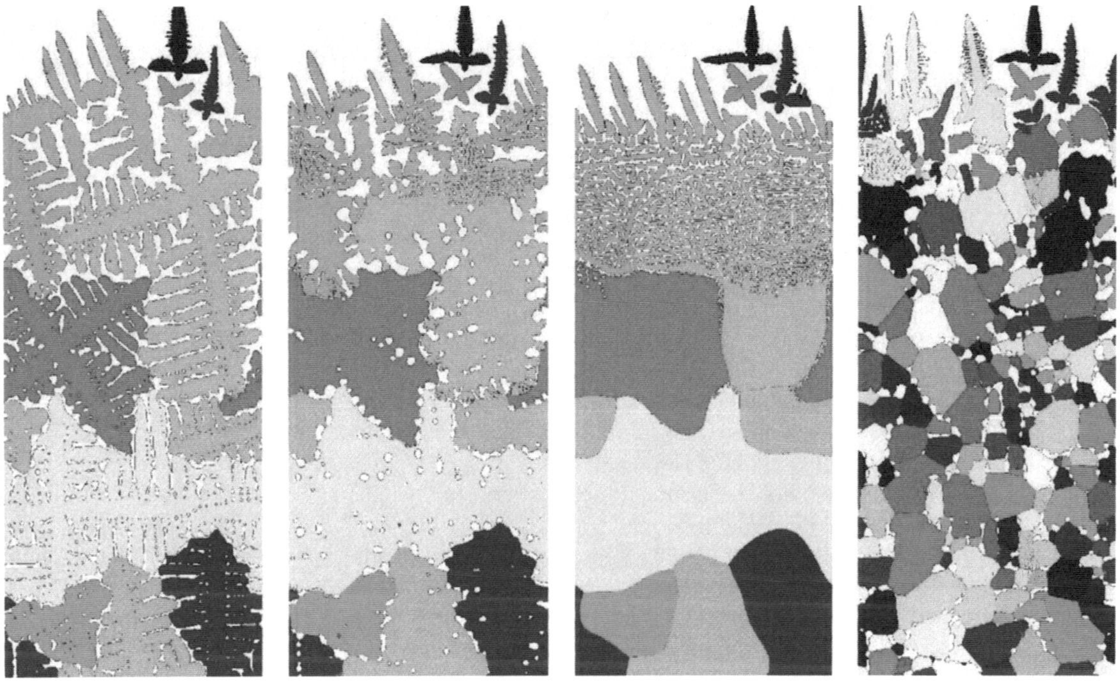

Figure 4. Melting and re-solidification of the four different initial microstructures (zoomed view #1 through #4; from left to right, width of each domain is approx. 200 μm) during a laser type process. Time step and process conditions are identical for all four microstructures. Depicted are the grain orientations respective grains.

4. Conclusions

In summary, modelling of melt based joining processes needs information about microstructures - especially segregation- which originates from the casting process and further evolved along the process chain. Modelling of segregation and its evolution along the entire process chain is highly important for both the properties of the materials but also for their further processing. Different segregation patterns lead to qualitative differences in the melting behaviour during thermal joining processes. Oversimplified assumptions - like e.g. a simple Voronoi type grain structure revealing a homogenous concentration of the alloying elements – will not be sufficient to correctly describe the melting process. In addition to modelling of solidification processes, the modelling of melting and dissolution processes of heterogeneous microstructures is expected to become important in the near future. "Processing-microstructure" relationships can be refined as "initial microstructure- processing-final microstructure" relationships in the future.

Acknowledgement
The present work has been supported by the DFG within the Cluster of Excellence "Integrative Production Technologies for High Wage Countries".

References

[1] National Research Council: Integrated Computational Materials Engineering: A Transformational Discipline for Improved Competitiveness and National Security; National Academic Press, Washington, D. C. (2008), ISBN: 0-309-12000-4.
[2] Schmitz G J and Prahl U (eds): Integrative Computational Materials Engineering- Concepts and applications of a modular simulation platform, Wiley VCH Verlag Weinheim, ISBN 978-3-527-33081-2 (2012).
[3] Schmitz G J and Prahl U: "ICMEg - the Integrated Computational Materials Engineering expert group - a new European Coordination action" Integrating Materials Manufacturing Innovation 2014, 3:2.
[4] www.ICMEg.eu – the "Integrated Computational Materials Engineering expert group" - accessed July 3rd 2014.
[5] MICRESS® - the MICRostructure Evolution Simulation Software (www.micress.de)
[6] Steinbach I, Pezzolla F, Nestler B, Seeßelberg M, Prieler R, Schmitz G J and Rezende J L L, A phase field concept for multiphase systems. *Physica* D **94**(1996), p.135-147.
[7] Eiken J, Böttger B, Steinbach I, Multiphase-Field approach for multicomponent alloys with extrapolation scheme for numerical application *Phys. Rev.* E **73** 066122 (2006).
[8] Databases available e.g. from Thermo-Calc Software www.thermocalc.se, Accessed Jul 3rd 2014.
[9] "COST 507: Thermochemical Database for Light Metal Alloys"; I. Ansara (editor) European Commission, Directorate-General XII, Science, Research and Development, 1994 ISBN2872631569, 9782872631568.
[10] Thermo-Calc Software www.thermocalc.se, accessed Jul 3rd 2014.
[11] MICRESS® Manual (Vol. 2): download from www.micress.de, Accessed Jul 3rd 2014.
[12] Eiken J: "A Phase Field Model for Technical Alloy Solidification", PhD Thesis RWTH Aachen 2009, Shaker Verlag, Aachen Band 62 (2010) ISBN 978-3-8322-9010-8.
[13] Böttger B, Eiken J, Apel M, Phase-field simulation of microstructure formation in technical castings – A self-consistent homoenthalpic approach to the micro–macro problem, *J. Comput. Phys.* **228** (2009), 6784-6795.
[14] Böttger B, Eiken J and Steinbach I, Phase-field simulation of equiaxed solidification in technical alloys *Acta Mater.* **54** (2006), 2697-2704.

IOP Conf. Series: Materials Science and Engineering **117** (2016) 012042 doi:10.1088/1757-899X/117/1/012042

Studies on synthesis of in-situ Al-TiC metal matrix composites

R N Rai[1,a], S C Saha[2], G L Datta[3] and M Chakraborty[4]
[1] Prod. Engg. Dept., NIT Agartala, India
[2] Mech. Engg. Dept., NIT Agartala, India
[3] K. L. University, Vijaywada, India
[4] Indian Institute of Technology, Bhubneswar, India

E-mail: [a]nareshray@yahoo.co.in

Abstract. In the present research work, synthesis and characterization of in-situ Al-TiC composites reinforced with ceramic phases was carried out. The formation of undesirable $TiAl_3$ particles could be avoided justifying the correct procedural requirement adopted while preparing Al-TiC composites. It was observed that distributions of reinforced particles were uniform along the grain boundaries. It was also observed that the average size of the TiC particles was of 0.5 μm. It was also noted that the presence of TiC particles in the composite enhances the yield strength and hardness substantially.

1. Introduction

Al-TiC belongs to the particulate type reinforced aluminium alloy based metal matrix composites. Aluminium based metal matrix composites have drawn the attention of many researchers due to their low density, excellent wear resistance, high specific strength and high specific modulus [1-2]. Particle reinforced metal matrix composites are likely to find high volume of commercial application due to their low cost, ease of fabrication and improved properties [1-5]. The practical applications of Al-TiC metal matrix composites are in the aerospace, automobile and structural industries [6-7]. Several processing techniques [4, 5] have been developed to disperse discontinuous (particulates, whiskers, platelets) reinforcements uniformly in the matrix. However, each fabrication technique has its own advantages and limitations. Some of the problems encountered relate to: uneven distribution of reinforcement, control of the matrix-reinforcement interface, scaling up of the process for industrial utilization and processing cost. To overcome some of the inherent problems that are associated with conventional methods, a new processing technique called, in-situ processing, has been chosen, which is cost effective and the interfaces produced are relatively stable and impurity-free [1, 6]. Interfacial bonding between the matrix and the reinforcement has been reported to be good [1].

The Al-TiC composite system has been studied by a number of researchers and it has been reported that during synthesis of Al-TiC composites using the Al-Ti-C system, fine TiC particulates are created in the Al matrix. However, the uniform distribution of the reinforcement may be hindered by the presence of $TiAl_3$ (a brittle compound) that degrades the mechanical properties of the composite [6-10]. $TiAl_3$ exhibits high strain-rate sensitivity and its flakes have been found to initiate and propagate cracks and low ductility in the composite

[11-12]. Elimination of this compound is, therefore, a primary concern for successful synthesis of Al-TiC composites.

In the present work Al-TiC composites have been developed via in-situ route. The microstructure and mechanical properties have been studied and the results are presented.

2. Experimental procedure

The in-situ process involves introducing carbon bearing activated charcoal into an Al-Ti melt, thereby forming TiC particles in the melt. In these experiments master alloys of Al-10Ti were prepared in an induction furnace at a temperature of 1200°C and poured into graphite moulds to solidify. Thereafter, the same Al-10Ti master alloys were re-melted in a pit furnace at different temperatures such as 1200°C, 1250°C and 1300°C and activated charcoal was added into the melt to meet the stoichiometry of TiC (Ti:C) to complete the reaction for the formation of TiC particles inside the melt. The reaction time was varied from 20 min to 60 min. The degasser hexachloroethane was used to remove the dissolved hydrogen gas from the melt. A small amount of potassium fluoride and sodium fluoride was added as a flux cover to remove the oxide film from the molten metal surface and to act as a protective barrier to gas absorption and for facilitating spontaneous incorporation of the particles into the melt [10]. Afterwards the melt was cast into rod form (15 mm in diameter and 150 mm in length) in metallic moulds to complete the process. Table 1 gives the processing conditions used for the synthesis of Al-TiC metal matrix composites. Al-TiC composites were prepared with varying weight percentages of TiC reinforced particles such as Al-3TiC, Al-5TiC, and Al-10TiC. The experimental setup for fabricating in-situ Al-TiC composites is shown schematically in figure1.

Table 1. Processing conditions used for the synthesis of Al-TiC composite.

Sl. No.	Nominal Composition	Processing condition	
		Reaction Temp (°C)	Reaction Time (Min)
1	Al-TiC	1200	45
2	Al-TiC	1200	60
3	Al-TiC	1250	20
4	Al-TiC	1250	30
5	Al-TiC	1250	45
6	Al-TiC	1250	60
7	Al-TiC	1300	20
8	Al-TiC	1300	30
9	Al-TiC	1300	45
10	Al-TiC	1300	60

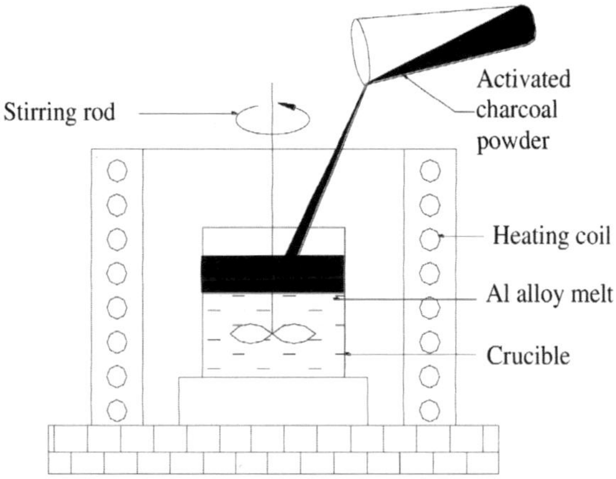

Figure 1. Schematic diagram of the apparatus for fabricating in-situ Al-TiC composites.

Microstructures of Al-TiC composites in the as-cast condition and of fractured surfaces of the tensile specimens were examined using a JEOL JSM-5800LV scanning electron microscope (SEM). Tensile tests were carried out at room temperature using an Instron 8516 machine with the loading speed controlled at 0.5 mm/min. The hardness of as-cast composites was measured on the HV scale using a Vicker's hardness tester at an applied load of 5 kg. A X-ray diffractometer (PW 1840) operating at 40 kV and 20 mA was used to identify the various phases present in the specimens. In order to understand the mechanisms responsible for the formation of TiC via this process, a mixture of Al-10Ti (as-cast) alloy powders and high purity activated charcoal in an arbitrary weight ratio of 1:1 was prepared and analysed by differential thermal analysis (DTA). This analysis was carried out in an argon atmosphere from room temperature to 1300°C with incremental rate of 10°C/min in an alumina crucible.

3. Results and discussions

3.1 Microstructure analysis

The SEM micrographs of the Al-TiC composites are shown in figure 2 (a-h). It was observed that Al-TiC composites prepared at 1200°C with 45 min and 1200°C for 60 min of holding time have both $TiAl_3$ and TiC particles, which are shown in figure 2 (a-b). It is felt that the temperature was not sufficient to dissolve all the $TiAl_3$ particles inside the melt. In the next set of experiments, composites were prepared at 1250°C and 1300°C with 20 min holding time and it was observed that in this case also $TiAl_3$ and TiC particles were co-existing as shown in figure 2 (c-d). When the temperature was maintained at 1250°C and 1300°C but with higher holding time of 30 min, 45 min and 60 min, the composites, thus prepared, had only TiC reinforced particles and no $TiAl_3$ particles. This observation has been conformed by EDX, XRD, DTA and SEM analysis. The SEM microstructure of Al-TiC composites prepared at 1250°C and 1300°C with 30 min holding time are shown in figure 2 (e-f). The figures show that at 1250°C with 30 min holding or higher temperature and holding time, all the $TiAl_3$ particles had been dissolved and formed stable TiC particles. The present results confirm the earlier reports of Rapp and Zhang [13] who reported that carbon normally reacts with titanium and aluminium to form titanium carbide (TiC) and aluminium carbide (Al_4C_3) and that at higher temperature TiC particles are more stable than Al_4C_3 particles. They suggested the following possible reactions and their accompanying free-energy changes:

Reaction 1 : $Ti+3Al \rightarrow TiAl_3$, $\Delta E_1 = -52503 + 21.483T$
Reaction 2 : $3/4Al + C \rightarrow 1/3\ Al_4C_3$, $\Delta G_2 = -89611 + 32.841T$
Reaction 3 : $Ti+C \rightarrow TiC$, $\Delta G_3 = -163382+80.347T+0.460x10^{-3}T^2+(3.094\ x\ 10^5)/T-0.962T\ ln\ T$

where ΔG = Gibb's free-energy (J mol^{-1}), T = Temperature (K). From reactions 2 and 3, it is revealed that the formation of TiC is more favorable than that of Al_4C_3. As Sahoo and Koczak [11] had presented the values ΔG for formation of TiC and Al_4C_3 are negative, indicating that the reactions can proceed spontaneously. In general, the more negative the standard free energy change of formation, the more stable the carbide is. They have showed that titanium is a very strong carbide former relative to Al_4C_3.

In the present experiments the average size of the TiC particles were observed to be 0.5 μm and those particles were found to be preferentially segregated at the grain boundaries. In addition, large clusters of carbides were also observed in the microstructure. Figure 2 (g-h), taken at higher magnifications, show clusters of TiC particles formed by the coalescence of several particles having faceted/polyhedral morphology. The particles observed in the microstructures were identified as TiC with the help of XRD and EDX analysis, the results of which are shown in figure 3 (a-b). figure 3 (a) represents the XRD of as-cast Al-TiC composite and extracted TiC particles. Figure 3 (b) represents the EDX of TiC particle.

IOP Conf. Series: Materials Science and Engineering **117** (2016) 012042 doi:10.1088/1757-899X/117/1/012042

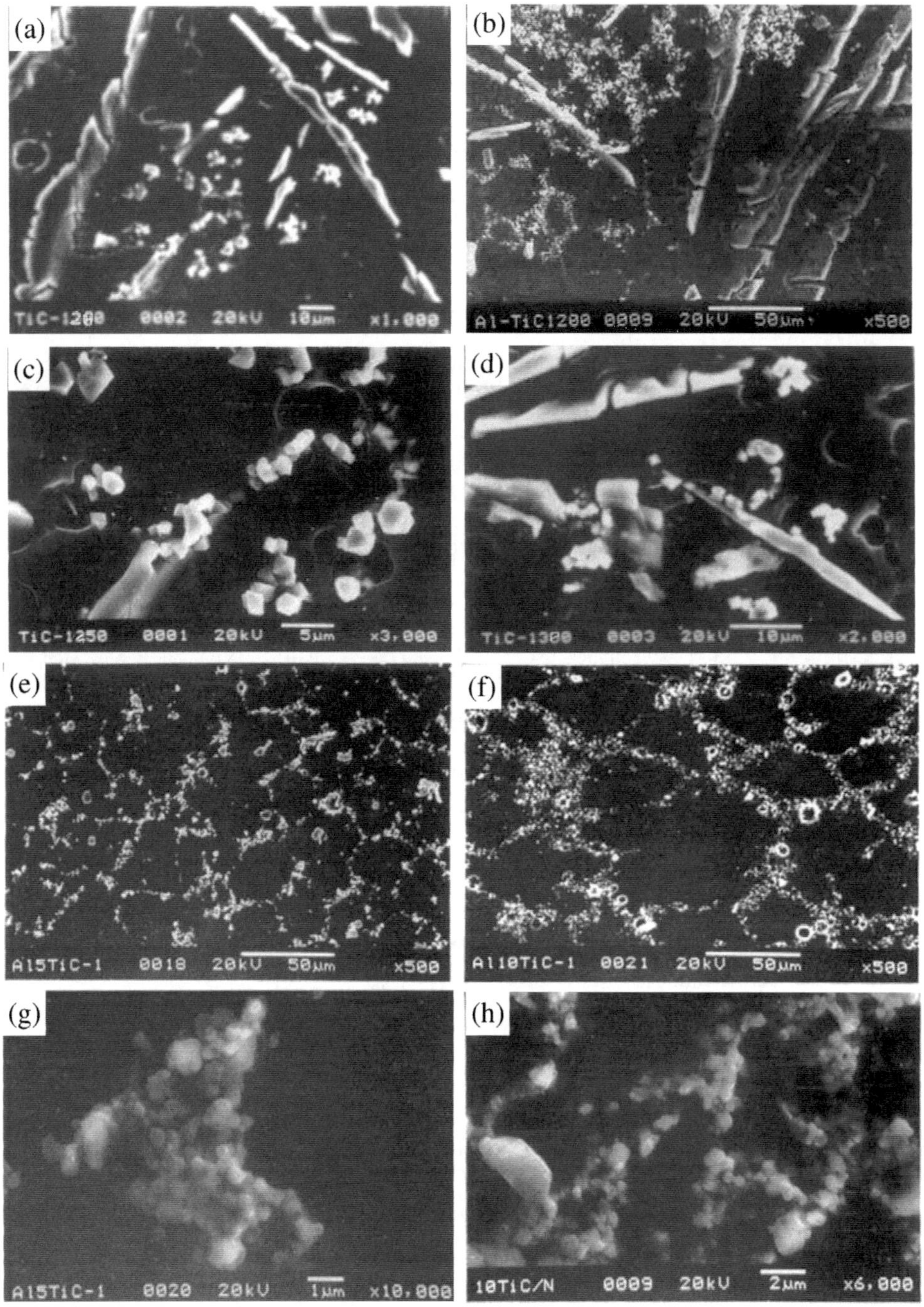

Figure 2. SEM micrographs of Al-TiC composites: (a) 1200°C with 45 min, (b) 1200°C with 60 min, (c) 1250°C with 20 min, (d) 1300°C with 20 min, (e) 1250°C with 30 min, (f) 1300°C with 30 min and (g) 1250°C with 30 min, (h) 1300°C with 30 min reaction time at higher magnification.

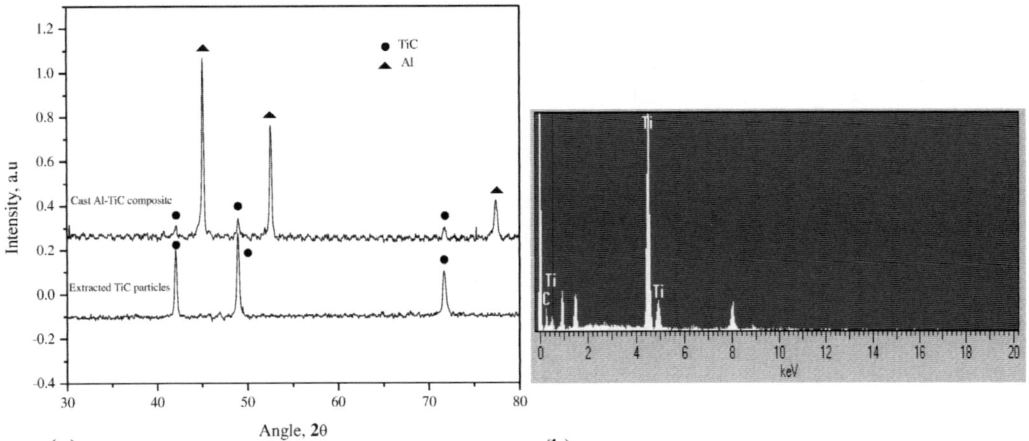

(a) (b)

Figure 3. XRD peak intensity of: (a) Extracted TiC particles and as-cast Al-TiC composites and (b) EDX of TiC particle.

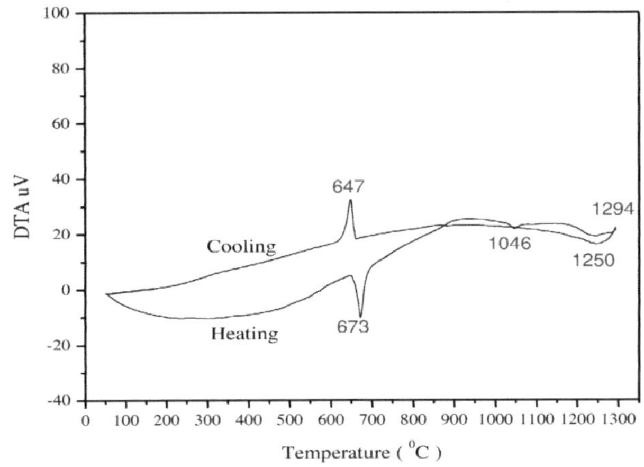

Figure 4. DTA analysis of Al-TiC composite during heating and cooling.

3.2 Thermal analysis

The DTA analysis of Al-TiC composite helps to understand the possible mechanism for the formation of TiC during the in-situ process. DTA results during heating and cooling are presented in figure 4. Three reaction peaks were found at 673°C, 1046°C and 1250°C, during the heating cycle. The reaction peak at 673°C represents the melting of aluminium and the peak at 1046°C represents formation of $TiAl_3$ particles. The third peak at 1250°C appears to be an exothermic reaction indicating the formation of TiC particles. There are two peaks occurring during cooling. The first peak at 1250°C represents the solidification of TiC particles and another at 647°C represents the freezing point of pure aluminium. This shows that all the Ti in the (Al-Ti) alloy melt had completely reacted with the carbon dispersed in the melt during heating. Microstructures of samples reveal many interesting results. Firstly, TiC particles within the matrix are finer than the activated charcoal powders used in the melt. Secondly and most importantly, the fine TiC particles are not only polyhedral in shape but also uniform in size. This strongly suggests that the formation of TiC is achieved via a mechanism of nucleation and growth during isothermal holding. Nucleation would require the dissolution and rapid diffusion of carbon from the large size (20 μm) of the activated charcoal powders into the melt, while the growth of the carbides would be controlled by the diffusion of the lower diffusivity element, Ti. Diffusion of carbon in molten aluminium is sufficiently fast for nucleation of TiC. The present results also confirm the findings of Nukami and Flemings [14] who had suggested the following in-situ reactions for synthesizing the TiC particulates reinforcement in the Al-Ti-C system based on DTA and XRD analysis:

Reaction 4: Ti +3Al→ $TiAl_3$
Reaction 5: $3TiAl_3$+ Al_4C_3→ 3TiC+ 13Al
Reaction 6: $TiAl_3$+ C→ TiC+ 3Al

Table 2. Tensile results and hardness of pure Al and the Al-TiC composites.

Sl. No.	Material	Yield Strength (MPa)	% Elongation	Hardness (HV_5)
1	Al	80	30	18.13
2	Al-3TiC	123	16	32.92
3	Al-5TiC	142	12	37.83
4	Al-10TiC	189	5	44.17

3.3 Tensile properties

The results of the tensile tests are presented in table 2. The effect of the TiC particles in increasing the yield strength of the Al-TiC composites is illustrated by comparing properties with those of commercially pure Al. The composites containing TiC particles have higher yield strength. The fine submicron TiC particles that are present in these composites are believed to contribute to strengthening. From table 2 it is observed that with increases in the amount of TiC particles in the Al-TiC composites the tensile strength increases whereas the elongation decreases. The increasing amount of TiC reduces the grain size of α-Al (because the α-Al nucleates on the surfaces of TiC particles) and hence there is a possibility of improving strength. According to Sahoo et al [11], a decrease in grain size from 300 to 60 μm would result in an improvement in yield strength of approximately 13 MPa. The improvement in the proof stress corresponding to this grain size reduction (for 18 vol. % particle additions) was about 50 MPa, indicating that the TiC particles contribute strengthening to the composite. As compared with pure Al, the Al-3TiC composite contains a small amount of TiC particles leading to increasing yield strength by 1.5 times, i.e. from 80 MPa to 123 MPa; and in the Al-5TiC and Al-10TiC composites the yield strengths increased by 1.75 and 2.3 times respectively. The hardness values of pure Al and the Al-TiC composites are also shown in table 3. The hardness of Al-TiC improved due to the formation of reinforcement particles. From the table 2 it can be concluded that with increases in the amount of TiC particles, the hardness increased substantially. It can thus be said that with the increase in the value of hardness there is a corresponding increase in tensile strength. Fractured surfaces of tensile specimens are shown in figure 5 (a-b), which reveals that the Al-3TiC composites had ductile fractures. The Al-10TiC (figure 5 (b)) composite exhibited smaller fracture dimples compared with those of Al-3TiC (figure 5 (a)). It can be said that materials having dimples of smaller sizes require more force for deformation/failure, which indicates that the Al-10TiC composite has still better strength.

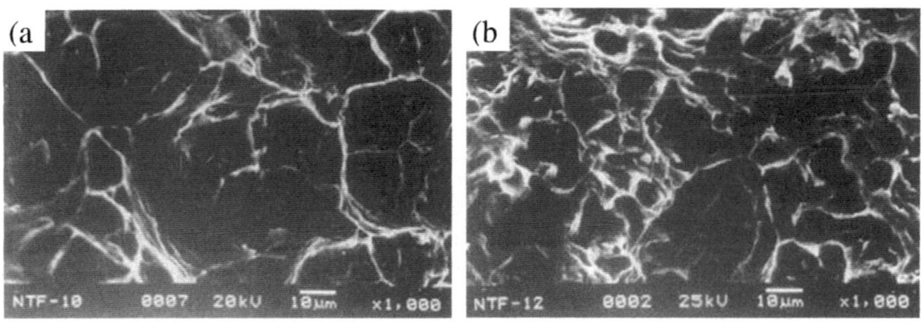

Figure 5. Fracture surface of: (a) Al-3TiC and (b) Al-10TiC composite.

4. Conclusions

Al-TiC metal matrix composites have been prepared successfully at a processing temperature of 1250°C and holding time of 30 minutes by an in-situ route which eliminates the incidence of unwanted brittle $TiAl_3$ particles and the average size of the TiC particles was observed to be about 0.5 µm. Microstructures of the composites showed uniform distributions of TiC particles along the grain boundaries. These particles have a polyhedral shape and are uniform in size. The presence of TiC particles in the composite increased the yield strength and hardness with substantial decrease in elongation.

References

[1] Tjong S. C and Ma Z. Y, Microstructural and mechanical characteristics of in-situ metal matrix composites. Materials Science and Engineering: R: Reports, 2000; 29: 49-113.

[2] Lloyd D. J, Particle reinforced aluminium and magnesium matrix composites. International Materials Reviews, 1994; 39: 1-23.

[3] Clyne T. W and Withers P. J, An introduction to metal matrix composites. Cambridge University Press, U. K, 1993.

[4] Everest R. K and Arsenault R. J, Metal Matrix Composites: Processing and interfaces, Academic Press, U. S. A., 1991

[5] Chawla K. K, Composites Materials: Science and engineering, Springer-Verlag, New Yark, 1987.

[6] Kennedy A. R and Wyatt S. M, Characterizing particle-matrix interfacial bonding in particulate Al-TiC MMCs produced by different methods. Composite Part A: Applied Science and Manufacturing, 2001; 32: 555-559.

[7] Selcuk C. and Kennedy A.R., Al–TiC composite made by the addition of master alloys pellets synthesized from reacted elemental powders. Materials Letters, 2006; 60: 3364-3366.

[8] Ma Z. Y and Tjong S. C, In situ ceramic particles-reinforced aluminum matrix composites fabricated by reaction pressing in the TiO_2(Ti)-Al-B(B_2O_3) system. Metallurgical and Materials Transactions: A, 1997; 28A: 1931-1942.

[9] Yong Z, Zheng M. S., Hitoshi H and Shuji T., Synthesis of high-purity polycrystalline Ti_3AlC_2 through pulse discharge sintering Ti/Al/TiC powders. Journal of Alloys and Compounds, 2008; 456: 456–460.

[10] Karantzalis A, Wyatt E. S and Kennedy A. R., The mechanical properties of Al-TiC metal matrix composites fabricated by a flux-casting technique. Materials Science and Engineering: A, 1997; 237: 200–206.

[11] Sahoo P and Koczak M. J, Analysis of in situ formation of titanium carbide in aluminum alloys, Materials Science and Engineering: A, 1991; 144: 37-44.

[12] Xiaoguang Ma, Xiangfa Liu and Haimin Ding, A united refinement technology for commercial pure Al by Al–10Ti and Al–Ti–C master alloys. Journal of Alloys and Compounds, 2009; 471: 56–59

[13] Rapp R. A and Zheng X, Thermodynamic consideration of grain refinement of aluminum alloys by titanium and carbon. Metallurgical and Materials Transactions: A, 1991; 22A: 3071-3075.

[14] Nukami T and Flemings M. C, In situ synthesis of TiC particles-reinforced aluminum matrix composite. Metallurgical and Materials Transactions: A, 1995; 26A: 1877- 1884.

The effect of rotating magnetic field on the microstructure of in situ TiB_2/Cu composites

C Zou, H Kang, R Li, M Li, W Wang, Z Chen and T Wang[a]
Key Laboratory of Materials Modification (Ministry of Education), School of Materials Science and Engineering, Dalian University of Technology, Dalian, 116024, China

E-mail: [a] tmwang@dlut.edu.cn

Abstract. Nano ceramic particulate reinforced metal matrix composites are confronted with the problem of particle aggregation emerging in the process of solidification. It sharply deteriorates the mechanical properties of the composites. In order to improve the microstructure and particle distribution, in situ TiB_2/Cu composites were prepared using Ti and Cu-B master alloys in a vacuum medium frequency induction furnace equipped with a rotating magnetic field (RMF). The effect of RMF magnetic field intensity employed on the microstructure and particles distribution of the TiB_2/Cu composites were investigated. The results show that with the applied RMF, TiB_2 particles are homogeneously distributed in the copper matrix, which significantly improves the mechanical properties of TiB_2/Cu composites. The mechanism of RMF may be ascribed to the following two aspects. On the one hand, the electromagnetic body force generated by appropriate RMF drives forced convection in the equatorial plane of composite melt during solidification. On the other hand, a secondary flow in the meridional plane is engendered by a radial pressure gradient, thus making a strong agitation in the melt. These two effects result in a homogenous dispersion of TiB_2 particles in the copper matrix, and hence excellent properties of TiB_2/Cu composites were obtained.

1. Introduction

Due to the good mechanical properties, high electrical conductivity and thermal conductivities in copper and copper alloy, it has been widely used as rail transit line contacts, integrated circuit lead frame, electrodes and heat exchangers [1]. The materials require both high conductivity, well thermal conductivity, and high strength. The major strengthening method of traditional copper alloy is solution strengthening. In those alloy, alloying elements solubilize in copper, and lead to lattice distortion the copper, which result in a big drop of conductivity in copper [2]. Therefore, the copper matrix composite with higher mechanical properties and thermo stability become the hotspot of current research.

Copper matrix composites can be fabricated via many routes, such as power metallurgy, self-propagating high-temperature synthesis (SHS), mechanical alloying, internal oxidation, etc [3-7]. Due to good wetting between ceramic particles and copper matrix by in situ reaction synthesis, preferable integrated performance (i.e. tensile strength and electrical conductivity) is more facilely obtained than the conventional methods [1, 8]. Among various in situ routes, the casting process is of particular interest owing to its lower cost and potential for mass production. Second phase particles dispersion strengthening copper composites are a good way to strengthen copper. Comparing to other ceramic particles, TiB_2 particles possess not only good conductivity, thermal conductivity, high hardness, but also excellent chemical stability, corrosion resistance and wear resistance. Moreover, the standard Gibbs free energy of TiB_2 is low, and could synthesise from boric acid potassium fluoride and fluoride potassium titanate compound salt, or directly generated by titanium and boron at high temperature. However, the traditional methods, such as directly added enhance phase particles to the matrix, or B, Ti in situ reaction, face the same problem of TiB_2 particles aggregation, which will seriously deteriorate the mechanical properties of composites.

Rotating magnetic field (RMF) has been widely used in materials processing to overcome this shortcoming [9, 10]. The liquid metal in the RMF is subjected to electromagnetic stirring, which has the advantages of refining the internal structures of the ingot, increasing the fraction of equiaxed grains and reducing the segregation and shrinkage cavity [11]. It is also inferred from the literature that using electromagnetic stirring in copper melt, the TiB_2 particles may exhibit a homogeneous distribution in the copper matrix. However, until now, the effects of RMF on the particle distribution in copper matrix composites have not been reported.

In this study, the in situ synthesis of TiB_2/Cu composite from Cu-4.6B and Ti with different RMF implied during solidification were prepared. The effect of RMF on the microstructure and mechanical properties of in situ TiB_2/Cu composite were investigated.

2. Experimental procedure

The experiment was performed in a vacuum medium frequency induction melting furnace with a rotating magnetic field. Figure 1 shows a schematic diagram of the experimental apparatus. The raw materials used were electrolytic copper (99.96% purity), titanium sponge (98.5% purity) and Cu-4.6wt%B master alloy.

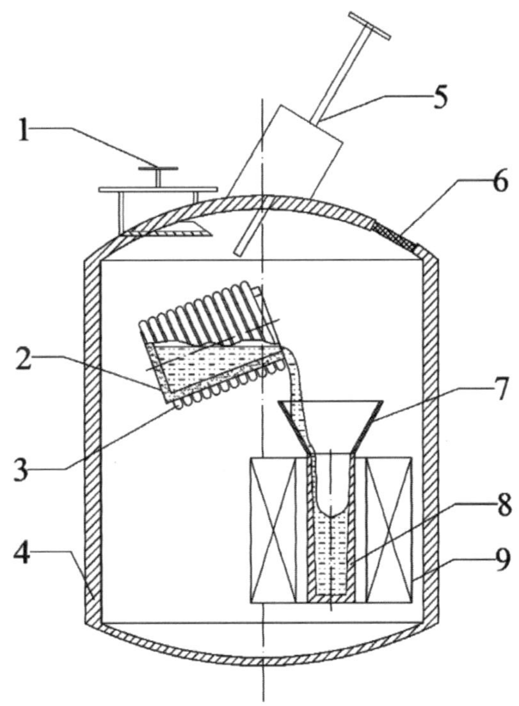

Figure 1. Schematic diagram of a vacuum induction heating furnace with RMF: (1) Hopper, (2) Graphite crucible, (3) Induction coil, (4) Vacuum furnace, (5) Material pounding rod, (6) Hagioscope, (7) Ceramic funnel, (8) Graphitic casting mould, (9) Rotating magnetic field generator.

Pure Cu was melted in a vacuum medium frequency induction melting furnace in an argon atmosphere, when the Cu melt in the furnace reached 1300 °C, Ti and Cu-B master alloys were incorporated into the pure Cu sequentially. After 5 minutes of holding, TiB_2 particles were formed via chemical reactions between Ti and B. Subsequently, the Cu-TiB_2 melt was poured into a cylindrical graphite mould (45 mm in diameter and 220 mm in height), which was inserted in a RMF generator for electromagnetic stirring. The RMF was charged with a frequency of 50 Hz shortly after pouring. The intensity of RMF was controlled by adjusting the alternating current. Four Cu-1wt.%TiB^2 samples were prepared with different current intensities (20 and 40 A, corresponding to the magnetic field intensity of 13 mT and 30 mT, respectively). The intensity of RMF was measured by Gauss/Teslameter Model 7030, Sypris Test & Measurement. For comparison, a reference sample without RMF was also prepared.

The as-cast samples were mechanically ground, polished and etched in a corrosive agent (3g $FeCl_3$, 2 mL HCl and 95 mL C_2H_5OH) for metallographic examination. The microstructure and TiB_2 distribution in the copper matrix were observed under a scanning electron microscopy (SEM, Zeiss supra 55) operated at secondary electron mode with an accelerating voltage of 15 kV. The tensile specimens, with a dimension of 40 mm gage length, 6 mm diameter, were machined from samples according to the GB/T 228-2002 standard. Tensile tests with cross head speed of 2 mm•min^{-1} were conducted at room temperature. Vickers hardness tests were performed under a load of 100 g for 10 s using a Vickers hardness tester (MH-6L). For each test, five measurements were performed and average experimental data was reported.

3. Results and discussion

3.1. Thermomechanical analysis

In order to determine the reaction in the composite melt, the Gibbs free energy of feasible chemical reactions in molten copper was calculated according to the thermodynamic data from thermodynamics manual [12]:

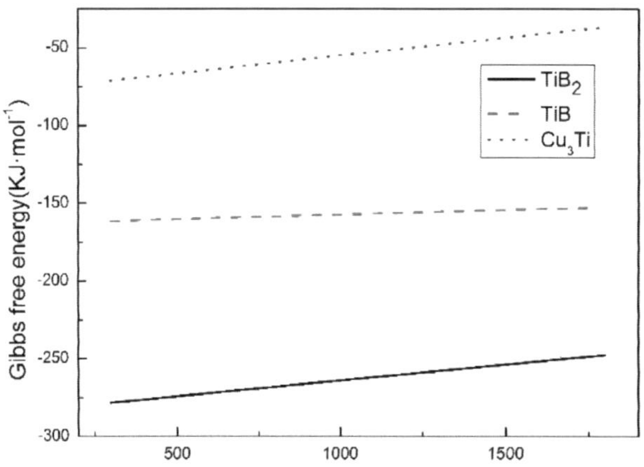

Figure 2. Variation of standard Gibbs free energy of possible reactions vs temperature in Cu-Ti-B melt.

The probable reactions that may exist in melt are as follows:

$$Ti+2[B]Cu \rightarrow TiB_2 \quad (1)$$

$$Ti+[B]Cu \rightarrow TiB \quad (2)$$

$$3Cu+Ti \rightarrow Cu_3Ti \quad (3)$$

By thermodynamics data of elements and compounds, the corresponding equations of Gibbs free energy change with temperature as follows:

$$\Delta G_1^o = -284500 + 20.5T \tag{4}$$

$$\Delta G_2^o = -163200 + 5.9T \tag{5}$$

$$\Delta G_3^o = -78200 + 23.2T \tag{6}$$

The thermodynamic calculation results are shown in figure 2. Due to the lowest Gibbs free energy of TiB_2 among the possible products of TiB, TiB_2, Cu_3Ti, the formation of TiB_2 in molten copper is more susceptible than that of other components. Therefore, only in situ formation of TiB_2 occurred in the copper melt. As shown in figure 3, only TiB_2 phase is determined in the composite.

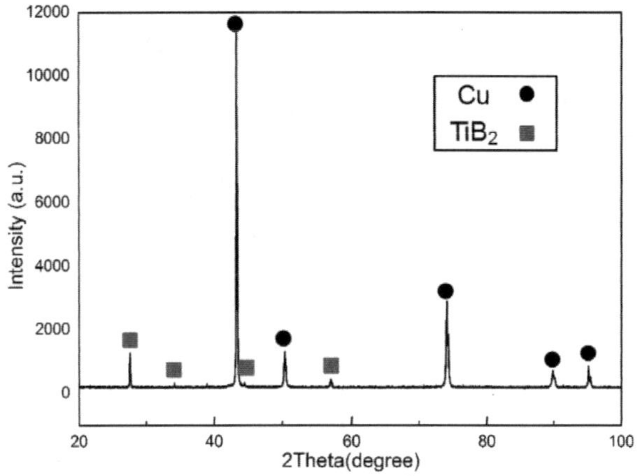

Figure 3. XRD pattern of Cu–1 wt.% TiB_2 composite.

3.2. Microstructure and TiB_2 distribution of Cu/TiB_2 composites

Figure 4 shows the microstructure the SEM micrographs of the distribution of TiB_2 particles in the Cu-1wt.% TiB_2 composites prepared with different field currents ranging 0, 20 A and 40 A.

It is well known that the distribution, size and spacing of reinforced particles in a copper matrix composite are the predominant factors influencing its mechanical properties. The aggregation of coarse TiB_2 particle can be clearly observed in the copper matrix figure 4(a). However, when a 20 A field current was applied, the distribution of TiB_2 in the copper matrix has been greatly improved in the mass. This implies that the RMF can be beneficial to the particle distribution of metal matrix composites. The improved distribution of TiB_2 is attributed to the electromagnetic stirring, which is produced inside the melt by RMF. It should be noted that some aggregates are still observed in this composite in certain local regions, as shown in figures 4(b). With field currents reach to 40 A, an increase in the field current leads to an improvement in the distribution of TiB_2. When a 40 A field current was applied, thanks to the stirring effect of RMF, the initial TiB_2 aggregates in the untreated composite were modified into a more homogeneously distribution throughout the copper matrix (figure 4(c)).

3.3. Mechanical properties

The effect of RMF on the tensile strength, elongation and hardness are shown in figure 5 and table 1. As shown in the figure, the implication of RMF on composite makes an obvious improvement of tensile strength and hardness. The strength increases from 178 MPa, 102HV to 200 MPa, 120.5HV respectively. However, the elongation of Cu-1 wt.% TiB_2 composites decreased a little after the application of RMF. As is known that the particles are always the crack initiation source in particle reinforced metal matrix composites. Therefore, the presence of TiB_2 particles will limit the ductility of the in situ composites [13]. Without RMF, the aggregation of TiB_2 particles results in more grains with fewer reinforcing particles, which may contribute a lot to the elongation.

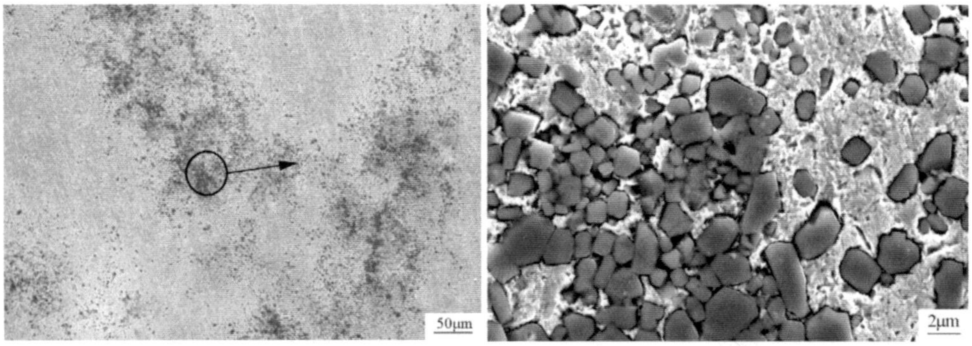

(a) Cu-1TiB2 0 A

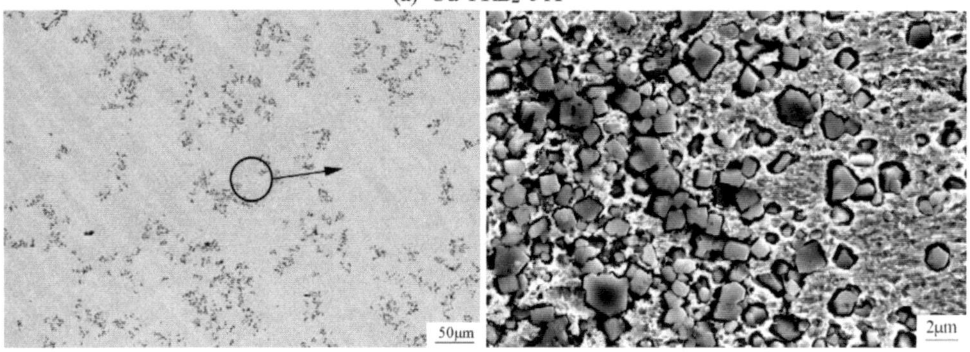

(b) Cu-1TiB2 20 A

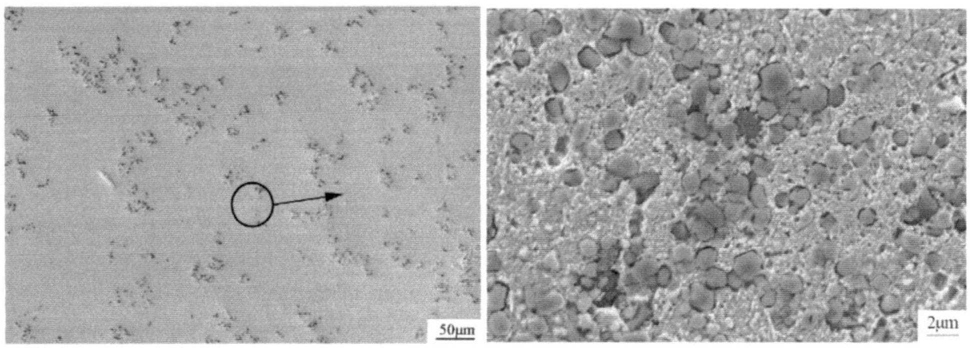

(c) Cu-1TiB2 40 A

Figure 4. SEM micrographs of Cu-1wt.% TiB_2 composite shows distribution of TiB_2 particles in the matrix under different field currents: (a) without, (b) 20 A, (c) 40 A.

Table 1. Hardness of Cu-1TiB_2 under different RMF.

ω% TiB_2/Cu	Field Current (A)	Hardness (HV)
	0	102.0±6.4
1	20	108.5±4.3
	40	120.5±3.5

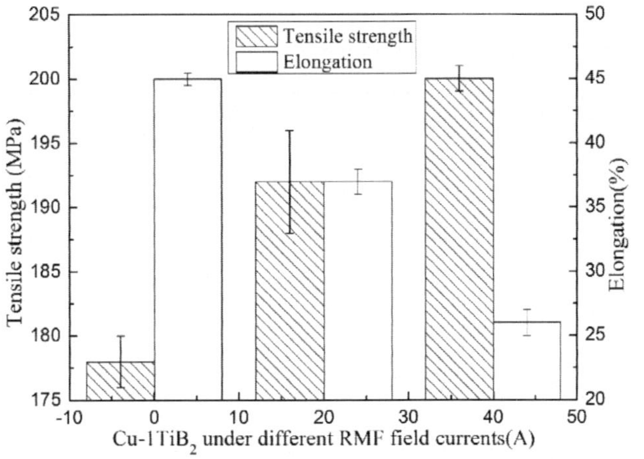

Figure 5. Tensile strength and elongation of Cu-1wt.%TiB_2 composites samples under different field currents.

The improvement of mechanical properties of Cu-TiB_2 is attribute to the more dispersive distribution of TiB_2 in copper matrix. The electromagnetic body force generated by RMF drives forced convection in composite melt. Besides the rotating, a secondary flow in the meridional plane engenders by a radial pressure gradient. The driven flow makes a strong agitation in composite, resulting in a homogenous dispersion of TiB_2 particles in the copper matrix, so the well properties Cu-TiB_2 composites were acquired.

4. Conclusions

In situ Cu-1TiB_2 composites were synthesised under different RMF. The RMF, imposed during the solidification of Cu-TiB_2 composite, facilitates a more homogenous distribution of TiB_2 particles in the copper matrix and significantly increased tensile strength and hardness. That because the driven flow generated by RMF separate the aggregation TiB_2 particles. The homogenous dispersion of TiB_2 particles in the copper matrix resulting in well mechanical properties of Cu-TiB_2 composites.

Acknowledgements

The authors gratefully acknowledge the supports of National Natural Science Foundation of China (No's. 51274054, U1332115, 51271042, 51375070, and 51401044), the Keygrant Project of Chinese Ministry of Education (No. 313011), the Science and Technology Planning Project of Dalian (No. 2013A16GX110), the China Postdoctoral Science Foundation, and the Fundamental Research Funds for the Central Universities.

References

[1] Tjong S C and Ma Z Y, 2000, Microstructural and mechanical characteristics of in situ metal matrix composites, *Materials Science and Engineering: R: Reports*, **29**(3–4), 49-113.

[2] Bozic D, Stasic J, Ruzic J, Vilotijevic M, and Rajkovic V Synthesis and properties of a Cu-Ti-TiB2 composite hardened by multiple mechanisms, *Mater. Sci. Eng. A-Struct. Mater. Prop. Microstruct. Process.*, 2011, **528**(28), 8139-8144.

[3] Biselli C, Morris D G, and Randall N, 1994, Mechanical alloying of high-strength copper-alloys containing TiB2 and AL2O3 dispersoid particles, *Scr. Metall. Materialia*, **30**(10), 1327-1332.

[4] Li H P, Bhaduri S B, and Sekhar J A, 1992, Metal-ceramic composites based on the Ti-B-Cu porosity system', *Metallurgical Transactions a-Physical Metallurgy And Materials Science*, **23**(1), 251-261.

[5] Guo M, Shen K, and Wang M, 2009,: Relationship between microstructure, properties and reaction conditions for Cu-TiB_2 alloys prepared by in situ reaction, *Acta Materialia*, **57**(15), 4568-4579.

[6] Lee A K, Sanchezcaldera L E, Oktay S T, and Suh N P, 1992,: Liquid-metal mixing process tailors mmc microstructures, *Advanced Materials & Processes*, **142**(2), 31-34.

[7] Kim J H, Yun J H, Park Y H, Cho K M, Choi I D, Park I M, 2007,: Manufacturing of Cu-TiB2 composites by turbulent in situ mixing process, *Mater. Sci. Eng. A-Struct. Mater. Prop. Microstruct. Process.*, **449**, 1018-1021.

[8] Ibrahim I A, Mohamed F A, and Lavernia E J, 1991,: Particulate reinforced metal matrix composites — a review, *Journal of Materials Science*, **26**(5), 1137-1156.

[9] Yan Z M, Chen M L, Yang L, Yang J, Gao H, and Li T J, 2013, Influence of rotating magnetic field on horizontal continuous casting of CuNi10Fe1Mn alloy hollow billet: from solidification structures to segregation of Ni and growth morphologies of Ni rich phase, *International Journal Of Cast Metals Research*, **26**(1), 37-42.

[10] Dold P and Benz K W, Rotating magnetic fields, 1999,: Fluid flow and crystal growth applications, *Progress In Crystal Growth And Characterization Of Materials*, **38**(1-4), 7-38.

[11] Willers B, Eckert S, Michel U, Haase I, and Zouhar G, 2005, The columnar-to-equiaxed transition in Pb-Sn alloys affected by electromagnetically driven convection, *Mater. Sci. Eng. A-Struct. Mater. Prop. Microstruct. Process.*, **402**(1-2), 55-65.

[12] Binnewies M M E, Thermochemical data of elements and compounds, 2002, Wienheim: Wiley-VCH.

[13] Manoharan M and Lewandowski J J, 1990, Crack initiation and growth toughness of an aluminum metal matrix composite, *Acta Metall. Mater.*, **38**(3), 489-496.

IOP Conf. Series: Materials Science and Engineering **117** (2016) 012044 doi:10.1088/1757-899X/117/1/012044

Numerical simulation on level fluctuation in bloom casting mold with electromagnetic stirring

H Zhang [1,a], H W Ni[1,b], Y Li[2,c] and Z F Zhao[2,d]
[1] Key Laboratory for Ferrous Metallurgy and Resources Utilization of Ministry of Education, Wuhan University of Science and Technology, Wuhan, 430081, China.
[2] Wuhan Iron and Steel Co Ltd, Wuhan, 430081, China

E-mail: [a]huazhang@wust.edu.cn, [b]nihongwei320@sohu.com, [c]136935077@qq.com, [d]zhaozhongfu73@yahoo.com.cn

Abstract. Based on a 380mm × 280mm bloom caster mold, the level fluctuation of steel-slag interface in the mold was simulated by the VOF model of commercial software Fluent. The effects of current intensity and frequency of EMS (electromagnetic stirring) on the level fluctuation in the mold were studied. The results show that whether or not with EMS, the maximum level fluctuation site of the mold occurs in the vicinity of the submerged entry nozzle. Compared with casting without EMS, molten steel flows horizontally rotatably under the action of the electromagnetic force by electromagnetic stirring, so the impact depth of molten steel decreases, then the level fluctuation slightly reduces, and the maximum level fluctuation value in the wide direction and the narrow direction of the mold, reduce from 4.24mm and 4.14mm to 4.04mm and 3.73mm respectively. With increasing intensity and frequency of current, the mold level fluctuation rises and the distribution uniformity of the level fluctuating amplitude worsens. The maximum level fluctuation enlarges by 0.18mm with raising the current intensity from 450A to 550A, but it enlarges by 0.79mm with 600A current intensity. The maximum level fluctuation enlarges by 0.15mm with raising the current frequency from 1.5Hz to 2.0Hz, but it quickly enlarges by 0.78mm with 2.5Hz current frequency. When the current strength and frequency are not more than 550A and 2.0Hz, level fluctuations are 4.00mm or less, which can meet requirements for controlling the bloom surface quality.

1. Introduction

Electromagnetic stirring (EMS) is widely used for its metallurgical effect on inclusion removal, the elimination of subsurface bubbles, reducing central segregation and improvement of central equiaxed grains. In recent years, the EMS technology has been researched by a number of metallurgical scholars. Liang Zhou et al [1-3] have tested the magnetic field of M-EMS by using a Gauss meter, qualitatively analysed the characteristics of the electromagnetic field, and the relationships of current intensity with stirring speed, molten steel flow, fluid velocity distribution were discussed. Miaoyong Zhu et al [4] have simulated the three-dimensional flow field and temperature field of a bloom mold, researched the influence of stirring intensity on the flow field and temperature field of molten steel, but dealt with the mold free surface as a plane of symmetry or no shear stress free wall. Furthermore, research on the level fluctuation was not done. At present, studies of the free surface mostly focuses on slab molds [5,6], and research on bloom molds is rarely reported [7,8]. Jiaquan Zhang et al [7] have studied the level fluctuation of the free surface under different nozzle hole angles by building a steel-slag-gas three phase free surface, but the influence of EMS was not considered. In the present work, the steel-slag-gas three phase free surface model was built by VOF multiphase flow model. The influence of EMS parameters on level fluctuation of the mold was studied. Thus the proper EMS parameters could be found to guide production.

2. Calculation methods and conditions

2.1. Assumptions

The actual fluid flow in mold is a complicated physical and chemical process. The principal assumptions are made in the modelling as follows:

- Molten steel are treated as incompressible viscous fluids with the physical properties as constants.
- A solidified shell doesn't exist in mold.
- The cover slag is regarded totally as liquid in mold.
- The influence of mold oscillation on the flow field in mold is neglected.

2.2. Governing equations

Basic conservation equations for molten steel flow include the mass conservation equation and the momentum conservation equation. The two-equation *k-ε* model is adopted as a turbulence model [7, 9, 10].

The SOLA-VOF method was used to simulate the mold free surface (air-powder-steel interface), and the geometric reconstruction method was applied to track the free surface fluctuations. The volume function of the free surface boundary F(x, y, z, t) is defined as follows [11, 12]:

$$F = \begin{cases} 1 & \text{volume is fully filled} \\ 0 \to 1 & \text{volume is filled in part} \\ 0 & \text{volume is empty} \end{cases} \tag{1}$$

F (x, y, z, t) satisfies the Volume function equation:

$$\frac{\partial F_i}{\partial t} + \left[\frac{\partial}{\partial x_i} (F_i \cdot u_i) \right] = 0 \tag{2}$$

where, u_i is the velocity components in x, y and z direction (m/s); t is time (s).

2.3. Boundary conditions

The boundary conditions of the model can be determined as follows:

- Inlet boundary condition: The top cross section of the submerged entry nozzle is set as the inlet. Molten steel injection rate is calculated by the casting speed.
- Outlet boundary condition: The bottom cross section of the mold is set as the outlet, which is the speed outlet boundary.
- Mold wall: The velocity component vertical to the wall is zero. Velocity component, pressure, k and ε parallel to the wall are set to the no-slip boundary condition. In the boundary layer nearby the mold wall, the fluid parallel to the wall is calculated by the wall function.
- The top surface of the mold is set to the atmospheric pressure inlet. The material of the inlet is air.
- Each phase initialization: the space below 100mm from top of the mold is the molten steel. The molten steel is covered by a 10mm thick layer powder. The space above the powder is air.

2.4. Calculation conditions and solving method

The size of mold was 380mm×280mm×800mm, and the computational domain height was 1800mm for eliminating the effect of the lower mold exports on flow field. A submerged entry nozzle is set at the centre of the mold. The calculation and operating conditions of the caster are listed in table 1.

The time averaged electromagnetic force data F_X, F_Y, which were simulated by electromagnetic stirring, were loaded on the momentum equation in X, Y directions using the UDF program, so that the molten steel in the mold produced a cross-sectional tangential velocity V_X, V_Y in order to simulate the mold by EMS, as shown in figure 1.

Table 1. Main process parameters.

Physical property	Value
The type of nozzle	Four-hole
Nozzle diameter, mm	81
Insertion depth of nozzle, mm	210
Casting speed, m/min	0.6
Molten steel density, kg/m^3	7100
Molten steel viscosity, Pa·s	0.0055
Molten Slag density, kg/m^3	2700
Molten Slag viscosity, Pa·s	0.41
Air density, kg/m^3	1.225
Air viscosity, Pa·s	1.79×10^{-5}

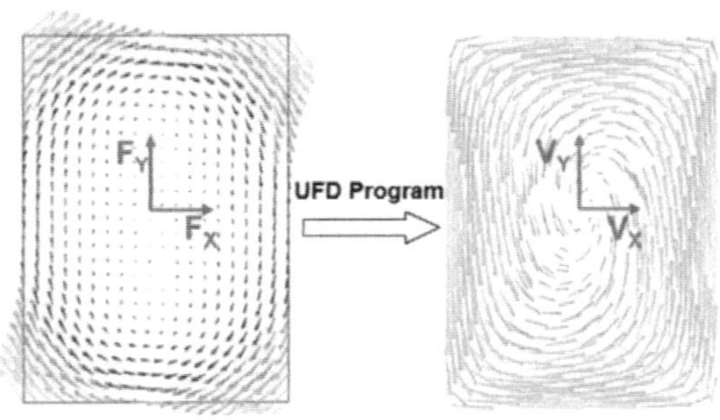

Figure 1. Electromagnetic force vector field (a) and flow field vector field (b) in the cross-section of the mold.

3. Results and Discussions

3.1. Comparison of level fluctuation with EMS and without EMS

The comparison of three dimensional flow field distributions of strand flow from the nozzle in the mold with EMS and without EMS is shown in figure 2.

As can be seen in figure 2: without EMS, the impact depth of molten steel is larger after the outlet of the molten steel, and part of the steel flows upwards to shock the powder and cause disturbance of the powder layer. After adding EMS, most of the steel spirals downwards along the mold axis under the electromagnetic force, then the impact depth of the molten steel is greatly reduced, and the small part of the steel that moves upward was weakened by the rotating electromagnetic force, so that longitudinal impact on the powder layer was decreased.

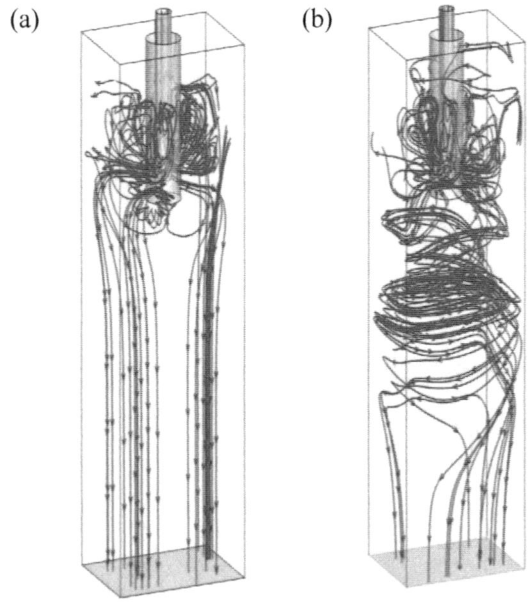

Figure 2. The comparison of three dimensional flow field distribution of mold steel flow without (a) and with (b) EMS.

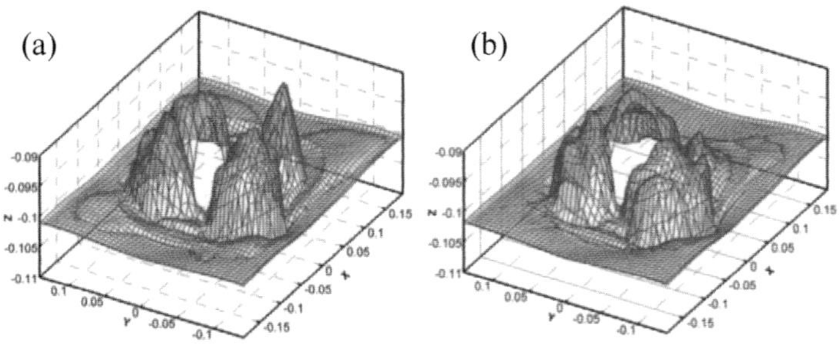

Figure 3. The comparison of three dimensional level fluctuation in the mold without (a) and with (b) EMS.

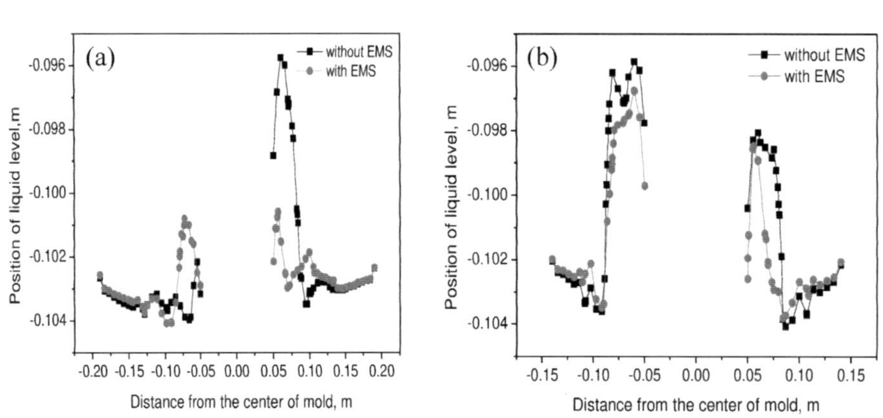

Figure 4. The level fluctuation value distributions in the wide direction (a) and the narrow direction (b) of the mold with and without EMS.

The comparison of three-dimensional time-averaged level fluctuation with and without EMS is presented in figure 3. The comparison of free surface level fluctuation value distributions in the wide direction and the narrow direction of the mold with and without EMS is shown in figure 4.

As can be seen in figure 3 and figure 4: whether with or without EMS, level fluctuation around the mold nozzle is apparently higher than elsewhere. The maximum fluctuation value in the wide direction and the narrow direction of the mold is 4.24mm and 4.14mm without EMS. The range of level fluctuation value was slightly reduced when the EMS was applied. The maximum fluctuation value in the wide direction and the narrow direction of the mold is 4.08mm and 3.73mm. Due to the application of electromagnetic stirring, the molten steel in mold moves spirally along the mold axis by the electromagnetic force (figure 2(b)). Then, the impact force of strand flow on the molten steel has deceased and the disturbance of strand flow on the upper surface of the mold has also been weakened.

3.2. The effects of current on fluctuation of liquid level

Figure 5 shows the three-dimensional level fluctuations under various current intensity when the stirring frequency is 2Hz. Figure 6 lists the level fluctuation value distributions in the wide direction and the narrow direction of the mold. It is noted from figure 5 and figure 6 that the fluctuation range of the liquid level in the mold rises with the increase of current intensity and that the maximum site of the level fluctuation occurs near the submerged entrance nozzle (within the scale of 0.5 to 0.1m in the radial direction) and presents as a convex shape. When the current is 450A, 550A or 600A, the largest fluctuated values in the wide direction and the narrow direction of the mold are 3.11mm, 3.29mm or 4.08mm and 3.26mm, 3.27mm or 3.80mm, respectively. The current intensity should not be too big under the condition of electromagnetic stirring. Thus, the fluctuated value of the liquid level must be closely controlled within 4mm when the current is less than 600A. Meanwhile, the powder in the vicinity of the nozzle should be evenly distributed and be supplemented timely in order to protect the surface of the molten steel from being exposed when faced with large liquid fluctuation.

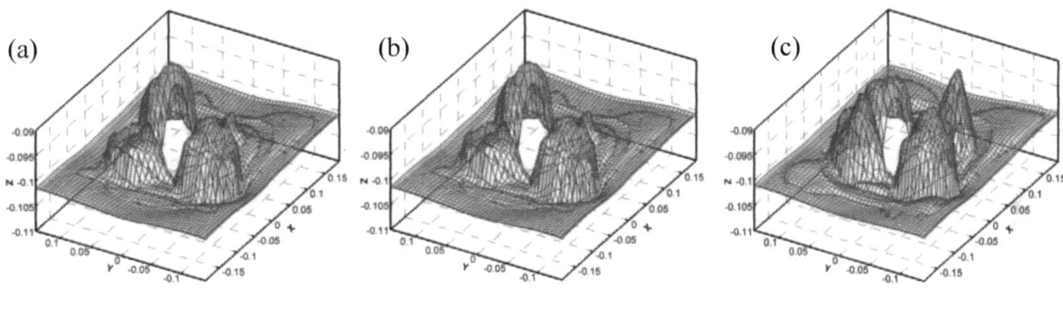

Figure 5. The three-dimensional liquid level fluctuations under 450A (a), 500A(b) and 600A(c) stirring current.

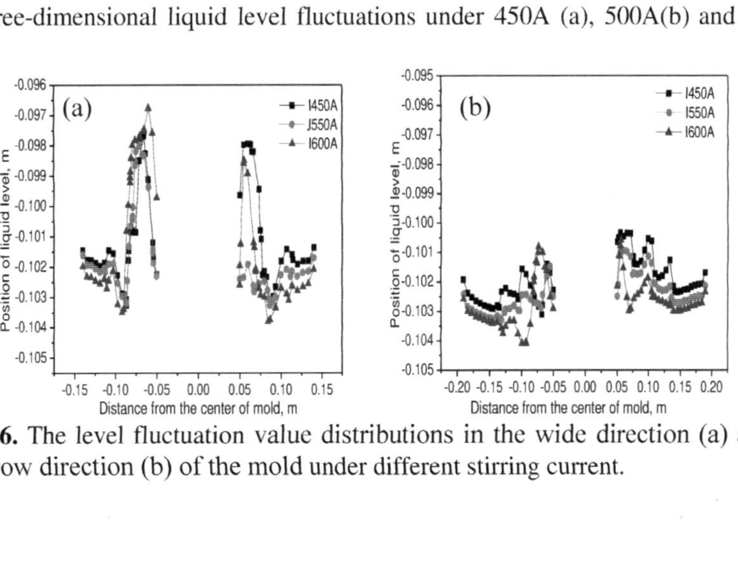

Figure 6. The level fluctuation value distributions in the wide direction (a) and the narrow direction (b) of the mold under different stirring current.

3.3. The influence of frequency on the level fluctuation

Figure 7 shows the three-dimensional level fluctuation in the mold under different stirring frequencies with the current intensity at 600A. The level fluctuation value distributions in the wide direction and the narrow direction of the mold are shown in figure 8.

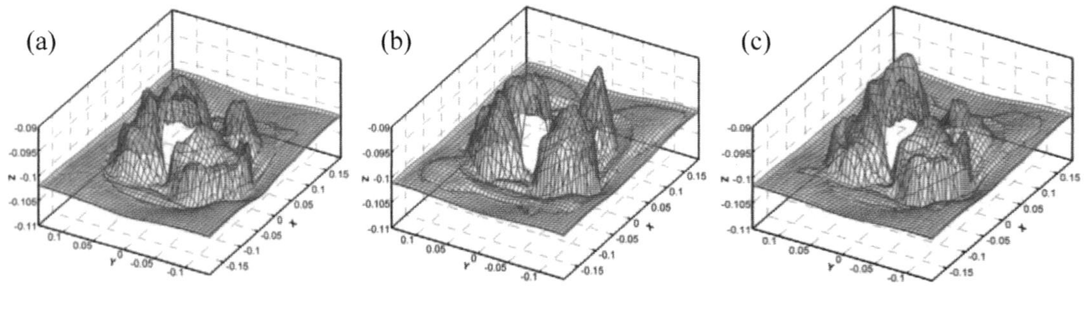

Figure 7. Three-dimensional diagrams of level fluctuation under 1.5Hz (a), 2.0Hz (b) and 2.5Hz (c) stirring frequency.

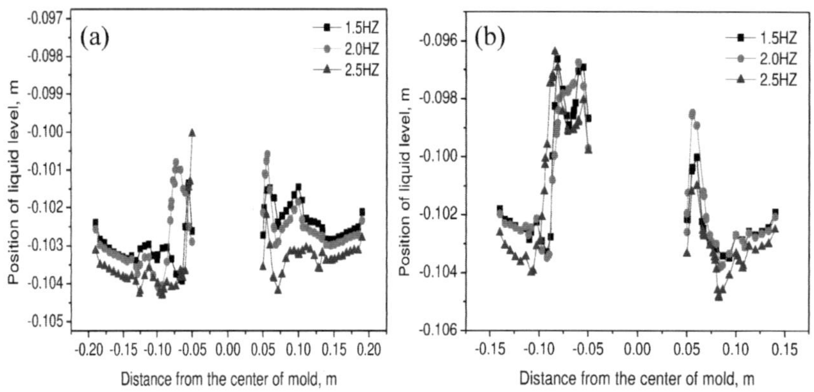

Figure 8. The level fluctuation value distributions in the wide direction (a) and narrow direction (b) of the mold under different stirring frequencies.

As can be seen in figure 7 and figure 8: with the increase of current frequency, the amplitude of liquid level fluctuation in the mold increased accordingly, the waveform gradually steepened, and the maximum fluctuation value all appeared nearby the nozzle (radial distance within 0.5m~1m). When the current frequencies are 1.5Hz, 2.0Hz and 2.5Hz, the maximum level fluctuation in the wide direction and the narrow direction of the mold are 3.93mm, 4.08mm and 4.29mm, and 3.49mm, 3.80mm and 4.86mm, respectively. When EMS is applied, the stirring frequency should not be too high. When the frequency was under 2.0Hz, the liquid fluctuation value was under 4mm.

4. Conclusions

- Compared with the bloom casting mold without electromagnetic stirring, the molten steel in the mold with EMS moves in a circular way under the rotating electromagnetic force, so that the fluctuations of the free liquid level decrease slightly and the maximum level fluctuation value decreases from 4.24mm to 4.04mm.
- The level fluctuation in the mold is enhanced with increasing current intensity and frequency by using electromagnetic stirring. The level fluctuation values in the mold are under 4mm when the current is less than 600A and the frequency is lower than 2Hz.
- With electromagnetic stirring, the maximum site of the liquid level occurs in the vicinity of the submerged entrance nozzle. In the practical continuous casting process, much more

attention should be paid to the thickness of the covering slag to protect the surface of the molten steel from being exposed in air caused by fierce fluctuations.

References

[1] Liu X D, Yang X D and Zhu M.Y 2007 *Journal of Iron and Steel Research* **14** 6
[2] Li Y, Ni H.W and Zhang H 2012 *Special Steel* **33** 10
[3] Zhou L, Tang P and Li Y X 2006 *Iron Steel Vanadium Titanium* **27** 10
[4] Ren B Z, Zhu M Y and Wang H D 2008 *Acta Metallurgical Sinica* **44** 507
[5] Liu H P and Wang Z Y 2002 *Research on Iron & Steel* **4** 47
[6] Wang Y F and Zhang L F 2010 *ISIJ International* **40** 1777
[7] Wang W W, Zhang J Q and Chen S Q 2007 *Journal of university of science and technology Beijing* **29** 816
[8] Ren Z P and Jiang M F 2006 *Steelmaking* **22** 16
[9] Yue Q, Zhou L and Wang J J 2011 *The Chinese Journal of Process Engineering* **11** 187
[10] Choi S L, Ryu K J and Park H S 2002 *Metals and Material International* **6** 527
[11] Tan LJ, Shen H.F and Liu B C 2003 *Acta Metallurgical Sinica* **39** 435
[12] Zhang Q Y and Wang X H 2010 *Journal of Iron and Steel Research* **17** 15

IOP Conf. Series: Materials Science and Engineering **117** (2016) 012045 doi:10.1088/1757-899X/117/1/012045

Influence of forced convection on solidification and remelting in the developing mushy zone

M Wu[1,2,a], A Vakhrushev[1], A Ludwig[2] and A Kharicha[1,2]
[1]Christian-Doppler Lab for Adv. Process Simulation of Solidification & Melting, University of Leoben, Franz-Josef-Str. 18, A-8700 Leoben, Austria
[2]Chair of Simulation and Modeling of Metallurgical Processes, University of Leoben, Franz-Josef-Str. 18, A-8700 Leoben, Austria

Email: [a]menghuai.wu@unileoben.ac.at

Abstract. The mushy zone and solid shell formed during solidification of a continuous casting are mostly uneven, and this unevenness of shell growth might lead to surface defects or breakout. One known example is the unevenness of shell growth at the impingement point between the jet flow (coming from submerged entry nozzle) and the solidification front. This phenomenon is primarily understood as the local remelting caused by the superheat of the melt, which is continuously brought by the jet flow towards the solidification front. A recent study of the authors [Metall. Mater. Trans. B, 2014, in press] hinted that, in addition to the aforementioned superheat-induced local remelting (1), two other factors also affect the shell growth. They are (2) the advection of latent heat in the semi-solid mushy zone and (3) the enhanced dissipation rate of energy by turbulence in the bulk-mush transition region. This paper is going to perform a detailed numerical analysis to gain an insight into the flow-solidification interaction phenomena. Contributions of each of the above factors to the shell formation are compared.

1. Introduction

The industry practice of continuous casting shows that the mushy zone and solid shell formed during solidification are mostly uneven, and this unevenness of shell growth might lead to surface defects or breakout in the worst case. Reasons for this are diverse: (1) the non-uniform heat flux from the casting to the mould [1-2]; (2) fragmentation of dendrites in the partially solidified shell [3-4]; (3) the dynamic flow-solidification interaction which leads to suppression of solidification or remelting locally [5-7]. The topic of non-uniform heat flux has drawn the significant attention of researchers [8-10]. For the fragmentation of dendrites due to forced convection, some preliminary knowledge [11-13] is also available. The current paper is going to perform numerical studies to gain an insight into the flow-solidification interaction which leads to suppression of solidification or remelting.

One known example of the unevenness of shell growth during continuous casting is the flow-solidification interaction at the impingement point between the jet flow (coming from the submerged entry nozzle) and the solidification front. According to B.G. Thomas [5-6], this can be primarily explained by the local remelting due to the superheat of the melt, which is continuously brought by the jet flow. A recent study of the authors [7] hinted that, in addition to the aforementioned superheat-induced local remelting; two other factors also affect the shell growth. They are: the advection of latent heat by the melt flow which is co-related to the motion/deformation of the semi-solid mushy zone and the enhanced dissipation rate of energy by turbulence in the bulk-mush transition region.

2. Key features of the numerical model and benchmark settings

An enthalpy based model [14-18] was used and extended by the current authors for thin slab casting [7]. Deformation of the partially solidified strand shell which affects the interdendritic melt flow is taken into account. We do not repeat all conservation equations, only the enthalpy equation

$$\frac{\partial}{\partial t}(\rho h)+\nabla\cdot(\rho\vec{u}h)=\nabla\cdot(\lambda_{\text{eff}}\nabla T)+S_{\text{e}}, \quad (1)$$

where

$$S_{\text{e}}=\rho L\frac{\partial f_{\text{s}}}{\partial t}+\rho L\cdot\nabla(\vec{u}_{\text{s}}f_{\text{s}}).$$

This equation is used to calculate the sensible enthalpy of the liquid-solid mixture $h_{\text{mixture}}^{\text{sensible}}$ (written as h hereafter), which represents the volume average of the sensible enthalpies of both liquid and solid phases: $h = f_\ell h_\ell^{\text{sensible}} + f_{\text{s}} h_{\text{s}}^{\text{sensible}}$. A further model simplification is made by assuming that the liquid and solid have the same sensible enthalpy ($h_\ell^{\text{sensible}} = h_{\text{s}}^{\text{sensible}} = h$). During solidification the latent heat (L) is treated separately by considering two source terms, as described in equation (1). In this study we set both liquid and solid the same density (ρ). Thermal conductivity (λ) is also calculated by phase volume averaging, and for the turbulent flow an effective thermal conductivity λ_{eff} ($=\lambda+\lambda_{\text{t}}$) is used to account for the turbulence enhanced energy dissipation.

A critical point of equation (1) is that it includes three different velocities: the intrinsic liquid velocity $\vec{u}_\ell$, solid velocity $\vec{u}_{\text{s}}$ and mixture velocity $\vec{u}$. They are related by

$$\vec{u}=f_\ell\vec{u}_\ell+f_{\text{s}}\vec{u}_{\text{s}}. \quad (2)$$

The mixture velocity is the solution of the momentum equation. The solid velocity is calculated with an incompressible rigid viscoplastic model on the basis of an assumed moving boundary condition [6]. From equation (2) we can derive the intrinsic liquid velocity ($\vec{u}_\ell$). Mass and volume conservations apply, i.e. $\nabla\cdot\vec{u}=0$ and $f_\ell + f_{\text{s}} = 1$. The evolution of the solid phase (f_{s}) is determined by the temperature according to the $f_{\text{s}}-T$ function (here the lever rule is used).

Focus of the paper is on the flow-solidification interaction. Three factors affecting solidification are investigated: the superheat (casting temperature over liquidus) induced suppression of solidification or remelting, the advection of latent heat by the interdendritic flow ($-\rho L\nabla\cdot(f_\ell\vec{u}_\ell)$, i.e. $\rho L\nabla\cdot(f_{\text{s}}\vec{u}_{\text{s}})$) and the enhanced dissipation rate of energy by turbulence (λ_{t}). A 2D benchmark as defined in the previous publication [7] is used, as shown in figure 1. A constant turbulence, kinetic energy (2.5 x 10^{-4} m^2 s^{-2}) and a constant dissipation rate (9.3 x 10^{-4} m^2 s^{-3}) are applied at the pressure inlet. A binary alloy (Fe-0.34 wt.%C) is considered, and its physical properties are listed in table 1. Four simulations are performed as listed in table 2.

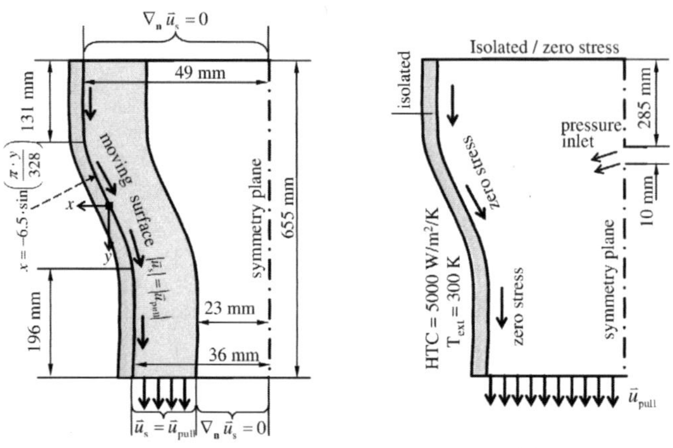

a) b)

Figure 1. Configuration of a 2D benchmark (a) for calculation of the solid velocity and (b) for the flow-solidification. The geometry in the vertical direction is scaled-down by 1/8. Reproduced from [7].

Table 1. Properties and parameters for the calculations of the 2D benchmark (Fe-0.34 wt.%C).

Thermal physical properties	Thermodynamic data
$c_p = 808.25 \ \ \mathrm{J \cdot kg^{-1} \cdot K^{-1}}$	$c_0 = 0.34$ wt.% C
$\lambda = 33.94 \ \ \mathrm{W \cdot m^{-1} \cdot K^{-1}}$	$T_{liquidus} = 1781.8$ K
$\rho = 7027 \ \ \mathrm{kg \cdot m^{-3}}$	$T_{solidus} = 1710.8$ K
$\mu_\ell = 5.6 \times 10^{-3} \ \ \mathrm{kg \cdot m^{-1} \cdot s^{-1}}$	$f_s - T$ function with the level rule.
$L = 2.5 \times 10^5 \ \ \mathrm{J \cdot kg^{-1}}$	
Process parameters	
$T_{inlet} = 1850$ K	$\vec{u}_{pull} = 0.07 \ \ \mathrm{m \cdot s^{-1}}$

Table 2. Parameter study of the flow-solidification interaction.

	superheat (K)	source term of latent heat	$f_s^{integral}$ ***
Case I	68.2	S_e	0.0841
Case II	1	S_e	0.1111 (+32%)
Case III*	68.2	disregarding $\rho L \nabla \cdot (f_s \vec{u}_s)$ in S_e	0.1331 (+58%)
Case IV**	68.2	S_e	0.0955 (+14%)

* Identical to case I, but $\rho L \nabla \cdot (f_s \vec{u}_s)$ in S_e is ignored.

** Identical to case I, but in the energy conservation equation the contribution of the turbulence to the effective thermal conductivity (λ_t) is ignored.

*** $f_s^{integral}$: volume average of f_s over the whole calculation domain at steady state.

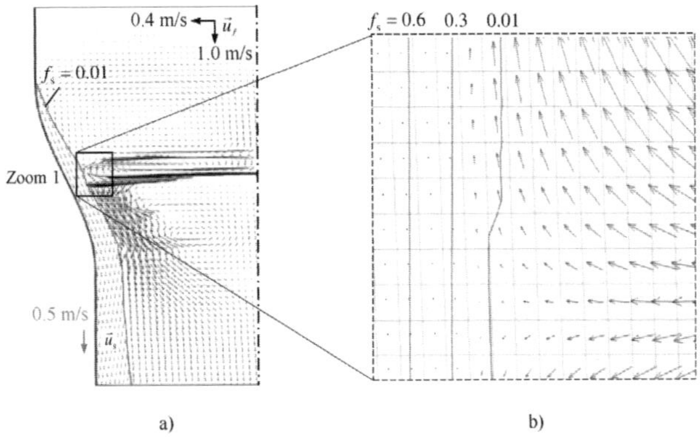

Figure 2. Result of flow pattern in Case I. a) $\vec{u}_s + \vec{u}_\ell$ (geometry in the vertical direction is scaled-down by 1/8); b) Zoom 1, ($\vec{u}_\ell - \vec{u}_s$) (geometry 1/1 scale). Some isolines of f_s are marked in figure.

3. Results and discussions

Figure 2 shows the calculated velocity fields of case I. The solid velocity is almost parallel to the curved mould surface. A jet flow coming from the inlet impinges on the solidification front, and the solidification front is slightly concaved at the impingement point. Figure 2(b) shows the details of the flow near and in the mushy zone. The flow can penetrate into the mush, but the interdendritic flow is significantly 'dampened' in the vicinity of the solidification front.

3.1 Influence of superheat

Comparison of the steady state solidification profiles between case I and II was made in figure 3: the former with a superheat of 68.2 K, the latter with a superheat of 1 K. It is clear that the higher superheat slows down the solidification. Taking the integral of f_s over the whole calculation domain, we get 32% solid phase with case II more than that with case I (table 2). With the current benchmark configuration, no notable remelting is evidenced.

3.2 Importance of advection of latent heat

Comparison of the steady state solidification profiles between case I and III was made in figure 4: the former with full source term S_e, the latter ignores the part $\rho L \nabla \cdot (f_s \vec{u}_s)$ from S_e. Overestimation of the shell growth by case III when the advection of latent heat due to the motion of solid phase in the mushy zone is ignored is huge: about 58% (table 2). The latent heat advection term, $\rho L \nabla \cdot (f_s \vec{u}_s)$, can also be written as $\rho L \vec{u}_s \cdot \nabla f_s$, as $\nabla \cdot \vec{u}_s \equiv 0$. The sign of this term can be '+' or '-', depending on the scalar product of the solid velocity vector $\vec{u}_s$ and the solid fraction gradient ∇f_s.

- A '+' sign indicates that this positive source term in the enthalpy equation will lead to an increase of the sensible enthalpy. The temperature might increase, with a corresponding decrease in the local cooling process. The consequence of this is that the solidification will slow down and even remelting could occur.
- A '-' sign indicates that this negative source term in the enthalpy equation will lead to a decrease of the sensible enthalpy. The temperature might decrease, with a corresponding increase in the local cooling process. The consequence of this is that the solidification will speed up.

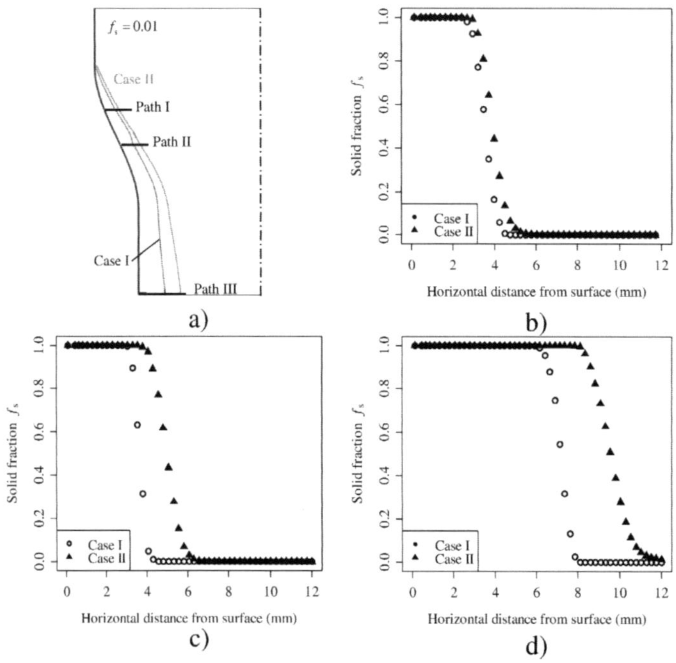

Figure 3. Influence of superheat on the shell growth: comparison of f_s distribution between case I and II. (a) Schematic of solidification front (f_s=0.01); f_s profiles along path I (b), path II (c), path III (d). The paths are 266, 323, 649 mm from top surface, correspondingly.

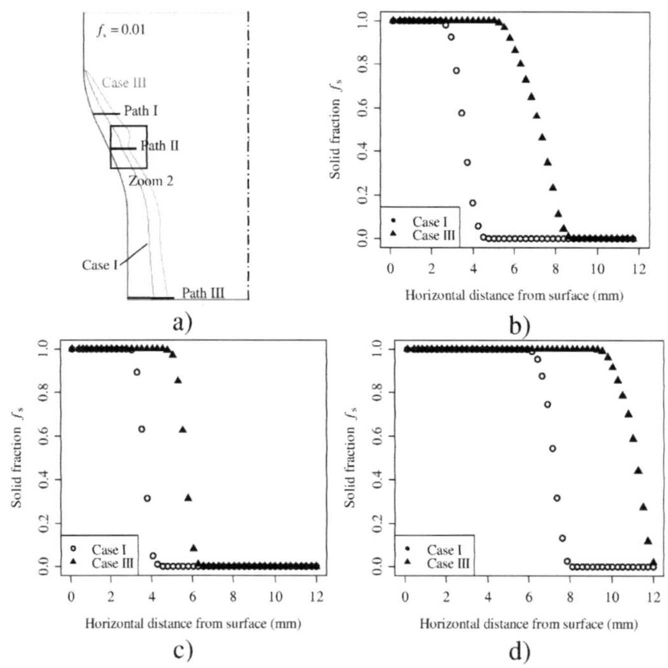

Figure 4. Influence of latent heat advection on the shell growth: comparison of f_s distribution between case I and III. (a) Schematic of solidification front (f_s=0.01); f_s profiles along path I (b), path II (c), path III (d). The paths are 266, 323, 649 mm from top surface, correspondingly.

The calculated $\rho L\vec{u}_s \cdot \nabla f_s$ for case III is shown by the colour scale in figure 5. The term of $\rho L\vec{u}_s \cdot \nabla f_s$ was actually disregarded from the energy equation in this case. It means that in the region with a '+' sign, a depression of solidification or remelting should occur due to this term, but this suppression of solidification or remelting is ignored; therefore it leads to an overestimation of the solidification. In contrast, in the region with a '-' sign, a speed-up of solidification should occur, but this speed-up of solidification is ignored by case III; therefore it leads to an underestimation of the solidification.

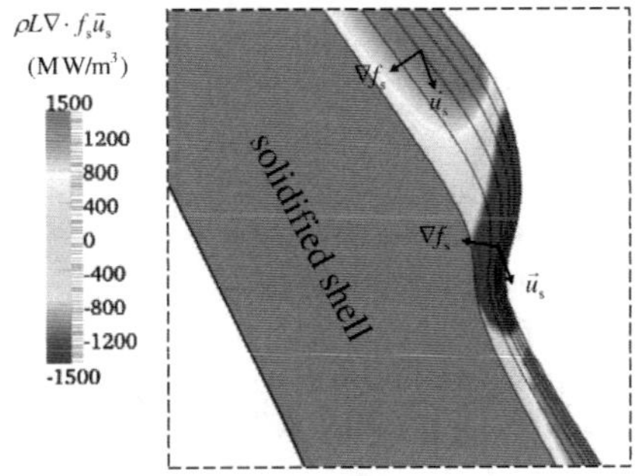

Figure 5. Alignment of the solid volume fraction gradient and the solid velocity vector in the mush zone (case III) in the Zoom 2, as marked in figure 4(a). Isolines of f_s = 0.01, 0.2, 0.4, 0.6, 0.8, 0.99 are shown. The $\rho L\vec{u}_s \cdot \nabla f_s$ term, which is actually ignored in case III, is shown by the colour scale.

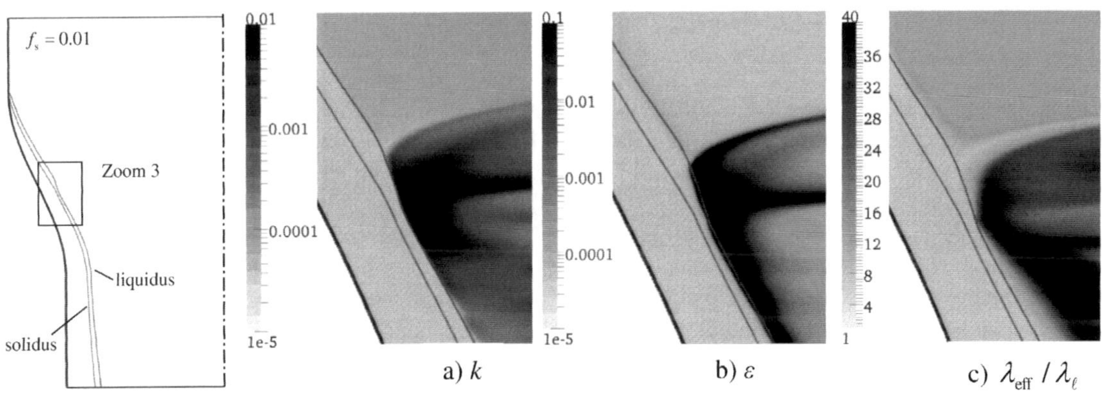

Figure 6. Calculated distributions of (a) turbulence kinetic energy, (b) dissipation rate and (c) normalized effective thermal conductivity in zoom 3 of case I.

3.3 Influence of turbulence

Case IV is identical to case I, but in the energy conservation equation of case IV the contribution of the turbulence (λ_t) to the effective thermal conductivity (λ_{eff}) is ignored. Generally, the turbulence-induced energy dissipation suppresses the solidification, which is observed especially in the lower domain below the jet impingement point. If we ignored this turbulence-induced energy dissipation, the solidification could be overestimated by about 14% (table 2). Details of the turbulence quantities for case I are shown in figure 6. The predicted effective thermal conductivity in the bulk melt region and near the jet impingement point can reach as much as 37 times higher than the physical value of the thermal conductivity.

4. Summary discussion

- A numerical parameter study using an enthalpy based solidification model gives an insight into the interaction between the forced convection and solidification in the developing mushy zone. Three influencing factors are analyzed: superheat induced suppression of solidification or remelting, advection of latent heat due to interdendritic flow which is coupled with the motion/deformation of semi-solid mushy zone and turbulence enhanced energy dissipation.
- A 2D benchmark is designed for this study:
- Turbulence suppresses the solidification due to the fact that the turbulence dissipation enhances the diffusive energy transfer. An ignorance of this part of energy transfer would lead to an overestimation of the solidification by 14%.
- Superheat (casting temperature over the liquidus) suppresses the evolution of the mushy zone (growth of the semi-solid shell). For example, a 68.2 K of superheat coming from inlet slows down the solidification by 32% if compared to 1 K of superheat.
- The advection of latent heat in the mushy zone plays a very important role in the formation of the mushy zone (shell). With the motion/deformation of the semi-solid shell, the advection of latent heat in the mushy zone can cause acceleration or depression of the solidification, depending on the interdendritic flow direction and the gradient of the volume fraction solid.
- People have used an "equivalent heat capacity model" [19] for solidification, in which the advection of latent heat due to the interdendritic flow is ignored. This model was successfully applied in many casting processes where the solidified phase is stationary, i.e. ingot castings and shape castings. We learnt that this kind of model cannot be used for continuous casting where the semisolid shell moves.

Acknowledgment
The financial support by RHI AG, the Austrian Federal Ministry of Economy, Family and Youth and the National Foundation for Research, Technology and Development is gratefully acknowledged.

References

[1] Savage J 1962 *J. Iron Steel Institute* **200** 1962 41-51
[2] Meng Y and Thomas BG 2003 *Metall. Mater. Trans.* **34B** 685-705
[3] Lesoult G, Combeau H and Moukassi M 1993 *Journal de Physique IV* **3** 813-822
[4] Hellawell A, Liu S and Lu SZ 1997 *JOM* **49(3)** 18-20
[5] Koric A, Thomas BG and Voller VR 2010 *Numer. Heat Transf. B Fundam.*, **57** 396-413
[6] Zhao B, Thomas BG, Vanka SP and O'Malley RJ 2005 *Metall. Mater. Trans.* **36B** 801-823
[7] Vakhrushev A, Wu M, Ludwig A, Tang Y, Hackl G and Nitzl G 2014 *Metall. Mater. Trans.* **45B** in press, DOI: 10.1007/s11663-014-0030-2
[8] Cho J, Shibata H, Emi T and Suzuki M 1998 *ISIJ Int.* **38** 440-446
[9] Camporredondo-S JE, Castillejos-E AH, Acosta-G FA, Gutierrez-M EP and Herrera-G MA 2004 *Metall. Mater. Trans.* **35B** 443-457
[10] O'Connor TG and Dantzig JA 1994 *Metall. Mater. Trans.* **25B** 541-560
[11] Yamazaki M, Natzume Y, Harada H and Ohsasa K 2006 *ISIJ Int.* **46** 903-908
[12] Campanella T, Charbo C and Rappaz M 2004 *Metall. Mater. Trans.* **35A** 3201-3210
[13] Su Z, Chen J, Nakajima K and He J 2009 *Steel Res Int.* **80** 824-833
[14] Prescott PJ and Incropera FP 1994 *J. Heat Transfer* **116** 735-741
[15] Prescott PJ, Incropera FP and Gaskell DR 1994 *Trans. ASME* **116** 742 – 749
[16] Prescott PJ and Incropera FP 1995 *Trans. ASME* **117** 716 – 724
[17] Voller VR and Prakash C 1987 *Int. J. Heat Mass Transfer* **30** 1709 – 1719
[18] Voller VR, Brent AD and Prakash C 1989 *Int. J. Heat Mass Transfer* **32** 1719 – 1737
[19] Hsiao JS 1985 *Numer. Heat Transf.* **8** 653-666

IOP Conf. Series: Materials Science and Engineering **117** (2016) 012046 doi:10.1088/1757-899X/117/1/012046

Control of solidification microstructure using programmable electro-magnetic pulses

T Manuwong[a], W Zhang[b], P L Kazinczi[c] , J Mi[d]
School of Engineering, University of Hull, Hull, HU6 7RX, East Yorkshire, UK

Email: [a]t.manuwong@2009.hull.ac.uk,.[b]w.zhang@2012.hull.ac.uk, [c]p.lobo-kazinczi@hull.ac.uk, [d]j.mi@hull.ac.uk

Abstract. A programmable electro-magnetic pulse (PEMP) apparatus has been designed and made, and used to study the effect of different pulse amplitude and frequencies on the solidification microstructure. The measured magnetic fluxes agree well with the simulated fluxes using finite element method. The solidification microstructure of a Sn-18%wtPb was studied using different pulse amplitudes and frequencies. It was found that higher pulse amplitude and frequency lead to more grain refinement.

1. Introduction

The mechanical properties of cast alloys, such as strength and ductility, are mainly determined by the grain structure and defect formed during solidification processes. How to achieve an optimal morphology, size and distribution of the grains and those of the defects (porosity and/or inclusions) to maximise the alloy's mechanical properties are the main driving force for metal alloy research for many years. Conventional methods of adding external grain refiners to refine the solidification structure of metal alloys have been used in industry for many decades [1, 2]. Those methods have inherent problems. Firstly different alloys need grain refiners of different chemistry, such as Al-Ti-B for Al-based alloys, Zr for Mg-based alloys, etc. [2, 3]. Secondly, different operation practices are needed for different alloys systems, and each of the operation inevitably add extra cost to the manufacturing of the alloys and most likely an adverse impact to the environment. To reduce the environment impact and develop a sustainable and green manufacturing strategy, in the past twenty years or so, many researches have been carried out to develop novel grain refining technologies. Among them, electrical or magnetic field based techniques have attracted a great deal of attention in research communities and industry. The electrical pulse current and electromagnetic stirring [4-7], for example, have shown to be effective techniques for grain refinement. The practical challenge is how to apply those current or field uniformly in a large volume of melt, and therefore to refine a large volume of melt without consuming a huge amount of energy with dangerously high electrical current or voltage [7, 8].

Currently, extensive worldwide researches have been conducted in this field in university laboratories in the UK, USA, Europe, Japan and China [9-13]. However, most of the research has been done by applying the electromagnetic pulses into stationary crucibles and there are very limited reports on applying the electro-magnetic pulse into a continuous casting operation. In this research, we designed a programmable electromagnetic pulse (PEMP) device, and a small scale continuous casting apparatus to mimic the operation in DC casting. The effects of electro-magnetic pulse of different voltages on the resulting solidification microstructure were studied in details.

2. The design of the PEMP device

We designed a novel device to generate programmable electro-magnetic pulses (PEMP). The amplitude, frequency and duration can be programmed to suit different alloys and processing conditions. The principle and design of the apparatus is briefly described below.

The PEMP device consists of two main units: the energy charging unit and pulse generating unit. In the charging unit, a capacity bank is charged using a long-duration but low-power energy. The voltage charged into and the energy stored within the capacitor can be calculated by:

$$V = V_c(1 - e^{(\frac{-t}{RC})}) \quad (1)$$

$$E = 0.5CV^2 \quad (2)$$

where V_c is the voltage applied to the circuit, *t* is time and *E* is the energy stored in the capacitor bank.

In the pulse generating unit, the energy stored in the capacitor bank is discharged in a controlled short period of time to send high power pulse to the working unit (the induction coil in this case) as showed in figure 1(a). The novelty of this device is that a Thyrister pulse trigger is designed to control the release of the pulse in term of time and duration and repetition. In this way, programmable high energy pulses with tuneable amplitudes, frequencies and duration can be generated to suit different alloys systems for process optimisation purpose. Examples of the charging process of the capacitor bank, and the pulse generated to the coil are shown in figure 1a and b. Through the induction coil (45 turns and 120 mm height), pulsed Lorentz forces are generated into the alloy during solidification process as schematically shown in figure 1(b) [14-16], and the Lorentz force vector points towards the centre of the coil.

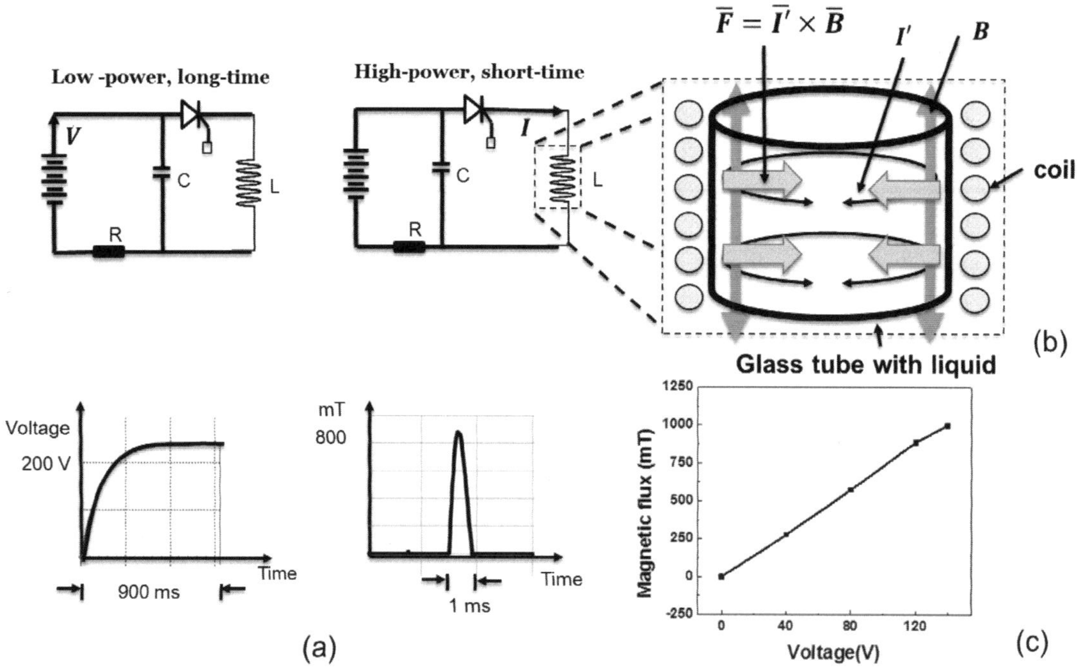

Figure 1. (a) Schematics to show the energy charging and pulse generating units of the PEMP device, (b) the Lorentz force created by the coil, (c) the peak magnetic flux as a function of the input voltages.

3. Experiment procedure

We also designed a small-scale casting apparatus. It consists of a furnace with close-loop temperature control unit, a sample stage to hold glass tubes (as a crucible) with metal, and a linear actuator that can release a controlled sample withdrawing out of the furnace with very high accuracy of travel (the travel step resolution is 0.49 μm). The apparatus can be used to mimic the semi-continuous DC casting operation.

3.1 Casting of alloy bars

Low melting point binary alloy (Sn-18%wt.Pb) was made by melting together pure Sn (purity of 99.75%) and pure Pb (purity of 99.97%) with the designed weight ratio in a Carbolite furnace. The alloys were cast into 130 mm long borosilicate glass tubes with the inner diameters of 9 mm by using a custom-made small counter-gravity casting apparatus. The controlled negative pressure draws the liquid metal uphill and fill into the glass tubes in a quiescent manner, avoiding the turbulence of the liquid metal during the filling process and therefore any entrainment of air bubbles, porosity or oxide films during casting.

3.2 Solidification experiment under pulse magnetic field

Borosilicate glass tubes were again used (130 mm long with 9 mm inner diameter) as the casting moulds. The alloy ingots were placed into the glass tubes, and at the bottom end, the glass tubes were sealed by Duratec ceramic rods, which were placed and linked to a linear actuator with a programmable travel speed of 100 μm/s. Five K-type thermocouples were positioned inside the glass tubes at 1, 3, 5, 7 and 9 cm from the bottom of the sample to measure the temperatures at those locations during the experiment and recorded using the TC-08 Thermocouple Data Logger, and PicoLog for Windows (PLW) recorder software.

The glass tubes with alloy bars inside were heated in the furnace and held above the alloy melting point at ~230 °C for 30 minutes to homogenise the melt temperature. Then the glass tubes were withdrawn out of the furnace at a constant speed of 100 μm/s.

Two groups of samples were cast, one with PEMP, one without. The experimental parameters used are listed in table 1, and the PEMP was applied continuously after the glass tubes were withdrawn out of the furnace until the melt completely solidified.

Table 1. The PEMP parameters applied into the samples during solidification

Discharged Pulse Frequency (Hz)	Discharge voltage (V_d)
1	40
	80
	120
10	40
	80
	120

After solidification, the as-cast alloy samples were taken out of the glass tubes, and were sectioned longitudinally into two halves. The sectioned surfaces were ground using SiC grinding papers of 600, 1200, 2500 grid size, and then polishing using diamond suspension of 6 and 1 μm size using Motopol 12 grinder and polisher operated at 150 rpm. Each grinding and polishing steps takes 20~30 minutes.

4. Modelling

In order to quantify the magnetic flux density induced by the voltage pulse discharged into the coil, we used COMSOL Multiphysics to calculate the induced magnetic flux density. Figure 2 shows the 2D symmetric geometry used in the simulation and the mesh (triangular mesh of 94,262 elements). The induction coil is defined as the Single-Turn Coil domain. The Biot–Savart law (equation 3 and 4) [17] was used to calculate the magnetic flux density:

$$0 = \nabla \cdot B \tag{3}$$

$$B = \frac{\mu_0 I}{4\pi} \int \frac{dl}{r^2} \tag{4}$$

where B is the magnitude of the magnetic field (the magnetic flux density), and is measured by Tesla (Wb/m^2), μ_0 is permeability of free space, r is the distance between the location of $d\ell$ and the location at which the magnetic field is being calculated.

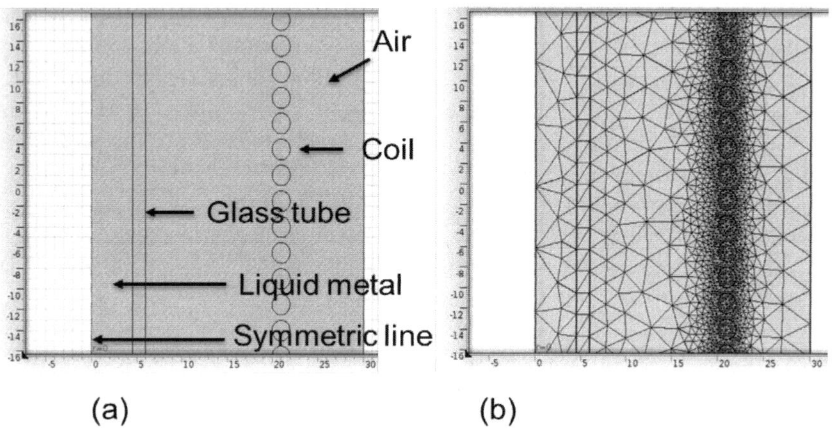

Figure 2. (a) the 2D symmetrical geometry for the simulation, and (b) the mesh structure.

5. Results and discussions

5.1 The effect of PEMP on the grain growth and dendrite morphology

Figure 3a shows the microstructure of the alloys cast without PEMP. The grains grew unidirectionally against the direction of withdrawing and formed very long columnar dendrites. Figure 3b-g show the microstructure formed under the influence of PEMP with the corresponding parameters applied listed in table 1. Apparently, when the PEMP was applied, the long columnar dendrites typically shown in figure 3a were disrupted, forming a series of dendrite clusters along the growth direction. We used ImageJ to segment the dendrite clusters of those showed in figure 3b-g where different discharging voltages and pulse frequencies applied. It was found that the average dendrite clusters become smaller and smaller as the discharge voltages and frequencies increase.

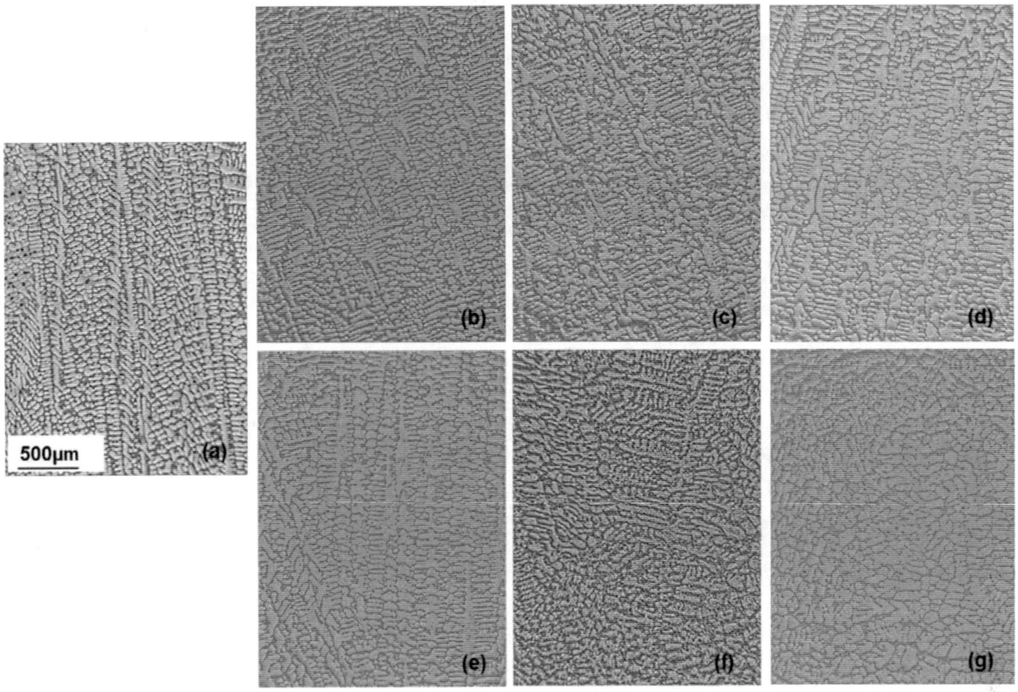

Figure 3. The solidified microstructure of the 9 mm diameter samples: (a) without PEMP, (b) PEMP with 40 V and 1 Hz, (c) 80 V and 1 Hz, (d) 120 V and 1 Hz, (e) 40 V and 10 Hz, (f) 80 V and 10 Hz, (g) 120 V and 10 Hz.

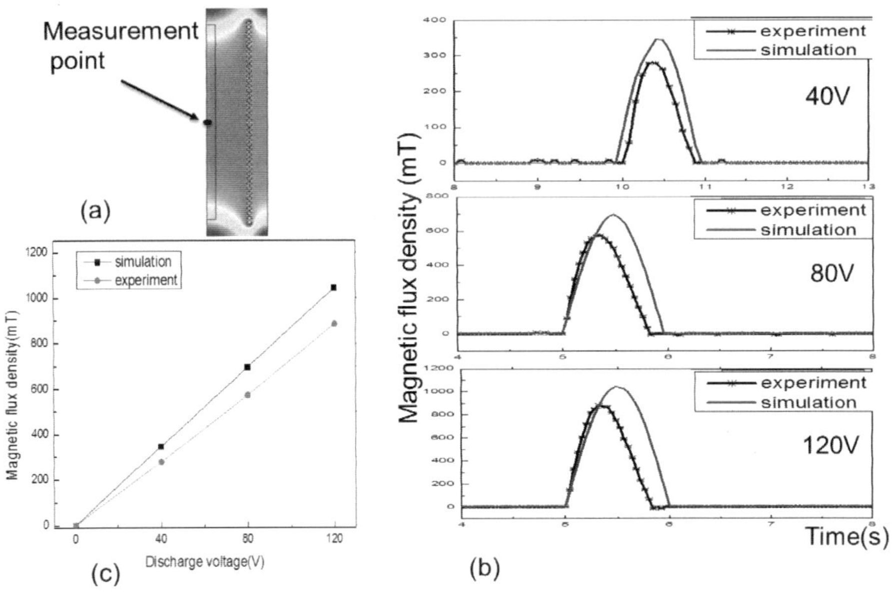

Figure 4. (a) The point where the magnetic flux density was measured, (b) the comparison of measured and simulated pulses of magnetic flux density, (c) the magnetic flux density as a function of the discharged voltage and frequency.

5.2 FEM simulation

Figure 4a shows the point where the maximum magnetic flux density is measured using Gaussmeter. Figure 4b shows the comparison between the measured magnetic flux density and those simulated using FEM method. In general, the measured and simulated shapes of the pulses for all different discharging voltages are very similar. While the peak values of the simulated pulses are 10~15% higher than those of the measured ones, indicating that, in the modelling, the energy loss in the discharging circuit and the coil is not considered, and therefore the simulated values are slightly higher than the measured values. It is interesting to find that the peak values of the magnetic pulses increase linearly with the increase of discharging voltages as shown in figure 4c, and the higher the discharging frequency, the higher the peak value of the magnetic pulse.

6. Conclusion

The peak of the magnetic flux density pulse has a linear relationship with the discharging voltages, and the higher the discharging voltage/frequency, the higher the peak value of the pulse. The magnetic pulse applied to the liquid metal during solidification can effectively disrupt the growth of the dendrite, and change a long columnar dendrite to many dendrite clusters. The higher the pulse density, the smaller the dendrite cluster created.

Acknowledgement

T. Manuwong would like to acknowledge the financial support from the Royal Thai Government for his PhD Studentship.

References

[1] Changjiang S, Qingyou H, Qijie Z. 2009 Review of grain refi nement methods for as-cast microstructure of magnesium alloy. *China foundry*. **6**:93-103.
[2] Kashyapkt Chandrashekar T 2001 Effects and mechanisms of grain refinement in aluminium alloys Bull *Mater Sci.* **24**:345-53.
[3] Qian M, StJohn D H, Frost M T 2003 Zirconium Alloying And Grain Refinement of Magnesium Alloys. *The Minerals, Metals & Materials Society.*
[4] Haifang S, Weiqiang Z 2003 Effects of electromagnetic stirring on microstructure of solidified aluminum alloys. *TransNonferrous MetSoc China*. **13**:1171-4.
[5] Ma J, Li J, Gao Y, Jia L, Li Z, Zhai Q 2009 Improving the Carbon Macrosegregation in High-Carbon Steel by an Electric Current Pulse. *Met Mater Int.* **15**:603-8.
[6] Nakada M, Shiohara Y, Flemings M C. 1990 Modification of Solidification Structures by Pulse Electric Discharging. *ISIT International.* **30**:27-33.
[7] Liao X, Zhai Q, Luo J, Chen W, Gong Y 2007 Refining mechanism of the electric current pulse on the solidification structure of pure aluminum. *Acta Materialia*. **55** 3103-9.
[8] Lie L, Jian-hong M, Chang-jiang S, Zhi-jun L, Yu-lai G, Qi-jie Z 2009 Columnar to Equiaxed Transition During Solidification of Small Ingot by Using Electric Current Pulse. *Journal of Iron and Steel Research, International*. **16**:7-12.
[9] Takagit, Iwai K, Asai S. 2003 Solidified Structure of Al Alloys by a Local Imposition of an Electromagnetic Oscillationg Force. *ISIJ International*. **43**:842-8.
[10] Zhen-Xing Y, Gong Y-Y, Li B, Cheng Y-F, Liang D, Zhai Q-J 2012 Refining of pure aluminum cast structure by surface pulsed magneto-oscillation. *Journal of Materials Processing Technology*. **212**:2629-34.
[11] Willers B, Eckert's, Nikrityuk P A, Räbiger D, Dong J, Eckert K, et al. 2008 Efficient Melt Stirring Using Pulse Sequences of a Rotating Magnetic Field: Part II. Application to Solidification of Al-Si Alloys. *Metallurgical And Materials Transactions B*. **39B**:304-16.
[12] Eckert S, Willers B, Michel U, Zouhar G, Nikrityuk PA, Eckert K. Directional Solidification Of PB-SN Alloys Affected By a Rotating Magnetic Field XXI ICTAM. Warsaw, Poland 2004.
[13] Wu XQ, Yang YS, Zhang JS, Jia GL, Hu ZQ 1999. Structure Characteristics in Industrially Centrifugally Cast 25Cr20Ni Stainless Steel Tubes Solidified under Different

Electromagnetic Field Intensity. *Journal of Materials Engineering and Performance*. **8**:525-30.

[14] Szajnar J, Stawarz M, Wróbel T, Sebzda W 2009. Influence of electromagnetic field on pure metals and alloys structure. *Journal of Achievements in Materials and Manufacturing Engineering*. **34**.

[15] Yoshikawa T, Morita K 2005. Refining of Si by the Solidification of Si-Al Melt with Electromagnetic Force. *ISIT International*. **45**:967-71.

[16] Asai S 2000. Recent development and prospect of electromagnetic processing of materials. *Science and Technology of Advanced Materials*. **1**:191-200.

[17] Meessen KJ, Paulides JJH, Lomonova EA 2013. Force Calculations in 3D Cylindrical Structures Using Fourier Analysis and the Maxwell Stress Tensor. *Magnetics IEEE Transactions* **49**:536 - 45.

Effect of electromagnetic vibration on the microstructure of direct chill cast Al-Zn-Mg-Cu alloy

Y Zuo[a], X Fu[b], Q Zhu[c], L Li, P Wang and J Cui[c]
Key Laboratory of Electromagnetic Processing of Materials, Ministry of Education, Northeastern University, Shenyang, Liaoning, 110819, China

Email: [a] zuoyubo@epm.neu.edu.cn, [b] fuxingwww@sina.com, [c] jzcui@epm.neu.edu.cn

Abstract. An electromagnetic vibration was achieved by the combined application of an alternating magnetic field and a stationary magnetic field during direct chill (DC) casting process. The ingots with 200 mm in diameter were prepared under the influence of electromagnetic vibration. The effect of electromagnetic vibration on the microstructure of an Al-Zn-Mg-Cu alloy was studied. The results showed that electromagnetic vibration has a significant effect on the solidification behaviour, under the influence of electromagnetic vibration during DC casting process, the microstructure is significantly refined and the uniformity of microstructure is evidently improved. This paper introduces the DC casting technology with the application of electromagnetic vibration, presents these results and gives corresponding discussions.

1. Introduction

Al-Zn-Mg-Cu alloys with ultra-high strength and good comprehensive performance are widely used in many industrial sectors including aerospace, defence, and sports equipment. DC casting process for aluminum alloys has been developed for about eighty years since it was invented in 1930s and is still the important method to produce ingot of high strength Al-Zn-Mg-Cu alloys. However, it is difficult to produce high quality ingots of Al-Zn-Mg-Cu alloys because casting defects such as hot tearing, cold crack, macrosegregation, porosity often occurs during casting process due to their high content of alloy elements, high strength and low ductility [1]. It is accepted that fine and uniform equiaxed grain structure of DC cast billet can improve mechanical properties, refine cast defects, decrease the sensitivity to hot tearing, result in a good surface finish, and enhance deformability and castability. Intensive development work on developing DC casting process has taken place over the last 30 years [2] to refine microstructure, reduce casting defects.

Grain refinement by grain refiners is an important and effective approach for refining microstructure [3, 4]. Although grain refinement of aluminium alloys has attracted intensive research the search for new and effective methods for grain refinement still continues. The application of external physical fields including intensive melt shearing [5] and electromagnetic field [6] during solidification process have attracted more intensive attention.

Fan and his co-workers [7] have found that intensive melt shearing provided by a twin-screw mechanism has a significant grain refining effect on both aluminium and magnesium alloys. Based on the similar principle, a new technology with simpler equipment (a rotor-stator high shear device) for intensive melt shearing was developed [8] and was used in the DC casting process [9].

The application of electromagnetic field in the DC casting process is also an effective method to control microstructure. Electromagnetic casting [10] developed in the early 1970s is perhaps the earliest application of electromagnetic field in DC casting of aluminium alloys. Refined microstructure and improved mechanical properties of 2024 aluminium alloy has been achieved by EMC process [11, 12]. Based on EMC, a CREM process was developed by Vives [13], which slightly modified the conventional DC casting mould and applied the industrial frequency alternating current to generate alternating magnetic field through a coil around the mould. Different electromagnetic fields such as high magnetic field [14, 15], rotating electromagnetic field [16, 17], travelling magnetic field [18] have been studied to refine microstructure.

Based on EMC and CREM, a low frequency electromagnetic casting (LFEC) process was developed recently by Cui and his colleagues [19, 20] in which the low frequency electromagnetic field was used to control the fluid flow and temperature field. Recently, based on the work of Vives [21], a low frequency electromagnetic vibration casting (LFEVC) was achieved by the combined application of an alternating magnetic field and a stationary magnetic field during direct chill (DC) casting process. The stationary magnetic field was achieved by a coil with application of direct current. The alternating magnetic field was achieved by a coil with application of alternating current.

In the present work, the electromagnetic vibration in the DC casting process was introduced and its effect on the microstructure of an Al-Zn-Mg-Cu alloy was studied.

2. Experimental procedure

Al-(9.5-10)Zn-(2.0-2.5)Mg-(2.0-2.3)Cu-(0.12-0.15)Zr (all in wt.%) aluminium alloy was used in the present work. The measured solidus and liquidus of the alloy are 470 °C and 632 °C, respectively. The pure aluminum was melted in a 500 kW medium frequency induction furnace. Pure Cu, Zn, Mg and Al-Zr master alloy were added to the melted aluminum at 750 °C. Then the alloy melt was transferred into a resistance furnace. The melt was degassed by C_2Cl_6 in the resistance furnace, and then was poured into DC casting mould through a launder. The experimental equipment for the LFEVC process is schematically illustrated in figure 1.

A 100 turn induction coil was arranged outside the hot-top to provide the stationary magnetic field. An 80 turn induction coil was arranged outside the mould to provide the alternating magnetic field. Billets in diameter of 200 mm were cast by the conventional DC casting and the LFEVC process, respectively, under following conditions: casting temperature (melt temperature in the furnace), 730-740 °C; casting speed, 80 mm/min; cooling water flow rate, 70 L/min.

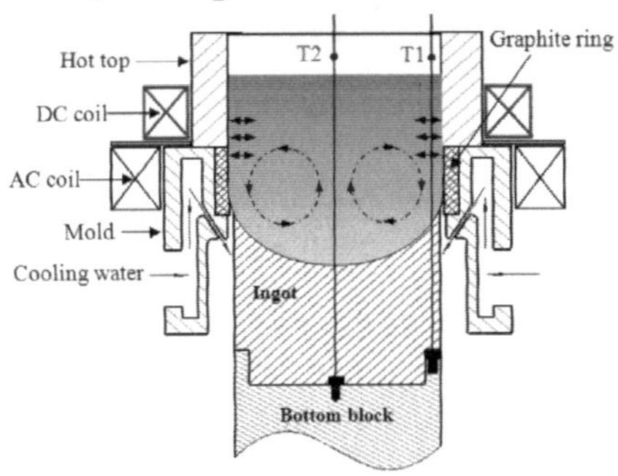

Figure 1. Schematic of LFEVC process.

During casting process, thermocouples (type K) were used to measure and record the temperatures, the positions of which were schematically shown in figure 1. The thermocouples T2 and T1 were fixed on a stainless steel rod which was connected to the bottom block. Thermocouples therefore can move with the billet to record the temperature variation from the liquid surface to the billet.

Samples for metallographic examinations were ground, polished and etched using a standard metallographic technique. Microstructures of these samples were analyzed under optical microscope (Leica DMR).

3. Results and discussions

3.1 Macrostructure

The ingots were prepared by the conventional DC casting process and LFEVC (DC current 100A, AC current 150A, 25Hz) process, respectively. The macrostructure of the ingot are shown in figure 2.

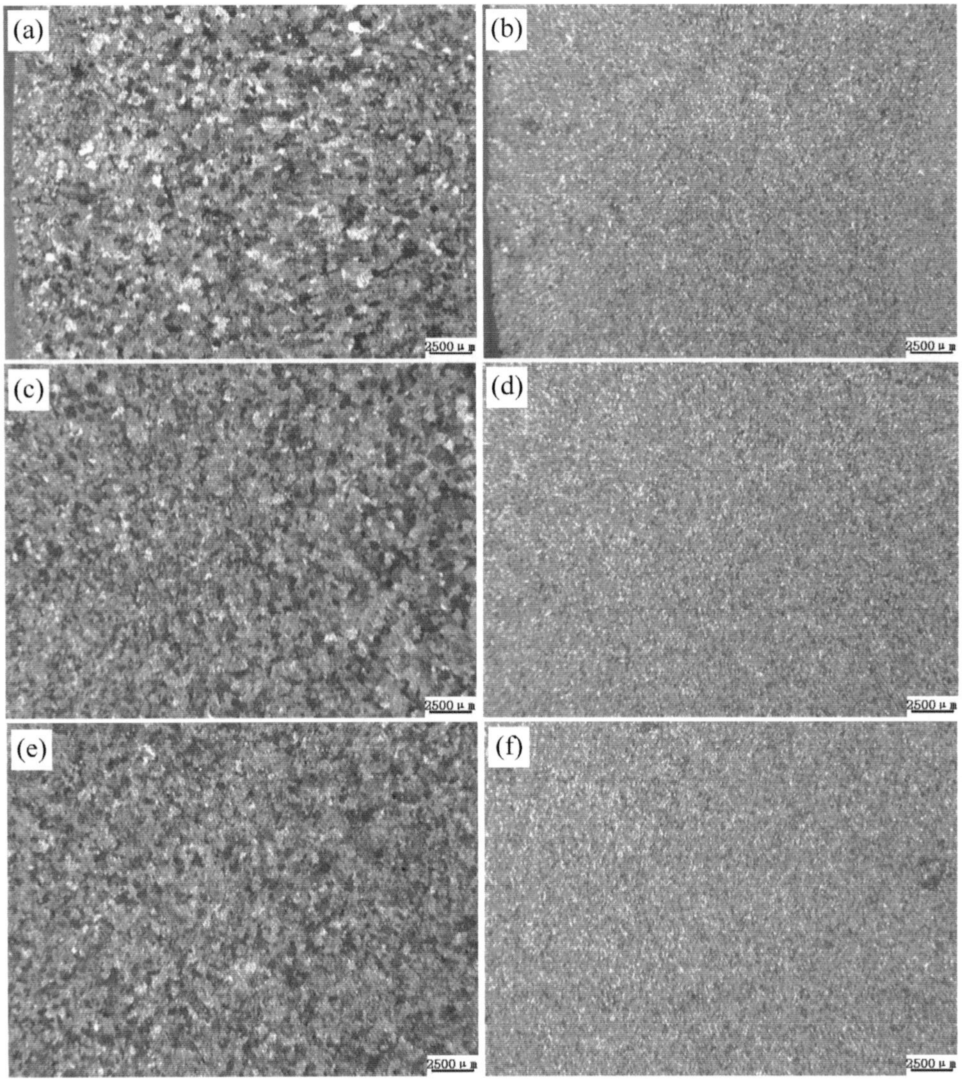

Figure 2. Macrostructure of the ingots prepared by DC and LFEVC (a), (b) border; (c), (d) 1/2 radius; (e), (f) centre; (a),(c),(e) DC; (b),(d),(f) LFEVC.

In general, the ingot prepared by the conventional DC casting process shows coarse macrostructure from the ingot surface to the centre of the ingot. As shown in figure 2a, there is a thin layer (about 2 mm) of fine macrostructure. Then the macrostructure shows a transition zone from fine macrostructure to coarse equiaxed structure then to coarse columnar structure and then to coarse equiaxed structure again. The ingot in both the 1/2 radius, figure 2c, and in the centre, figure 2e, shows a coarse equiaxed structure. The macrostructure of the LFEVC ingots is much finer than that of DC ingots. The macrostructure of the ingot from the surface, figure 2b, to the centre, figure 2f, shows fine and uniform equiaxed structure.

3.2 Microstructure

The microstructure of the LFEVC ingot is much finer than the coarse microstructure of DC ingot. Figure 3 a, c, e show dendritic microstructures of DC cast ingot, with the increase of the dendrite arm space from the surface to the centre. With the application of electromagnetic vibration, figure 3 b, d, f show fine and uniform equiaxed microstructure.

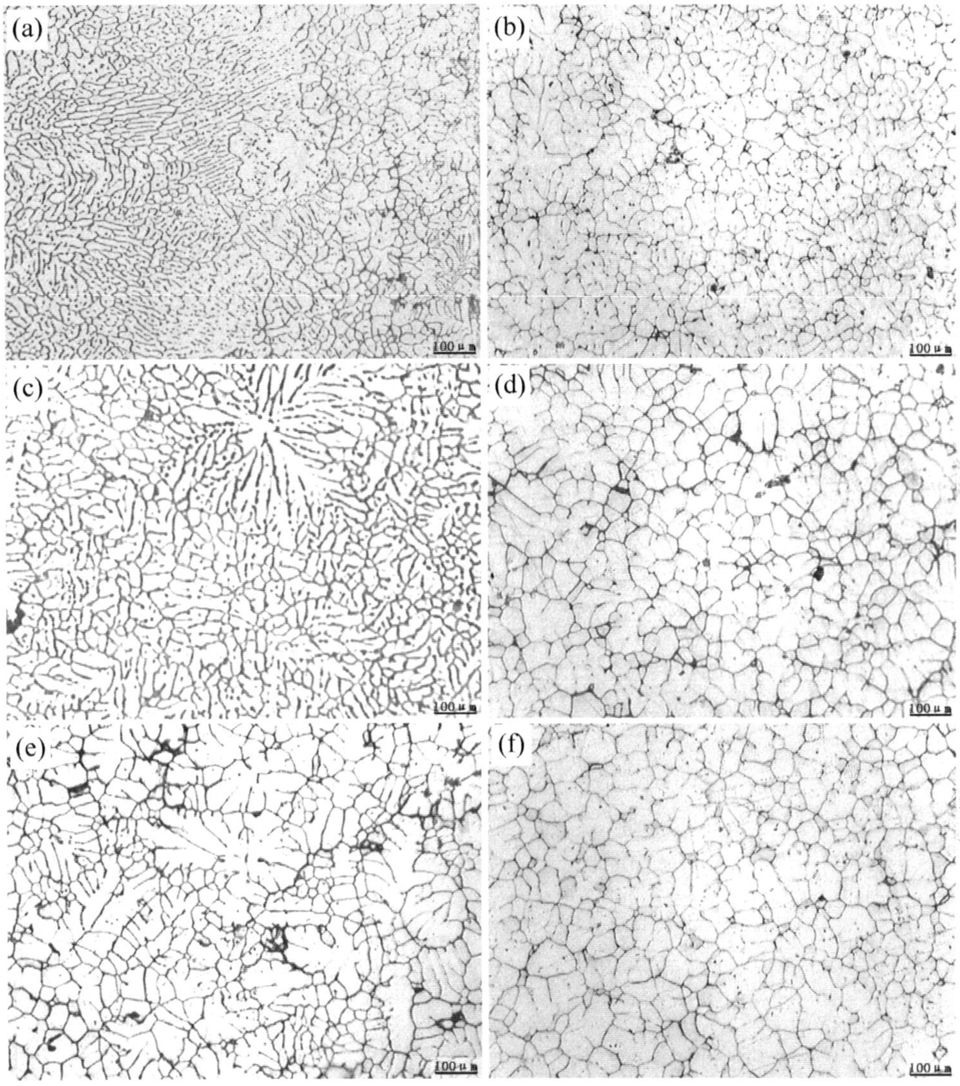

Figure 3. Microstructure of the ingots prepared by DC and LFEVC (a), (b) border; (c), (d) 1/2 radius; (e), (f) centre; (a),(c),(e) DC; (b),(d),(f) LFEVC.

3.3 Temperature curves

The application of electromagnetic vibration results in a significant grain refining effect on the Al-Zn-Mg-Cu alloy. In order to understand and analyze the mechanism of grain refinement, the temperature curves were measured during casting process. As schematically shown in figure 1, thermocouple 1 (T1) and thermocouple 2 (T2) were fixed on the stainless rods connected to the bottom block. Therefore, T1 and T2 can be used to record the thermal history from liquid to the solid and analyze the thermal gradient of the melt and/or ingot.

As shown in figure 4, during conventional DC casting process, the temperature near to the ingot surface is close to the liquidus of the alloy while in the centre it is much higher than the liquidus, indicating a radial thermal gradient. During LFEVC casting process, the measured temperature both in the centre and near to the surface are close to the liquidus. The temperature field during LFEVC casting process is more uniform than that of conventional DC casting process, making a significant contribution to the grain refinement.

During LFEVC casting process, an alternating magnetic field, B(t), and a stationary magnetic field, B_0, are generated in the melt by imposing an alternating current (frequency f) into AC induction coil and a direct current into the DC coil as shown in figure 1. The alternating magnetic field, B(t), gives rise to an induced current, J, in melt and/or ingot. The interaction of stationary magnetic field, B_0, and induced current J engenders a vibrating force: $F_1 = J \times B_0$ with a frequency of f, which is schematically shown in figure 1 by dual direction arrows. Therefore an electromagnetic vibration is generated in the melt and/or ingot, resulting in an alternation of high pressure and low pressure in a local region. The electromagnetic vibration could increase the nucleation on oxide particles, intermetallics or the wall of the mould.

The interaction of the alternating magnetic field, B(t), and its induced current, J, can also generate a Lorentz force, $F_2 = J \times B(t)$. The Lorentz force, F_2, is composed a potential force and a rotational force. The potential electromagnetic force is balanced by the stationary pressure of the melt, resulting in a decrease in the contacting pressure on the mold (soft contact). The rotational force can induce a forced convection in the melt, which is also schematically shown in figure 1 by circles and arrows under the liquid surface. The forced convection can improve the uniformity of temperature field in the melt as shown in figure 4 and promote the separation and movement of nuclei resulting in the fine and uniform microstructure, as shown in figure 2 and figure 3.

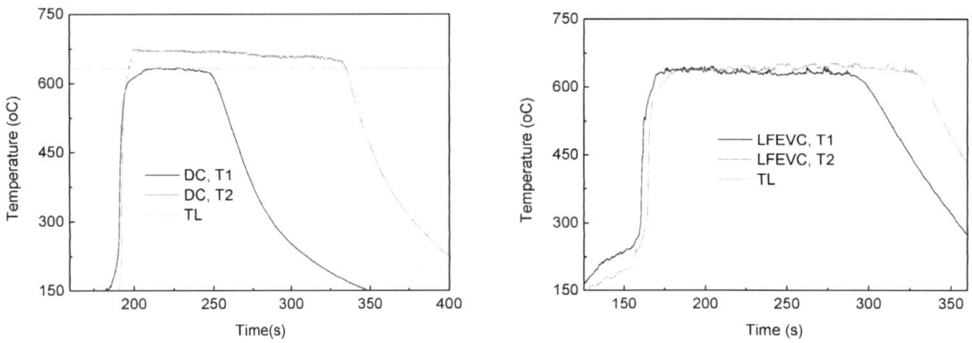

DC LFEVC

Figure 4. Measured temperature curves during conventional (left) DC casting and (right) LFEVC process.

3.4 Ingot surface

The ingot surface was observed to further understand the LFEVC casting process. As shown in figure 5a the conventional DC cast ingot shows a shining surface with cold shuts; while the ingot prepared by the LFEVC process shows smoother gray surface without shining. The gray colour of the surface could be due to the electromagnetic vibration and some graphite was adhered to the surface of the ingot.

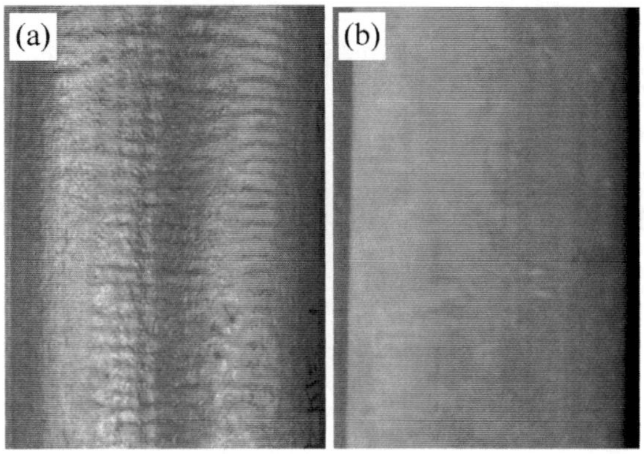

Figure 5. Surface of ingots prepared by DC and LFEVC (a) DC ingot; (b) LFEVC ingot.

4. Summary

An electromagnetic vibration was achieved by the combined application of an alternating magnetic field and a stationary magnetic field during direct chill (DC) casting process. The ingots of an Al-Zn-Mg-Cu alloy with 200 mm in diameter were prepared under the influence of electromagnetic vibration. The results showed that electromagnetic vibration has a significant grain refining effect on the Al-Zn-Mg-Cu alloy. Under the influence of electromagnetic vibration during DC casting process, the uniformity of microstructure is evidently improved.

Acknowledgement

This work is supported by the National Natural Science Foundation of China (51374067), the outstanding young scholars growth plan in the colleges and universities of Liaoning Province (LJQ2014032), Fundamental Research Funds for the Central Universities (N120409002) and the National Basic Research Program of China (2012CB619506).

References

[1] Cui J, Zhang H, Zuo Y 2013, *Materials Science Forum* **765** 165-169.
[2] Schneider W 2002, Light Metals 953-960.
[3] Eskin D G 2008, *Physical metallurgy of direct chill casting of aluminum alloys*, CRC Press, Boca Raton.
[4] Murty B S, Kori S A, Chakraborty M 2002, *International Materials Reviews* **47** 3-29.
[5] Li H T, Wang Y, Fan Z 2012, *Acta Materialia* **60** 1528-1537.
[6] Zuo Y, Cui J, Zhao Z, Zhang H, Li L, Zhu Q 2012 *Journal of Materials Science* **47** 5501-5508.
[7] Fan Z, Wang Y, Xia M, Arumuganathar S 2009 *Acta Materialia* **57** 4891-4901.
[8] Fan Z, Zuo Y B , Jiang B, 2010 Apparatus and method for liquid metals treatment, in: UK.
[9] Zuo Y B, Jiang B, Zhang Y, Fan Z 2011 Grain refinement of DC cast magnesium alloys with intensive melt shearing, in: 3rd International Conference on Advances in Solidification Processes, ICASP-3, June 7, 2011 - June 10, 2011, vol 27, *Institute of Physics Publishing*, Aachen, Netherlands.
[10] Getselev Z N 1971, *J. Met.* **10** 38-43.
[11] Cao Z, Jia F, Zhang X, Hao H, Jin J 2002, *Materials Science and Engineering A* **327** 133-137.
[12] Hao H, Zhang X G, Park J P, Kim H Y, Jin J Z 2003, *Journal of Materials Processing Technology* **142** 526-531.
[13] Vives C 1989 *Metallurgical and Materials Transactions B* **20** 623-629.
[14] Li M, Tamura T, Miwa K 2007, *Acta Materialia* **55** 4635-4643.
[15] Li X, Ren Z M, Fautrelle Y 2006, *Journal of Crystal Growth* **290** 571-575.

[16] Li X, Guo Z, Zhao X, Wei B, Chen F, Li T 2007, *Materials Science and Engineering: A* 460-461 648-651.

[17] Bai F, Sha M, Li T, Lu L 2011, *Journal of Alloys and Compounds* **509** 4835-4838.

[18] Metan V, Eigenfeld K, Rabiger D, Leonhardt M, Eckert S 2009, *Journal of Alloys and Compounds* **487** 163-172.

[19] Dong J, Zhao Z, Cui J, Yu F, Ban C 2004, *Metallurgical and Materials Transactions A* **35** 2487-2494.

[20] Zhang B, Cui J, Lu G 2003, *Materials Letters* **57** 1707-1711.

[21] Vives C 1996, *Journal of Crystal Growth* **158** 118-127.

IOP Conf. Series: Materials Science and Engineering **117** (2016) 012048 doi:10.1088/1757-899X/117/1/012048

Microstructure refinement of commercial 7xxx aluminium alloys solidified by the electromagnetic vibration technique

M Li [a], T Tamura, N Omura, Y Murakami and S Tada
The National Institute of Advanced Industrial Science and Technology (AIST), Materials Research Institute for Sustainable Development, Nagoya, Aichi 463-8560 Japan

E-mail: [a]li.mingjun@aist.go.jp

Abstract. This paper examines the microstructure refinement of commercial 7xxx aluminium alloys solidified by the electromagnetic vibration technique (EMV) as a function of vibration frequency, f. The microstructure evolution reveals that at the low frequency of $f = 62.5$ Hz, the solidified microstructure is coarse and with the increase of vibration frequency to $f = 500$ Hz, the grain size becomes the finest and further increase of frequency to $f = 2000$ Hz results in coarsening of microstructures. The refinement mechanism is clarified when considering the significant difference in electrical resistivities of the solid and the liquid in mushy zone, in which both phases coexist and subject to vibration. The frequency-dependent refinement behaviour is revealed when the displacement of the mobile solid and sluggish liquid is taken into account during solidification. In contrast to 3xxx aluminium alloys, no giant compounds have been discerned in the present 7xxx alloy regardless of the solidification condition. The formation of crystalline twin is briefly discussed when considering the vibration condition.

1. Introduction

The 7xxx alloy is one of the mostly widely used aluminium alloys in aerospace engineering. To improve the performance of the alloy during service, it is always desired to produce an ingot with grain-refined microstructures that are of great significance in obtaining enhanced mechanical properties according to the Hall-Petch relation. For the refinement of an aluminium alloy billet, the most widely used technique is to inoculate the melt using TiB_2 compound that has been confirmed to be quite effective for aluminium-based alloys [1]. The problem is that both titanium (Ti) and boron (B) are rare elements with an unstable supply market, which makes the billet expensive and less competitive. Moreover, the addition of the external inoculant may deteriorate deformation properties, in particular, for the deep drawing or press processing and in addition, can increase the cost of recycling process. Hence, it is desirable to develop a novel technique, by which no external additives are required.

The electromagnetic vibration (EMV) technique is based on the Fleming left hand rule, by which a conductor is driven to move periodically upon its initial equilibrium position when an alternating current passes perpendicularly to the direction of the magnetic field. The pioneering work by using the EMV technique to the solidification field may be dated back to Vives [2,3] who investigated a series of aluminium alloys and confirmed that the EMV was effective in grain refinement when EMV was imposed at a certain frequency range. After this great work, some other researchers [4-6] improved and modified the technique and examined the microstructure formation in various metals [7,8] and alloys [9,10]. The present authors [11-13] employed this technique to solidify a series of magnesium-based alloys and revealed that grain refinement could take place when EMV was imposed at the medium frequency interval.

In this paper, we examine the solidification of a commercial 7xxx aluminium alloy when EMV is added at various vibration frequencies. The solidified microstructure is qualitatively observed and quantitatively measured using an Image Pro software package. The refinement mechanism of the microstructure is clarified when considering the pronounced difference of electrical resistivity between the primary solid and the remaining liquid. The formation of crystal twins during solidification is briefly discussed upon the imposition of EMV processing.

2. Experimental procedures

The 7xxx billet was cast into a fine rod with 6 mm in diameter and then encapsulated into an Al_2O_3 tube. The rod was then stabilized by two graphite electrodes, from which two copper electrodes were connected. A removable arc-shaped carbon heater was covered onto the tube to heat the sample. To monitor the thermal profile during the heating and cooling process, a K-type thermocouple with a sampling rate of 10 scans per second was inserted into an as-prepared hole in the central position of the ceramic tube. More detailed procedures can be found elsewhere [14].

To diagnose the influence of vibration frequency, f, the other two parameters, i.e., magnetic flux density, B_0, and effective alternating current, J_e, were kept as constants of $B_0 = 10$ T and $J_e = 60$ A. In each EMV run, the experimental operation was conducted with the same procedure as follows; after the sample was melted, overheated, and then held at 1023 K (750 °C) for about 120 seconds, EMV was imposed and simultaneously, the heating power was turned off and the carbon heater was pulled up to cool the sample. When the sample temperature reached 773 K (500°C), the alternating current was switched off and thus EMV was terminated to complete the vibration process.

After the specimen was solidified, it was cut, ground, polished and then etched using the traditional metallographic method. The polarized optical microstructures were observed and characterized quantitatively in order to reveal the structure-frequency dependent relation.

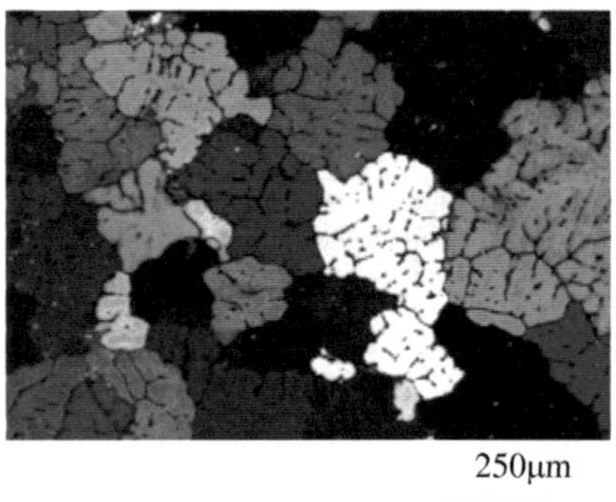

Figure 1. Polarized optical structure of a 7xxx alloy solidified at the casting state without EMV.

3. Results

In order to elucidate the influence of EMV on microstructure formation, the microstructure solidified under a usual casting state without EMV is included in figure 1, which consists of coarse dendrites with well-developed trunks (for easy description, hereafter $f = 0$ Hz). Here it should be supplemented that the sample without EMV is solidified at the same cooling rate as that with EMV, which ensures the microstructures to be comparable.

Figure 2 depicts the polarized optical microstructures of the sample solidified at various frequencies of (a) $f = 62.5$ Hz, (b) $f = 125$ Hz, (c) $f = 250$ Hz, (d) $f = 500$ Hz, (e) $f = 1000$ Hz and (f) $f = 2000$ Hz, respectively. Note that both figure 1 and figure 2 are observed at the same magnification. Qualitatively, one can see that in comparison with the structure formed at $f = 0$ Hz, the microstructure yielded at

f = 62.5 Hz is much fine. Overall, the size of equiaxed dendrites decreases and the number of fine grains increases. When the vibration frequency is further increased to f = 125 Hz and f = 250 Hz, the structures become finer and finer and more and more independent equiaxed grains can be observed. In contrast, there are only a few premature dendrites, as highlighted by a circle in figure 2(c). When the alloy is solidified at f = 500 Hz, the microstructure consists of uniform and equiaxed round grains, as shown in figure 2(d). With the increase of the vibration frequency to f = 1000 Hz, some undeveloped dendrites once more can be discerned, as circled in figure 2(e). As EMV is carried out at f = 2000 Hz, the microstructure becomes coarse with a number of developed dendrites, as depicted in figure 2(f). The general microstructure evolution versus the vibration frequency exhibits a V-shaped tendency with a minimum average grain size at medium frequency of f = 500 Hz and comparatively large grain sizes at both low and high frequency regimes.

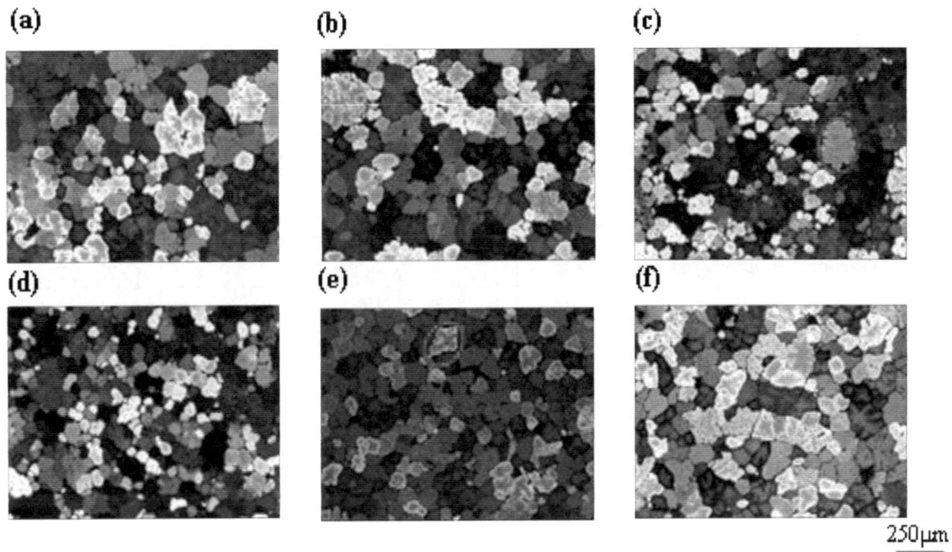

Figure 2. Polarized optical microstructures of the 7xxx aluminium alloy solidified at the vibration frequency of (a) f = 62.5 Hz, (b) f = 125 Hz, (c) f = 250 Hz, (d) f = 500 Hz, (e) f = 1000 Hz and (f) f = 2000 Hz, respectively.

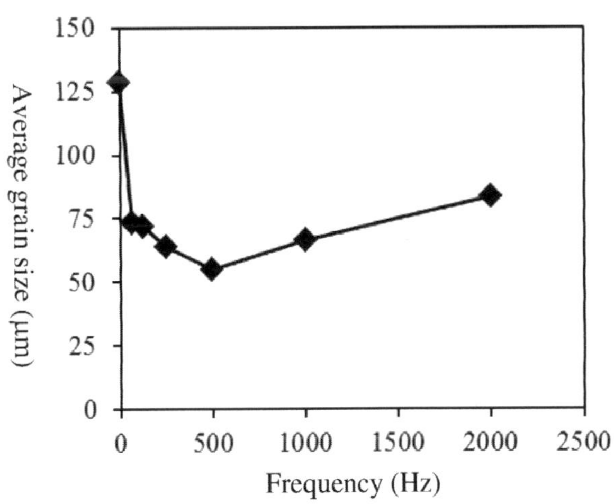

Figure 3. The measured average grain size of solidified alloys as a function of vibration frequency for the present 7xxx alloys.

To characterize the microstructure frequency-dependent relation quantitatively, the crystal grain size is measured using an Image Pro software package. Figure 3 indicates the average grain size as a function of vibration frequency, which exhibits a similar variation tendency as qualitatively stated previously. The average grain size is rather large of about 128 μm in diameter at f = 0 Hz and decreases drastically to about 73 μm at f = 62.5 Hz. When vibration frequency is increased from f = 125 Hz to f = 500 Hz, the average grain size continuously decreases from around 72 μm to the minimum of 54 μm. Further increase the frequency to f = 1000 Hz will result in an increase in average grain size to about 66 μm.

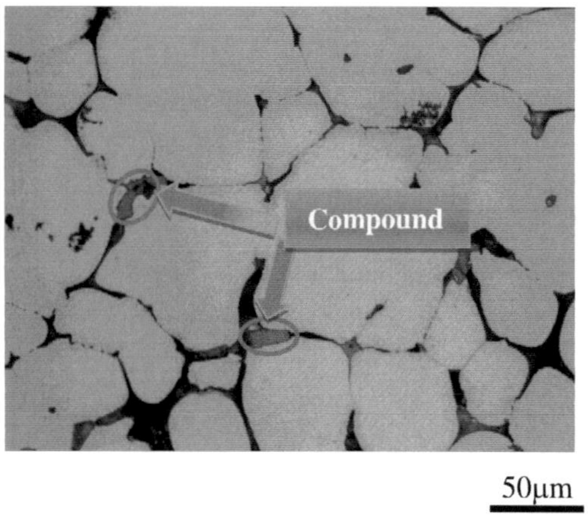

Figure 4. An optical microstructure showing the morphology of intermetallic compounds in the present 7xxx alloy

When the alloy is solidified at the present maximum frequency of f = 2000 Hz, the average grain size becomes as large as 83 μm. From the presentation above, it is clear that the optimum frequency for achieving grain-refined microstructures in the present 7xxx aluminium alloy is around 500 Hz when the alloy is solidified at B_0 = 10 T and J_e = 60 A. Actually, this kind of "V-shaped" transition behaviour has been confirmed in magnesium alloys [11-13].

In addition to the microstructure formation in the 7xxx aluminium alloy, whether or not the coarse intermetallic compounds can be yielded is another great concern in engineering, because the giant compounds are detrimental to make the alloy brittle during deformation processing. Yoshikawa et al. [15] summarized the conditions for the intermetallic compound formation upon crystallization in the 7075 alloy. We calculated the equivalent solute amount for compound formation of the present alloy and found that the total equivalent solute amount of the alloy is still less than the critical boundary for compound formation, indicating that it is hard to yield giant compounds. However, this does follow that no compound can be found in the alloy. On the contrary, intermetallic compounds can be identified at grain boundaries when the microstructure is observed under a large magnification. Figure 4 shows an optical microstructure with a high magnification. Very fine compound particles with the size of about tens micrometres can be identified, as highlighted in red ellipse in figure 4. In comparison with the considerable amount of Mn in 3xxx alloys, the entire volume of intermetallic compounds in the present 7xxx alloy is rather small and negligible. As the present authors [16] have examined the formation and refinement of compounds in 3xxx alloys in detail, here we omit to repeat the topic.

4. Discussion

From the experimental results presented above, one can readily raise two questions as follows; i.e., (1) why can EMV produce grain refined microstructures compared with that formed at the usual casting state and (2) why the microstructure evolution is frequency-dependent in the present 7xxx alloy, that is, why can the "V-shaped" transition behaviour be identified?

The first question deals with the refinement mechanism that has been a great concern for decades and one of the most striking mechanisms was that cavitation occurred during vibration and thus resulted in grain refinement [17]. Here it should be mentioned that by using a high speed digital imaging system, Shu et al. [18] have confirmed that cavitation could indeed take place in liquid at ultrasonic frequency band and can induce breakup of crystals from dendrite arms, which can lead to multiplication of grains and thus result in refinement. However, in terms of microstructure refinement using the EMV technique, the most effective frequency regime is usually at sonic frequency range of hundreds of hertz, e.g., in the present 7xxx alloy and other magnesium-based alloys [11-13], which are two orders of magnitude lower than the well-accepted definition for ultrasonic frequency, i.e., higher than 16000 Hz in physics. So it is doubtful whether or not it is suitable to transplant the phenomenon in ultrasonic band to sonic regime. Furthermore, there have been no reports that cavitation can really occur during sonic vibration.

With this discrepancy in mind, the present authors propose a new model when considering the electrical properties of the solid and liquid in mushy zone. The electrical resistivities of pure Al melt at the melting temperature is 242 nΩ·m, whereas the value is 110 nΩ·m [19] for the solid at the same temperature, i.e., only about half in quantity as that of liquid. For an Al alloy, the overall electrical resistivity of the liquid is estimated to be about twice of that solid. As the solid and the liquid in the mushy zone are to be a parallel circuit, the electric current in the solid, $J_e(\mathrm{s})$, is about twice that in the liquid, $J_e(\mathrm{l})$, according to the Ohm's law, i.e., $J_e(\mathrm{s}) \approx 2J_e(\mathrm{l})$. Hence, according to Newton's second law of motion and electromagnetic theory, the solid in the mushy zone is subject to twice Lorentz force of that liquid and thus the solid is driven to move with twice acceleration of that liquid due to their similar densities. Consequently, the solid moves faster with a higher velocity and thus results in a longer displacement in comparison with that of the liquid within half vibration cycle, which have been quantified elsewhere in detail [11].

Based on the analysis stated above, it is concluded that uncoupled motion occurs between the mobile solid and the surrounding sluggish liquid. Considering that the mushy zone is confined within an alumina tube rather than in free space, one can readily tell that melt flow can be initiated to take place during EMV processing, which has been demonstrated to be effective in promoting grain multiplication during solidification [20].

As far as the second question is concerned, we should consider the melt flow intensity that is closely related to vibration frequency. The non-synchronic motion between the primary mobile solid and the sluggish liquid produces the relative leading displacement, Δs, which decreases per a square relation with the increase of vibration frequency [11]. At $f = 50$ Hz, the relative displacement of the leading solid spans about 64 mm, which is much larger than the diameter with 6 mm for the Al_2O_3 tube. Under an extreme condition that a solid particle is driven to move across the diameter, the time duration for movement is about $\Delta t = 6/(60 \cdot 2f) = 0.001$ s within a half cycle, i.e., only 10% of the half cycle interval of 0.01 s. This indicates that within nearly 90% duration time of a half cycle, solid particle are impeded to stop around the tube wall and kept to be still as the sample is encased in the sealed container. This means that at this vibration frequency, the total moving time for solid is rather short while the dormant time is quite long. Therefore, only weak melt flow can be induced at this low frequency and thus microstructure is relatively coarse.

With the increase of vibration frequency, the relative displacement decreases. When f = 500 Hz, Δs is about 570 μm that is larger than the thickness of the boundary layer and thus it is sufficient to move out of the regime of the solute boundary layer. In this case, a stable solute distribution boundary cannot be established ahead of advancing front and hence, refinement can be effectively produced without dormant time. When frequency is further increased to f = 2000 Hz, Δs is decreased to about 35 μm that is much less than the thickness of the layer. In this case, although uncoupled movement occurs between solid and liquid, this short distance is still within the boundary regime and thus it cannot break out the hierarchy of solute distribution during solidification. This is why the microstructure once more becomes coarse. Considering the relation between the relative displacement and vibration frequency one can readily tell the origin of the "V-shaped" microstructure evolution during EMV processing.

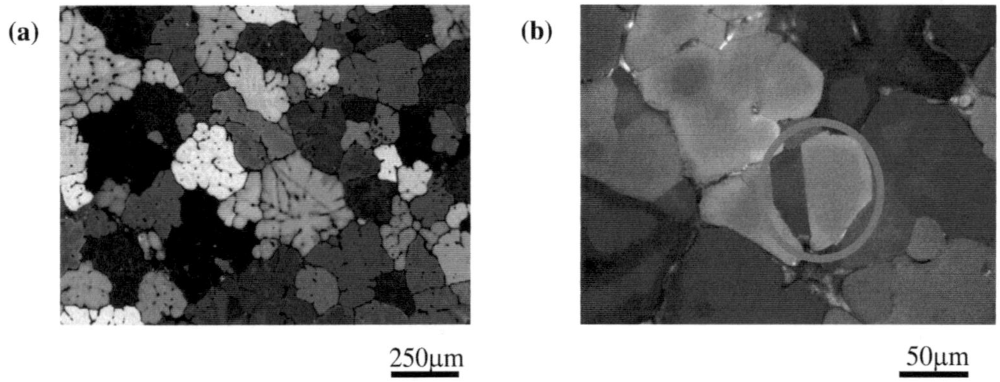

Figure 5. The microstructure solidified (a) at the usual casting state without EMV showing no crystal twins and (b) with EMV indicating deformation twins in the present face-centred cubic aluminium alloys due to the imposition of Lorentz force

The impact of Lorentz force, for one aspect, can produce refined structures during EMV processing, and for the other aspect, can result in the formation of crystal twins. Figure 5 depicts two microstructures solidified (a) without EMV at the usual casting state and (b) with EMV. For the aluminium alloy with a face-centred cubic structure, there are virtually no twins in the as-cast alloy [21], which can be verified in figure 5(a). When EMV is imposed, the Lorentz body force is added to the specimen. The Lorentz force is directional, while crystallographic orientation of grains is random, as the polarized microstructures reveal in figure 2. In this case, deformation twins may be produced when the Lorentz force is higher than the critical resolved shear stress for twin formation for those grains, in which their easy deformation crystallographic directions are consistent with that of the Lorentz force. This is why twins can be observed in certain grains, as highlighted in figure 5(b) in red circle while no twins can be discerned in other grains.

5. Conclusions

We examined the microstructure formation of the 7xxx aluminium alloy subject to EMV processing. The microstructure is coarse at the low vibration frequency and the average grain size decreases with the increase of frequency and eventually reaches the minimum of 54 μm at f = 500 Hz and then increases gradually when frequency is increased up to 2000 Hz in the present experiment. The refinement should be attributed to the uncoupled movement between the mobile solid and sluggish liquid because of a significant difference in electrical resistivity for the solid and liquid at high temperature. As the relative leading distance is strongly dependent on frequency, the "V-shaped" microstructure evolution behaviour can be achieved. No giant intermetallic compounds have been found regardless of imposition condition of vibration while, in terms of deformation twins, they can be yielded only in certain grains due to the directional feature of Lorentz force when EMV is imposed during solidification.

Acknowledgements
We appreciate the UACJ Corporation for supplying 7xxx ingots for the present experiment.

References

[1] Quested T and Greer A L 2004 *Acta mater.* **52** 3859
[2] Vives C 1989 *Metall. Mater. Trans. B* **20** 623

[3] Vives C 1996 *Metall. Mater. Trans. B* **27** 445
[4] Asai S 2004 *Modell. Simul. Mater. Sci. Eng.* **12** R1
[5] Dong J, Cui J, Yu F, Ban C and Zhao Z 2004 *Metall. Mater. Trans. A* **35** 2487
[6] Radjai A, Miwa K and Nishio T 1998 *Metall. Mater. Trans. A* **29** 1477
[7] Mizutani Y, Tamura T and Miwa K 2005 *Mater. Sci. Eng. A* **413-414** 205
[8] Mizutani Y, Miwa K, Tamura T, Nakai Y and Otsuka Y 2006 *Mater. Trans.* **47** 1793
[9] Radjai A and Miwa K 2001 *Metall. Mater. Trans. A* **33** 3025
[10] Mizutani Y, Kawata J, Miwa K, Yasue K, Tamura T and Sakaguchi Y 2004 *J. Mater. Res.* **19** 2997
[11] Li M, Tamura T and Miwa K 2007 *Acta mater.* **55** 4635
[12] Li M, Tamura T and Miwa K 2009 *J. Mater. Res.* **24** 145
[13] Li M, Tamura T, Omura N and Miwa K 2010 *J. Alloy Compd.* **494** 116
[14] Li M, Tamura T and Miwa K 2009 *Metall. Mater. Trans. A* **40** 1422
[15] Yoshikawa K, Sakamoto T, Furukawa T and Mitsuta Y 1995 *Light Metals* **45** 76 (in Japanese)
[16] Li M, Tamura T, Omura N, Murakami Y, Tada S, Miwa K and Takahashi K 2014 *J. Alloy Compd.* **610** 606
[17] Vives C 1998 *JOM-e* **50** http://www.tms.org/pubs/journals/JOM/9802/Vives/Vives-9802.html
[18] Shu D, Sun B, Mi J and Grant P S 2012 *Metall. Mater. Trans. A* **43** 3755
[19] Davis J R 1993 in ASM Specialty handbook aluminium and aluminium alloys, The Materials International Society, Materials Park OH, p. 641
[20] For a review, see Hellawell A, Liu S and Lu S Z 1997 *JOM* **49(3)** 18
[21] Cahn R W 1954 *Philos. Mag.* **3** 363

IOP Conf. Series: Materials Science and Engineering **117** (2016) 012050 doi:10.1088/1757-899X/117/1/012050

Grain Refinement of an Al-2 wt%Cu Alloy by Al3Ti1B Master Alloy and Ultrasonic Treatment

E Q Wang[1], G Wang[1,a], M S Dargusch[1], M Qian[2], D G Eskin[3] and D H StJohn[1]
[1]Centre of Advanced Materials Processing and Manufacturing (AMPAM), The University of Queensland, School of Mechanical and Mining Engineering, St Lucia, QLD 4072 Australia
[2]RMIT University, School of Aerospace, Mechanical and Manufacturing Engineering, Centre for Additive Manufacturing, GPO Box 2476, VIC 3083 Australia
[3]BCAST, Brunel University, Uxbridge UB3 PH, United Kingdom

Email: [a]gui.wang@uq.edu.au

Abstract. Both inoculation by AlTiB master alloys and Ultrasonic Treatment (UT) are effective methods of refining the grain size of aluminium alloys. The present study investigates the influence of UT on the grain refinement of an Al-2 wt% Cu alloy with a range of Al3Ti1B master alloy additions. When the alloy contains the smallest amount of added master alloy, UT caused significant additional grain refinement compared with that provided by the master alloy only. However, the influence of UT on grain size reduces with increasing addition of the master alloy. Plotting the grain size data versus the inverse of the growth restriction factor (Q) reveals that the application of UT causes both an increase in the number of potentially active nuclei and a decrease in the size of the nucleation free zone due to a reduction in the temperature gradient throughout the melt. Both these factors promote the formation of a fine equiaxed grain structure.

1. Introduction

Microstructural grain refinement of cast metallic alloys improves the mechanical properties of as-cast components and subsequent thermo-mechanical processing behaviour of semi-fabricated products such as billet and slab materials [1, 2]. The addition of master alloys containing inoculant particles is standard grain-refining practice for most commercial cast aluminium (Al) alloys where the most common grain refiners are based on Al-Ti-B and Al-Ti-C master alloys [3-6]. Alternatively, Ultrasonic Treatment (UT) has proved to be effective for the grain refinement of various Al alloys [7-12]. The effect of UT on microstructural evolution is derived from its physical influence on the molten melt, such as cavitation, acoustic streaming and radiation pressure [13]. Although much research has been undertaken over many decades, the mechanism of ultrasonic grain refinement is still a controversial issue. The current proposed mechanisms are mainly associated with the role of cavitation, including cavitation-induced dendrite fragmentation and cavitation-assisted heterogeneous nucleation [7-9, 13]. The present research investigates the combined effect of Al3Ti1B master alloy and UT on the grain refinement of an Al-2 wt% Cu alloy. The grain refining efficiency is evaluated for a range of master alloy additions with and without the application of UT.

2. Experimental procedure

The Al-2 wt% Cu alloy was prepared from commercial purity Al ingot (99.7%) and pure copper (Cu) (99.9%) using an electric furnace in a 4 kg batch. The equilibrium liquidus and solidus temperatures of the alloy as calculated using ThermoCalc software are 655 and 620°C, respectively. The ultrasonic device consists of a 2-kW commercial ultrasound generator, an air cooled 20-kHz piezoelectric transducer and a sonotrode made of a molybdenum alloy with an 18-mm diameter tip. About 1 kg of the alloy was melted and preheated to 720±3°C inside a graphite-clay crucible with 90 mm top diameter, 60 mm bottom diameter and 120 mm in height. For samples without the addition of master alloy, the crucible containing the molten alloy was removed after reaching 720±3°C and transferred to the UT platform [14]. For samples with the addition of master alloy, the equivalent amounts of Al3Ti1B master alloy (in wt%) containing 50, 100, 500, 1000 and 2000 ppm Ti were introduced into the molten melt at 720±3°C, which then was mechanically stirred and held for 5 minutes before being transferred to the UT platform. The UT sonotrode was switched-on without preheating and then immersed 15 mm below the top surface of the melt. Two K-type thermocouples were inserted into the melt beside the sonotrode: one adjacent to the wall of the crucible and the other being placed 12.5 mm apart. Both thermocouples were positioned 45 mm above the bottom of the crucible. The temperature data was collected by a data-acquisition system with a sampling rate of four readings per second. The cooling rates from the measurement were 0.25±0.01 °C/sec without UT and 0.54±0.01 °C/sec with UT. The sonotrode was not inserted into the melt for the experiments without UT.

The UT experiments were conducted with a fixed power input of 1 kW at an amplitude of 20 μm from 695 °C which is 40 °C above the liquidus temperature, for 4 minutes during subsequent solidification and ceasing at 653°C. A previous investigation [14] has revealed that an appropriate amount of superheat of the liquid metal and continued application of UT for a period of time below the liquidus temperature for ultrasonic grain refinement are essential when an unpreheated sonotrode is introduced into the melt. Superheat of 40°C above the liquidus temperature can insure that there is no formation of a strong solidified layer on the sonotrode. This solidified layer could dampen the UT effect in the surrounding liquid which may result in a significant reduction in cavitation and acoustic streaming thus preventing enhanced nucleation and rapid grain transport while continued application of UT for a period of time below the liquidus temperature promotes nucleation therefore refinement can be achieved.

Metallographic samples were sectioned along the centre symmetrical axis and mechanically ground and polished for observation. In order to measure the grain size, small samples were cut at 45 mm (the same height as the thermocouples and 25 mm below the sonotrode) from the bottom of the sectioned piece. Micrographs were obtained by a Leica Polyvar microscope with polarized light after anodizing using a 0.5% HBF_4 solution for about 20 seconds at 30 V. The grain sizes were measured using the linear intercept method (ASTM E112-10). The value of the growth restriction factor, Q, was calculated using the Al-Cu-Ti ternary phase diagram generated by ThermoCalc.

3. Results

3.1 Ultrasonic refinement of Al-2Cu alloy

Solidification of the Al-2 wt% Cu alloy without UT and without the addition of master alloy produces a coarse dendritic macrostructure (figure 1a). Figure 2a shows the typical equiaxed dendritic grain structure obtained from the centre of the ingot. The average dendritic grain size is about 1500 μm.

With UT applied from 40°C above the liquidus temperature of the Al-2 wt% Cu alloy for 4 minutes during subsequent cooling, the macrostructure presents a uniform fine grained structure througout the whole ingot (figure 1b). The anodized microstructure of samples from the central part is shown in figure 2b, and the average grain size is about 150 μm, indicating that significant grain refinement results from the application of UT.

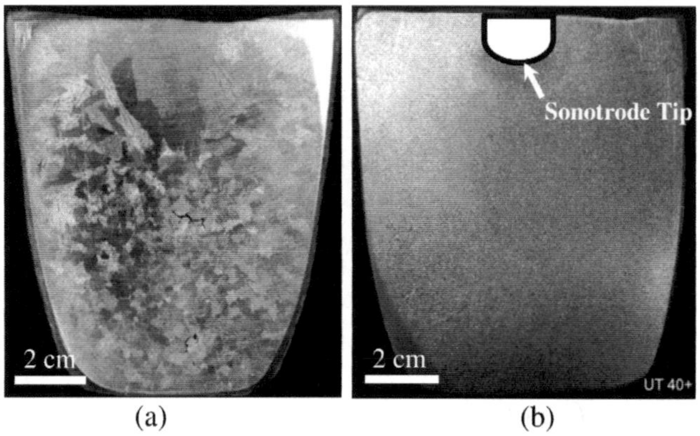

Figure 1. The macrostructure of Al-2% Cu alloy ingot samples (a) without UT and (b) with UT applied at 40 °C above the liquidus temperature for 4 minutes.

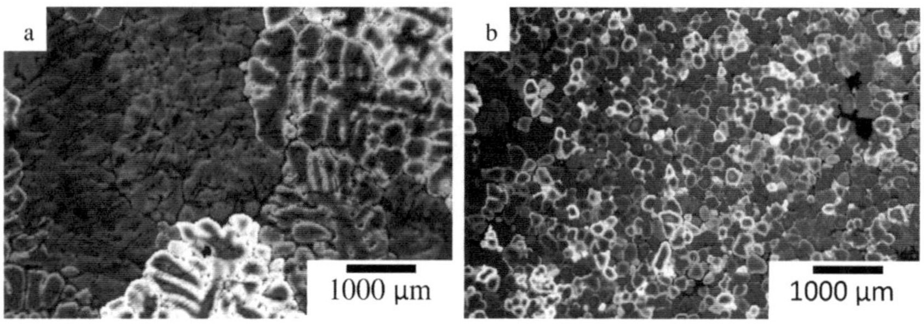

Figure 2. The microstructure of anodized samples from the central part of the Al 2Cu alloy ingot samples: (a) without UT and (b) with UT.

3.2 The influence of Al3Ti1B master alloy on grain refinement

Figures 3 a, b and c show the effect of adding (a) 50 ppm Ti, (b) 100 ppm Ti, (c) 1000 ppm Ti on the microstructure near the centre of the ingot. Figure 4a plots the measured grain sizes versus Ti content up to 2000 ppm Ti where the grain size decreases from about 230 microns to 125 microns. The diminishing effect of further high levels of master alloy is typically observed as there is a cube root relationship between the number of particles nucleating grains and the grain size [15].

3.3 The influence of UT on the refinement of Al-2 wt% Cu alloy with additions of master alloy

Figures 3 d, e and f show the microstructures of the Al-2Cu alloy ingot samples with (d) 50 ppm Ti, (e) 100 ppm Ti and (f) 1000 ppm Ti after the application of UT. Figure 4a shows that the grain size is consistently smaller than when UT is not applied and ranges from about 130 microns to just under 100 microns.

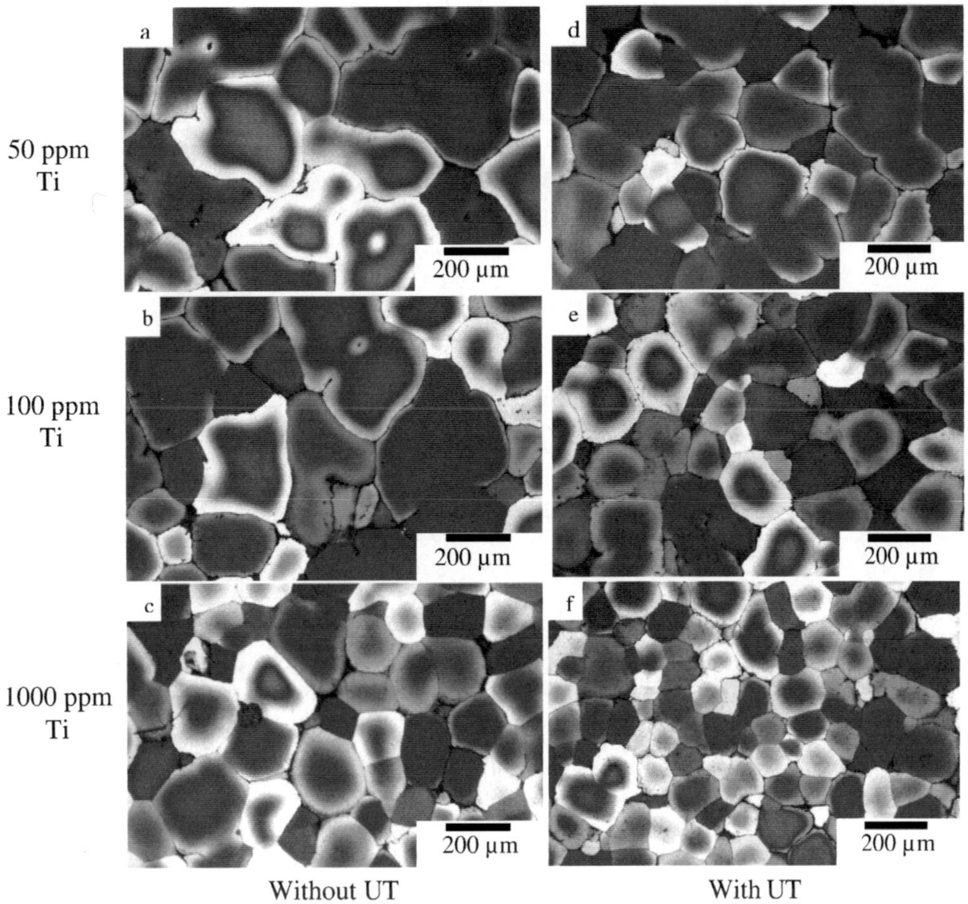

Figure 3. The microstructure of Al-2Cu alloy ingot samples with equivalent amounts of 50 ppm Ti, 100 ppm Ti and 1000 ppm Ti where (a), (b) and (c) are without UT and (d), (e) and (f) are with UT.

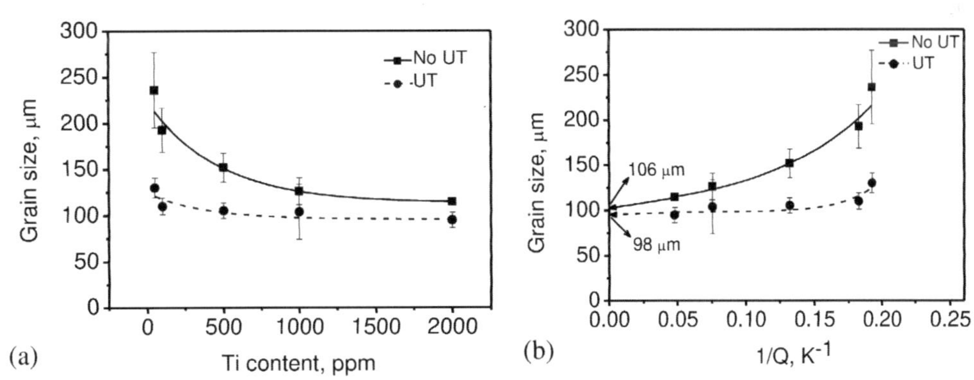

Figure 4. (a) The relationship between grain size and Ti content with and without UT application. (b) The relationship between grain size and inverse growth restriction factor Q.

4. Discussion

The average grain size of ingots with and without the application of UT is plotted against the equivalent content of Ti in figure 4a. There is a significant decrease in grain size from 50 to 500 ppm Ti content in both master alloy addition alone and the combined master alloy and UT cases, and the grain size decreases slowly with further increases in the Ti content. However, while plotting grain size against composition shows a diminishing effect, the number density of active particles may in fact be increasing at the same rate in both cases. Figure 5 converts grain size into grain density [16] which is directly related to the number of active nucleant particles. It should be noted that the rate of increase in grain density diminishes as the Ti content increases. This curvature may be due to the fact that the increasing amount of Ti solute added by the master alloy is increasing the growth restriction factor, Q, which has a greater effect on grain size reduction at low levels of addition [15].

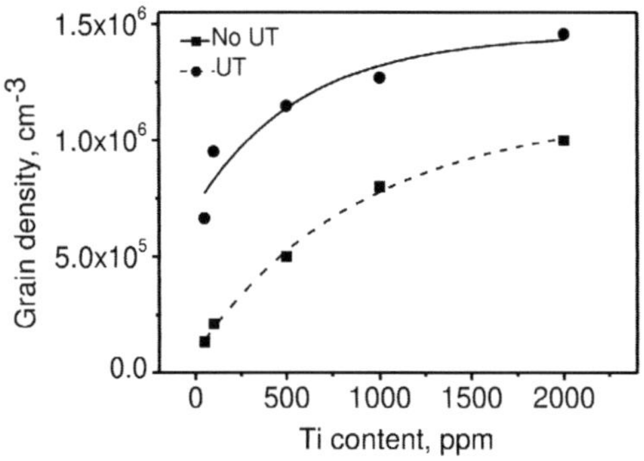

Figure 5. The relationship between grain density and Ti content with and without the application of UT.

Other effects of UT could be that cavitation and acoustic streaming promote a finer distribution of TiB_2 particles and/or deagglomeration of TiB_2 clusters creating more sites for nucleation [16]. In addition, the collapse of cavitation bubbles may enhance the wetting of particles which may improve their efficiency for grain refinement [7].

By calculating Q for the alloy composition (2 wt% Cu plus the amount of Ti solute), the relationship between grain size and Q with and without UT is presented in figure 4b. Note that the curves of best fit are not straight lines as observed when the particle number density is kept constant [15]. Curves are observed when both the particle number density and Q increase simultaneously [17]. Figure 4b clearly shows that UT decreases the gradient of the curve of best fit which may indicate that the nucleant potency of the refining particles has increased. Atamanenko et al. [9] found that UT alters the potency of the nucleant particles in Al-Zr–Ti alloy. However, in this case the effect of UT is likely to be related to the refinement of the particles. In the present study the nucleant particles are only TiB_2, and though the number of active particle may change with UT due to the deagglomeration and better dispersion the change in slope may be due to other factors as well. The gradient of the curves in figure 4b can be significantly influenced by change in the size of the Nucleation-Free Zone (x_{NFZ}) surrounding each nucleated grain with change in the value of Q [17] according to [5]

$$x_{NFZ} = 5.6\left(\frac{D \cdot z\Delta T_n}{vQ}\right) \tag{1}$$

where D is the diffusion coefficient, v is the rate of growth of the solid-liquid (S-L) interface, and $z\Delta T_n$ is the incremental amount of undercooling required to activate the next nucleation event as the temperature gradient moves towards the thermal centre of the casting. x_{NFZ} contributes to the final grain size. The other factor contributing to the grain size is the distance between the most potent particles in the melt which is denoted x_{Sd} [18]. The sum of x_{NFZ} and x_{Sd} equals the grain size.

The non-linear relationship shown in figure 4b derives mainly from the non-liner dependency of x_{Sd} on $1/Q$ in this study.

Therefore, a change in any of the parameters in equation (1) will change the gradient. The parameter z in equation (1) represents the effect of temperature gradient on the size of x_{NFZ}. Because the temperature gradient decreases due to UT treatment [14] the value of z decreases thus reducing x_{NFZ} and, therefore, the gradient of the UT curve in figure 4b. The other point to note is that the intercept with the y-axis in figure 4b which defines x_{Sd} for an infinite number of active particles, decreases when UT is applied. By extrapolating the curves to the y-axis a value of 106 is obtained without UT and with UT the intercept is 98. Although this appears to be a small difference, it represents an increase in grain density of about 20% implying the number of activatable particles has increased by a similar amount with the application of UT. In addition, for a given composition a reduction in x_{NFZ} also enables more TiB_2 particles to nucleate grains [18].

5. Summary

The application of UT to an Al-2 wt% Cu alloy without master alloy additions resulted in significant grain refinement when applied from 40 °C above the liquidus temperature for 4 minutes during subsequent cooling. The separate application of UT and the addition of Al3Ti1B master alloy to an Al-2 wt% Cu alloy both individually produced considerable grain refinement. Further refinement was achieved by applying both methods together. The relationship between grain size and the inverse growth restriction factor indicates that UT changes the nucleation efficiency of the master alloy by lowering the temperature gradient before and during the nucleation stage in the melt which decreases the size of the nucleation-free zone, and by increasing the proportion of activatable TiB_2 particles along with the addition of the UT induced nucleation by about 20% through the nucleation stage. Both of these factors lead to a reduction in grain size.

Acknowledgements

Mr. Matthew Travers and Mr. Jayden Smith are acknowledged for their help during sample preparation. The authors gratefully acknowledge support from the ARC Centre of Excellence for Design in Light Metals, ARC Discovery Project DP140100702, and the ExoMet Project co-funded by the European Commission's 7th Framework Programme (contract FP7-NMP3-LA-2012-280421), by the European Space Agency and by the individual partner organisations. We also thank AMG Aluminium for the provision of Al3Ti1B master alloy.

References

[1] Wang F, Qiu D, Liu Z-L, Taylor J A, Easton M A, Zhang M-X, *Acta Materialia*, **61** (2013) 5636-5645.

[2] Chandrashekar T, Muralidhara M K, Kashyap K T, Rao P R, *Int J Adv Manuf Technol*, **40** (2009) 234-241.

[3] Greer A L, Cooper P S, Meredith M W, Schneider W, Schumacher P, Spittle J A, Tronche A, *Advanced Engineering Materials*, **5** (2003) 81-91.

[4] Qian M, Cao P, Easton M A, McDonald S D, StJohn D H, *Acta Materialia*, **58** (2010) 3262-3270.

[5] StJohn D H, Easton M A, Qian M, Taylor J A, *Metall and Mat Trans A*, **44** (2013) 2935-2949.

[6] Stjohn D H, Cao P, Qian M, Easton M A, *Advanced Engineering Materials*, **9** (2007) 739-746.

[7] Eskin G I, *Ultrasonic Treatment of Light Alloy Melts*, Taylor & Francis, 1998.

[8] Abramov O V, *Ultrasound in liquid and solid metals*, CRC Press, 1994.

[9] Atamanenko T V, Eskin D G, Zhang L, Katgerman L, *Metall and Mat Trans A*, 41 (2010) 2056-2066.

[10] Han Y, Li K, Wang J, Shu D, Sun B, *Materials Science and Engineering: A*, **405** (2005) 306-312.

[11] Jian X, Meek T T, Han Q, *Scripta Materialia*, **54** (2006) 893-896.

[12] Jian X, Xu H, Meek T T, Han Q, *Materials Letters*, **59** (2005) 190-193.
[13] Abramov O V, *Ultrasonics*, **25** (1987) 73-82.
[14] Wang G, Dargusch Ms, Qian M, Eskin D G, StJohn D H, *Journal of Crystal Growth*, **408** (2014) 119-124.
[15] Easton M, StJohn D, *Metall and Mat Trans A*, **36** (2005) 1911-1920.
[16] Underwood E, Quantitative Stereology for Microstructural Analysis, in: J. McCall, W. Mueller (Eds.) *Microstructural Analysis*, Springer US, 1973, pp. 35-66.
[17] Easton M A, StJohn D H, Prasad A, Grain Refinement of Aluminium Alloys: Recent Developments in Predicting the As-Cast Grain Size of Alloys Refined by Al-Ti-B Master Alloys, in: *Light Metals 2014*, John Wiley & Sons, Inc., 2014, pp. 937-944.
[18] StJohn D H, Qian M, Easton M A, Cao P, *Acta Materialia*, **59** (2011) 4907-4921.

IOP Conf. Series: Materials Science and Engineering **117** (2016) 012051 doi:10.1088/1757-899X/117/1/012051

Convection and solidification influenced by thermo-electric effect

J Wang[1,2], G S Abou-Jaoude[3,4], O Budenkova[1], G Reinhart[3,4], N Mangelinck[3,4], X Li[2], H Nguyen-Thi[3,4], Z-M Ren[2], and Y Fautrelle[1,a]
[1] SIMAP/EPM/Grenoble Institute of Technology/CNRS, BP 75, 38402 Saint Martin d'Hères, France
[2]Shanghai University, 149 YanChang Road, Shanghai 200072, P.R. China
[3]Aix-Marseille Université
[4]CNRS, IM2NP, UMR 7334 Campus Saint-Jérôme, Case 142, 13397 Marseille Cedex 20, France

Email : [a]yves.fautrelle@simap.grenoble-inp.fr

Abstract. We analyse the thermo-electric-magnetic (TEM) forces both on the liquid metal and immersed solid particles which mimics equiaxed grains. Firstly, we provide a summary of analytical works performed in a previous paper. Then, we present experimental works on in situ and real time visualization of the directional solidification of Al-10wt%Cu alloy by means of synchrotron X-ray radiography during directional solidification under a transverse 0.08 T permanent magnetic field. The experiments bring definite evidence of the existence of various phenomena which could be related to the thermoelectric effect acting directly on equiaxed grains. Comparisons between the predicted and observed horizontal deviations of the particles are in fairly good agreement.

1. Introduction

Convection induced by magnetic field is now widely used in metallurgy to master liquid metal flows during solidification and thus the final properties of the materials. Usually, AC magnetic fields are commonly used to generate fluid flows whereas DC fields are supposed to create an anisotropic damping of liquid motions [1-3]. However, the presence of temperature gradient may lead to the generation of thermo-electric currents both in the liquid and in the solid phases near the solidification interface. The interaction of non-parallel thermo-electric current and the applied magnetic field creates electromagnetic forces which act both on the liquid and on the solid [4]. In previous works, experimental and numerical analyses of thermo-electric magnetic convection (TEMC) for different length scales under various magnetic fields up to 10T have been investigated, and it has been found that besides TEMC in the melt thermoelectric magnetic (TEM) force on the solid was important as well [5-8]. In general, all materials during solidification have thermal gradient at the interface and different material properties between the liquid and the solid, including thermal and electric conductivities as well as absolute thermo-electric powers. The existence of temperature gradient having a non-zero component along the interface (e.g. the liquid-solid front) may cause the appearance of internal electric current along the interface in accordance with the well-known Seebeck effect, associated with the Thomson effect. Thermo-electric effect may be encountered in various situations as in solidification processes of electrically conducting materials. When static magnetic fields are superimposed there appears Lorentz forces (TEM forces) which may be responsible for thermo-electric hydromagnetic flows, the so-called TEMHD [4]. Various types of effects may be expected according to the magnetic field amplitude and direction as well as the physical properties of both the liquid and solid. The existence of electrical currents depends on the thermo-electric power difference $S_l - S_s$ [4, 8] between the liquid and solid. Furthermore, the TEM forces are able to drive a liquid motion only if $\vec{\nabla} S_l \times \vec{\nabla} T \neq 0$ [4].

In the present paper we focus our attention on the TEM forces and motions on equiaxed grains in a solidifying Al-10wt.%Cu alloy. Firstly, we summarize the theoretical results obtained in a previous works [9]. Then, we present experimental results related to in situ and real time observations of the equiaxed grain motion during directional solidification of Al – 10 wt.% Cu under static magnetic field which were carried out by means of synchrotron X-ray radiography. The latter work was published in [10].

2. Theory of the motion of a spherical particle under thermo-electric effect

We consider now an isolated solid spherical particle immersed in a liquid metal submitted both to a constant vertical temperature gradient and a static magnetic field. The geometry is illustrated in figure 1. Details of the theoretical analysis may be found in [9]. We only provide here a summary of the results. The thermo-electric current has a dipole-type distribution [9]. The interaction of the TEM electric current density $\vec{j}_i$ and a superimposed static magnetic field $\vec{B}_0$ produces a TEM body force $\vec{F}_i$. In the case of a transverse magnetic field, the orientation of the TEM force acting on the sphere is illustrated in figure 1. The force is perpendicular both to the temperature gradient and the magnetic field. The analytical expression of the body force is $\vec{F}_i = \vec{j}_i \times \vec{B}_0$, $i = l$ or s respectively in liquid and in solid. In spherical coordinates (r, θ, ϕ) the expressions of the electric current both in the liquid and the solid are [9]

$$\vec{j}_s = -\frac{2\sigma_s \sigma_l}{(\sigma_s + 2\sigma_l)}(S_s - S_l) G \vec{i}_z \quad (1)$$

$$\vec{j}_l = \frac{\sigma_s \sigma_l}{(\sigma_s + 2\sigma_l)}(S_s - S_l) G R^3 \left(\frac{2}{r^3}\cos\theta\, \vec{i}_r + \frac{1}{r^3}\sin\theta\, \vec{i}_\theta\right) \quad (2)$$

The symbols R, G, σ_i, S_i respectively denote the sphere radius, the applied constant temperature gradient, the electrical conductivity and the absolute thermo-electric power of each phases.

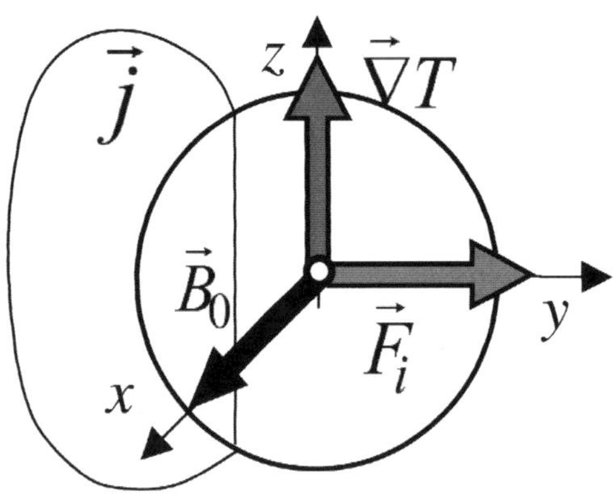

Figure 1. Sketch of the geometry and illustration of the thermo-electric current distribution both in the solid and in the liquid as well as the corresponding TEM force under a transverse magnetic field.

Note from equations (1) and (2) that the TEM force acts both on the liquid and the solid. The TEM force on the solid tends to generate a deviation of the particle along the y-direction in the present geometrical conditions. Neglecting the influence of the surrounding flow, an estimate of **the particle**

TEM velocity U may be obtained from the balance between the TEM force on the solid and the Stokes drag. We then obtained the following expression:

$$U = \frac{2R^2}{9\mu} F_s , \qquad (3)$$

μ denoting the dynamic viscosity of the fluid.

The latter phenomenon will be confirmed by the in-situ experiments shown in the next paragraph. The TEM force also generates fluid flows around the particle. Figure 2 provide an example of computed flow obtained by introducing the analytical force expressions (1) and (2) in ANSYS/FLUENT.

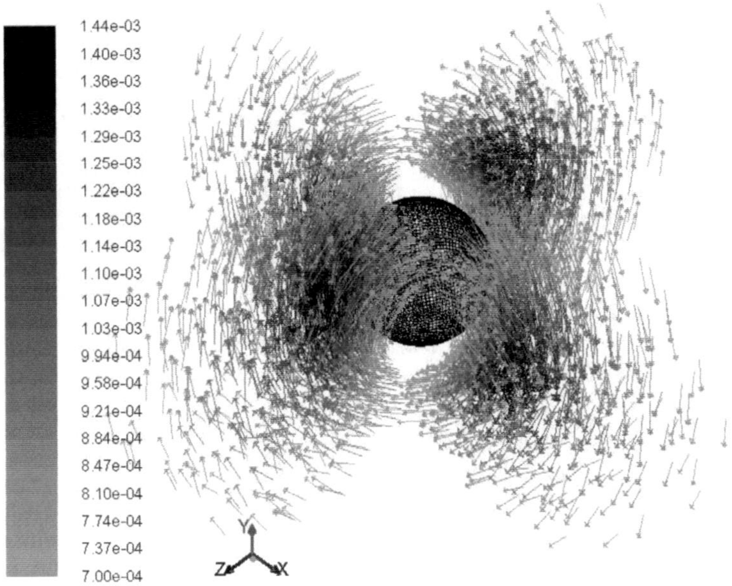

Figure 2. View of the flow pattern around the spherical particle in the case of the transverse magnetic field. The parameter values are $G = 3000$ K/m, sphere radius $R = 0.05$ mm, $B = 0.08$ T. The velocity scales is in m/s. $U_{max} = 0.00143$ m/s.

3. In situ observations of the thermo-electric effect on equiaxed grains

In situ experiments were carried out at the European Synchrotron Radiation Facility (ESRF, Grenoble, France). Directional solidification experiments were performed inside a Bridgman furnace, which has been described in detail previously [11]. The thermal gradient G (in the z-direction) was imposed by two separated heating elements and the solidification was controlled by applying the same cooling rate CR on the two heating elements. The temperature adjustment was monitored using two embedded K-type thermocouples, one at the top and one at the bottom on either side of the sample, 2 cm away from each other. A series of solidification experiments were carried out with a same cooling rate $CR = 2$ K/mn, the same magnetic field value, and various temperature gradients G decreasing from 20 K/cm to nearly zero. The thin samples studied were 35 mm in length, 5 mm to 6 mm in width and 150 µm to 200 µm in thickness. To minimize grain sinking phenomena, solidifications were conducted on Al – 10 wt.% Cu alloy for which the grain density is weakly larger than the surrounding liquid [12]. The static magnetic field was generated by a cubic neodymium magnet, which was fixed close to the Bridgman furnace and imposed a permanent magnetic field B_0 equal to 0.08 Tesla. The direction of the field was set before the experiment parallel (x-direction) to the beam, and its intensity was almost uniform over the sample. Some details of the experiments may be found in [10, 11]. The solid-liquid interface was visualized by X-ray radiography; the main surface of the sample (y-z plane) was set perpendicular to the incident monochromatic X-ray beam.

3.1. Experimental evidence of the TEM force

Figure 3 shows the time evolution of the solidification microstructure for a temperature gradient $G = 10$ K/cm. The presence of an axial temperature gradient in the molten sample led to a non-homogeneous nucleation phase, from the bottom to the top of the sample. More specifically, first grains nucleated and showed up earlier in the cooler bottom part of the sample. As discussed in a previous paper [13], it can be reasonably assumed that the grain nucleation takes place from heterogeneous nucleation events at the crucible wall/oxide skin at liquidus temperature, or negligibly below. After the nucleation, few grains remained stuck on the oxide layer of the sample but the great majority of the grains moved roughly from the right side to the left side of the sample as indicated by the white arrows in figure 3a to figure 3d. It is worth noticing that the grains continued to grow during their motion. The global movement of the grains caused the equiaxed grains to form a tightly packed dendritic matrix on the left side of the sample, which, gradually moved towards the right side of the sample. This effect is more easily visible in the videos from which figure 3 are taken because grains were forced to move towards one side of the sample during the equiaxed growth. The number of grains across the transverse direction of the sample was non-uniform, as was their size. The fact that all free grains moved from the right side toward the left side of the sample definitely demonstrates that the grains were pushed by a horizontal TEM force in y-direction. Furthermore, the experiments showed that the deviation of the particles with respect to gravity increased with the temperature gradient in agreement with the theoretical analysis.

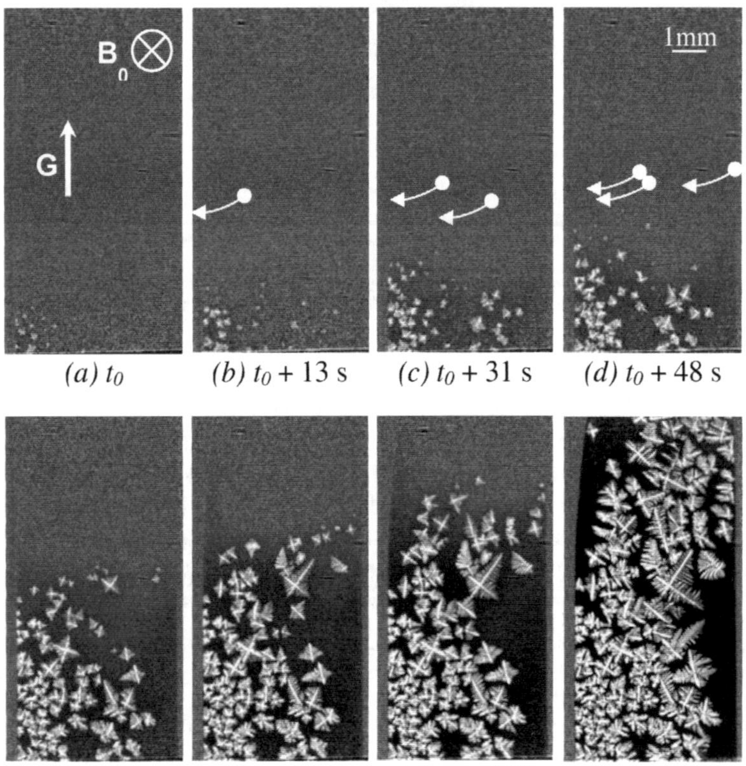

Figure 3. Sequence of radiographs showing the equiaxed solidification experiment of Al-10wt%Cu (CR = 2K/mn ; G = 10K/cm) under a 0.08T static magnetic field. Equiaxed grains nucleated from the bottom to the top, according to the temperature gradient direction, and then moved from right to left in the sample as indicated by few arrows in figure 3a to figure 3d.

3.2. Analysis of grain velocity as a function of grain size

Measurements of the grain surface and the grain velocity were carried out at each step of the solidification experiments between two successive radiographs. The shape of a dendritic grain is obviously very complex and far from a spherical solid in three-dimensions. For a non-spherical particle, an "equivalent surface diameter" d_S can be defined as the diameter of the disc that would have the same area as the projected image of the irregular dendritic grain. This definition does not take into account phenomena which can take place in the third dimension (sample thickness direction). Figure 4 shows the evolution of the grain velocity U for seven grains as a function of the equivalent diameter d_S. The grains were selected in the same area of the sample to prevent as most as possible any influence of sample thickness, temperature gradient or solute segregation. Despite these precautions, a large dispersion was observed for each value of diameter. In our experiments, the maximum Reynolds number $Re = Ud_S/\nu$ (ν being the kinematic viscosity) is much lower than unity, typically about 10^{-2}. Therefore, in a first approximation we may assume to be in Stokes regime and the velocity U can be deduced from equation (3). The effect of the thermo-electric flow around the particle is neglected following Baltaretu et al. [17].

A comparison of the Stokes velocity (dotted line in figure 4) with experimental data showed a large discrepancy, although the magnitudes are of the same order. Several effects might be responsible for the aforementioned discrepancy, such as:

(i) A confinement effect due to the lateral wall for large size particles,
(ii) The actual morphology equiaxed grains which are different from that of spheres,
(iii) The non-zero permeability of the equiaxed grains.

Those phenomena have an impact not only on the flow field but also on the electric current distribution, hence the TEM forces. To illustrate the previous discussion, some corrections of the Stokes law were introduced to account for the wall interference, that is to say the morphology and permeability of the particles [14]-[16]. Notes that all the corrections tend to decrease the particle velocity. The results illustrated in figure 4 show that owing to the corrections it is possible to obtain a better agreement.

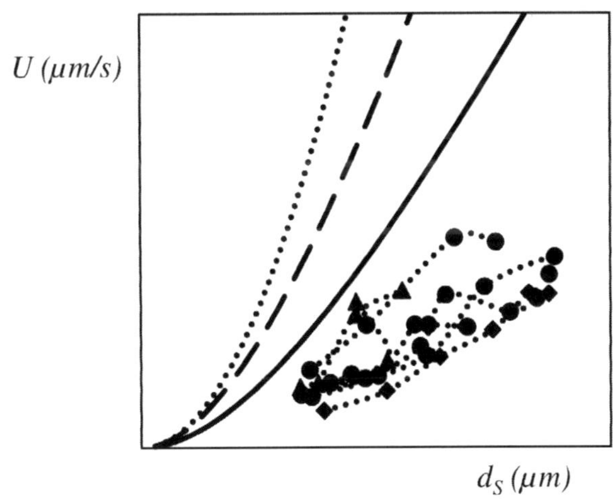

Figure 4. Plot of the velocity as a function of the equiaxed grain diameter. The symbols represent experimental measurements of seven grains, nucleating almost at the same position in the sample. The Stokes model is indicated by the dotted line, the model adding the wall correction to Stokes law is plotted by a dashed line and the model taking into account both wall confinement and grain morphology is plotted in full line.

4. Conclusions

We have analysed the thermo-electric-magnetic (TEM) effect on solid electrically conducting particles in liquid metal under a static magnetic field. That situation is applied to equiaxed solidification in liquid metals. Firstly, a heuristic theoretical model was developed. It was aimed at determining the TEM body forces acting both in the particle and on the surrounding liquid. It was shown that the particle is submitted to a transverse force, which tends to deflect the grain from the vertical direction. We also performed in situ and real time observations of equiaxed grain motion during directional solidification of Al-10 wt% Cu alloy under a weak static magnetic field (0.08 T). The experimental results show that equiaxed grains were moved approximately along a direction perpendicular to both the temperature gradient and the magnetic field direction, which confirms that the grains are forced to move by gravity force and TEM force. The deviation angle in the early stages of growth for the different temperature gradients were determined and compared to the prediction achieved by using an analytical model for a spherical particle. A quite good agreement was obtained. The variation of the grain velocity as a function of the grains diameter was then measured from the radiographs and compared with predictions of analytical models which take into the account the wall confinement and the grain morphology. Despite the corrections, it appears that the drag force in our experiments is much larger, probably due to grain behaviour (tilting and rotation) during the motion, long-distance interaction with other equiaxed grains and local variation of liquid composition related to solute rejection during solidification and modifications of the electric current distribution.

References

[1] Fautrelle Y, Ernst R and Moreau R 2009 *Int. J. on Materials Research* **100** 1389-1398
[2] Moreau R 1990 *Magnetohydrodynamics* (Kluwer Publications, Dordrecht, The Netherlands)
[3] Hurle DTG 1994 *R.W. Series, Use of a magnetic field in melt growth* (North Holland, Amsterdam ed.)
[4] Shercliff JA 1979 *J. Fluid Mech.* **91** 231
[5] Lehmann P, Moreau R, Camel D, Bolcato R 1998 *Acta Materialia* **46** 4067
[6] Li X, Fautrelle Y and Ren ZM 2007 *Acta Materialia* **55** 5333
[7] Li X, Fautrelle Y and Ren ZM 2007 *Acta Materialia* **55** 3803
[8] Wang J, Fautrelle Y, Ren ZM, Nguyen-Thi H, Abou Jaoude GS, Reinhart G, Mangelinck-Noel N, Li X, Zhong YB, Kaldre I, Bojarevics A and Buligins L 2012 *Appl. Phys. Lett.* **101** 251904
[9] Humbert E, Wang J, Budenkova O, Ren ZM and Fautrelle Y 2014 *Proc. Int. Conf. on Modelling for Electromagnetic Processing* (Leibniz University, Hannover Germany)
[10] Abou Jaoude GS, Wang J, Abou Khalil L, Reinhart G, Ren ZM, Mangelinck-Noel N, Li X, Fautrelle Y and Nguyen-Thi H 2014 *J. Crystal Growth,* to appear
[11] Schenk T, Nguyen Thi H, Gastaldi J, Reinhart G, Cristiglio V, Mangelinck-Noël N, Klein H, Härtwig J, Grushko B, Billia B, Baruchel J 2005 *J. Crystal Growth* **275** 201
[12] Nguyen-Thi H, Bogno A, Reinhart G, Billia B, Mathiesen RH, Zimmermann G, Houltz Y, Löth K, Voss D, Verga A, Pascale FD 2011 *Journal of Physics: Conference Series* **327** 012012
[13] Bogno A, Nguyen-Thi H, Reinhart G, Billia B, Baruchel J 2013 *Acta Materialia* **61** 1303
[14] Lindgren ER 1999 *Phys. Scr.* **60** 97
[15] Delaleau P, Mathiesen RH, Schaffer PL, Arnberg L, Beckermann C 2009 *Proc. TMS Int. Conf.* (Minerals, Metals & Materials Soc, Warrendale PA, USA)
[16] Zakhem R, Weidman PD, Degroh HC 1992 *Metallurgical Transactions A-Physical Metallurgy and Materials Science* **23** 2169
[17] Baltaretu F, Wang J, Letout S, Ren ZM, Li X, Budenkova O and Fautrelle Y 2015 *Magnetohydrodynamics*, 51 3

A multiphysics and multiscale model for low frequency electromagnetic direct-chill casting

N Košnik[1,2,4,a], A Z Guštin[2b], B Mavrič[2c], and B Šarler[2,3,d]

[1] Jožef Stefan Institute, Jamova 39, Ljubljana, Slovenia
[2] Institute of Metals and Technology, Lepi pot 11, Ljubljana, Slovenia
[3] University of Nova Gorica, Vipavska 13, Nova Gorica, Slovenia
[4] Department of Physics, University of Ljubljana, Jadranska 19, Ljubljana, Slovenia

E-mail: [a]nejc.kosnik@imt.si, [b]agnieszkazuzanna.gustin@imt.si, [c]bostjan.mavric@imt.si, [d]bozidar.sarler@imt.si

Abstract. Simulation and control of macrosegregation, deformation and grain size in low frequency electromagnetic (EM) direct-chill casting (LFEMC) is important for downstream processing. Respectively, a multiphysics and multiscale model is developed for solution of Lorentz force, temperature, velocity, concentration, deformation and grain structure of LFEMC processed aluminum alloys, with focus on axisymmetric billets. The mixture equations with lever rule, linearized phase diagram, and stationary thermoelastic solid phase are assumed, together with EM induction equation for the field imposed by the coil. Explicit diffuse approximate meshless solution procedure [1] is used for solving the EM field, and the explicit local radial basis function collocation method [2] is used for solving the coupled transport phenomena and thermomechanics fields. Pressure-velocity coupling is performed by the fractional step method [3]. The point automata method with modified KGT model is used to estimate the grain structure [4] in a post-processing mode. Thermal, mechanical, EM and grain structure outcomes of the model are demonstrated. A systematic study of the complicated influences of the process parameters can be investigated by the model, including intensity and frequency of the electromagnetic field. The meshless solution framework, with the implemented simplest physical models, will be further extended by including more sophisticated microsegregation and grain structure models, as well as a more realistic solid and solid-liquid phase rheology.

1. Introduction

Direct-Chill (DC) casting [5] turns out to be an efficient technology for production of aluminium semi-products. The process is robust and relatively simple, however it can induce a spectra of defects in the ingots. The control of the process, depending on the strand geometry and alloy composition, basically relies on adjusting the melt level, the casting speed, the melt temperature, and the intensity of the water jets that chill the ingot. These control parameters indirectly influence the melt flow and thus the solidification characteristics. An additional handle that can directly affect the flow is offered by alternating EM fields [6], applied to the DC process by means of coils placed around the casting system. The fields induce the eddy currents in the melt, which cause Lorentz force and thus provide a stirring force on the melt. A pioneering study of the application of low frequency electromagnetic field to the casting process was presented in [7], while a recent magnetohydrodynamic study of this problem was done by our group in [8]. The main aim of the present paper is the extension of this model to include macrosegregation, mechanical effects and grain structure and thus forming a reasonably complete basic multiphysics and multiscale model of the process for the axisymmetric billets.

2. Macroscopic models

2.1. Thermofluid equations

We express all the governing macroscopic transport equations in cylindrical coordinates (r,z), with base vectors $\hat{\mathbf{e}}_r$ and $\hat{\mathbf{e}}_z$ with z coordinate opposite to the casting direction. We limit ourselves to the cases where velocity vector points in the $r-z$ plane. The mixture continuum concept [9], adjusted to DC casting with columnar only solid phase [10] is adopted. The momentum conservation equation accounts for the Darcy drag term, the volume force, which is decomposed into thermal and solutal Boussinesq buoyancy term and the Lorentz force term

$$\rho_m \frac{\partial}{\partial t}(\mathbf{v}_m)+\rho_m \nabla\cdot(\mathbf{v}_m\mathbf{v}_m)=-\nabla P_l+\nabla\cdot\left(\mu_l\frac{\rho_m}{\rho_l}\nabla\mathbf{v}_m\right)-\frac{\mu_l}{K}\frac{\rho_m}{\rho_l}(\mathbf{v}_m-\mathbf{v}_{\text{cast}})+\mathbf{b}, \tag{1}$$

$$\mathbf{b}=-\rho_l\left[\beta_{T,l}(T-T_0)+\sum_{i=1}^{n}\beta_{C,l}(C_l^i-C_0^i)\right]\mathbf{g}+\langle\mathbf{b}_{\text{EM}}\rangle. \tag{2}$$

The latter is denoted by $\langle\mathbf{b}_{\text{EM}}\rangle$ and will be detailed in the following section. The solid phase velocity $\mathbf{v}_S$ is assumed to be constant and equal to the casting velocity $\mathbf{v}_S=\mathbf{v}_{\text{cast}}$, that is parallel to the z coordinate. The mixture density and mixture velocity are defined as $\rho_m=f_l\rho_l+f_s\rho_l$, $\mathbf{v}_m=(f_l\rho_l\mathbf{v}_l+f_s\rho_s\mathbf{v}_s)/\rho_m$, with f_l and f_s standing for the volume fraction of the liquid and solid phase, respectively. Mass conservation thus reads $\nabla\cdot\mathbf{v}_m=0$. The permeability constant of the Darcy term, relevant in the mushy zone is modelled as $K=K_0f_L^3(1-f_L)^{-2}$, and is set to 0 below the consolidation temperature, corresponding to $f_l^{\lim}$, at which the liquid phase becomes trapped within the dendrites. The heat and species transfer is formulated with the mixture convection-diffusion equations

$$\frac{\partial}{\partial t}(\rho_m h_m)+\nabla\cdot(\rho_m h_m\mathbf{v}_m)=\nabla\cdot(\lambda_m\nabla T)-\nabla\cdot\left[\rho_m(h_l-h_m)(\mathbf{v}_m-\mathbf{v}_{\text{cast}})\right], \tag{3}$$

$$\begin{aligned}&\frac{\partial}{\partial t}(\rho_m C_m^i)+\nabla\cdot(\rho_m C_m^i\mathbf{v}_m)=\\&=\nabla\cdot(\rho_m f_l D_l^i\nabla C_m^i)+\nabla\cdot\left[\rho_m f_l D_l^i\nabla(C_l^i-C_m^i)\right]-\nabla\cdot\left[\rho_m(C_l^i-C_m^i)(\mathbf{v}_m-\mathbf{v}_{\text{cast}})\right].\end{aligned} \tag{4}$$

The entalphy of the mixture is defined as $h_m=f_lh_l+f_sh_s$, where the constitutive relations are given as $h_S(\mathrm{T})=\rho_s c_{ps}T$ and $h_l(T)=\rho_l\left[c_{ps}T+c_{pl}(T-T_{sol})\right]+h_M$, with h_M denoting melting enthalpy. We assume that f_l rises consistent with the Lever rule in the mushy zone, $T>T_{liq}(C_m^i): f_l=1,\ C_l^i=C_m^i$; $T_{liq}(C_m^i)\geq T\geq T_{sol}(C_m^i): C_m^i=(1-f_l)k^iC_l^i,\ \ T=T_{liq}(C_l^i)$; $T<T_{sol}(C_m^i): f_l=0,\ C_l^i=C_m^i/k^i$ and that the heat transfer coefficient is a weighted average of the phase coefficients $\lambda=f_l\lambda_l+f_s\lambda_s$. On the inlet boundary (north) of the axisymmetric domain we impose the fully developed Poisseuille flow profile with known casting temperature and nominal concentration of the alloying elements C_0^i

$$\mathbf{v}_S = \mathbf{v}_L = -\hat{\mathbf{e}}_z v_{cast} \frac{2R^2}{R_{inlet}^2}\left(1-\frac{r^2}{R_{inlet}^2}\right), \quad T = T_{cast}, \tag{5}$$

where R_{inlet} and R are the inlet and mould radii, respectively. On the top free surface we set the temperature to the casting temperature and concentration to the nominal alloy concentration, and for the velocity, we set $\partial v_r / \partial z = 0,\ v_z = 0$. On the left (symmetry) boundary, we prescribe $v_r = 0,\ \partial v_z / \partial r = 0,\ \partial T / \partial r = 0,\ C_l^i = C_0^i$. On the right boundary (cooling side), the sticking boundary condition in the moving system is $\mathbf{v} = -v_{\text{cast}} \hat{\mathbf{e}}_z$. The cooling side temperature boundary condition is modelled with the Robin boundary condition $-k\partial T / \partial r = h(z)(T - T_{env})$, where $h(z)$ takes into account variation of the heat transfer coefficient with coordinate z. It vanishes in the hot-top region. In the mould chill region we set it to a constant value h_{MC} whereas in the direct chill region it starts at value h_{DC} just below the mould chill and linearly rises to $3h_{DC}$ on the bottom of the direct chill region. Finally, on the bottom boundary we require constant field derivatives in the casting direction $\partial T / \partial z = 0,\ \partial v_z / \partial z = 0,\ v_r = 0,\ \partial C_l^i / \partial z = 0$.

2.2. *Electromagnetic field equations*

The dynamical pair of Maxwell equation is

$$\nabla \times \mathbf{E} = -\frac{\partial \mathbf{B}}{\partial t}, \quad \nabla \times \mathbf{B} = \mu_0 \mathbf{J} + \frac{1}{c^2}\frac{\partial \mathbf{E}}{\partial t}, \quad \left(\mu_0 = 4\pi \times 10^{-7}\,\text{Vs/Am}\right), \tag{6}$$

where we assume $c = \infty$ and work in the quasi-static approximation. Electric $(\mathbf{E})$ and magnetic $(\mathbf{B})$ fields are both divergence-free, i.e., solenoidal. In addition, the Ohm's law in the moving frame relates fields with induced eddy currents, $\mathbf{J} = \sigma(\mathbf{E} + \mathbf{v} \times \mathbf{B}) + \mathbf{J}_{ext}$, where σ is conductivity that is nonzero only in the conductor, whereas $\mathbf{J}_{ext}$ is nonzero only in the coil and can be viewed as a boundary condition. We work with the electromagnetic potential $\mathbf{A}$ defined through $\mathbf{E} = -\nabla\phi - \frac{\partial \mathbf{A}}{\partial t}$, $\mathbf{B} = \nabla \times \mathbf{A}$ and impose the Coulomb gauge, $\nabla \cdot \mathbf{A} = 0$, for the vector potential. Maxwell equations thus reduce to the induction equation for $\mathbf{A}$

$$\nabla^2 \mathbf{A} = \mu_0 \sigma \left(\frac{\partial \mathbf{A}}{\partial t} - \mathbf{v} \times \nabla \times \mathbf{A}\right) - \mu_0 \mathbf{J}_{\text{ext}}, \quad \phi = 0. \tag{7}$$

We set $\mathbf{v} = \mathbf{0}$ on the right-hand side since the effect of velocity on the EM field is small in regime of small magnetic Reynolds number. Details of the equations in axisymmetry are given in [8].

2.3. *Solid mechanics equations*

The solid mechanics model is solved by assuming the calculated temperature field from the thermal model. The assumed thermoelastic governing equation is

$$G\nabla^2\mathbf{u} + (G+\lambda)\nabla\nabla\cdot\mathbf{u} + \nabla\lambda\nabla\cdot\mathbf{u} + \nabla G\left[\nabla\mathbf{u} + (\nabla\mathbf{u})^T\right] = \nabla\left[\beta_T\left(T - T_{ref}\right)\right] - \mathbf{f}, \tag{8}$$

where G stands for shear modulus, λ for Lamé parameter and $\mathbf{f}$ is the body force. The coupling with the temperature field is described by the coefficient β defined as $\beta = (3\lambda + 2G)\alpha$ which is composition dependent. Another free parameter is the reference temperature T_{ref}, which denotes the temperature at which the thermal expansion is considered to be zero.

3. Microscopic models

The microscopic model is solved in a post-processing mode. The model inputs temperature and concentration data from the macroscopic model. The representative input data are taken from the billet center, middle and surface streamlines in Lagrangean sense. The time for microscopic model is calculated from the velocity and length of the streamline. A typical 3x3mm representative traveling area is considered on each of the three considered streamlines.

3.1. Nucleation kinetics

To describe the nucleation event the continuous log normal nucleation model is adopted. The variation of the grain distribution in each time step depends on the undercooling temperature in the bulk of the casting. The density of grains $n(\Delta T_{nuc})$ activated at the given degree of undercooling ΔT_{nuc} is given by the integral of nucleation density distribution

$$\frac{dn}{d\Delta T_{nuc}} = \frac{n_{\max}}{\sqrt{2\pi}\Delta T_\sigma} \frac{1}{\Delta T_{nuc}} \exp\left[-\frac{1}{2}\left(\frac{\ln \Delta T_{nuc} - \ln \Delta T_0}{\Delta T_\sigma} \right)^2 \right], \tag{9}$$

where ΔT_0, ΔT_σ, $n_{\max}$ represent the mean nucleation undercooling, the standard deviation and the maximum density of nuclei that can form in the melt, respectively. These three standard nucleation parameters are determined experimentally for each alloy and grain refiner by measuring the grain size.

3.2. Grain growth kinetics

A modified KGT model is used to predict the grain growth velocities. The growth process is driven [11] by the local undercooling, approximated from the linearized phase diagram of a multicomponent alloy $\Delta T = T_{liq} - T(t) + \sum_{i=1}^{n} m_i \left(C_l^i - C_0^i \right)$, where T_{liq}, T, m_i, C_0^i, C_l^i represent the liquidus temperature, the temperature recalculated from the macroscopic model, the liquidus slope of species i, the initial concentration of species i and the liquid concentration of species i, recalculated from the macroscopic model. In order to calculate the grain velocity V, the transport equations are solved for each species i.

The solution relates the nondimensional supersaturations for each alloying element i, to the corresponding growth Péclet number $\Omega_i = F(P_i)$, $P_i = rV/2D_l^i$ where r is a tip radius. The supersaturation Ω_i is expressed by the Ivantsov function for the purely diffusive growth regime $F(P_i) = \mathrm{P}_i \exp(\mathrm{P}_i) E_1(\mathrm{P}_i)$ where E_1 is the integral exponential function. The total undercooling can be expressed as

$$\Delta T = \sum_{i=1}^{n} m_i C_0^i \left(1 - \frac{1}{1-(1-k_i)F(P_i)} \right), \tag{10}$$

where m_i is the liquidus slope of species i. The marginally stable plane front solution is applied in order to select the solution $V(\Delta T)$, with $r = \frac{1}{\sqrt{V}} \sqrt{4\pi^2\Gamma / \sum_{i=1}^{n} m_i (k_i - 1) C_l^i / D_l^i}$, and Γ standing for the Gibbs-Thomson coefficient.

4. Solution procedures

The time-averaged Lorentz force is calculated by discretizing the computational domain in local domains with 9 nodes with the approximation function expressed as a minimum least squares fit of 6 monomials $(1, x, y, xy, x^2, y^2)$. The weight function in the least squares problem is chosen to be

Gaussian, $e^{-|\mathbf{r}|^2/\xi}$. ξ is an average squared distance from the center node to all nodes in the local domain. The sparse linear system is solved using the sparse system solver [12]. The EM field calculation domain extends $1.5H$ beyond the casting system in vertical direction and is terminated at $3R$ in the radial direction (see figure 2). Unstructured node arrangement is used for the EM field calculation that has embedded the nodes for the thermofluid equations in a way that in the mould the EM and thermofluid nodes coincide. The transport equations for mass, momentum, energy, and concentration are solved by collocation with multiquadrics radial basis functions with 5-noded subdomains in case of the thermal model (as described in [13, 14]) and 6-noded subdomains in case of the mechanical model as elaborated in the present proceedings [15]. The phase change kinetics equations are solved by the PA method. The representative microscopic calculation domain with a dimension 3x3 mm is discretized into 360 000 randomly located nodes. The post processing calculations of temperature and concentration for this domain are taken from the macro level of calculations. The details of the PA method are elaborated in [16]. The potential nucleation grains are randomly located in this domain in each time step. The new grain represents a new family and gets a randomly chosen orientation (color) at the time when it appears. As grains nucleate, they start to grow with respect to the PA neighborhood configuration which is associated with the position of the neighboring PA nodes which fall into a circle with assumed radius $R_h = 0.5\,\mu\text{m}$. Due to the random arrangement of the nodes the neighborhoods are different. For the given values of the physical and thermodynamic properties m_i, k_i, D_l^i of each alloy element i, the variable grain growth velocity V for a specified undercooling temperature ΔT can be calculated. Independent calculations of the growth velocity for a given set of these values are carried out in each time step by using an iterative method. This requires the micro model to be provided with values of T and C_l^i, taken from the macro model at every 0.05 s.

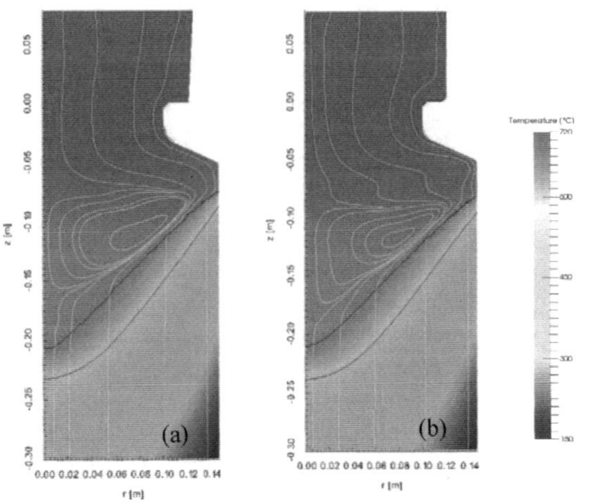

Figure 1. Temperature and velocity profiles for casting speed 80 mm/min without (a) and with (b) I=10A, f=10Hz EM field.

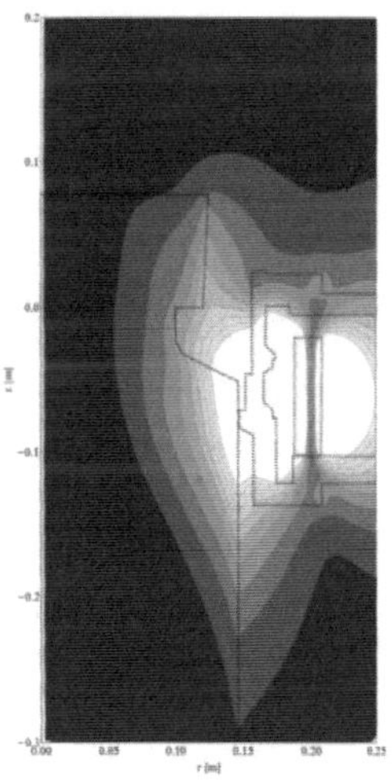

Figure 2. Computed magnitude of the magnetic field I=10A, f=10Hz.

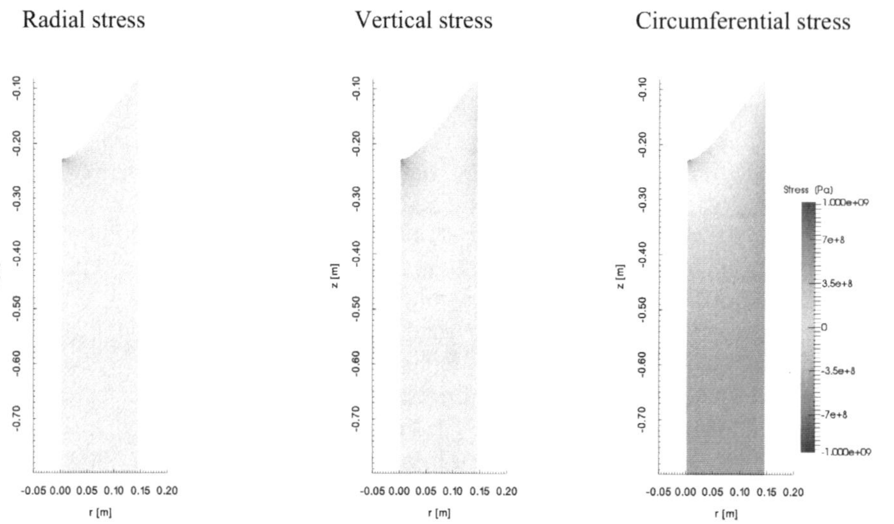

Figure 3. Stress calculations from the thermoelastic model at conditions from figure 1 (a).

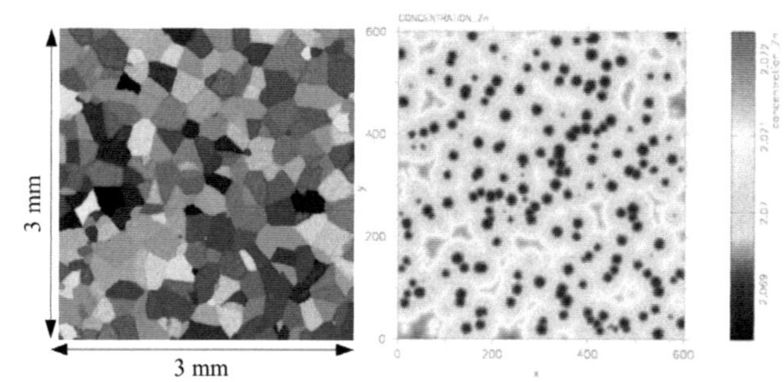

Figure 4. Results from the grain structure model. Left: grain structure. Right: Zn concentration for Al- Zn-5.3% Mg-2.35% Cr-1.5% Cu-0.5% alloy at conditions from figure 1(a).

5. Implementation of the models

The capabilities of the model are demonstrated in figures 1- 4 at various process parameters. Figure 1 shows the temperature field without and with EM field. Figure 2 shows computed magnitude of the magnetic field. Figure 3 shows stress calculations from the thermoelastic model for radial, vertical and circumferential stress. Figure 4 presents grain structures and Zn concentration distribution.

6. Conclusions

A reasonably complete multiphysics and multiscale model of the low frequency electromagnetic casting has been developed. The electromagnetic field is calculated first, afterwards the momentum equation, pressure velocity-coupling, energy equation, and species equations. The mechanical calculations are performed without feedback, based on the thermal calculations. The microstructure calculations are coupled in Lagrangean sense to the calculated velocity and concentration fields in a post-processing mode. The models are on the macro level solved by the diffuse approximate method and on the micro-level by the point automata method. This novel approach is completely meshless and no polygonisation is needed. The considered physical models incorporate only simple, basic elements that obviously need further improvements. In the field of macrosegregation, the movement of the solid phase and influence of the grain refiner represent the next necessary steps [17]. In the field of solid mechanics, the introduction of viscoelastic behaviour and feedback to the thermal model (mould heat transfer) seems important. In the field of microstructure modelling, the inclusions of the secondary phases in the model are necessary. It should be pointed out that the extension of the present numerical approach to three dimensions seems straightforward, since the coding of the present axisymmetric meshless formulation is equivalent in higher dimensions.

Acknowledgements
Support from the Slovenian Grant Agency in the framework of the project L2-6775 and IMPOL Aluminium Industry is kindly acknowledged.

References

[1] Vertnik R, Založnik M and Šarler B 2006 *Eng.Anal.Bound.Elem.* **30** 847-55
[2] Šarler B and Vertnik R 2006 *Comput. Math. Appl.* **51** 1269-82
[3] Chorin A J 1967 *J. Comput. Phys.* **2** 12-26
[4] Lorbiecka A Z and Šarler B 2014 *Materials Science Forum* **790-791** 115-20
[5] Eskin D G 2008 *Physical Metallurgy of Direct Chill Casting of Aluminium Alloys*, (Boca Raton: CRC Press/Taylor & Francis)
[6] Davidson P A 2001 *An Introduction to Magnetohydrodynamics*, (Cambridge: Cambridge University Press)
[7] Dong J and Cui J Z 2004 *Metall. Mater. Trans.* **35A** 2487-95
[8] Košnik N, Vertnik R and Šarler B 2004 *Materials Science Forum* **790-791** 390-395
[9] Bennon W D and Incropera F P 1988 *Numer. Heat Transf.* **13A** 277-96
[10] Du Q, Eskin D G and Katgerman L 2007 *Metall. Mater. Trans.* **38A** 180-89
[11] Appolaire B, Combeau H and Lesoult G 2007 *Mater. Sci. Eng. A* **487** 33-45
[12] Schenk O and Gärtner K 2004 *Journal of Future Generation Computer Systems* **20** 475-487
[13] Kosec G Založnik M, Šarler B and Combeau H 2011 *Comput. Mater. Cont.* **22** 169-59
[14] Kosec G and Šarler B 2014 *Eng. Anal. Bound. Elem.* **45** 36-44
[15] Mavrič B and Šarler B (in present proceedings)
[16] Lorbiecka A Z and Šarler B 2010 *CMC* **18** 69-104
[17] Nadella R, Eskin D G, Du Q and Katgerman L 2008 *Progress in Materials Science* **53** 421-480

Rapidly solidified Ag-Cu eutectics: A comparative study using drop-tube and melt fluxing techniques

Y Yu[a], A M Mullis[b] and R F Cochrane[c]
Institute for Materials Research, University of Leeds, Leeds, LS2 9JT, UK.

Email: [a]pmysy@leeds.ac.uk, [b]A.M.Mullis@leeds.ac.uk, [c]r.f.cochrane@leeds.ac.uk

Abstract. A comparative study of rapid solidification of Ag-Cu eutectic alloy processed via melt fluxing and drop-tube techniques is presented. A computational model is used to estimate the cooling rate and undercooling of the free fall droplets as this cannot be determined directly. SEM micrographs show that both materials consist of lamellar and anomalous eutectic structures. However, below the critical undercooling the morphologies of each are different in respect of the distribution and volume of anomalous eutectic. The anomalous eutectic in flux-undercooled samples preferentially forms at cell boundaries around the lamellar eutectic in the cell body. In drop-tube processed samples it tends to distribute randomly inside the droplets and at much smaller volume fractions. That the formation of the anomalous eutectic can, at least in part, be suppressed in the drop-tube is strongly suggestive that the formation of anomalous eutectic occurs via remelting process, which is suppressed by rapid cooling during solidification.

1. Introduction

Ag-Cu eutectic alloys have been widely studied for their remarkable strength and conductivity under both near-equilibrium and rapid solidification conditions. As one of the simplest binary eutectic systems, with the same face-centred cubic crystal structure in both the Cu-rich and Ag-rich phases, it is ideally suited for the study of rapid solidification behaviour. Various containerless processing techniques, such as splat quenching, drop tube processing, electromagnetic levitation and melt fluxing have been used to achieve high undercooling in Ag-Cu droplets [1]. In melt fluxing, high undercoolings may be achieved after thermal cycling to remove possible impurities attached on the melt surface, but during post-recalescence solidification the highest cooling rate achieved is typically no more than 10 K s^{-1} [2], leading to significant post-recalescence modification of the original solidification structure [3]. Conversely, in drop-tube processing, a range of cooling rates from 10^3 to 10^6 K s^{-1} is achieved via the production of a fine spray of free falling droplets of varying sizes [4]. However, determination of the nucleation temperature, and hence the undercooling within individual droplets is not possible. Therefore, an indirect method of estimating these quantities from the size of the as-solidified droplet may be employed, details of which are given in [1, 5].

The mechanism of anomalous eutectic formation during rapid solidification has been the subject of controversy for many years. In Ag-Cu, most such studies have mainly concentrated on the condition of low undercooling. The highest undercooling in which the formation of anomalous eutectic is observed in Ag-Cu has been variously reported as 70 K [6], 75 K [3] and 76 K [7], with higher undercoolings resulting in the formation of supersaturated single-phase dendrites instead. Wang et al [8] also estimated a maximum undercooling of 70 K in drop-tube processed samples, in agreement with the results of direct undercooling studies. At intermediate undercoolings, less than the critical undercooling [9] and typically < 60 K [10], it has been widely conjectured that the initial planar eutectic front breaks down into several eutectic dendritic tips ahead of S/L interface. The rapid release of latent heat associated with this rapid growth may then remelt the initially formed eutectic dendrite giving the observed anomalous eutectic.

In this paper, the microstructure of Ag-Cu alloy droplets that have been rapidly solidified by drop-tube processing will be compared with the results of our previous investigation [6] in which melt fluxing was used to undercool an alloy of the same eutectic composition. The cooling rate and undercoolings of droplets formed during drop-tube processing are calculated as a function of particle size which may in turn provide new insights into the mechanism of anomalous eutectic formation.

2. Experimental methods

Rapid solidification of Ag-Cu eutectic alloy was studied in a 6.5 m drop-tube. In order to achieve high undercoolings, a thin liquid jet fell uncontained through the tube in an atmosphere of dried, oxygen-free N_2 gas to solidify as small droplets at the bottom of the tube.

A high-purity ingot of Ag-28 wt. %Cu eutectic alloy was prepared from 99.99% purity silver and copper shot. The Ag-Cu eutectic sample was placed in a 10 mm diameter and 20 mm long alumina crucible with 3 laser drilled holes in the base, each of which was 300 μm in diameter. This was then fixed at the top of the drop-tube surrounded by the RF coils. Prior to backfilling with purified nitrogen gas at 400 kPa, the tube was evacuated to a pressure of < 1 Pa by an oil-sealed rotary-vane pump and then to 10^{-4} Pa using a turbo-molecular pump. Following evacuation and backfilling the ingot was melted by RF induction heating of a graphite susceptor. When the desired superheat was obtained the melt was forced through the holes in the crucible base using 0.4 MPa N_2 gas. The various sizes of droplets were collected at the bottom of the drop-tube once they had cooled to ambient temperature. Further details of the experimental procedure are given in [4].

The droplets were sieved into the following size ranges; 850, 700, 500, 300, 212, 150, 106 μm, following which they were hot mounted and ground on a series of progressively finer SiC papers and polished using diamond paste. The morphologies of well-polished sample were subsequently characterized by means of SEM backscattered-electron imaging and EDX mapping.

3. Computational calculation

Due to the absence of an effective method for the direct observation of the formation of droplets in flight, computational methods are used to predict the process of droplet solidification and to provide a better physical understanding of the effect of experimental parameters upon the solidification morphology. The equation for the cooling rate, $\dot{T}$, as a function of volume, v_s, of droplet is given in equation (1) [11], where ρ_l is the mass density of the melt, C_P^l the specific heat of the liquid, and h_m the heat transfer coefficient. For the prediction of undercooling ΔT, as a function of diameter D we use the heat transfer model proposed by Lee and Ahn [5] as given equation (2). In order to estimate the wetting angle factor, $f(\theta)$, as a function of the droplet size, we use the empirical expression given by equation (3) [12]. All thermal physical parameters of Ag-Cu alloy and nitrogen gas have been obtained from [4, 5] respectively.

$$\dot{T} = \frac{1}{v_s \rho_l C_P^l \delta T} \int_{T_o}^{T_n} \left[h_m (T - T_R) + \varepsilon \sigma_{SB} (T^4 - T_R^4) \cdot dT \right] \tag{1}$$

$$\frac{\Psi(\theta)}{T_N \cdot \Delta T^2} = \ln \Phi(T_N, \theta, D) \tag{2}$$

$$f(\theta) = -0.0307 + \frac{35.5212}{D} \tag{3}$$

The calculated cooling rate and undercooling against droplet diameter is plotted in figure 1. It can be seen that with increasing droplet size there is a decline in the level of both undercooling and cooling rate, as expected [1]. We note however that the cooling rates calculated here are significantly lower than those given in [4]. This is due to the different inert gas chosen as a protective environment. As the specific heats of nitrogen and helium are 1039 J kg^{-1} K^{-1} and 5195 J kg^{-1} K^{-1} respectively, the heat transfer from the sample to the environment is much more rapid in helium than in nitrogen.

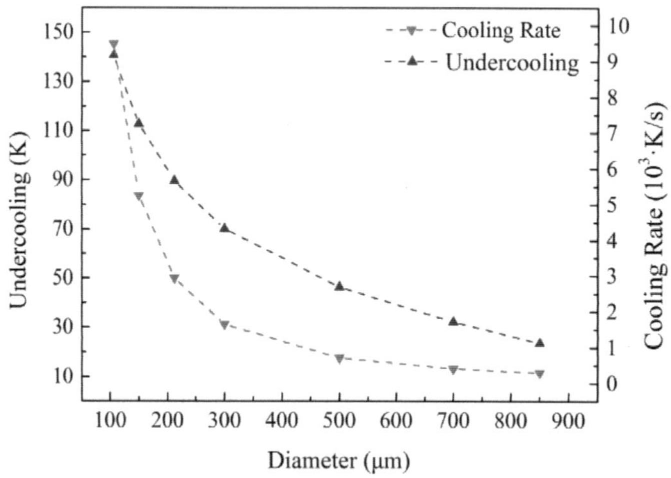

Figure 1. Calculated relationship between diameter, cooling rate (left Y-axis) and maximum undercooling (right Y-axis). Interpolation of the curves is only provided to guide the eye. The estimated cooling rate decreases from 9.50 × 10^3 K s^{-1} to 0.30 × 10^3 K s^{-1} over the range of droplet size produced in the drop-tube.

4. Results

In samples processed by melt fluxing there is extensive evidence for anomalous eutectic formation, as shown in figure 2(a), in which the microstructure is typically cellular, with cells consisting of lamellar eutectic in the cell body and anomalous eutectic occurring at the cell boundaries. In contrast, the anomalous eutectic formed in drop tube processed samples has a less clear distinction between these morphologies. With reference to figure 2(b), we note that the formation of colonies of lamellar eutectic, in the size range 10 ~ 20 μm, are favoured inside droplets. More interestingly, there are numerous 2 ~ 5 μm Ag-rich regions formed at, or near, the colony boundaries, which appear as white spots in the micrograph due to Ag having a higher atomic number than Cu. These are not observed in the flux undercooled samples (figure 2(a)), despite each experiment using material from the same, fully homogenized, ingot of master alloy. In bulk undercooled alloys it is normally accepted that volume fraction of anomalous eutectic is a monotonically increasing function of the undercooling [3, 9]. In contrast, in drop-tube processed samples the volume fraction of anomalous eutectic shows no clear trend with particle size, which is a good proxy for both cooling rate and undercooling.

With decreasing droplet size it is much more likely that individual droplets will achieve higher undercooling during drop-tube processing due to the melt sub-division effect. This is clearly illustrated in figures 2(c), (d) and (e). It can be seen that the formation of single-phase dendrites becomes increasingly intense with decreasing droplet size, which is clearly indicative of increasing undercooling. In comparison with the morphology of figure 2(b), the microstructure displayed in figures 2(d) and (e) is clearly indicative of a different mechanism for microstructure formation above a critical undercooling of order of 70 ~ 90 K. Similar findings for melt-fluxed samples are presented in [3, 6, 7, 9], in which the supersaturated single-phase dendrite is formed as the primary phase in samples undercooled above the critical undercooling of ~70 K.

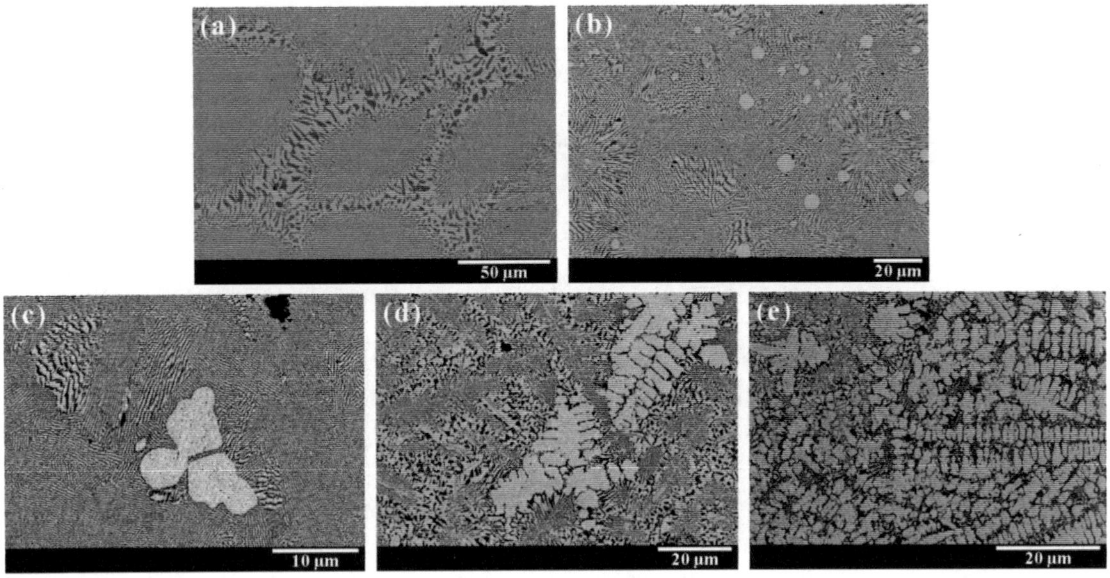

Figure 2. Series of micrographs showing the different morphologies via melt fluxing technique at (a) with undercooling 40K and drop-tube technique at (b) with the quantitative calculated undercooling in the droplet of the order of 30 K. With the decrease of droplet size, the Ag-rich single dendrite at figure (c) becomes increasingly dominant from figure (d) to figure (e).

5. Discussion

Above the critical undercooling samples produced by both melt fluxing and drop-tube processing clearly display single-phase dendrites as the primary solidification morphology, followed by the nucleation and growth of lamellar eutectic to fill the interdendritic space. Depending on the subsequent liquid temperature, anomalous eutectic formation might occur due to the remelting of either primary dendrites, or the subsequent interdendritic eutectic. In the case of drop-tube formed samples, the high cooling rate accompanying high undercooling appears favourable to the original solidification structure being retained, as shown in figure 2(d) and (e).

Conversely, the mechanism for anomalous eutectic formation for undercoolings less than 60 K via melt fluxing (or other bulk undercooling techniques) is most plausibly understood via the formation, and subsequent remelting, of eutectic cells or eutectic dendrites [6], with lamellar eutectic growing to infill the residual intercellular or interdendritic spaces. For similar undercoolings the microstructure obtained via drop-tube processing shows considerable difference to that obtained by melt fluxing, as can be seen by comparing figure 2(b) with figure 2(a). In particular, the eutectic grows in clearly defined colonies with numerous Ag-rich inclusions at or near the boundaries of the eutectic colonies. A reasonable explanation of the Ag-rich region formation should therefore play a central role in better understanding the mechanism of anomalous eutectic formation.

Figure 3(a) shows that the Ag-rich inclusions appear to be intimately related to the anomalous eutectic, merging seamlessly with the Ag-rich portion of the anomalous structure (figure 3(b) shows the EDX map for the same region, confirming that the inclusion is indeed Ag-rich). With reference to the metastable Ag-Cu phase diagram [14] we note that, for a eutectic formed from its undercooled parent melt, the Ag-rich phase will be supersaturated in Cu, but that the Cu-rich phase will still be close to, or slightly below, its equilibrium concentration of Ag. Upon reheating to the eutectic temperature, the Ag-rich phase will partially remelt while the Cu-rich phase will remain fully solid, but may of course be subject to coarsening. With the subsequent nucleation and growth of lamellar eutectic along the two-phase dendrite interface, the residual liquid will be Ag-rich, relative to the eutectic composition, plausibly giving rise to the observed Ag-rich inclusions at the colony boundaries.

The measurement of lamellar spacing permits an estimation of the conditions under which lamellar eutectic growth occurred in the post-recalescence period. Our past work [10] relying on the Jackson-Hunt model strongly suggests that for flux undercooled samples the post-recalescence lamellar growth is independent of both the bulk undercooling of the sample and position within the sample. It was found that growth occurred within a very narrow range of lamellar spacing and with an indicative undercooling of around 4 K. This can be seen in the central cellular structure in figure 2(a), where the lamellar eutectic has an almost constant spacing of ~0.28 μm. In contrast, figure 2(b) shows that the lamellar eutectic spacing in drop-tube processed samples is dependent upon their location within the colony. A comparison of lamellar eutectic spacing in the different colonies by means of image analysis indicates that the lamellar spacing varies from around 0.56 μm to the nano-scale. This difference in lamellar spacing between the melt fluxing and the drop-tube techniques is suggestive that in the free falling droplets less time is available for the lamellar growth phase, which is consistent with the expected high cooling rate. We note with reference to figure 3(a) that the coarsest lamellar eutectic is either near the anomalous eutectic or attached to the Ag-rich spot, with a gradual decrease in lamellar spacing with distance away from each of those two areas.

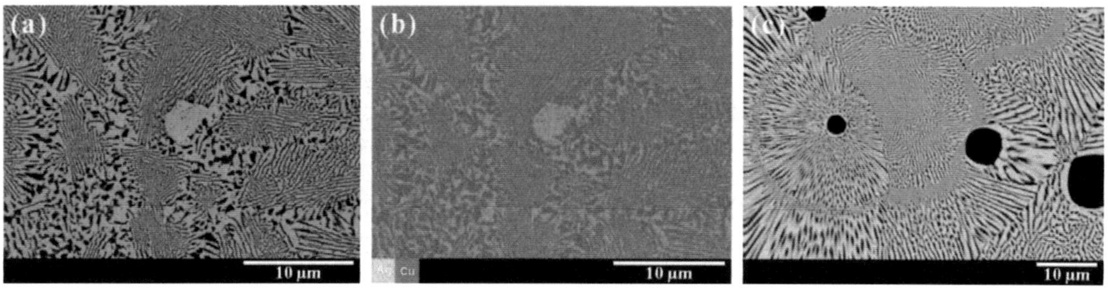

Figure 3. Micrographs showing different morphologies between droplets with fully molten liquid prior to free fall (a) and those with partially molten liquid prior to free fall (c). In order to clearly determine the composition distribution in the selected area (a), EDX mapping (b) is used as extensive evidence of droplet morphology, where green indicates Ag-rich regions and red Cu-rich regions.

Finally, we consider the possibility that the inclusions did not form from the melt, but instead were already present in the melt in the form of small unmelted fragments. To test this we performed a drop-tube run in which the master alloy was not fully melted prior to ejection. Figure 3(c) displays an example microstructure in which there was residual solid in the melt upon ejection during the drop-tube experiment. Firstly, we note that Cu-rich, rather than Ag-rich, inclusions are observed, a result of this being the high melting point phase. Secondly, it can be seen from figure 3(c) that the eutectic colonies radiate out from the solid Cu-rich inclusions which act as nuclei for solidification. This is contrary to the Ag-rich inclusion case, where the inclusions tend to cluster at colony boundaries, indicating they were one of the last features to solidify. Additionally, in this experiment no high undercooling (single phase dendrite) morphologies were observed, due to the strong heterogeneous nucleation effect within the droplets.

6. Conclusion

A comparative study on the solidification of Ag-Cu eutectic alloy via drop-tube and melt fluxing techniques has been undertaken and the results have been shown to yield useful insights into the formation of the anomalous eutectic morphology. Further work is in progress to determine the composition of Ag-rich region and to determine whether this correlates with the estimated undercooling within the droplets.

References

[1] Herlach D M, Cochrane, R F, Egry I, Fecht H J, Greer A L 1993 *International Materials Reviews* **38** 273-347.
[2] Çadirli E, Herlach D M, Volkmann T 2010 *Journal of Non-Crystalline Solids* **356** 461-466.
[3] Walder S, Ryder P L 1993 *Journal of Applied Physics* **73** 1965-1970.
[4] Erol M, Böyük U, Volkmann T, Herlach D M 2013 *Journal of Alloys and Compounds* **575** 96-103.
[5] Lee E S, Ahn S 1994 *Acta Metallurgica Et Materialia* **42** 3231-3243.
[6] Clopet C R, Cochrane R F, Mullis A M 2013 *Applied Physics Letters* **102** 031906.
[7] Walder S, Ryder P L 1993 *Journal of Applied Physics* 73 1965.
[8] Wang N, Cao C D, Wei B 1999 *Advances in Space Research,* **24** 1257-1261.
[9] Zhao S, Li J, Liu L, Zhou Y 478 *Journal of Alloys and Compounds* **478** 252-256.
[10] Clopet C R, Cochrane R F, Mullis A M 2012 *Acta Materialia* **61** 6894-6902.
[11] Herlach D M, Galenko P P K, Holland-Moritz D, Metastable solids from undercooled melts, PERGAMON Press, 2007.
[12] Wang N, Wei B –B 2004 *Chinese Physics Letters* **21** 1120.
[13] Schwarz M, Karma A, Eckler K, Herlach D M 1994 *Physical Review Letters* **73** 1380-1383.
[14] Murray J L 1984 *Metallurgical Transactions a-Physical Metallurgy and Materials Science* **15** 261-268.

The origins of spontaneous grain refinement in deeply undercooled metallic melts

A M Mullis[a], E G Castle[b] and R F Cochrane[c]
Institute for Materials Research, University of Leeds, Leeds LS2-9JT, UK

E-mail: [a] a.m.mullis@leeds.ac.uk, [b] pm07egc@leeds.ac.uk, [c] r.f.cochrane@leeds.ac.uk

Abstract. Spontaneous grain refinement in undercooled metallic melts has been a topic of enduring interest within the solidification community since its discovery more than 50 years ago. Here we present a comparative study of the solidification microstructures and velocity-undercooling behaviour in two dilute Cu-Ni alloys (3.98 & 8.90 wt.% Ni), which have been undercooled by a melt encasement (fluxing) method. Cu-3.98 wt.% Ni shows grain refinement at both low and high undercooling, with a dendritic growth regime separating the two grain refined regions. Within the grain refined region dendritic fragments are clearly evident in the centres of the refined grains and on the surface of the undercooled droplet, suggesting a dendritic fragmentation mechanism. Cu-8.90 wt.% Ni displays also grain refinement at both high and low undercoolings. In the low undercooling grain refined region the samples display curved grain boundaries with a dendritic substructure that extends across grains, indicative of a recovery and recrystallisation mechanism. Conversely, prior to the onset of the high undercooling grain refinement transition extensive regions of dendritic seaweed are observed, suggesting that it is remelting of a dendritic seaweed that gives rise to this structure. Consequently, in two closely related Cu-based systems we have strong microstructural evidence for the operation of all three mechanisms currently considered to give rise to grain refinement. This may help to resolve the grain refinement controversy, although it remains to be determined what factors determine which mechanism operates in any given system.

1. Introduction

A long-standing problem with regard to the dendritic solidification of metals has been that of spontaneous grain refinement in undercooled melts, first reported in pure Ni by Walker [1]. At a well defined undercooling, ΔT^*, Walker observed an abrupt transition from a coarse columnar, to a fine equiaxed, structure with a reduction in grain size of at least one order of magnitude. This effect has subsequently been identified in other pure metals [2, 3, 4] and in a range of alloy systems [5, 6, 7, 8, 9], in which a more complex evolutionary sequence is often observed. At low undercooling an initially columnar growth pattern is observed which gives way to an equiaxed grain structure as the undercooling is increased beyond a critical value, ΔT_1^*. At yet higher undercooling a second region of columnar growth is observed which, in most of systems, is replaced by a second region of equiaxed growth at still higher undercooling, ΔT_2^*. In many systems the onset of spontaneous grain refinement is observed to be simultaneous with a discontinuity in the velocity-undercooling curve [2].

The origins of the effect are controversial. Early theories suggested a range of mechanisms including nucleation ahead of the solidification front induced by the pressure pulse associated with solidification [1], recrystallisation, either during or immediately after solidification [10, 11], or the role

IOP Conf. Series: Materials Science and Engineering **117** (2016) 012054 doi:10.1088/1757-899X/117/1/012054

of minor solute additions [12, 13, 14]. However, more recent theories have tended to invoke fragmentation of the primary crystal, either during [15, 16], or immediately post-recalescence [17, 18]. The model that has perhaps achieved the most currency is that of Schwarz *et al.* [17, 18]. The model postulates that two characteristic timescales can be defined. The first is the break-up time, τ_{bu}, which is the time required for fragmentation of side branches due to remelting and Rayleigh instability. This is a monotonic function of the dendrite tip radius, with small radii giving short breakup times. The second timescale is the plateau time, τ_{pl}, which is the time the melt remains at, or around, the melting temperature during recalescence and is determined by the macroscopic heat extraction rate. The theory postulates that grain refinement occurs when $\tau_{bu} < \tau_{pl}$, which corresponds to the tip radius, ρ, being below some critical value, ρ^*, determined by the heat extraction rate. The original analysis was performed in the framework of the LKT model [19] for prediction of ρ, wherein the model appears to offer a natural explanation for grain refinement in both pure metals and alloys, although similar results can now be obtained using the more physically justifiable microscopic solvability theory [20].

An alternative mechanism for grain-refinement has been postulated in which a tip-splitting instability at the dendrite tip leads to the growth of an unstable 'dendritic seaweed' structure, which subsequently remelts to give the observed grain-refinement. This is supported the observation, in deeply undercooled ultra-high purity Cu, of a 'frozen in' seaweed morphology [21] above a critical undercooling, which resulted in a discontinuity in the velocity-undercooling curve identical in nature to that observed in spontaneous grain refinement [22].

In this paper we present the results of a microstructural investigation into the solidification of two closely related alloys (Cu- 3.98 wt.% Ni and Cu- 8.90 wt.% Ni) from their deeply undercooled melt. Both alloys show a variety of growth morphologies, including grain refinement, but subtle variations in the morphologies may yield important insights into the origin of spontaneous grain refinement.

2. Experimental method

High purity, 99.999% metals basis, Cu and Ni powders, of 100 and 120 mesh respectively, were weighed, mixed and compacted into pellets at nominal compositions close to Cu- 5 wt.% Ni and Cu-10 wt.% Ni. These were subsequently arc-melted together under an inert argon atmosphere. X-ray diffraction (XRD) patterns obtained from the alloys confirmed that a sufficient degree of homogeneity had been achieved and indicated the presence of a small amount of NiO. The composition of the as-solidified master alloys, accounting for any mass loss during preparation, was confirmed by Inductively Coupled Plasma – Optical Emission Spectroscopy (ICP-OES) analysis as Cu- 3.98 wt.% Ni and Cu-8.90 wt.% Ni. Prior to the conduct of undercooling experiments 5-7 mm pieces of the master alloy were ultrasonically cleaned in a warm aqueous solution of ammonium peroxodisulphate.

Undercooling experiments were carried out within a stainless steel vacuum chamber, evacuated to 10^{-5} mbar and backfilled to 500 mbar with N_2 gas. Samples were heated in fused silica crucibles by induction heating of a graphite susceptor, contained within an alumina radiation shield. In order to reduce the number of potential heterogeneous nucleation sites, the droplets were isolated from the crucible walls by suspension in a dewatered 40% Na_2SiO_3 – 60% B_2O_3 glass flux. The sample was visible throughout the experiment via a glass viewing window in the vacuum chamber and slots cut into the susceptor and alumina shield. High speed digital imaging (up to 15 000 fps) of the sample was used to make *in-situ* growth velocity measurements of the recalescence front.

Samples were subjected to repeated heating and cooling cycles, in which they were superheated to around 130 K above the liquidus temperature and held at this temperature for 15 minutes, ensuring complete melting and encasement of the sample and removal of gas bubbles from the flux. Samples were then slowly cooled through the liquidus at an average rate of 10 K/min. Temperature was monitored by means of an r-type thermocouple positioned at the base of the crucible. Solidification was allowed to nucleate spontaneously as this was found to occur naturally over a wide range of undercoolings and ensures that solidification proceeds from a single nucleation point. Further details of the experimental procedure are given in [23].

Samples, undercooled by varying amounts, were sectioned and mounted in Bakelite. Since growth proceeds radially outwards from the nucleation point, droplets with evident nucleation points could be sectioned with a well-defined orientation relative to their growth direction. To assess the interior microstructure of the droplets they were then polished and etched using an aqueous solution of ammonium persulphate. Microstructural analysis was by means optical microscopy (Olympus BX51 microscope), SEM secondary-electron imaging (Carl Zeiss EVO MA15 SEM) and XRD pole figure analysis (Phillips/PanAlytical diffractometer).

3. Results

The measured velocity undercooling behaviour for the two alloys is shown in figure 1. The same general trend is apparent in both cases. For low undercooling there is a relatively slow increase in growth velocity In this region a power-law fit to the data (not shown) indicates that, to a good approximation, the velocity increases as $(\Delta T)^2$. Beyond a critical undercooling, for which our best estimate is $\Delta T = 192$ K ($v \approx 31$ m s^{-1}) for Cu- 3.98 wt.% Ni and $\Delta T = 204$ K ($v \approx 29$ m s^{-1}) for Cu-8.90 wt.% Ni, there is a discontinuity in the velocity-undercooling curve, with the velocity increasing much more rapidly at higher undercooling. Where this has been observed previously it is normally indicative of either a change in dendrite growth direction, which in other Cu alloys has been from <100> to <111> [24], or a transition from growth controlled by solute diffusion to growth controlled by thermal diffusion as a result of solute trapping (see e.g. [25, 26] relating to Ni-B and Ni-Zr respectively). The microstructure determined from the resulting as-solidified samples is also summarised in figure 1.

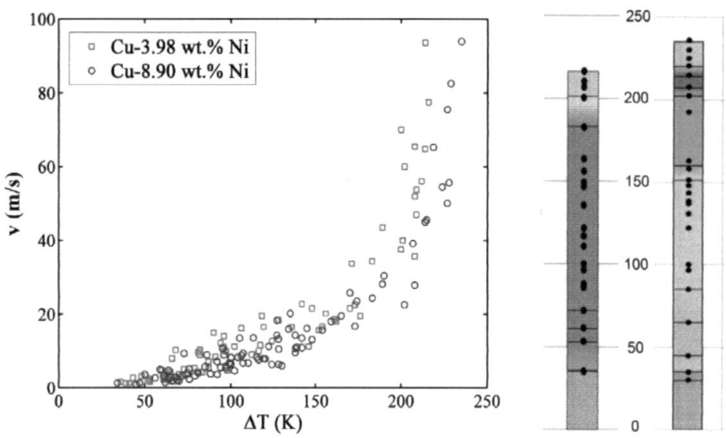

Figure 1. Velocity ΔT curves for the dilute Cu-Ni alloys investigated in this study. Also shown summary of the structures observed for the 3.98 (left) and 8.90 (right) alloys – Grain refined, low ΔT (pink); dendritic, low ΔT (light blue); mixed orientation (khaki); 8-fold dendritic (pale green); 6-fold dendritic (purple); dendritic (dark blue), grain refined (orange).

For $\Delta T < 35$ K it is difficult to nucleate solidification of the flux encased droplet and consequently microstructure determination has not been possible. At $\Delta T = 35$ K the Cu- 3.98 wt.% Ni already displays a grain refined microstructure, although on the assumption that growth at equilibrium is dendritic we assume that there is a low undercooling dendritic → grain refined transition but that this is located in the inaccessible region of very low undercoolings. Conversely, in the Cu-8.90 wt.% Ni alloy a coarse grained dendritic structure is observed up to undercoolings of 45 K, with a grain refined structure being obtained thereafter. These low undercooling grain refined regions persist up to undercoolings of 53 K and 85 K for the low and high Ni alloys respectively.

In both alloys this is then followed by an extended region in which dendrites with a variety of growth orientations are observed. In Cu- 3.98 wt.% Ni this begins with a region in which the dominant microstructure is for the droplet to comprise a single grain formed from a <100>-oriented dendrite with a strong orthogonality being observed between the primary and secondary arms. However, above $\Delta T = 72$ K, and up to at least $\Delta T = 148$ K, this gives way to a region in which the dendrites exhibit features of mixed-orientation growth; with secondary arms growing at both orthogonal and non-orthogonal angles to the primary trunk, some of which also appear to show twinning about the {111} plane. At undercoolings above 184 K this is replaced by a grain refined structure.

In contrast, in the Cu-8.90 wt.% Ni alloy, a much clearer transition in growth direction occurs. Above 85 K undercooling the grain refined structure gives way to dendritic growth with a clear 8-fold structure. This persists up to ΔT = 155 K, beyond which, and up to ΔT = 205 K, the dendrites adopt a 6-fold symmetry. Intermixed with these dendritic structures are regions of 'dendritic seaweed'. These 'seaweed' regions are bound by the dendrites and evolve from a diverging split primary branch, whose inner secondary branches have undergone multiple tip-splitting. In some cases, orthogonal outer secondary branches are observed to have grown from the split primary branches. In these cases, the seaweed growth is observed to terminate in an array of parallel <100> type dendrites. Finally, at very high undercooling a region of coarse, 4-fold symmetric, dendritic growth is observed up to undercoolings of 220 K, which finally gives way to a grain refined structure which is observed up to the highest undercoolings attained.

There are also significant differences in the morphologies observed in the grain refined regions. Figure 2a shows the typical microstructure observed in the Cu- 3.98 wt.% Ni alloy at low undercooling. Within the centre of the droplet, a large grain structure consisting of small subgrains is observed. Fragmented dendrite 'cores' are evident, with the low-angle subgrain boundaries most likely formed due to the slight misorientations between them. A surface micrograph (not shown) reveals that dendrite fragments which appear to exhibit secondary arms at 90° to the primary; suggesting that normal, <100>-oriented growth took place initially. The {200} pole figure taken from this sample indicates the presence of at least three unrelated grains, confirming the discontinuity of the substructure between them. The low undercooling region of grain refinement in this alloy appears therefore to take place via the fragmentation of <100>-oriented dendrites.

In contrast, in the low undercooling grain refined region in the Cu- 9.80 wt.% Ni alloy there appear to be large parent grains consisting of numerous small sub-grains. This is indicated by the difference in contrast between large areas, as shown in the top image of figure 2b - an optical micrograph of a sectioned, polished and etched sample which has been undercooled by ΔT = 65 K prior to solidification. The existence of curved grain boundaries suggests that grain boundary migration has taken place, whilst the DIC micrograph of the same sample (figure 2b bottom) appears to show that the coarse dendritic segregation pattern extends across the grain boundaries. The {200} pole figure plot taken from a randomly-oriented section of this sample exhibits only three sharp poles related to one another through 90° angles, confirming that the dendritic substructure is continuous throughout. This is consistent with the recrystallisation mechanism associated with the low undercooling region of grain refinement observed in many alloys [10, 11].

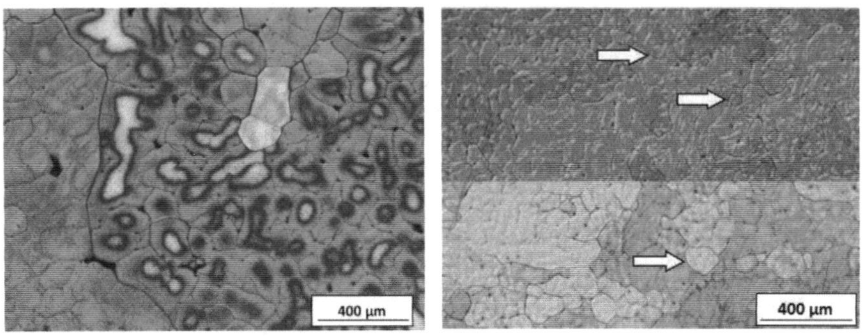

Figure 2. Grain refined microstructure of (a-left) Cu-3.98 wt.% Ni alloy (b-right) Cu-8.90 wt.% Ni alloy in the low undercooling regime. (a) shows clear evidence of dendrite fragmentation, (b) shows evidence of recrystallisation, indicated by curved grain boundaries and a substructure which is continuous across grains. This is particularly evident under DIC (top part of image).

For the Cu- 3.98 wt.% Ni alloy, grain refinement in the high undercooling regime also appears to proceed via a dendrite fragmentation mechanism. A surface microstructure, obtained from a droplet undercooled by ΔT = 216 K is shown in figure 3a. A fragmented dendrite structure can be seen; with each fragment constituting a single grain. The interior microstructure (not shown) appears to show a dendritic substructure that traverse some, but not all, of the grain boundaries. The {111} pole figure displays numerous random, unrelated sets of poles, confirming the presence of a large number of randomly oriented grains. The morphology of these dendrites is unclear, making it difficult to determine the character of the original growth structure. However, the velocity data, and in particular the fact that grain refinement occurs only on the high velocity side of the discontinuity in the velocity-undercooling curve, suggests a <111> orientation. It therefore appears that the high undercooling region of grain refinement in this alloy arises from the fragmentation of dendrites, possibly with a <111> growth direction.

A different morphology is observed in the Cu- 8.90 wt.% Ni alloy at the highest undercoolings achieved, wherein the grain structure transitions from large elongated grains to fine equiaxed grains. This is confirmed by the {111} pole figure (figure 3b) which contains a large number of poles which are mostly confined to one half of the plot, indicating the presence of a number of individual grains, yet with overall directionality still existing. The small equiaxed grains appear devoid of any substructure. However, the substructure in areas which are not grain refined (where large elongated grains are observed) is distinctly dendritic seaweed-like in character, as shown in figure 3b. Here, we appear to be observing an intermediate stage in the spontaneous grain refinement process, in which it looks like the growth of dendritic seaweed is the precursor structure.

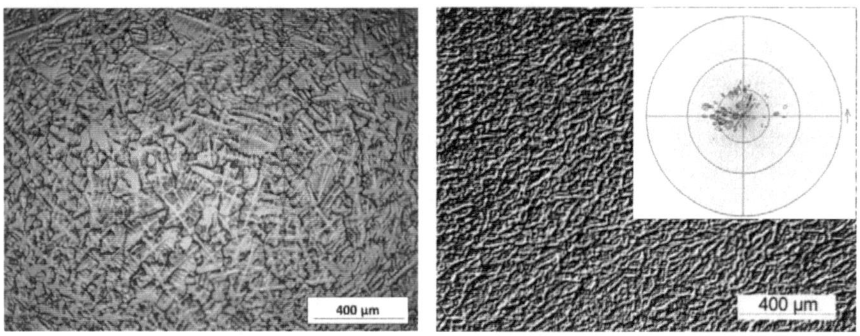

Figure 3. Grain refined microstructure of (a-left) Cu-3.98 wt.% Ni alloy (b-right) Cu-8.90 wt.% Ni alloy in the high undercooling regime. (a) shows clear evidence of dendrite fragmentation in the through focused image of the droplet surface, (b) shows extensive evidence of seaweed growth just prior to full grain refinement with {111} pole figure inset.

4. Discussion

Three distinct grain refinement mechanisms have been identified in the two closely related alloys studied here. Firstly, spontaneous grain refinement in the low-Ni alloy appears to be consistent with the model of Schwarz et al. [17], since two grain refinement transitions are observed to arise from the fragmentation of dendrites. At low undercooling grain refinement appears to occur from the fragmentation of refined <100> dendrites and, following an orientation transition, from the fragmentation of refined <111> dendrites at high undercooling.

A second grain refinement mechanism of recrystallisation has then been observed at low undercooling in the high-Ni alloy, consistent with a number of other reports [10, 11]. The difference between the fragmentation mechanism observed in the low-Ni alloy and the recrystallisation mechanism observed in the high-Ni alloy appears to arise from the difference in the original growth structure. In the low-Ni alloy, stable <100>-oriented growth is observed well beyond the first grain refinement transition, suggesting that the precursor in this case is <100> dendrites; in which case the greatest driving force for grain refinement seems to be the fragmentation of these dendrites following a gradual scale refinement. However, in the high-Ni alloy, there is an abrupt transition from the grain refined to the mixed-orientation regime, suggesting that the two regimes may coincide. Thus, the precursor to grain refinement in this instance would be mixed orientation dendrites; the presence of which most likely indicates the existence of a high dislocation density, thereby providing sufficient driving force for a recovery/recrystallisation process in place of refinement and fragmentation.

Finally, the remelting and fragmentation of a seaweed precursor appears to be the most likely mechanism for spontaneous grain refinement at high undercooling in the high-Ni alloy; in line with that proposed by Mullis et al. [16]. In order to understand why seaweed is observed for the high Ni alloy only, a consideration of the effect of Ni content is required. Since the addition of more Ni to Cu shifts the mixed orientation regime to lower undercoolings (with respect to the first grain refinement transition), and since this facilitates a transition to seaweed growth, it is suggested that the addition of Ni changes the balance between the capillary and kinetic anisotropies, an idea supported by the work of Haxhimali et al. [27] and Dantzig et al. [28]. In particular, orientation selection maps produced by Haxhimali et al. showed both Cu and Ni being very close to the <100>-<110> orientation boundary.

5. Conclusions

Spontaneous grain refinement has been a long standing and contentious issue in the solidification of deeply undercooled melts. A number of mechanisms have been proposed which plausibly account for this phenomenon, including recovery and recrystallisation, fragmentation of dendrites and growth and subsequent fragmentation of a dendritic seaweed. A comparative study on the solidification of two closely related Cu-Ni alloys has revealed that under varying conditions all three mechanisms appear to operate, perhaps explaining the long standing confusion over the origins of the phenomenon.

6. References

[1] Walker J L 1959 *The Physical Chemistry of Process Metallurgy* (New York: Interscience).
[2] Wilnecker R, Herlach D M and Feuerbacher B 1989 *Phys. Rev. Lett.* **62** 2707.
[3] Battersby S E, Cochrane R F and Mullis A M 1999 *J. Mater. Sci.* **34** 2049.
[4] Dragnevski K, Cochrane R F and Mullis A M 2004 *Mater. Sci. Eng. A*, **375-377** 479.
[5] Battersby S E, Cochrane R F & Mullis A M 2000 *J. Mater. Sci.* **35** 1365.
[6] Zhang T, Liu F, Wang H F, Yang G C 2010 *Scripta Mater.* **63** 43.
[7] Li J F, Zhou Y H and Yang G C 2000 *Mater. Sci. Eng. A* **277** 161.
[8] Li J F, Jie W Q, Y and G C and Zhou Y H 2002 *Acta Mater.* **50** 1797.
[9] Wang N, Gao J R and Wei B 1999 *Scripta Mater.* **41** 959.
[10] Southin R T and Weston G M 1973 *J. Aust. Inst. Met.* **18** 74.
[11] Ovsiyenko D E, Maslov V V and Dneprenko V N 1979 *Russ. Met.* **6** 124.
[12] Jones B L and Weston G M 1970 *J. Aust. Inst. Met.* **15** 167.
[13] Jones B L and Weston G M 1970 *J. Aust. Inst. Met.* **15** 189.
[14] Powell G L F and Hogan L M 1969 *Trans. AIME* **245** 407.
[15] Mullis A M and Cochrane R F 1997 *J. Appl. Phys.* **82** 3783.
[16] Mullis A M and Cochrane R F 1998 *J. Appl. Phys.* **84** 4905.
[17] Schwarz M, Karma A, Eckler K and Herlach D M 1994 *Phys. Rev. Lett.* **73** 1380.
[18] Karma A 1998 *Int. J. Non-equilibrium Processing* **11** 201.
[19] Lipton J, Kurz W and Trivedi R 1987 *Acta Metall.* **35** 957.
[20] Alexandrov D V and Galenko P K 2013 *Phys. Rev. E* **87** 062403.
[21] Dragnevski K, Cochrane R F and Mullis A M 2002 *Phys. Rev. Lett.* **89** 215502.
[22] Mullis A M and Cochrane R F 2001 *Acta Mater.* **49** 2205.
[23] Castle E G, Mullis A M and Cochrane R F 2014. *Acta Mater.* **66** 378.
[24] Dragnevski K, Mullis A M and Cochrane R F 2004 *Metall. Mater. Trans. A* **35A** 3211.
[25] Eckler K, Herlach D M and Aziz M J 1994 *Acta Metall. Mater.* **42** 975.
[26] Galenko P K, Reutzel S, Herlach D M, Danilov D and Nestler B 2007 *Acta Mater.* **55** 6834.
[27] Haxhimali T, Karma A, Gonzales F, Rappaz M 2006 *Nat. Mater.* **5** 660.
[28] Dantzig J A, Napoli P D, Friedli J, Rappaz M. 2013 *Metall. Mater. Trans. A* **44** 5532.

Dendrite growth morphologies in rapidly solidified Al-4.5wt.%Cu droplets

M Bedel[1,a], G Reinhart[1], A-A Bogno[2], H Nguyen-Thi[1], E Boller[3], Ch-A Gandin[4] and H Henein[2]
[1] Aix-Marseille University & CNRS, IM2NP UMR 7334, Campus Saint-Jérôme, Case 142, 13397 Marseille Cedex 20, France
[2] AMPL, Department of Chemical and Materials Engineering, University of Alberta, Canada
[3] ESRF – The European Synchrotron, CS 40220, 38043 Grenoble Cedex 9, France
[4] MINES ParisTech & CNRS, CEMEF UMR 7635, 06904 Sophia Antipolis, France

E-mail: [a] marie.bedel@im2np.fr

Abstract. The impulse atomization process developed at the University of Alberta (Canada) enables metallic powders to be solidified with controlled process parameters and improved properties. In order to investigate the microstructure morphologies in droplets of Al-4.5wt.%Cu alloys, three-dimensional reconstructions of several droplets are obtained by using synchrotron X-ray micro-tomography, allowing a visualization of the inner microstructure in three dimensions. The analysis of the reconstructed volumes reveals that a wide range of morphology, from highly branched to "finger-bundle", can be obtained for different droplets of similar diameter and produced in the same batch. Unexpectedly for this alloy, microstructural features also indicate that the development of the dendrite arms (primary and of higher orders) occurs in most droplets along <111> crystallographic axes, instead of the usual <100> directions observed in conventional casting technologies.

1. Introduction

The increase in strength and ductility of aluminium alloys is an ongoing challenge for automotive and aerospace applications. The improvement of these properties is possible through solidification microstructure refinement, obtained for instance by increasing the solidification velocity [1]. In addition, rapid solidification techniques are known to impact the solidification structure through morphology change [3], extended solid solubility [4] and non-equilibrium phase formation [4].

Different rapid solidification techniques have been developed through the years, which differ by the means used to form small liquid volume and to extract heat. The atomization techniques are the most common as they enable the formation of metallic powders which can later be packed onto the desired object by sintering or by cold or hot pressing [5]. In these techniques, the liquid metal jet is destabilized and disperses into fine droplets in a much colder medium where they solidify rapidly.

The goal of the present work is to present results obtained by synchrotron X-ray micro-tomography carried out at ESRF (European Synchrotron Radiation Facility, Grenoble, France) of Al-4.5wt.%Cu droplets formed by the atomization technique named impulse atomization (IA) developed by the AMPL (Advanced Materials & Processing Laboratory), Edmonton, Canada [6]. Three-dimensional reconstructions have been achieved for a large number of droplets and thus enable for the first time the inner microstructures to be analysed statistically in detail. Moreover, the impact on the microstructure morphology of parameters acting on the cooling rate, such as the droplet size or the nature of the cooling gas, is considered. From this analysis, an interpretation of the observed morphology evolution is proposed.

2. Experimental details

The widely used Al-4.5wt.%Cu alloy is chosen for this study as it is a model alloy for solidification studies. Its thermo-physical properties are well defined and former powder atomization studies were carried out [7]. The investigated droplets were produced by the IA technique [6]. This is one of a number of techniques termed single fluid atomization methods. In IA, the alloy is melted in a furnace and then pushed through a nozzle plate by impulsions, forming liquid jets that break into small droplets due to the Plateau-Rayleigh instability. The droplets lose their heat while falling through a stagnant gas atmosphere of either argon or helium. They are fully solidified before reaching an oil quench bath placed 4 meters under the atomizing nozzle. The droplet diameter can vary from less than 100 μm to more than 1 mm in the same batch. They are then sieved into different sizes based on the technique described in [8]. Then, several droplets of a given size range are introduced into a cylindrical glass capillary filled with epoxy resin to prevent them from moving during the image acquisition.

Synchrotron X-ray micro-tomography [9] is used to investigate the microstructure morphology of the droplets. Image acquisition and volume reconstruction were performed at ESRF on the ID19 beamline. This technique provides a three-dimensional representation of the droplets inside the glass capillary as shown in figure 1a. Two resolutions were used: a high resolution of 0.18 μm/pixel (field of view of 369 μm) to analyse in detail the fine microstructure of small droplets, and a medium resolution of 0.56 μm/pixel (field of view of 1146 μm) to scan either the bigger droplets, or several small droplets at the same time, which allows statistical studies. Two kinds of images are obtained depending on the post-processing technique used for the volume reconstruction: the standard filtered back-projection algorithm that provides images based on the difference in X-ray attenuation coefficient between the different phases in the droplet, and a reconstruction algorithm developed more recently from the work of Paganin et al. [10] based on phase retrieval. In both cases, the difference in grey level in the images is mainly due to the difference in density between the primary phase (aluminium, dark grey in figure 1c) and the eutectic (Al + Al_2Cu, light grey in figure 1c). Thereafter, the "Paganin" images will be used because they showed a better contrast between the two phases. Finally, the reconstructed volumes are analysed by using ImageJ software [11] to separate the droplets from each other. As the structure is highly complex due to three dimensional competition between dendrite arms, we focus on determining the position of planes showing characteristic morphologies, such as the dendritic plane as shown in figure 1b and figure 1c.

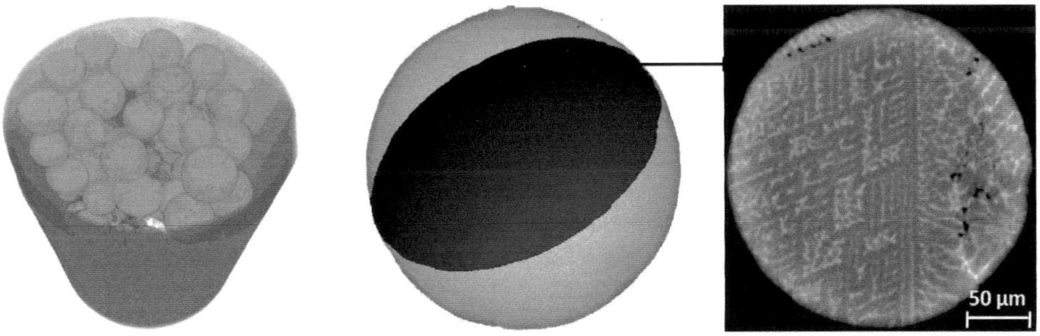

(a) (b) (c)

Figure 1. Typical X-ray micro-tomography result: (a) 3D-view of a capillary containing droplets, (b) position of a specific plane in a droplet and (c) corresponding image showing a dendritic structure.

3. Results

3.1. Dendrite morphologies

We have chosen to focus in the present article on the forty Al-4.5wt.%Cu alloy droplets cooled in argon, whose diameter is between 250 and 300 μm. By identifying characteristic planes in each droplet, four typical morphologies are distinguished. We do not intend here to extensively develop the characteristics of each of these morphologies but simply to present their main characteristics. The whole forty droplets are observed to be formed of a single grain. The first morphology shown in figure 2a is characterized by a highly branched microstructure. No clear nucleation position could be determined but it is worth noting that a rotation of 90° vertically (along OA) or horizontally (perpendicular to OA) from one characteristic plane lead to the observation of planes with similar morphology, suggesting that the dendritic microstructure developed along the usual <100> direction. For the second morphology figure 2b, the nucleation site (white dots noted O) as well as primary arms (white arrows noted OA and OB) can be distinguished and the morphology becomes highly branched (as in figure 2a) farther from the nucleation centre.

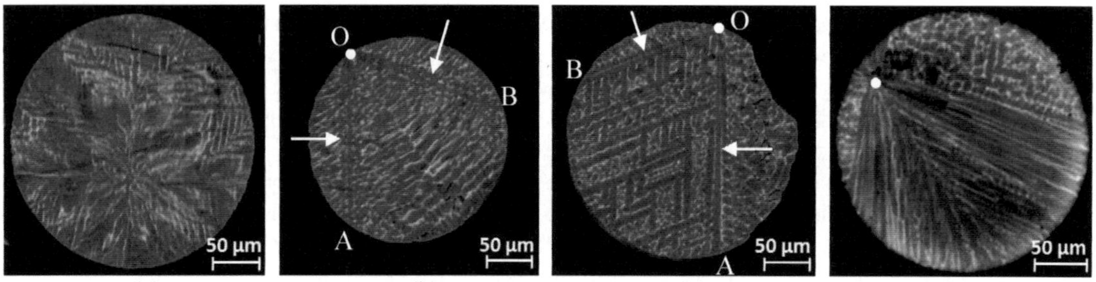

(a) (b) (c) (d)

Figure 2. The four morphologies identified in the Al-4.5wt.%Cu droplets of diameter between 250 and 300 μm: (a) highly branched morphology, (b) highly branched morphology with visible primary arms, (c) dendritic morphology and (d) "finger bundle" morphology. The nucleation position noted O is shown by a white dot and the primary arms noted OA and OB by white arrows.

The third morphology is shown in Figure **2**2c. The nucleation site can also be detected and the microstructure is dendritic with primary, secondary and tertiary arms clearly visible. It is worth noting that the secondary and tertiary arms developed in the same plane containing the two primary arms, more precisely on the left of the vertical primary arm OA and below the primary arm developing to the left OB. Finally, figure 2d shows an example of the fourth morphology. The nucleation site can still be found but characteristic planes are more difficult to determine because several arms with few branching expand from the nucleation centre and grow in an almost parallel manner. To our knowledge, such a microstructure has not been reported before and we termed this morphology "finger-bundle". Then, the morphology becomes dendritic as in figure 2c farther in the droplet.

3.2. Evidence for <111> dendritic growth

One striking feature of the dendritic microstructure in figure 2c is that the primary, secondary and tertiary arms are developing with an angle smaller than 90°, which is not expected for face-centered cubic structures in general [12]. Figure 3 shows characteristic dendritic planes found in another droplet. The two primary arms highlighted in red in figure 3a make an angle of approximately 70°. Similarly, the secondary arm highlighted in green makes a 70° angle with the primary arms and so on for higher arms orders. In addition, by rotating around the vertical primary arm noted OA in figure 3a, only two other similar dendritic planes are found: the plane in figure 3b by rotating of -120° and the plane figure 3c by rotating from +120°.

The angle values between the arms being characteristic of the growth orientation of the dendrites, they can help us identifying it. The angles measured on the dendritic microstructure in figure 3 are actually coherent with a <111> growth orientation. Indeed, with this growth orientation, the theoretical angle between two primary arms is equal to 70.5° and the angle between planes containing primary arms is equal to 120°, in a perfect agreement with the previous measured angles. The measurements made for the presented dendritic planes can be easily extended to the eight primary arms and six dendritic planes identified in the droplet. Finally, by placing the nucleation centre of the dendrite at the centre of a cube, we could determine that all the primary arms are pointing towards the cube corners (figure 4), confirming that the eight primary arms are directed in the eight <111> directions.

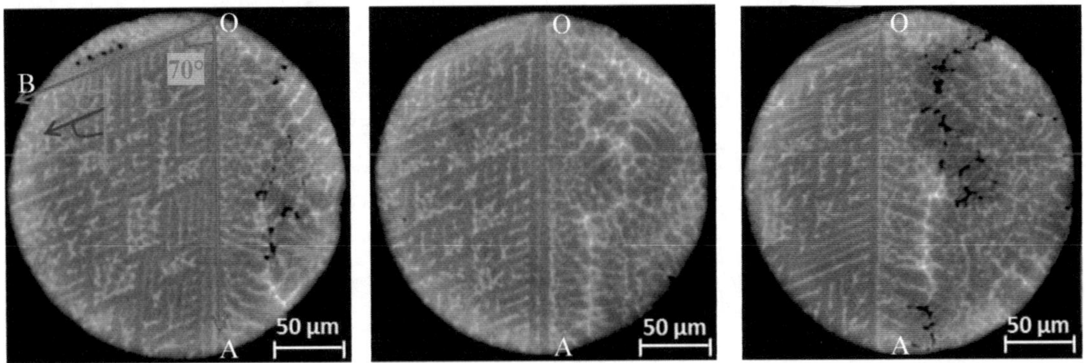

(a) $\theta = 0°$ (b) $\theta = -120°$ (c) $\theta = +120°$

Figure 3. Three of the characteristic planes identified in a dendritic droplet. From the plane shown in (a), we obtained the one shown in (b) by rotating from -120° along the OA, and the plane shown in (c) by rotating from +120°. The arrows highlight the primary, secondary and tertiary dendritic arms.

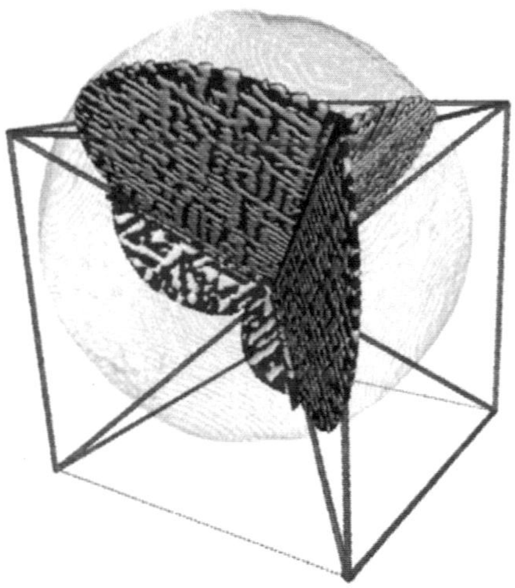

Figure 4. 3D-representation of the droplet with its surface in transparency and 6 dendritic planes in black. The droplet is placed in a cube with the nucleation centre at the centre and the primary arms correspond to the cube diagonals, illustrating the <111> growth orientation of the microstructure.

4. Discussion

As previously mentioned, the <111> growth direction (actually found in 85% of the 40 studied droplets, as can be seen in figure 5) was not initially expected. Indeed, the aluminium-copper alloy is commonly considered to grow in <100> directions [13], as for the face-centered cubic (fcc) structures in general [12]. However, growth direction changes in fcc crystals have already been reported experimentally. Kahlweit et al. highlighted the growth orientation change of NH_4Cl-H_2O crystals from <100> to <110> and then <111> with increasing growth velocity [14]. The authors investigated the growth orientation by varying the nucleation undercooling through solute composition and solidification temperature. Chan et al. suggested that, at low undercooling, the growth orientation is imposed by the anisotropy of interfacial free energy, while at high undercooling it is controlled by the "anisotropy of rate constant" [15]. By directionally solidifying this alloy, oscillations between these different growth modes were also seen by Gudgel et al. [16]. On the other hand, the investigation of Haxhimali et al. [17] on the dendrite growth orientation of directionally solidified Al-Zn alloys showed a change from <100> to <110> for a range of solute compositions. The latter result being independent of the solidification velocity, the authors concluded that the continuous orientation change is due to anisotropy variation of the surface energy and not to attachment kinetics anisotropy for this dendrite orientation transition. Using EBSD analysis, Chen et al. showed that with Al-0.6wt.%Fe and Al-1.9wt.%Fe atomized using IA, a textured structure showing a preference in the <111> direction was observed [18] . In the same way, Castle et al. recently reported a number of morphological transitions with increasing undercooling for a Cu-Ni alloy, interpreted as an extended transition between fully <100> and <111> growth orientations [19]. All these experiments suggest that the variation of interfacial energy anisotropy can induce a <100> to <110> transition, while the attachment kinetics anisotropy variation would explain the <111> orientation outbreak, with the transition to <111> linked to the solidification rate.

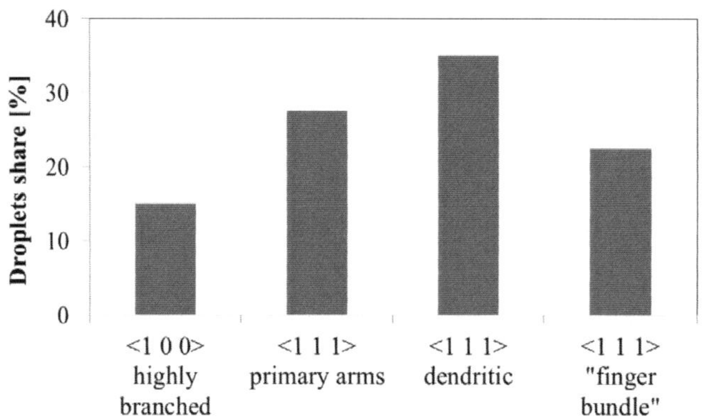

Figure 5. Droplet share of the four morphologies for the 40 droplets solidified in argon which diameter is between 250 and 300 μm.

This interpretation is coherent with our observations. Indeed, the solute composition and the process parameters being imposed, a distribution of the solidification velocity of the droplets can be the reason for the growth morphology distribution. As described in paragraph 3.1, the growth orientation seems to be <100> in highly-branched droplets (figure 2a) and we therefore can suppose that they formed at the slowest velocity. Then, for higher solidification velocity, the primary arms would start growing in <111> directions but the higher level arms that solidify later and more slowly would grow in <100> directions. This seems to correspond to the second morphology shown figure 2b. Then, when the velocity is even higher, the growth orientation remains < 111> in the whole droplet as illustrated figure 2c and figure 3. Finally, for the fastest velocities, the competition between several primary arms growing in the <111> direction leads to the kind of "finger bundle" morphology observed in figure 2d, then the dendritic structure is found back in regions of the droplet with slower solidification velocity.

Thus a range of solidification rates may be obtained in the IA production of droplets, even for imposed process conditions and for a given droplet size. The origin of the solidification velocity range can be the local thermal interactions between the droplets or a distribution of nucleation undercooling. This distribution would be large enough to induce a morphology distribution, from highly-branched to "finger-bundle", passing by highly branched with visible primary arms and dendritic morphologies.

5. Conclusion

Synchrotron X-ray micro-tomography has been used to characterize Al-4.5wt.%Cu droplets formed by IA process. This characterization technique is non-destructive and enables to statistically analyse droplets in three-dimensions. The determination of characteristic planes by studying a large number of droplets of roughly the same size and formed under the same process conditions leads to the identification of four different morphologies. The morphologies were shortly described and a <111> growth orientation was highlighted. The existence of the four morphologies was attributed to a range of solidification velocity for droplets of the same batch. This work will be continued by analysing more precisely the three other morphologies. Moreover, droplet solidification modelling will be used to improve the understanding of the solidification velocity impact on kinetic anisotropy during dendritic growth.

Acknowledgements
This study has been performed in the framework of MIMOSA project funded by the French National Research Agency (ANR) in collaboration with the Natural Sciences and Engineering Research Council of Canada and Novelis. In addition, we would like to thank Barbara Fayard and Olivier Guiraud from Novitom for their technical support.

References
[1] Hatch J E 1984 (ASM International)
[2] Nussbaum G, Sainfort P, Regazzoni G and Gjestland H 1989 *Scr. Metall.* **23** 1079–84
[3] Boettinger W J, Bendersky L and Early J G 1986 *Metall. Trans. A* **17** 781–90
[4] Sudarshan T S and Srivatsan T S 1993 (CRC Press)
[5] Prasad A 2006 *PhD Thesis* 70
[6] Henein H 2002 *Mater. Sci. Eng. A* **326** 92–100
[7] Wiskel J B, Henein H and Maire E 2002 *Can. Metall. Q.* **41** 97–110
[8] Federation M P I 2012 (S.l.: Metal Powder Industry)
[9] Nguyen-Thi H, Salvo L, Mathiesen R H, Arnberg L, Billia B, Suery M and Reinhart G 2012 *Comptes Rendus Phys.* **13** 237–45
[10] Nugent K A, Gureyev T E, Cookson D F, Paganin D and Barnea Z 1996 *Phys. Rev. Lett.* **77** 2961–4
[11] Abramoff M D, Magalhães P J and Ram S J 2004 *Biophotonics Int.* **11** 36–42
[12] Dantzig J A and Rappaz M 2009 (EPFL Press)
[13] Kato H and Umeda T 1978 *Metall. Trans. A* **9** 1795–800
[14] Kahlweit M 1970 *J. Cryst. Growth* **7** 74–8
[15] Chan S-K, Reimer H-H and Kahlweit M 1976 *J. Cryst. Growth* **32** 303–15
[16] Gudgel K A and Jackson K A 2001 *J. Cryst. Growth* **225** 264–7
[17] Haxhimali T, Karma A, Gonzales F and Rappaz M 2006 *Nat. Mater.* **5** 660–4
[18] Chen J, Dahlborg U, Bao C M, Calvo-Dahlborg M and Henein H 2011 *Metall. Mater. Trans. B* 1–11
[19] Castle E G, Mullis A M and Cochrane R F 2014 *Acta Mater.* **66** 378–87

IOP Conf. Series: Materials Science and Engineering **117** (2016) 012056 doi:10.1088/1757-899X/117/1/012056

Phase field crystal modelling of the order-to-disordered atomistic structure transition of metallic glasses

W Zhang[a] and J Mi[b]
School of Engineering, University of Hull, Hull, East Yorkshire, HU6 7RX, UK

Email: [a]w.zhang@2012.hull.ac.uk, [b]j.mi@hull.ac.uk

Abstract. Bulk metallic glass composites are a new class of metallic alloy systems that have very high tensile strength, ductility and fracture toughness. This unique combination of mechanical properties is largely determined by the presence of crystalline phases uniformly distributed within the glassy matrix. However, there have been very limited reports on how the crystalline phases are nucleated in the super-cooled liquid and their growth dynamics, especially lack of information on the order-to-disordered atomistic structure transition across the crystalline-amorphous interface. In this paper, we use phase field crystal (PFC) method to study the nucleation and growth of the crystalline phases and the glass formation of the super cooled liquid of a binary alloy. The study is focused on understanding the order-to-disordered transition of atomistic configuration across the interface between the crystalline phases and amorphous matrix of different chemical compositions at different thermal conditions. The capability of using PFC to simulate the order-to-disorder atomistic transition in the bulk material or across the interface is discussed in details.

1. Introduction

Bulk metallic glasses have attracted great attention because of their unique combination of mechanical and functional properties, such as very high strength, large elastic strain limit, exceptional corrosion and wear resistance capability [1]. However their very low ductility prevents them from being used as structural materials in wide industrial applications. Recently, a new class of alloy systems, i.e. bulk metallic glass composites (BMGC) has been developed. In these BMGC, the alloy composition is designed to allow ductile crystalline phase nucleate and grow in-situ inside the glassy matrix during solidification. By tailoring alloy compositions and controlling the thermal conditions during solidification, the volume fraction, morphology and size of the crystalline phases can be manipulated to provide a good combination of strength and ductility. However, the fundamental understanding on how the crystalline phases are nucleated in the super-cooled alloy matrix and their growth dynamics is still not fully understood, especially the order-to-disordered atomistic structure transition across the crystalline-amorphous interface.

A number of theoretical analyses have been made in the past to understand the fundamentals of order-to-disordered structure transition in glassy materials. For example, in the condense physics community, mode coupling theory and molecular dynamics have provided insight into the initial stage of atom slowing down in a supercooled liquid due to the increase of viscosity. The dynamic density function theory (DDFT) method [2-4] with a conserved form of equation of motion has also been proved to be a useful modeling method to resolve the order-to-disorder transition in glass formation [5,

6]. However, as explained by Jaatinen [7], the difficulty of DDFT in resolving the sharp peaks in density field causes serious problems in numerical computing. Therefore current research in physics and materials science communities is aiming to develop a general modeling framework to understand and describe glass transition.

Elder, *et al* [8, 9] firstly proposed the phase field crystal (PFC) methodology in 2002. Since then, extensive research and model development have been made in physics and materials science communities concerning PFC, and recently, dynamic density functional theory and thermodynamic concepts have been also linked to PFC to extend the modelling framework. PFC model operates at atomic length scale (from nm to μm) and diffusive time scales (in μs range), and is a computationally efficient alternative to molecular dynamics simulation methods. Berry [10, 11] argued that PFC is a promising modelling method for the studies of glass transition and crystallization in super-cooled liquid. In this study, we used PFC model to study the atomistic order-to-disordered transition of a binary glass at different thermal conditions, providing more quantitative insight on understanding the order-to-disordered transition across the crystalline and amorphous interface.

2. Phase field crystal model

We used the conserved form of motion equation of a binary alloy PFC model first proposed by Elder, et al [9], and the modified highly nonlinear free energy functional from the Vacancy Phase Field Crystal model (VPFC) proposed by Chan et al [12] in our research. The dimensionless free energy functional of a binary alloy is:

$$F = \int d\vec{r}[f_{AA} + f_{BB} + f_{AB}] \tag{1}$$

where f_{ii} and f_{AB} is the free energy density function, from VPFC [12], a specific form of the free energy function is adopted as below:

$$f_{ii} = \frac{n_i}{2}[r_i + \left(q_i^2 + \nabla^2\right)^2]n_i - \frac{w_i}{3}n_i^3 + \frac{u_i}{4}n_i^4 + H_i(|n_i|^3 - n_i^3) \tag{2}$$

and

$$f_{AB} = \frac{n_A}{2}\left(q_{AB}^2 + \nabla^2\right)^2 n_B + \frac{r_{AB}}{2}n_A^2 n_B^2 \tag{3}$$

where f_{ii} is the free energy density for atom i, and f_{AB} is the free energy density for A and B atom mixture. n_i is the scaled time averaged number density of i atom, r_i is related to the liquid modulus, q_i sets the equilibrium distance between atoms of the same species, q_{AB} sets that between A and B, and H_i and r_{AB} are constants. The terms multiplied by H_i penalize n_i when $n_i < 0$, and are the unique feature of the VPFC model. w_i and u_i are derived from the nonlinear terms of DDFT free energy functional [9].

The conserved equation of motion, the so-called model B [13] is adopted to describe the dynamics of the density fields, diffusive relaxations can be described through a direct minimization of the local free energy, and all dynamic processes on shorter time scales can be absorbed into a stochastic noise term, given by

$$\frac{\partial n_i}{\partial t} = \nabla^2 \frac{\delta F}{\delta n_i} + A\xi \tag{4}$$

where, ξ is the conserved noise satisfying the fluctuation-dissipation theorems:

$$\left\langle \xi(\vec{x}_1, \vec{t}_1)\xi(\vec{x}_2, \vec{t}_2) \right\rangle = \eta\nabla^2\delta(\vec{x}_1 - \vec{x}_2)\delta(\vec{t}_1 - \vec{t}_2)$$

We can rewrite the equation as

$$\frac{\partial n_i}{\partial t} = \nabla^2 \frac{\delta F}{\delta n_i} + A \nabla \eta \tag{5}$$

where η is a n-D vector whose components are independent Gaussian white noise, which can be computed by the Box-Muller transformation, and A is the amplitude of stochastic thermal noise, which can be considered as temperature [11] in the simulation.

3. Numerical methods

In equation (5), there are several high-order, even-powered gradient terms, making the implicit Euler numerical method more difficult to be derived for the gradients terms. We therefore adopted the semi-implicit pseudo-spectral method (PSM) [13-15] to solve equation (5). Using PSM, all the even-powered gradient terms are converted into even-powered algebraic expressions of the wave vector in the frequency space. PSM is particularly powerful for solving equations that exhibit periodic solutions such as those in this PFC model. After Applying Fourier transformation on both sides of equation (5), and using the forward finite difference method for the time marching, the PSM discretion of equation (5) without the thermal noise term can be written as:

$$\frac{\hat{n}_A^{t+dt} - \hat{n}_A^t}{dt} = \Delta_k{}^2 C_{AA} \hat{n}_A^{t+dt} + r_{AB} \hat{n}_B{}^2 \hat{n}_A^t + \Delta_k{}^2 N(\hat{n}_A^t) + \frac{1}{2} \Delta_k{}^2 C_{AB} \hat{n}_B^t \tag{6}$$

where $C_{ii} = r_i + \left(q_i^2 + \Delta_k{}^2\right)^2$, $C_{AB} = q_{AB}^2 + \Delta_k{}^2$, $\Delta_k{}^2$ is the discrete Fourier space representation of the ∇^2 for a finite size system, and $N(\hat{n}_A) = -w_A \hat{n}_A{}^2 + u_A \hat{n}_A{}^3 + 3H_A(|\hat{n}_A| - \hat{n}_A)\hat{n}_A$. In addition, the thermal stochastic noise term can be discretised as:

$$\vec{\nabla} \vec{\eta}_{i,j} = \sum_{z_1, z_2} \left(\frac{\eta_{i+1,j+1} + 2\eta_{i+1,j} - \eta_{i-1,j-1} - \eta_{i-1,j+1} - 2\eta_{i-1,j} + \eta_{i+1,j-1}}{8\Delta x} \right) \tag{7}$$

where, i, j denote the grid indices in the computational domain and Δx is the spatial step size. Because there is no data exchange required between the current one and adjacent grids, we used parallel computing technique to solve the discretised equation (6). The Message Passing Interface (MPI) [16, 17] parallel computing has been implemented together with PSM method.

4. Results and discussion

A series of simulations have been conducted to study the atomistic order-to-disordered transition. All simulations were carried out using a 64-core computing server (Dell PowerEdge C6145) available in our research group. Unless specified otherwise, all parameters defined in the model are referenced from [18] as: n_A=0.15, n_B=0.15, q_A=0.9, q_B=0.9, q_{AB}=0.9, r_A=-0.9, r_B=-0.9, H_A=1500, H_B=1500, u_A=1, u_B=1, w_A=0, w_B=0, dt=0.001, dx= π/4, N_x=256, N_y=256.

Since the PFC free energy is derived through Ryoichi Yamamoto functional [19], a representative of DFT functional in solidification. All the parameters entering the PFC model is temperature dependent. However the stochastic thermal noise defined by equation (7) is used to represent the changes of temperature [12, 18] (*T* is mimicked by the amplitude *A*), and the cooling rate is defined by *dA*/*dt*. In subsequent simulation, the different thermal conditions were achieved by varying the amplitude *A*, and the total computing time steps.

4.1. *Liquid-glass-crystalline phase transition*

As the super-cooled liquid cools from a high temperature to a low one, there are two possible phase changes: liquid to crystalline or liquid to glass, mainly depending on the compositions and cooling rates.

In a particle system, the radial distribution function (RDF) [20, 21], *G(r)* is normally used to describe the probability of the location of particles relative to the reference point, and to quantify the number density or co-ordination number of a particular group of particles relative to the reference point. RDF is widely used to describe the characteristics of disordered molecular systems, such as liquids and amorphous alloys. To quantify the structure of the disordered amorphous alloys, we used the partial radial distribution function (PRDF) to describe how atoms are radially packed relative to the reference atom of the same species.

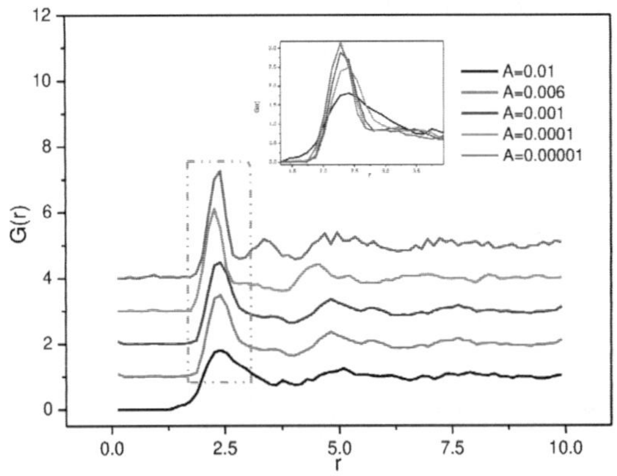

Figure 1. A series of $G(r)$ of atom A at different temperatures, which is offset vertically by $1*n$ with $n = 0,1,..,4$.The inset shows the zoom-in first peak of all curves without offset.

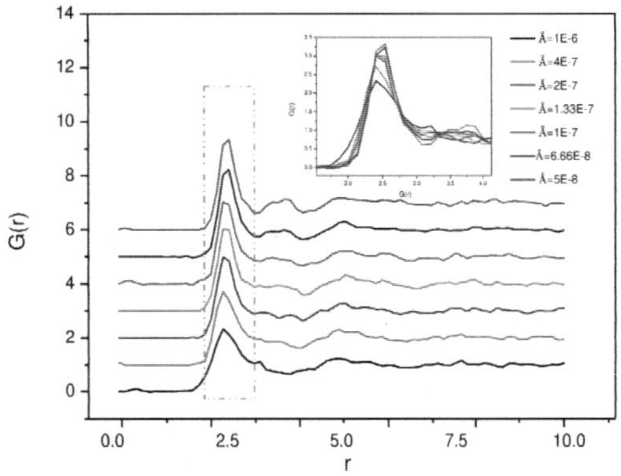

Figure 3. A series of $G(r)$ of atom A at different cooling rates, which is offset vertically by $1*n$ with $n = 0, 1,.. 6$. The inset shows the zoom-in first peak of all curves without offset.

Figure 1 shows the partial radial distribution function of atom A at different isothermal temperatures. By comparing the characteristics of the first peaks, mainly the sharpness of the peaks, and the full-width-at-half-maximum (FWHM). It clearly shows that from $A = 0.01$ to $A = 0.00001$, the first peaks becomes shaper and shaper, and their corresponding FWHM become smaller and smaller, especially at $A = 0.00001$, the second minor peak start to emerge, showing a typical crystalline feature. In order to quantify the ordered, and disordered atomic structures, we define that the structure with H<2.0 and FWHM>1.2 as pure random liquid structure, that with 2.5<H< 3.0 and 0.5<FWHM<1.2 as glassy (short plus medium range ordered) structure and that with H> 3.0 and FWHM<0.5 as crystalline structure.

4.1.1. Order to disorder transition at different cooling rates. Based on the above analysis and definition, in the simulation, we "quench" the liquid with purely random atomic structure simulated at A=0.01 (high temperature) to A=0.00001 (low temperature) with a series of different cooling rates, i.e. dA/dt =1.00E-06 to dA/dt =5.00E-08.

Figure 2 shows the simulated patterns for the atomic configuration at different cooling rates, and the corresponding G(r) curves for each pattern are shown in figure 3. Clearly, the maximum of first peaks of G(r) changes from 2.31 to 3.32 and the FWHM from 0.78 to 0.45 with the decrease of the cooling rates, and the corresponding patterns become more and more orderly arranged. Compared to the pattern shown in figure 2a, those in figure 2b, c have more medium range ordered clusters with the decrease of the cooling rates. Further decrease in cooling rates give enough time for the random atoms in liquid to grow into crystal-like clusters as showed in figure 2d, e, f. It is interesting to note that a small peak on the immediate right side of the first peaks start to emerge with the decrease of cooling rates, especially at dA/dt=1.33-E-07 to dA/dt=5.0E-08.

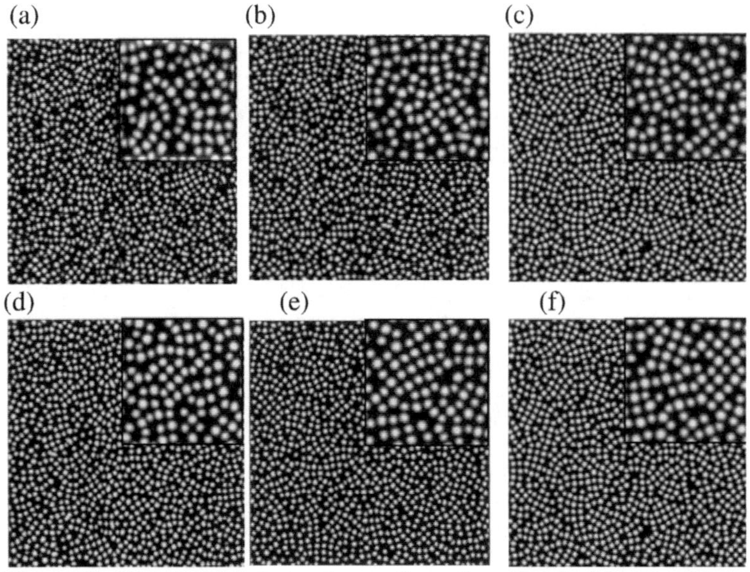

Figure 2. The simulated atomic configurations for the binary alloy using different cooling rates: (a) ***dA/dt*** =4.0E-07, (b) ***dA/dt*** =2.0E-07, (c) ***dA/dt*** =1.33-E-07, (d) ***dA/dt*** =1.0E-07, (e) ***dA/dt*** =6.66E-08, (f) ***dA/dt*** =5.0E-08.

4.2. Order-to-disordered transition across the amorphous-crystalline interface

In order to study the atomistic order to disordered transition across the interface between the crystalline and amorphous phase, the simulation domain is purposely enlarged as N_x=512, N_y=256, and treated as two equal sized domain (each has Nx=256, Ny=256), in which different cooling rates (see section 4.2.1), and also chemical compositions represented by the ratio of number density (see section 4.2.2), were used to produce a diffuse interface at the interface between the two domains.

4.2.1. The effects of cooling rates. In the simulation, the left half-domain is calculated using low cooling rate, dA/dt=5.0E-08, and the right one with high cooling rate dA/dt=2.0E-07. The whole domain is quenched from super-cooled liquid with a high temperature (A =0.01).

Figure 4a-d show the supercooled liquid state at A=0.01, from which two different cooling rates are applied on left and right part of computation domain respectively to create a diffuse interface. The left part is quenched with a slow cooling rate, and therefore the atoms have enough time to diffuse and grow into crystalline-like clusters. While the right part experiences a slowing down of the atom movement because of fast cooling rate applied (right half of figure 4b). As time evolves, the left part completely freezes into crystalline solid (figure 4d), while the right parts starts to nucleate and form median range ordered crystals.

Figure 4b and c show a clear interface because of the different cooling rate generated different atomic configuration, and the interface separates the two domains that contain different size and orientation of crystals. Although the simulated medium range-ordered structure can relax quickly into crystalline-like structure, this model shows that this transition can be captured elegantly during the simulation.

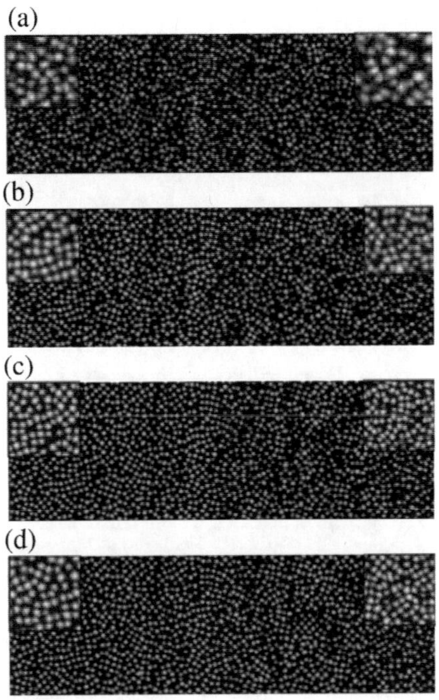

Figure 4. The interface across the domains simulated using different cooling rates at the simulation time steps of, (a)t=1.1E5 dt, (b) 1.3E5 dt, (c) 1.4E5 dt, (d)1.5E5 dt.

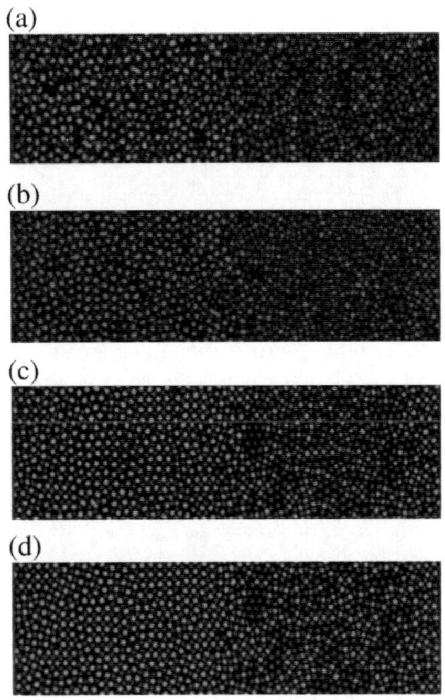

Figure 5. The interface across the domains interface between the two phases with different number density. The example atomistic configurations is shown, (a)t=1.0E6 dt, (b) 5.0E6 dt, (c) 1.0E7 dt, (d)1.5E7 dt.

4.2.2 The effect of number density (composition). As discussed in [12], the composition in at % increases linearly with the increment of number density if $n_i \leq 1.5$. Therefore, variations of number density can be used to represent the chemical composition changes in real material system.

The effect of different compositions on the order-to-disordered atomistic structure transition is studied by varying the ratio of number density, n_A and n_B. Similar to section 4.2.1, the computational domain consists of left and right halves. $n_A = 0.1$(denoted by red particles in figure 5), n_B=0.2(denoted by green particles in figure 5) is given to the left part, and n_A=0.2, n_B=0.1 for the right part. The whole domain is again "quenched" from A=0.01 to A=0.00001 in t=2.0E7 × dt time steps to allow enough time for the diffusion of atoms to occur. Figure 5 shows that there is a significant difference in atomistic structure in the two halves at different time. For example, starting from t=1.0E7 ×dt, small cluster of crystals nucleate and appear in the super-cooled liquid in the left half, and grow into medium range ordered crystals as showed in figure 5d. However, the right half remains in a randomly distributed amorphous state throughout. Clearly, the effect of chemical compositions on the transition of amorphous to crystalline structure is again captured elegantly by the PFC model.

5. Conclusion

Phase field crystal model was used to simulate the evolution of atomistic structure of a binary alloy. The atomic structure changes due to the variation of chemical composition and different cooling rates of the alloy system are elegantly simulated by the phase field crystal model. Further validation of the modelled atomistic structures and pair distribution functions against the diffraction data obtained from real metallic glass and composite alloys are under way to establish the link between the dimensionless model parameters employed in simulation and the real alloy composition and thermal conditions during solidification.

Acknowledgements

I acknowledge the joint CSC-UOH scholarship from The Chinese Scholarship Council and University of Hull for my PhD research. And also many thanks to the sponsorship by LIME and BCAST for me to attend the ICASP-4 conference as a UK-based student.

References

[1] Wang W-H, Dong C and Shek C Bulk metallic glasses 2004. *Materials Science and Engineering: R: Reports*. **44**:45-89.
[2] Jin Y M and Khachaturyan A G 2006. Atomic density function theory and modeling of microstructure evolution at the atomic scale. *Journal of applied physics*. **100**:013519.
[3] Löwen H, Melting, freezing and colloidal suspensions 1994. *Physics Reports*. **237**:249-324.
[4] Evans R 1979 The nature of the liquid-vapour interface and other topics in the statistical mechanics of non-uniform, classical fluids. *Advances in Physics*. **28**:143-200.
[5] Archer A J, Robbins M J, Thiele U, Knobloch E 2012. Solidification fronts in supercooled liquids: How rapid fronts can lead to disordered glassy solids. *Physical Review E*. **86**:031603.
[6] Singh Y 1991. Density-functional theory of freezing and properties of the ordered phase. *Physics Reports*. **207**:351-444.
[7] Jaatinen A, Achim C, Elder K, Ala-Nissila T 2009, Thermodynamics of bcc metals in phase-field-crystal models. *Physical Review E*. **80**:031602.
[8] Elder K, Katakowski M, Haataja M and Grant M 2002, Modeling elasticity in crystal growth. *Physical review letters*. **88**:245701.
[9] Elder K, Provatas N, Berry J, Stefanovic P and Grant M 2007, Phase-field crystal modeling and classical density functional theory of freezing. *Physical Review B*. **75**:064107.
[10] Berry J, Elder K and Grant M 2008, Simulation of an atomistic dynamic field theory for monatomic liquids: Freezing and glass formation. *Physical Review E*. **77**:061506.
[11] Berry J and Grant M 2014, Phase-field-crystal modeling of glass-forming liquids: Spanning time scales during vitrification, aging, and deformation. *Physical Review E*. **89**:062303.
[12] Chan P Y, Goldenfeld N and Dantzig J 2009, Molecular dynamics on diffusive time scales from the phase-field-crystal equation. *Physical Review E*. **79**:035701.
[13] Provatas N and Elder K 2011, Phase-field methods in materials science and engineering: Wiley-Vch.
[14] Tegze G, Bansel G, Tóth GI, Pusztai T, Fan Z and Gránásy L 2009, Advanced operator splitting-based semi-implicit spectral method to solve the binary phase-field crystal equations with variable coefficients. *Journal of Computational Physics*. **228**:1612-23.
[15] Berry J M 2011 Liquid-Solid Systems out of Equilibrium: Phase-Field Crystal Studies of Solidification, Melting, and Plasticity.
[16] Grama A 2003, Introduction to parallel computing: Pearson Education.
[17] Guo Z, Mi J and Grant P 2012. An implicit parallel multigrid computing scheme to solve coupled thermal-solute phase-field equations for dendrite evolution. *Journal of Computational Physics*. **231**:1781-96.
[18] Berry J and Grant M 2011, Modeling multiple time scales during glass formation with phase-field crystals. *Physical review letters*. **106**:175702.
[19] Yamamoto R and Onuki A 1998, Dynamics of highly supercooled liquids: Heterogeneity, rheology, and diffusion. *Physical Review E*. **58**:3515.
[20] McQuarrie D 1976, *Statistical Mechanics*, Harper Collins, New York.
[21] Wang J, Hodgson P D, Zhang J, Yan W and Yang C 2009, Effects of quenching rate on amorphousstructures of $Cu_{46}Zr_{54}$ metallic glass. *Journal of Materials Processing Technology*. **209**:4601-6.

IOP Conf. Series: Materials Science and Engineering **117** (2016) 012057 doi:10.1088/1757-899X/117/1/012057

Solidification of single droplets under combined cooling conditions

N Ellendt[1,a], N Ciftci[1], C Goodreau[2], V Uhlenwinkel[1] and L Mädler[1]
[1]Foundation Institute for Materials Science, Process & Chemical Engineering
Badgasteiner Str. 3, 28359 Bremen, Germany
[2]University of Alberta, Department of Chemical and Materials Engineering, T6G2V4
Edmonton, Alberta, Canada

E-mail: [a] ellendt@iwt.uni-bremen.de

Abstract. In this study, a pneumatic high-temperature droplet generator was used to generate individual droplets (diameter range: 350 – 1200 μm) which were cooled under combined cooling conditions. The individual droplets were cooled in a nitrogen atmosphere after ejection and during subsequent free fall. After a defined falling distance, the particles were quenched in either oil or water to further increase their cooling rate. Two alloys in different temperature ranges were used to study the effect of different cooling conditions quantitatively by the analysis of different microstructural features. To show the working range of the droplet generator, a metallic glass FeCo35.1Nb7.7B4.3Si2.8 (liquidus: 1210 °C) was used as a high-temperature alloy, and its resulting amorphous fraction was quantified as an indicator for different cooling conditions. Furthermore, the aluminium alloy AlCu4.5 (liquidus: 650 °C) was solidified under different conditions and the subsequent secondary dendrite arm spacing (SDAS) measurements were analyzed.

1. Introduction

Droplet generators can generate single droplets of a defined size [1-3] and are common tools to study phenomena such as droplet impact [4, 5] or droplet collisions [6, 7] under defined conditions. While droplet generation is simple to achieve with cold fluids, specific requirements for the droplet generators provide more challenges for the droplet generation of liquid metals with increasing melt temperature. Major requirements include the implementation of a suitable heating system, adequate refractory materials for the crucible, the orifice production in this material, and the necessity of an inert gas atmosphere. After ejection, the droplets cool down in a defined way in the inert atmosphere [8] which depends mainly on the particle size. However, if this cooling step is combined with subsequent quenching, the microstructure of one certain particle size can be adjusted by changing either the falling distance before quenching or the type of quenchant. The aim of the present work is to show how a high-temperature drop-on-demand droplet generator can be used to adjust the microstructure of generated particles exemplified with an Fe-based metallic glass and the conventional aluminium alloy AlCu4.5.

2. Experiments

The droplet generator consists of a crucible with a melt volume of 30 ml in a small pressure vessel (figure 1). A solenoid valve, which is connected to an overpressure reservoir, is opened for a short time (2 - 6 ms, depending on droplet size) allowing the pressure wave to travel through tubing (figure 2), the vessel, and subsequently through the melt. This introduces an overpressure and pushes the melt out. After the wave is reflected from the bottom of the crucible, the pressure drops and starts to retract the melt. If the pressure drop is fast enough, the melt volume that was pushed out of the nozzle can separate and form a single droplet. The pressure wave is relieved through a vent hole at the top of the vessel. After the pressure wave has completely vanished, a new droplet can be generated. A high speed camera was used to ensure the parameters were correctly adjusted to achieve a drop-on-demand mode, generating exactly one droplet for each pressure pulse.

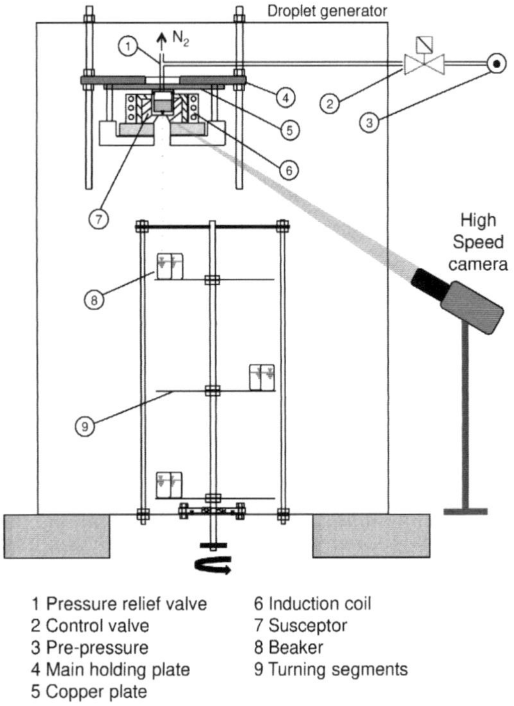

Figure 1. Schematic of experimental setup.

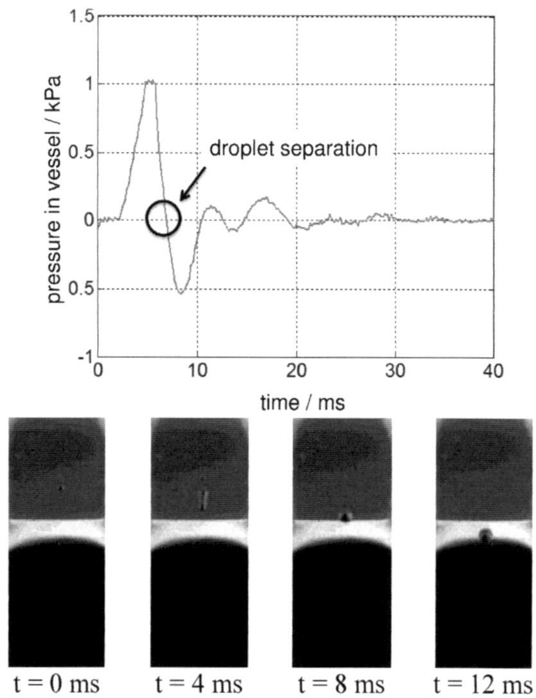

Figure 2. Pressure curve in the droplet generator and ejection of a droplet.

The entire system is placed in a chamber which is purged with nitrogen to avoid oxidation. The oxygen level under the droplet generator was measured to be below 50 ppm. However, the oxygen content in the vicinity of the nozzle is assumed to be much lower due to hot graphite parts. Induction heating is used to either heat a graphite crucible or a graphite susceptor that heats a boron nitride crucible by conduction. Temperatures of up to 1600 °C can be reached, allowing the use of a variety of technical alloys in the system. Particles can be collected in a quenchant in separate beakers after adjustable falling distances from the nozzle.

2.1 Experiments with FeCo35.1Nb7.7B4.3Si2.8 (metallic glass)
To show the effect of different combined cooling conditions, particles of the iron-based metallic glass FeCo35.1Nb7.7B4.3Si2.8 were collected after four different falling distances for a droplet diameter of 360 μm. If cooled in nitrogen only during free-fall, the particles crystallized fully [9]. Figure 3 shows schematically the cooling of a single droplet of a glass former alloy in a CCT diagram during free-fall in a gas atmosphere and subsequently quenched in a liquid after different falling times. If the particle is not quenched (condition a), it crosses the crystallization area slowly and crystallizes. If it is quenched early enough (condition b), crystallization is suppressed and a fully amorphous particle can be generated. In condition c, the particle is quenched shortly after it has entered the crystallization zone, causing partial crystallization. In condition d, the particle is quenched too late where crystallization was already complete. In the experiments, the falling distance was varied between 0.22 and 0.82 m for one particle size (table 1).The critical cooling rate of the alloy is approx. 5000 K/s [9].

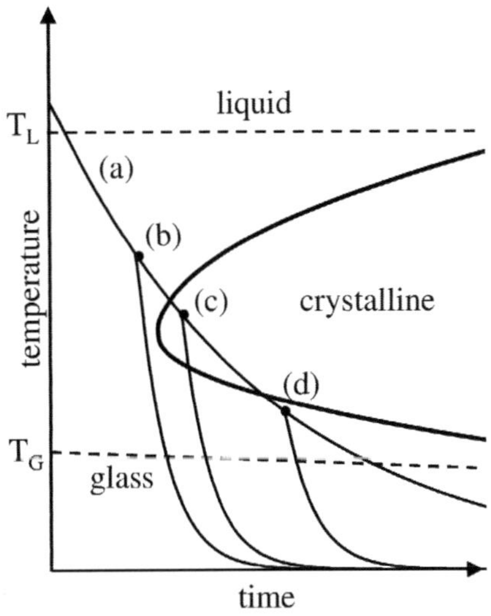

Figure 3. Different quenching conditions in the CCT-diagram, (a) gas cooling to achieve fully crystalline material, (b-d) with subsequent liquid quenching to achieve fully amorphous, partially amorphous and fully crystalline material, respectively.

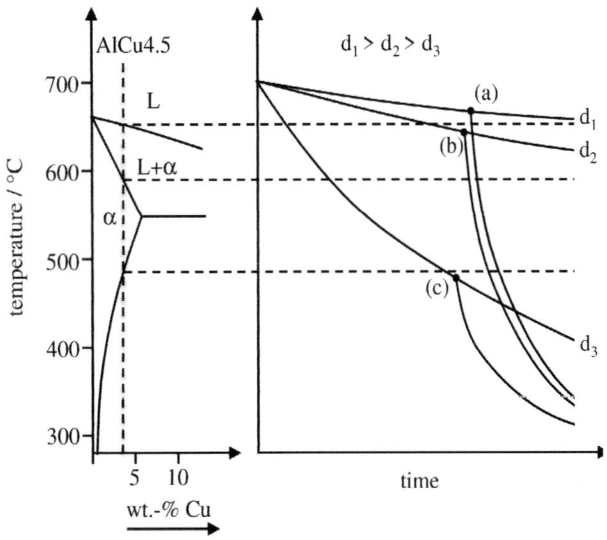

Figure 4. Quenching conditions for AlCu4.5 experiments for different droplet diameters. Condition (a) affects the whole microstructure, condition (b) partially affects dendrite growth and (c) can only affect microstructure growth in the solid.

2.2 Experiments with AlCu4.5

Figure 4 shows possible quenching conditions for droplets of different diameters from AlCu4.5 with regard to its phase diagram. For the largest droplet (condition a) the droplet is quenched in a fully liquid condition, so that its entire volume will cool down rapidly resulting in a very fine microstructure. However, if it is quenched below liquidus during dendrite growth (condition b), the remaining dendrite growth will yield smaller dendrites compared to the dendrite growth during gas cooling. Finally, if the particle is quenched below solidus (condition c), microstructural growth in the solid can still be avoided. In this study, the droplet diameter was changed for a constant quenching distance of 0.22 m (table 1).

Table 1. Experimental parameters

	FeCo35.1Nb7.7B4.3Si2.8	AlCu4.5
Melt temperature / °C	1250	750
Inert gas	nitrogen	nitrogen
Nozzle diameter / µm	350	250, 350
Falling distance / m	0.22, 0.35, 0.48, 0.82	0.22, 0.48, 0.82
quenchant	oil	distilled water

3. Analyses

3.1. Particle size

Image analyses of the size of the spherical particles were performed using the Malvern Morphologi G3. For the metallic glass, a tight particle size distribution with a number median diameter $d_{50,0}$ = 360µm geometric standard deviation of $d_{84,0} / d_{50,0} = 1.08$ was achieved (figure 5).

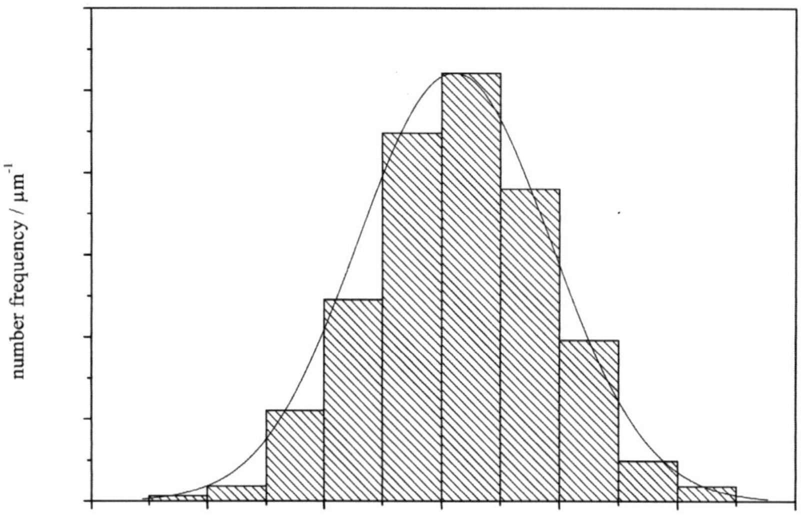

Figure 5. Unsieved particle size distribution for the Fe-based metallic glass

Since semisolid particles can deform as they strike the quenchant, image analysis is not possible and a sphere-equivalent average particle diameter was determined from the mass of 20 particles. Due to the lower surface tension and viscosity of AlCu4.5, a tighter particle size distribution was achieved with a geometric standard deviation <1.05 for all particles.

3.2. *Microstructure*

Micrographs of particles were prepared from both alloys for different process conditions to show the effect of cooling conditions on the microstructure. While the metallic glass was etched in aqua regia for 3 s, the aluminium alloy was etched in NaOH 5% for 30 s. Light microscopy was used for evaluation. The amorphous fractions of the metallic glass samples were measured using differential scanning calorimetry (Perkin Elmer Pyris 1) as proposed by Ciftci et al. [9]. The AlCu4.5 alloy was evaluated using secondary dendrite arm spacing (SDAS). The average dendrite arm spacing and its standard deviation were measured by image analysis of micrographs.

4. Results and discussion

4.1. *Metallic glass*

Figure 6a shows micrographs of the metallic glass for a falling distance of 0.22 m, 0.48 m and 0.82 m (corresponding to a falling time of 190, 300 and 410 ms). While the particles appear featureless for the shortest falling distance of 0.22 m, crystalline fractions increase as the falling distance increases. This shows that the cooling rate resulting from gas cooling is insufficient for glass formation. It is so low that the shortest falling time is still too long to avoid the crystallization area as denoted in figure 3.

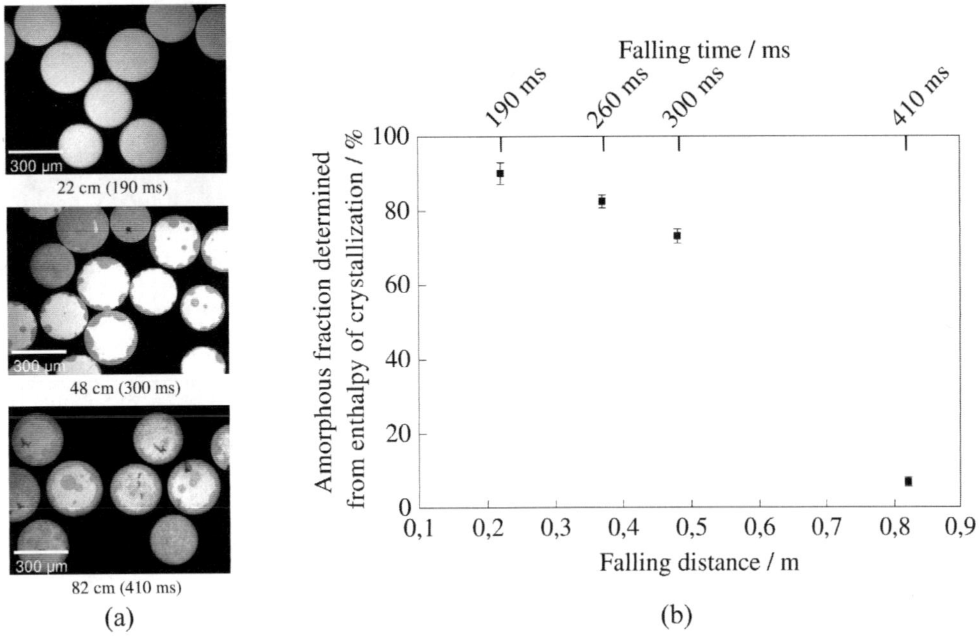

Figure 6. (a) Micrographs of metallic glass particles quenched after different falling distances, average particle size: 360 μm (apparent variations of particle size can result from different cutting planes), (b) Particle size distribution of the metallic glass measured by image analysis.

To achieve a fully amorphous sample, either the quenching distance could be reduced or the melt temperature could be raised, so that the particles are quenched at a higher temperature. However, both options might result in deformed particles due to the hotter impingement condition. Figure 6b shows the amorphous fraction as determined from DSC analysis. An amorphous fraction of 90% was achieved for the shortest falling distance and decreases to 4 % with increasing falling distance. The standard deviation of less than ± 3 % indicates a high reproducibility for each measured sample.

4.2. AlCu4.5

Figure 7 shows micrographs of AlCu4.5 droplets with different droplet diameters for different cooling conditions. For the smallest droplet diameter, 350 μm, two different sizes of dendrites were observed where the majority of the dendrites were larger. This reflects cooling condition (c) from figure 4 where the larger dendrites form during gas cooling, while the remaining dendrites are formed during quenching. If the droplet diameter is increased to 820 μm, the dendrites size greatly decreases. This could be due to cooling condition (b), where the particle cooling is so slow that it is quenched fully or almost liquid. If the droplet diameter is increased further to a diameter of 1065 μm, the droplet still impinges in a state that refers to condition (b). A possible reason for the increased dendrite size is the occurrence of the Leidenfrost effect. A maximum cooling rate of the droplet is achieved if the heat released by instantaneous evaporation of the quenchant during the impingement is high enough for full solidification; if there is liquid left when an insulating vapour film builds up, this will greatly reduce the cooling rate. This effect heightened as the droplet diameter is increased to 1600 μm.

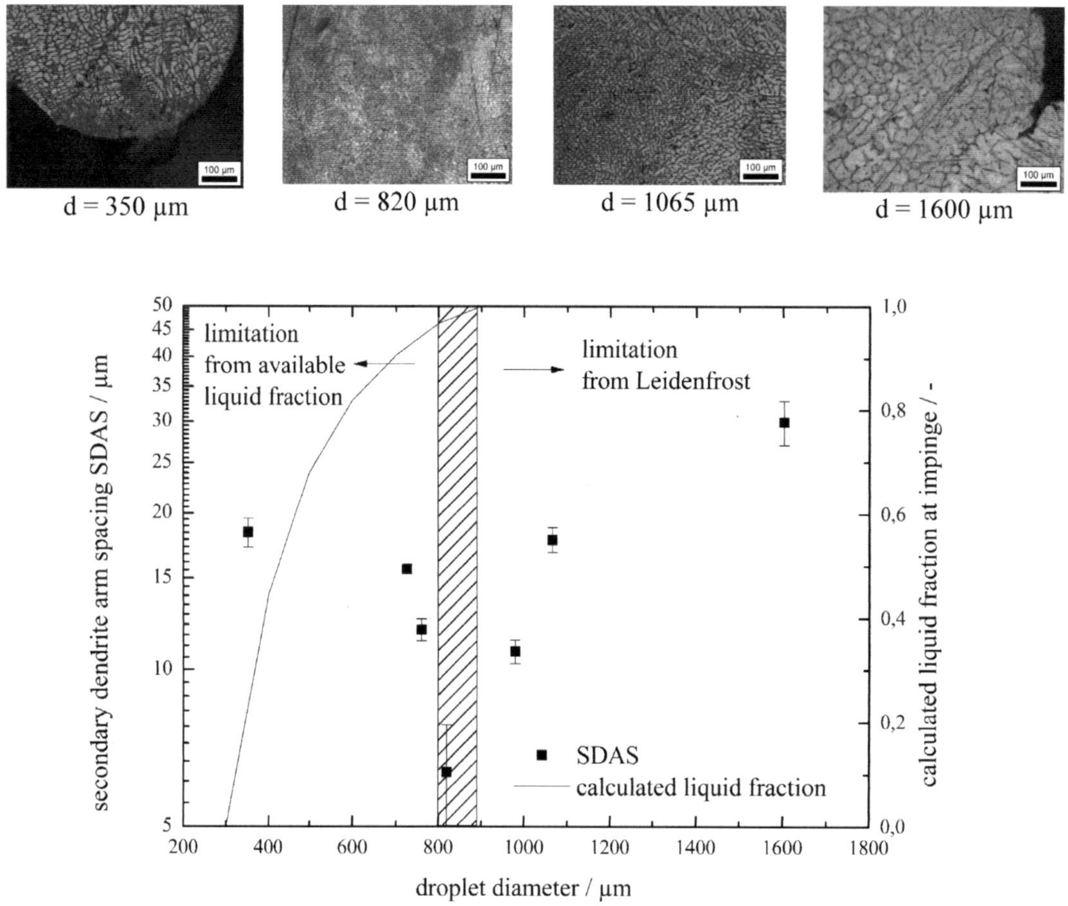

Figure 7. Micrographs of AlCu4.5 droplets of different diameter (falling distance: 0.22 m) and measured secondary dendrite arm spacing (SDAS)

Figure 7 also shows the results of the quantitative secondary dendrite arm spacing. The qualitative trend discussed previously can also be found here. A minimum dendrite size is achieved with a droplet diameter of approx. 820 μm. At this size, the droplet has not cooled below liquidus and has a small enough mass that the dendrites can be formed during impingement. The liquid fraction of the impinging droplets was calculated by the method proposed by Whiskel et al. [8]. Droplets with a diameter smaller then 300 μm are predicted to impinge fully solid, while droplets above 850 μm are predicted to impinge fully liquid. This fits well with experimental results described previously.

5. Conclusion

Through of the experimentation of an iron based metallic glass and the alloy AlCu4.5, it is evident that drop-on-demand droplet generation with subsequent quenching is a proper technique to alter the microstructure of the resulting particles. The amorphous fraction of the metallic glass could be adjusted between 4 and 90 % for an average particle diameter of 360 μm and falling distances between 22 and 82 mm. For the aluminium alloy, the droplet diameter was changed to evaluate the influence of the quenchant. It was found that the effect on the SDAS is limited to a range of droplet diameters: small droplets are almost solidified when they are quenched while large droplets contain too much mass to allow a quenching without the Leidenfrost effect. The droplet diameter with the strongest quenching effect was found to be in the range of 820 μm for a falling distance of 0.22 m.

IOP Conf. Series: Materials Science and Engineering **117** (2016) 012057 doi:10.1088/1757-899X/117/1/012057

6. References

[1] Miura A, Dong W, Fukue M, Yodoshi N, Takagi K, and Kawasaki A 2011 J. of *Alloys and Compounds* **509** 5581-5586

[2] Ellendt N, Uhlenwinkel V, Ahmetovic A, Boddez L, Henein H 2008 Proc. PM2008

[3] Amirzadeh A, Chandra S 2010 *Exp. Thermal and Fluid Science* **34** 1488-1497.

[4] Fang M, Chandra S, Park C B 2007 J. *Manuf. Sci. Eng.-Trans. ASME,* **129** 311-318.

[5] Azar R G, Yang Z, Chandra S, Mostaghimi J 2005 *Int. J. Heat and Fluid Flow* **26** 334-347.

[6] Willis K, Orme M **2003** *Exp Fluids*, **34** 28-41

[7] Gao S, Fritsching U 2010 *Materialwissenschaft und Werkstofftechnik*, **41** 547-554.

[8] Whiskel J B, Henein H, Maire E 2002 *Canadian Metallurgical Quarterly* **41** 97-110.

[9] Ciftci N, Ellendt N, von Bargen R, Henein H, Mädler L, Uhlenwinkel V 2014 J. of *non-cryst. solids* **394-395** 36-42

Effect of convection on the dendrite growth kinetics in undercooled melts of D2 tool steels

J Valloton[1,2,a], D M Herlach[2] and H Henein[1]
[1]Advanced Materials and Processing Laboratory, University of Alberta, Edmonton, Canada
[2]Institut für Materialphysik im Weltraum, DLR, 51170 Cologne, Germany

Email: [a]valloton@ualberta.ca

Abstract Rapid solidification of D2 tool steel is investigated experimentally using the electromagnetic levitation technique under terrestrial and reduced gravity conditions. The microstructures of samples covering a broad range of undercoolings (40 K $\leq \Delta T \leq$ 280 K) are analysed. At low undercooling coarse grained dendritic microstructure is observed, while at higher undercoolings this dendritic feature disappears in favour of a grain refined equiaxed structure. In the latter case, the eutectic carbides are more evenly dispersed throughout the microstructure. The sample solidified in microgravity during parabolic flight experiment exhibits only a few very large grains with twinning relationship. This highlights the effect of convection on grain refinement in this system.

1. Introduction

D2 tool steels are widely used in industry because of their good wear and abrasion properties [1]. This is due to the high volume fraction of carbides that precipitate during the eutectic reaction. The wear resistance can be further improved by reducing the size and evenly distributing carbides [4]. Therefore, a rapid solidification technique is used on D2 tool steel to refine the microstructure. In addition to the cooling rate, the undercooling ΔT prior to solidification has a significant effect on the microstructure during solidification of metals. However, systematic experimental work on the rapid solidification of D2 tool steels is still lacking.

In this study, electromagnetic levitation (EML) is applied to droplets of D2 tool steel. This containerless technique enables direct measurement of the undercooling prior to solidification and of the velocity of the growth front. Comparison between results obtained under terrestrial and reduced gravity conditions allows the investigation of convection effect in dendrite growth dynamics and in the evolution of the microstructure.

2. Experimental

The D2 tool steel is a high carbon, high chromium ferrous alloy with Fe-1.55C-11.8Cr-0.40Mn-0.80Mo-0.80V-0.30Si (all in wt.%). The corresponding pseudo-binary phase diagram is shown in Figure 1 (calculated using ThermoCalc software and the TCFE6 data bank), with the dotted line highlighting the carbon content of D2 steel. Upon cooling, solidification starts with the crystallization of austenite followed by the eutectic decomposition of the interdendritic liquid into austenite and carbide ($L \rightarrow \gamma + (Cr,Fe)_7C_3$). At room temperature, the microstructure consists of ferrite and $(Cr,Fe)_7C_3$. However, during rapid solidification of tool steels metastable supersaturated austenite can be retained in the microstructure [3].

For *in-situ* studies of deeply undercooled D2 tool steel melts the electromagnetic levitation technique (EML) is used as schematically shown in Figure 2. This containerless technique provides a large range of undercoolings prior to solidification and allows direct observation of the solidification process. The release of latent heat during rapid solidification leads to a visible contrast between the solid and the undercooled liquid (the former appearing brighter than the latter). This enables the recording of the solidification front by a high-speed video camera. The temperature is measured contactless by an infrared-pyrometer. High purity 6N He is used to cool down the sample. Detailed information on the EML technique can be found in [4].

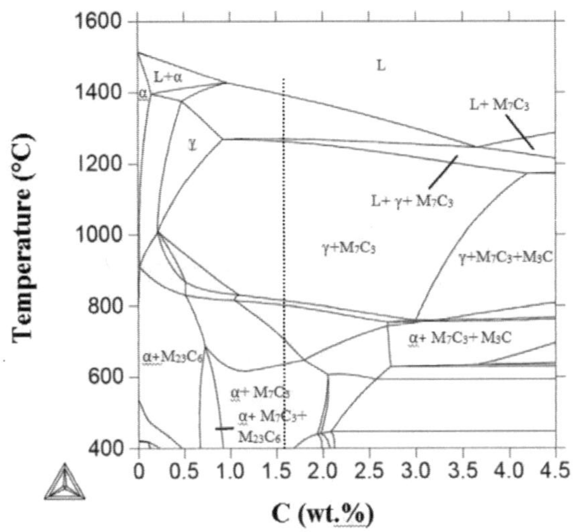

Figure 1. Calculated phase diagram for D2 tool steel from TCFE6 database using ThermoCalc software.

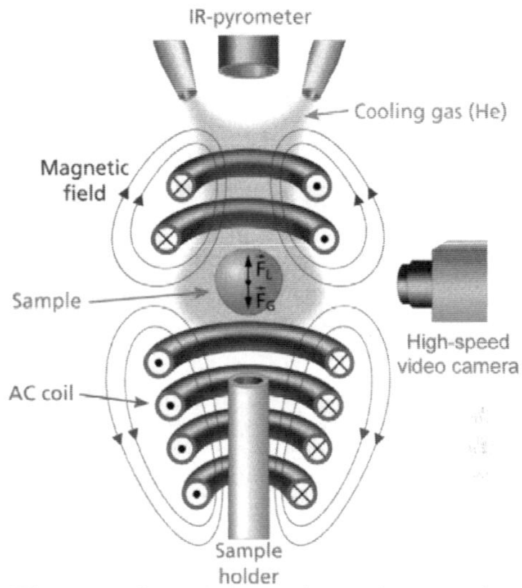

Figure 2. Schematic view of an electromagnetic levitation apparatus.

3. Results and discussion

Figure 3 shows high-speed video snapshots of electromagnetic levitated samples solidified at different undercoolings ΔT. At low ΔT a single coarse equiaxed dendrite is growing through the undercooled sample (a). At higher ΔT several fine dendrites originating from a single nucleation point can be seen (b). At very high ΔT the front becomes spherical with no observable dendritic features (c). As expected, the growth velocity of the front increases with the undercooling.

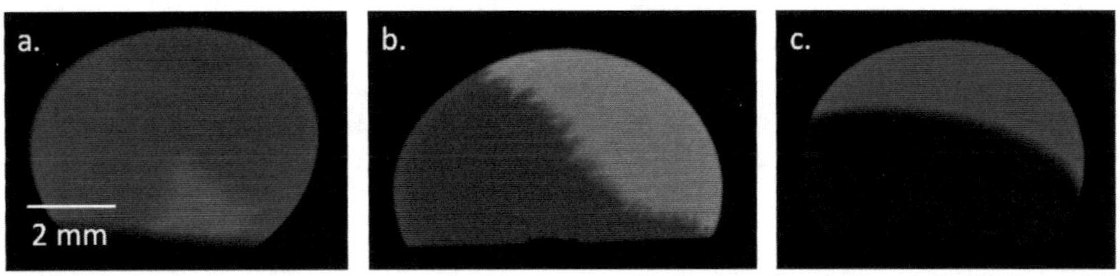

Figure 3. High-speed video images of electromagnetic levitated samples solidified at different undercoolings ΔT. The dark grey area is the undercooled liquid. The light grey region corresponds to the growing solid, which appears brighter due to the release of latent heat during recalescence. a. $\Delta T = 45$ K, v = 0.03 m/s. b. $\Delta T = 179$ K, v = 1.3 m/s. c. $\Delta T = 272$ K, v = 8.4 m/s. A transition from coarse to fine dendrites to spherical front is observed.

Micrographs of the microstructures of these three samples are presented in Figure 4. In all cases, primary austenite and interdendritic austenite + carbide eutectic can be identified. Rapid solidification of D2 in EML completely suppresses the formation of ferrite and martensite. At $\Delta T = 45$ K a classic coarse grained dendritic microstructure is observed, while at higher undercoolings this dendritic feature disappears in favour of a grain refined equiaxed structure. It is assumed that grain refined microstructures were developed by the faster growth dendritic velocity at higher undercoolings and possibly by dendrite fragmentation after recalescence [5]. High undercoolings give rise to much finer microstructures, with the cell spacing decreasing from about 10-30 μm at $\Delta T = 179$ K to 5-10 μm at $\Delta T = 272$ K. This results in a better dispersion of the $(Cr,Fe)_7C_3$ carbides throughout the microstructure.

Comparing the micrographs of samples b and c with the corresponding high speed video recordings of solidification, there appears to be no notable change in microstructure with regards to the solidification front morphology apart from the cell size. This suggests that the spherical front depicted in Figure 3c is composed of very fine dendrites whose features are too small to resolve on the recording.

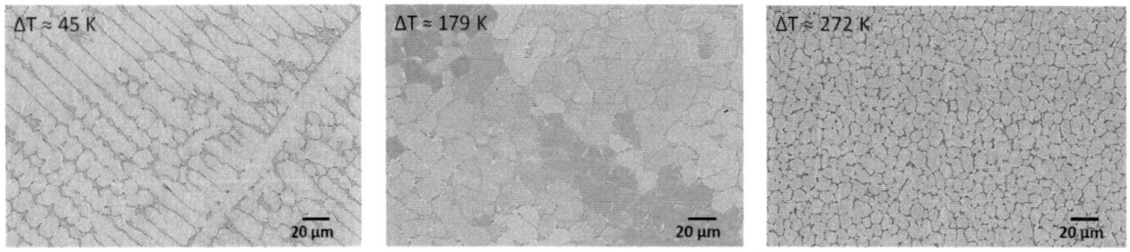

Figure 4. SEM micrographs of cross sections of D2 samples solidified at $\Delta T = 45$K, 179 K, and 272 K. Dendrites or cells of austenite can be observed, surrounded by the eutectic.

Figure 5 shows electron back-scattered diffraction maps (EBSD) and the corresponding pole figures for samples b and c. Both samples exhibit an equiaxed polycrystalline microstructure. At $\Delta T = 179$ K the grain size ranges from 100 to 500 μm while at $\Delta T = 272$ K the range is 10-50 μm, i.e. the highly undercooled sample shows a stronger grain refinement. As the measured growth velocity is high, the dendrites are expected to be very slim. Thus, fragmentation of the primary trunks or secondary arms would be easier to achieve.

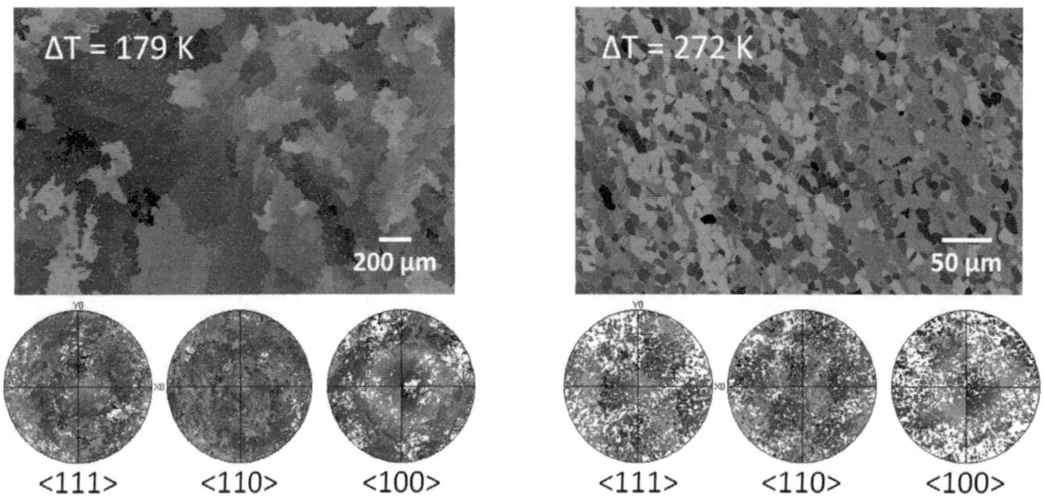

Figure 5. EBSD maps and corresponding pole figures of cross sections of D2 samples solidified at $\Delta T = 179$ K and 272 K respectively.

In order to observe the possible influence of convection on the solidification of levitated D2 tool steel, parabolic flight experiments were performed under reduced gravity conditions using the TEMPUS facility [6]. In µg-EML the fluid flow is about 0.05 m/s, compared with 0.3 m/s during ground experiments [7]. Figure 6 shows the micrograph of a cross section of a sample solidified in such conditions at an undercooling of $\Delta T = 164$ K. At first glance the microstructure appears similar to that of the sample solidified under 1g conditions at $\Delta T = 179$ K. Both the primary austenite and the austenite + $(Cr,Fe)_7C_3$ carbide eutectic are observed. Furthermore, the cell size of the primary phase ranges also from 10 to 30 µm. However, diffraction results are strikingly different. The EBSD map in Figure 7 shows only two different grains, one large grain covering the majority of the cross section (about 1 mm in size) and one small grain in the lower left part. Interestingly, these two grains have a twin relationship with respect to each other. The common <111> direction and the trace of the {111} twin plane containing the three common <110> directions are highlighted on the pole figures. It is not known if solidification started with one single grain with subsequent stacking faults or if twinning occurred at nucleation. EBSD analysis performed on a cross section deeper into the sample (not shown here) revealed three grains with twin relationships, two of them being the same grains as in Figure 7. The size of the largest grain is about 2.3 mm but is probably larger as the EBSD map only covered a portion of the cross section. Such large grains have never been observed in D2 samples solidified in 1g conditions at undercoolings of the same magnitude. This clearly indicates that convection in the undercooled liquid during EML solidification plays a major role in grain refinement in this system. Whether fluid flow promotes fragmentation of the dendrites via mechanical shear or local remelting due to solute redistribution remains however unclear.

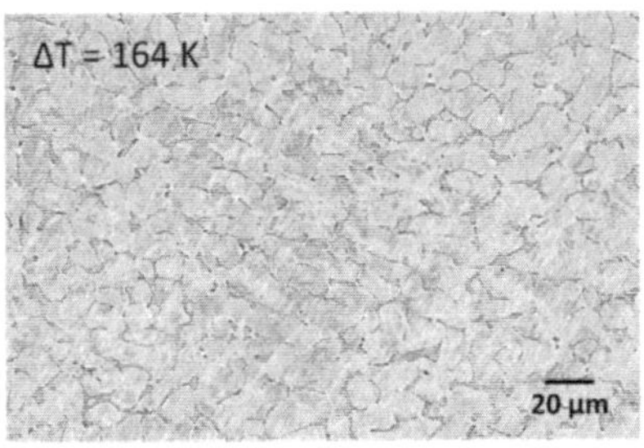

Figure 6. SEM micrograph of a cross section of a D2 sample solidified at $\Delta T = 164$ K in µg.

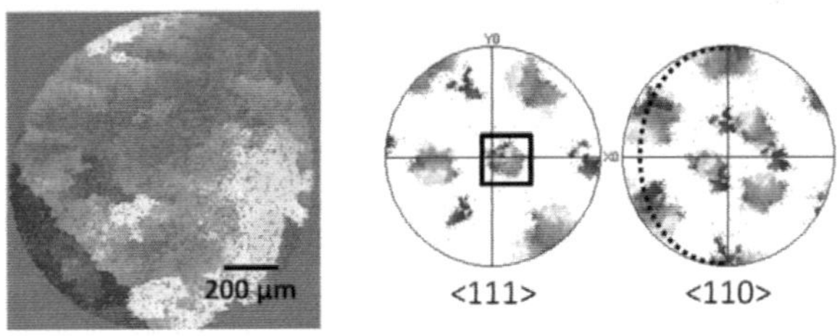

Figure 7. EBSD map and pole figures of a cross section of a D2 sample solidified at $\Delta T = 164$ K in µg. The common <111> direction and the trace of the {111} twin plane are highlighted.

4. Conclusion

In this study, droplets of industrial D2 tool steel were solidified in EML over a wide range of undercoolings. Samples solidified at high undercooling exhibit a fine microstructure with strong grain refinement. This results in a better dispersion of the eutectic carbides as at lower undercoolings. Rapid solidification should thus prove to be a good way to improve the mechanical properties of this alloy.

EBSD analysis showed large differences in the microstructure between droplets solidified on the ground and in microgravity conditions. While the former ones are polycrystalline with many different grains, the sample solidified in µg shows very few much larger grains with twinning relationships. This indicates that fluid flow has a strong influence on grain refinement in this system.

Acknowledgments

The authors are grateful to the TEMPUS crew performing the experiment during parabolic flights. Special thanks are extended to M. Kolbe, T. Volkmann, J. Gegner, and C. Karrasch for support and fruitful discussions. Financial support from the Canadian Space Agency FAST program and the European Space Agency within the CCEMLCC project under contract #4200020277 is acknowledged.

References

[1] Roberts G, Krauss G and Kennedy R 1998 *Tool Steels* (Materials Park: ASM International)
[2] Luan Y, Song N, Bai Y, Kang X and Li D 2010 *J. Mater. Process. Technolog.* **210** pp 536–541
[3] Bhargava A K and Tiwari A N 1992 *Int. J. Rapid Solidification* **7** pp 51–66
[4] Herlach D M 2012 *Solidification of Containerless Undercooled Melts*, ed D M Herlach and D M Matson (Weinheim: Wiley-VCH), pp 9–16
[5] Schwarz M, Karma A, Eckler K and Herlach D M 1994 *Phys. Rev. Let.* **73** pp 1380–83
[6] Lohöfer G and Piller G 2002 *Proc. 40th AIAA (Reno)*
[7] Lee J, Matson D M, Binder S, Kolbe M, Herlach D M and Hyers R W 2014 *Metal. Mater. Trans. B* **45** pp 1018–23

IOP Conf. Series: Materials Science and Engineering **117** (2016) 012059 doi:10.1088/1757-899X/117/1/012059

Instability of nanoscale metallic particles under electron irradiation in TEM

X Y Chen[a], S G Zhang[b], M X Xia[c] and J G Li[d]
School of Materials Science and Engineering, Shanghai Jiao Tong University, Shanghai, 200240, China

Email: [a]jennychenxy@sjtu.edu.cn, [b]sgzhang@sjtu.edu.cn, [c]mingxu.xia@sjtu.edu.cn and [d]lijg@sjtu.edu.cn

Abstract. The stability of nano metallic glass under electron beam in transmission electron microscope (TEM) was investigated. The most common voltage of TEM used in metallic materials characterization was either 200 kV or 300 kV. Both situations were investigated in this work. An amorphous metallic particle with a dimension of a few hundred nanometers was tested under 300 keV electron irradiation. New phase decomposed from the parent phase was observed. Moreover, a crystal particle with the same composition and dimension was tested under 200 keV irradiation. Decomposition process also occurred in this situation. Besides, crystal orientation modification was observed during irradiation. These results proved that the electron beam in TEM have an effect on the stability of nanoscale samples during long time irradiation. Atomic displacement was induced and diffusion was enhanced by electron irradiation. Thus, artifacts would be induced when a nanoscale metallic sample was characterized in TEM.

1. Introduction

Amorphous alloys (metallic glasses) are non-equilibrium materials, which will process phase transformations when heated or mechanically forced. Crystallization is the most important phase transition of amorphous alloys. Crystallization is not only a question related to the service life of glass materials, but also a scientific question including glass transition process and nucleation and growth in glass state. There is another phase transition which is frequently found and gets more attention in metallic glasses, that is phase separation [1]. Phase separation is originated from the positive mixing enthalpy between atoms [1, 2]. It is found that metallic glasses show obvious different mechanical properties after phase separation occurs [3, 4].

In-situ observation is favored in phase transformation investigation. As to now, in-situ observations of metallic glasses are mainly based on small angle scattering (SAS) [5] and transmission electron microscope (TEM) [6-8]. To induce phase transition, samples are reheated during SAS examination, while in TEM observation the interaction between the electron beam and the sample could stimulate or accelerate phase transitions. Moreover, TEM can provide direct morphology observation. Thus, TEM is more convenient and useful than SAS. However, the energy of the electron beam should be more than mega-electron-volt (MeV) to induce crystallization in metallic glass [6-8]. On the contrary, it seems that 200~300 keV electron beam is safe for metallic samples to be characterized without extra artifacts.

However, considering the size effect, small samples with large surface to volume (S/V) value may be unstable even under 200~300 keV electron irradiation. The unstable phenomenon is confirmed in this work. Not only the metallic glass particle, but also the crystal particle process transformations during electron irradiation.

2. Experiments

$(Fe_{62.6}Mn_{12.5}Cr_{9.3}Ni_{4.6}Si_{11})_{90}C_{10}$ alloy with low glass forming ability (GFA) and phase separation potential was chosen. Low GFA was aimed to limit the dimension of the glass phase in the composited samples as prepared. The phase separation potential was expected on Fe-Mn, Fe-Cr, Fe-Ni and Mn-Cr atomic pairs, whose mixing enthalpy are either positive or close to zero [9].

The master alloy was prepared by induction-melting in an argon atmosphere and injected into copper mold. To obtain nano particles, chemical collection process was used. Slices cut from the cast sample were firstly polished to 50 μm thick. Then they were soaked in a mixed solution of H2O:H2O2:HF (10:50:3) to etch for 6 hours. After that the solution was diluted with alcohol to ease corrosion. In this way, the non-corrosive solution containing metallic particles was obtained. At last, the solution was dropped onto the copper grid to make the particles attach to the carbon film on the copper grid for TEM (JEOL JEM-2100, 200kV and JEM-2100F, 300 kV) observation. During the continuous irradiation and graphing, each bright field (BF) image was taken under the same condition. Each selected area diffraction (SAD) pattern was taken 1.5 min after the corresponding BF image taken. The exposure time for each SAD pattern was kept the same.

3. Results and discussion

3.1 Amorphous particle under electron irradiation

Figure 1 shows the evolution of an amorphous particle under electron irradiation in-situ in TEM (JEM-2100F, 300 kV). Figure 1 (a, b, c) present the BF images of the particle during irradiation, while (d, e, f) show the SAD patterns corresponding to (a, b, c), respectively. The particle is made up of several small pieces, and the thickness is not homogeneous. Around the particle, new phases with round shape and dark contrast come up during irradiation. The new phase concentration areas are marked by arrows. Even in the middle of the particle (the arrow place), the contrast changed. However, the figurations of the new phases in the middle area are not as clear as the outside ones. It may be caused by the overlapping of the small piece samples with new phases. Each SAD pattern in (d, e, f) consists of 4 halo rings, indicating the particle keeps amorphous structure during irradiation. However, 4 halo rings are not regularly seen in SADs of amorphous alloys. The 4 halo rings here imply the amorphous structure is partly ordered. R_1 is the brightest one, while the others show lower brightness. The radiuses of R_1~R_4 are almost the same from (d) to (f), but the brightness is increasing. Halo rings in SAD represent an amorphous microstructure, while brightness increasing indicates the local clusters taking part in diffraction are coming up during irradiation. That is to say, electron irradiation promotes the formation of short-range-ordered clusters. Thus, it is concluded that the microstructure of the particle is kept in an amorphous state under irradiation, but the local atomic structure is ordering.

The new dark phases decomposed from the parent phase are supposedly related to phase separation. These dark phases nucleate and grow up as (b, c) shown. While phase separation occurs, SAD is suspected to present different rings since the new phase has different local atomic distances from the parent phase. However, SAD rings in (d, e, f) present the same radiuses. This can be explained that the diffraction intensity of the new phase is not strong enough. To clarify this question, energy dispersive spectrometer (EDS) exam is needed. If it is phase separation, the composition of the dark phase will be rich of one or more elements and the light phase will be rich of the other elements. However, the electron beam sport is larger than the decomposed phase. Also, the magnetism of the sample lead drifting of electron imaging, which will disturb the beam spot position. These facts make it unreliable for EDS in TEM to reflect the real composition difference of the decomposed phase and the matrix phase.

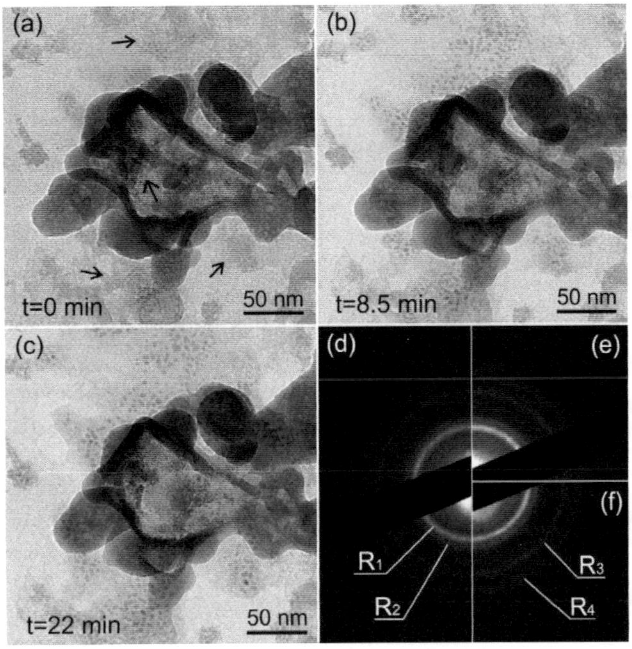

Figure 1. In-situ evolution of amorphous particle under electron irradiation (300 keV). (a, b, c) are bright field (BF) images. (d, e, f) are the SAD patterns corresponding to (a, c, e), respectively. Shot time of each picture is marked in the bottom-left corner. The radius of the SAD halo rings are R_1 3.58 nm^{-1}, R_2 4.07 nm^{-1}, R_3 5.80 nm^{-1} and R_4 6.88 nm^{-1}.

3.2 Crystal particle under electron irradiation

Figure 2 shows the evolution of a crystal particle of the same composition with the amorphous particle under electron irradiation in-situ in TEM (JEM-2010, 200 kV). Figure 2 (a, b, c) present the BF images of the particle during irradiation, while figure 2 (d, e, f) shows the SAD patterns corresponding to figure 2 (a, b, c) respectively. The particle is gathered of 3 small particles, and the microstructure is non-homogeneous. Around the particle, new phases appear as same as figure 1 shows. Moreover, the round dark new phases occupy the whole particle in figure 2 (c). The 3 small particles grow up into one particle. The SAD in figure 2 (d) shows some diffraction spots located in several rings, indicating the particle consists of nanocrystals. R_1 and R_2 are the main rings marked by arrows and their radiuses are 3.39, 4.86 nm^{-1}, respectively. The interplanar spacing of R_1 is 0.295 nm, of R_2 is 0.206 nm. SAD in figure 2 (e) is almost the same with that in figure 2 (d), except the brightness decreased a little. In figure 2 (f), there is one diffraction ring R_2 left and less few spots at further radius. R_1 disappeared and only R_2 remained with an increased brightness, indicating crystal lattice orientation modification is performed in the crystal particle under electron irradiation.

Phase separation not only occurs in the amorphous particle, but also in the crystal particle under electron irradiation. Nevertheless the SADs in figure 2 can't reflect the decomposition process either. The mechanism of crystals modifying orientation under electron irradiation can be crystal rotation or re-crystallization. Re-crystallization is preferred to take charge of the orientation adjusting process because the brightness of SAD pattern decreased in (e) and increased in (f). If rotation is the main mechanism, the brightness of R_2 will increase all the way.

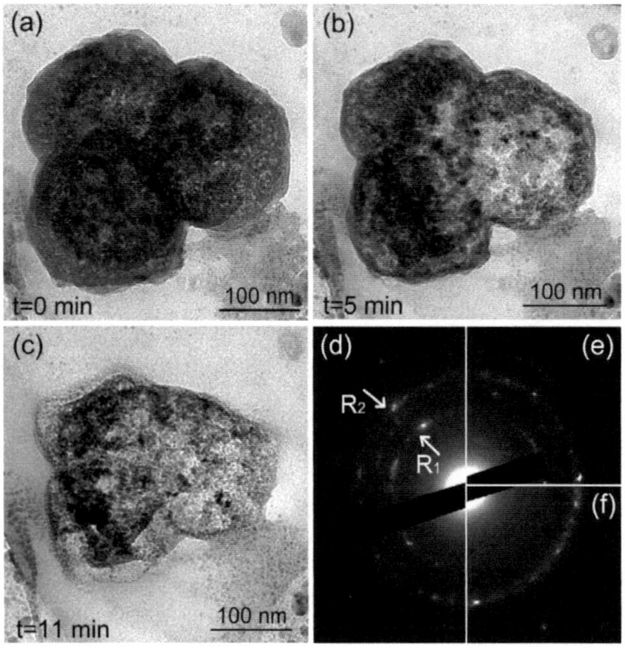

Figure 2. In-situ evolution of crystal particle under electron irradiation (200 keV). (a, b, c) are bright field (BF) images. (d, e, f) are the SAD patterns corresponding to (a, b, c), respectively. Shot time of each picture is marked in the bottom-left corner. (d) and (e) show similar diffraction patterns, while (f) shows less diffraction rings and spots than (d, e). The radius of the main SAD rings are R_1 3.39 nm^{-1} and R_2 4.86 nm^{-1}.

3.3 Effect of electron irradiation on the samples

The interaction between the incident electron and the metallic sample includes electron excitation and knock-on displacement. The former effect is between electron and electron, while the latter is between electron and nucleus. Electron excitation can release very fast in metallic materials thus it doesn't cause significant influence. Knock-on displacement means the incident electron hit a nucleus to move. To achieve knock-on displacement, the electron beam requires high energy to satisfy the threshold energy of atoms [6-8]. The threshold voltage of electron beam to stimulate knock-on displacement of the elements composed in the particles are Fe (370 kV), Mn (360 kV), Cr (440 kV), Ni (440 kV), C (200 kV) and Si (140 kV) [7]. Thus, only carbon atoms and silicon atoms are displaced directly by electron irradiation of 200~300 keV.

It is noticeable that the shape of particles in figure 1 and figure 2 are changing with irradiation as figure 3 shows. The shape change route of the particle is 1a-1b-1c or 1a-2b-2c. The areas of the particles are decreasing. It implies that the atoms have moved a long distance during transitions under electron irradiation. Atomic diffusion is the key of the transitions under electron irradiation. The atomic radius of carbon (0.077 nm) is smaller than the other atoms, whereas the atomic radius of silicon (0.117 nm) is similar to the metallic elements (0.112~0.125 nm) [9]. That is to say, the vacancies induced by the displacement of Si can be taken up by the other metallic elements. Thus, the diffusion of all the atoms in the sample is enhanced by low energy electron beam.

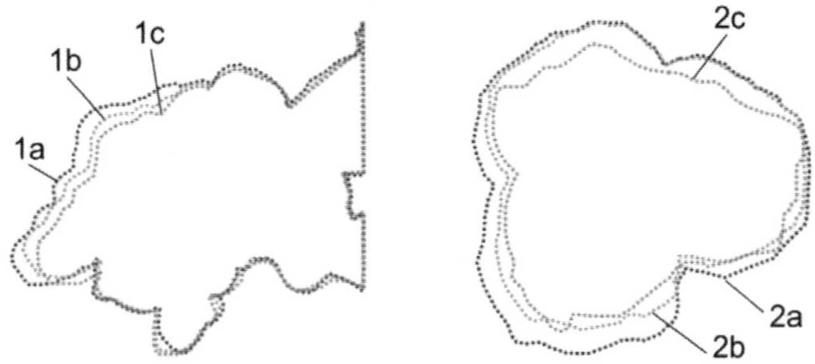

Figure 3. The shape change of the particles during electron irradiation is painted by dashed lines. The left one is the particle in figure 1, while the right one is the particle in figure 2. The shapes of particles corresponding to each image are marked. For example, '1a' means corresponding to figure 1 (a).

Another factor improving atomic diffusion is the large S/V value of the nanoscale particles. As know to all, atoms at surface are inherent of higher energy than the inside atoms. A large S/V value indicates a large amount of atoms in the particle are carrying high energy. These atoms are active and are easy to diffuse. It is concluded that, transitions are induced by electron irradiation through atoms diffusion.

Radiation-induced diffusion provides a means for the atoms to rearrange to a more stable form of nanoscale metallic particles. The voltage of the electron beam and the size of the particle both have an effect on the stability of metallic particle inside a TEM.

4. Conclusions

Decomposition occurred in the amorphous alloy particle under 300 keV electron irradiation and in the crystal particle under 200 keV irradiation. Moreover, orientation modification occurred in the crystal particle. It proved that 200 or 300 kV TEM would introduce artifacts during long time observation.

The atomic displacement was induced by electron irradiation even with low energy as 200 keV. It is suggested that the electron irradiation should be reduced when nanoscale amorphous alloys was characterized.

References

[1] Kim D H, Kim W T, Park E S, Mattern N and Eckert J 2013 *Prog. Mater. Sci.* **58** 1103
[2] Park E S, Na J H and Kim D H, 2010 *J. Appl. Phys.* **108** 053515
[3] Du X H, Huang J C, Hsieh K C, Jang J S C, Liaw P K, Chen H M, Chou H S and Lai Y H 2009 *Adv. Eng. Mater.* **11** 387
[4] Yao K F and Zhang C Q 2007 *Appl. Phys. Lett.* **90** 061901
[5] Wang X L, Almer J, Liu C T, Wang Y D, Zhao J K, Stoica A D, Haeffner D R and Wang W H 2003 *Phys. Rev. Lett.* **91** 265501
[6] Nagase T, Sanda T, Nino A, Qin W, Yasuda H, Mori H, Umakoshi Y and Szpunar J A, 2012 *J. Non-Cryst. Solids* **358** 502
[7] Nagase T, Nino A, Umakoshi Y 2011 *J. Alloys and Comp.* **509S** S202
[8] Nagase T, Nino A, Hosokawa T and Umakoshi Y 2007 *Mater. Trans.* **48** 1651
[9] Takeuchi A and Inoue A 2005 *Mater. Trans.* **46** 2817

IOP Conf. Series: Materials Science and Engineering **117** (2016) 012060 doi:10.1088/1757-899X/117/1/012060

Criterion function for predicting freckles in CMSX-4 during directional solidification

B Pustal[1a], D Ma[1], N Warnken[3], E Subasic[2], J Jakumeit[2] and A Bührig-Polaczek[1]
[1]Foundry Institute at RWTH Aachen University, Intzestr. 5, 52072 Aachen, Germany
[2]ACCESS e.V. at RWTH Aachen University, Intzestr. 5, 52072 Aachen, Germany
[3]University of Birmingham, School of Metallurgy and Materials, Edgbaston, Birmingham, B15 2TT, United Kingdom

E-mail: [a]b.pustal@gi.rwth-aachen.de

Abstract. In the present work the impact of curvature effects on freckle formation in CMSX-4 during directional solidification was modelled and simulated. Modelling work is based on a four-step criterion function for predicting freckles. First, only surface elements are taken into account. Second, critical cross-sectional areas are identified where freckle formation is possible. As a new aspect, third, curvature effects are taken into account for predicting freckles in radiation shadow. Fourth, the Rayleigh number is calculated based on a permeability function of dendrite arm spacing and fraction of liquid. In the thus identified areas sufficient thermo-solutal convection takes place to initiate freckle formation. With the new model freckles can only appear in areas where the thermal temperature profile is concave, that is, at locations where the temperature gradient is positive against the surface normal. The present model was tested simulating step samples. For that, the individual parts of our industrial Bridgman furnace were input and joined in one FEM model. The thermal results clearly show concave solidus isothermal at the shadow faces of the step sample in the cluster. Freckles are predicted at the corresponding locations. When comparing simulated freckle formation with experimental findings, good correlation was found. Furthermore, we could predict both the transition length that occurs when the cross-sectional area increases stepwise and the run length when the cross-sectional area is reduced.

1. Introduction

Directionally solidified or single crystal castings of superalloys are characterized by their excellent high temperature properties. However, the production of defect-free components is more problematic with increasing component size. Typical structural defects are particularly freckles, which are observed as aligned chain-like small defect grains predominantly at the outer surface of thick-walled casting areas. Because of the negative influence of freckles on the life of the components, this structure defect is usually a discard criterion and should therefore be avoided. The solidification of nickel based superalloys is characterized in that the interdendritic residual melt is enriched with the light elements Ti and Al, while the heavy elements W and Re are depleted. This lower density results in interdendritic upward convection in the mushy zone. From broken and melted off dendrites the chain-like freckles are formed. So far, three important impact factors of freckle formation are known: (1) the chemical composition, that is, alloys with high contents of T, Al, W, and Re show increasing probability of freckles; (2) the size of the castings, freckles occur primarily in thick-walled castings rather than in thin-walled; (3) the solidification parameters, a low temperature gradient and a slow rate of solidification promote freckles formation.

2. State of the art

Freckles appear under certain conditions, primarily at the surface of directionally solidified castings on what is referred to as the *surface effect*. Copley et al. [1] report that in directional solidification virtually all freckles found occur at the external surface of the castings. Thus when it comes to freckle formation, a critical cross-sectional size of the component must be achieved, which refers to the so-called *size effect*. In thin sections freckle formation does not occur. For this purpose a minimum cross-sectional area A_0 and a diameter D_0 is defined. From the local geometry the equivalent diameter $D_{eq} = 4\ A\ /\ U$ is calculated, wherein A is the local cross-sectional area and U is the circumference. From this, the critical diameter criterion for freckles formation derives as follows:

$$I_D = D_{eq} / D_0 \tag{1}$$

Ramirez and Beckermann [2] investigated correlations of freckle formation depending on material and process based on different Rayleigh number definitions. Auburtin [3] examined freckle formation in directionally solidified rods made of the three superalloys (Waspaloy, Mar-M247, CMSX-4). In his research, he developed a modified Rayleigh based criterion that incorporates the anisotropy of the mushy zone, permeability and gravity to describe the influence of process parameters, and the angle of the growth front on freckle formation. This is known as the *process and material effect*. He assumes that the thermo-solutal convection takes place either perpendicular or parallel to the primary dendrites:

$$Ra_x = F_x /(\eta D / K_x^2) \qquad \text{and} \qquad Ra_z = F_t /(\eta D / K_\alpha^2). \tag{2}$$

The driving force for freckle formation is given by:

$$F_x = g\ (d\rho / dT)\ G_F \sin(\alpha) \qquad \text{and} \qquad F_Z = g\ (d\rho / dT)\ G_F. \tag{3}$$

For the modified Rayleigh criterion of Auburtin the parameter set in table 1 was used. Viscosity and thermal diffusion constant were assumed to be concentration- and temperature-independent in the solidification range. The critical specific change in density that results in the formation of freckles was determined experimentally with 30 kg/m^3/K. By microprobe measurements on quenched samples the concentration profiles could be determined and transformed into density profiles by weighted averaging. By inserting these parameters in equations 2 and 3, the Rayleigh numbers can be simplified as follows:

$$Ra_x = 4.62 \cdot 10^{-15} G_F (G_L V)^{-2.75} \sin(\alpha) \qquad \text{and} \qquad Ra_Z = 3.64 \cdot 10^{-14} G_F (G_L V)^{-1.32}. \tag{4}$$

If one of the thus determined Rayleigh numbers exceeds the critical value Ra^*(CMSX-4) = 10^{-8} according to Auburtin [3], freckles are formed. The criterion for freckle formation corresponds to:

$$I_{Ra} = Ra / Ra^*, \tag{5}$$

with Ra = max(Ra_x, Ra_z). By combination of the criterion for the critical cross-section and the Rayleigh number, a generalized criterion can be formulated as follows:

$$I = I_D \cdot I_{Ra} \tag{6}$$

Table 1. Parameters and physical data for the calculation of Rayleigh numbers.

parameter	value and unit	Description	Ref.
g	$= 9.81$ [m/s^2]	acceleration due to gravity	
D_T	$= 9*10^{-6}$ [m^2/s]	thermal diffusivity	
η	$= 4*10^3$ [kg/ms]	dynamic viscosity	
f_L	$= 0.5$ at the location of freckles	fraction of liquid	
$d\rho/dT$	$= 30$ [kg/m^3/K]	specific density change	
G_L	process parameter [K/cm]	gradient at the growth front	
G_S	process parameter [K/cm]	gradient at solidus	
G_F	$= 0.8\ G_L + 0.2\ G_S$ [K/cm]	gradient at location of freckles formation	
K_x	$= 3.62 * 10^3 f_L^{3.34} \lambda_1^{0.699} \lambda_2^{2.73}$ [m^2]	perpendicular permeability	[4]
Kz	$= 3.75 * 10^{-4} f_L^2 \lambda_1^2$ [m^2]	parallel permeability	[4]
λ_1	$= 0.0015/(G_L * V)^{0.33}$	distance between primary dendrites	[5]
λ_2	$= 0.0004/(G_L * V)^{0.42}$	dendrite arm spacing	[5]
V	= process parameter [m/s]	solidification rate	
α	= process parameter [°]	angle at the growth front	

3. Work carried out

The modified Rayleigh criterion of Auburtin [3] considers only the growth angle of the solidification front, the curvature has been neglected. In fact, the concave curvature and the associated *curvature effect* is a basic requirement for freckle formation on the surface, as shown by Ma et al. [6]. In the present study the curvature requirement was included in the modified criterion by a critical angle β which was defined between the surface normal of the cast component and the growth front. If $\beta > 90°$, we obtain a concave solid-liquid interface and freckle formation will occur. The overall model with the extension to the curvature requirement of the solidification front has been implemented as shown in figure 1.

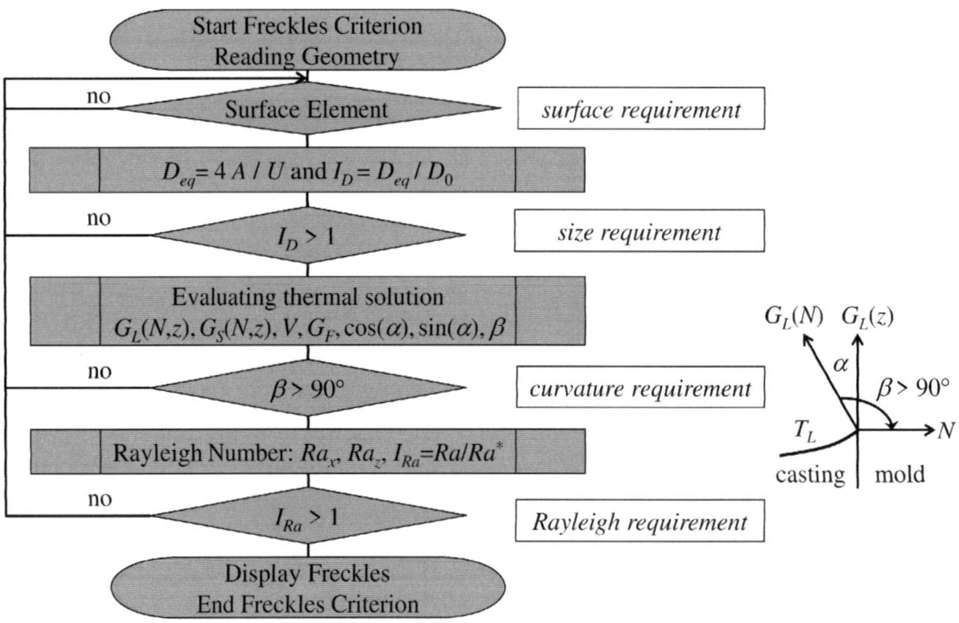

Figure 1. Flow chart for calculating the freckles criterion including curvature requirement.

The additional effect of the curvature now allows also the prediction of shadow effects in freckle formation. For this purpose, the geometry of a cluster mold was created linking six step samples. Also the industrial Bridgman furnace, in which the experiments were carried out, was entered in its geometry and the individual components were linked together. The overall model consists of chill plate, cast metal, shell, lower / upper heater, inner / outer baffle, cooling, insulation and wall material. The materials used have been provided with the appropriate thermo-physical data and the initial (cast metal and shell 1430°C), boundary (upper heater 1480°C, lower heater 1500°C, chill plate, and cooling 20°C) and run-time conditions (velocity of withdrawal 1mm/min) adapted to the experiments carried out. In the simulation model CASTS, that Laschet et al. [7] described in detail, the cast component is withdrawn in the cooling zone with a fixed withdrawal velocity, calculating the view factor matrix for each lowering step. Through the view factor matrix the radiation exchange among surfaces is calculated taking account of the variation due to the withdrawal. In this way the temperature profile impacted by shadow effects can be taken into account. The temperature field determined from this simulation is evaluated in the extended criterion function for freckle formation regarding temperature gradients and solidification conditions.

4. Results

The basis for the prediction of freckles in the present model is the geometry of the sample and the temperature result through processing the sample in the Bridgman furnace. For studying the influence of geometry changes on freckle formation sample geometries were used with varying cross sections. The gradation of the cross-section is 1.5mm and the step length 25mm. Figure 2 shows one example in a sectional view along the centre line of the xz-plane of the temperature field of a cluster of step samples after 5010s at a constant drawing rate of 1mm/min. The isothermal lines for liquidus are drawn in green and for solidus temperature in blue. It is clearly seen that under the chosen process conditions the solidification front is far ahead of the baffle. For each withdrawal step, the view factor matrix is recalculated. This shows the prediction of radiation shadows as they exist in the centre of the cluster. This can be recognized on the slope of the solidus isothermal lines towards the middle of cluster. For the outer step samples the concave curvature of the solidus isotherms into the direction of the centre of the cluster can be seen. We can also see that the liquidus isothermal line spreads at the corners where the cross-sectional area increases. Due to the Reynolds and the new curvature requirement no freckles will appear at corresponding locations.

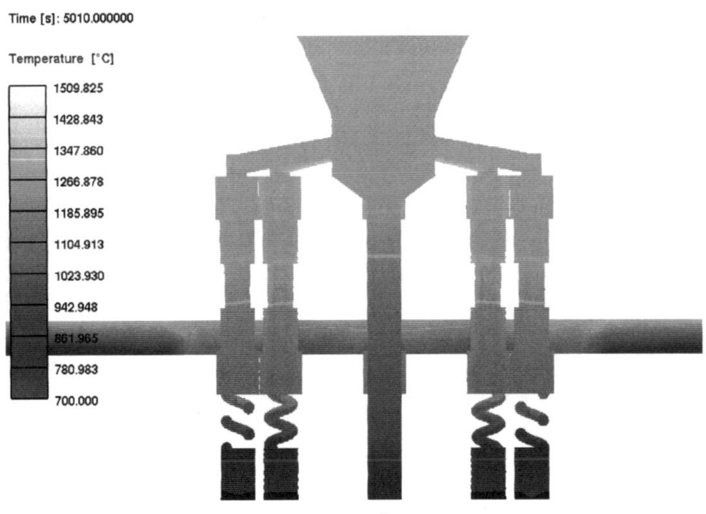

Figure 2. Temperature field during the withdrawal of a cluster of samples with levels for liquidus and solidus isothermal lines.

To validate the simulation results, we come back to the results published by D. Ma et al. [6]. Freckles result, as mentioned, preferably on the sample surface. This work showed (cf. Figure 3d), that when an abrupt change in cross section occurs, the continuity of the surface is broken, and freckle formation supportive conditions have to rebuild on the upper level. In a cross-sectional widening the freckles on the lower end can arise only when a sufficient "support zone" was established. In this support zone are sufficient segregation and density inversion to allow convection, which then leads to freckle formation above the transition length. In the case of a cross-sectional reduction there exists already a large interdendritic mushy zone below the new surface. Freckles may appear on the new interface immediately.

The freckles criterion indicates the relative percent for exceeding the critical Rayleigh number and diameter in calculations. In the case of the CMSX-4 alloy used, the critical Rayleigh number is 10^{-8}, which is exceeded by up to 25% according to the prediction in figure 3. Well predicted is the effect of the transition length, which can be seen on the steps with cross-sectional enlargement. Conversely, the results also show clearly that a run length exist before the necking. The model revised to curvature effects clearly shows the maxima of the freckle formation in good agreement compared to the experiments. Before the introduction of the curvature requirement for freckle formation, freckles were predicted over the entire circumference of the sample (cf. figure 3a), whereas now only in the shadow zone a freckle probability is predicted (b-c), which is much more consistent with the experimental observations (d). The discrepancies in the level of freckle probability before (a) and after implementation of the curvature requirement (b-c) are due to minor errors corrected alongside.

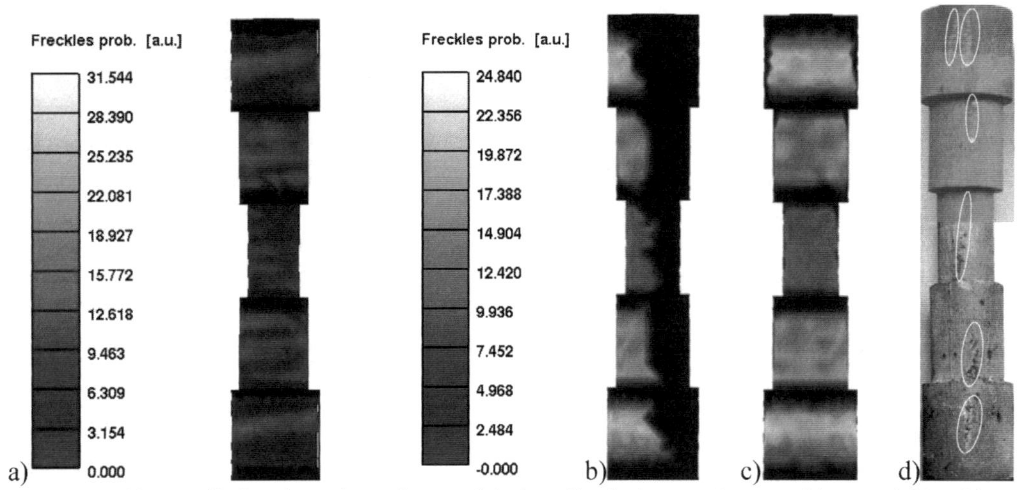

Figure 3. Freckle prediction based on the model described above, a) not taking and b) taking into account the new curvature requirement showing transition lengths with increasing or decreasing cross sections, c) view of the shadow side and d) comparison with experiment by Ma et al. [6].

5. Summary and conclusions

The proposed modified model for freckle formation of Auburtin was implemented numerically and extended to curvature effects in this work. Thus, there is a four-step approach to freckle formation: surface requirement, size requirement, curvature requirement, and the Rayleigh requirement. The model was compared with results of experimental castings. With the developed model, it was possible to predict the freckles that occur exclusively on the shadow side. Furthermore, the already known transition lengths when widening the cross-section as well as the run lengths when reducing the cross-section of the specimen have been predicted. The run lengths can also be seen in the experimental results to some extent. If the material-specific characteristics of an alloy are well known, critical zones in the cast component can be predicted with the extended model on the basis of the process parameters without much computational effort. It thus provides a practical approach for predicting freckles. As a new aspect, Ma et al. [6] observed freckles also at edges of samples with rectangular cross sections. This together with internal freckles that have also been seen in their study

IOP Conf. Series: Materials Science and Engineering **117** (2016) 012060 doi:10.1088/1757-899X/117/1/012060

should be integrated with the prediction. Also surface structures have been varied and quantified with respect to freckle formation by Ma et al. [8], which is a point of future interest for modeling and simulation.

Acknowledgment
This work was funded by the German Research Foundation DFG within the project BU1072 / 15-2. In particular, we would like to take this opportunity to thank M. Schallmo for the implementation of a previous model to freckle formation.

References
[1] Copley S M, Giamei A F, Johnson S M and Hornbecker M F 1970 *Met. Trans* **1** 2193
[2] Ramirez J C and Beckermann C 2003 *Met. Mat. Trans* **34A** 1525
[3] Auburtin P 1996 PhD Thesis, U. of British Columbia, Canada
[4] Poirier D R 1987 *Met. Mat. Trans.* **18B** 245
[5] Bouse G K and Mihalisin J R 1989 Superalloys, Supercomposites and Superceramics; Tien J K et al. Ed. 99
[6] Ma D, Qiang W and Bührig-Polaczek A 2011 *Met. Mat. Trans.* **43B** 344
[7] Laschet G, Neises J and Steinbach I 1998 Lecture Notes of 4iéme Ecole d' été du CNRS, Porquerolles, France C8 1-42
[8] Ma D, Mathes M, Zhou B and Bührig-Polaczek A 2011 *Adv. Mat. Res* **278** 114

Numerical study of influence of inclusion movement on channel segregation in Fe- 0.21 wt% C- 0.1 wt% S alloy

D R Liu[a], Z P Yang, Q Y Sun, L P Wang, and B X Ma
School of Materials Science and Engineering, Harbin University of Science and Technology, No.4 Lin Yuan Road, Xiang Fang District, Harbin 150040, China

E-mail: [a] nacylake@126.com

Abstract. A two-dimensional continuum model on solute and heat transport, and fluid flow is developed to numerically investigate the influence of inclusion movement on the development of channel segregation. A trajectory model is used to track the moving path of inclusion particles. Inclusion movement affects the flow field in simulation by means of interfacial friction coefficient. Simulations are performed on the Hebditch-Hunt casting. A parametric study is carried out to study the effects of with and without inclusions, and diameter (5×10^{-6} m, 10×10^{-6} m, 20×10^{-6} m and 40×10^{-6} m) of inclusions on channel segregation. It is found that the channel segregation is strengthened with the consideration of inclusion movement. Compared to other diameters, inclusions with diameter 20×10^{-6} m are found to enhance the channel segregation. This is because the larger inclusions (40×10^{-6} m) present a faster floating velocity that reduces the interaction time between inclusion upward movement and the development of solidification front, and then lessens the disturbance to solidification front that is important to the initialization of channel segregation. The upward movement of smaller inclusions (5×10^{-6} m and 10×10^{-6} m) cannot greatly increase the upward velocity of fluid flow. Therefore, the formation of channel segregation is less affected.

1. Introduction

One particularly striking macrosegregation form during solidification of casting is channel segregation which is considered to be the most harmful defect, due to the fact that it cannot be removed in the manufacturing process [1]. Channel segregation typically occurs in alloys experiencing a slow solidification rate. They are finger-like A-shape or V-shape segregation bands [2]. Those bands are characterized to have a higher segregation than that of the nearby solid regions. Elements with the low melting point cluster here, corresponding to detrimental properties [2].

In order to deeply understand the fundamental origins of channel segregation, numerical modelling of channel segregation during solidification of metallic alloy has received great interest. Flemings believed that local dissolution characterized as a decrease of solid fraction and the resulting flow instability lead to channel segregates [3]. A series of research work of Založnik and Combeau showed that channel segregation stems from a perturbation to the advancement of the mushy zone front induced by the interdendritic liquid [1,4]. The local dissolution (decrease of solid fraction) does not necessarily occur to initiate a channel segregate. Li et al. developed a two-phase columnar solidification model and numerically investigated the formation of channel segregation [5, 6]. The simulations have justified that Rayleigh number is a qualitative indicator for characterizing the origin of channel segregates. Our previous work investigated the influence of increasing the initial concentration of sulphur on the formation of channel segregation during solidification of iron-carbon-sulphur ternary system [7]. Simulations showed that channel segregation occurs accompanied by the local dissolution at the mushy zone front, when a high initial concentration of sulphur is considered. For a low initial concentration, however, the local dissolution does not necessarily to induce channel segregation.

Although a lot of work has been done to understand channel segregation formation, the influence of inclusion transport has not been considered until now. Inclusions exist as a kind of impurity in the casting process, whose movement was believed to affect the flow field and then change the segregation pattern [8]. Thus the goal of this study is to analyze the impact of inclusion motion on the channel segregation. Parametric studies regarding different inclusion sizes are performed, in some degree, to quantify the conditions necessary for inclusions to strength channel segregation.

2. Model description

The model for the solidification of ternary alloys is an extension of the model for binary alloys developed by Pardeshi and Voller [9]. A set of equations of conservation of mass, energy and solute concentration is solved in conjunction with the momentum equation. A Boussinesq approximation is used to account for buoyancy driven flow. The mushy zone is treated as a porous medium of variable porosity. The multiple species on the transport phenomena manifests itself through some coupled effects [10]; the differences in the solutal expansion coefficients for the elements in the liquid phase, and the differences in the partition coefficients for the elements as well as the effects of the various species on the liquidus temperature. The continuum mixture model is used as a starting point for the simulation of macrosegregation.

The trajectory model is employed to investigate inclusion movement [11]. In order to simplify the computation of inclusion moving, a few assumptions are made; (1) inclusions are spherical, (2) every inclusion is considered to move independently and collision between inclusions is not included, (3) once inclusions reach the casting walls, they will be captured and stop moving.

The translational motion of one inclusion of velocity, v_{sx} and v_{sy}, diameter, d, and density, ρ_s, can be described using Newton's law [11]:

$$\frac{dv_{sy}}{dt} = -\frac{3}{4}\frac{\rho_l}{\rho_s d}\left(v_{sy} - v_{ly}\right)\left|v_{sy} - v_{ly}\right| C_D + \left(1 - \frac{\rho_l}{\rho_s}\right) g \qquad (1)$$

$$\frac{dv_{sx}}{dt} = -\frac{3}{4}\frac{\rho_l}{\rho_s d}\left(v_{sx} - v_{lx}\right)\left|v_{sx} - v_{lx}\right| C_D \qquad (2)$$

where v_{lx} and v_{ly} and ρ_l are the velocity and density of the liquid, respectively; C_D is a drag coefficient; g is the gravitational acceleration; and t is time. The first term on the right-hand side of equation (1) is the drag force exerted by the liquid, and the second is the buoyancy force.

The movement of inclusions in the molten metal is assumed to affect the flow streamlines. The interfacial friction coefficient is calculated by the Gidaspow correlation [12]:

$$K_{ls} = \frac{3}{4} C_D \frac{F_{si} f_l \rho_l}{d} \left|v_{sx} - v_{lx}\right| f_l^{-2.65} \qquad (3)$$

where K_{ls} is the interphase friction coefficient; f_l the volume fraction of liquid; F_{si} the volume fraction of inclusion in the mesh. When the local volume fraction of solid exceeds 0.2, the inclusion is thought to be captured by the dendritic network and the velocity is abruptly set to zero.

The general governing conservation equation is discretized using a finite volume method. A staggered grid is employed for the discretization of the momentum equation. A power law scheme is adopted to estimate the convection-diffusion flux at the control volume faces. Equations (1) and (2) are discretized implicitly. The movement of inclusions does not step into the pressure correction equation. The resolution of the velocity-pressure coupling is performed by the single-phase solution algorithm SIMPLER.

3. Results and discussion

The model for simulating macrosegregation was verified using a numerical benchmark proposed by Založnik et al. [13]. The test was concerned with solidification of a binary Sn- 5wt%Pb alloy in a Hebditch-Hunt casting. Full details of validation concerning mesosegregation and macrosegregation patterns are available in a published article [7]. The test with an acceptable outcome gives us confidence that the self-developed computer codes are reasonable for the following analysis. In the present contribution, the previously verified model is extended to include the inclusion movement.

The study configuration in [7] is used in this research. A Hebditch-Hunt casting is a two-dimensional rectangular mould cavity of 0.06 m height and 0.1 m width. The mould cavity is thermally insulated on the left, top and bottom walls and chilled by water at the right wall. The coolant temperature keeps constant at 25 °C. For the following simulations, uniform grids 100×60 with grid size 0.001 m are employed. The model alloy is Fe- 0.21 wt% C- 0.1 wt% S. The 500 inclusions are initially introduced into the liquid at random positions. Inclusions have the same diameter and their size will not change during solidification. Main thermophysical properties and boundary conditions used in the computations are listed in table 1 [13]. In order to study the impact of inclusion movement on the channel segregation formation, two cases are defined. The main characteristics are given in table 2. The first case is used as the reference case. For Case 2, the movement of inclusions is switched off. For Case 3 and Case 4, the influences of inclusion size on channel segregation are investigated.

Table 1. Thermophysical data and boundary conditions used in simulation.

	Fe - 0.21 C - 0.1 S	Inclusion
Density (kg m^{-3})	6910.0	3640.0
Binary partition coefficient (C)	0.169	—
Binary partition coefficient (S)	0.024	—
Liquidus slope (C, °C $wt\%^{-1}$)	-118.41	
Liquidus slope (S, °C $wt\%^{-1}$)	-30.4	
Thermal expansion coefficient (K^{-1})	6.0×10^{-5}	—
Solutal expansion coefficient (C, $wt\%^{-1}$)	0.011	—
Solutal expansion coefficient (S, $wt\%^{-1}$)	0.0123	—
Secondary dendrite arm spacing (m)	500×10^{-6}	—
Coolant temperature (°C)	25	—
Heat transfer coefficient at the chill (W m^{-2} s^{-1})	20	—

Table 2. Cases definition

	Inclusion movement	Diameter (m)
Case 1	Yes	20×10^{-6}
Case 2	No	—
Case 3	Yes	5×10^{-6}
Case 4	Yes	10×10^{-6}
Case 5	Yes	40×10^{-6}

3.1. Influence of inclusion movement (comparison between Case 1 and Case 2)

Comparisons between simulated macrosegregation maps in Case 1 and Case 2 are shown in figure 1. Only the distribution of carbon concentration is presented. The global macrosegregation pattern of sulphur is very similar to that of carbon, as the two elements both have partition coefficients smaller than 1 and show the same sign in the solutal expansion coefficients. Their corresponding thermal expansion coefficients are small enough ensuring that the effects of solutal buoyancy outweigh the effects of thermal buoyancy.

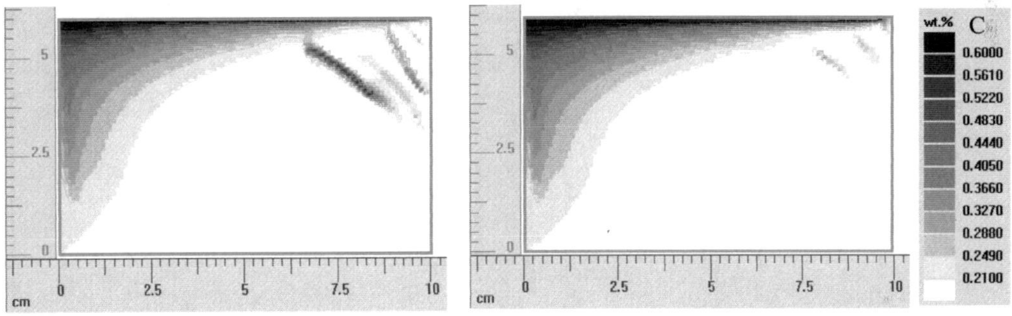

(a) (b)

Figure 1. Maps of macrosegregation of carbon in (a) Case 1 and (b) Case 2.

The channel segregation (solute-rich strip) is intensive in Case 1. Since inclusions have a lower density than that of liquid, they tend to float to the top of the casting cavity. 51 inclusions among 500 inclusions are randomly chosen. The moving trajectories are illustrated in figure 2. With diameter 20×10^{-6} m, all these inclusions float up within 100 s and finally are captured by the top boundary of casting. The more penetrated channels are the direct consequence of a relatively higher velocity induced by inclusion movement at the initial stage of solidification (figure 3a). For the melt solidifying from liquidus temperature (zero superheat), a porous mushy zone quickly forms and the solutal buoyancy plays a dominant role from the onset of solidification. The solutal configuration creates an upward flow in the melt at the cooled side across the phase changing front and a downward flow in the bulk liquid. At some locations, the upward fluid flow ahead of the solidification front is subject to drag forces from the floating inclusions. Such stronger upward flow easily triggers instability at the solidification front in the regions where it leaves the mushy zone. Such instability that the continuously advancing solidification front faces leads to channel segregation [4].

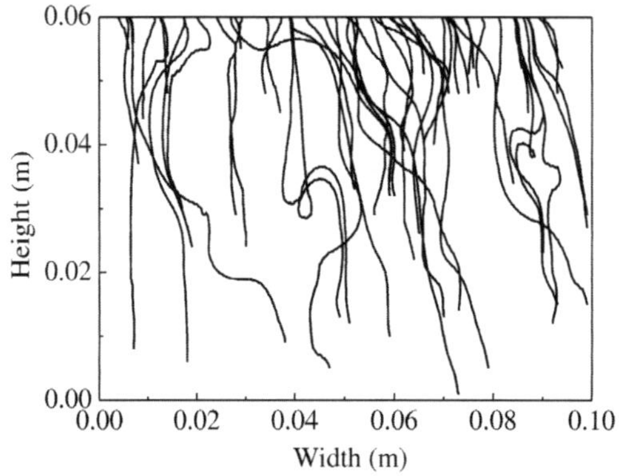

Figure 2. Moving trajectories of 51 inclusions randomly selected among the 500 inclusions.

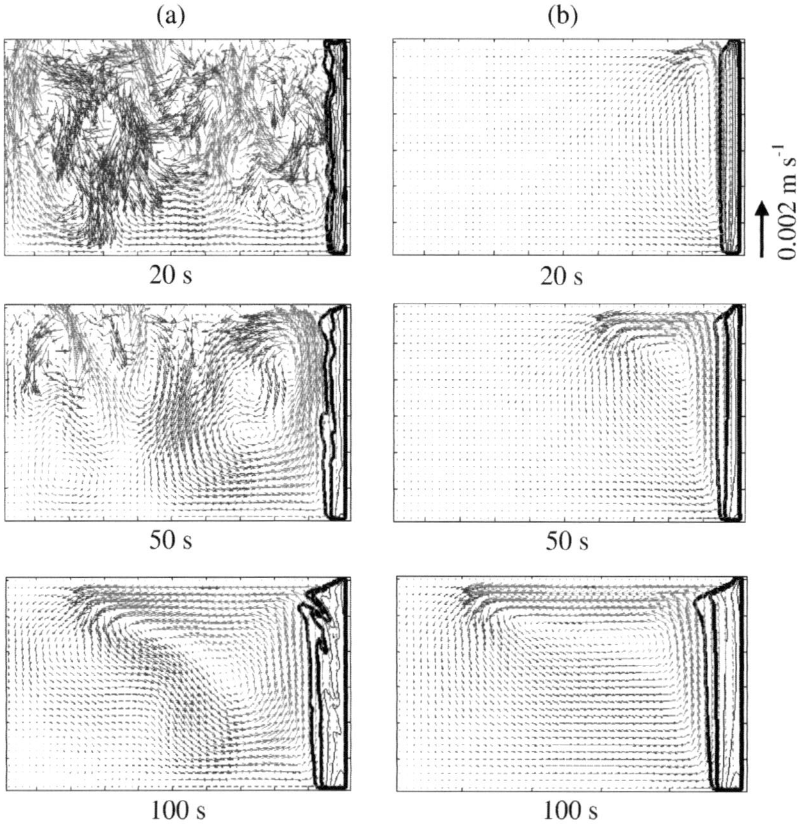

Figure 3. Flow field and iso-fractions of solid at different times: (a) Case 1 and (b) Case 2

When comparing figure 3a and figure 3b, we can see that without inclusion movement (Case 2), during solidification, the flow field is clearly structured as a counter-clockwise cell and the magnitude of fluid flow is relatively weak. The oscillations on iso-fractions of solid are less pronounced. It is why figure 1b shows segregated channels of low intensity.

3.2. Influence of inclusion size

The formation of mesosegregation when changing inclusion diameters is illustrated in figure 4. The intensity of channel segregation is low in Case 5 with a larger inclusion size (40×10^{-6} m, figure 4 d). Compared with Case 5, segregated channels with a stronger intensity in Case 1 deeply diffuse into the mushy zone (20×10^{-6} m, figure 4 c). The degree of channel segregation is comparable in Case 3 and Case 4 (figures 4a and b). According to Stokes law, the larger the inclusions, the faster the floating velocity. The quick upward movement of large inclusions slightly coincides with the development of solidification front. The up-movement of inclusions cannot greatly strengthen the upward flow velocity of interdendritic liquid. Therefore, the perturbation from the interdendritic fluid flow is reduced. In comparison with large inclusions, small inclusions (5×10^{-6} m and 10×10^{-6} m) are more easily entrained by the fluid flow. They do not present a strong tendency of floating up. Under such condition, the channel segregates are, on some degree, suppressed.

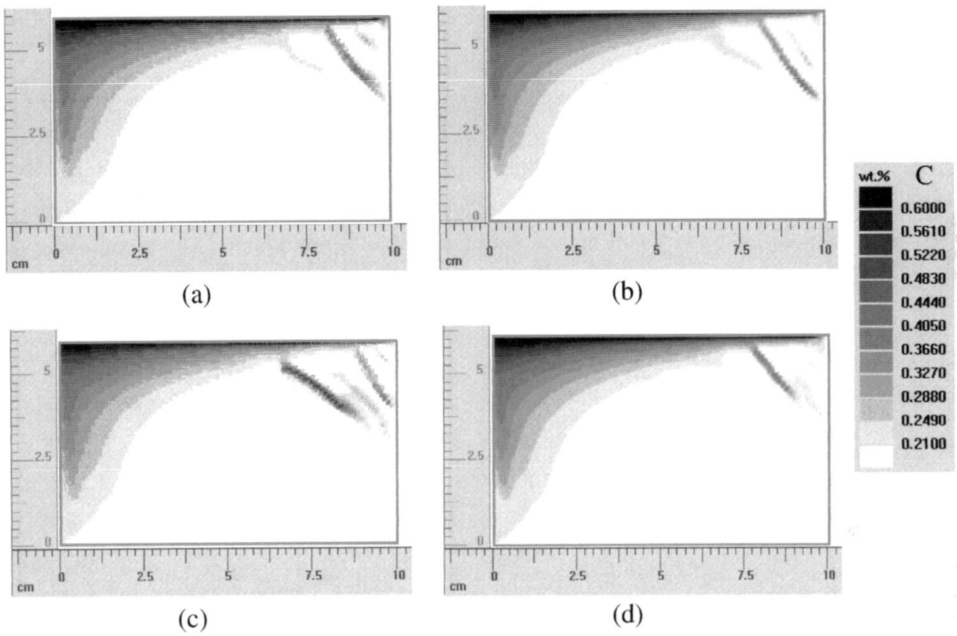

Figure 4. Maps of macrosegregation of carbon: (a) Case 3, (b) Case 4, (c) Case 1, (d) Case 5.

4. Conclusion

A mathematical model was developed to simulate the transport of inclusions and the formation of mesosegregation. Inclusion motion has an impact on increasing the intensity of channel segregation. A relatively serious channel segregation is found in the case with inclusion movement and with inclusion diameter 20×10^{-6} m. With increasing the size to 40×10^{-6} m or decreasing the size to 5×10^{-6} m, the degree of channel segregation is reduced. All the above simulations are performed with melt superheat zero. With inclusion movement, the parametric study indicates that there exists a relatively small window of initial conditions that result in a serious channel segregation. Although the calculations provide some insight into the impact of inclusion movement on channel segregation, an experimental validation is needed for model improvement.

Acknowledgements
Authors acknowledge financial supports from the National Natural Science Foundation of China (Grant No. 51101045) and from the Research and Development and Application of Technology Project for young researcher of Harbin (Grant No. 2014RFQXJ026).

References
[1] Combeau H, Založnik M, Hans S, Richy PE 2009 *Metall. Mater. Trans. B* **40** 289
[2] Bennon W D, Incropera FP 1987 *Metall. Trans. B* **18** 611
[3] Flemings M C 2000 *ISIJ Int.* **40** 833
[4] Založnik M, Combeau H 2010 *Int. J.Therm. Sci.* **49** 1500
[5] Li J, Wu M, Hao J, Ludwig A 2012 *Comput. Mater. Sci.* **55** 407
[6] Li J, Wu M, Hao J, Kharicha A, Ludwig A 2012 *Comput. Mater. Sci.* **55** 419
[7] Liu DR, Kang X H, Li D Z 2012 *IOP Conf. Ser. Mater. Sci. Eng.* **33** 012093
[8] Li D, Chen X, Fu P, Ma X, Liu H, Chen Y, Luan Y, Li Y 2013 *Mat. Sci.* Submitted http://arxiv.org/pdf/1308.3344.pdf
[9] Pardeshi R, Voller V R, Singh A K, Dutta P 2008 *Int. J. Heat Mass Transfer* **46** 1115
[10] Schneider MC, Beckermann C 1995 *Metall. Mater. Trans. A* **26** 2373
[11] Gu J P, Beckermann C, Giamei A F 1997 *Metall. Mater. Trans. A* **28** 1533
[12] Li W, Shen B, Shen H, Liu B 2012 *IOP Conf. Ser. Mater. Sci. Eng.* **33** 012090
[13] Založnik M, Kumar A, Combeau H 2010 *Comput. Mater. Sci.* **48** 11

Simple metrics for verification and validation of macrosegregation model predictions

I Vušanović[1a] and V R Voller[2]

[1]Faculty of Mechanical Engineering, University of Montenegro, George Washington St. bb 81000 Podgorica, Montenegro
[2]Department of Civil, Environmental, and Geo- Engineering, University of Minnesota, Minneapolis, MN 55455, USA

Email: [a]igor.vusanovic@ac.me

Abstract. While the numerical simulation of macrosegregation is now a common place activity efforts can still be enhanced by developing quantitative measures of the results. Here, on treating the nodal field of concentration predictions from a macrosegregation simulation as a sample from a statistical distribution, we demonstrate how statistical measures can be used in verification and validation. The first set of such measures is simply the central moments of the distribution, i.e., the mean, the standard deviation, and the skewness; measurements that provide quantitative checks of mass balance and grid convergence. In addition, building on recently reported work [1], we also demonstrate how to construct and use a cumulative distribution function (CDF) of the nodal concentration field; a measure that can be used to determine the fraction of the casting volume concentrations less than a specified value. We show how the CDF can be used to compare the influence of various process conditions and phenomena related to domain size, cooling rate, permeability, and micro-segregation.

1. Introduction

Current macrosegregation simulations are advanced and when coupled with sophisticated graphical outputs lead to detailed predictions of the distribution of solute in cast products. Ultimately our objective with these simulations is twofold. In the first place we might like to compare differences in process settings such as cooling conditions, and geometry. Secondly, we may be interested in understanding the basic phenomena underlying the simulations, e.g., microsegregation, and mushy region flow behavior. Clearly, graphical output provides an immediate qualitative evaluation of changing process settings or model phenomena. But this powerful tool is lacking in obvious quantitative measures that can be used in case to case comparisons. One way of obtaining quantitative measures is to treat the predicted nodal values of solute concentrations from a macrosegregation calculation as a statistical distribution. In this way measures of central moments and derived distribution functions [1] can be used as quantitative representations of macrosegregation behavior. The objective of this paper is to introduce these measures and, through a number of example macrosegregation simulations, show how they can be used to evaluate process settings and phenomenological models.

2. Statistical measures

Usually, the output from macrosegregation simulations essentially provides a list, C_i (i=1, 2,..N) of solute concentration values taken from (N) specified discrete points (nodes) in the casting. Often these are more conveniently stored as a set of macrosegregation levels $M_i = C_i / C_0$, where C_0 is the initial solute concentration in the melt. When this data is treated as a sample from a statistical distribution, central moments can be immediately calculated. Each of the first three central moments of such a distribution provides useful quantitative measures.

The first moment is the mean, defined as:

$$\overline{M} = \frac{\sum_{i}^{N} M_i}{N} \tag{1}$$

This will return the average macrosegregation level which can be used as a check on the mass balance of the simulation, i.e., if our simulation is correctly conservative we would expect that $\overline{M} = 1$. The second central moment is the sample standard deviation

$$s = \sqrt{\frac{1}{N-1}\sum_{i=1}^{N}\left(M_i - \overline{M}\right)^2} \tag{2}$$

This provides a measure of the spread of the data. Previously introduced as the macrosegregation number [2] it provides a basic measure of the level of macrosegregation. The third moment is the sample skewness, which based on the SKEW function found in spreadsheet software, can be defined as:

$$g = \frac{N}{(N-1)(N-2)}\sum_{i=1}^{N}\left(\frac{M_i - \overline{M}}{s}\right)^3 \tag{3}$$

This measure tells us how much the solute distribution is biased to the positive or negative side of the mean (the initial concentration). For example, a positive value of skewness (g) indicates that positive segregation values have a wider range than negative segregated values.

In addition to measures of central moments we could also construct a distribution function of our data. In recent work Voller and Vušanović [1] have proposed a simple way of constructing a so called Cumulative Distribution Function (CDF). This involves (i) listing the nodal values M_i in descending order, (ii) assigning a rank number j = 1, 2,3, …, N to each item in the list, and then (iii) associating with each entry an inverse Weibul number

$$F_j = \frac{j}{N+1} \tag{4}$$

Essentially the value of F_j is the volume of the solidified casting which has a macrosegregation level greater than or equal to that of $M_j = C_j / C_0$. The method is best suited to cases where the data acquired from experimental or numerical simulations is over a uniform grid. In cases where this is not case, e.g., simulation results obtained with an unstructured or cloud grid, care may need to be taken in constructing and interpreting the CDF.

Voller and Vušanović have used this CDF approach to investigate the grid dependence of macro-segregation simulations. A result, that when used in conjunction with experimental measurements for a Sn-10%Bi alloy [3] suggests that the positive segregated region may have a power-law tail. Thus indicating that as numerical or measurement resolution is refined one will, up to the formation of a second phase, encounter higher and higher values of solute concentration. Here we will extend the use of the (CDF) recently proposed in [1] and demonstrate how it can also be used to evaluate changes in process and model conditions.

3. A baseline case

As a baseline case, to compare the effects of process and phenomena conditions, we consider the solidification in a side cooled two-dimensional cavity containing an Al-4.5%Cu like alloy, with a fixed columnar microstructure initially at the liquidus temperature. The governing equations are the standard macrosegregation mixture model of heat, solute, and momentum conservation introduced by Bennon and Incropera [4], full details of the equations and solution strategy can be found in [5,6]. We note that our principal intention here is not to detail the model and solution used but rather to illustrate how the statistical measures, introduced above, can be used to systematically study the effects of process and phenomena on macrosegregation, in particular variations in the domain size, cooling conditions, microsegregation treatment, and permeability model. The base case consist of a 20mm x 20mm domain size cooled by applying a constant heat flux $Q=1.0\text{MW/m}^2$ on the left vertical side. Key factors in the model include (i) the construction of a microsegregation model to determine the nodal liquid solute concentration field from the predicted mixture concentration C_i, and (ii) a devise to suppress the fluid flow as a nodal solid fraction f_i in the mushy region approaches 1 (full solidification). In the base model we use a lever rule which assumes, within the micro scale of the mushy region (a secondary arm space), complete solute diffusion in both liquid and solid phases. Thus allowing us to determine the liquid concentration from a simple mass balance $C_i = fkC_l + (1-f)C_l$. The standard model to suppress fluid flow, as full solidification is reached, is to add a Darcy like sink term, scaling as ~ (inverse permeability x velocity), to the momentum equation. In the base case, the permeability is modeled as $K=\kappa_0 f^3/(1-f)^2$, where the constant $\kappa_0=1.0\text{x}10^{-10}\text{m}^2$. In this way as we approach full solidification $f\to 1$ the inverse permeability $K^{-1}\to\infty$, effectively suppressing the flow velocity in the mush.

The first panel in figure 1 shows the predicted final macrosegregation pattern for the base case using a grid size of 40x40 control volumes and a time step $\Delta t = 2.0\text{x}10^{-3}$sec. The second panel shows the CDF in log – log scale, for four different grid sizes. In support of figure 1, table 1 shows the mean, standard deviation, and skewness change with grid size.

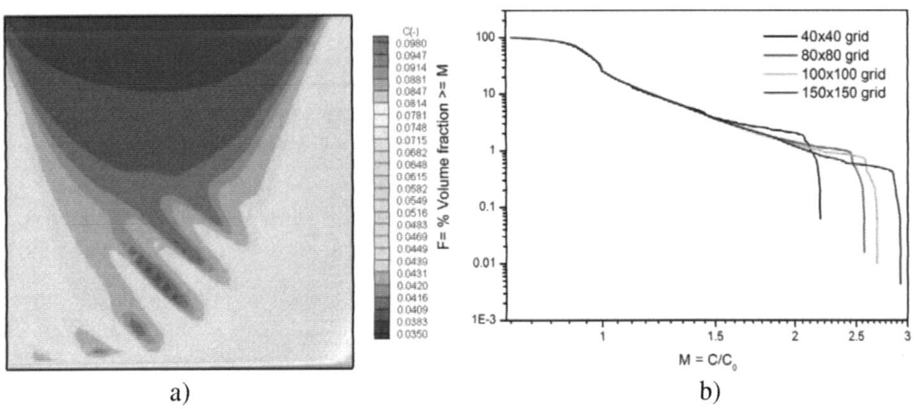

a) b)

Figure 1. Macrosegregation profile for baseline case (a), and (CDF) for the baseline case with different grid sizes (b).

Matching the findings in Voller and Vušanović [1] the main observation to take from figure 1b), is that as the grid is refined, the positive segregation region extends to the right and the predicted maximum truncation value of (M) increases. Essentially, a finer resolution, be it from simulation or experiments, will continue, up to some physical limit (e.g. the appearance of another solid phase), to uncover additional volumes with higher concentrations. This behavior is also seen in the values of table 1, where both the standard deviation (macrosegregation number) and skewness increase linearly with decreasing grid spacing. This suggests that care needs to be taken when increasing the grid size in computer models, and at the same time indicates why it is often so difficult to economically arrive at grid resolved macro-segregation simulation predictions.

Table 1. Statistical measures of macrosegregation results.

Grid spacing (mm)	Mean $\overline{M}$	Standard deviation (s)	Skewness (g)
0.50	1.0	0.05019	3.15733
0.25	1.0	0.05533	3.74400
0.2	1.0	0.05634	3.87929
0.134	1.0	0.05786	4.08245

4. Applications of (CDF) approach for parametric study of macrosegregation

To demonstrate how to use the *C*umulative *D*istribution *F*unction (CDF) in a parametric study we conduct a number of simulation runs where we change a single process or phenomena setting in our base line simulation. The changes are: (1) reductions in the cooling flux Q, (2) an increase in the permeability constant (corresponding to an increase in the assumed secondary arm space), and (3) the use of a Scheil microsegregation model where a zero solid state solute diffusion is assumed. These deviations from the base case are summarized in table 2 and the resulting CDF's (plotted in both linear and log-log scales) are shown in figures 2-4. Note that all runs are done with the relatively coarse grid size of (0.5 mm)

Table 2. Variations of process parameter for CDF analysis of macrosegregation results.

Case	Domain size	Cooling rate	Micro model	Permeability
Baseline case: 20x20mm, Q=1MW/m^2, Lever, $\kappa_0=1.0\text{x}10^{-10}\,m^2$				
1.1	20x20mm	Q=0.5 MW/m^2	Lever	$\kappa_0=1.0\text{x}10^{-10}\,m^2$
1.2	20x20mm	Q=0.25 MW/m^2	Lever	$\kappa_0=1.0\text{x}10^{-10}\,m^2$
2.1	20x20mm	Q=1MW/m^2	Scheil	$\kappa_0=1.0\text{x}10^{-10}\,m^2$
2.2	20x20mm	Q=1MW/m^2	Lever	$\kappa_0=1.0\text{x}10^{-9}\,m^2$
2.3	20x20mm	Q=1MW/m^2	Lever	$\kappa_0=1.0\text{x}10^{-11}\,m^2$

With reference to the linear CDF plot (figure 2b) we see that a fast cooling rate, implying a faster solidification rate, leads to less segregation; this is indicated by the reduction in the spread of the CDF as the cooling rate is increased. In contrast, a comparison of the CDF's obtained with a Scheil and lever microsegregation model (figure 3) indicates that for the majority of the casting, the limit microsegregation treatments have little impact on the solute distribution. In the last 5% of the casting, however, as clearly indicated in the log-log plot (figure 3a), there is a pronounced difference. This is driven by the fact that, in the later stages of solidification, the Scheil model leads to significantly higher values of the liquid solute concentration. Finally, as might be expected, changes in the permeability, figure 4, result in the most dramatic changes in the CDF. With a high value, the flow in the mushy region is more intense and can continue to quite high solid fractions. The result is a much wider solute distribution, terminating with a larger level of macrosegregation *M*. In contrast, a low value effectively cuts the flow in the mushy region providing limited opportunity for macrosegregation and a much tighter CDF.

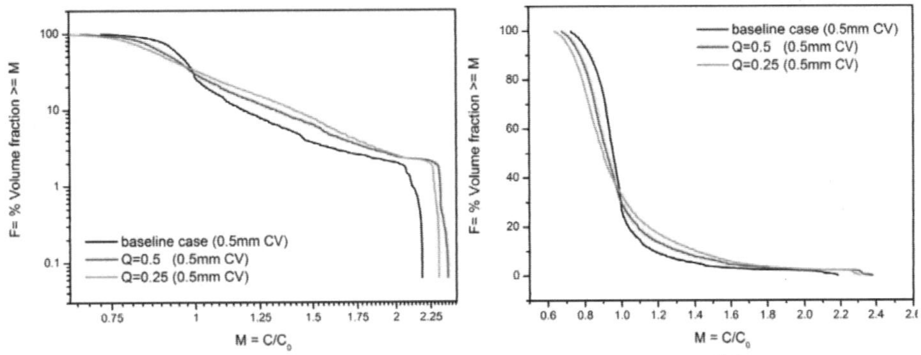

(a) (b)

Figure 2. Influence of cooling rate (a) log-log scale, (b) linear scale.

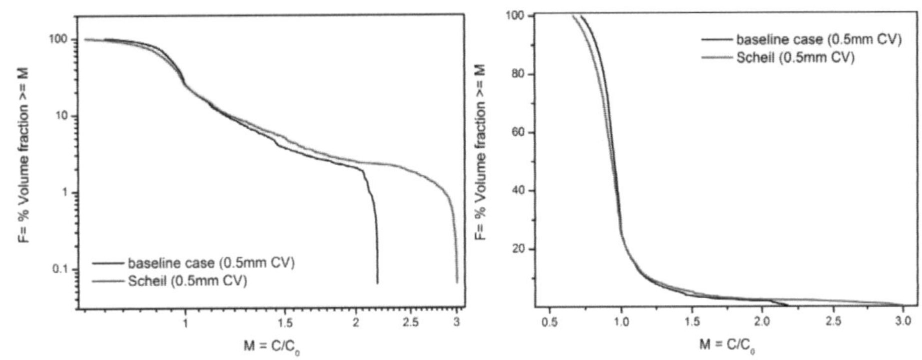

(a) (b)

Figure 3. Influence of microsegregation model (a) log-log scale, (b) linear scale.

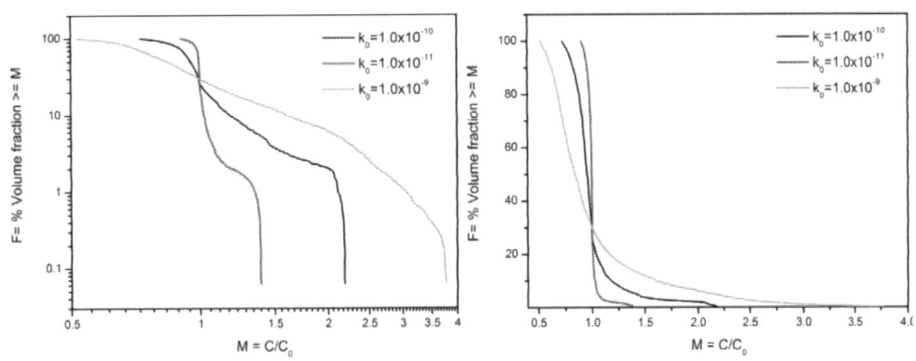

(a) (b)

Figure 4. Influence of different permeability in Darcy law (a) log-log scale, (b) linear scale.

5. Remarks and conclusions

Here we have demonstrated how simple statistical measures can be used to guide us in constructing reliable macrosegregation simulations. Even given the accuracy limitations of the grid size and the physical shortcomings in the mixture model, these statistical measures, in particular the Cumulative Distribution Function (CDF), provide two important insights. The first, supporting the initial work presented in Voller and Vušanović [1], is that the power-law tail in the CDF indicates why it is often so difficult to grid converge macrosegregation simulation. The second, is that the shape and spread of the CDF is extremely sensitive to change in the parameters that control the fluid flow in the mushy region. In practice there are many factors that can control the fluid flow in the mushy region, in particular coarsening of the arm spaces (leading to a transient permeability) and the movement of solid grains (non-fixed microstructure). As a result, although it may be obvious to state, we believe that it is in the modeling treatment of the mushy region flow that drives the observed divergence between macrosegregation code predictions observed in benchmark studies [2]. Therefore, in moving towards simulations that can consistently and faithfully match those seen in experiment or plant, it is key to focus on the development of robust physical models of the flow in the mushy region. Use of the statistical approaches demonstrated here, in particular the Cumulative Distribution Function [1], will be a useful tool in developing and tuning such models.

References

[1] Voller R V and Vušanović I 2014, *Int. J. Heat Mass Transfer* **79** 468
[2] H. Combeau, et al. 2012 *IOP Conf. Series: Materials Science and Engineering* **33** 012086.
[3] Quillet G Ciobanas A Lehmann P and Fautrelle Y 2007 *Int. J. Heat Mass Transfer*, **50** 654.
[4] Bennon W D and Incropera F P 1987 *Int. J. Heat Mass Transfer* **30** 2161
[5] VušanovićI and Voller R V 2014 *Mat. Sci. Forum*, **790-791** 74.
[6] Voller V R, Mouchmov A and Cross M, 2004 *App. Math. Model.* **28**, 79.

IOP Conf. Series: Materials Science and Engineering **117** (2016) 012063 doi:10.1088/1757-899X/117/1/012063

Effect of the dendritic morphology on hot tearing of carbon steels

M R Ridolfi
Centro Sviluppo Materiali S.p.A., Via di Castel Romano, 100, Rome, Italy

E-mail: m.ridolfi@c-s-m.it

Abstract. Hot tears form during solidification in the brittle region of the dendritic front. Most hot tearing criteria are based on solid and fluid mechanics, being the phenomenon strictly depending on the solid resistance to applied strains and on the liquid capability of filling the void spaces. Modelling both mechanisms implies the precise description of the dendritic morphology. To this scope, the theory of coalescence of the dendritic arms at grain boundaries of Rappaz et al. has been applied, in this work, to the columnar growth of carbon steels by means of a simple mathematical model. Depending on the alloy composition, solid bridging starts at solid fractions down to about 0.8 and up to above 0.995 (very low carbon). The morphology of the brittle region changes drastically with increasing carbon and adding other solutes. In particular, ferritic dendrites, typical of low carbon steels, tend to offer short and wide interdendritic spaces to the surrounding liquid making possible their complete filling, and few solid bridges; peritectic steels show the rise of austenite growing and bridging rapidly in the interdendritic spaces, preventing void formation; austenitic dendrites form long and narrow interdendritic spaces difficult to reach for the liquid and with a lot of solid bridges. Sulphur addition mainly acts in delaying the coalescence end, more markedly in ferritic dendrites.

1. Introduction

Hot tearing is a common defect, which develops during solidification because of the application of tensile strains to the material with poor ductility due to the presence of small liquid fractions.

The description of the physical mechanism of the hot tearing phenomenon as well as most of the hot tearing criteria [1-7] are based on the mechanical evolution of the solid bridges that counteract the applied strains till reaching a possible rupture, and the fluid-dynamics of the liquid that is sucked into the interdendritic spaces and is not always able to completely fill the void spaces opened by the strain action. Both the mechanical and the fluid-dynamic aspects require to know the morphology of the dendritic frame to be properly described.

Exploring more specifically the field of steel solidification, for which hot tearing is commonly recognized as a main problem [8-9], the microstructure of reference is undoubtedly the columnar dendritic one. The description of the morphology of the dendritic framework can be approached by first explaining the main characteristics of the columnar microstructure; strains form because of thermal gradients on average directed parallel to the dendrite growth direction and to the average columnar grain direction; as the applied strains tend to separate the columnar grains, the material develops mechanical resistance only at the onset of coalescence of adjacent grains.

Grain boundaries play an important role as they are weak sites because of the coalescence delay caused by the dendrites misorientation. Therefore they both determine the onset of mechanical resistance and are sites of strain accumulation and hot tearing formation.

2. Columnar dendrites

The first step to set up a model of the columnar dendritic morphology is to define a simplified geometry of the dendrites. At this scope, an upper paraboloid shape and a lower square parallelepiped one (figure 1) are assumed. The paraboloidal growth arrests when adjacent dendrites come in touch as soon as their diameter reaches the value of the primary dendrite arm spacing (*PDAS*). From that time on, the solidification proceeds towards the corners until they are fully solidified. The solid fraction at the first contact between adjacent dendrites is easily obtained in an approximated way, by dividing the area of the circle of diameter *PDAS* by that of the square of side length *PDAS*:

$$f_{s-touch} = \pi\left(\frac{PDAS^2}{4}\right)(PDAS^2)^{-1} = \frac{\pi}{4} = 0.785 \tag{1}$$

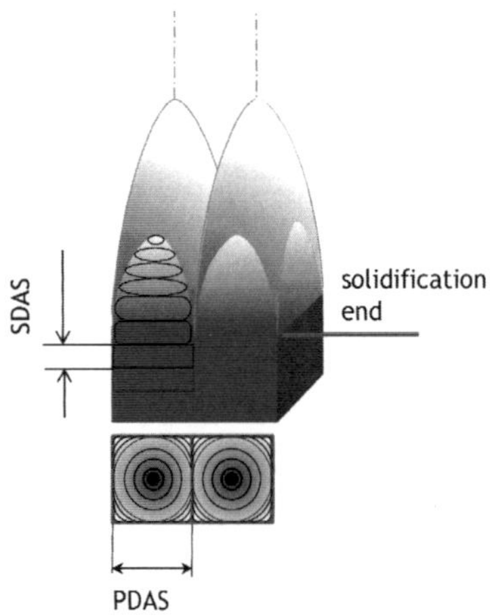

Figure 1. Simplified scheme of the columnar dendrite growth (PDAS: primary dendrite arm spacing, SDAS: secondary dendrite arm spacing).

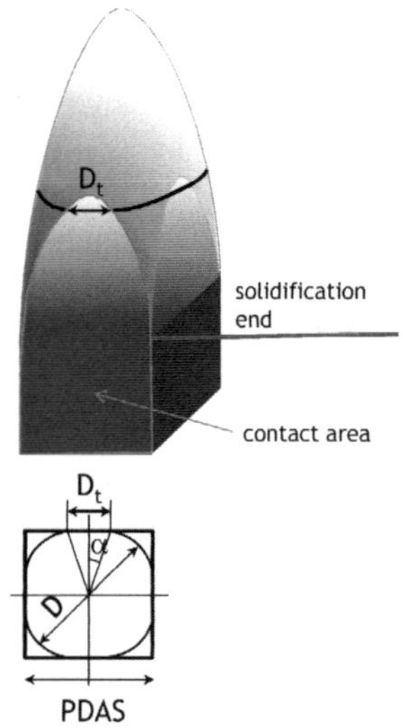

Figure 2. Scheme for the calculation of the progression of the contact area.

The solid fraction given by equation (1) is better approximated for cells than for dendrites since, when the side branches of these last touch, there is still liquid within the dendritic frame, consequently the actual solid fraction should be lower. It is worth noting that the obtained value is very close to commonly accepted values of solid fraction of zero strength.

To describe the dendrite morphology according to this simplified scheme, it is needed to express the extent of the contact area between adjacent dendrites, given along the dendrite transverse section by the geometrical parameter D_t, of figure 2, given by:

$$\frac{D_t}{2} = \frac{D}{2}\sin\alpha \tag{2}$$

with:

$$\alpha = \arccos\left(\frac{PDAS}{D}\right) \tag{3}$$

The correlation between the geometrical parameter D (related to the curvature of the rounded corners of the transverse section as shown in figure 2, and the solid fraction f_s, is obtained equating this last to the ratio between the area of the transverse section at a given height and that of the dendrite square base:

$$D = \left(\frac{f_s}{\pi}\right)^{\frac{1}{2}} 2PDAS \quad (4)$$

3. Coalescence of the columnar dendrites

Contacting side branches of adjacent parallel dendrites belonging to the same grain coalesce immediately as soon as their mutual distance is of the same order of the interface thickness. Dendrites at the grain boundaries coalesce if the grain boundary energy is lower than the solid-liquid interface energy of the two initial surfaces [10]. For tilt grain boundaries it is required that the interface undergoes solidification undercooling (ΔT_b = bridging undercooling) to balance the difference between grain boundary and interfaces free energies, before coalescence can start. ΔT_b, for pure materials, is derived balancing the sum of the free energies of the two solid/liquid interfaces and that of the interposed undercooled liquid, with the grain boundary free energy, per unit area [10]:

$$\Delta T_b = \frac{\gamma_{gb} - 2\gamma_{sl}}{\Delta s_f} \frac{1}{\delta} \quad (5)$$

with: γ_{gb}=grain boundary energy; γ_{sl}=solid-liquid interface energy; Δs_f=fusion entropy for unit volume and δ=solid-liquid interface thickness.

The resulting expression relates the undercooling to the entropy of fusion of the liquid entrapped between the two grains. ΔT_b increases as the difference between the grain boundary energy and the interface energy increases, and also as the entropy of fusion of the residual liquid decreases.

The grain boundary energy is function of the misorientation angle θ according to the Read-Shockley relationship [11]:

$$\gamma_{gb} = \gamma_{gbo}(1 - \ln\theta) \quad (6)$$

The solid/liquid interface free energy has been derived for pure materials (Spaepen-Thompson) [12]:

$$\gamma_{sl} = \frac{\alpha_m \Delta s_f}{(NV^2)^{\frac{1}{3}}} \quad (7)$$

where: α_m=0.71 for bcc e α_m=0.86 for fcc and hcp; N=Avogadro's number; V=molar volume of the solid at the interface; and Δs_f=molar fusion entropy.

Body centred cubic and fcc or hcp lattices have different coefficients. The bcc lattice tends to develop a smaller solid/liquid interface free energy than the other two ones.

In the case of multi-component alloys the fusion entropy at each stage of the solidification accounts for the composition at the interface resulting from microsegregation and diffusion.

Modeling of the microsegregation and diffusion in the solid during the multi-component dendrite growth is used to derive the variables (values referred to the interface) needed for calculating γ_{sl}: entropy of fusion from enthalpy of fusion, molar mass and volume. The density of the solid at the interface is derived through empirical relationships available on the literature as function of temperature and composition.

The model used in this work consists of two separate models, for microsegregation and coalescence; the first one provides input data to the second one. In this way, the reduction of undercooling due to back diffusion of solutes from the last liquid in the interface, is neglected.

Grain boundaries are modelled as they are sites where preferentially hot tearing forms because of the low mechanical resistance induced by the longer permanence of the liquid. The main features of the grain boundary morphology (figure 3) are provided by the bridging undercooling and the primary and secondary arm spacing as they influence the hot tearing tendency by determining shape and extension of the coalesced areas, and shape and extension of the liquid channels respectively. The former determine the solid resistance to applied strains, while the latter the volumes where the liquid will not be able to flow. The possibility of tearing is higher for thin solid bridges forming at low solid fraction and small primary and secondary arm spacing. Therefore, tearing is favoured when bridging starts with low undercooling. The last liquid forms channels mainly located at the corners of the primary trunks. The critical region for hot tearing formation is where the local pressure drops below a minimum value. This region generates hot tearing porosity only above a minimum volume below which it becomes invisible (figure 4).

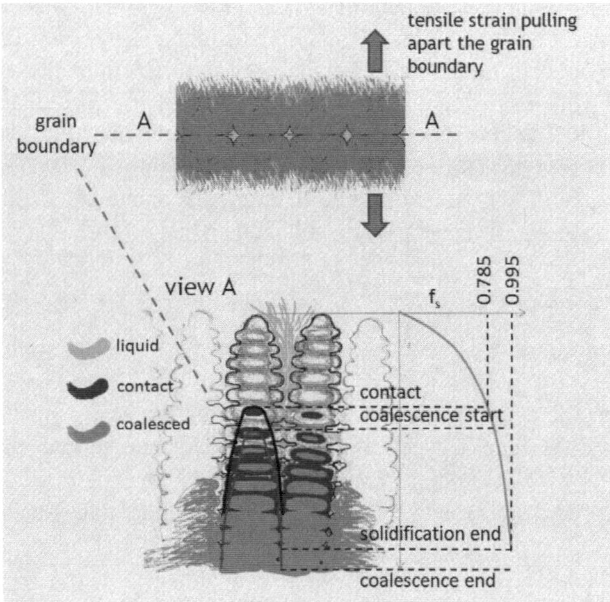

Figure 3. Drawn sketch of a grain boundary showing the locations of the first contact between dendrites, the start and end of coalescence and their position with respect to the solidus temperature, assumed to correspond to solid fraction 0.995. The position of the residual liquid volumes at the corners is also shown.

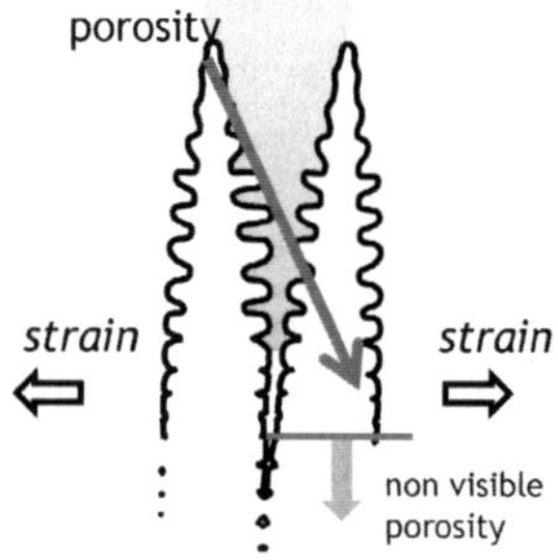

Figure 4. Drawn sketch of the region of the liquid channel where hot tearing forms.

4. Calculated grain boundary features for the Fe-C system

The microsegregation and coalescence models have been applied to the binary Fe-C system for C ranging between 0.002 and 1 wt%, considering grain boundaries with fixed tilt angle and a fixed cooling rate of 1 K/s. Figure 5 shows where the lines start and end of coalescence lay with respect to the solidus and liquidus ones in the phase diagram and, for some compositions, the schematic representations of the dendrites in the style of figure 3, where the shapes of the dendrites and of their contacting and coalesced areas can be observed. It is noted how solid bridging, and the associated mechanical resistance, start at solid fractions as high as ~1 for low carbon content, going down to 0.8 for carbon content higher than about 0.3%. This behaviour can be mainly ascribed to the lower interface energy of ferrite with respect to austenite. The morphology of the brittle region changes drastically with increasing carbon, in particular: ferritic dendrites (low carbon) form short and wide interdendritic spaces easily filled by liquid with few and thick solid bridges; peritectic steels show the rise of austenite growing and bridging rapidly in the interdendritic spaces, preventing void formations in a way similar to eutectics growth; austenitic dendrites form long and narrow interdendritic spaces difficult to reach for the liquid and with a lot of weak solid bridges. Narrowing and elongating of the residual liquid volumes of austenitic dendrites can be partly attributed to the primary arm spacing decrease with increasing carbon content.

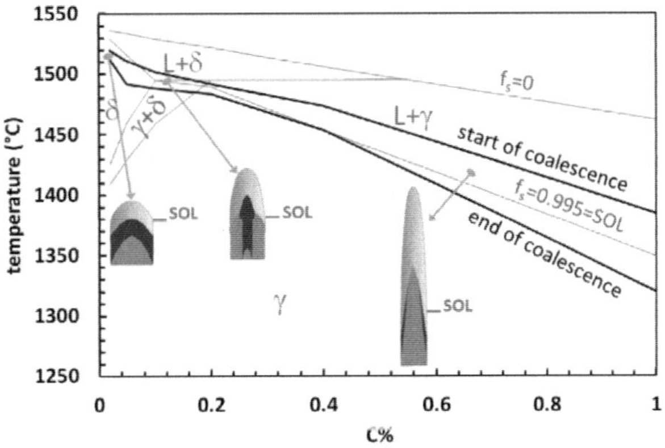

Figure 5. Fe-C phase diagram with the lines of start and end of coalescence and the sketches of the dendrite morphology drawn for reference compositions (SOL is equal to solid fraction 0.995).

The different tendency shown by low and high carbon steels is explained in terms of physical behaviour at the grain boundaries (figure 6). Strains applied to low carbon steel dendrites accumulate in the disconnected grain boundary, which open forming wide spaces filled by liquid. Thick solid bridges, formed through sudden coalescence near to the solidus, break only when subjected to large strains. Differently, strains applied to high carbon steel dendrites are shared between grains and solid bridges in the grain boundaries, liquid flows in narrow channels and stops at solid fraction less than solidus, where thinner bridges can be easily broken at small strains.

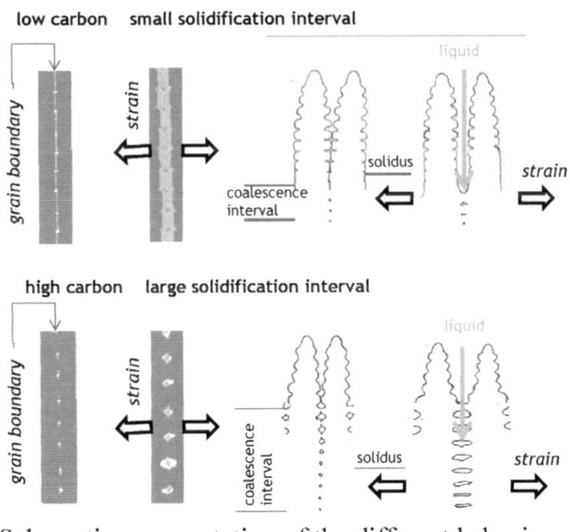

Figure 6. Schematic representation of the different behaviour of low and high carbon steel grain boundaries after application of tensile strain.

The addition of other alloying elements changes both the solidification and the coalescence temperature ranges. Figure 7 shows how the addition of 0.02% of sulphur to the binary Fe-C alloy causes coalescence of ferritic dendrites starting at lower solid fractions, and lowering of the coalescence end temperature for all carbon contents, with a more marked effect for C<~0.2%. The deleterious effects of sulphur addition consist in favouring the formation of thin bridges and in generating narrow liquid channels. The first is mainly due to the drop of the solidus line, the second one to the drop of the end of coalescence line. These effects are more marked in low carbon steels. The addition of other alloying elements changes both the solidification and the coalescence temperature ranges.

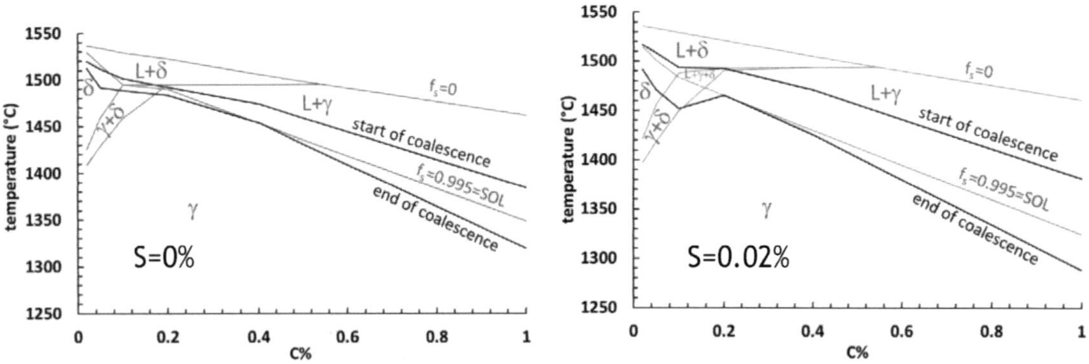

Figure 7. Comparison of the position of the start and end of coalescence lines with respect to the corresponding phase diagrams between the binary Fe-C system and after addition of 0.02% of sulphur.

5. Conclusions

The different tendency to form hot tears shown by Fe-C alloys is basically explained by the features of the dendrite solidification mainly expressed by the extension of the solidification range and the primary and secondary arm spacing. The effect of bridging undercooling is to weaken the grain boundary concentrating tensile strains in it. Bridging undercooling overlaps its effects to the basic cracking tendency simply due to solidification, introducing important differences depending on the lattice structure and the solutes concentration. Bridging undercooling embrittling action is mainly due to the elongation of the boundary characterized by thin solid bridges with low solid fraction, associated to low permeability to the liquid flow. No embrittling effect can be adduced only to the delay of the start of coalescence, since the possible gap formed by separating the two grains can be easily filled by the liquid. It can be concluded that the variable tendency to hot tearing of multicomponent steels is caused by the effects of the alloying elements on both solidification path and coalescence evolution.

References

[1] Won Y M, Yeo T-J, Seol D J, Oh K H 2000 *Metall. Mater. Trans.* B **31B** 779
[2] Bellet M, Cerri O, Bobadilla M, Chastel Y 2009 *Metall. Mater. Trans.* A **40A** 2705
[3] Rappaz M, Drezet J M, Gremaud M 1999 *Metall. Mater. Trans.* A **30A** 449
[4] Monroe C, Beckermann C 2005 *Mater. Sci. Eng.* A **30A** 449
[5] M'Hamdi M, Mo A, Martin C L 2002 *Metall. Mater. Trans.* A **33A** 2081
[6] Mathier V, Drezet J-M, Rappaz M 2007 *Model. Simulat. Mater. Sci. Eng.* **15** 121
[7] Ludwig O, Drezet J-M, Martin C L, Suery M 2005 *Metall. Mater. Trans.* A **36A** 1525
[8] Santillana B, Boom R, Eskin D G, Mizukami H, Hanao M, Kawamoto M 2012 *Metall. Mater. Trans.* A **34A** 5048
[9] Mehrara H, Eskin D G, Petrov R H, Lalpoor M, Katgerman L 2014 *Metall. Mater. Trans.* A **45A** 1445
[10] Rappaz M, Jacot A, Boettinger W J 2003 *Metall. Mater. Trans.* A **34A** 467
[11] Read W T, Shockley W 1950 *Physical Review* **78** 275
[12] Thompson C V, Spaepen F 1983 *Acta metall* **31** 2021

IOP Conf. Series: Materials Science and Engineering **117** (2016) 012064 doi:10.1088/1757-899X/117/1/012064

Effect of superheat on macrostructure and macrosegregation in continuous cast low-alloy steel slabs

T Pikkarainen[1,2,a], **V Vuorenmaa**[2], **I Rentola**[2], **M Leinonen**[2] **and D Porter**[1]
[1]Centre for Advanced Steels Research, University of Oulu, P.O. Box 4200, FI-90014, Oulu, Finland
[2]Ruukki Metals Oy, Rautaruukintie 155, Raahe, Finland

E-mail: [a] teppo.pikkarainen@ruukki.com

Abstract. The effect of superheat on grain sedimentation and macrosegregation has been investigated using experimental castings of a low-alloyed steel grade. With a high superheat of ~ 40 °C, the central equiaxed parts of the slabs consisted of randomly oriented fine dendrites but with a low superheat of ~ 10 °C, coarse globular structures formed. The mean carbon content measured with optical emission spectroscopy was of the order of 15 % smaller with coarse globular structures than in fine equiaxed dendritic structures. Electron probe microanalysis of other alloying elements indicates that the negative segregation in the slab central zones is caused by sedimentation of globulites. With superheat in the range ~ 20 - 40 °C, the equiaxed zone is bordered by a columnar to equiaxed transition (CET) zone. In this region a positive macrosegregation of carbon and other alloying elements was observed. These phenomena are important when considering the through-thickness properties of the slabs and final products.

1. Introduction

The as-cast structure and its effect on macrosegregation have been studied widely under recent decades. In continuous casting of steel, macrosegregation has been found to be caused by the motion of solute-enriched liquid and solute-lean solid [1]. In ingots, many researchers have found sedimentation of equiaxed grains to be the main cause of negative segregation in the bottom part of the ingot [2-6]. Some researchers have also noted the influence of grain sedimentation on macrosegregation in continuous cast steel slabs [7, 8].

Slab macrostructure is largely related to casting temperature (superheat), strand dimensions, cooling and steel composition. Usually the slabs have a fine equiaxed zone near the surfaces, which is formed during the rapid start of solidification in the mold. Depending on casting parameters, the central area usually solidifies as a combination of columnar dendrites from the chill zone towards slab center and equiaxed dendrites in the central region of the slabs. With a very high superheat, the equiaxed zone can be absent so that the structure is fully columnar dendritic. With a very low superheat, the slabs may solidify with a fully equiaxed dendritic structure. Some researchers [7] have pointed out that the equiaxed zone may consist of randomly orientated dendrites or globulites. When the macrostructure consists of columnar and equiaxed grains, a columnar-to-equiaxed transition (CET) occurs. In slab casting, the CET is often associated with negative segregation, so called white bands, which may occur when using electromagnetic stirring [2]. Positive segregation at the CET has also been reported in the case of continuously cast steel billets [9].

In this study, the effect of casting temperature on slab macrostructure has been studied and the different macrostructures caused by different casting superheat are evaluated. The effect of sedimentation of globulites on macrosegregation in the slabs is considered in detail and the instances of intermediate superheat, with a positive macrosegregation at the CET are investigated.

2. Experimental procedure

2.1 Materials and sampling

Experimental castings of a low-alloyed steel grade were performed with a different superheat. Composition of the studied steel grade is given in table 1.

Table 1. Chemical composition of the studied steel grade.

C	Si	Mn	Al	Cu	Cr	Ni	Mo	Ti	B
0.16	0.60	1.10	0.04	0.01	0.60	0.20	0.40	0.02	0.001

The dimensions of the casting strand were 270 mm x 1975 mm (thickness x width). Slab samples of 100 mm x 270 mm x 20 mm were cut for studying the as-cast structure from the 1/2 and 1/4 width positions of the strand. Smaller samples of 50 mm x 40 mm x 20 mm were cut from the above mentioned samples for optical emission spectroscopy (OES) and electron probe microanalysis (EPMA) and more detailed characterization of slab macrostructures, see figure 1.

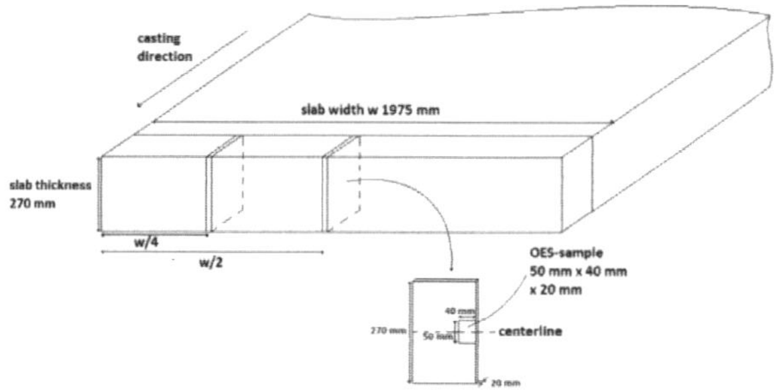

Figure 1. Sampling of the investigated steels.

2.2 Metallography and chemical analysis

Slab samples of 270 mm x 100 mm x 20 mm were ground and etched with HCl to reveal the primary structure of the slabs. Smaller samples of 50 mm x 40 mm were polished and etched with Oberhoffer's reagent [10] to reveal the primary structure in detail. An optical emission spectrometer (OBLF QSG 750, OBLF, Witten, Germany) was used for macrosegregation analyses of the slab samples. The sample size was 50 mm x 40 mm (height x width). An analysis matrix of 30 analysis points was used as shown in figure 2.

Microsegregation of Mn, Cr and Si was measured on polished surfaces with an electron probe microanalyzer (JEOL JXA-8200, JEOL Ltd., Tokyo, Japan) using line analyses with a step size of 40 μm, an accelerating voltage of 15 kV and a beam current of 15 nA. Because of their low diffusion coefficients their concentration provides information of the as-cast structure of the slabs.

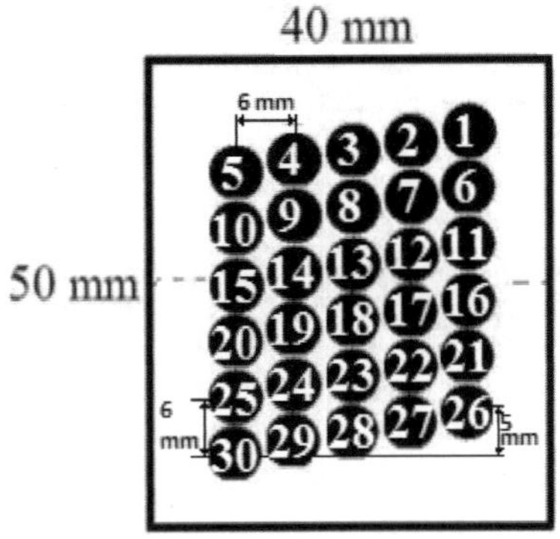

Figure 2. OES measurement matrix.

3. Results

3.1 Slab macrostructure

The macrostructure of the investigated slab samples was divided to three types: columnar dendritic, equiaxed dendritic and globular. The different macrostructures are shown in figure 3.

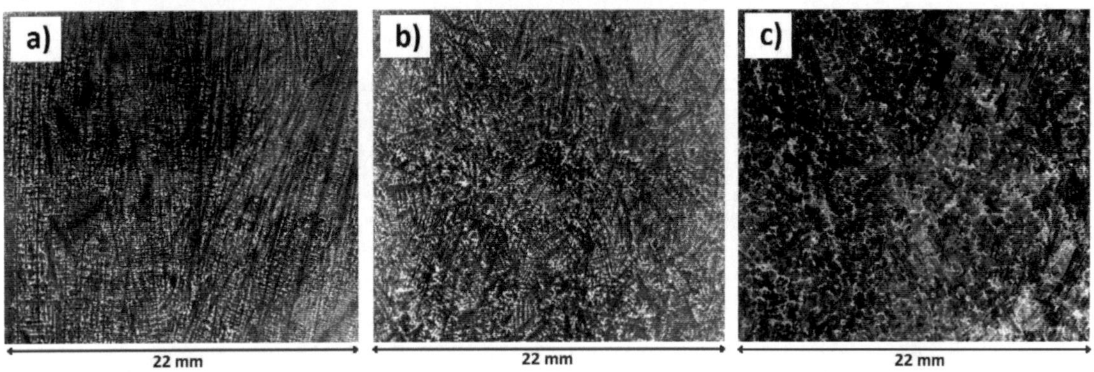

Figure 3. Slab macrostructures in the studied steels. a) columnar dendritic, b) equiaxed dendritic and c) globular. Etched with picral acid, solute enriched areas appear white.

All macrostructures of the studied slabs had a fine equiaxed chill zone near the surfaces. With a high superheat (~40 °C above liquidus) the rest of the solidification structure consisted of columnar dendrites (figure 3a) and an equiaxed dendritic zone (figure 3b) in the central parts of the slabs. With an extremely low superheat (~10 °C above liquidus), the columnar structure was replaced by an equiaxed dendritic structure, and the central areas consisted of a globular structure (figure 3c). The fraction of different solidification structures is given in table 2.

Table 2. Fractions of solidification structures.

	Sample A	Sample B	Sample C
Superheat (°C)	44	22	7
Chill zone (%)	7	7	7
Columnar dendritic (%)	75	70	0
Equiaxed dendritic (%)	18	4	44
Globular (%)	0	19	49

A dramatic difference in macrostructures is seen between samples A and C, shown in figure 4. The central area of the slab sample cast with high superheat (44 °C) consists of a columnar dendritic zone and an equiaxed dendritic zone (figure 4a). The sample cast with a low superheat consists of a totally globular structure in the central area (figure 4b).

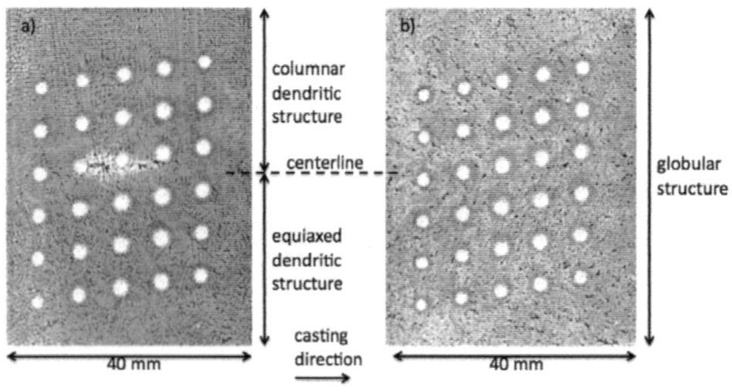

Figure 4. Macrostructures of central areas of slab samples a) A (superheat 44 °C) and b) C (superheat 7 °C). Etched with Oberhoffer's reagent, solute enriched areas appear dark (the bright spots are remnants of the OES analyses).

With an intermediate superheat (22 °C) the slabs consisted of a combination of columnar zones and a globular and equiaxed dendritic zone in the center part of the slabs. In some instances an abrupt CET was observed and a positive macrosegregation was seen at the location of the CET. In figure 5, the appearance of the slab structure cast with an intermediate superheat is shown. As can be seen from the image of the sample etched with HCl (figure 5a) there is a positive macrosegregation at the centerline of the sample but also at the location of the CET.

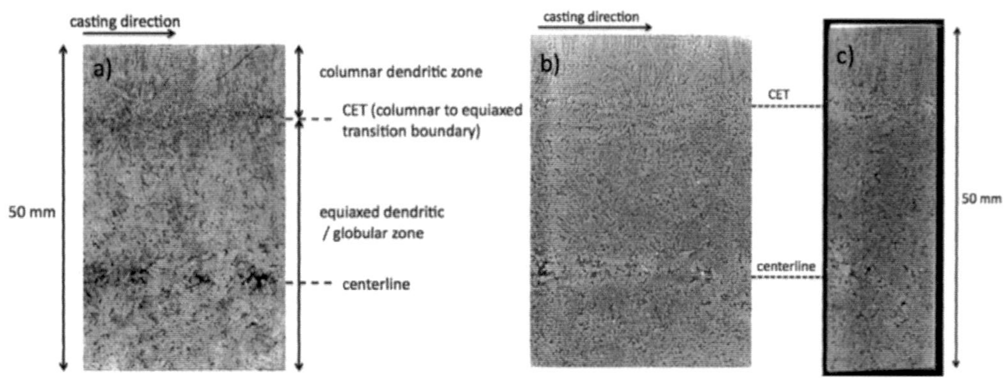

Figure 5. Macrostructure of central area of a slab sample with an intermediate superheat (22 °C). a) etched with HCl, b) section parallel to the casting direction, etched with Oberhoffer's reagent, and c) section perpendicular to the casting direction, etched with Oberhoffer's reagent.

3.2 OES analyses

OES analyses were made to measure composition changes in the slab samples in a macroscopical scale. Results are presented as segregation coefficients, i.e. local concentration in the measured area relative to the nominal bulk concentration in the steel (C/Cnom). Each measurement is an average over a spot size of approximately 6 mm. The mean, maximum and minimum values of segregation coefficients of C, Si, Mn, Cr, Ni and Mo are given in table 3 for samples A and C. A segregation coefficient for the carbon equivalent CEV is also given.

Table 3. Segregation coefficients from OES analyses of samples A and C.

Sample		Segregation coefficients C/Cnom						
		C	Si	Mn	Cr	Ni	Mo	CEV
Sample A	Maximum	1.50	1.09	1.07	1.05	1.12	1.14	1.19
	Average	1.03	1.02	1.00	1.01	1.05	1.02	1.01
	Minimum	0.89	0.96	0.95	0.97	1.02	0.98	0.95
Sample C	Maximum	0.93	1.03	1.00	1.02	0.97	1.05	0.96
	Average	0.81	0.98	0.94	0.99	0.92	0.98	0.91
	Minimum	0.72	0.95	0.91	0.96	0.89	0.93	0.87

It can be seen that carbon is the element with the highest and lowest segregation coefficient of the elements considered. In sample A, the maximum carbon segregation coefficient 1.5 was measured at the centerline segregation. The average values C/Cnom of carbon in samples A and C are 1.03 and 0.81, respectively. Considering other elements, the difference between samples A and C is small. Thus, the average values of C/Cnom for carbon equivalent CEV in samples A and C are 1.01 and 0.91, respectively. In a dataset of 120 OES measurements C/Cnom for carbon was 15 % smaller in samples cast with a low superheat (~10 °C) than in samples cast with a high superheat (~40 °C). The OES analysis curves are shown in figure 6a. An example of macrosegregation at the location of the CET is given in figure 6b.

It can be seen from figure 6a, that both samples A and C have macrosegregation peaks at the slab centerline, and areas of negative segregation adjacent to centerline. In sample A the centerline segregation is higher and in general the carbon segregation coefficient and CEV values are smaller in sample C. In figure 6b, one sample shows a positive macrosegregation at the CET, and in the other sample, the composition stays nearly constant. The macrosegregation peak at the CET is wider than at centerline in sample A (figure 6a), but the maximum segregation coefficient is smaller at the CET than the centerline.

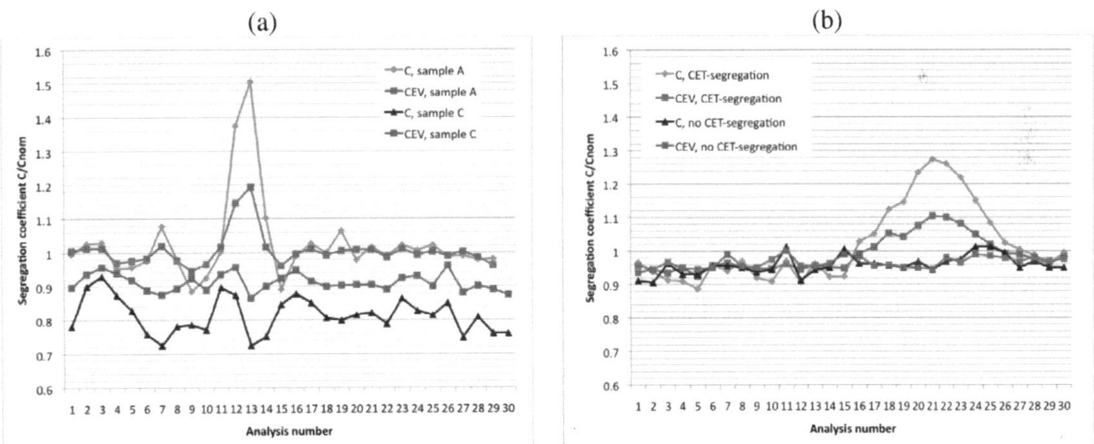

Figure 6. OES measurements from slab samples. (a) A (superheat 44 °C) and C (superheat 7 °C). (b) Samples with and without CET segregation (superheat ~20 °C). (Analysis numbers are spot numbers in a matrix like those shown in figure 4. Increasing number means increasing distance from the top surface of the slab.)

3.3 Microsegregation analysis

EPMA line analyses of Si, Cr and Mn in different cast structures are given in figure 7. From figure 7, it can be seen that the distance between microsegregations is smaller in the equiaxed dendritic zone (figure 7c) than the globular zone (figure 7d). Segregation intensities are slightly higher in the globular zone (figure 7b). The composition stays more stable in the equiaxed dendritic zone (figure 7a), as in the maxima compared to the minima of the compositions are higher in the globular zones.

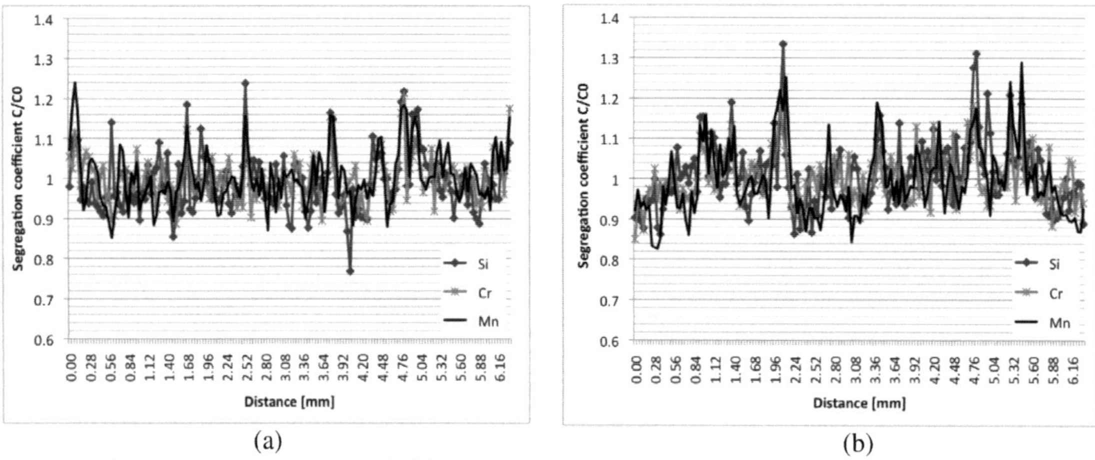

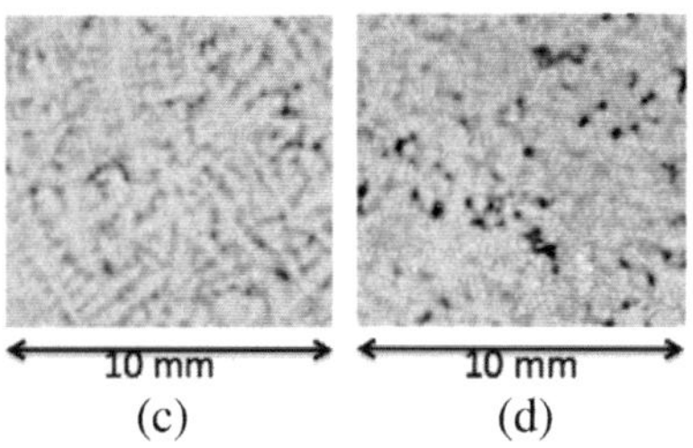

Figure 7. Segregation coefficients determined from EPMA line analyses (a), (b) and macrostructures (c, d). (a) and (c) equiaxed, (b) and (d) globular structures.

4. Discussion

The slab samples cast with a very low superheat were seen to be depleted of carbon and other solutes in the central areas. This is consistent with the sedimentation of globulites as suggested by Wunnenberg and Jacobi [7]. To produce a homogeneous concentration profile through the slab thickness, a columnar dendritic, or a combination of columnar dendritic and equiaxed dendritic structures would therefore be desirable. According to [7] there is a certain temperature below which the globular structure starts to dominate the casting structure. The present results suggest that, in the case studied, that temperature is of the order of 20 °C. Therefore, casting should be performed at high temperatures in applications where negative segregation is undesirable. Casting at an intermediate superheat causes the formation of a CET. Sometimes, positive macrosegregation was observed at the CET boundaries. The segregation coefficient for carbon was smaller at the CET than at the centerline, which is in agreement with the study of Choudhary and Ghosh [9] concerning continuously cast steel billets. According to Combeau et al. [5] the fraction of solid is larger in the packed globular zone than in the equiaxed dendritic zone. In continuous casting when the globulites sink in the strand in front of the columnar dendrites, the remaining solute-enriched liquid may result in positive CET segregation at the boundaries of columnar and the tightly packed globular zones. With regard to spot-like segregations, some researchers have pointed out the importance of minimizing local segregation width for improved inner slab quality [1, 11, 12]. This work has shown that the width of the segregation spots is larger in globular structures than in randomly orientated dendritic structures. The EPMA analyses also revealed that intensity of the segregation was higher in the segregation spots in globular zones than in equiaxed zones. Overall, therefore, globular structures in the central parts of the slab are undesirable if uniform element concentration distributions are desired, i.e. narrow low-intensity spot-like segregation and absence of positive CET macrosegregation.

5. Conclusions

This study has shown that continuous casting with very low superheat causes sedimentation of globulites and negative segregation in the slab central areas. In addition to centreline segregation, with intermediate superheats positive macrosegregation may occur at the columnar to equiaxed transition boundaries. Spot-like segregation widths are smaller with equiaxed dendrites than globular structures. The importance of these phenomena will depend on the intended final product and requirements with regard to the homogeneity of the chemical composition.

Acknowledgements

This work has been done within the DEMAPP programme of the Finnish Metals and Engineering Competence Cluster FIMECC Ltd. We gratefully acknowledge the financial support from the Finnish funding agency for innovation Tekes and the companies participating in the programme.

References

[1] Ghosh A 2001 *Segregation in cast products* Sadhana **26** 5
[2] Lesoult G 2005 *Mater. Sci. Eng. A* **413–414** 19
[3] Flemings M 2000 *ISIJ Int.* **40** 833
[4] Pickering E 2013 *ISIJ Int.* **53** 935
[5] Combeau H, Zaloznik M, Hans S and Richy P 2009 *Metall. Mater. Trans. B* **40B** 289
[6] Ludwig A and Wu M 2002 *Metall. Mater. Trans. A* **33A** 3673
[7] Wunnenberg K and Jacobi H 1983 *Arch. Eisenhüttenwes.* **54** 217
[8] Schwerdtfeger K 1992 *Metallurgie des stranggießens* (Düsseldorf: Verlag Stahleisen)
[9] Choudhary S and Ghosh A 1994 *ISIJ Int.* **34** 338
[10] Vander Voort G 1984 *Metallography: Principles and Practice* (New york: McGraw- Hill)
[11] Matsumiya T 2006 *ISIJ Int.* **46** 1800
[12] Shima S, Maruki Y, Nakashima J, Yamamura H and Oka Y 2004 *Scanmet II - 2nd International Conference on Process Development in Iron and Steelmaking, Luleå,* **2** 19

The behaviour of entrainment defects formed in commercial purity Mg alloy cast under a cover gas of SF_6

T Li[a] and W D Griffiths[b]

School of Metallurgy and Material, College of Engineering and Physical Science, University of Birmingham, Birmingham, UK, B15 2TT

Email: [a]TXL802@bham.ac.uk, [b]w.d.griffiths@bham.ac.uk

Abstract. In the casting of light alloys, the oxidised film on the melt surface can be folded due to surface turbulence, thus forming entrainment defects that have a significant negative effect on the mechanical properties of castings. Previous researchers reported that the surface film of Mg alloys formed in an atmosphere containing SF_6 had a complicated structure composed of MgO and MgF_2. The work reported here aims to investigate the behaviour of entrainment defects formed in magnesium alloys protected by SF_6-containing atmospheres. Tensile test bars of commercial purity Mg were cast in an unsealed environment under a cover gas of pure SF_6. 34Scanning electron microscopy (SEM) of the fracture surface of the test bars indicated entrainment defects that consisted of symmetrical films containing MgO, but also sulphur and fluorine. The results of these examinations of the symmetrical films were used to infer the potential formation and development of entrainment defects in commercial purity Mg alloy.

1. Introduction

Due to the rapid oxidation of liquid magnesium in air, a cover gas is used to protect Mg alloys in the casting process. The melt surface is generally covered by a continuous solid protective film formed due to the reaction between the molten Mg and the cover gas. Surface turbulence, (e.g. a splash or a vortex), which is common in casting operations, can lead to a folding action of the surface film, entraining a small amount of the surrounding atmosphere. When the folded structure is submerged into the bulk liquid, it becomes an entrainment defect. This kind of defect has been widely recognized as a major factor in the poor quality and lack of reproducibility of the properties of both Al and Mg alloy castings [1, 2].

Currently sulphur hexafluoride (SF_6) is one of the most widely used cover gases in industry [3], but its protective mechanism was not clear for decades. Cashion et al. [4] suggested that MgO particles could form a cohesive "raft" on the Mg melt surface in a SF_6 atmosphere, halting the rapid oxidation. Pettersen et al. [5] partly supported Cashion's process, and suggested that SF_6 could dissociate to provide atomic fluorine, which would diffuse into the porous MgO layer, forming MgF_2. Xiong and Liu [6] found MgF_2 particles between the film and the Mg alloy substrate, giving support to Pettersen's process. The sulphur contents in the surface films were reported to be very low (0.01%~0.02%) [5], or not detected [4].

So far, research into entrainment defects in Mg alloys has been limited [1]. The work reported here was aimed at trying to understand the behaviour of entrainment defects by investigating the surface films of entrainment defects formed in a SF_6 atmosphere.

2. Experiment procedure

Castings were produced using PEPSET resin-bonded sand mold, cast with commercial purity magnesium. The sand contained 2% Na_2SiF_6 as an inhibitor. The casting process was carried out in an unsealed environment with a cover gas of pure SF_6.

After solidification, the castings were machined into test bars, (of 37mm gauge length and 6.75mm diameter in the gauge length), and tested using a Zwick 1484 tensile test machine, at a strain rate of 1mm/min. Scanning Electron Microscopy (SEM) and Energy Dispersive X-ray Spectroscopy (EDS) were then carried out, using a Philips JEOL6060 SEM and Philips JEOL7000 SEM with Oxford INCA EDS, to investigate the nature of the defect found on the fracture surfaces of the broken test bars.

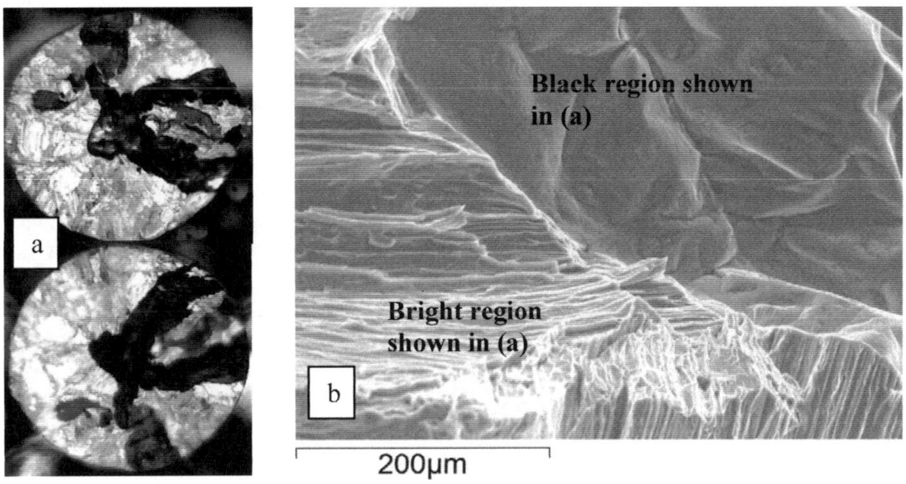

Figure 1. (a) Fracture surfaces of a broken tensile test bar (the diameter of the test bar was 6.75 mm); (b) The boundary between the black and the bright regions shown in (a).

Table 1. EDS results of in both regions shown in figure 1, compared by Student's T test.

		C	O	Mg	Si	F	S
	Sample size	15	15	15	15	15	15
Bright area	Mean	1.784	0.476	97.655	0.0727	Nd.	Nd.
(weight %)	Standard Deviation	0.821	0.159	0.967	0.0562	Nd.	Nd.
	Sample size	15	15	15	15	15	15
Black area	Mean	5.08	10.36	83.13	0.25	0.49	0.09
(weight %)	Standard Deviation	1.94	3.62	5.13	0.10	0.15	0.06
P value		<0.0001	<0.0001	-	<0.0001	-	-

Nd. = Not detected

3. Results

3.1. Entrainment defects formed in pure Mg under a cover gas of pure SF_6

Three different types of entrainment defects, formed under the same conditions, (casting of commercial purity Mg protected by pure SF_6), were found.

Figure 1(a) shows the fracture surfaces of a broken tensile test bar. Two symmetrical macroscopic black areas were seen on both fracture surfaces. Figure 1(b) shows a SEM image of the boundary between the black and bright areas shown in figure 1(a). Slip planes were seen in the bright area, and the black region had a comparatively smooth surface.

Table 1 shows the EDS results corresponding to both regions and it can be seen that sulphur and fluorine were only detected in the black region. The oxygen content of the black region was significantly higher than that of the bright area, and indicated the oxygen in the black region was not produced by oxidation after tensile testing, and that the black regions were entrainment defects with a surface film that contained oxygen, fluorine, and sulphur.

Figure 2(a) shows another entrainment defect formed in the same casting, (commercial purity Mg cast under pure SF_6). EDS analysis (see figure 2(b)), corresponding to the black region shown in figure 2(a), revealed that the surface film contained oxygen and fluorine, but no sulphur.

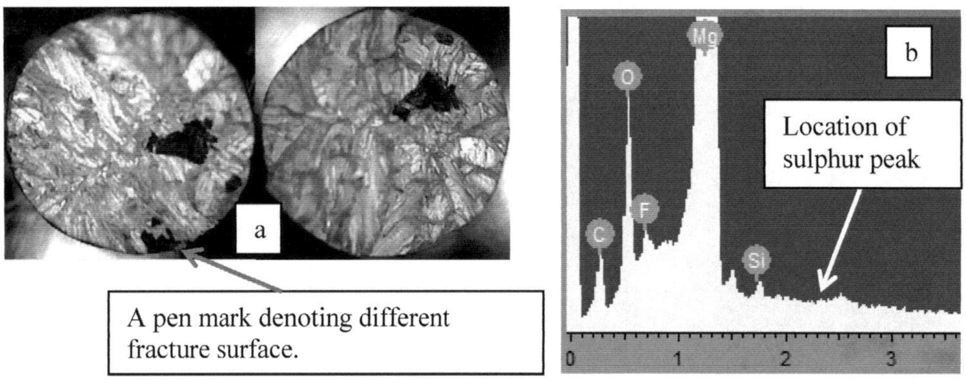

Figure 2. (a) The surface film of another entrainment defect formed in pure Mg/SF_6, the diameter of the test bar is 6.75 mm; (b) EDS spectrum corresponding to the black area shown in (a).

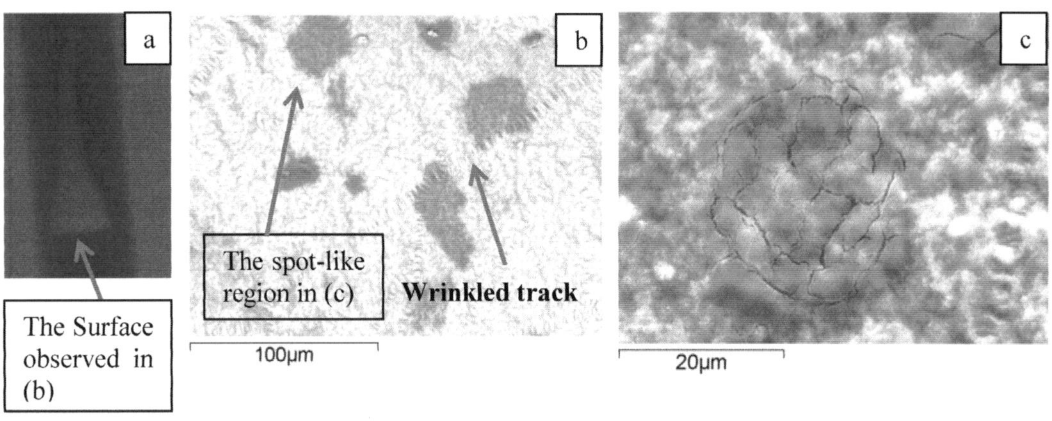

Figure 3. (a) A Real Time X-ray image of a trapped bubble formed in a tensile test bar of pure Mg/SF_6, the diameter of the bar is 10 mm. (b) The surface film of the bubble, using a backscatter electron beam; (c) a closer observation of the spot-like black region shown in (b), indicating cracks on the spot surface.

Figure 3(a) shows an entrained bubble (around 0.48 cm^3) captured in a tensile test bar. A SEM examination, using a backscattered electron beam, was carried out to observe the surface film of the trapped bubble (figure 3(b)), indicating dark spot-like regions distributed on the film surface. In addition, a wrinkled track was also found. An observation at higher magnification (figure 3(c)) showed cracks in the surface of the dark region. The EDS results (table 2) revealed that the dark spots had higher contents of sulphur and oxygen than the surrounding white areas. However, fluorine could not be clearly recognized by EDS in both regions, indicating that MgS or $MgSO_4$ might be the active species making the film protective. Thus it can be suggested that $S_2(g)$ or SO_2 also existed in the final residual entrained gas of the bubble.

Table 2. A comparison of the EDS results of the element contents in both regions shown in figure 3(b), by Student's T test.

		C	O	Mg	S
	Sample size	15	15	15	15
The dark spot	Mean	36.6	14.24	46.6	1.04
(weight %)	Standard Deviation	10.4	5.54	17.5	1.04
	Sample size	15	15	15	15
White area	Mean	8.96	4.05	86.61	0.07
(weight %)	Standard Deviation	3.51	1.54	4.89	0.08
P value		<0.001	<0.001	-	0.001

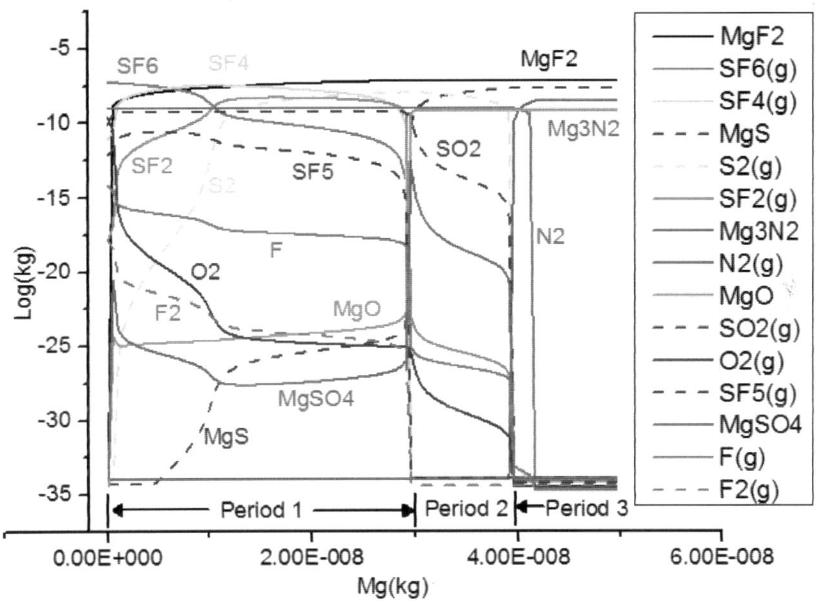

Figure 4. Equilibrium diagram for the reaction between 1e-8 kg 90%SF_6/Air and different amounts of Mg. To make it clear, the amounts of residual reactants in the cover gas and products present by logarithm (base 10) in Y-axis.

3.2. Thermodynamic calculation to approximate the evolution of gas occurring inside entrainment defects

A hypothetical thermodynamic calculation (figure 4) was carried out, using HSC software from Outokumpu, HSC Chemistry for Windows (http://www.hsc-chemistry.net/), in order to explore what might occur to the atmosphere inside an entrainment defect as the entrained gas reacted with the surrounding melt. This software can suggest which products are most likely to occur, based on thermodynamic calculations. Since the casting process was carried out in an unsealed environment, the cover gas inevitably contained some air. The entrained gas for the calculation was assumed to be 1e-08 kg 90%SF_6/Air ($\approx$ 1.67 mm^3) in contact with sufficient Mg for any reaction to go to completion. The temperature was set to 700 °C, the pressure to 1 atm, and the decomposition products of SF_6 were assumed to be SF_5, SF_4, SF_2, F_2, and F(g) [7, 8]. In addition, according to the EDS results shown in table 2, S_2(g) and SO_2 were assumed to also exist in the entrained gas.

In figure 4, the horizontal X-axis denotes the amount of liquid magnesium in contact with the entrained gas, and the vertical Y-axis denotes the logarithm (base 10) of the amounts of residual reactants and products. The whole process could be divided into three periods: (i) in period 1, liquid Mg preferentially consumes fluorides in the cover gas, forming MgF_2, while S_2(g) accumulates in the residual gas. A portion of S_2(g) reacts with oxygen, forming SO_2. MgO, MgS, and $MgSO_4$ are theoretically produced, but they can be neglected in a practical process since their amounts are too small (less than 10^{-20} kg, while MgF_2 occurs at more than 10^{-10} kg); (ii) in period 2, nearly all fluorides in the entrained gas have been consumed, and liquid Mg reacts with S_2(g) and SO_2, forming MgO and MgS. The existence of $MgSO_4$ can also be practically neglected, due to its small amount (less than 10^{-15} kg, while both MgO and MgS occurs at more than 10^{-10} kg); (iii) in period 3, S_2(g) and SO_2 have been depleted, and liquid Mg reacts with nitrogen, forming Mg_3N_2.

Therefore, the detected sulphur shown in table 1 and 2 might mainly exist as MgS, consistent with the suggestions of [8, 9], and the existence of MgS in the surface film can be suggested to depend on the F:S ratio in the cover gas.

4. Discussion

Sulphur and fluorine were not found in all of the films formed in the commercial purity Mg castings protected by pure SF_6; sulphur was only found in two films and fluorine in two other different films. The thermodynamic calculation shown in figure 4 suggested that the existence of MgF_2 and MgS in the surface film was determined by the F:S ratio in the cover gas. Therefore, the evolution process for an entrainment defect can be suggested to be as shown in figure 5.

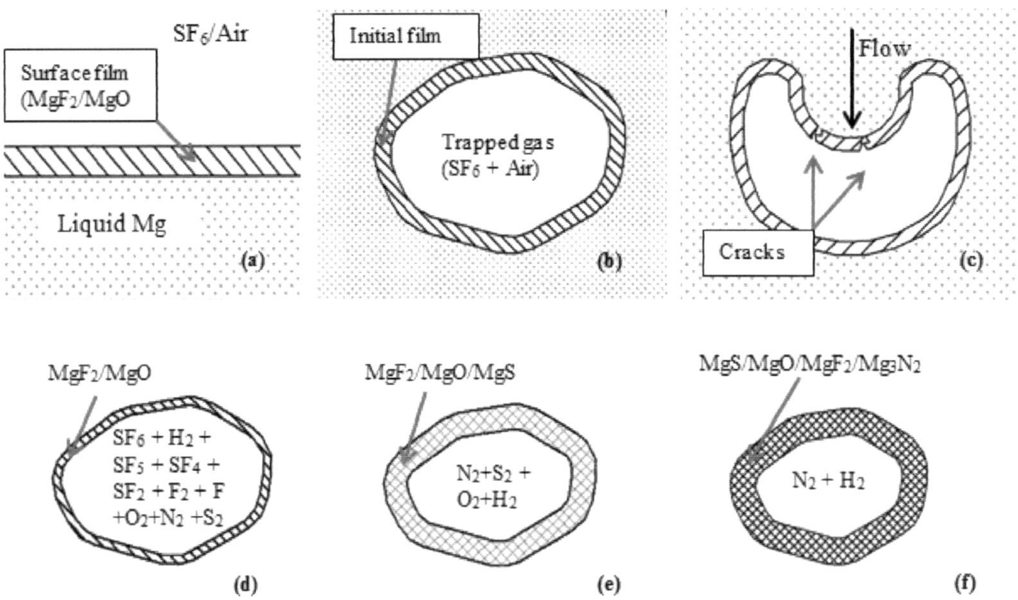

Figure 5. Formation and evolution process of the observed entrainment defects. (a) a surface film formation; (b) original entrainment defect; (c) cracks caused by melt flow; (d ~ f) final entrainment defects.

In stage 1, figure 5(a), a protective film has been formed on the melt surface due to the reaction between liquid Mg and the cover gas. The protective film might contain MgF_2 but not MgS, based on [4, 5] and the observation shown in figure 2. Moreover, according to Xiong and Wang [10], who found that the film formed on the surface of Mg melt had a high content of oxygen under a cover gas of 0.1%SF_6/air, MgO can be contained in the surface film shown in figure 5(a), resulting from $2Mg + O_2 = MgO$, when the surrounding atmosphere of the Mg melt has a high concentration of air, especially during the pouring process.

In stage 2, figure 5(b), the protective film can be folded over onto itself due to surface turbulence, forming entrainment defects. The original entrainment defect would be covered by the initial surface film, consisting of MgF_2 and MgO. The entrained gas may originally consist of SF_6 and air.

In stage 3, figure 5(c), the initial surface film can be cracked or even partly removed, by the movement of the melt. Thus fresh liquid Mg can come into contact with the interior of the defect through the cracks, reacting with the trapped gas and forming new surface films. In conjunction with the calculation shown in figure 4, the formation of the new surface film can reduce the total fluorine content in the trapped gas, but the sulphur content might not decrease. Simultaneously, hydrogen might diffuse into the entrained gas [2], through the cracks and the permeable film.

Therefore in stage 4, three types of final entrainment defects might exist, as shown in figure 5(d) ~ (f), and the different observed defects on the fracture surfaces have given evidence of different types of entrainment defects:

Firstly, if the amount of the liquid Mg that comes into contact with the entrained gas, was not enough to consume the fluorides in the trapped gas, corresponding to period 1 shown in figure 4, the surface film can only contain MgF_2 and MgO, as shown in figure 5(d), consistent with the results shown in figure 2. The residual entrained gas might contain SF_6, hydrogen, air and the decomposition products of SF_6.

Secondly, if the liquid Mg, in contact with the entrained gas, consumed nearly all of the fluorides in the trapped gas, but did not deplete sulphur, corresponding to the period 2 shown in figure 4, the final surface film can contain MgS, MgF_2, and MgO, as shown in figure 5(e), consistent with the observation shown in figure 1. The residual entrained gas therefore mainly contains sulphur, air and hydrogen at this stage. In addition, if the initial surface film containing MgF_2 was removed, the final surface film would only contain MgS and MgO as shown in figure 3 and table 2.

Finally, if all fluorides and sulphur in the entrained gas were consumed by the Mg melt, corresponding to period 3 shown in figure 4, the surface film might contain MgS, MgO, MgF_2 and Mg_3N_2, as shown in figure 5(f). The residual entrained gas might only contain N_2 and H_2. If the process continued further, making more liquid Mg come into contact with the entrained gas, N_2 can be depleted, so that the final entrained gas might only contain H_2. However, Mg_3N_2 has not been found in the samples of the experiment. Thus there may not be enough time for the entrained gas to react with liquid Mg, forming Mg_3N_2, before solidification.

The evolution process and results revealed that entrainment defects can exist with a trapped gas in Mg castings, acting as a crack initiator, similar to the entrainment defects in Al castings [2]. In addition, the evolution process also indicated that all the components of the initial entrained gas can be consumed by reaction with liquid Mg. Thus the effect of entrainment defects in Mg casting might be effectively reduced, by increasing the holding time before solidification [13], so that the reproducibility of the properties of Mg castings can be improved. Compared with aluminum alloy, which cannot efficiently consume nitrogen (78% volume of the initial trapped gas) in the entrained gas [11, 12], Mg alloy castings might achieve a better reproducibility through this technique.

5. Conclusion

The surface films of three typical entrainment defects formed in commercial purity Mg cast under a cover gas of pure SF_6 have been observed. The surface films contained different combinations of compounds: (1) MgS and MgO; (2) MgF_2 and MgO; (3) MgS, MgF_2 and MgO. In addition, a thermodynamic prediction revealed that fluorides in the entrained gas could be preferentially consumed compared to air. An evolution process of the entrainment defects formed in pure Mg/pure SF_6 was suggested, and the composition of the surface film depended on which stage the defect was in.

Acknowledgements
The authors acknowledge funding from the EPSRC LiME grant EP/H026177/1, and the technical assistance of Mr. Peter Cranmer and Mr. Adrian Caden of the University of Birmingham.

References

[1] Griffiths W D and Lai N W 2007 Double oxide film defects in cast magnesium alloy *Metall Mater Trans A-Phys Metall Mater Sci* **38A** 190-196
[2] Campbell J 2004 *Castings* (Oxford: Butterworth-Heinemann)
[3] Polmear I J 2006 *Light alloys: metallurgy of the light metals* (Oxford: Butterworth-Heinemann)
[4] Cashion S P, Ricketts N J and Hayes P C 2002 The mechanism of protection of molten magnesium by cover gas mixtures containing sulphur hexafluoride *Journal of Light Metals* **2** 43-47
[5] Pettersen G, Ovrelid E, Tranell G, Fenstad J and Gjestland H 2002 Characterisation of the surface films formed on molten magnesium in different protective atmospheres *Mater Sci Eng A-Struct Mater Prop Microstruct Process* **332** 285-294
[6] Xiong S M and Liu X L 2007 Microstructure, composition, and depth analysis of surface films formed on molten AZ91D alloy under protection of SF6 mixtures *Metall Mater Trans A-Phys Metall Mater Sci* **38A** 428-434
[7] Kubaschewski O and Hesselemam K 1991 *Thermo-chemical properties of Inorganic Substances* (Belin: Springer-Verlag)
[8] Aarstad K 2004 Protective Films on Molten Magnesium PhD Thesis Norwegian University of Science and Technology
[9] Shih T S, Liu J B and Wei P S 2007 Oxide films on magnesium and magnesium alloys *Mater Chem Phys* **104** 497-504
[10] Xiong S M and Wang X F 2010 Protection behavior of fluorine-containing cover gases on molten magnesium alloys *Trans Nonferrous Met Soc China* **20** 1228-1234
[11] Kumari S S S, Pillai U T S and Pai B C 2011 Synthesis and characterization of in situ Al-AlN composite by nitrogen gas bubbling method *J Alloy Compd* **509** 2503-2509
[12] Swaminathan S, Rao B S and Jayaram V 2002 The influence of oxygen impurities on the formation of AlN-Al composites by infiltration of molten Al-Mg *Mater Sci Eng A-Struct Mater Prop Microstruct Process* **337** 134-139

The effect of alloy elements on the density variation of steel melt at the interdendritic region during solidification

Y F Cao, Y Chen[a], X P Ma, P X Fu, X H Kang, H W Liu and D Z Li
Shenyang National Laboratory for Materials Science, Institute of Metal Research, Chinese Academy of Sciences, Shenyang, Liaoning, 110016 China

Email: [a]chenyun@imr.ac.cn

Abstract. Alloying elements in steels have essential effects on the formation of macrosegregation by inducing a density difference between the mushy zone and the bulk melt, and even by the alteration of the solidified microstructure. Hence, in terms of the thermodynamic laws for solidification of an idealized dilute solution, a systematic theoretical analysis on the effect of elements on the density variation of the interdendritic melt is presented for common Fe-based binary alloys. It shows that density variation closely associates with three crucial parameters: the microsegregation parameter λ, the temp-comp-expansion parameter β and the initial composition C_L^0. With these derived parameters, a simple analytical model is proposed to assess the influence of alloy elements on density change. The theoretical analysis indicates that compared to silicon, the effect of carbon on the density variation cannot be ignored, which is remarkably different from the previous recognition of these two elements. The macrosegregation induced by carbon only is experimentally validated by two dissected 500 kg ingots with different carbon contents. Furthermore, to directly validate the results of theoretical analysis a macrosegregation model with two different microsegregation laws (lever rule and Scheil equation) is established. Then simulations of the convection during solidification in the experimental 500 kg reference ingot are performed for Fe-C and Fe-Si alloys, respectively. It demonstrates that numerically simulated effects of carbon and silicon on the interdendritic convection induced by density inversion are fairly consistent with the analytical predictions.

1. Introduction

Macrosegregation of solute elements in steels is usually a serious inhomogeneity problem impairing the quality of final products. Since the mid-1960s, huge amounts of research has been performed on the formation mechanism of macrosegregation [1-6]. It has been well recognized that the composition of an alloy often plays an essential role in macrosegregation formation by changing the dendrite morphology and magnifying the density variation of liquid melt in the mushy zone, which leads to intense interdendritic flow and finally results in the macrosegregation of elements in a solidified product. Specifically, reducing the content of Si and increasing the content of Mo [6] has been employed frequently as an effective approach to reduce channel segregates in steel ingots.

NOMENCLATURE

C	concentration, wt.%	P	pressure, Pa
f	volume fraction	k	partition coefficient
H	enthalpy, J kg^{-1}	m	liquidus slope, K (wt.%)$^{-1}$
T	temperature, K	M	atomic weight
β	temp-comp-expansion parameter, kg m^{-3} (wt.%)$^{-1}$	w	mole fraction
λ	microsegregation parameter	ΔH	heat of fusion, J kg^{-1}
$\Delta_{fus}H_m$	molar enthalpy of fusion, J mol^{-1}	R	gas constant, J K^{-1} mol^{-1}
K	permeability, m^2	t	time, s
$\vec{U}$	superficial velocity, m s^{-1}	$\vec{g}$	gravity acceleration, m s^{-2}
u	velocity component in x direction, m s^{-1}	μ_L	liquid dynamic viscosity, pa·s
v	velocity component in y direction, m s^{-1}	ρ	density, kg m^{-3}
c_p	specific heat, J kg^{-1} K^{-1}	α	heat conductivity, W m^{-1} K^{-1}

Subscripts and Superscripts

δ, γ	δ and γ phases	m	melting point
S, L	solid and liquid phases	ρ	density
T, C	thermal and solutal effects	o	initial concentration
$1, 2$	solvent and solute	i, j	the ith and jth solute elements
ref	reference		

However, evaluating the density change or the convection intensity in the mushy zone from experimental results is difficult since the thermal and solute fields during experiments may not be accurately known. Commercial software packages, such as JMatPro, Thermo-Calc, ProCAST and Magma, are able to deliver some parameters in steels, but outputs from them are limited in their current calculations [7]. In addition, it has been evidenced by some previous reports [8-9] that in carbon steels, silicon has the most important influence on the change of liquid density, whereas carbon plays a negligible role. Nevertheless, experimental investigations on the macrosegregation in steel ingots have demonstrated that more serious segregation exists in high carbon steels compared with that in low carbon steels, which indicates that the effect of carbon content on the macrosegregation formation should not be neglected. Therefore, in order to clearly understand the influence of alloying elements on the formation of macrosegregation in steels, further research oriented on the relationship between alloying elements and the density difference is described in the current paper. Firstly, a theoretical analysis of the effect of alloying elements on the density change is carried out. Then two experimental steel ingots are dissected to verify the analytical results. Finally, the flow fields in the Fe-C and Fe-Si systems induced only by density difference during solidification of the 500 kg reference ingot are simulated numerically based on our theoretical analysis.

2. Theoretical analysis

Based on microsegregation models of the lever rule and the Scheil equation, for a dilute multicomponent steel melt the density change can be expressed as following [8],

$$\Delta\rho_L = \sum_i C_{Li}^o \times (\beta_T^\rho \times m_i + \beta_{Ci}^\rho) \times (f_L^{k_i - 1} - 1)$$
$$+ \sum_j C_{Lj}^o \times (\beta_T^\rho \times m_j + \beta_{Cj}^\rho) \times (\frac{(1-f_L)\times(1-k_j)}{f_L \times (1-k_j) + k_j})$$
$$= \sum_i (C_{Li}^o \times \beta_i \times \lambda_i) + \sum_j (C_{Lj}^o \times \beta_j \times \lambda_j) \qquad (1)$$

where β_i and β_j represent the density expansion caused by temperature and composition simultaneously for substitutional element i and interstitial element j, respectively, and they are defined as the "temp-comp-expansion parameter". λ is defined as the "microsegregation parameter" and is determined by microsegregation law. Then, the effect of elements on the density in the mushy zone via the initial concentration C_L^o, expansion parameter β and microsegregation parameter λ, will be discussed in the following.

2.1. *Temp-comp-expansion parameter β*

In the derived temp-comp-expansion parameter β, the values of β_C^ρ are almost the same in most of the published papers for various alloying elements and the value of β_T^ρ of iron can be obtained using the commercial software, such as the Thermo-Calc software. Therefore, the only controversial parameter is the liquidus slope m for each alloying element, which can be derived within the thermodynamic laws for a dilute solution. According to the dilute solution approximation, if both the solid and liquid conform to Raoult's law, the relation between the concentration of an alloying element and the melting point of iron T_m (1809 K) can be written as [10]

$$\ln \frac{w_1^L}{w_1^S} = \frac{\Delta_{fus} H_m}{R} (\frac{1}{T_m} - \frac{1}{T}) \qquad (2)$$

where w_1^L and w_1^S are the mole fractions of solvent in the liquid and solid respectively (subscript:1 solvent, 2-solute); the molar enthalpy of fusion $\Delta_{fus}H_m$ of iron is equal to 15490 J mol^{-1}; R is the gas constant (8.314 J K^{-1} mol^{-1}). As the concentration of alloy element is small and w_1^L = 1- w_2^L, w_1^S = 1- w_2^S, equation (2) can be approximated to:

$$w_2^L - w_2^S = \frac{\Delta_{fus} H_m}{RT_m^2} (T_m - T) \qquad (3)$$

where $w_2^S = k \cdot w_2^L$. Then for an element with atomic weight M the absolute value of the liquidus slope m can be obtained using equation (3), where the mole fraction of solute element has been changed to mass percentage,

$$|m| = \frac{1000(1-k)}{M} \qquad (4)$$

The calculated results of m using equation (4) for some common elements during phase transformation from liquid to δ-iron or γ-iron, together with the previous calculation [8-11], are listed in table 1 for comparison. It denotes that the present calculated values for all the selected elements are rather close to the referred values.

Table 1. Comparisons between the calculated results of m using the present model and references.

Elements	C	Si	Mn	P	S	Mo
m_δ, K (wt.%)$^{-1}$	-67	-6.1	-1.8	-28	-31	-1.5
m_γ, K (wt.%)$^{-1}$	-58	-18	-4.5	-30	-30	-4.2
$m^{[8]}$, K (wt.%)$^{-1}$	-55	-13	-4.8	-30	-30	-1.9
$m^{[11]}$, K (wt.%)$^{-1}$	-76	-10	-4	-31	-34	-3

Based on respective invoked parameters, table 2 shows the calculation results of β. Here, in our calculations β_T^ρ = -0.635 kg m^{-3} K^{-1}, which is obtained from Thermo-Calc software. The partition coefficients k_δ and k_γ and β_C^ρ are cited from Ref. [8], and are similar to that in Ref. [11]. It should be noticed that when we checked the values of β as expressed by $\frac{\partial \rho_L}{\partial T}\frac{\partial T_L}{\partial C_L}+\frac{\partial \rho_L}{\partial C_L}$ in Ref. [8], the values listed in that paper are different from the calculated results using the corresponding input parameters. This is possibly because of a mistake in the units in the paper, which leads to the calculated values β^b deviating two orders of magnitude from β^c. Thus we imposed these two sets of data together in table 2. It shows that the obvious discrepancy between the present calculation and the previous reported results are the values of β for C element. In the present calculation its value is in the same order of magnitude as those of Si, S and P elements, whereas, in both references, this value is so small that it can be ignored. This difference might be mostly ascribed to the different choices of m and β_T^ρ during the calculations of β. In the previous calculations the values of m and β_T^ρ referred to very early experimental data and varied with experimental methods and conditions. For example, the value of β_T^ρ cited is different in some papers [8-9, 11], while in current calculations, it is obtained from Thermo-Calc software, which should be more reliable. For the calculation of C element, small differences of m and β_T^ρ will result in sensitive changes of β because of the different signs between the product of them and β_C^ρ.

Table 2. The calculated results of temp-comp-expansion parameter β in different research.

Elements	C	Si	Mn	P	S	Mo
β_δ^a, kg m^{-3} (wt.%)$^{-1}$	-38	-83	-13	-66	-71	15
β_γ^a, kg m^{-3} (wt.%)$^{-1}$	-43	-76	-11	-65	-71	17
β^b, kg m^{-3} (wt.%)$^{-1}$	-25	-6800	-700	-4100	-4700	1700
β^c, kg m^{-3} (wt.%)$^{-1}$	-0.25	-68	-7	-41	-47	17
β^d, kg m^{-3} (wt.%)$^{-1}$	-4	-77	-10	-53	-56	17

[a] Current calculation results.

[b] The results in Ref. [8].

[c] The modified results from Ref. [8], and all the parameters are the same with the reference.

[d] The results in Ref. [11].

2.2. Microsegregation parameter λ

Furthermore, to precisely understand the effect of an alloying element on macrosegregation, the value of $\beta\cdot\lambda$ should actually be used rather than just β according to equation (1). The calculated values of β indicate that S and P will have strong effects on the density change, but their contents usually are limited within dozens of ppm during the steelmaking process. Hence, we will mainly focus on the effects of C and Si. Equations (5) and (6) show the calculated ratios of $\beta\cdot\lambda$ between carbon and silicon for δ-iron and γ-iron respectively. It indicates that the values of $\beta\cdot\lambda$ of C and Si are rather close and so carbon will have a comparable effect on the density change to that of silicon, which is different from the conclusions in Refs. [8] and [11].

$$\left(\beta_C \cdot \lambda_C\right)_\delta / \left(\beta_{Si} \cdot \lambda_{Si}\right)_\delta \in (2.14, 2.56), f_L \in (0.05, 0.99) \tag{5}$$

$$\left(\beta_{Si} \cdot \lambda_{Si}\right)_\gamma / \left(\beta_C \cdot \lambda_C\right)_\gamma \in (1.26, 3.08), f_L \in (0.05, 0.99) \tag{6}$$

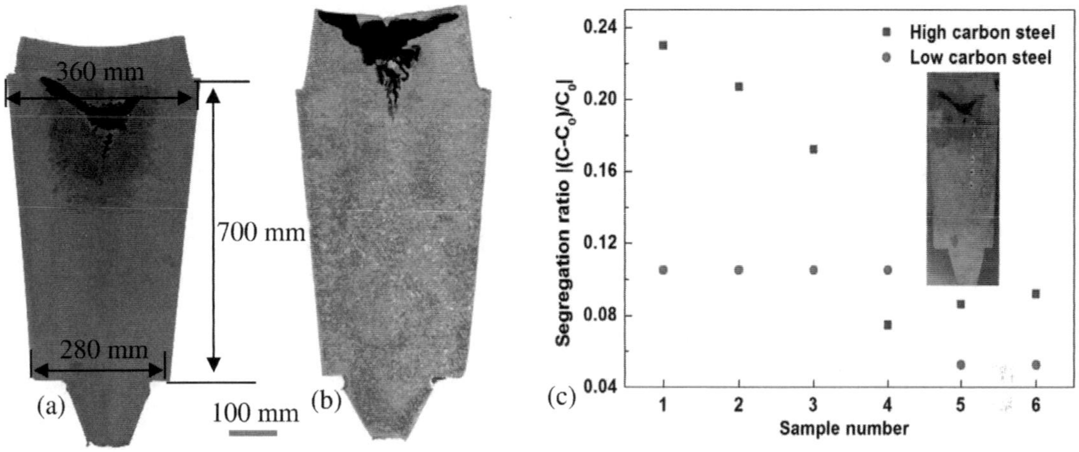

Figure 1. The etched surfaces and carbon segregation ratio of two vacuum poured 500 kg ingots cut half along the axle plane. (a) High carbon (1.74 wt.%) ingot and its body dimensions; (b) low carbon (0.019 wt.%) ingot; (c) the segregation ratio of measured C concentration in the two ingots. The inset figure shows the positions of six drilled samples.

In addition, the current calculation results can also explain the noticeable phenomenon in the industrial process that macrosegregation of carbon in high carbon steels is more serious, relative to that in low carbon steels, as shown in figures 1(a) and (b). A sand mould was used for two round 500 kg ingots (other alloying element contents are the same for the produced ingots and are limited to a very low level during the refining process) and the sectioned surfaces were ground, polished and etched using a 20% HNO_3-5% H_2SO_4-H_2O solution. Six samples which were located in the same positions of each ingot were drilled for chemical analysis. Figure 1(a) indicates that in the high carbon steel both positive segregation (black region) at the upper part and the negative segregation at the bottom (white region) along the centreline are distinct, but in the low carbon steel (figure 1(b)) the segregation of carbon is not so obvious. Moreover, further measurements of carbon in the drilled samples (figure 1(c)) quantitatively show that the segregation ratio is generally larger in high carbon steel than that in low carbon steel. These compared results indicate that the effect of C element on the macrosegregation along the centreline in steel ingots should not be ignored and it will become more serious with increasing C content. These results are in accord once with the above theoretical analysis.

3. Numerical implementations

Furthermore, in order to elucidate the importance of carbon in driving natural convection and even macrosegregation, the flow field caused only by density inversion during solidification, neglecting solidification shrinkage and grain settlement, is simulated in binary Fe-0.36 wt.% C and Fe-0.36 wt.% Si systems respectively. The details of the applied model can be referred to our previous work [12], and here some key conservation equations are listed.

Conservation of mass: $$\nabla\cdot\left(\vec{U}\right)=0 \tag{7}$$

Conservation of energy: $$\frac{\partial[\rho H]}{\partial t}+\nabla\cdot\left(\rho c_p\vec{U}T+\rho\vec{U}\Delta H\right)=\nabla\cdot\left(\alpha\nabla T\right) \tag{8}$$

Conservation of solute: $$\frac{\partial[\rho C]}{\partial t}+\nabla\cdot\left(\rho\vec{U}C_L\right)=0 \tag{9}$$

Conservation of momentum (X direction): $$\frac{\partial\left(\rho u\right)}{\partial t}+\nabla\cdot\left(\rho\vec{U}u\right)=-\frac{\partial P}{\partial x}-\frac{\mu_L}{K}u+\nabla\cdot\left(\mu_L\nabla u\right) \tag{10}$$

Conservation of momentum (Y direction):

$$\frac{\partial\left(\rho v\right)}{\partial t}+\nabla\cdot\left(\rho\vec{U}v\right)=-\frac{\partial P}{\partial y}-\frac{\mu_L}{K}v+\nabla\cdot\left(\mu_L\nabla v\right)+\rho\vec{g}\left[\beta_T\left(T-T_{ref}\right)+\beta_C\left(C_L-C_{ref}\right)\right] \tag{11}$$

Table 3. The initial and thermal boundary conditions in current simulations.

Material/Property	ρ (kg m^{-3})	C_P (J kg^{-1} K^{-1})	α (W m^{-1} K^{-1})
Sand mould	1370	1070	0.73
Insulation sleeve	600	1130	0.9
Covering flux	210	400	0.18
Interfacial heat transfer coefficient (W m^{-2} K^{-1})	Ingot and covering flux: 50 Ingot and sand mould: 200 Ingot and insulation sleeve: 50 Covering flux and the outside: $5 + 4 \times 5.67 \times 10^{-8} \times \varepsilon \times T^3$, $\varepsilon = 0.5$ Sand mould and the outside: $7.5 + 4 \times 5.67 \times 10^{-8} \times \varepsilon \times T^3$, $\varepsilon = 0.9$		

It should be noted that in the current model two different microsegregation models are applied: Lever and Scheil laws for C and Si respectively. As a reference, the experimental 500 kg ingot (see figure 1(a)) is numerically studied. The necessary boundary and initial conditions are also shown in table 3, and physical parameters of elements C and Si are referred to in the above analytical results. The other alloy properties such as density, specific heat, conductivity and secondary dendrite arm spacing and so on, are referred to in paper [13]. The grid size is 0.5 cm × 0.5 cm. The initial superheat is set to zero.

4. Numerical simulation results and discussion

Figure 2 shows the velocity and solid fraction fields when the solidification front nearly reaches half of the ingots for Fe-0.36 wt.% C and Fe-0.36 wt.% Si alloys. It denotes that in both alloys the solutal buoyancy gradually overwhelms thermal buoyancy with the development of solidification, and the flow is ascending along the solidification front and descending along the centreline of the ingot. The convection finally dominated by solutal buoyancy can be well explained by the values of the temp-comp-expansion parameter β in table 2, which are negative and mean that $|m\cdot\beta_T^\rho| < |\beta_C^\rho|$. In addition, it is worth noting that the mushy zone in the Fe-C system is wider than that in the Fe-Si system due to the larger liquidus slope induced by element carbon. Quantitatively, figure 3 shows that the average and maximum velocity magnitudes in the mushy zone where solid fraction varies from 0.15 to 0.45 are higher in the Fe-C alloy than those in the Fe-Si alloy, which indicates the prediction by equation (5) that the driving force induced by density inversion in the Fe-C system is slightly larger compared with the Fe-Si system. Thus, the simulation results again demonstrate that the C element plays a comparable role on the evolution of thermosolutal convection in steels and are consistent with the theoretical analysis in section 2.

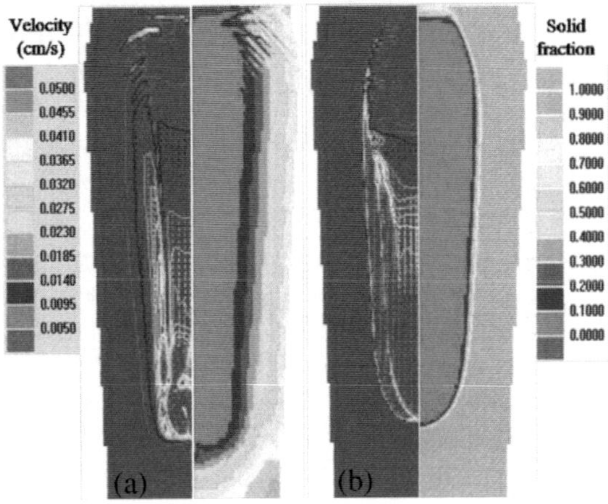

Figure 2. Predicted flow and solid fraction fields. (a) Fe-0.36 wt.% C; (b) Fe-0.36 wt.% Si.

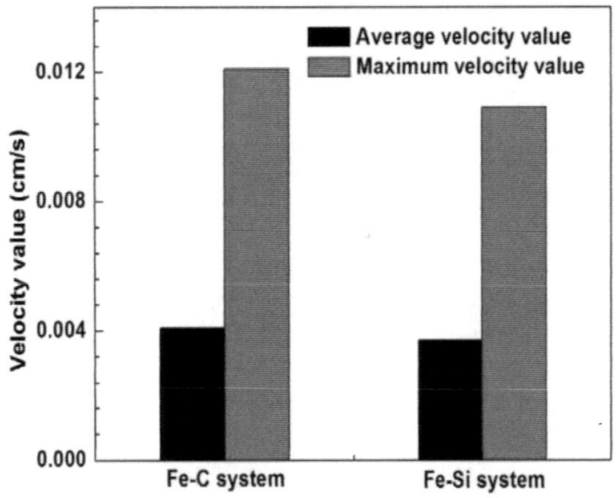

Figure 3. The average and maximum values of velocity in the mushy zone of the two systems.

5. Conclusions

To summarize, the effect of alloying elements on the density change in the mushy zone in steel melts has been studied systematically by theoretical analysis, experiments and numerical simulations, and some main conclusions are yielded as follows:

(1) A theoretical analysis with consideration of microsegregation parameter λ and expansion parameter β has been performed to assess the effect of elements on the density variation during solidification of steels. Guidance to composition design of steels can be offered to reduce the macrosegregation within this theoretical framework.

(2) Theoretical analysis results have been verified by two dissected experimental ingots and the corresponding numerical simulations. They demonstrate that the effect of carbon on the density variation or even macrosegregation can be comparable with silicon, which is dramatically different from the previous recognition of these two elements that the contribution of carbon on the thermosolutal buoyancy can be neglected when compared to silicon.

References

[1] Flemings M C 1974 *Metall. Trans.* **5** 2121
[2] Mehrabian R, Keane M, and Flemings M C 1970 *Metall. Trans.* **1** 1209
[3] Flemings M C 1974 *Solidification Processing* (New York: McGraw-Hill Book Co.) p 244
[4] Ishida H, Sakamoto K, Nohara M, Oki H and Ohsasa K 2009 *Tetsu-to-Hagnné.* **95** 26
[5] Fredriksson H and Nilsson S O 1978 *Metall. Trans.* B **9** 111
[6] Suzuki K, Miyamoto T 1981 *Trans. Iron Steel Inst. Jpn.* **21** 235
[7] Pickering E J 2013 *ISIJ Int.* **53** 935
[8] Fujii T, Poirier D R, and Flemings M C 1979 *Metall. Trans.* B **10** 331
[9] Jie W Q and Zhou Y H 1989 *Metall. Trans.* B **20** 723
[10] Derge G 1964 *Basic Open Health steelmaking* (New York: AIMMPE) p 643
[11] Mori N and ŌGI K 1991 *Metall. Trans.* A **22** 1663
[12] Liu D R, Sang B G., Kang X H and Li D Z 2011 *Metall. Mater. Trans.* B **42** 210
[13] Combeau H, Založnik M, Hans S and Richy P E 2009 *Metall. Mater. Trans.* B **40** 289

IOP Conf. Series: Materials Science and Engineering **117** (2016) 012067 doi:10.1088/1757-899X/117/1/012067

In situ X-ray observations of gas porosity interactions with dendritic microstructures during solidification of Al-based alloys

A G Murphy[1a], D J Browne[1], Y Houltz[2] and R H Mathiesen[3]
[1]School of Mechanical and Materials Engineering, University College Dublin, Belfield, Ireland
[2]Swedish Space Corporation, Science Services Division PO Box 4208, SE-17104 Solna, Sweden
[3]Department of Physics, NTNU, N-7491 Trondheim, Norway

Email: [a]andrew.murphy@ucd.ie

Abstract. *In situ* X-radiography solidification experiments were performed on Al-based alloys, using both synchrotron and laboratory-based X-ray sources, in conjunction with a gradient furnace and a newly developed isothermal furnace, respectively. The effect of gas porosity nucleation and growth within the semi-solid mush during both columnar and equiaxed solidification was thereby observed. In all experimental cases examined, gas porosity was observed to nucleate and grow within the field-of-view (FOV) causing various levels of distortion to the semi-solid mush, and thereafter disappearing from the sample leaving no permanent voids within the solidified microstructure. During columnar growth, a single bubble caused severe remelting and destruction of primary trunks leading to secondary fragmentation and evidence of blocking of the columnar front. Equiaxed solidification was performed under microgravity-like conditions with restricted grain motion in the FOV. The degree to which the nucleated gas bubbles affected the surrounding grain structure increased with increasing solid fraction. However, bubble sphericity remained unaffected by apparent solid fraction or grain coherency.

1. Introduction

Defects in cast components, such as micro and shrinkage porosity, can be detrimental to service-life performance, causing a reduction in ductility and fatigue life [1,2]. While shrinkage porosity is largely due to restricted liquid transport in the semi-solid mush during the late stages of solidification, thereby causing voids in the as-cast microstructure, gas porosity is due to the effervescence of dissolved gas within the liquid melt as a consequence of lower gas solubility in the solid than in the liquid, e.g. hydrogen in aluminium alloys [3]. Traditionally, the method used to characterise porosity distribution in castings has been through post-mortem metallurgical examination of the as-cast microstructure. These examinations, however, provide little information on porosity-microstructural interactions during the early stages of alloy solidification. Real-time *in situ* X-radiography has emerged as a powerful tool in solidification science, providing invaluable and unique insights into early and transient solidification dynamics [4], e.g. nucleation, growth kinetics, porosity formation [5,6], etc.

Some brief *in situ* X-ray observations of gas porosity nucleation, interaction with dendritic microstructure during solidification, and subsequent disappearance from the sample are presented and discussed. This work is the result of both laboratory and synchrotron-based *in situ* X-radiography solidification experiments.

2. Equipment and experiment description

The results presented were obtained from two independent experimental campaigns. The first involved the use of a Bridgman-type solidification furnace and synchrotron-based X-radiography, and the second used a newly developed prototype isothermal solidification furnace and laboratory-based microfocus X-radiography. Figure 1 shows the general arrangement of the *in situ* X-ray diagnostics setup used in both campaigns. The incident X-ray beam (arrow, figure 1) intersects the alloy sample at the location of the adiabatic zone inside the solidification furnace. The transmitted beam then projects an image onto the scintillator/CCD (charge-coupled device), showing the contrasting solid and liquid phases as light and dark zones in the field of view (FOV), respectively. The differing intensities at the scintillator/CCD are as a result of differing X-ray attenuation coefficients of the independent solid and liquid phases within the FOV giving rise to absorption contrast X-radiography.

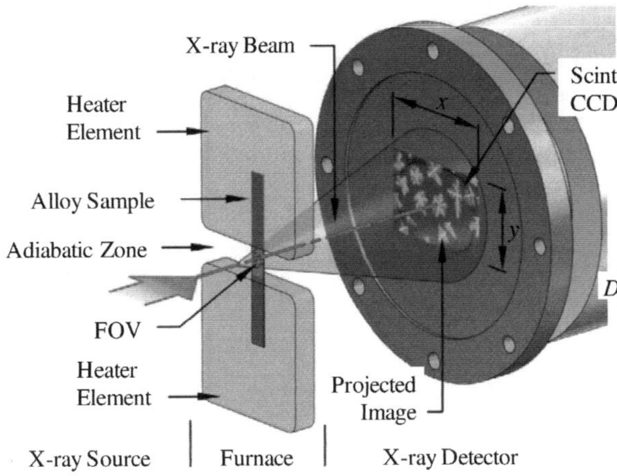

Figure 1. Schematic illustration of *in situ* X-ray diagnostic setup. The X-ray beam passes through the aluminium sample and projects an image onto the scintillator/CCD. Dimensions x and y indicate the size of the projected FOV.

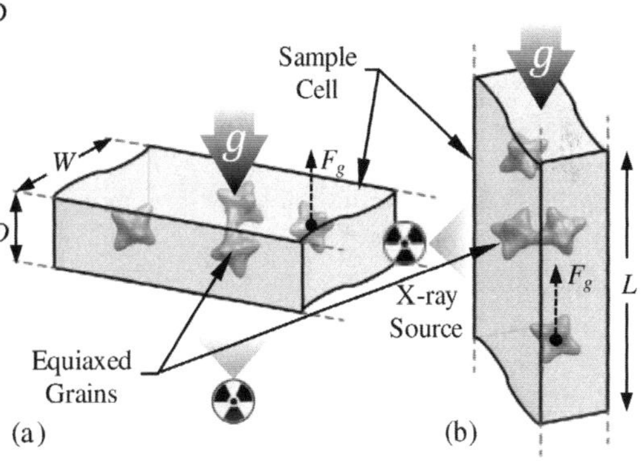

Figure 2. Sample orientation relative to the force of gravity. Dimensions D, W, and L represent the sample thickness, width, and length, respectively. F_g denotes the direction of buoyant grain motion, w.r.t. gravity (g) for the alloys used.

The synchrotron experiments were performed at the European Synchrotron Radiation Facility (ESRF) in Grenoble, France, on beamline ID15A, using a high energy white beam (30 → 500 keV). Mathiesen *et al.* [6] described in detail the relevant characteristics of the furnace, along with the method of sample manufacture and encapsulation used. The alloy material used was Al-12wt%Ge inoculated with 0.15wt%Al-Ti-B (5/1) master alloy, of dimensions 30 × 15 × 0.2 mm ($L \times W \times D$, figure 2). Solidification was promoted by pulling the sample through a constant thermal gradient (G_T) applied across the adiabatic zone, at a constant velocity (V_p). Spatiotemporal resolutions of 2.54 μm/px (pixel) and 0.045 s were achieved, resulting in a FOV measuring 1.3 × 1.3 mm on a Sarnoff fast-readout 512 × 512 px CCD.

The laboratory-based *in situ* X-radiography equipment used incorporated a Hamamatsu microfocus X-ray tube with a 7 μm focus spot operating at 6 W, outputting a 39° conically divergent white beam. The sample material was an Al-20wt%Cu sample inoculated with 0.1wt%Al-Ti-B (5/1) master alloy. Optimal settings for the X-ray tube with the Al-20wt%Cu system used were 55 keV × 120 μA. The X-ray detector comprised a Vosskuhler 11000XR CCD visible light camera fitted with a Scint-X structured scintillator [7]. The camera was operated in 2 × 2 binning mode providing a 2012 × 1340 px CCD, with a physical pixel dimension of 18 μm. As a consequence of the high beam divergence (green cone, figure 1) a magnification effect was produced at the CCD, which is the ratio of the source-detector to the source-sample distance. Magnification in this case was measured at ~ 5.8 times, giving a virtual pixel size of 3.1 μm. The actual spatial resolution, however, is a function of both source size and scintillator pitch, thus the real spatial resolution was closer to 10 μm [7]. Images were recorded every 1 second.

Unlike the synchrotron experiments discussed previously, the solidification furnace used in this work was a newly developed prototype isothermal design. While a fully detailed description is beyond the scope of this article, the furnace comprised a circular alloy sample, measuring diameter, Ø 23 × 0.2 mm, surrounded by eight independently controlled heater elements arranged concentrically about the sample axis. The X-ray FOV was coincident with the centre of the sample. Solidification was initiated by reducing all heater temperatures at a constant cooling rate. The sample manufacture and preparation techniques used for the circular sample were largely similar to those reported for similar rectangular samples, previously [8].

The modular and compact nature of laboratory-based *in situ* X-ray equipment allowed the entire setup to be oriented in any direction w.r.t. the force of gravity. Figure 2 demonstrates this aspect from the perspective of the sample geometry. Figure 2 (a) shows the sample in the horizontal orientation, with the X-ray beam projection anti-parallel to gravity, *g*. In this configuration gravity acts through the thin dimension of the sample, *D*, resulting in significantly reduced equiaxed grain buoyancy in the FOV. However, in the vertical orientation (figure 2 (b)), which was the case for the synchrotron-based experiments, significant grain buoyancy was unavoidable, thereby adding significant complexity to characterisation of equiaxed nucleation and growth. Thus, to provide an environment to minimise equiaxed grain movement during solidification the X-ray diagnostics in this work were oriented such that the sample was horizontal, (figure 2 (a)).

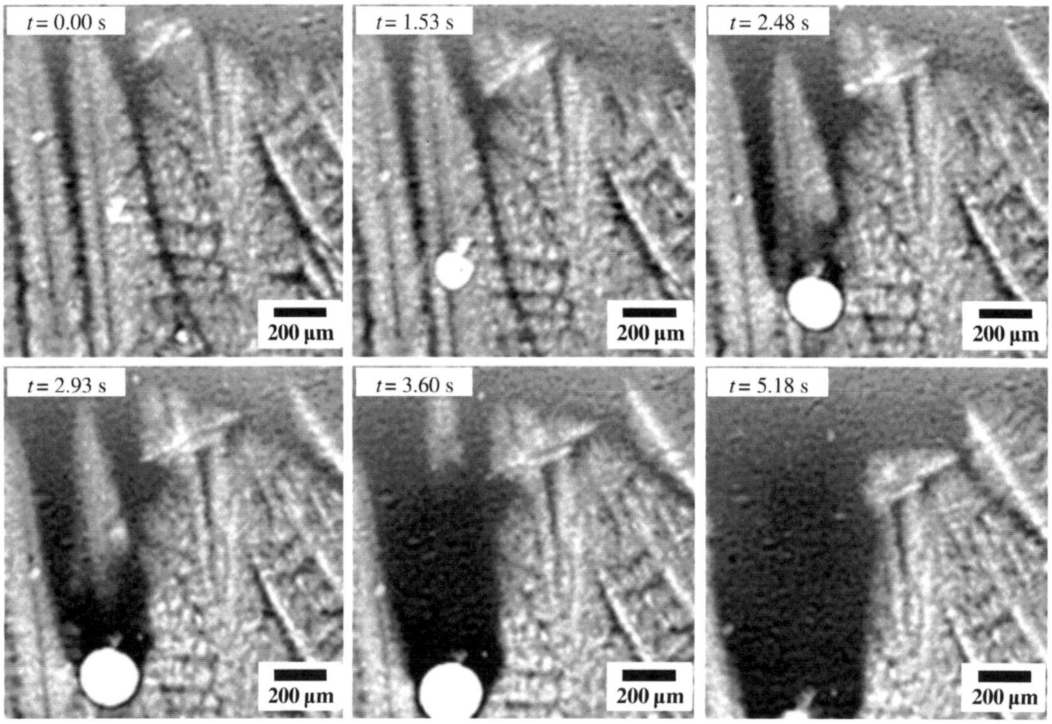

Figure 3. Directionally solidified Al-12wt%Ge alloy showing porosity nucleation within the interdendritic mush, causing remelting and destruction of the dendritic network. V_p = 85 μm/s↓; G_T = 79 K/mm; g↓.

3. Results and discussion

3.1. Solidification experiments using synchrotron X-radiography

Figure 3 demonstrates the effect of porosity nucleation within the interdendritic mush on the evolving microstructure. Columnar growth was initiated by pulling the sample, at a constant speed V_p = 85 μm/s, through a relatively high thermal gradient, G_T = 79 K/mm, similarly described in Ref. [6]. The indicated image time, t = 0 s, represents a datum time prior to the observed bubble nucleation. At $t \approx 1.53$ s, a bubble was observed nucleating deep within the interdendritic mush and, thereafter, relocated to a position coincident with a primary dendrite trunk. As the bubble continued to grow, remelting of the microstructure was observed at the bubble location, causing the top portion of the primary columnar dendrite to buoyantly rise out of the mushy zone, consequently melting as the fragment approached the super-heated liquid. Concurrently the surrounding columnar dendrites also experienced remelting, developing a large interdendritic pocket, as a consequence of the bubble development. The pore achieved a maximum diameter ≈ 200 μm, i.e. of the order of the sample thickness, and, thereafter, disappeared from the FOV. The cause of the microstructure remelting was likely the bubble acting as a thermal baffle within the mushy zone. Heat flow from the top of the FOV to the bottom, as a consequence of the imposed thermal gradient, was, therefore, restricted, causing a build-up of heat within the mushy zone ahead of the pore, thus leading to the remelting observed.

3.2. Solidification experiments using laboratory microfocus X-radiography

Figure 4 shows the equiaxed dendritic mush apparent in the FOV starting from ~ 76 seconds after the first α-solid nucleation event was observed. Note, image time, t, was set to zero at ~ 8 seconds prior to the onset of equiaxed nucleation, while the sample was fully liquid. The temperature indicated on each of the images in figure 4 was defined based on the assumption that the sample temperature at the instance of first equiaxed nucleation was equal to the liquidus temperature of the alloy, i.e. T_L = 602.6 °C at t = 8 s. Solidification was initiated by application of a constant cooling rate of $\dot{T}$ = 0.084 K/s near-isothermally across the entire sample. The total FOV dimension shown in figure 4 was limited to 3.17 × 3.17 mm to save space as well as to remove some CCD edge effects. Analysis of the real-time image data showed the temperature difference between nucleation commencement and nucleation cessation within the FOV as $\Delta T \approx 1.5$ K, with no significant aberrant temperature gradient evident.

The image sequence in figure 4 reveals the effects of hydrogen porosity nucleation and growth on equiaxed grains during solidification, with the additional complexity that bubbles, after reaching a diameter approximately equal to the sample thickness (200 μm), then disappeared from the FOV. The arrows in figure 4 (a) show the position in the FOV where bubble nucleation was first observed for a solid fraction of g_S = 0.13. In both cases highlighted, bubbles were observed to nucleate in approximately the same location as an equiaxed grain. Consequently, the bubbles, remaining stationary, increased in diameter and pushed the equiaxed grains from their original nucleation positions. Comparing the highlighted grains in figure 4 (a-iii) with the original grain positions (figure 4 (a-i)) it is clear that prior to mechanical grain coherency the effect of bubble nucleation was localised to grains within the immediate bubble vicinity. Finally, however, as shown in figure 4 (a-iv), both bubbles eventually disappear from the FOV leaving no permanent visible trace in the semi-solid microstructure. Maximum diameters measured were $d \approx 210$ μm and $d \approx 225$ μm for the top and bottom bubbles in figure 4 (a-iii), respectively. Figure 4 (b) shows the equiaxed dendritic microstructure at an apparent solid fraction of, $g_S \approx 0.47$, approximately 384 s after figure 4 (a-iv). At this stage during solidification the equiaxed microstructure was more densely packed, with mechanical grain impingement clearly apparent, figure 4 (b-i). The arrows in 4 (b-i) indicate the position of a single pore nucleation event within the semi-solid mush. With such a high solid fraction, however, the distortion of the dendritic mush was more far-reaching, as illustrated by the highlighted regions, figure 4 (b-i) - (b-iii)). Interestingly, the maximum diameter achieved by the bubble was measured at $d \approx 198$ μm, which suggests the coherent microstructure was insufficiently rigid to cause any distortion to the pore shape. Further, as observed earlier during solidification, once the nucleated bubble reached a maximum diameter, it disappeared from the FOV this time leaving a relatively large intergranular spacing in the microstructure, (arrow in figure 4 (b-iv)).

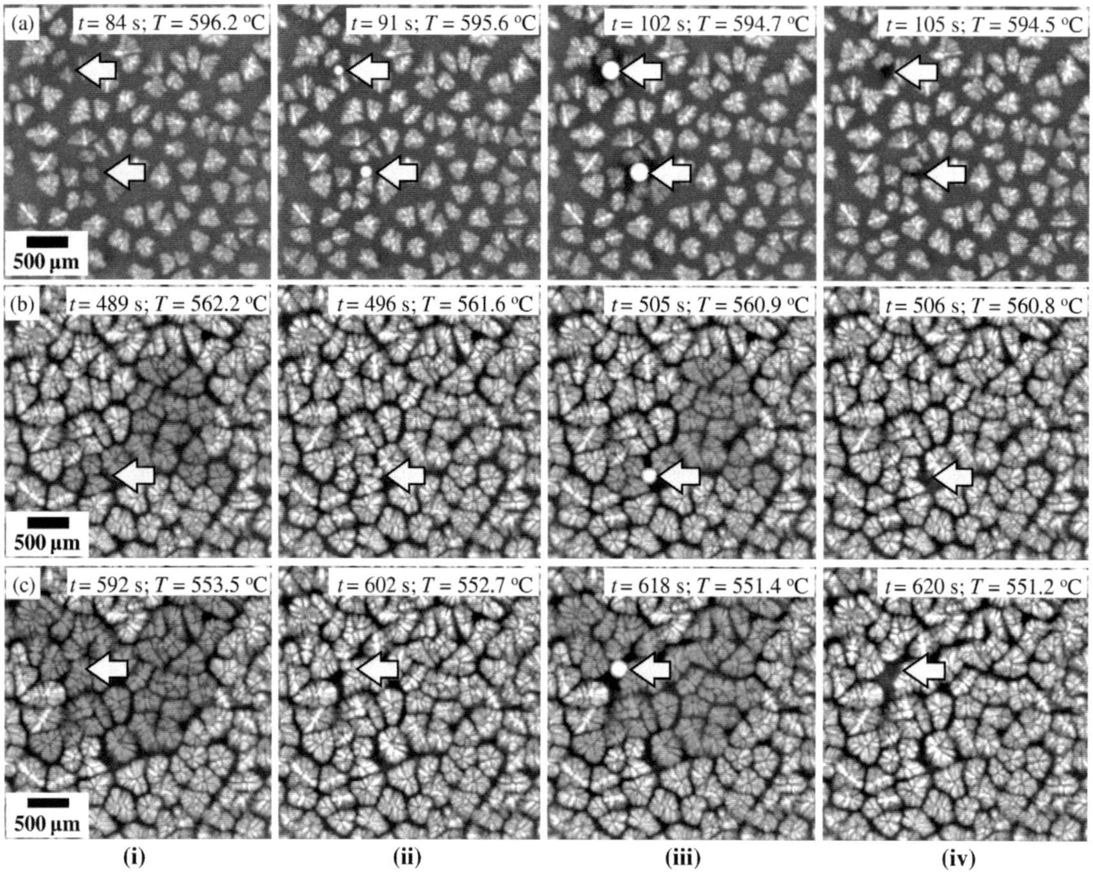

Figure 4. Near-isothermally solidified grain refined Al-20wt%Cu sample. Cooling rate, $\dot{T}$ = 0.084 K/s. Rows (a), (b), and (c) show the apparent solid fractions (g_S) 0.13, 0.47, and 0.52, respectively. Columns (i), (ii), (iii), and (iv) show time increasing from left to right. Arrows indicate the location of porosity nucleation. Colourisation shows which grains were affected by the nucleation and growth of pores; blue (column (i)) represent pre-pore nucleation position, red (column (iii)) represents the maximum mush distortion observed.

Figure 4 (c) shows a similar case to figure 4 (b) with a further increase in the apparent solid fraction, $g_S \approx 0.52$. The remaining liquid was close to the eutectic composition and thus the sample was almost completely solidified. Once again a gas bubble nucleated in the intergranular liquid, pushing and distorting the microstructure further than previously observed (highlighted regions, figure 4 (c)). As with the previous case, however, no distortion was observed in bubble morphology, maintaining relatively uniform sphericity at a high apparent solid fraction, and reaching a maximum diameter, $d \approx 218$ µm. As before, when the bubble disappeared from the FOV a large liquid void remained in its place. Approximately 29 seconds after figure 4 (c-iv) was captured, the eutectic transformation was observed crossing the FOV from the bottom-left corner towards the top-right corner. Figure 4 (c-iv), however, represents the final as-cast microstructure observed, showing no detectable signs of porosity in contrast to the real-time imaging which demonstrated bubble nucleation continuously during solidification.

4. Summary, conclusion, and outlook

This article presents a brief analysis of gas porosity nucleation and growth observed at various stages during both columnar (directional) and equiaxed (isothermal) *in situ* X-ray solidification experiments. In all cases examined the maximum pore diameter appeared to be limited by the sample thickness, i.e. 200 μm, suggesting that once the pore came into contact with the free surface oxide layers the gas was expelled from the sample. After the pore disappeared from the FOV no significant trace remained within the semisolid mush, giving a further demonstration of the importance of real-time *in situ* studies. During columnar growth the nucleation and growth of a gas bubble deep within the dendritic mush caused rapid remelting of the dendritic network and significant fragmentation of the remaining primary and secondary dendrite arms. These fragments continued to grow, both ahead and within the dendritic mush, causing a CET-like (Columnar-to-Equiaxed Transition) phenomenon. While the experimental evidence is somewhat limited, it would seem that gas bubbles nucleating in the dendritic mush and causing remelting/fragmentation may be possible as a further mechanism of CET formation. Pore growth velocity measurements for the near-isothermal equiaxed (microgravity-like) solidification experiments showed no significant dependence on temperature. However, a slight decrease in pore growth velocity was observed as the solid fraction increased. This was related to the reduction in the physical space available for the pore to grow, thereby requiring the gas volume to push against larger and larger regions of interconnected equiaxed dendrites. Most significant was the lack of any observed distortion in sphericity of the gas bubble as a consequence of interaction with the dendritic structure suggesting relatively high internal gas pressure.

Finally, the near-isothermal solidification experiments and analysis presented in this work are currently in their infancy. Development and testing of the prototype isothermal furnace for *in situ* X-radiography solidification studies of Al-based alloys is in the final stages. The furnace is scheduled to fly on board the MASER 13 sounding rocket in spring 2015, investigating near-isothermal equiaxed solidification under microgravity conditions.

Acknowledgements

The authors wish to acknowledge the financial support of the European Space Agency (ESA) under the PRODEX program, (contract numbers. 90392 and 4000110414). This work is part of the ESA-MAP (Microgravity Applications Promotion) project XRMON. The authors would also like to thank Dr Marco Di Michiel (Expt. Division, ESRF) for his assistance with the synchrotron experiment.

References

[1] Samuel A M and Samuel F H 1995 *Metall Mater Trans A* **26** 2359–72
[2] Skallerud B, Iveland T and Härkegård G 1993 *Eng Fract Mech* **44** 857–74
[3] San-Martin A and Manchester F D 1992 *Journal of Phase Equilibria* **13** 17–21
[4] Mathiesen R, Arnberg L, Nguyen-Thi H and Billia B 2012 *JOM* **64** 76–82
[5] Lee P D and Hunt J D 1997 *Acta Mater* **45** 4155–69
[6] Mathiesen R H, Arnberg L, Ramsøskar K, Weitkamp T, Rau C and Snigirev A 2002 *Metall Mater Trans B* **33** 613–23
[7] Rakete C, Baumbach C, Goldschmidt A, Samberg D, Schroer C G, Breede F, Stenzel C, Zimmermann G, Pickmann C, Houltz Y, Lockowandt C, Svenonius O, Wiklund P and Mathiesen R H 2011 *Rev Sci Instrum* **82** 105108 –105108–10
[8] Murphy A G, Browne D J, Mirihanage W U and Mathiesen R H 2013 *Acta Mater* **61** 4559–71

Analysis and simulation of non-metallic inclusions in spheroidal graphite iron

B Pustal[a], B Schelnberger and A Bührig-Polaczek
Foundry Institute at RWTH Aachen University, Intzestr. 5, 52072 Aachen, Germany

E-mail: [a]b.pustal@gi.rwth-aachen.de

Abstract. Non-metallic inclusions in spheroidal cast iron (SGI) reduce fatigue strength and yield strength. This type of inclusion usually accumulates at grain boundaries. Papers addressing this topic show the overall impact of both the fraction of so-called white (carbides) and black (non-metallic) inclusions on mechanical properties. In the present work we focus on the origin and the formation conditions of black Mg-bearing inclusions, further distinguishing between Si-bearing and non-Si-bearing Mg inclusions. The formation was simulated applying thermodynamic approaches. Moreover, appropriate experiments have been carried out and a large number of particles have been studied applying innovative feature analysis with regard to shape, size, and composition. Magnesium silicates are predicted at elevated oxygen concentrations, whereas at low levels of oxygen sulphides and carbides appear at a late stage of solidification. Experiments with three consecutive flow obstacles show that the amount of magnesium silicates decrease after each of the three obstacles, whereas the fraction of non-Si-bearing inclusions remains approximately constant. The size of inclusions divides in halves over the flow path and the number of particles increases accordingly. We point out that based on feature analysis Mg-O-C bearing inclusion show disadvantageous form factors for which reason this kind of inclusions may be extremely harmful in terms of crack initiation. All results obtained indicate that magnesium silicates are entrapped on mould filling, whereas Mg-(O, C, S, P, N) bearing particles are precipitates at late stages of solidification. Consequently, the only avoidance strategy is setting up optimum retained magnesium content.

1. Introduction

Non-metallic inclusions decrease fatigue strength and yield strength of SGI, since they accumulate near grain boundaries. This weakening notch effect in the matrix structure is severe if caused by a jagged or worm-like morphology of these inclusions. Papers addressing this topic show the overall impact of both the fraction of so-called white (carbides) and black (non-metallic) inclusions on mechanical properties. The cause for the formation of these inclusions and the associated morphology is not yet sufficiently known. After the magnesium treatment of the cast iron melt for the production of spheroidal graphite iron (SGI) small amounts of oxygen, sulphur, magnesium, and other trace elements will remain in the melt. These elements form silicates on the one hand, and on the other hand the solubility of these elements in the austenite is usually very low, which causes microsegregation. At a sufficiently high concentration of these elements in the residual melt, Mg-bearing precipitates are expected. One aim of the present study was to determine the thermodynamic conditions under which such inclusions occur. Furthermore, two hypotheses were followed: the inclusion hypothesis after which such inclusions form before or during mold filling, and the precipitation hypothesis that non-metallic precipitations occur towards the end of solidification in the residual melt areas. Supplementary experiments were conducted to support these theses, or to refute them.

2. State of the art

Sofue et al. [1] investigated the influence of grain size, graphite shape, amount of non-metallic inclusions, matrix structure and matrix hardness on the fatigue strength. Their results show that the graphite roundness increases steadily with the remaining magnesium content, but also the number of non-metallic inclusions. The weakening of the grain boundaries through non-metallic inclusions exceeds the effect of the spheroid roundness at about 0.6 wt.- % retained magnesium. Beyond this point, elongation and fatigue strength decrease again. Sofue et al. created rules for the production of high quality SGI with improved fatigue strength on the basis of their results. They divided non-metallic inclusions into white (Ti, Cr carbides) and black Mg-bearing inclusions. They noted that the detailed composition of these inclusions has not been adequately characterized. Using electrolytic etching and EDX, Hiroshi et al. [2] investigated the three-dimensional shape and composition of various inclusions, thereby distinguishing titanium phosphides, Ti (C, N), and non-metallic inclusions from the Mg-Al-Si-O and Ti-C-Si-P-S systems. In their work with high silicon cast iron, Lin et al. [3] pointed out, that the presence of magnesium inclusions initiates cracks at the cell boundaries under cyclic load.

Thermodynamic calculations on the phase structure of steel materials are state of the art for many applications [4] and [5]. In optimization of cast iron materials, however, such a methodology is still the exception [6]. The software Thermo-Calc [7] or FactSage [8] are capable of performing thermodynamic calculations with multicomponent systems. From the thermodynamic data, such as enthalpy, entropy, and heat capacity, the solidification behaviour can also be derived [7]. With this method, effects of individual elements on the amount of non-metallic inclusions depending on the temperature were determined [9]. The main kinetic factors influencing cell distances based on the nucleation or diffusion in the austenite shell cannot be calculated with thermodynamic methods alone. This can only be achieved by means of kinetic models such as the microsegregation model of Pustal et al. [10]. The continued segregation of the alloying elements on the basis of thermodynamic and mobility data and the formation of related precipitates is numerically calculated. In this model simulation of the solidification and diffusion occurs along an axis that reflects a characteristic length, depending on the assumed morphology. In dendritic growth this length corresponds to half the dendrite arm spacing. In a eutectic cell, this axis corresponds to the half distance between the graphite nuclei. The species transport problem of all alloying elements can be solved in several solid phases, and for the thermodynamic equilibrium calculations at the phase boundary, the tq-interface of Thermo-Calc [7] is used. The model takes into account the mobility of the elements as well as their interaction during segregation. For example, up-hill diffusion of carbon can be predicted. These effects influence the solidification behavior of SGI considerably.

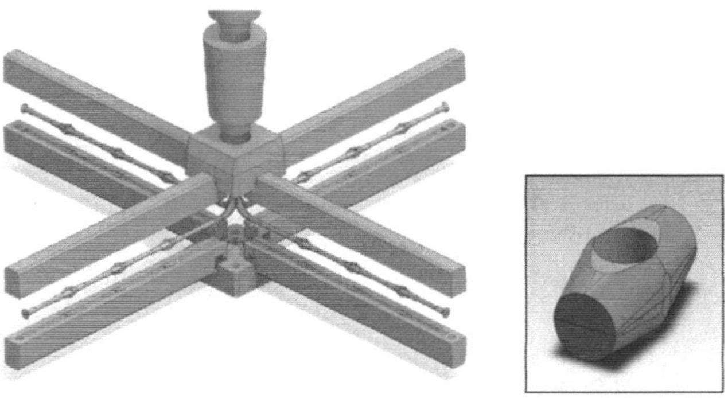

Figure 1. Quadruple mold with arrangement of three flow obstacles in each of the four arms (left) and cylindrical obstacle geometry (right).

3. Work carried out

3.1. Casting

To quantify the influence of mold filling on the amount and type of non-metallic inclusions, an experiment was set up, that allows generating such inclusions in flow obstacles. The mold in figure 1, designated as "quadrupel test", has been successfully used for the selective generation of oxide inclusions in magnesium casting.

The modular mold is assembled of individual cold-box cores, the down sprue through which the pouring height can be varied, the gating element at the base, and four horizontal arms, which represent the actual casting. In each arm up to three flow obstacles can be inserted at regular intervals. During mold filling eddies form in the flow shadow of the flow obstacle. These in turn lead to increased contact between melt and atmosphere, so that an increased possibility for the formation of oxides exists, where appropriate reactants are present. If turbulent mold filling has significant influence on the formation of Mg inclusions, then their amount should increase in the casting behind the obstacles along the flow path.

We further subdivided the inclusions in Si-bearing and non-Si-bearing Mg inclusions. Key research questions of casting trials were whether an increase of Mg-bearing inclusions along the flow path is taking place, and whether the proportion of finely divided non-Si-bearing inclusion behaves in the same manner as the amount of Si-bearing particles. If the amounts of non-Si-bearing inclusions along the flow path develop in the same or similar manner as the amounts of Si-bearing inclusions, then this would be an indication of an influence of the mold filling conditions on non-Si-bearing particle formation. But, if the non-Si-bearing inclusions are equally distributed, while magnesium silicates increase, then the influence of mold filling on the non-Si-bearing inclusions is secondary and non-Si-bearing inclusions are precipitates. For this purpose, a sample was taken after each flow obstacle, ground, and polished for analysis.

3.2. Analysis

For identification of inclusions or precipitates, scanning electron microscopy (SEM) was used. Due to the small size of the particles a so-called feature analysis was applied instead of conventional image analysis. In this method, image analysis is conducted on the basis of automatically generated SEM images over a large sample area. Subsequently, all characteristic particles are automatically checked with EDX measurement. In this way, a record was created for each particle to the entire measurement area of about 0.38 mm^2 containing size, composition, shape, and surface area. Because of this highly time-consuming procedure, this process could be performed for only three samples at one arm.

3.3. Thermodynamic / thermodynamic-kinetic simulations

For the thermodynamic calculations the software FactSage [8] along with the databases FStel and Ftoxid [8] were applied. Calculations were performed at thermodynamic equilibrium and according to the Gulliver-Scheil model, that is, no diffusion in the solid state, infinite rapid diffusion in the liquid phase. As a starting point for variations of oxygen the composition C3.45 Si2.19 Mn0.14 P0.018 S0.006 Mg0.028 wt.-% was selected. The oxygen content was varied between 19 ppm and 0.1 wt.-% and the effect of the presence or absence of sulphur in the melt during the development of Mg-bearing precipitates was taken into account. The results of this study are given in Section 4.2.

Applying the microsegregation software microphase along with the database Iron-01c, both described in Pustal et al. [10], simulations were performed with the initial compositions Fe C3.62 Si2.4 Mn0.08 wt.-% and variation of the retained Mg: 0.02, 0.04, 0.08 wt.-%. In each such case the resulting residual melt concentration at 1 wt.-% liquid was calculated as: C3.96 Mn0.29 Si1.21 wt.-% and Mg: 0.21, 0.43, 0.87 wt.-%. These concentrations represent the conditions in the intercellular residual melt areas where non-metallic inclusions were found. These conditions were then applied in FactSage as initial concentration for calculating compounds at thermodynamic equilibrium. The calculations were carried out at temperatures between 1000 °C and 1400 °C with the addition of 19 ppm of oxygen and the final results of this procedure are shown in Section 4.2.

4. Results

4.1. Results of analysis

Feature analysis was conducted at three samples taken behind each of the three flow obstacles of one arm, to determine the statistical change in the number of particles, the particle size, and the area ratio for the different kinds of particles bearing Mg with and without Si. Figure 2 (a) shows that both types of particles are present in roughly equal amounts in the microstructure. The area ratio of all Mg-bearing particles decreases after each flow obstacle. This is mainly due to the decrease of magnesium silicates, since the amount of Mg (O, C, S, P, N)-bearing particles without Si remains approximately constant. The discrepancies at sample position 2 are due to porosity at this location which impacts the acquisition of the values measured. When neglecting sample position 2, the drawn trend of decrease or increase remains. Figure 2 (b) shows that the particles behind the first flow obstacle are large with an average of about 1.5 μm^2, whereas behind the third module these particles are small; the mean area is only about 0.65 μm^2. Along with the particle size, the particle density (n-particles/100μm^2) decreases. The form factor decreases strongly, that is, the particle roundness improved considerably over the sample position. This means, that after the first flow obstacle a few large and rather jagged particles occur and behind the last module, there are many more small rounded particles. In a further differentiation, it was found that the non-Si-bearing particles decrease in size more noticeably, whereas the form factor improvement is more pronounced in the Si-bearing particles. It can be concluded that the non-Si-bearing particles due to their size and disadvantagous shape may impact the mechanical properties more than the Si-bearing inclusions.

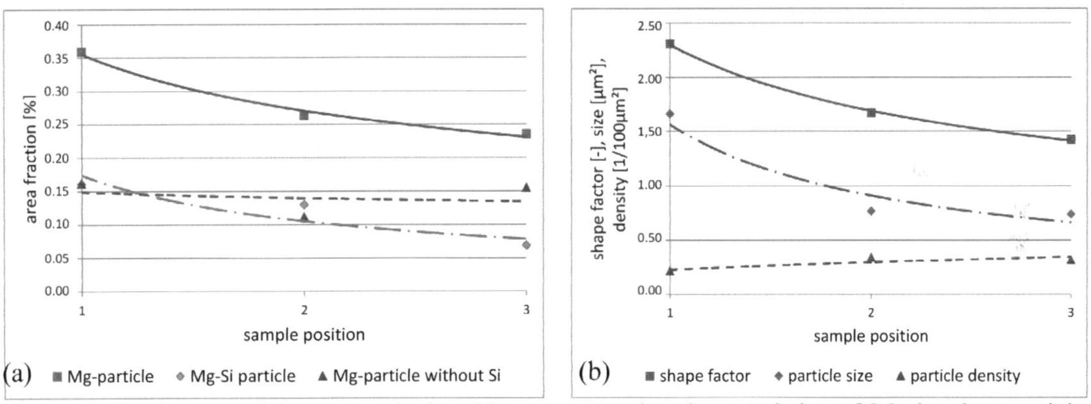

Figure 2. Evaluation of feature analysis with respect to the characteristics of Mg-bearing particles with and without Si along the sample arm.

4.2. Results of simulations

The results of the thermodynamic calculations in figure 3 show that for cast-iron relevant temperatures MgO or MgS may form directly, as soon as Mg, O, and S are in the melt or at the free surface. The amounts of MgO at about 1780 °C and MgS at about 1700 °C arise due to the assumed oxygen or sulphur content of the melt. If the locally available oxygen and sulphur are depleted by reaction with magnesium, further reaction products such as Mg_2C_3 can occur at lower temperatures. A further study varying the oxygen content shows that in the multi-component system Fe-C-Mg-Si-O, the oxygen content determines which magnesium compounds are stable before solidification starts. At a temperature of 1250 °C and oxygen content up to 0.03 wt.-% in the full equilibrium liquid magnesium and magnesium oxides are expected.

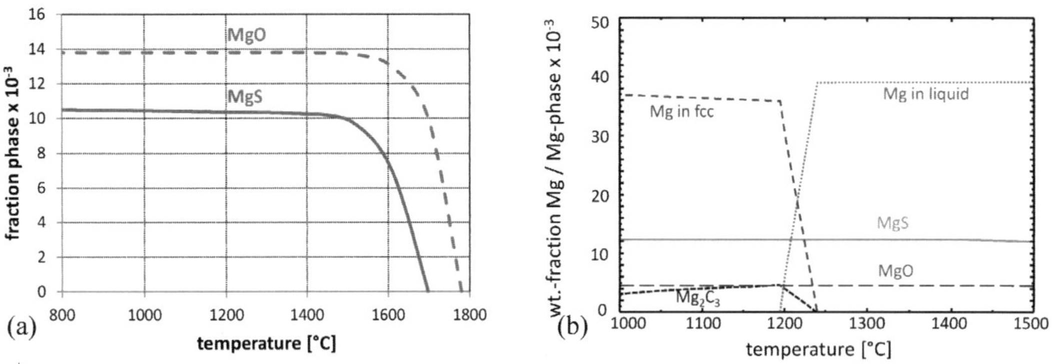

Figure 3. Weight fractions of MgO and MgS calculated with FactSage (a) and of Mg-bearing phases based the composition given in Section 3.1 without Mn and P (b).

The microsegregation simulation showed a significant enrichment of magnesium in the melt on solidification. Calculations were performed based on the same melt concentration varying the retained magnesium content: 0.02, 0.04, 0.08 wt.-%. These calculations resulted in the following concentrations at 1 wt.-% liquid phase: Fe C3.96 Si1.2 Mn0.29 and Mg 0.21, 0.43, 0.87 wt.-% respectively.

With these concentrations subsequent equilibrium calculations with FactSage [8] were performed. The results of these calculations with respect to the formation of MgO and Mg_2C_3 are shown in figure 4. After the residual oxygen is consumed, Mg_2C_3 may form at the given composition at about 1260°C. The amount of Mg_2C_3 scales linearly with the concentration of the residual liquid, which the results of Sofue et al. [1] confirm. In summary, the formation of silicate inclusions is to be expected at higher oxygen concentrations such as surface contact, and at lower oxygen concentrations MgO, MgS, or Mg_2C_3 can be formed. This does not exclude the formation of complex compounds that were not included in the databases or due to other reactants such as P, S, Ca, Ce.

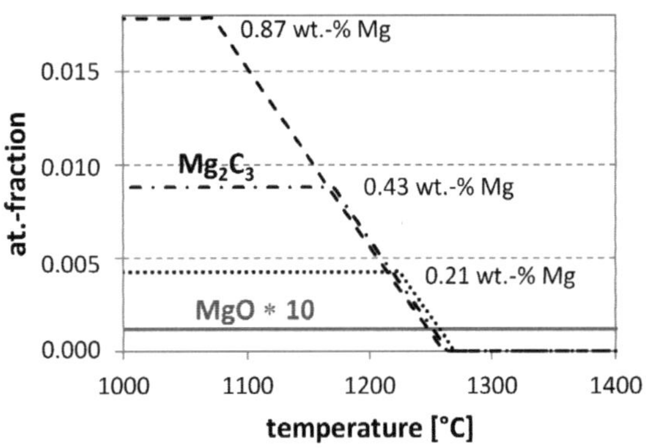

Figure 4. FactSage calculations with the databases Iron-01c [10] and FToxid [8]. Phase fractions for varying magnesium concentrations of the residual liquid: 0.21, 0.43, 0.87 wt.-%, and 19 ppm oxygen. The initial retained magnesium content of the melt was 0.02, 0.04, and 0.08 wt.-%.

5. Summary and conclusions

The thermodynamic results indicate that a reaction of magnesium and oxygen already takes place above the liquidus temperature leading to the formation of MgO and, at higher concentrations of oxygen, also to the formation of magnesium silicates. Smaller magnesium silicates can thus arise, for example during mold filling in surface contact and then pass into the cast component. The remaining magnesium segregates and reacts at a sufficient concentration with the segregating elements oxygen and sulphur, until they are fully consumed. According to the calculations Mg_2C_3 is following. The influence of the oxygen concentration on the reaction equilibria indicates that during the Mg treatment at first magnesium silicates form.

The amount of the non-Si-bearing precipitates remained approximately constant in each sample, whereas particle number and size varied greatly. This observation supports the precipitation hypotheses for non-Si-bearing Mg inclusions. All results and analyses of the present study suggest that non-metallic inclusions, such as Mg (O, C, S, N, P) only form during solidification by segregation as precipitates. Astonishingly, the amount of Si-bearing inclusions decreased. This observation contradicts the inclusion hypotheses because more surface contact due to vortices developing behind the three flow obstacles does not increase the amount of magnesium silicates as expected. This means that no significant amount of Si-bearing particles is formed on mold filling; these particles must be pre-existing. The pre-existing particles may be trapped in the flow shadow of the flow obstacles and decrease therefore. The overall decrease in the size and proportion of Mg-bearing particles as well as the improvement of the form factor lead to the conclusion that vortex formation may have a positive impact on fatigue strength. This has to be further investigated.

The literature review shows that all particles near the grain boundaries weaken the grain boundaries under stress, even with rounded shape, which is why these are undesirable together with carbides also caused by segregation. The avoidance of Mg-bearing precipitates is possible in SGI only in so far as that the optimal retained Mg content of 0.06 wt.-% should not be exceeded. At higher retained Mg contents, the effect of grain boundaries weakened by non-metallic precipitates outweighs the effect of improved particle roundness. Sofue et al. [1] show the relationships between retained Mg content, nonmetallic inclusions, graphite mold and matrix hardness on the fatigue strength. On this basis, they developed criteria for optimal properties.

Acknowledgment

The research project 16698 N / 1 of the Research Association for Foundry Technology e.V. (FVG) was sponsored by the AiF under the programme for the promotion of industrial research and development (IGF) by the Federal Ministry of Economics and Technology based on a decision of the German Bundestag and professionally supervised by the Federal Association of German Foundry Industry (BDG). For the support of this work we thank the agencies involved cordially. Also, we thank the participants of the accompanying working group, in particular the DOSSMANN GmbH, Wallduern-Rippberg and SHW Casting Technologies GmbH, Aalen-Wasseralfingen.

References

[1] Sofue M, Okada S and Sasaki T 1979 *AFS Trans.* **87** 173
[2] Horie H, Kowata T, Onisawa H and Shindo Y 1987 J. Japan Foundrymen's Society **59** 302
[3] Lin H M, Lui T S and Chen L H 2003 *Mat. Trans.* **44** 1209
[4] Schneider A 2006 Salzgitter-Mannesmann Aktuell 2
[5] Wiener J, Gigacher G, Penz S and Bernhard C 2009 *Berg- und Hüttenmännische Monatshefte* **154** 33
[6] Obermaier S, Kleiner S, Zeipper L and Kauffmann M 2009 *Gießtechnik im Motorenbau.* VDI- Bericht 2061
[7] Andersson J, O Helander T, Höglund L, Shi P F and Sundman B 2002 *Calphad* **26** 273
[8] Bale CW, Bélisle E, Chartrand P, Decterov SA, Eriksson G, Hack K, Jung IH, Kang YB, Melançon J, Pelton DA, Robelin C and Petersen S 2009 *Calphad* **33** 295
[9] Gagne M, Paquin M P and Cabanne P M 2008 Ductile Iron Society T&O Meeting 142
[10] Pustal B, Subasic E, Hallstedt B, Schäfer W, Bartels C, Siebert H, Schneider J M and Bührig-Polaczek A 2010 *Adv. Eng. Mat.* **12** 158

IOP Conf. Series: Materials Science and Engineering **117** (2016) 012069 doi:10.1088/1757-899X/117/1/012069

Three dimensional simulation of macrosegregation in steel billets by a meshless method

R Vertnik[1,2,a] and B Šarler[2,3,b]
[1] Štore Steel d.o.o., Research, Železarska 3, SI-3220 Štore, Slovenia
[2] Laboratory for Simulation of Materials and Processes, Institute of Metals and Technology, Lepi pot 11, SI-1000 Ljubljana, Slovenia
[3] Laboratory for Multiphase Processes, University of Nova Gorica, Vipavska 13, SI-5000 Nova Gorica, Slovenia

E-mail: [a] robert.vertnik@store-steel.si, [b] bozidar.sarler@imt.si

Abstract. There is a continuously developing need for benchmarking of solidification simulations - from the theoretical as well as from the applied points of view. The history of related benchmarking shows differences of the results between different numerical methods, and differences in comparison with the experiments when solving even quite simple solidification situations. The present benchmark test proposes a three dimensional version of the recently developed two dimensional solidification benchmark of the continuous casting with turbulent fluid flow and solidification with macrosegregation. The macroscopic transport equations for mass, momentum, energy, species, turbulent kinetic energy and dissipation rate are considered. Turbulent effects are incorporated through the solution of a low-Re turbulence model. The solidification system is treated by the mixture-continuum model, where the mushy zone is modeled as a Darcy porous media with Kozeny-Karman permeability relation and columnar solid phase moving with the casting velocity. The Fe-C binary phase diagram and the lever rule microsegregation model are used to obtain the carbon segregation in the billet. The transport equations are solved by our newly developed and efficient meshless numerical technique, based on local collocation with radial basis functions and fractional step velocity-pressure coupling. The presented results represent first simulations of a three-dimensional solidification problem by a meshless method.

1. Introduction

Continuous casting [1] is the most common process in the production of steel. The process starts by pouring the molten metal into the water cooled mold, where cooling intensity is high enough to solidify steel around the inner surface of the mold and generates the solid shell with molten metal in the center of strand. After several minutes, the strand is pulled into the secondary cooling system, which contains water spray systems with much smaller cooling intensity as in the water channels in the mold, and rollers, which support and guide the strand up to the end of the casting machine. The quality of the casted product (round or square billets, blooms or slabs) depends mainly on the process parameters in the mold region, where complex physical phenomena occur. The liquid metal, poured with a high velocity from the submerged entry nozzle (SEM) into the mold, produces turbulent flow with several re-circulating zones. Large heat fluxes, extracted from the mold, are a consequence of very high flow rates of the cooling water in the mold channels. It is impossible to measure temperature, velocity field and macrosegregation [2] inside the strand due to the very high temperatures of steel and inaccessibility of the region during the process. The numerical models [3, 4] help us to better understand the casting behavior and to further improve and optimize the process, particularly in such experimentally sophisticated situations. Various numerical methods have already been used to simulate the described spectra of problems [5, 6, 7, 8, 9]. However, until now, there is a lack of benchmark tests that would enable verification of numerical methods for solving transport phenomena in continuous casting of steel in an ordered way. We recently tried to fill this gap with proposing a first such benchmark [10, 11]. In these papers, the two dimensional turbulent heat transfer, fluid flow, and species transfer solidification discussion in [12] is upgraded to three dimensions. The main aim of the test is to get new reference results and to be able to further compare the results of different numerical methods for the same physical model. The presented physical model considers turbulent fluid flow and macrosegregation of carbon steel. Turbulent flow is modeled by the two-equation eddy-viscosity model with the low-Re corrections. The numerical results are shown for the case with lever microsegregation model of 0.55 wt%C steel. The novel meshless Local Radial Basis Function Collocation Method (LRBFCM) is employed for solving the coupled system of six partial differential equations in three dimension. The objective of the present paper is to produce a first reference numerical solution of the given problem. LRBFC was first developed in [13] for diffusion problems, to convection diffusion problems in [14], and to driven cavity and natural convection problems in [15, 16]. The turbulence was tackled in [17, 18]. The melting benchmark [19] was successfully solved in [20] and the macrosegregation benchmark [3] in [21, 22]. The magneto-hydrodynamic effects were tackled in [23] for natural convection in a cavity and in [24] for backward facing step situation.

2. Governing equations

The system of equations, describing the coupled transport phenomena in the continuous casting of steel, is derived based on the mixture continuum formulation by Bennon and Incropera [25] and Reynolds time-averaging approach for modelling incompressible turbulent flow [26]. The time-averaged transport equations for mass, energy, momentum and solute conservation are

$$\nabla\cdot\mathbf{u}=0, \tag{1}$$

$$\rho\frac{\partial\mathbf{u}}{\partial t}+\rho\nabla\cdot(\mathbf{uu})=-\nabla P+\nabla\cdot\left[2(\mu+\mu_t)\mathbf{S}\right]-\frac{2}{3}\rho\nabla k-\mu_L\frac{K_0\left(1-f_L^2\right)}{f_L^3}(\mathbf{u}-\mathbf{u}_S)+\mathbf{F}_b, \tag{2}$$

$$\rho\frac{\partial h}{\partial t}+\rho\nabla\cdot(\mathbf{u}h)=\nabla\cdot(\lambda\nabla T)+\rho\nabla\cdot\left[f_S(h_L-h_S)(\mathbf{u}-\mathbf{u}_S)\right]+\nabla\cdot\left(f_L\frac{\mu_t}{\sigma_T}\nabla h_L\right), \tag{3}$$

$$\rho\frac{\partial C}{\partial t}+\rho\nabla\cdot(\mathbf{u}C)=\nabla\cdot(f_SD_S\nabla C_S+f_LD_L\nabla C_L)+\rho\nabla\cdot\left[(\mathbf{u}-\mathbf{u}_S)(C_L-C)\right]+\nabla\cdot\left(\frac{f_L\cdot\mu_t}{\sigma_C}\nabla C_L\right) \tag{4}$$

with $\mathbf{u}$, P, h, T and C for velocity, pressure, enthalpy, temperature and solute concentration, respectively, and ρ, μ, μ_t, f, λ, D are standing for density, molecular dynamic viscosity, turbulent dynamic viscosity, volume phase fraction, thermal conductivity, and diffusivity of the material. In above equations, subscripts L and S denote liquid and solid phase. $\mathbf{S}$ stands for the strain-rate tensor. The fourth term on the right hand side of equation (2) represents the Darcy term, where K_0 is the morphology constant of the porous medium and $\mathbf{u}_S$ is velocity of the solid phase. The fifth term on the right hand side of equation (2) represents the buoyancy forces, defined as

$$\mathbf{F}_b=\rho\left[\beta_T\mathbf{g}(T-T_{ref})+\beta_C\mathbf{g}(C_L-C_{ref})\right] \tag{5}$$

where β_T, β_C, T_{ref} and C_{ref} stand for thermal expansion coefficient, solute expansion coefficient, reference temperature and reference concentration, respectively. The turbulent dynamic viscosity is defined as $\mu_t = \rho c_\mu f_\mu k^2 / \varepsilon$, where k and ε are kinetic energy and dissipation rate, respectively. They are calculated by the following transport equations

$$\rho\frac{\partial k}{\partial t} + \rho\nabla\cdot(\mathbf{u}k) = \nabla\cdot\left[\left(\mu_L + \frac{\mu_t}{\sigma_k}\right)\nabla k\right] + P_k - \rho\varepsilon + \rho D + \mu_L\frac{K_0\left(1-f_L^2\right)}{f_L^3}k, \quad (6)$$

$$\rho\frac{\partial \varepsilon}{\partial t} + \rho\nabla\cdot(\mathbf{u}\varepsilon) = \nabla\cdot\left[\left(\mu_L + \frac{\mu_t}{\sigma_\varepsilon}\right)\nabla \varepsilon\right] + \rho\left(c_{1\varepsilon}f_1 - c_{2\varepsilon}f_2\varepsilon\right)\frac{\varepsilon}{k} + \rho E + \mu_L\frac{K_0\left(1-f_L^2\right)}{f_L^3}\varepsilon. \quad (7)$$

In equations (3,4,6,7) c_μ, f_μ, $c_{1\varepsilon}$, f_1, $c_{2\varepsilon}$, f_2, σ_T, σ_C, σ_k and σ_ε are the closure coefficients, while D and E are additional source terms of the low-Reynolds turbulent model [26]. The constitutive temperature-enthalpy relationship h_S and h_L in equation (3) are defined as

$$h_S = \int_{T_{ref}}^{T} c_{pS}dT, \quad h_L = \int_{T_{ref}}^{T_S} c_{pS}dT + \int_{T_S}^{T} c_{pL}dT + h_m = h_S(T) + \int_{T_S}^{T}\left(c_{pL} - c_{pS}\right)dT + h_m, \quad (8,9)$$

respectively. The liquid fraction f_L, concentration of solute in liquid phase C_L, partition ratio k_p, and liquidus temperature T_L are determined based on the level rule microsegregation model, i.e.

$$f_L = 1 - \frac{1}{1-k_p}\frac{T-T_L}{T-T_m}, \quad C_L = \frac{C}{1+f_S\left(k_p-1\right)}, \quad k_p = \frac{C_S}{C_L}, \quad T_L = T_m + \left(T_e - T_m\right)\frac{C}{C_e}, \quad (10,11,12,13)$$

where T_m, T_e and C_e are melting temperature, eutectic temperature and solute concentration in eutectic alloy, respectively. They are obtained from the phase diagram for the iron-carbon system.

3. Solution procedure

Our solution procedure is based on the novel meshless technique that follows the following steps. We seek the solution of the equations for the velocity, pressure, temperature, concentration, turbulent kinetic energy, and dissipation at time $t+t_0$ by assuming known fields $\mathbf{u}$, p, T, C, k, and ε at time t_0 and known boundary conditions at time $t > t_0$. The coupled set of mass conservation equation (1) and momentum conservation equations (2) are solved by the fractional step method [27], where the continuity of the mass (1) is considered by constructing the pressure Poisson's equation. The governing equations are discretized by using the explicit time discretization. This leads to the following algorithm, summarized through the following 7 steps: (I) The intermediate velocity components are calculated without the pressure gradient, (II) the pressure Poisson equation is treated, (III) the velocity components are corrected by the pressure gradient, (IV), after the solution of the velocity field, the transport equation (3) for energy is solved, (V) the temperature field is calculated from the enthalpy field, calculated in previous step, using the inverse of the constitutive temperature-enthalpy relationships (8,9), i.e. $T = T(h)$, (VI) species concentration (4), turbulent kinetic energy (6) and dissipation rate (7) fields are calculated, and finally (VII) the turbulent viscosity, solute concentrations (11,12) and liquidus temperature (13), are updated for each node and the solution is set ready for the next time step. All derivatives of the transport equations are calculated by the LRBFCM [12], where the collocation is made locally on overlapped sub-domains. On each sub-domain, the scalar function Φ (standing for temperature, species concentration, velocity components, pressure, turbulent kinetic energy, and turbulent dissipation rate) is represented over a set of (in general) nonequally spaced nodes $\mathbf{p}_n$; $n = 1,2,...,{}_l N$ in the following way $\Phi(\mathbf{p}) \approx \sum_{k=1}^{K} \psi_k(\mathbf{p})\alpha_k$, $\psi_k(\mathbf{p}) = \left[r_k^2 + c^2\right]^{1/2}$, where ψ_k stands for the multi-quadric radial basis shape functions [28], α_k for the coefficients of the shape functions, K the number of the shape functions, c the shape parameter, and r_k the radial distance between two points in the sub-domain. The detailed procedure of calculating the derivatives with the LRBFCM can be found in [3]. Seven subdomain nodes are used in three dimensions.

4. Description of CC case

Geometry: The geometry represents a simplified 3D model of the continuous casting process. The geometry of the simplified casting machine is shown in figure 1. A billet with a square cross-section is assumed. The computational domain represents a quarter of the longitudinal section of the billet with length 1.15 m, shown in figure 2. The dimensions of the casting geometry are billet width $a_b = 0.18\text{m}$, diameter of the SEN $d_2 = 0.035\text{m}$ and mold height $l_m = 0.8$ m, respectively.

Material properties: Generally, the material properties of steel are temperature dependent. However, in this work, constant values are used for each phase. They are given in table 1.

Table 1. Material properties.

ρ (kg m-3)	c_p (J kg^{-1}K^{-1})	λ (W m^{-2}K^{-1})	μ (kg/m s)	T_{liq} (K)	T_{sol} (K)	h_m (J/kg)	K_0 (m^{-2})	k_p (/)
7200	700	30	0.006	1760	1680	250000	1.6×10^{8}	0.48
D_S (m^{2}s^{-1})	D_L (m^{2}s^{-1})	β_T (K^{-1})	β_C (/)					
1.6×10^{-11}	1.0×10^{-8}	1.0×10^{-4}	4.0×10^{-3}					

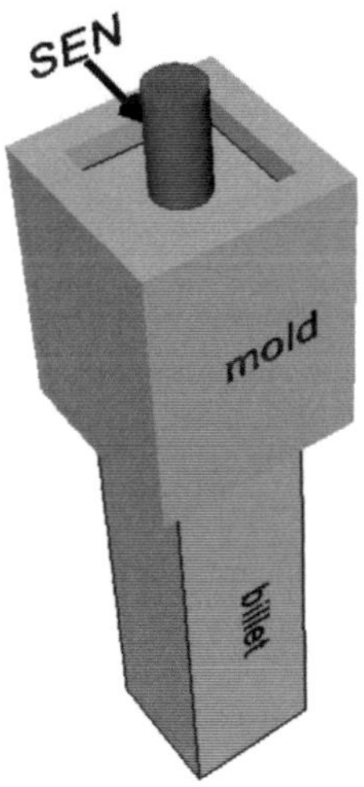

Figure 1. Scheme of geometry.

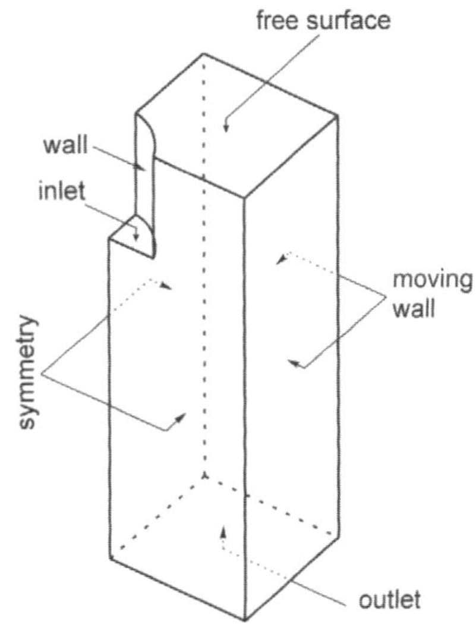

Figure 2. Computational geometry (a quarter of the billet)

Boundary conditions: The boundary conditions for the velocity, temperature, turbulent kinetic energy, dissipation rate, and species concentration are set as follows

Inlet: Constant values are prescribed for all variables: $u_{inlet,x}=0$, $u_{inlet,y}=a_b u_{cast}/d_2$, $k_{inlet}=1.5\left(I_t u_{inlet,y}\right)^2$, $\varepsilon_{inlet}=\dfrac{c_\mu^{0.75}k_{inlet}^{1.5}}{0.07d_2}$, $T_{inlet}=T_0$, $C_{inlet}=C_0$, where u_{cast} is the casting velocity in the casting direction, and I_t is the turbulent intensity calculated from the following relation

$$I_t=0.16\,\mathrm{Re}^{-1/8},\ \mathrm{Re}=\rho u_{inlet,y}d_2/\mu. \tag{14,15}$$

Normal gradient of pressure is set to zero, i.e. $\dfrac{\partial P}{\partial y}=0$.

Outlet: The pressure outlet is prescribed. The derivative in vertical direction of all other variables is set to zero

$$P=0,\frac{\partial u_x}{\partial y}=0,\frac{\partial u_y}{\partial y}=0,\frac{\partial k}{\partial y}=0,\frac{\partial \varepsilon}{\partial y}=0,\frac{\partial T}{\partial y}=0,\frac{\partial C}{\partial y}=0. \tag{16}$$

Top surface (meniscus): At the meniscus, the symmetry boundary conditions are used (free surface flow). The normal derivative of all variables is set to zero, except the vertical velocity u_y, which is set to zero.

Symmetry: The normal derivative of all variables is set to zero

$$\frac{\partial P}{\partial \mathbf{n}}=0,\frac{\partial u_x}{\partial \mathbf{n}}=0,\frac{\partial u_y}{\partial \mathbf{n}}=0,\frac{\partial k}{\partial \mathbf{n}}=0,\frac{\partial \varepsilon}{\partial \mathbf{n}}=0,\frac{\partial T}{\partial \mathbf{n}}=0,\frac{\partial C}{\partial \mathbf{n}}=0. \tag{17}$$

Moving walls: The walls with the solidified steel move with casting velocity along the casting direction. At the walls, where the liquid phase exists, the no-slip boundary conditions for the velocity are set. In the mold, the Robin boundary condition is used, with the surface heat transfer coefficient equal to $2000\ \mathrm{Wm^{-2}K^{-1}}$. Below the mold, at the secondary cooling system, the heat transfer coefficient is equal to $800\ \mathrm{Wm^{-2}K^{-1}}$. The normal derivative of the solute concentration is set to zero.

5. Numerical results

In this conference paper, the preliminary numerical results are obtained, without considering inverse segregation and shrinkage-induced flow. The low-Re turbulence model by Abe-Kondoh-Nagano (AKN) [29] is selected, where the normal distance in closure coefficients is calculated up to the wall. The simulation was performed on the node arrangement with 643891 non-uniformly positioned nodes. The steady-state solution shown in figures is approached by a false transient calculation using a fixed time-step of $7.5\cdot10^{-4}$ s. The following process parameters were used: $u_{cast}=1.1$ m/min, $T_0=1807$ K and $C_0=0.55$ wt%.

Figure 3 represents the following fields: velocity magnitude, temperature and carbon concentration. The same fields at cross-section $y=-0.8$ m are presented in figure 4. The liquid metal is pouring from the nozzle into the mold, where the flow is separated into upward and downward direction. In the upward direction, liquid flows around SEN up to the meniscus (top surface), and then turn into the mold wall. In the upper half of the mold, the flow in the mushy zone is oriented upward, which results in rejecting the carbon out from the mushy zone in to the liquid pool. We get negative segregation in the mushy zone and positive segregation in the liquid pool.

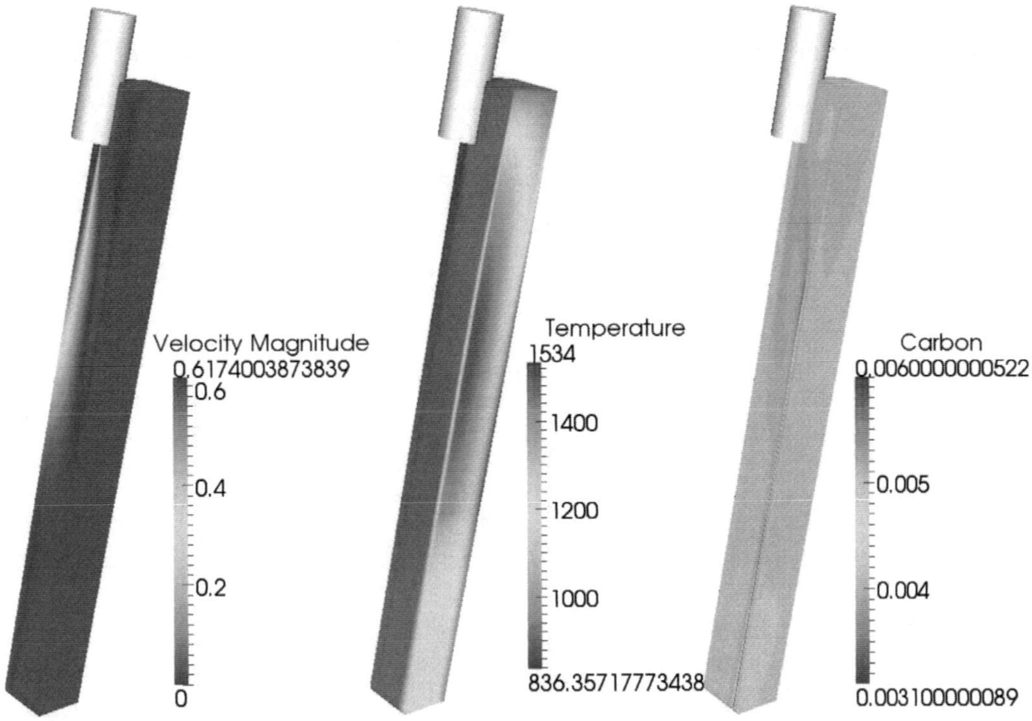

Figure 3. Left: velocity magnitude field. Center: temperature field. Right: carbon segregation field.

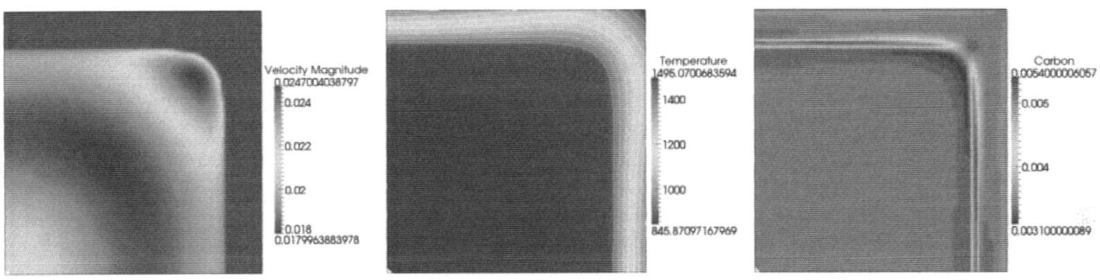

Figure 4. Left: velocity magnitude field. Center: temperature field. Right: carbon segregation field.

6. Conclusions

This paper represents a continuation of numerical benchmarking [10] proposed for the continuous casting of steel. The geometry and material properties of the continuous casting process are simplified in order to enable straightforward performance comparisons with various numerical methods. In the paper, our in-house developed meshless numerical method, based on the local radial basis function collocation, is used to get the steady state solution of the turbulent fluid flow with heat transfer, solidification, and macrosegregation of carbon in three dimensions. A symmetric quarter of the billet was considered. The results were obtained without electromagnetic forces in the momentum equation, and with a lever microsegregation model. The benchmark will be further upgraded and adapted, based on the feedback from the colleges in the field of the numerical modelling of solidification processes. A journal paper with tabulated results of the related simulations, similar as [11], is underway. A web page of the benchmark results with systematically tabulated results of different numerical methods will be established, similar to [3].

Acknowledgement
The financial support of the Slovenian Grant Agency under grants L2-6775 and P2-0379 is kindly acknowledged.

References

[1] Irwing W R 1993 *Continuous Casting of Steel*, (London: The Institute of Materials)
[2] Lesoult G 2005 *Mat. Sci. Eng. A-Struct* **413-414** 19-29
[3] Bellet M, Combeau H, Fautrelle Y, Gobin D, Rady M, Arquis E, Budenkova O, Dussoubs B, Duterrail Y, Kumar A, Gandin C A, Goyeau B, Mosbah S and Založnik M 2009 *Int. J. Therm. Sci.* **48** 2013-16
[4] Thomas B G 2002 *Modeling of the Continuous Casting of Steel - Past, Present Future*, www.brimacombecourse.org/pdf/2001_Thomas.pdf
[5] Kang K G, Ryou H S and Hur N K 2005 *Numer. Heat Transf. A-Appl.* **48** 461-81
[6] Lei H, Zhang H W and He J C 2009 *Chem. Eng. Technol.* **32** No.6 991-1002
[7] Combeau H, Založnik M, Hans S and Richy P E 2009 *Metall. Mater. Trans. A* **40** 289-304
[8] Li J, Wu M, Ludwig A and Kharicha A 2014 *Int. J. Heat Mass Trans.* **72** 668-679
[9] Kajitani T, Drezet J M and Rappaz M 2001 *Metall. Mater. Trans. A* **32** 1479-1491
[10] Šarler B, Vertnik R and Mramor K 2012 MCWASP XIII, IOP Conf.Series: Mat.Sci.Eng. 33 012012
[11] Vertnik R and Šarler B 2014 *Eng. Anal. Bound. Elem.* **45** 45-61
[12] Šarler B and Vertnik R 2014 *Materials Science Forum* **790-91** 279-284
[13] Šarler B and Vertnik R 2006 *Comput. Math. Appl.* **51** 1269-82
[14] Vertnik R and Šarler B 2006 *Int. J. Numer. Method Heat Fluid Flow* **16** 617-640
[15] Mramor K, Vertnik R and Šarler B 2013 *Comput. Mater. Cont.* **36** 1-21
[16] Kosec G and Šarler B 2009 *Int. J. Numer. Method Heat Fluid Flow* **18** 868-82
[17] Vertnik R and Šarler B 2009 *Comput. Model. Engin. Sci.* **44** 65-95
[18] Vertnik R and Šarler B 2011 *Advances in Applied Mathematics and Mechanics* **3** 259-279
[19] Gobin D and Le Quéré P 2000 *Comput. Assist. Mech. Eng. Sci.* **7** 289-306
[20] Kosec G and Šarler B 2009 *Comput. Model. Eng. Sci.* **47** 191-216
[21] Kosec G Založnik M, Šarler B and Combeau H 2011 *Comput. Mater. Cont.* **22** 169-59
[22] Kosec G and Šarler B 2014 *Eng. Anal. Bound. Elem.* **45** 36-44
[23] Mramor K, Vertnik R and Šarler B 2013 *Comput. Model. Eng. Sci.* **92** 327-52
[24] Mramor K, Vertnik R and Šarler B 2014 *Eng Anal. Bound. Elem.* doi: 10.1016/j.enganabound.2014.04.013
[25] Bennon W D and Incropera F P 1988 *Numer. Heat Transf. A-Appl.* **13** 277-96
[26] Wilcox, D C 1993 *Turbulence modeling for CFD* (California: DCW Industries, Inc.)
[27] Chorin A J 1967 *J. Comput. Phys.* **2** 12-26
[28] Buhmann M D 2003 *Radial Basis Function: Theory and Implementations* (Cambridge: Cambridge University Press)
[29] Abe K, Kondoh T and Nagano Y 1994 *Int. J. Heat Mass Transf.* **37** 139-51

Oxide film defects in Al alloys and the formation of hydrogen-related porosity

W D Griffiths[1a], A J Gerrard[2] and Y Yue[1]

[1]School of Metallurgy and Materials, College of Engineering and Physical Sciences, University of Birmingham, Edgbaston, Birmingham, B43 7BA, UK
[2]Alcoa Europe, Kitts Green, Birmingham, B33 9QR, UK

Email: [a]W.D.Griffiths@bham.ac.uk

Abstract. Double oxide film defects have also been held responsible for the origins of hydrogen porosity, where hydrogen dissolved in the Al melt passes into the interior atmosphere of the double oxide film defect causing it to inflate. However, this is in opposition to long-established evidence that H cannot readily diffuse through aluminium oxide. To investigate this further, samples of commercial purity Al were first degassed to remove their initial H content, and then heated to above their melting point and held in atmospheres of air and nitrogen respectively, to determine any differences in H pick-up. The experiment showed that samples held in an oxidising atmosphere, and having an oxide skin, picked up significantly less H than when the samples were held in a nitrogen atmosphere, which resulted in the formation of AlN in cracks in the oxide skin of the sample. It is suggested that double oxide film defects can give rise to hydrogen-related porosity, but this occurs more quickly when the oxygen in the original oxide film defect has been consumed by reaction with the surrounding melt and nitrogen reacts to form AlN, which is more permeable to H than alumina, more easily allowing the oxide film defect to give rise to a hydrogen pore. This is used to interpret results from an earlier synchrotron experiment, in which a small pore was seen to grow into a larger pore, while an adjacent large pore remained at a constant size.

1. Introduction

Double oxide film defects occur in Al alloy castings when, during mould filling or other metal transfer operations, the oxidised surface of the liquid metal folds over onto itself and is entrained in the liquid metal. Since the oxidised surfaces are at a temperature thought to be too low for sintering, the defect consists of unbonded oxide surfaces between which is trapped a thin layer of the local atmosphere, trapped when the liquid metal surfaces folded over onto themselves. The thickness of the defects is probably just a few μm's, but their area can be several mm^2. These gas-filled crevices with an area of several mm^2, present in the solidified casting as defects, can bring about a considerable reduction in mechanical properties. The randomness of the fluid flow effects occurring during their formation, and their subsequent dispersion and orientation, means that oxide film defects have a considerable effect on the reproducibility of cast Al alloy properties [1-3]. Figure 1 shows a sketch of the formation of a double oxide film defect, and figure 2 shows an example of such a defect on the fracture surface of a tensile test bar.

Double oxide film defects in Al alloys have also been associated with the formation of other defects, such as shrinkage porosity, hydrogen gas porosity, hot tears and the nucleation of low melting point phases such as Fe-rich intermetallics [2]. It has been claimed that the nucleation of porosity during solidification, both shrinkage porosity and gas porosity, is best explained by the presence of double oxide film defects, as their internal atmosphere acts as a precursor pore, avoiding the problem of accounting for the initial nucleation of a pore [1-3].

Experimental evidence by Griffiths and co-workers supports the possibility of hydrogen diffusing through the oxide layer of a double oxide film defect into the interior atmosphere [4,5]. However, obstacles to this interpretation are to be found in the reports by other researchers of the difficulty for hydrogen to diffuse through an oxide layer. For example, alumina, in layers as little as 20 nm thick, has been proposed as a barrier to the permeation of hydrogen through steel [6,7]. Furthermore, in early experiments to measure hydrogen dissolved in solid Al, considerable problems were experienced due to the presence of an alumina layer on the sample surface [8]. Most interesting were experiments reported by Campbell [1], in which water was placed on the surface of an Al alloy melt while the hydrogen content was being continuously measured. No change in hydrogen content was observed.

The results presented here derive from an experiment in which samples of commercial purity (CP) Al alloy were heated in either an oxygen or a nitrogen atmosphere, to change their surface characteristics, and then held in a hydrogen atmosphere, followed by measurement of hydrogen pick-up. The results were used to interpret the behaviour of pores observed by synchrotron X-ray tomography in small samples of Al held in the liquid state [9].

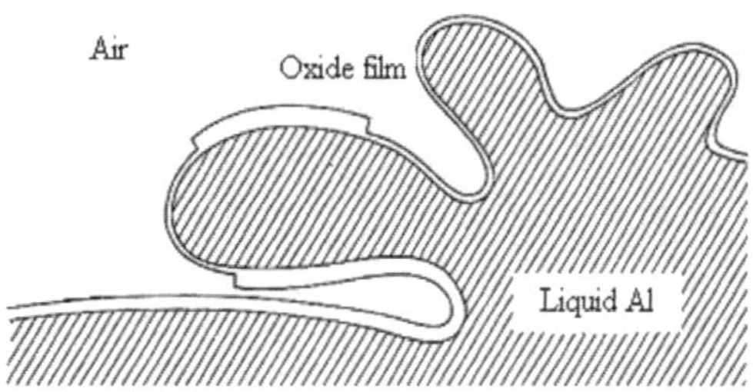

Figure 1. Sketch showing the formation of a double oxide film defect in liquid metal [3].

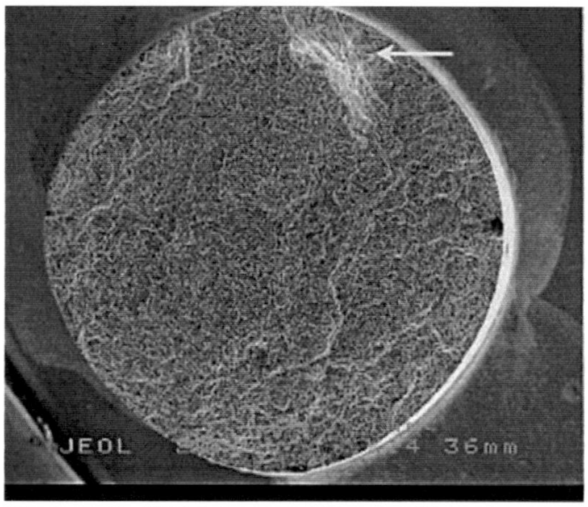

Figure 2. SEM image of one side of a double oxide film defect on the fracture surface of a tensile test bar, (arrowed). The diameter of the test bar is 6 mm.

2. Experimental procedure

Samples of commercial purity (CP) Al alloy of dimensions 8 mm in diameter and 50 mm in length were used in the experiment, as they had suitable dimensions for measurement of hydrogen in a LECO instrument. (The LECO instrument heats up the sample and extracts out dissolved hydrogen into a nitrogen carrier gas; the thermal conductivity of the N_2/H_2 gas mixture is measured in order to determine the hydrogen content of the Al alloy sample).

Initially samples of CP Al alloy were measured in the LECO instrument three times, in order to reduce the H content to nil, (or less than the measurement accuracy of the instrument, (± 0.005 ppm). The H-free samples were then placed in a sealed glass bottle to minimise H pick-up from the environment, and then placed in a furnace with a controllable atmosphere.

In one experiment the furnace was heated at 40 $Kmin^{-1}$ to 700°C with the samples placed in an atmosphere of air. As the samples of CP Al were heated and expanded, it would be expected that their surface alumina film would crack, and be resealed by the formation of fresh alumina, by reaction between the Al and the oxygen in the air. The samples were held at 700°C for 15 minutes, after which the furnace was evacuated to a pressure of 1 Pa, and then filled with an atmosphere of H_2 at a pressure of 150 kPa, and held for 1 hour. The furnace was then cooled at a rate of 20 $Kmin^{-1}$ to 50°C, at which point the hydrogen atmosphere was removed, and the samples taken out for measurement of their hydrogen pick-up by LECO. Two samples were treated in this way.

In a second experiment, two samples of CP Al were similarly completely degassed, by repeated LECO measurement, and then placed into the furnace at room temperature, together with 50 g of Hf turnings, (to act as an oxygen getter). The furnace atmosphere was evacuated to 1 Pa and back-filled with Ar (99.999% purity) to 507 kPa, (5 bar), to minimise the oxygen content in the furnace atmosphere. The furnace was then filled with pure N_2, (99.998% purity), to a pressure of 150 kPa, and then heated to 700°C at the same heating rate, and held for 15 minutes in an atmosphere of N_2. Again it would be expected that heating of the samples would cause them to expand and their oxide film crack, as before, but the cracks in the oxide film should be resealed with AlN, due to reaction between Al and the N_2 atmosphere. The furnace was then evacuated and again back-filled with hydrogen to a pressure of 150 kPa and held at this pressure for 1 hour. The furnace was then cooled and the H content of the samples measured to determine the H pick-up when AlN was present on the sample surface.

In unrelated work, the porosity of samples of 2L99 Al alloy containing porosity was examined using the synchrotron source at the Swiss Light Source. In this work samples of 3 mm in diameter and 6 mm in length, were melted and heated to 727°C in the TOMCAT beamline, and synchrotron X-ray microtomography used to construct 3D images of the porosity and its change in morphology with time. The results of the previous holding experiments were used to interpret the results obtained from the synchrotron imaging. Further experimental details are given in reference [9].

3. Results

3.1 Results of the hydrogen pick-up experiments.

Figure 3 shows the results of the H pick-up experiments. The mean hydrogen pick-up for the two samples heated in air was 0.04 ppm, while for the two samples heated in the nitrogen atmosphere it was 0.08 ppm. Figure 3 also shows the hydrogen pick-up from samples of 5083 alloy, (prepared in the same way), which has a higher Mg content, (4-5wt.%), and therefore should form a surface of porous MgO, as well as having a higher solubility for hydrogen. In this case the hydrogen pick-up was much greater, at 0.28 ppm.

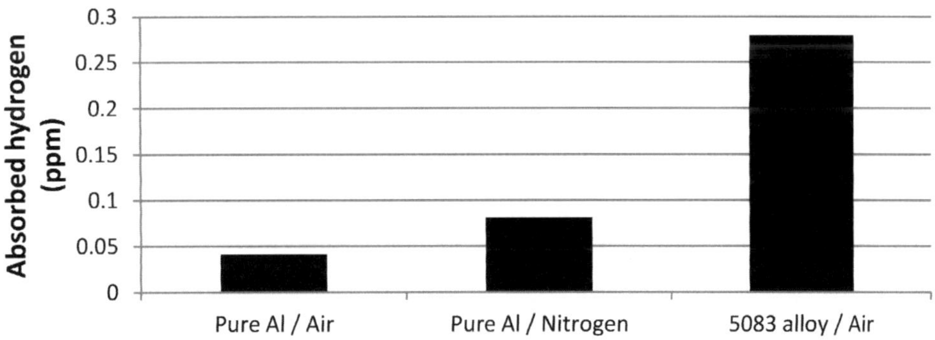

Figure 3. Bar chart showing the relative amounts of hydrogen pick-up in the different experiments.

Figure 4 shows SEM images of the surfaces of the samples heated in the nitrogen atmosphere. At higher magnification, (see figure 4(b)), coupled with EDX analysis, (see figure 4(c)), it was found that the cracks in the surface film contained AlN. The image in figure 4(b) suggested that the structure of AlN would be more permeable than that of alumina. (This type of structure for AlN has been demonstrated in earlier work [4]).

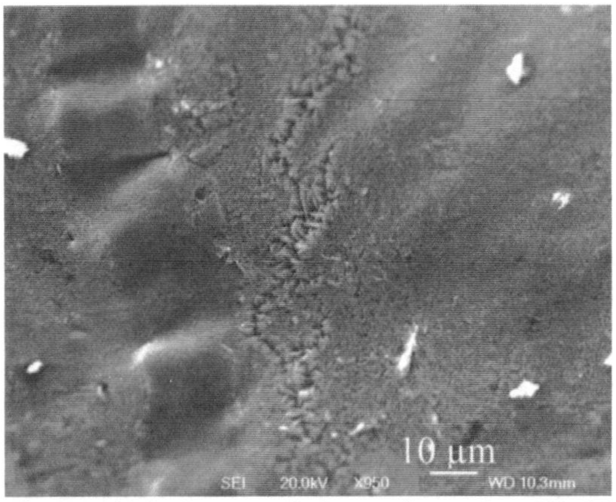

Figure 4(a). Image of the surface of a CP Al alloy heated in N_2.

Figure 4(b). Higher magnification image of the surface of a CP Al alloy heated in N_2.

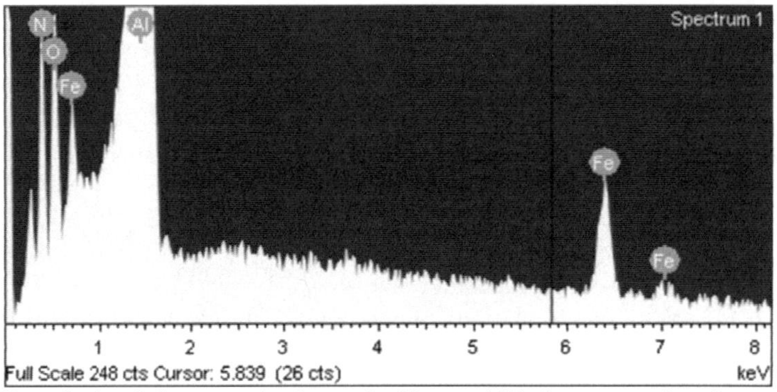

Figure 4(c). EDX analysis of the crystals shown in Figure 4(b) demonstrating the occurrence of AlN.

3.2 Results of synchrotron X-ray tomographic microscopy of pores in 2L99 Al alloy.

Figure 5 shows four images obtained in the TOMCAT beamline at the Swiss Light Source, and show the growth of a pore inside a sample of 2L99 Al alloy, held in the liquid state at 727°C for a period of more than 27 minutes. Figure 5(a) shows a pore in the solid state, about 1 mm in diameter, before the experiment began. During heating to the liquid state and holding at 727°C for 16 minutes the pore collapsed to a thinner, irregular form, (see figure 5(b)). Figure 5(c) shows that the pore appeared similar after holding for a further 11 minutes, (a total of 27 minutes). However, the formation of a small spherical pore was noted, (arrowed in figures 5(b-c), adjacent to the large irregular pore. It might be expected that, if this was a hydrogen pore, that the hydrogen would preferentially diffuse into the large irregular pore, whereas growth of the small spherical pore immediately nearby seems to have been preferred. Figure 5(d) shows the same two pores after solidification.

4. Discussion

The results shown in figure 3 and figure 4 therefore suggest that AlN is more permeable to H than aluminium oxide. Aluminium oxide is evidently not completely impervious to the diffusion of hydrogen, but the more open structure of the AlN would be expected to have the higher diffusion rate, as the results of the H pick-up experiments confirmed. The SEM images of the cracks in the oxide films in which the AlN grew, and the EDX analysis of the region, suggests a complex relationship between alumina and AlN, in which the latter may have grown in cracks in the former, but alternatively, the AlN may have grown on an underlying substrate of alumina. This requires further investigation. The preferential passage of H through AlN compared to alumina may be used to explain the different behaviour of the pores studied in the synchrotron experiment shown in figure 5.

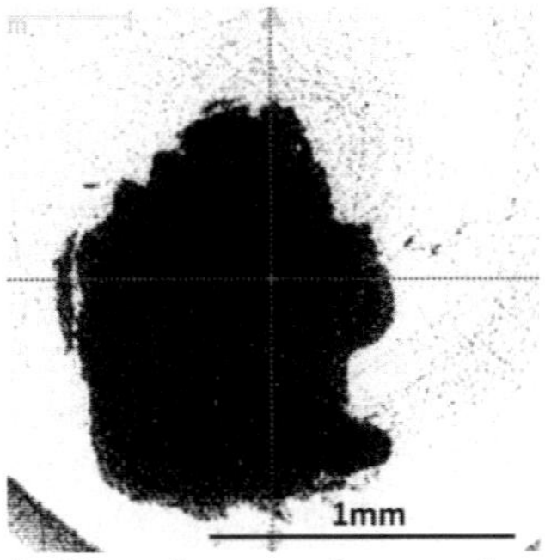

Figure 5(a). X-ray image of a pore in a sample of CP Al at 25°C.

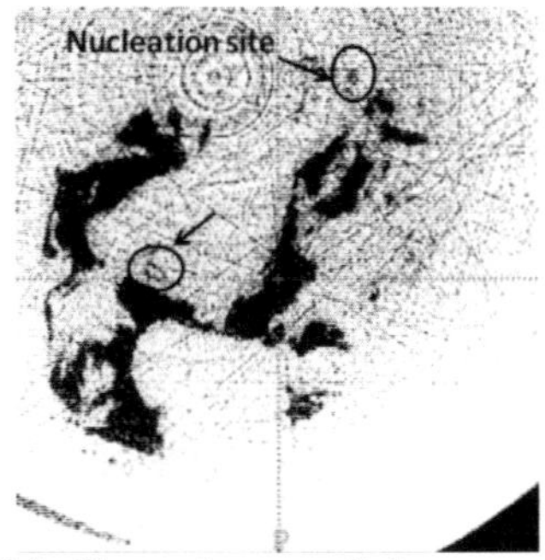

Figure 5(b). X-ray image of the same pore after holding in the liquid state at 727°C for 1000 s.

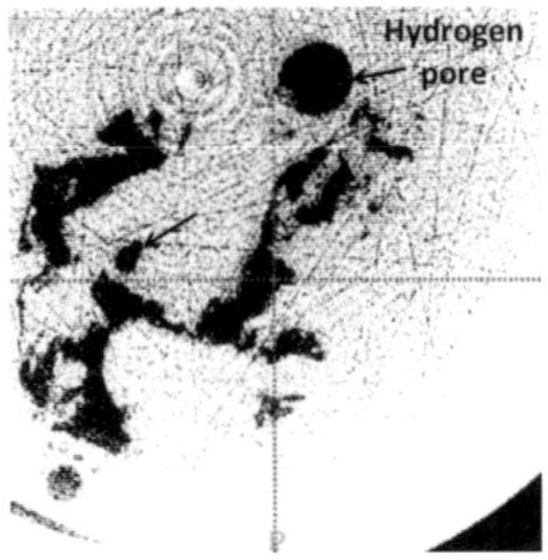

Figure 5(c). X-ray image of the same pore after holding in the liquid state at 727°C for 1640 s.

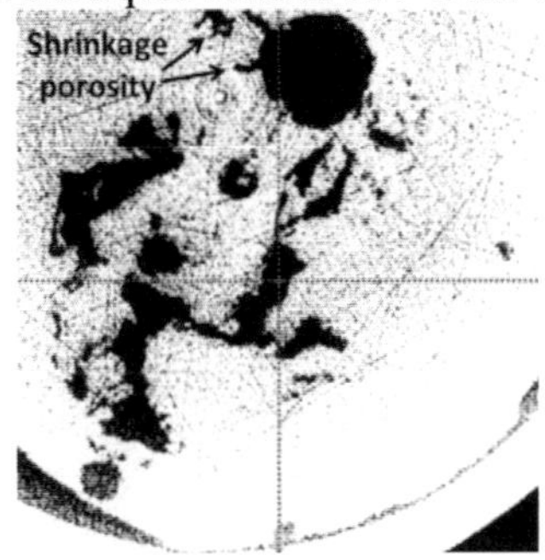

Figure 5(d). X-ray image of the same pore after solidification and cooling to 25°C.

The solid sample was obtained from a bar cast in such a way as to contain an entrainment defect, in this case a large, mm-sized pore, shown in figure 5(a). This pore presumably therefore originally contained a mixture of air, (but depleted to some extent of oxygen), arising from the entrainment event, and also some hydrogen, arising from the hydrogen dissolved in the metal, which had passed into the pore during solidification. Upon heating of the sample to the liquid state the pore collapsed, either during melting, or during holding in the liquid state. The extent of the collapse suggests that all of the oxygen contained in the pore would have reacted with the surrounding melt, and the pore atmosphere would now perhaps consist of a mixture of N_2 and H_2, (strongly suggested by Pore Gas Analysis measurements on air bubbles held in liquid Al alloys [5]. Between 16 and 27 minutes its morphology was constant, (see figures 5(b) and (c)), and the interior atmosphere was perhaps unchanged, and still consisted of nitrogen and hydrogen. However in the same time period, between 16 and 27 minutes, a spherical pore was formed, which then grew slightly to become a spherical pore in the solid state which, if seen in a metallographic section, would probably have been attributed to hydrogen porosity (see figure 5(d)).

It was presumed that the small pore, since it appears to grow independently of the large pore, was not connected. The hydrogen dissolved in the liquid metal appears to have preferentially entered the smaller pore, rather than the larger, tangled pore. The large pore shown in figure 5 had progressed to an age where it should be expected that AlN would have been formed at the pore surface, by reaction between the surrounding Al melt and the nitrogen in the pore. The observation that AlN provides an easy diffusion path for hydrogen into the interior of a double oxide film defect, based on the LECO experiments reported in figure 3, suggests that, since the smaller pore expanded, hydrogen dissolved in the melt passed into the smaller pore because the latter offered the least barrier to diffusion.

The origin of the small pore may or may not have been an entrainment defect. If it originated in an oxide film defect, then the same arguments might apply, the presence of AlN would assist the diffusion of H into the pore. If the pore origin did not lie with an entrainment defect, no interfacial barrier to diffusion of H would be expected.

These results therefore suggest that the formation of hydrogen porosity during solidification is complex. In these results three different mechanisms have been identified. Firstly, passage of hydrogen through an alumina layer into a double oxide film defect can evidently occur, although this occurs slowly, (figure 3). Secondly, passage of hydrogen through an AlN layer can occur, more quickly than through alumina, (see figure 3). Finally, formation of hydrogen porosity either spontaneously or through the presence of pre-existing pores could be a third mechanism, shown in figure 5. The small pore observed in figures 5(b) and (c) may have been a residual pore from a prior entrainment event, although in this case one might have expected the adjacent irregular pore to behave similarly, so perhaps this may have been a new pore coming into existence.

5. Conclusions

1. Hydrogen can diffuse through AlN at a faster rate than through alumina, suggesting that older oxide films, which have formed AlN at their interface, may be a better candidate for the origins of hydrogen porosity in Al alloys than newer double oxide film defects which still retain oxygen.
2. Synchrotron X-ray tomography has suggested that pore formation in Al alloys can arise for multiple reasons, and a monocausal effect should not be sought.

Acknowledgements

The authors acknowledge the support of the EPSRC LiME grant EP/H026177/1, and the financial support of the School of Metallurgy and Materials at the University of Birmingham.

References

[1] Campbell, J., (2nd ed.), "Castings" Butterworth-Heineman, Oxford, 2003.

[2] Campbell, J., "An overview of the effects of bifilms on the structure and properties of cast alloys," *Met. and Mat. Trans. B*, vol. 37 B, pp. 857-863, 2006.

[3] Campbell, J., "Complete Castings Handbook", Butterworth-Heineman, Oxford, 2013.

[4] Raeiszadeh, R. and Griffiths, W. D., "A method to study the history of a double oxide film defect in liquid aluminium alloys", *Met. Mat. Trans. B*, Dec. 2006, vol. 37, 865-871.

[5] El-Sayed, M. A., Salem, H. A. G., Kandeil A. Y. and Griffiths, W. D., "Determination of the lifetime of a double oxide film in Al castings", *Met. and Mat. Trans. B*, in press.

[6] Yamada, T. H., "Hydrogen permeation barrier performance characterization of vapor deposited amorphous aluminum oxide films using coloration of tungsten oxide," *Surface and Coatings Technology*, Vols. 153(2-3), pp. 114-118, 2002.

[7] Aggarwal, S., "Use of amorphous aluminium oxide on a capacitor sidewall for use as a hydrogen barrier," Patent , no. US6876021 B2, 2005.

[8] Ransley, C. E. and Neufeld H, "The solubility of hydrogen in liquid and solid aluminium", *J. Inst. Met.*, (Aug. 1948), pp. 599-620.

[9] Yue, Y., Griffiths, W. D., Fife, J. L. and Green, M. R., "In-Situ Characterization of Entrainment Defects in Liquid Al-Si-Mg Alloy," proc. conf., *1st Int. Conf. 3D Materials Science*, Seven Springs Mountains, PA, 2012.